NOMINA ANATOMICA AVIUM

An Annotated Anatomical Dictionary of Birds

3

NOMINA ANATOMICA AVIUM

An Annotated Anatomical Dictionary of Birds

Edited by

JULIAN J. BAUMEL

and

ANTHONY S. KING, ALFRED M. LUCAS, JAMES E. BREAZILE and HOWARD E. EVANS

Consultant for Taxonomy: RICHARD L. ZUSI

Consultant for Classical Languages: LUBOMIR MALINOVSKÝ

Prepared by the International Committee on Avian Anatomical Nomenclature, a committee of the World Association of Veterinary Anatomists. Supported in part by the U.S. National Institutes of Health (NIH Grant LM02420 from the National Library of Medicine)

1979

ACADEMIC PRESS

London New York Toronto Sydney San Francisco

A Subsidiary of Harcourt Brace Jovanovich, Publishers

ACADEMIC PRESS INC. (LONDON) LTD
24/28 Oval Road
London NW1

United States Edition published by
ACADEMIC PRESS INC.
111 Fifth Avenue
New York, New York 10003

British Library Cataloguing in Publication Data

Nomina anatomica avium.
1. Birds - Anatomy - Terminology
I. Baumel, Julian J II. International
Committee on Avian Anatomical
Nomenclature
598.2′4′03 QL697 78-67890

ISBN 0–12–083150–3

Printed in Great Britain by
Page Bros (Norwich) Ltd. Mile Cross Lane, Norwich NR6 65A

GENERAL INTRODUCTION

HISTORICAL BACKGROUND

The first comprehensive study of avian structure was the treatise on hawking by Emperor Frederick the Second in the thirteenth century, in which Aristotle's errors on the homologies of the leg bones were corrected (Cole, 1944). These homologies were analyzed further in 1555 by the French physician Belon, a pioneer of the comparative method in anatomy, but it was the Dutch physician Coiter who was the first to study comparative anatomy on a large scale with many admirable observations on birds in his works of 1572 and 1575. Coiter may justly be regarded as the father of avian anatomy. Notable contributions were made during the sixteenth century at Padua by Fabricus and Casserius, and, in the seventeenth century, by their distinguished pupil, William Harvey. The anatomists of the Amsterdam School (Blasius); of the peripatetic German Collegium Academiae Naturae Curiosorum (Murait); of the Copenhagen School (Jacobsen, Steno); and particularly of Paris (Perrault, Duverney) all published penetrating observations during the seventeenth century on the anatomy of birds.

All of these early comparative anatomists were in fact physicians (Cole, 1944). The advancement of human medicine was the main objective of their comparative studies in the formative years. From here onwards a different objective appeared. Avian and other branches of comparative anatomy began to be studied for their own sakes. From the onset of the eighteenth century to the present there have been two broad streams of avian anatomists, one concerned primarily with the medical sciences, and the other with comparative anatomy as a discipline in its own right. Today one finds avian anatomists in the departments of anatomy, physiology, biochemistry, pathology, and microbiology of medical and veterinary schools and government research institutes, and many more in departments of biology, zoology, paleontology, etc. of the universities and museums.

What is the effect of this historical background on current problems of avian anatomical nomenclature? During the formative years, the terminology of comparative anatomy was firmly based on that of human anatomy. Terms like M. triceps brachii and M. semitendinosus, more or less descriptively meaningful in man, have served also in birds and other vertebrates in which they are often not descriptively accurate. Some such mammalian terms cause no difficulty, but others such as jejunum and ileum, which are obscure even in man, pose further problems. Attempts to apply mammalian names to avian structures have contributed to some of the nomenclatural confusion in the avian anatomical literature. The homologies of structures based on the relatively close relationship of birds to the reptiles have often been ignored in naming avian anatomical structures. For example, the sauropsid M. depressor mandibulae is commonly (erroneously) designated as M. masseter. M. depressor mandibulae is innervated by N. facialis, and depresses the lower jaw; whereas M. masseter in mammals belongs to the trigeminal nerve field, and functions to elevate the mandible. For those numerous basic structures shared by the higher vertebrates, the foundation of medical anatomical terminology has indeed proved a valuable unifying nomenclature which is common to vertebrates in general. On the other hand, numerous other structures are unique to birds. Heretofore the lack of a uniform anatomical nomenclature for birds often has been a severe hindrance; for instance, there were at least 13 different names for the clavicular air sac before the year 1900. Such proliferation of synonyms in the widely scattered literature of avian anatomy has created considerable difficulty.

Meanwhile the pace of research on birds has sharply increased in recent years; for example, the *Zoological Record* contained about 1500 avian entries in 1950, 2000 in 1954, and 7500 in 1969. Much of this work has been done on the domestic fowl in the search for cheap animal protein; the fowl is an important biological model as in immunological studies, and a mainstay in experimental embryological research. Still other diverse areas such as bone and fat metabolism, neurophysiology, temperature regulation, extrarenal excretion, pulmonary function, flight energetics, and circadian rhythms are being actively explored in birds of many different species. Moreover, the so-called information explosion requires the use of an internationally acceptable terminology to facilitate the storage and immediate retrieval of anatomical information about birds by means of computers.

The analysis of anatomical nomenclature has always been contentious; surprisingly passionate emotions can be generated over the names of quite insignificant structures. In the three centuries after Vesalius, repeated attempts were made to introduce order into the chaotic mixture of medieval and classical Latin, Greek, and Arabic terms which had persisted in human anatomy. But the prestige of those who wished to reform the nomenclature was inadequate to enforce their views. It required the authority of Hyrtl to establish the *Basle Nomina Anatomica* (*BNA*) in 1895, which eliminated several thousand names and condensed the remainder to about 4500 of the original total (Donath and Crawford, 1969).

The problems of avian anatomical nomenclature today are simpler in some respects than those of mammalian anatomy before the *BNA*, since so many of the *BNA* terms apply to all higher vertebrates including birds. However, considered from another aspect, compilation of the avian nomenclature presents complications not faced to an appreciable extent in the formulation of the mammalian nomenclatures. Whereas the *BNA* and subsequent mammalian nomenclatures treated only man and several domestic species representing three orders of Class Mammalia, the avian terminology is constructed as a system of anatomical names applicable to birds generally (some 25 orders of Class Aves representing about 8500 species).

Inasmuch as the literature of avian anatomy is ambiguous and difficult to use on account of the diversity of terminologies, the publication of a standard set of unambiguous anatomical terms is expected to be of substantial benefit to avian science. For this reason it has been essential to take into account the requirements of all those most concerned with avian anatomy, especially the ornithologists, comparative and veterinary anatomists, paleontologists, comparative physiologists, and poultry scientists. *It is now important that the official Latin terms should be used in scientific articles and books, on the first occasion (at least) when a term appears in the body of the text, and in the title and key words supplied for data-retrieval systems.*

Although the work on this first edition of *Nomina Anatomica Avium* has already extended over a dozen years (see below, History of ICAAN), there will never be an end to the effort. Knowledge of the anatomy of birds is already vast enough to exercise an army of anatomists for many years over problems of nomenclature. Even though the *Nomina Anatomica Avium* has been compiled with all possible attention to anatomical accuracy and completeness, it is scarcely to be expected that the First Edition will emerge unscathed from criticism. Naturally the errors that inevitably will be disclosed will be eliminated in subsequent editions, just as the nomenclature of human anatomy has undergone repeated revisions since the *Basle Nomina Anatomica* of 1895. See below and Introduction to Fourth Edition of the *Nomina Anatomica* (*IANC*, 1977).

History of ICAAN

Two major revisions of the *BNA* took place in the first half of this century. These were the *British Revision (BR)* in 1933 and the *Jena Nomina Anatomica*, largely a German effort, in 1936. The *JNA* terminology was based on the quadrupedal anatomical position for man. At the VIth International Congress of Anatomists in 1955 the *Paris Nomina Anatomica (PNA)* was adopted. Unfortunately it reintroduced terms of orientation related to man in the upright, standing position; consequently it was opposed by veterinary anatomists. In 1957 the International Association of Veterinary Anatomists set up the International Committee on Veterinary Anatomical Nomenclature (ICVAN), with the purpose of preparing a nomenclature for veterinary anatomy based on the *PNA*. In 1961 the International Association of Veterinary Anatomists had become the World Association of Veterinary Anatomists (WAVA). It reorganized ICVAN into twelve subcommittees, eleven being concerned with the various body systems of mammals and the twelfth, with the title Anatomica Avium, concerned solely with birds.

The mammalian subcommittees completed their work with the publication in 1968 of the first edition of the *Nomina Anatomica Veterinaria* (ICVAN, 1968); a second edition of *NAV* followed in 1973. Avian anatomy was not included in the *NAV* but the subcommittee on avian anatomy was expanded in 1965 to include J. J. Baumel, H. E. Evans, A. S. King, V. Komárek, A. M. Lucas, V. Simić, and H. Yasuda, under the Chairmanship of P. C. Blin. A tentative list of avian terms had been drafted by P. C. Blin and V. Komárek in the early 1960s, but was not adopted since there was a general feeling that a very extensive survey would eventually be required. In 1971 WAVA dissolved the subcommittee on Anatomica Avium, and replaced it by the International Committee on Avian Anatomical Nomenclature (ICAAN) with A. M. Lucas as chairman. Under his active leadership, the membership was rapidly augmented to include about 70 international authorities representing all branches of avian morphology. Within two years ICAAN had prepared Draft Lists for all the systems of the body. These lists were examined in detail at the First General Meeting of ICAAN at Omaha, Nebraska, USA in August 1973. At this meeting A. S. King became chairman. Provisional Lists were considered at the Second General Meeting of ICAAN at Liverpool England, in August 1974, and the Tentative Final List was presented to WAVA in July 1975 at Thessaloniki, Greece. J. J. Baumel was elected chairman in 1974 and has held this office to the present date. At Liverpool, in June 1976 and at Ithaca, New York in August 1977 the Editorial Committee of ICAAN prepared the final revisions. The first edition of the *Nomina Anatomica Avium (NAA)* was finally completed for publication in early 1979.*

Initially, the great bulk of this heavy load of work was done by the Subcommittee Chairmen, with the support of their Sub-Committee Members. In the later stages, the labor fell more and more heavily on the Editorial Committee and in particular on the Chairman of the Editorial Committee. Suggestions for improvement were received from members of WAVA who were not members of ICAAN, notably F. Preuss (who surveyed the whole of the Tentative Final List) and R. Habel.

GENERAL OBJECTIVES OF THE *NOMINA ANATOMICA AVIUM*

The ultimate objective of any standardized nomenclature must be the promotion of international communication by establishing an agreed list of

* This publication of *Nomina Anatomica Avium* was supported in part by the U.S. National Institutes of Health (NIH Grant LM 02420 from the National Library of Medicine).

terms in a universally acceptable language. One impediment is the lack of formal rules for anatomical nomenclature comparable to those which regulate the naming of species and higher taxa of mammals (see below, Selection of Terms). Codifying of prevailing usage is undoubtedly the principal means by which the *NAA* (and the mammalian terminologies) have developed. In practice this has often amounted to selecting one term from among several alternatives, all in more or less general use, e.g. choosing between uterus and shell gland. In the *NAA* important synonyms have not been totally discarded, but are mentioned in annotations and listed in the index; others have been retained as official alternative terms (see below, Format, *Bracketed terms*). Well-established terms devoid of equally well-established rivals have been preserved whenever possible.

In all anatomical nomenclatures it seems to have been easier to defend the *status quo* than to attack it. The Introduction to the second edition of the human *Nomina Anatomica* (IANC, 1964) stated as its first principle: "As few changes as possible should be made in anatomical nomenclature . . .", and added, "We changed nothing unless the term was obviously inconsistent or inaccurate". Admittedly the *Nomina Anatomica* was designed for use by the medical profession which may tend to be generally conservative. Equally stiff opposition to change can be encountered from zoologists. Nevertheless the *NAA* has eliminated some substantial errors in the prevailing usage, particularly those founded on inaccurate anatomical beliefs or incorrect homologies.

An inescapable commitment of the *NAA* has been to produce Latin names for major structures which have been described but not named, or named only in the vernacular of the author; in developing Latin names, extensive use has been made of the anatomical and biological dictionaries of Donath and Crawford (1969) and Kenneth (1966). These broad objectives connected with usage can be summarized as follows: (1) To select one term where two or more are in general use for the same structure; (2) to replace terms which are grossly defective; (3) to produce new terms in Latin where none is available; and (4) within the limitations of the above, to codify prevailing usage.

The *Nomina Anatomica Avium* has always had one other general objective of an entirely different character, namely the advancement of anatomical knowledge of birds. To meet this objective the *NAA* contains extensive annotations explaining structure, homology, synonymy, and species variation. Numerous illustrations are included for further clarification of terminology. Finally there is a bibliography showing the authority for the anatomical facts and the selection of the names. Thus the *Nomina Anatomica Avium* is more than a list of terms; it is a major source-book of avian anatomical information, containing many original observations previously unpublished.

SELECTION OF TERMS

The following principles, which in general are common to the *Nomina Anatomica Avium* and *Nomina Anatomica Veterinaria*, have received full attention:

1. As few changes as possible should made made in well-established terms.
2. With a few exceptions, each structure should be named by only one term.
3. Each term should be in Latin.
4. Each term should be short and simple as possible.
5. Each term should be easy to remember and should, above all, be informative.
6. Structures which are closely related topographically should have similar names.
7. Contrasting adjectives should generally be opposites, such as major and minor.
8. Eponyms should not be used.
9. No attempt should be made to name very minor structure every discovered.

These principles, notably the first, have led in many instances to the immediate adoption by the *NAA* of the same terms as those listed in the *Nomina Anatomica*, *Nomina Anatomica Veterinaria*, and *Nomina Histologica* (IANC, 1975). However, certain conflicts are apparent between these precepts, notably between items 1 on the one hand and items 4 and 5 on the other. Since international communication is the ultimate objective, informativeness, simplicity, and familiarity have in general been given precedence. The more closely the Latin term resembles its vernacular equivalent, the more likely it is to be immediately understood and freely used by scientists with little or no knowledge of Latin—the overwhelming majority. A number of important terms have therefore been rendered into Latin forms which closely resemble the terms in English and the Romance languages (these being used as a second language by many biologists outside those language groups). For example, Cavum nasi (*NA* and *NAV*) has been changed to Cavitas nasalis, which more closely resembles "nasal cavity", "cavité nasale", "cavità nasale", and "cavidad nasal". As with Cavitas nasalis, adjectival forms have in general been adopted rather than the genitives of nouns. In common with anatomists since the time of Vesalius, members of ICAAN have also not been afraid to evolve Latin neologisms in instances where these are informative and simple and no orthodox term is available. The emphasis on informativeness, simplicity, and familiarity is expected to facilitate the widest possible use of the *Nomina Anatomica Avium*, not only by avian anatomists but by avian scientists in many disciplines. It is important that the *Nomina* should not be neglected simply because of the unfamiliarity of Latin. It bears repeating that the official Latin Terms should be used in scientific articles and books in order to enhance international scientific communication.

The principles behind these arguments have been well-stated by R. Warwick in his Introduction to the Fourth Edition of the *Nomina Anatomica* (IANC, 1977) and his article of 1978. The Fourth Edition of the *NA* is a combination for the first time of *Nomina Anatomica*, *Nomina Histologia*, and *Nomina Embryologica*; therefore the application of these principles to the Fourth Edition is likely to have far-reaching effects on anatomical nomenclature. The Fourth Edition also reflects a new attitude to the rôle of Latin in anatomical nomenclature, namely that Latin is to be the servant, not the master of anatomists. The *Nomina Anatomica Avium* supports this reform. It is indeed unfortunate that although there has been consultation between the Honorary Secretary of IANC (R. Warwick) and the present and past chairmen of ICAAN (J. J. Baumel and A. S. King), the *NA* and the *NAA* have been prepared simultaneously without the opportunity for comparisons of the two lists; it is probable that in view of their common principles, even more agreement between the two lists could have been achieved than there exists.

It was decided from the outset that emphasis should be placed mainly on the macroscopic and mesoscopic anatomical structures, and the *NAA* is intended to be comprehensive in these areas (subject to item 9 of the principles listed above). However, the line between the mesoscopic and microscopic levels of structure is often too indistinct to enable the microscopic structures to be systematically excluded (a difficulty which is not new, as shown for example by the inclusion in the *NA* and *NAV* of the Vas afferens and Vas efferens of interlobular arteries of the kidney; although it is impossible to see these vessels without a microscope). Furthermore the understanding of structure requires the integration of all aspects of structure. Therefore the *Nomina Anatomica Avium* includes a number of microscopic and submicroscopic structures. These are restricted mainly to viscera and sense organs, and no attempt has been made to cover systematically the more general aspects of histological and cytological nomenclature.

The long list of Termini Generales which precedes the Third Edition of the *NA* and Second Edition of the *NAV* has been omitted from the *NAA* except for relatively short lists that occur at the beginning of certain Chapters such as Osteologia, Arthrologia, and Myologia. A long, preliminary list appears to have negligible value, since most of the terms so listed already occur in special terms in various parts of the *Nomina*. Furthermore, as pointed out by the Honorary Secretary of *IANC* in the Introduction to the Fourth Edition of the *NA*, there is a danger that a long preliminary list of Termini

Generales may acquire the status of an official source of preferred Latin beyond which one may not go. The *NAA* has followed the conventional rejection of dipthongs and eponyms, but hyphens have been employed, particularly in Arthrologia and Systema nervosum centrale (see **Arthr**. Intro).

ORGANIZATION OF ICAAN

General. The International Committee for Avian Anatomical Nomenclature (ICAAN) is formed as follows: General Chairman, Vice-Chairman, Secretary, Chairman of Sub-Committees, and a Consultant for Classical Languages. The Committee is organized into 15 Sub-Committees related to the body systems, each Sub-Committee having its own Chairman. The Executive Committee consists of the Chairman and Vice-Chairman of ICAAN, Sub-Committee Chairmen, and the Editorial Committee.

At the time of publication of the *NAA* these offices were held as follows: General Chairman, J. J. Baumel; Vice-Chairman, M. Yasuda; Secretary, J. McLelland; Consultant for Classical Languages, L. Malinovský. The Chairman of the Editorial Committee is J. J. Baumel; its members are J. E. Breazile, H. E. Evans, A. S. King, A. M. Lucas and R. L. Zusi.

The Sub-Committee Chairmen are as follows:

J. J. Baumel (Osteologia, Arthrologia, Systema cardiovasculare)
J. E. Breazile (Systema nervosum centrale)
J. E. Breazile and M. Yasuda (Systema nervosum peripheriale)
H. E. Evans (Organa sensoria)
A. B. Gilbert (Glandulae endocrinae)
A. S. King (Systema respiratorium, Systema urogenitale)
A. M. Lucas (Integumentum commune, Anatomia topographica externa)
A. M. Lucas and V. Komárek (Termini situm et directionem)
J. McLelland (Systema digestorium)
L. N. Payne (Systema lymphaticum)
J. C. Vanden Berge (Myologia).

Functions of General Chairman and Vice-Chairman. The General Chairman and Vice-Chairman work together to manage the general policy of ICAAN, to select Sub-Committee Chairman and Members of Sub-Committees, to organize General Meetings of ICAAN, meetings of the Executive Committee, and of the Editorial Committee, to arrange for the production of lists of terms,and to settle controversial terms (see below).

Although the General Chairman is responsible for the approval of additional Sub-Committee Members, any member of the Committee may propose new Sub-Committee Members. Such proposal is sent to the General Chairman. The General Chairman consults the Sub-Committee Chairman in the appropriate field. If the Sub-Committee Chairman approves of the proposal, the General Chairman obtains from the proposed new Sub-Committee Member a curriculum vitae and a list of his avian publications. If the General Chairman now approves of the proposal, he notifies the new member of his decision, and sends him the relevant documents.

Functions of Sub-Committee Chairmen and Sub-Committee Members. The Sub-Committees are responsible for the various anatomical systems. It is the duty of the Chairman of each Sub-Committee to prepare the list of terms for that system, with the help of his Sub-Committee Members.

Functions of the Executive Committee. The Executive Committee formulates and implements ICAAN policy (see above for composition of the Executive Committee).

Functions of the Editorial Committee. The Editorial Committee is responsible for the publication of the *Nomina Anatomica Avium*.

Controversial terms. In general, controversial terms are settled through discussion at General Meetings of ICAAN. However, in difficult cases the decision rests with those best qualified, i.e. the Sub-Committee Chairmen, after appropriate discussion at General Meetings. If all else fails, the decision is made by the General Chairman in consultation with the Vice-Chairman.

INFORMATION ON THE USE OF *NOMINA ANATOMICA AVIUM*

Terms of Orientation

Cranial and *caudal* are used throughout the *Nomina Anatomica Avium*, except that *rostral* replaces *cranial* within the head from the level of the occipital condyles. *Superior* and *inferior* are entirely avoided, but the prefixes *supra-* and *infra-* are retained for convenience and because of long usage, e.g. Nervus infraorbitalis. *Anterior* and *posterior* are used only in the eye (see **Org. sens.**), but again the prefix *post-* has occasionally been used in terms which are well established, such as Ligamentum postorbitale. The prefix *pro-* has generally been preferred to *pre-*, although both have been used. That some avian anatomists strongly prefer anterior and posterior as general alternatives to cranial and caudal is acknowledged as these are time-honored terms in the literature of vertebrate morphology. Worthy though they are, these terms are capable of confusion through their entirely different meaning in human anatomy, where they refer to ventral and dorsal. Anyone not convinced of this potential confusion should examine the anatomical writings of the eminent nineteenth century prosectors of the Zoological Society of London, W. A. Forbes, A. H. Garrod, and F. E. Beddard. Although writing as zoologists in a zoological journal, these authors commonly used anterior and posterior in the human sense. Sometimes the usage of human anatomy and the zoological usage were both employed in the same paper (e.g. Beddard, 1886). Even so, it is not realistic to expect the instant and universal adoption of cranial and caudal; nevertheless, it is hoped that in the long run this will be achieved.

Anatomical position of the bird. All terms of location and direction are applied to the bird specimen considered in the arbitrarily defined, standard reference position. In this so-called anatomical position, the bird is oriented with the wings completely abducted and extended as in flight. Thus *dorsal* and *ventral* apply to the upper and lower surfaces of the wing, and *cranial* and *caudal* apply to the leading and trailing edges of the wing. In the anatomical position the long axis of the ellipsoidal Facies articularis of the Caput humeri is nearly vertical as pointed out by Fürbringer (1888). The pelvic limbs are oriented in the standing position with the knees and ankle joints slightly flexed. Following the *Nomina Anatomica Veterinaria*, terms of direction and position are used in this manner: *cranial* and *caudal* apply to that part of the limb proximal to the intertarsal (tibiotarso-tarsometatarsal or ankle) joint, and *dorsal* and *plantar* apply to the part of the limb distal to the intertarsal joint, i.e. to the foot.

Taxonomic Nomenclature

Most of the anatomical structures that are listed are common to birds generally, but important structures or variations of features peculiar to individual species or larger taxa are included if the necessary anatomical facts are known. The scientific taxonomic nomenclature used in the *Nomina Anatomica Avium* is that of Morony, Bock and Farrand (1975); common names in English have been selected from widely used field guides and faunal references; the common names have been standardized throughout the volume (see Taxonomic List).

In certain instances both the common name and scientific names of birds are given; in other instances the common name alone, or the scientific name alone are used. The binomial specific names are employed in the annotations when reporting previously unpublished, original observations of the Sub-Committee members. When published works of authors are cited, often only the generic name is given; the full binomial name of the species (if known) may be obtained from the cited references. Scientific names of orders and families are often converted into their informal equivalents (e.g. gruids for Gruidae, or passeriforms for Passeriformes).

The common names of the ordinary laboratory and domestic forms always refer to the species of genera indicated below. It is recognized that the derivation of these birds has not, in fact, always been conclusively established, but for convenience, the relationships are assumed to be as stated:

Duck: forms of *Anas platyrhynchos.*
Goose: forms of *Anser anser.*

Pigeon:	forms of *Columba livia*
Turkey:	forms of *Meleagris gallopavo.*
Chicken:	forms of *Gallus gallus.*
Quail:	forms of the genus *Coturnix.*
Canary:	forms of *Serinus canarius.*
Guinea fowl:	forms of *Numida meleagris.*
Budgerigar:	forms of *Melopsittacus undulatus.*

Format

Annotations. Annotations are indicated by a superscript number placed after the term to which it refers, and are compiled at the end of each chapter. Annotations are employed to indicate essential anatomical facts, important synonyms, homologies, variation in structure between different groups of birds, reasons for choice of terms, etc.

Illustrations. Illustrations are included where confusion in terminology might otherwise occur.

Bracketed terms. Names in square brackets in the lists and the annotations are official alternative terms, secondary to the main term which is preferred.

Italicized terms. Italics are used for ontogenetic terms in the lists.

References. Literature references show the authority for the anatomical facts or concepts which have formed the basis for the selection of a term, especially if the facts, concepts, or choice are controversial. The references may also indicate the existence of important synonyms in the literature, and call attention to major reviews or monographs.

Index. The index includes not only the recommended anatomical names, but their important synonyms as well.

Abbreviations

For each of the General Abbreviations the first abbreviation is singular and the second, in parentheses, is plural. Thus "A". means the singular Arteria, and "Aa." means the plural Arteriae. Then follow the nominative and genitive singular forms of the term in full, e.g. Arteria, arteriae, followed by the nominative and genitive plural cases, e.g. arteriae, arteriarum. The abbreviations in the Key to Main Headings in bold type represent the titles of the major subdivisions of the terminology, e.g. **Myol.** represents Myologia.

General Abbreviations

A. (Aa.)	= Arteria, arteriae; arteriae, arteriarum
Annot.	= Annotation(s)
Ant.	= Anterior
Apt. (Aptt.)	= Apterium, apterii; apteria, apteriorum
Artc. (Artcc.)	= Articulatio, articulationis, articulationes, articulationum

Caud.	= Caudalis
Cran.	= Cranialis
G.(Gg.)	= Ganglion, ganglii; ganglia, gangliorum
Gl. (Gll.)	= Glandula, glandulae; glandulae, glandularum
Lat.	= Lateralis
Lig. (Ligg.)	= Ligamentum, ligamenti; ligamenta, ligamentorum
M. (Mm.)	= Musculus, musculi; musculi, musculorum
Maj.	= Major
Med.	= Medialis
Min.	= Minor
N. (Nn.)	= Nervus, nervi; nervi, nervorum
Nuc.	= Nucleus, nuclei
Pers. comm.	= Personal communication
Pers. obs.	= Personal observation
Post.	= Posterior
Pt. (Ptt.)	= Pteryla, pterylae; pterylae, pterylarum
Proc. (Procc.)	= Processus, processus; processus, processum
R. (Rr.)	= Ramus, rami; rami, ramorum
Rdx. (Rdxx.)	= Radix, radicis; radices, radicum
Rec. (Recc.)	= Rectrix, rectricis; retrices, rectricum
Reg. (Regg.)	= Regio, regionis; regiones, regionum
Rmx. (Rmxx.)	= Remex, remigis; remiges, remigium
Sp. (Spp.)	= Species of bird
Sut. (Sutt.)	= Sutura, suturae; suturae, suturarum
Synd. (Syndd.)	= Syndesmosis, syndesmosis; syndesmoses, syndesmosium
Synos. (Synoss.)	= Synostosis, synostosis; synostoses, synostosium
Tec. (Tecc.)	= Tectrix, tectricis; tectrices, tectricum
Tr. (Trr.)	= Tractus, tractus; tractus, tractuum
V. (Vv.)	= Vena, venae; venae, venarum
Vas. l. (Vasa l.)	= Vas lymphaticum, vasis lymphatici; vasa lymphatica, vasorum lymphaticorum

Key to Main Headings

Art.	= Arteriae
Arthr.	= Arthrologia
Cardvas.	= Systema cardiovasculare
Cloaca	= Cloaca
CNS	= Systema nervosum centrale
Cor	= Cor
Diges.	= Systema digestorium
Endoc.	= Glandulae endocrinae
Genit. fem.	= Organa genitalia feminina

Genit. masc. = Organa genitalia masculina
Integ. = Integumentum commune
Gen. Intro. = General Introduction
Lym. = Systema lymphaticum et Splen
Myol. = Myologia
Org. sens. = Organa sensoria
Osteo. = Osteologia
Pericar. = Pericardium, Pleura, et Peritoneum
PNS = Systema nervosum peripheriale
Resp. = Systema respiratorium
Term. sit. = Termini situm et directionem—Termini generales
Topog. anat. = Anatomia topographica externa
Urin. = Organa urinaria
Urogen. = Systema urogenitale
Ven. = Venae

Recommended Form of Citations for Use by Authors

1) Book citation in body of text:

   Baumel *et al.* (1979)

2) Chapter citation in body of text:

   Breazile, J. E. (1979)

3) Book citation in bibliography:

   Baumel, J. J., King, A. S., Lucas, A. M., Breazile, J. E. and Evans, H. E. (eds). (1979). "Nomina Anatomica Avium". Academic Press, London.

4) Chapter citation in bibliography:

   Evans, H. E. (1979). Organa sensoria. *In* "Nomina Anatomica Avium" (J. J. Baumel, A. S. King, A. M. Lucas, J. E. Breazile and H. E. Evans, eds). Academic Press, London.

LIST OF ICAAN MEMBERS

Dr R. N. C. Aitken,
Department of Histology and Embryology,
University of Glasgow Veterinary School,
Glasgow, G61 1QH, Scotland

Dr A. R. Akester,
Department of Anatomy,
University of Cambridge,
Cambridge, England

Dr Peter L. Ames,
150 South Wacker Drive,
Chicago, Illinois 60606, USA

Dr Peter Ballmann,
5 Köln 60,
Am Botanischen Garten 68,
West Germany

Dr Julian J. Baumel,
Department of Anatomy,
School of Medicine,
Creighton University,
Omaha, Nebraska 68178, USA

Dr Terence Bennett,
Department of Physiology,
The Medical School,
University of Nottingham NG7 2RD,
England

Dr Betsy G. Bang,
School of Hygiene and Public Health,
Johns Hopkins University,
Baltimore, Maryland 21205, USA

Dr Andrew J. Berger,
Department of Zoology,
University of Hawaii,
Honolulu, Hawaii 96822, USA

Dr J. L. Bhaduri,
Department of Zoology,
University of Calcutta,
Calcutta, India

Dr Biswama Biswas
Zoological Survey of India,
Indian Museum,
Calcutta 700013, India

Prof. Dr P.-C. Blin,
Service d'Anatomie,
École Nationale Veterinaire d'Alfort,
94-Maisons-Alfort, France

Dr Walter J. Bock,
Department of Biological Sciences,
Columbia University,
New York, New York 10027, USA

Dr Robert L. Boord,
Department of Biological Sciences,
University of Delaware,
Newark, Delaware 19711, USA

Dr James E. Breazile,
Department of Physiological Sciences,
Oklahoma State University,
Stillwater, Oklahoma 74074, USA

Dr Pierce Brodkorb,
Department of Zoology,
University of Florida,
Gainesville, Florida 32603, USA

Dr Paul Bühler,
Zoologisches Institut der Universität Stuttgart-Hohenheim,
7 Stuttgart 70, West Germany

Dr P. J. K. Burton,
British Museum (Natural History),
Sub-Department of Ornithology,
Tring,
Herts HP23 6AP, England

Dr Robert B. Chiasson,
Department of Biological Sciences,
University of Arizona,
Tucson, Arizona 85721, USA

Dr Joel Cracraft,
Department of Anatomy,
College of Medicine,
University of Illinois at the Medical Center,
Chicago, Illinois 60680, USA

Dr Jacob L. Dubbeldam,
Zoölogisch Laboratorium,
Rijksuniversiteit te Leiden,
Kaiserstraat 63,
Leiden, Nederland

Dr L. N. Das,
Orissa Veterinary College,
Department of Anatomy,
Bhubaneswar-3,
Orissa, India

Prof. Dr H.-R. Duncker,
Zentrum für Anatomie und Cytobiologie,
Justus-Liebig Universität Giessen,
Aulweg 123,
D 6300 Giessen, West Germany

Dr Howard E. Evans,
Department of Anatomy,
New York State Veterinary College,
Cornell University,
Ithaca, New York 14850, USA

Dr Donald S. Farner
Department of Zoology,
University of Washington,
Seattle, Washington 98105, USA

Dr M. Roger Fedde,
Department of Anatomy and Physiology
College of Veterinary Medicine
Kansas State University
Manhattan, Kansas 66502, USA

Dr Gy. Fehér,
Institut für Anatomie und Histologie der Veterinärmedizinischen Universität,
Budapest, Hungary

Prof. Brian K. Follett,
Department of Zoology,
University of North Wales,
Bangor, Caernarvonshire,
North Wales

Dr S. L. Freedman,
Department of Anatomy,
University of Vermont,
Burlington, Vermont 05401, USA

Dr B. M. Freeman,
Houghton Poultry Research Station,
Houghton, Huntingdon PE17 2DA,
England

Dr Toshitake Fujioka,
Laboratory of Veterinary Anatomy,
Nagoya University,
Chikusa-Ku Nagoya,
Japan

Dr A. B. Gilbert,
Reproductive Physiology Section,
Agricultural Research Council,
Poultry Research Centre,
Edinburgh, EH9 3JS, Scotland

Dr Kenneth J. Hill,
Unilever Research Laboratory,
Sharnbrook, Bedfordshire,
England

Dr R. D. Hodges,
Wye College,
University of London,
Near Ashford,
Kent TN25 5AH, England

Dr Heinrich Hoerschelmann,
Zoologisches Institut u.Museum,
Universität Hamburg, West Germany

Dr William Hodos,
Department of Psychology,
University of Maryland,
College Park Maryland 20742, USA

Prof. V. Ilyichev,
Academy of Sciences,
Institute of Evolutionary Animal Morphology and Ecology,
117312, Moscow W-312,
USSR

Prof. O. W. Johnson,
Biology Department,
Moorhead State College,
Moorhead, Minnesota 56560, USA

Dr Malcolm T. Jollie,
Department of Biological Sciences,
Northern Illinois University,
DeKalb, Illinois 60115, USA

Dr J. Kaman,
Department of Pathological Anatomy and Histology,
University of Veterinary Medicine,
612 42 Brno, Czechoslovakia

Dr Harvey J. Karten,
Department of Psychology,
Massachusetts Institute of Technology,
Cambridge, Massachusetts 02139, USA

Prof. Anthony S. King,
Department of Veterinary Anatomy,
University of Liverpool,
P.O. Box 147, Liverpool L69 3BX,
England

Dr R. L. Kitchell,
Department of Anatomy,
School of Veterinary Medicine,
University of California,
Davis, California 95616, USA

Dr A. Kjaerheim,
Department of Pathological Anatomy,
Faculty of Medicine and Dentistry,
The University, Oslo 3, Norway

Prof. MV Dr Vladimír Komárek,
Sídlištní ul. 204,
16500 Praha 6- Lysolaje,
Czechoslovakia

Dr E. N. Kourotchkin,
Palaeontological Museum Academy of Sciences,
Leninsky Prospect 16,
Moscow V-71, USSR

Dr P. E. Lake,
Reproduction Section,
Agricultural Research Council,
Poultry Research Centre,
Edinburgh EH9 3JS, Scotland

Flora E. F. Lindsay,
Department of Veterinary Anatomy,
University of Glasgow,
Glasgow G61 1QH, Scotland

Dr Alfred M. Lucas,
Department of Poultry Science,
Michigan State University,
East Lansing, Michigan 48824, USA

Dr John McLelland,
Department of Anatomy,
Royal (Dick) School of Veterinary Studies,
Edinburgh EH9 1QH,
Scotland

Prof. Dr med. Lubomir Malinovský,
Department of Anatomy,
Medical Faculty,
University J. E. Purkyněv Brně,
662 43 Brno,
Czechoslovakia

Prof. Dr G. Michel,
Karl-Marx-Universität Leipzig,
Sektion Tierproduktion und Veterinärmedizin,
Fachgruppe Anatomie, Histologie u. Embryologie,
710 Leipzig,
German Democratic Republic.

Dr. Takayoshi Miyaki,
Department of Anatomy,
Faculty of Medicine,
Kanazawa University,
Kanazawa, 920 Japan

Professor G. V. Morejohn,
Department of Biological Sciences,
School of Sciences,
California State University,
San Jose, California 95192, USA

Dr Timothv J. Neary,
Department of Anatomy,
School of Medicine,
Creighton University,
Omaha, Nebraska 68178, USA

Dr Takao Nishida,
Laboratory of Veterinary Anatomy,
Faculty of Agriculture,
University of Tokyo,
Tokyo 113, Japan

Prof. A. Oksche,
Zentrum für Anatomie und Cytobiologie,
Justus-Liebig Universität Giessen,
Aulweg 123,
D 6300 Giessen, West Germany

Dr Y. K. Paik,
Laboratory of Veterinary Anatomy,
Agricultural College,
Chon-puk University,
Chonju, Korea

Dr E. Pastea,
Department of Anatomy and Anat. Research,
Faculty of Veterinary Medicine,
Bucuresti, Roumania

Dr L. N. Payne,
Houghton Poultry Research Station,
Houghton, Huntingdon PE17 2DA,
England

Dr Ronald Pearson,
Department of Zoology,
University of Liverpool,
Liverpool, L69 3BX,
England

Dr T. P. Pessacq,
Servicio de Patologia,
Hospital Sur Roque de Gonnet,
Casilla de Correo 350,
La Plata, Argentina

Dr Robert J. Raikow,
Department of Life Sciences,
University of Pittsburgh,
Pittsburgh, Pennsylvania 15260, USA

Dr U. M. Rawal,
Department of Zoology,
School of Sciences,
Gujarat University,
Ahmedabad-9,
India

Prof. J. Sandoval,
Catedràtico de Anatomía y Embriologia,
Facultad de Veterinaria,
Cordoba, Spain

Prof. P. Sengel,
Laboratoire de Zoologie,
Universite Scientifique et Medicale de Grenoble,
38041 Grenoble Cedex, France

Dr Mary Jane Showers,
Hospital of Philadelphia College of Osteopathic Medicine,
Spruce Street at 48th,
Philadelphia, PA 19139, USA

Prof. Dr Vladeta Simić,
Institut d'Anatomie de la Faculté de Médecine Vetérinaire de l'Universitè de Belgrade,
Belgrade, Jugoslavia

Dr R. I. C. Spearman,
Dermatology Department,
University College Hospital Medical School,
London WC1, England

Dr Peter Stettenheim,
Meriden Road, Lebanon,
New Hampshire 03766, USA

Dr M. D. Tingari,
Department of Veterinary Anatomy,
Faculty of Veterinary Science,
Khartoum North,
Republic of the Sudan

Dr A. Tixier-Vidal,
Laboratoire de Biologie Moléculaire,
11 Place Marcelin Berthelot,
Paris V, France

Dr Elena Traciuc,
Institut de Biologie,
Academie de la R.S.R.,
Bucuresti, Roumania

Prof. R. Tucker,
Division of Veterinary Science,
University of Dar Es Salaam,
Morogoro, Tanzania

Prof. Walter S. Tyler,
National Center for Primate Biology,
Davis, California 95616, USA

Dr James C. Vanden Berge,
Northwest Regional Center for Medical Education,
Indiana University,
Gary, Indiana 46408, USA

Dr H. Völker,
Institut für Veterinärwesen,
University of Rostock, Rostock,
German Democratic Republic

Dr A. Waluszewska-Bubień
Akademia Rolnicza,
Wydzial Weterynaryjny,
Katedra Anatomii Zwierzat,
Wroclaw 12, ul. Kozuchowska 1/3,
Poland

Dr R. W. Warner,
76 Biggin Hill, Upper Norwood,
London SE19 3HU, England

Dr Tohru Watanabe,
Department of Veterinary Anatomy,
Faculty of Agriculture,
Nagoya University,
Furo-Cho, Chikusa-Ku,
Nagoya 464, Japan

Dr P. A. L. Wight,
Anatomy Section,
Poultry Research Centre,
Edinburgh EH9 3JS, Scotland

Dr K. G. Wingstrand,
Institut for Sammenlingnende Anatomi,
University of Copenhagen,
Denmark

Dr Mikio Yasuda,
Nagoya University Library,
Faculty of Agriculture,
Nagoya University,
Chikusa-Ku Nagoya,
Japan

Dr V. Ziswiler,
Zoologisches Museum der Universität.
8006 Zürich, Switzerland

Dr Richard L. Zusi,
Division of Birds,
Smithsonian Institution,
United States National Museum,
Washington, D.C. 20560, USA

ACKNOWLEDGEMENTS

Support of the ICAAN project. Most of the very substantial costs of preparing the many preliminary lists were borne from the beginning by the University or Institute Departments of the Sub-Committee Chairmen. Some members of the Committee contributed funds from personal resources. However, the later stages of completing the first edition were greatly eased by a most generous grant from the National Library of Medicine, USA, and it is doubtful whether the *NAA* would have been published without this financial assistance.*

For the past eight years many persons have contributed to the ICAAN project by assisting the members of ICAAN. These include secretaries, artists, photographers, laboratory technicians among others. Although most of them will remain anonymous, appreciation is nonetheless extended to all of them for their efforts on behalf of the ICAAN project.

Several of these individuals, however, deserve special commendation and expression of gratitude for their extraordinary contributions to the project. In addition to their regular secretarial duties, both Mrs Helen Chapell of the University of Liverpool and Mrs Edith Witt of Creighton University have performed over the years an enormous amount of painstaking typing and clerical work in the preparation of the several successive drafts of the terminology and the textual material of the manuscript of *Nomina Anatomica Avium*. A substantial portion of the art work was produced by Mr W. P. Hamilton IV, Cornell University, who illustrated the chapter on Osteologia, and Mr M. Lynn, University of Liverpool, who illustrated the chapter on Systema Nervosum Centrale. Fr Edward Sharp, S.J. and his staff of the Computer Center of Creighton University have been most helpful in the preparation of the index to this volume.

The skillful assistance of the Publisher, Academic Press, London is gratefully acknowledged, and we are especially indebted to R. Farrand of Academic Press for his enthusiastic support for the publication of this work.

One could not wish for a more capable, cooperative, and patient Sub-Editor than Ms Sheila M. Smith of Academic Press. I extend to her my gratitude and offer apologies for the delays and the multitude of corrections to the manuscript.

J.J.B.

* This publication of *Nomina Anatomica Avium* was supported in part by the U.S. National Institutes of Health (NIH Grant LM 02420 from the National Library of Medicine).

CONTENTS

TERMINI SITUM ET DIRECTIONEM TERMINI GENERALES

ALFRED M. LUCAS

U.S.D.A. Avian Anatomy Project,
Department of Poultry Science,
Michigan State University,
East Lansing,
Michigan 48824, USA

VLADIMÍR KOMÁREK

Sídlištní ul. 204,
16500 Praha 6-Lysolaje,
Czechoslovakia

With contributions from Sub-Committee Member: V. Simić, *Universitè de Belgrade*

Anatomical position of the bird (Fig. 1). In applying the various adjectives that indicate relative positions of body parts, their locations or relationships, or in referring to planes, axes or surfaces, the body of the bird should be considered in the standard anatomical position. The anatomical position is defined as one in which the bird stands erect with knees and ankles slightly flexed, with the wings outstretched laterally, and with the neck completely extended. (In the living bird the neck typically forms an S-shaped curve.)

The upper surface of the wing is designated as dorsal regardless of its position, even when folded against the side of the body; its lower surface is designated as ventral. In addition, the wing has cranial and caudal margins. In the pelvic limb caudal and cranial apply proximal to the tibiotarso-metatarsal joint (ankle) and dorsal and plantar are used in the foot distal to the ankle joint. (See Fig. 2; and **Gen. Intro.** Terms of orientation.)

TERMINI SITUM ET DIRECTIONEM

Aboralis
Afferens
Ascendens
Caudalis [Posterior][1] (Fig. 2)
Centralis
Centrifugalis
Centripitalis
Coronalis

TERMINI SITUM ET DIRECTIONEM—*continued*

Cranialis [Anterior][1] (Fig. 2)
Descendens (see **Cardvas.** Intro.)
Dexter
Distalis[2] (Fig. 2)
Dorsalis (Figs. 1, 2)
Efferens
Externus
Fibularis
Frontalis
Horizontalis
Inferior (see **Gen. Intro.** Terms of orientation)
Intermedius
Internus
Lateralis (Fig. 1)
Longitudinalis
Medialis[3]
Medianus; Medius[3]
Occipitalis
Oralis
Palmaris[4]
Plantaris[5] (Figs. 1, 2)
Periphericus, Peripheriale
Perpendicularis
Profundus
Proximalis[2] (Fig. 2)
Radialis
Rostralis[1] (Fig. 2)
Paramedianus [Sagittalis][6]
Sinister
Superficialis
Superior (see **Gen. Intro.** Terms of orientation)
Tibialis
Transversalis, Transversus
Ulnaris
Ventralis (Figs. 1, 2)
Verticalis

AXES, LINEAE ET PLANA

Axis rostrocaudalis[7]
Axis proximodistalis[8]

Linea mediana dorsalis[9]
Linea mediana ventralis[9]
Planum medianum[10]
Plana paramediana [sagittalia][6, 10]
Plana transversalia[11]
Plana dorsalia[12]

TERMINI GENERALES[13]

ANNOTATIONS

(1) **Caudalis [Posterior]; Cranialis [Anterior]; Rostralis.** Cranialis denotes toward the head; caudalis denotes toward the tail. Rostralis applies to structures situated between the occiput and the tip of the rostrum (beak). Anterior and posterior are used only in the eye; elsewhere cranialis, rostralis, and caudalis are preferred. (See **Gen. Intro.** Terms of orientation.)

(2) **Distalis; Proximalis.** These terms are used to indicate the relative position of anatomical features or parts from a site used for reference. For the limbs the trunk is the reference point; therefore, the elbow is proximal to the wrist, and the hand distal to the forearm. With arteries the heart is the ultimate site of reference. To indicate near and far relationships of peripheral nerves, the central nervous system is used for reference, but in applying the concept to the spinal cord its caudal end would be regarded as distal to parts nearer the brain (Davenport, 1966).

(3) **Medialis; Medianus; Medius.** Medialis means situated closer to the middle of the body or a part; opposite of lateral. Medianus and medius mean situated in the midline or middle of an object. Medius simply means middle (Donath, 1969).

(4) **Palmaris.** Synonymy: Volaris. This term is not generally used in avian anatomical literature because the manus is poorly defined and forms an integral part of a total plane composed of antebrachium, carpus, and manus. Thus it is more appropriate to use dorsal and ventral for the total wing as well as for each of its parts.

(5) **Plantaris.** Refers to the sole of the foot. In most birds the metatarsal part of the foot is elevated in the standing position, but for example in grebes (*Podiceps*), the entire plantar surface of the foot rests on the ground.

(6) **Paramedianus [Sagittalis]; Plana paramediana [sagittalia].** Synonymy: Parasagittalis. These terms refer to positions or planes lateral and parallel to the median plane of the body.

(7) **Axis rostrocaudalis.** The longitudinal axis of the body; a theoretically straight line from tip of rostrum to tip of tail.

(8) **Axis proximodistalis.** The longitudinal axis of a limb; a theoretically straight line from the center of the shoulder to the tip of Digitus major of the hand or from the center of the hip joint to the claw of Digitus III of the foot.

(9) **Lineae mediana dorsalis et ventralis.** Lines projected onto the dorsal and ventral surfaces of the body representing the surface edges of Planum medianum (see Annot. 10).

(10) **Planum medianum.** Synonymy: Midsagittal plane. The middle dorsoventral plane that bisects the body longitudinally into right and left halves.

(11) **Plana transversalia.** Planes that cut across the body from side to side, at right angles to the rostrocaudal axis of the body. Transverse planes are at right angles to proximodistal axes of limbs or other appendages.

(12) **Plana dorsalia.** Synonymy: Plana horizontalia; Plana frontalia. Planes parallel to the back or dorsal surface of the body that cut the body at right angles to both median and transverse planes. The term, Planum dorsale, is most appropriate for quadrupeds; frontal planes apply principally to the human, and horizontal planes depend on an orientation to features outside the body itself. See *Nomina Anatomica Veterinaria* (ICVAN, 1973).

(13) **Termini generales.** The reader is referred to the comprehensive lists of general terms to be found in the human *Nomina Anatomica* (IANC, 1966) and the *Nomina Anatomica Veterinaria* (ICVAN, 1973). Among these terms were word roots used to construct anatomical names for all vertebrates including birds. Several of the sections on avian organ systems in the present work are preceded by brief lists of general terms.

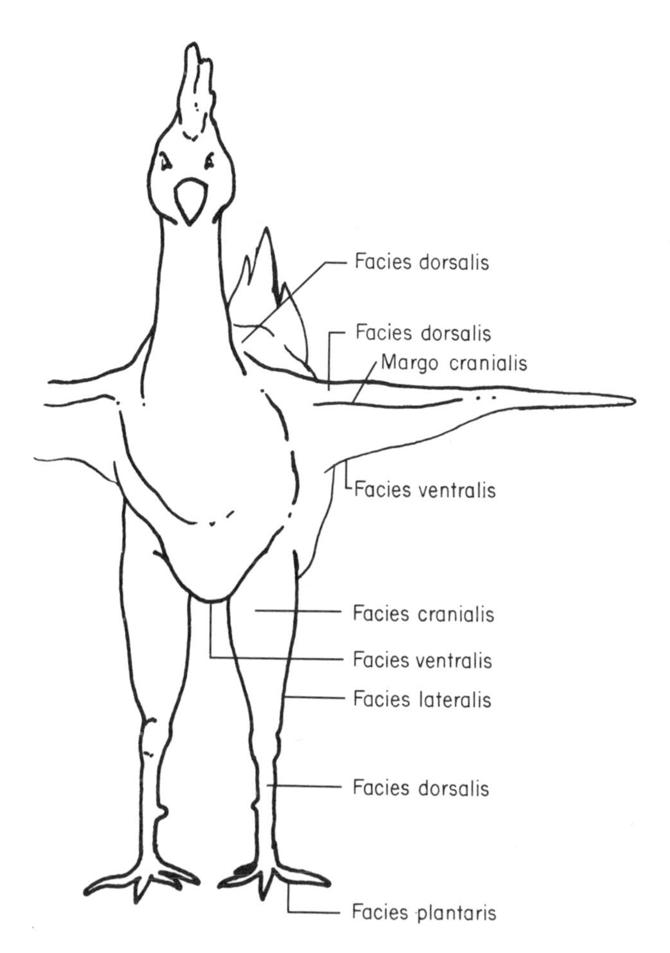

Fig. 1 The standard anatomical position of the bird. The wings are outstretched laterally as in gliding flight. The surfaces of the wing, therefore are dorsal and ventral rather than lateral and medial as in the folded position (see **Gen. Intro.**, Anatomical position of the bird).

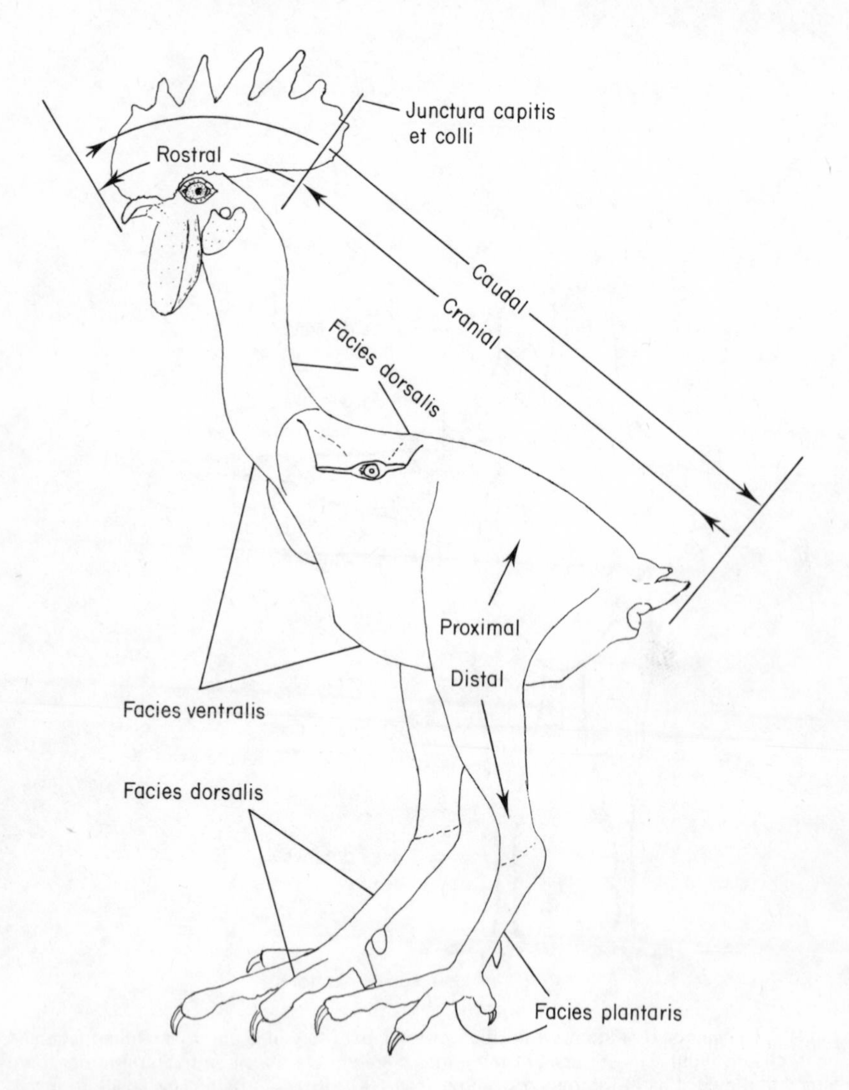

Fig. 2 Terms of position and direction in the bird. Note that rostral refers to the direction toward the tip of the beak, and is used in the head only (see **Gen. Intro.,** Terms of orientation).

ANATOMIA TOPOGRAPHICA EXTERNA

ALFRED M. LUCAS

U.S.D.A. Avian Anatomy Project,
Department of Poultry Project,
Michigan State University,
East Lansing, Michigan 48824, USA

With contributions from Sub-Committee Members: J. J. Baumel, *Creighton University*; A. J. Berger, *University of Hawaii*; W. J. Bock, *Columbia University*; M. T. Jollie, *Northern Illinois University*; Vl. Komárek, *16500 Praha 6- Lysolaje Czechoslovakia*; P. Sengel, *University Scientifique et Medical de Grenoble*; M. Yasuda, *Nagoya University*; R. L. Zusi, *U.S. National Museum*

Regional or topographic anatomy considers all the structures and the relationships of parts in a given region, for example, the axilla (Davenport, 1966). This section on external topographic anatomy provides names for only the superficial features, regions, surfaces, and parts. It represents an amalgamation of the material traditionally treated as two separate sections in other anatomical nomenclatures, i.e. Partes corporis and Regiones corporis. In those nomenclatures the same names are frequently used to identify parts of the body as well as regions of the body, although the meanings of the identical names differ. For example, the knee as a body part includes joint capsule, ligaments, menisci, blood vessels, nerves, etc., but the knee as an external topographical entity (region) is a surface area circumscribed by arbitrary boundaries.

The Regiones and Partes of the entire surface of the body of any given species may be precisely delineated as Komárek (1958) and Lucas and Stettenheim (1972) did for several different birds. Such a detailed topographical map may be useful for a particular species, however variations in the surface landmarks restrict its application to birds generally.

The terminology presented in this chapter is a moderately detailed compilation of names which the Sub-Committee Members believe will fulfill the ordinary needs of most avian biologists. The list of terms is designed to give *approximate* indications of the main landmarks on the surface of the body, providing a flexible vocabulary for the external anatomy of any bird for taxonomic and general descriptive purposes. When the list of terms is not sufficiently detailed, it is assumed that the terms can be readily extended either by adding the standard anatomical adjectives indicating direction and position or by devising regions as required. Thus the user may freely convert the name of a body part into a body region, for example, Genu (knee) is easily transformed into Regio genicularis or Regio genus. Those who do require a more elaborate and refined terminology of body parts and regions should consult the very comprehensive book of Lucas and Stettenheim (1972).

FACIES ET MARGINES[1]

(see **Term. sit.** Figs. 1, 2)

Facies [Margo] dorsalis
Facies [Margo] ventralis
Facies [Margo] lateralis
Facies [Margo] medialis
Facies [Margo] cranialis
Facies [Margo] caudalis
Facies [Margo] rostralis
Facies plantaris

CAPUT

Cranium

Regio frontalis
Cristae[2]
  Crista ossea[2]
  Crista carnosa[3]
  Crista pennarum[4]
  Corona
  Occiput
  Regio auricularis
    Apertura auris externae (see **Org. sens.** Annot. 66)
      Operculum auris[5]
    Lobus auricularis[6] (see **Integ.** Fig. 4)

Facies

Regio oralis
  Apertura oris[7]
  Rostrum[8]
    Rostrum maxillare (see Fig. 2; and **Osteo.** Annot. 53)
    Rostrum mandibulare (Fig. 2)

Rhamphotheca[9]
Culmen[10]
Gonys[11]
Tomium maxillare[12] (Fig. 2)
Tomium mandibulare[12] (Fig. 2)
Lamellae[13] (see **Digest.** Fig. 2)
Unguis maxillaris[14]
Unguis mandibularis[14]
Bucca
Rictus[15]
Pars maxillaris (Fig. 2)
Pars mandibularis (Fig. 2)
Mentum[16]
Regio interramalis[17]
Regio submalaris [gularis][17] (see **Integ.** Annot. 48)
Saccus gularis[17]
Palea[18] (Fig. 2)
Palear[19]

Regio nasalis
Naris (Fig. 2) (see **Resp.** Annot. 1)
Operculum nasale[20] (Fig. 2)
Cera[21]
Angulus craniofacialis [frontonasalis][22] (see **Arthr.** Annot. 46)
Lorum[23]

Regio orbitalis
Oculus
Palpebra dorsalis[24] (see **Org. sens.** Annot. 37)
Palpebra ventralis[24] (see **Org. sens.** Annot. 37)
Membrana nictitans [Palpebra tertia][25]

COLLUM[26]

Pars cranialis
Nucha
Palear[19]
Pars intermedia
Pars caudalis
Barba cervicalis[27]
Regio ingluvialis[28]
Patagium cervicale[29]
Diverticula subcutanea[30]

TRUNCUS[31]

(see **Osteo.** Cavitas thoracica)

Thorax[31]
Abdomen[31]
Pelvis[31]
Dorsum trunci
    Columna vertebralis
    Regio scapularis
    Omus[32]
    Regio interscapularis
    Regio synsacralis (see **Osteo.** Annot. 141)
    Pyga[33]
    Coxa (see **Art.** Annot. 70)
    Regio ilii preacetabularis
    Regio ilii postacetabularis
Latus trunci[34]
    Axilla
    Pectus
    Regio costalis
    Plica inguinalis[35]
    Regio abdominalis
Ventrum trunci
    Regio ingluvialis[28]
    Pectus
    Carina
    Regio abdominalis
        Area incubationis (see **Integ.** Annot. 17)
    Inguen[35]
    Ventus[32] (see **Cloaca** Annot. 20, 21)
    Promontorium cloacale (see **Genit. masc.** Annot. 26)
    Orificium venti

CAUDA

(see **Myol.** Fig. 4)

Dorsum caudae
    Eminentia uropygialis (see **Integ.** Fig. 5)
        Papilla uropygialis
Latus caudae
Ventrum caudae

ALA [MEMBRUM THORACICUM]

Patagia
  Propatagium[36]
  Postpatagium[37]
  Metapatagium[38]
  Patagium alulare[39]
Omus[32]
Axilla
Brachium
Cubitus
Antebrachium
Manus
  Carpus
  Metacarpus (see **Osteo.** Intro.)
  Digiti[40] (see **Osteo.** Intro.)
    Digitus alularis
    Digitus major
    Digitus minor
    Phalanges
      Ungues digitorum manus (see **Integ.** Annot. 70)

MEMBRUM PELVICUM

Coxa
Femur
Genu
  Regio patellaris
  Regio poplitea
Crus
  Sura (see **Art.** Annot. 74)
Pes[41]
  Regio tarsalis[42]
  Tarsometatarsus[43]
    Calcar metatarsale[44] (Fig. 1)
    Regio ossis metatarsalis I [hallucis][45] (Fig. 1)
    Pulvinus metatarsalis[46] (see Fig. 1; and **Integ.** Corpora adiposa)
    Plica metatarsalis[46] (Fig. 1)
  Digitus pedis I [Hallux] (Fig. 1)
  Digiti pedis II–IV (Fig. 1)
    Phalanges
      Pulvinus digitalis (Fig. 1)
      Area interpulvinaris (Fig. 1)

MEMBRUM PELVICUM—*continued*

Unguis digiti pedis (see Fig. 1; and **Integ.** for parts of Unguis)
Tela interdigitalis medialis[47]
Tela interdigitalis intermedia[47] (Fig. 1)
Tela interdigitalis lateralis[47] (Fig. 1)

ANNOTATIONS

(1) **Facies et Margines.** The terms that are listed apply to the surfaces and margins of the head, neck, trunk, limbs, and other appendages. This basic list of terms will serve adequately for most descriptive purposes. Several synonymous terms for surfaces and margins are established in conventional anatomical usage. For example, tibial and fibular (radial and ulnar) borders or surfaces of a limb, the popliteal surface of the knee joint, the occipital, frontal, or temporal surfaces of the head. A number of these synonyms are employed for parts of the body other than superficial features; e.g. the costal surface of the lung, the visceral surface of the sternum, the anterior surface of the eyeball, the tendinous surface of the muscular stomach, the parietal surface of the liver, etc.

(2) **Cristae.** Birds possess three sorts of crests: bony, fleshy and plumed, each showing modifications in different birds.

Crista ossea. Special names are often applied to these crests of bone (and corneous material), such as the (1) casques of the cassowary (*Casuarius casuarius*) and of the hornbills (e.g. *Buceros bicornis*); (2) frontal shields of the gallinule (*Porphyrula martinica*) and the Northern Jacana (*Jacana spinosa*); and (3) a slender, stiff cartilaginous spike extending forward in the Horned Screamer (*Anhima cornuta*).

(3) **Crista carnosa.** In the chicken these fleshy crests are referred to as combs, and display much variation in their shape (e.g. single, rose, pea, buttercup, silkie, strawberry, cushion, and V-shaped). Among wild birds the male Andean Condor (*Vultur gryphus*) exhibits a fleshy crest. Although not a crest, the frontal process (snood) of the turkey has been called the nasal comb (Schneider, 1931). See Lucas (1970); Lucas and Stettenheim (1972) for comparisons of histological structure of the comb of chickens and the frontal process and caruncles of head and neck of the turkey.

(4) **Crista pennarum.** Plumed crests occur commonly among wild and domestic birds such as the guineafowl (*Guttera plumifera*), the touraco (*Tauraco macrorhynchus*), cockatoo (e.g. *Cacatua leadbeateri*), the Hoatzin (*Opisthocomus hoazin*), and curassows (members of the genus *Crax*). Particularly striking crests are those of the Crowned Crane (*Balearica pavonina*) and the Crowned Pigeon (*Goura cristata*). Among chickens certain breeds such as the Silky, Polish, Houdans, Crevecoeurs, have feather crests as well as combs (Amer. Poultry Assoc., 1974).

(5) **Operculum auris.** A fleshy flap or fold along the rostral margin of the external acoustic meatus of some owls that lies over the orifice, and can be erected (Thomson, 1964). (See **Org. sens.** Annot. 69.)

(6) **Lobus auricularis.** Fleshy ear lobe of the chicken. See **Integ.** Annot. 19.

(7) **Apertura oris.** Synonymy: Rima oris (Komárek, 1958). The opening of the mouth is commonly called the "gape" in the ornithological literature. See Annot. 12.

(8) **Rostrum.** In avian anatomy this term designates the beak or bill; it includes parts of the skeleton of upper and lower jaws covered by the Rhamphotheca (see Annot. 9; and **Osteo.** Annot. 53.)

(9) **Rhamphotheca.** The horny covering of the beak; that for Rostrum maxillare has been designated the "rhinotheca" and "gnathotheca" for Rostrum mandibulare (Thomson, 1964). The Rhamphotheca in most birds is a coherent sheath that grows continuously. On the other hand, the rhamphotheca of procellariiform birds consists of segmented horny plates and in the puffins (*Alcidae*) the segmented rhamphotheca consists of 7 to 9 deciduous elements which are regularly molted (Coues, 1903, p. 1063).

(10) **Culmen.** The middorsal contour border of the maxillary rhamphotheca; the profile of the Rostrum maxillare. The culmen is usually curved ventrally near the tip of the beak, but may be straight as in the snipe (*Gallinago gallinago*) and in the stilts (*Himantopus*), or may even be curved dorsally as in the adult avocet (*Recurvirostra americana*) (Coues, 1903, p. 109).

(11) **Gonys.** That part of the Rhamphotheca investing the whole line of union of the mandibular rami; corresponds to Symphysis mandibularis (see **Arthr.** Annot. 21; **Osteo.** Annot. 40, 42. The Gonys therefore refers to the midventral profile or contour of the mandibular Rhamphotheca. According to Coues (1903) the Gonys is the opposite of Culmen (see Annot. 10). Coues (p. 109) reviewed the history of the term, and suggested that the proper term is Genys (meaning lower jaw or chin), but Gonys (meaning knee) has become so well established in the literature that it would be difficult to change the term.

(12) **Tomium maxillare; Tomium mandibulare.** The cutting edge of Rhamphotheca on each ramus of the upper and lower jaws. The Tomia partially bound Apertura oris (see Annot. 7, 15).

(13) **Lamellae.** The serrations along the edge of the Rhamphotheca in ducks and other anseriforms, used for straining food. The term also applies to the teethlike serrations of mergansers (e.g. *Mergus serrator*), useful for catching and holding fish.

(14) **Unguis maxillaris; Unguis mandibularis.** Thickenings of the Rhamphotheca at the tips of the maxillary and mandibular rostra often referred to as "nails"; occur commonly in anseriform birds.

(15) **Rictus.** The fleshy area of skin, approximately triangular in shape, at the angle of the mouth; continuous with the caudal parts of the maxillary and mandibular Rhamphothecae. The Rictus of some birds is brilliantly colored and distinctive. The ricti and tomia (Annot. 12) have the functions of lips and teeth of mammals (Coues, 1903, p. 105). The Rictus has maxillary and mandibular parts. See Fig. 2.

(16) **Mentum.** Literally this is the chin; the rostral part of the interramal region (Annot. 17) caudal to the Gonys (Annot. 11) according to Coues (1903) and Thomson (1964). The Mentum is continuous caudally with the submalar [gular] region (Annot. 17).

(17) **Regio interramalis.** Synonymy: Regio intermandibularis. The triangular area between the two mandibular rami.
Regio submalaris [gularis]. The caudal part of the interramal region forming the throat or floor of the pharynx, continuous with the Mentum rostrally and with the ventral neck caudally. See Thomson (1964).
Saccus gularis. A caudally directed diverticulum of the oral cavity. When filled with air, the skin of the neck protrudes as in frigate birds (Fregatidae) and bustards (Otididae) (Coues, 1903, p. 216; Thomson, 1964, p. 693). See **Digest.** Annot. 3.

(18) **Palea.** The wattles of chickens and other birds; paired structures beneath each mandibular ramus (see Fig. 2).

(19) **Palear.** The dewlap; a median, unpaired fleshy fold in the gular region and neck of the turkey, chicken (Bantam Silkie) and in both sexes of the so-called African goose (see Lucas and Stettenheim, 1972, p. 25).

(20) **Operculum nasale.** This plate-like structure forms a partial or complete covering flap over the Naris (external nasal aperture). In procellariiform birds the operculum is modified to form the characteristic nasal tube. See **Resp.** Annot. 2.

(21) **Cera.** The thickened portion of integument that straddles the base of the nasal region; may be feathered or bare. The cera is found among pigeons, parrots, owls, and other birds, and is often brightly colored.

(22) **Angulus craniofacialis [frontonasalis].** Birds possess a skeletal mechanism involving the jugal arches, the quadrate, pterygoid, and palatine bones whereby the upper jaw can be elevated. The transverse line of bending across the region of junction of the nasal and premaxillary bones with the frontal and ethmoid bones establishes the bending zone; the craniofacial angle is formed at the junction of the Culmen with the slope of the frontal region. The Great Horned Owl (*Bubo virginianus*) has an exaggerated angle of slightly more than 90 degrees. See **Arthr.** Annot. 46.

(23) **Lorum.** Area between the eye and the upper jaw (Coues, 1903, p. 103). The English terms are: lore (singular), lores (plural), and loral (adjective). The lore is often identified by its distinctive feathers. Often loral bristles are present (see Stettenheim, 1974). The lore lies dorsocaudal to the Rictus, and is superficial to most of the underlying nasal cavity.

(24) **Palpebra dorsalis; Palpebra ventralis.** In the domestic birds the ventral mobile portion of the retracted upper lid is folded under the fixed dorsal portion (also observed in the owls). In owls the mobile part of the upper eyelid drops to meet the lower mobile lid, and sometimes the lower lid rises to meet the upper (Lucas and Stettenheim, 1972, p. 31). Newton and Gadow (1896, p. 234) noted that in most birds only the lower eyelid is movable. See **Org. sens.** Annot. 37.

(25) **Membrana nictitans** [Palpebra tertia]. The nictitating membrane of non-mammalian vertebrates including birds is a protective translucent membrane that in closing is drawn toward the temporal angle of the eye from its folded position at the nasal angle of the eye. See **Org. sens.** Annot. 40; Simič and Jablan-Pantic (1959).

(26) **Collum.** Synonymy: Cervix. The necks of many birds are long with variable

numbers of vertebrae (see **Osteo.** Annot. 129). The neck is usually carried in an S-shaped curve. Caudally the root of the free neck is located between the shoulders. This latter part has been referred to as its "interscapular" portion. Functionally and morphologically the avian neck consists of three segments. The osteological and myological characteristics of these segments are treated in the works of Boas (1926) and Zusi (1962). The dorsal part of the neck adjacent to the Basis cranii is referred to as the Nucha.

(27) **Barba cervicalis.** The beard of the male turkey.

(28) **Regio ingluvialis.** In those avian species that possess a crop this term refers to the region of the crop at the thoracic inlet ventral to the root or base of the neck.

(29) **Patagium cervicale.** The triangular fold of skin that extends from the lateral aspect of the base of the neck laterally to the cranial margin of the shoulder.

(30) **Diverticula subcutanea.** Inflatable, subcutaneous evaginations of the oral cavity (**Digest.** Annot. 4), esophagus (**Digest.** Annot. 23), or respiratory tract (**Resp.** Annot. 21, 31). Coues (1903, p. 206) mentions the enormous development of subcutaneous air cells in the areolar tissue of pelicans and gannets. These are quite different from the large cavities originating from the upper digestive tract that lie in the loose connective tissue between muscles and skin.

(31) **Truncus; Thorax; Abdomen; Pelvis.** The trunk is all of the body between neck and tail. The thorax is that part of the trunk bounded by the rib cage, sternum, and vertebral column. Boundaries between abdomen and pelvis are indefinite, since the bony pelvis is open ventrally in most birds. In most birds the so-called thoracic viscera (heart, lungs, esophagus) occupy the cavity of the thorax; however the skeletal thorax of birds also encloses and protects some of the abdominal viscera (e.g. liver, proventriculus and ventriculus), as in mammals. See **Osteo.** Cavitas thoracica.

(32) **Omus.** A Greek word meaning shoulder, preferable to the Latin word, "Humerus" which also means shoulder. "Omo-" is used in **Nomina Anatomica** (IANC, 1966) and **Nomina Anatomica Veterinaria** (ICVAN, 1973) only as a combining form; however in neither of these works is the shoulder as a part of the body given a name from the classical languages.

(33) **Pyga.** The "rump" or caudal part of the synsacral region and base of tail.

(34) **Latus trunci.** The side of the trunk; composed of two parts, cranial and caudal. The division line between the two parts follows the cranial border of the preacetabular ilium and the cranial margin of the thigh. Most of the caudal part of the trunk is situated deep to the thigh that covers all except a small area adjacent to the tail region.

(35) **Plica inguinalis; Inguen.** When the thigh of the bird is extended (caudally) the skin forms a fold from the cranial aspect of the thigh to the trunk. Inguen is the term for groin.

(36) **Propatagium.** The triangular fold of skin that forms the leading or cranial edge of the wing from the shoulder to the wrist. (See **Integ.** Fig. 5.)

(37) **Postpatagium.** The fold of skin that extends from the elbow distally along the caudal margin of the antebrachium and hand; aligns the primary and secondary flight feathers which are implanted into the fold. (See **Integ.** Fig. 6.)

(38) **Metapatagium.** The triangular skin fold extending from the side of the trunk to the caudal margin of the brachium. (See **Integ.** Fig. 5.)

(39) **Patagium alulare.** The fold of skin connecting the Digitus alularis to the cranial border of Digitus major.

(40) **Digiti.** At present there is no agreement on homologies of the avian digits of the hand. In order to avoid the problem of arbitrarily numbering the digits either I, II, and III or II, III, and IV, the digits and their metacarpals are designated: alular. major and minor. (See **Osteo**. Intro.; and **Myol.** Intro.)

(41) **Pes.** The scutes of the feet and toes and the several terms applied to their different sizes, shapes, arrangements, locations, etc.; the variation in the number of toes, the relationships of toes to the Tarsometatarsus; the fusions, grasping and walking positions of the toes are summarized briefly by Newton and Gadow (1896, p. 972); Coues (1903, pp. 132–139); Van Tyne and Berger (1976, pp. 556–560).

(42) **Regio tarsalis.** The region of the ankle or hock joint. In birds this is an intertarsal joint rather than a tarsocrural joint as in mammals. (See **Arthr.** Annot. 167.)

(43) **Tarsometatarsus.** The body part corresponding to the fused metatarsal bones, usually covered with nonfeathered Podotheca. In much of the ornithological literature the tarsometatarsal segment of the foot is called "tarsus" (as in booted tarsus) rather than the more correct terms Tarsometatarsus or Metatarsus. (See **Arthr.** Annot. 167, 173; and **Osteo.** Annot. 288.)

(44) **Calcar metatarsale.** Synonymy: Calcar tarsometatarsale. The metatarsal spur found in galliform birds. (See **Integ.** Annot. 69; and **Osteo.** Annot. 296.)

(45) **Regio ossis metatarsalis I [hallucis].** The metatarsal bone of Digitus I supports the two phalanges of this digit; the terminal phalanx forms the core of the claw. Os metatarsale I does not fuse with the other metatarsal bones. (See **Arthr.** Annot. 174–177.)

(46) **Pulvinus metatarsalis.** The pad at the distal end of the Metatarsus and at the bases of digits II–IV. See Fig. 1; and **Arthr.** Annot. 179.
Plica metatarsalis. The narrow transverse fold situated between the first digit and the metatarsal pad.

(47) **Tela interdigitalis.** The web of skin between the toes. Tela interdigitalis medialis occurs only in the pelecaniforms (the so-called totipalmate birds) between the Hallux and Digitus II.

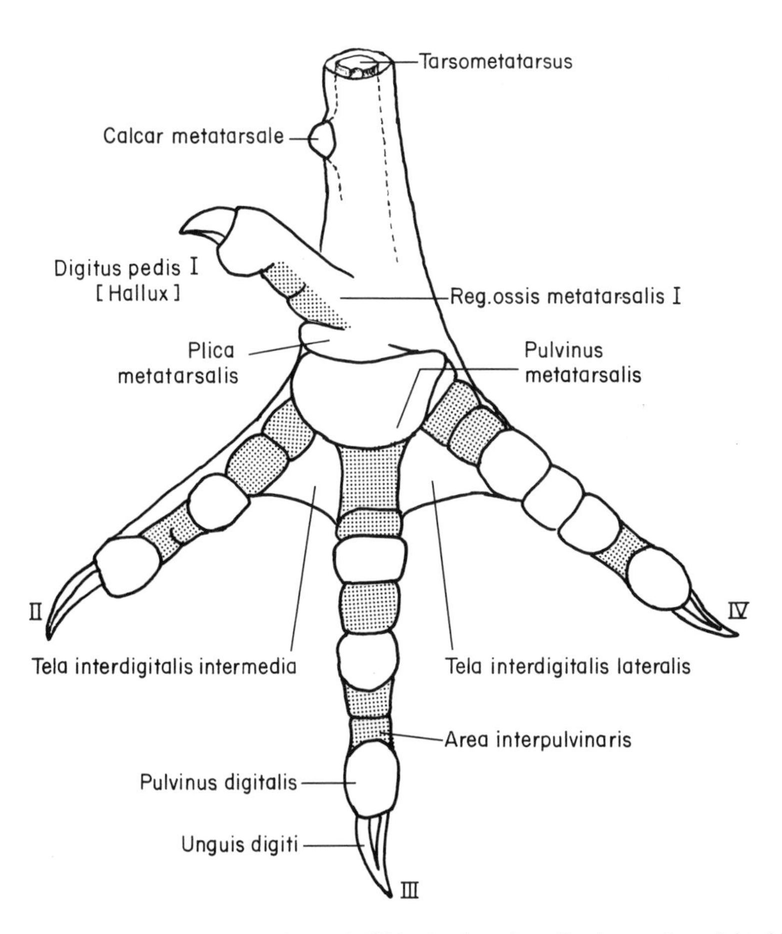

Fig. 1 Foot of the chicken (single comb White Leghorn breed); plantar view, right side.

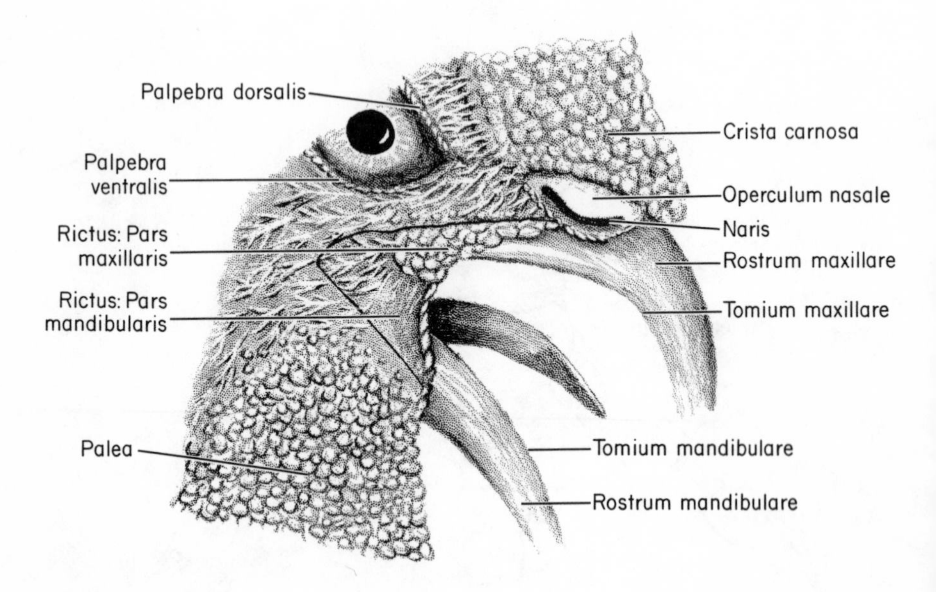

Fig. 2 Rostral part of the head of the chicken (single comb White Leghorn breed); right lateral view.

INTEGUMENTUM COMMUNE

ALFRED M. LUCAS

Department of Poultry Science, U.S.A. Avian Anatomy Project, Michigan State University, East Lansing, Michigan 48824, USA

With contributions from Sub-Committee Members: J. J. Baumel, *Creighton University*; Vl. Komárek, *Sídlištní ul. 204, 16500 Praha 6- Lysolaje, Czechoslovakia*; P. Sengel, *Universite Scientifique et Medicale de Grenoble*; R. I. C. Spearman, *University College Hospital Medical School, London*; P. Stettenheim, *Lebanon, New Hampshire, USA*

INTEGUMENTUM PROPRIUM

Cutis
- Sulcus cutis
- Plica cutis

Epidermis (Fig. 1)
- Stratum corneum[1]
- Stratum germinativum[2]
  - Stratum transitivum[3]
  - Stratum intermedium[4]
  - Stratum basale
- Membrana [Lamina] basalis[5]
- *Periderm*[6]

Dermis [Corium][7] (Figs. 1, 2)
- Stratum superficiale[8]
- Stratum profundum
  - Stratum compactum
  - Stratum laxum
- Lamina elastica[9]
- Mm. nonstriati dermatis
  - Mm. pennales[10]

Dermis [Corium]—*continued*
Mm. apteriales[11]
Tendo elasticus[11]

TERMINATIONES NERVORUM IN CUTE[12]

Tela subcutanea (Fig. 2)
Fascia superficialis
Corpora adiposa [Panniculus adiposus][13]
Corpus adiposum ophthalmicum
Corpus adiposum dorsocervicale
Corpus adiposum laterocervicale
Corpus adiposum tracheale
Corpus adiposum coracoclaviculare
Corpus adiposum spinale
Corpus adiposum synsacrale
Corpus adiposum thoracicum laterale
Corpus adiposum abdominale laterale
Corpus adiposum pectorale
Corpus adiposum abdominale transversum
Corpus adiposum abdominale medioventrale
Corpus adiposum ischiopudendale
Corpus adiposum subalare
Corpus adiposum fermorale craniale
Corpus adiposum femorale caudale
Corpus adiposum plantare profundum[13]
Corpus adiposum plantare superficiale[13]
Lipocytus multivesicularis[14]
Lipocytus univesicularis[14]
Fascia profunda[15]

Bursa sternalis[16]

Area incubationis[17]

APPENDICES INTEGUMENTI

Appendices carnosae
Crista carnosa[18]
Palea[18]
Rictus[18]
Lobus auricularis[19]
Processus frontalis[20]

Carunculae cutaneae[21]
Barba cervicalis[22]
Papilla barbae
Filamentum barbae

Glandulae cutaneae[23]

Gl. uropygialis (Fig. 3)
Lobus gl. uropygialis
Septum interlobare
Capsula gl. uropygialis
Papilla uropygialis
Ductus gl. uropygialis
Porus ductus uropygialis
Gll. auriculares[24]
Gll. venti (see **Cloaca** Annot. 25)

PENNAE

Plumae[25]

Circulus venti[50] (Fig. 6)
Plumae caudae[23]
Circulus uropygialis[23]
Plumae alae (Figs. 7–10)
Plumae antebrachii
Plumae manus
Plumae alulae
Pulviplumae[26]

Semiplumae[25]

Pennae contourae[25]

Remex (see **Arthr.** Ligg. pennarum)[53, 56]
Tectrix
Rectrix (see **Arthr.** Ligg. pennarum)
Penna contoura generalis[25]

PARTES PENNAE[26]

Scapus (Fig. 12)

Calamus (Fig. 12)
Paries
Umbilicus distalis [superior] (Fig. 12)
Umbilicus proximalis [inferior]

Scapus—*continued*

Galerus pulposus[27]

Rachis (Fig. 12)

Cortex (Fig. 11)

Plica corticalis

Medulla (Fig. 11)

Sulcus ventralis

Vexilla (Fig. 12)

Vexillum internum[28]

Vexillum externum[28]

Pars plumacea[29] (Fig. 12)

Pars pennacea[29] (Fig. 12)

Apex

Margo

Incisura vexilli[30]

Zona impendens[31]

Barba [Ramus] (Fig. 11)

Cortex (Fig. 11)

Medulla (Fig. 11)

Incisura barbae[32]

Petiolus[32]

Crista dorsalis[33] (Fig. 11)

Crista ventralis[33] (Fig. 11)

Tegmen[33]

Villi

Ruga proximalis[34] (Fig. 11)

Ruga distalis[34]

Vexilla barbae[35]

Vexillum barbae proximalis

Vexillum barbae distalis

Barbula [Radius][36] (Fig. 11)

Basis (Fig. 11)

Arcus dorsalis[37] (Fig. 11)

Dens ventralis[37] (Fig. 11)

Pennula (Fig. 11)

Nodus

Dens nodosus

Internodus

Annulus

Processus barbulae

Stilus dorsalis

Nodulus dentis

Cilium dorsale (Fig. 11)

Cilium ventrale (Fig. 11)
Hamulus (Fig. 11)
Spiculum
Flexura[38]

Hypopenna[39] (Fig. 12)
Hyporachis
Hypovexillum
Barbae umbilicales

Filopluma

Setae[40]
Cilia [Setae] palpebrarum
Setae nariales
Semisetae

Folliculus[41] (see **Arthr.** Annot. 189, 190)
Collum
Cavitas
Paries

PTERYLAE[42]

Pterylae capitales[43] (Fig. 4)
Pt. frontalis
Pt. coronalis
Pt. occipitalis
Pt. loralis
Pt. superciliaris
Pt. palpebralis
Pt. rictalis[43]
Pt. buccalis [suborbitalis][43]
Pt. temporalis
Pt. auricularis (see **Org. sens.** Annot. 68)
Pt. caudoauricularis
Pt. interramalis[43]

Pterylae spinales[44] (Fig. 5)
Pt. cervicalis dorsalis
Pt. interscapularis
Pt. dorsalis
Pt. pelvica

Pteryla trunci lateralis[45]

Pterylae ventrales[46] (Fig. 6)

Ptt. cervicales ventrales
  Pt. submalaris [gularis][48]
  Pt. cervicalis ventralis[49]
Pt. pectoralis
Pt. sternalis
Pt. abdominalis[50]

Pterylae caudae

Pt. dorsalis caudae (Fig. 5)
  Tectrices majores
  Tectrices intermediae
  Tectrices minores
Rectrices (see **Arthr.** Ligg. pennarum)
Pt. ventralis caudae[51] (Fig. 6)
  Tectrices majores
  Tectrices intermediae
  Tectrices minores

Pterylae alae

Ptt. brachiales
  Pt. humeralis[52] (Fig. 5)
  Pt. subhumeralis (Fig. 6)
  Pt. caudohumeralis (Figs. 5, 6)
Ptt. antebrachiales (Figs. 6, 9)
  Remiges secundarii[53] (see Figs. 7–10; and **Arthr.** Ligg. pennarum)
    Diastema[53]
  Tectrices secundariae dorsales[54] (Fig. 7)
    Tectrices majores
    Tectrices medianae
    Tectrices minores[54]
      Ordo primus
      Ordo secundus
  Tectrices secundariae ventrales (Fig. 9)
    Tectrices majores
    Tectrices medianae
    Tectrices minores
Tectrices propatagii
  Tectrices marginales dorsales propatagii (Fig. 7)
  Tectrices marginales ventrales propatagii (Fig. 9)
  Pt. antebrachialis ventralis (Fig. 9)

Ptt. carpales
Remex carpalis[55] (Figs. 7, 8)
Tectrix carpalis dorsalis (Figs. 7, 8)
Tectrix carpalis ventralis (Figs. 9, 10)
Ptt. manuales
Remiges primarii[56] (see Figs. 7–10; and **Arthr.** Ligg. pennarum)
Tectrices primariae dorsales (Figs. 7, 8)
Tectrices majores
Tectrices medianae
Tectrices minores
Tectrices primariae ventrales (Fig. 10)
Tectrices majores
Tectrices medianae[58]
Tectrices minores
Pt. manualis ventralis (Fig. 10)
Remiges alulares[57] (Figs. 7–10)
Tectrices marginales manuales (Figs. 8, 10)

Pterylae membri pelvici
Pt. femoralis (Figs. 5, 6)
Pt. cruralis[59] (Figs. 5, 6)
Pt. metatarsalis
Ptt. digitales

APTERIA[42]

Apteria capitalia[60]

Apteria spinalia[61]
Apt. cervicale dorsale
Apt. interscapulare
Apt. scapulare (Fig. 5)
Apt. dorsale
Apt. pelvicum medianum
Apt. pelvicum laterale (Fig. 5)

Apteria ventralia[62]
Apt. interramale
Apt. submalare [gulare]
Apt. cervicale ventrale (Fig. 6)
Apt. sternale (Fig. 6)
Apt. pectorale (Fig. 6)
Apt. abdominale medianum

Apteria ventralia—*continued*

Apt. abdominale laterale
Apt. abdominale venti (Fig. 6)

Apteria lateralia

Apt. cervicale laterale (Figs. 5, 6)
Apt. truncale laterale (Figs. 5, 6)

Apteria caudae[63]

Apt. laterale caudae
Apt. dorsale caudae
Apt. ventrale caudae

Apteria alae[64]

Aptt. alaria dorsalia
- Apt. humerale (Fig. 5)
- Apt. cubitale (Fig. 7)
- Apt. alulare (Figs. 7, 8)
- Apt. manuale (Figs. 7, 8)

Aptt. alaria ventralia[64] (Figs. 6, 9)
- Apt. subhumerale (Fig. 6)
- Apt. antebrachiale (Fig. 9)
- Apt. cubitale (Fig. 9)
- Apt. propatagiale (Fig. 9)
- Apt. alulare (Figs. 9, 10)
- Apt. manuale (Figs. 9, 10)

Apteria membri pelvici

Apt. crurale (Fig. 5)
Apt. intracrurale[64]
Apt. metatarsale craniale[65]
Apt. metatarsale caudale[65]

Podotheca[66]

Scuta[67]
Scutella
Pulvinus hypotarsalis[68]
Pulvinus metatarsalis (see **Topog. anat**. Annot. 46)
Pulvinus digitalis
Areae interpulvinares

Rhamphotheca (consult **Topog. anat.** for parts of Rhamphotheca)

Calcaria

Calcar craniale
Calcar carpale et metacarpale
Calcar metatarsale[69]

Ungues

Ungues digitorum manus[70]
Unguis digiti alularis (Figs. 7–10)
Unguis digiti majoris
Unguis digiti pedis
Scutum dorsale
Scutum plantare
Eponychium
Hyponychium

ANNOTATIONS

(1) **Stratum corneum.** The Stratum corneum is the product of the living cells that lie below it. This horny layer constitutes a protective barrier between the environment and the body. Adjacent to the layer of living cells, the laminae of the corneum are strongly bonded whereas when the laminae are pushed toward the free surface there may be a weakening of the adhesive substance with the spreading apart of the laminae, namely dehiscence. This process is followed by desquamation. In the bird there seems to be no distinct Stratum disjunctum [disjunctivum], but rather a general loosening of the outer laminae from one another.

Leblond (1951) and Giroud and Leblond (1951) introduced the terms "soft" and "hard" keratins. Soft keratin of birds is present in the soft skin, in grooves between scales, in Eponychium and Hyponychium, and on the plantar surface of claws. The laminae of soft keratin are desquamated. Hard keratin is present in feathers, claws, Rhamphotheca and some scales, and does not show dehiscence except in the proximal parts of the beaks and the plantar scutes of the claws.

(2) **Stratum germinativum.** After extensive consideration of a suitable term to include all of the living cells of the epidermis, it was finally agreed to use the name adopted by Bohm and von Davidoff (1910). The term "germinative" has as much validity and precedent for designating the property of a tissue to produce a new tissue layer, as it has to designate dividing cells (Spearman, pers. comm., 1968). The terminology for the layers of the avian skin has been extensively reviewed by Lucas and Stettenheim (1972, p. 485; see also Figs. 338, 367, 375 and 381). In recent years many publications have contributed to the components of the epidermis: Cane and Spearman (1967); Leblond *et al.* (1964); Matoltsy (1962, 1969); Matoltsy and Huszar (1972); Spearman (1966, 1969); and recently Sengel (1976).

(3) **Stratum transitivum.** This is a new term that designates vacuolated cells of the epidermis immediately beneath the corneous layer. It differs from the Stratum granulosum of mammals although it has the same position relative to the other layers of the Stratum germinativum.

(4) **Stratum intermedium.** It is preferable to use the term "Stratum intermedium" rather than "Stratum spinosum". Only in a thickened Stratum germinativum can one, with the light microscope, see clearly the so-called "intercellular bridges" that led originally to the use of the term "spiny" cells. In the epidermis of the bird, spiny cells may not be limited to the intermediate layer. The term "Stratum intermedium" is applied in mammalian epidermal histogenesis to denote a "middle" layer that later becomes the spiny layer (Sengel, 1976, p. 11).

(5) **Membrana [Lamina] basalis.** The basement membrane as viewed through the light microscope following Luxol fast blue and periodic acid Schiff reaction (PAS) is a clearly defined line at the junction of basal cell plasma membrane and dermis. The electron microscope reveals that it is composed of at least two layers and a space. Investigators in the area of ultrastructure generally use the term "dermo-epidermal boundary." Selby (1955, p. 436) used the term "dermal membrane" to distinguish it from the typical basement membrane of the classical histologist. The membrane is already present at day two of incubation, and appears attached to the epidermal cells before there is close contact with the mesoderm (Sengel, 1976, p. 31). The term "basement membrane" is still in common usage in student texts of histology and pathology and should be retained for the many who still use the light microscope.

(6) **Periderm.** This superficial layer of the epidermis in the mammalian embryo has the name "epitrichium", but this term is not appropriate for the bird. The Periderm is desquamated during several days preceding hatching (Hamilton, 1962).

(7) **Dermis [Corium].** The morphology of the avian dermis is so different from that of the mammal that new identifying names are presented, most of them from the avian literature. This set of names and the descriptions accompanying them were found to be applicable to skin in various parts of the avian body and to specialized avian integumentary structures. See Lucas and Stettenheim, 1972, Chap. 9.

(8) **Stratum superficiale.** This connective tissue layer varies in thickness in different parts of the body, and is generally less dense than the connective tissues of the Stratum compactum. The Stratum superficiale is especially clearly identifiable when it contains sinus capillaries. These give to the organ a strong red color; e.g. comb, wattles, face skin of older chickens, rictus, and red ear lobes. See Fig. 2.

(9) **Lamina elastica.** This layer of the dermis, when stained to reveal elastic connective tissue, is a reliable guide for the separation of dermis and subcutis. When made visible it clearly demonstrates that the entire feather follicle lies within the dermis, whereas the basal ends of hair follicles as generally described are located in the Tela subcutanea.
Mm. nonstriati dermatis. These smooth muscles are situated in the dermis layer of the skin and are of two types: feather muscles and apterial muscles. See **Myol.** Annot. 4, Mm. subcutanei.

(10) **Mm. pennales.** These nonstriated feather muscles are attached by way of elastic tendons to the connective tissues of the external follicle walls of contour, bristle, semiplume, and down feathers. Filoplumes lack feather muscles. Basically the feather muscles have a parallelogram arrangement, either complex or simple. The muscles may function as erectors, depressors, retractors, and rotators (Langley, 1904, p. 242). The numerous patterns of feather and apterial muscles and the complex

morphology of those attached to remiges, the upper tectrices and the rectrices are reported in detail by Lucas and Stettenheim (1972, Chap. 8).

(11) **Mm. apteriales.** In the chicken, apterial muscles (nonstriated) are most abundant in the caudal part of the scapular apterium, lateral trunk apterium, lateral pelvic apterium, lateral apterium of the tail, and pectoral apterium. The apterial muscles were designated the "Stratum musculo-elasticum" by Lange (1931), but the muscles are part of the loose layer (Stratum laxum), and do not constitute a separate layer of the dermis.
Tendo elasticus. Feather muscles and apterial muscles are joined to each other, end to end, by elastic bundles. The fibers are closely packed. Each fiber is thicker than the elastic fibers found among the collagenous fibers of dermal and subcutaneous layers.

(12) **Terminationes nervorum in cute.** The terminations of nerves may be found in epidermis and dermis. Free and specialized nerve endings are found in many other organs of the avian body, as well as in the integument. See nerve endings listed under **Org. sens.**

(13) **Corpora adiposa [Panniculus adiposus].** To our knowledge the publication of Liebelt and Eastlick (1954) is the only one which gives names to the avian subcutaneous fat bodies. Their English names have been translated into Latin; "sartorial fat body" and "femoral fat body" have been changed into "cranial femoral fat body" and "caudal femoral fat body" respectively.
Corpus adiposum plantare profundum; Corpus adiposum plantare superficiale. The plantar fat bodies in the sole of the foot were observed and named by J. J. Baumel (1974, pers. comm.). See **Arthr.** Annot. 179.

(14) **Lipocytus multivesicularis; Lipocytus univesicularis.** In birds one finds both multivesicular (plurivacuolar) and univesicular (univacuolar) fat cells (Clara, 1929).

(15) **Fascia profunda.** Generally a dense connective tissue adjacent to underlying skeletal muscles (Fig. 2) or adjacent to the periosteum of bone or the perichondrium of cartilage. Within the Tela subcutanea are located striated muscles to which the name Mm. subcutanei is applied (see **Myol.** Annot. 64).

(16) **Bursa sternalis.** The sternal bursa, often regarded in caged chickens and turkeys as a pathological entity when seen in older birds, has its origin as a normal histological development at about 4 weeks of age (Lucas and Stettenheim, 1972, pp. 514–531). The bursa lies ventral to the cranial end of the keel.

It originates as a flattened chamber (Bursa superficialis) traversed by strands of connective tissue, each a Filum intrabursarium. Stomata interbursaria may connect the Bursa superficialis with the Bursa profunda by penetrating the Lamina interbursaria. The superficial bursa contains paired triangular folds of vascular tissue (Plica lateralis triangularis) on each side. With increasing age the superficial and deep cavities may become confluent, giving the appearance of a single cavity. At such later stages there may be evidence of infection. Removal of the cranial tip of the sternum will prevent the development of a sternal bursa (Koonz *et al.*, 1963).

(17) **Area incubationis.** An incubation and brood patch is developed to varying degrees among many kinds of birds resulting in increased vascularity of the ventral

skin. The loss of ventral feathers brings the warmth of this modified skin into direct contact with the eggs. Beer (1964) mentioned that some kinds of birds fail to develop brood patches, e.g. cormorants (Phalacrocoracidae) and gannets (Sulidae). In the latter the incubating function of the brood patch seems to be performed by the highly vascularized webbed feet. Among some passerines sometimes males incubate the eggs, and in these cases may develop brood patches (see Van Tyne and Berger, p. 491). See **Art.** Annot. 44.

(18) **Crista carnosa; Palea; Rictus.** The comb, wattle and rictus have been grouped together because they are basically similar in their histological structure. They are brilliant red in the chicken due to the blood coursing through the sinus capillaries, and on this basis one could include the red face skin of an adult male chicken and the red earlobes of many breeds of chickens, male and female. The comb, earlobe and rictus have in common, a fibromucous layer which is absent in immature males and females but present throughout the life of the mature male. In the mature female it is present during production, but disappears when laying ceases. See the above organs in **Topog. anat.** for terms applicable to their gross morphology.

(19) **Lobus auricularis.** The earlobes of chickens are red or white and also may show a mixture of these two colors. The Silkies have a blue to purple earlobe due to a high concentration of melanin pigment overlying the red color produced by blood in the sinus capillaries. The pheasant (*Phasianus colchicus*) has flat, poorly defined earlobes. These have the red color of the face skin. The redness of the face skin is accentuated by short red feathers implanted almost at right angles to the skin surface.

(20) **Processus frontalis.** This organ of the turkey arises from the dorsal surface of the head between the levels of orbit and naris. When elongating, its diameter becomes less than in the contracted state. It contains no erectile tissue. Details of its internal structure may be found in Lucas (1970) and Lucas and Stettenheim (1972). Small bristle feathers are scattered over the surface of the frontal process as they are over the so-called bare skin of the head. See **Topog. anat.** Annot 3.

(21) **Carunculae cutaneae.** Found on the head and the cranial part of the neck of the turkey. Their histological structure is similar to that of the frontal process but with less capability to expand and contract.

(22) **Barba cervicalis.** This structure occurs in turkeys (Lucas and Stettenheim, p. 571, 1972). The beard develops from a papilla near the base of neck in the ventral midline of the male. Seldom do filaments develop in the female. Bulliard (1926) mentions certain species of condors, cassowaries, herons and storks that have a thickening of skin in the midline in the lower third of the neck. In the Marabou Stork (*Leptoptilos crumeniferus*), filaments may emerge from the dermal papilla. Bulliard identified and described three zones for the pulp along the length of a filament in the turkey: basal, median, and terminal, as well as a proliferation of the epidermis filling all of the spaces between filaments.

(23) **Glandulae cutaneae.** It has been suggested that the cells of the epidermis in the aggregate function as an extensive cutaneous gland that releases lipoid sebaceous secretion toward the corneum (Lucas, 1968; Lucas and Stettenheim, 1972, p. 627). Shah and Menon (1972) and Shah *et al.* (1977) confirmed the existence of lipoid spheres in the epidermis of the denuded capital tract of the Painted Stork and in the

skin of the pigeon. For this structural phenomenon found in the skin of birds they coined the term "holocrine epidermis". The term has merit, but probably final acceptance will depend upon a detailed morphological comparison between the cytomorphosis of secretory cells in the uropygial gland (which is the holocrine type) with the layers of the skin epidermis.

Plumae caudae; Circulus uropygialis. Down feathers are present on the dorsal and ventral surfaces of the tail; some of these are associated with the rectrices and others form a circle at the tip of the uropygial papilla. The Circulus uropygialis is present in the chicken and turkey, but is reported as absent in some species of birds (Chandler, 1916). The absence of uropygial down feathers may be due to abrasion of the papilla, which may give a bare appearance to its tip (often the case in the tom turkey). One feather of the Circulus is shown in Fig. 3. Burt (1929, p. 438) stated that the tip of the uropygial papilla was tufted in North American woodpeckers on the basis of his examination of 23 species. In the extinct starling (*Fregilupus varius*) the papilla is nude (Berger, 1957, p. 233) and nude in the Blue Coua (*Coua caerulea*) and other cuckoos (Berger, 1953, 1960).

(24) **Gll. auriculares.** These glands are located in the external wall of the acoustic meatus. Although, like the uropygial gland, they are holocrine glands they are otherwise entirely different, being composed of a number of sausage-shaped tubes, each opening independently into the meatus. The large cavity of each gland unit is filled with cell debris. Stettenheim (1972) quotes Glimstedt (1942) as stating that the gland produces wax, which is in agreement with the studies of Lucas and Stettenheim (1972) as well as Moser (1906), Plate (1918), Schumacher (1919) and Gomot (1958). Plate noted an enormous blood supply around the acinar units of the gland, as well as accumulations of lymphoid tissue.

(25) **Plumae; Semiplumae; Pennae contourae.** A plume or down feather has completely fluffy (plumaceous) vanes, with its Rachis shorter than its longest barbs. A semiplume feather possesses completely plumaceous vanes; the length of its Rachis exceeds that of its longest barbs. A contour feather is one with flat, closely knit (pennaceous) vanes, although the basal part of the vane may be variably plumulaceous (Fig. 12). The main types of contour feathers are the flight feathers of wing and tail with their coverts and the general contour feathers of the body, head, neck, and limbs. See Lucas and Stettenheim (1972, p. 263). Names have been applied only to the down feathers of tail and wing which show a high degree of orderly arrangement. Elsewhere the downs are more variable and less well arranged (see below). See Lucas and Stettenheim (1972, Figs. 119, 120).

Plumae. Natal down feathers are those seen on newly-hatched birds such as petrels, ducks, galliforms, shorebirds, gulls, pigeons, and passerines. Other birds, naked at hatching, show these feathers within a few days (e.g. loons, pelicans, cormorants, and frigate-birds). Adult or definitive down feathers occur at various places on the body as part of the immature and adult feather coats of most birds. Following are patterns of down feather distribution among birds: (1) Evenly distributed over entire body (e.g. penguins, ducks, pelicans, parrots); most abundant on water birds. (2) Sparsely or unevenly distributed, in apteria and tracts (e.g. shorebirds, gulls, owls, kingfishers). (3) Confined to feather tracts (tinamous). (4) Confined to apteria (e.g. herons, bustards, most galliforms, goatsuckers, swifts). (5) Sparsely distributed in apteria or entirely absent (e.g. ratites, pigeons, cuckoos, hummingbirds, trogons, coraciiforms, piciforms, and passerines). See Lucas and Stettenheim (1972, pp. 264–267).

The plumes of wing, tail, and vent demonstrate distinct, organized patterns; only the plumes of these parts have been assigned names.

(26) **Pulviplumae.** These are the powder down feathers. A fine, granular, waxy powder is produced by the specialized powder down feathers in some birds: pigeons (especially the Archangel, Levi, 1957), herons (Ardeidae), toucans (Ramphastidae), parrots (Psittacidae) and Bowerbirds (Ptilonorhynchidae) (Thomson, 1964, p. 641). Eiselen (1939) discussed the nature and development of the waxy substance (see also Lucas and Stettenheim, 1972, p. 396). Minute quantities of powder are produced in contour feathers according to Schuz (1927, p. 215). Powder downs may be concentrated in patches as in herons and in a frogmouth (*Podargus*); these may be two inches long (see "powder-downs" in Newton and Gadow, 1896, p. 738).
Partes pennae (Fig. 12). The structure of contour feathers, particularly the flight feathers, is much more complex than semiplume and down feathers. For illustrations of the detailed parts of feathers see Chaps. 5, 6 and 7 of Lucas and Stettenheim (1972).

(27) **Galerus pulposus.** Pulp cap. In the mature feather a series of pulp caps may be seen within the hollow core of the calamus of the feather. These form transverse partitions, separated by air-filled spaces. Each pulp cap consists of a plate (concave on its proximal aspect) and the segment of tube between it and the adjacent cap proximal to it. Consult Lucas and Stettenheim (1972, pp. 381–383) for details of the formation of the pulp caps.

(28) **Vexillum internum; Vexillum externum.** Vexilla are vanes projecting from each side of the Rachis of a contour feather. The terms "internum" and "externum" are related to the overlap of adjacent feathers; the vane that lies on top of another feather is external, if it lies beneath an adjacent feather it is internal. The direction of overlap is constant for remiges and rectrices, but may be variable for the wing coverts among different species of birds; these coverts may show either a contrary or a conforming overlap (Bates, 1918).

(29) **Pars plumacea; Pars pennacea** (Fig. 12). The fluffy or plumaceous part of a contour or semiplume feather lies near the basal end, and the pennaceous part consists of the distally placed flat vanes. The barbs of the Pars pennacea of each vane are united by the hooklets (Hamuli) of the barbules that maintain the continuous surfaces of the vanes (see Fig. 11).

(30) **Incisura vexilli.** The vane of some of the primary remiges may show an incisure or notch. This is slight in the chicken and involves primaries 4–8, the notch produced by a shift in the angle of the barbs to the rachis rather than a change in length. The incisure is deep in the primaries of vultures, eagles, some hawks, and some ducks and swans. The incisure, plus the spread of the primaries in flight, increases the size of the slot between the feather tips (Lucas and Stettenheim, 1972). The margination (notching) of the inner vanes of waterfowl is variable (Humphrey and Clark, 1964, p. 170). See Annot. 32, Incisura barbae.

(31) **Zona impendens.** This term refers to the friction zones of the vanes of the flight feathers of wing and tail (example, remiges of the pigeon). Modifications of the barbules in these areas of overlap with adjacent feathers assists in keeping the feathers in place against one another. See Lucas and Stettenheim (1972, p. 260) for details.

(32) **Incisura barbae; Petiolus.** The base of the Barba has been named petiole, and immediately distal to it the barb may be notched at the Incisura barbae.

(33) **Crista dorsalis; Crista ventralis** (Fig. 11). Synonymy: dorsal and ventral ridges (Lucas and Stettenheim, 1972). These are dorsal and ventral crests of the shaft of the barb. The Crista dorsalis arises above the level of the roots of the barbules. The ventral crest in some avian species may be expanded and curved to form a shelf, the **Tegmen**; from the free edge of the Tegmen arise many short stiff Villi. The ventral crests may be absent in some birds, but are moderately wide on the barbs in the primary feathers of pelicans, cormorants, herons, storks, vultures, hawks, sandpipers, plovers, and sandgrouse (Lucas and Stettenheim, 1972, pp. 258–259).

(34) **Ruga proximalis; Ruga distalis.** Proximal and distal ledges or ridges that are situated on each side of the shaft of the barb near the attachment of the barb to the Rachis.

(35) **Vexilla barbae.** Collectively the barbules projecting from each side of a barb form small vanes (vanules). **Vexillum barbae proximalis** is the vanule on the barb nearest the base of the feather itself; **Vexillum barbae distalis** is on the side of the barb toward the free tip of the feather. The barbules of the proximal vanule differ structurally from those of the distal barbule. Consult Lucas and Stettenheim (1972, pp. 239–252) for details.

(36) **Barbula [Radius]** (Fig. 11). The barbules have considerable numbers of shapes and forms; they may be pennaceous, plumaceous, a friction type (Barbula trita), rachidial barbules, and a stylet type. The rachidial barbules are those attached directly to the side of the Rachis, but in shape are similar to those at the proximal end of the barb.

In the literature two points of view have been presented: (1) that the barbule was part of a barb (e.g. Chandler, 1916, p. 251; and Sick, 1937, p. 211) and (2) that the barbule was an entity on a par with the barb (e.g. Rawles, 1960, p. 200). The latter point of view has been chosen for this list.

(37) **Arcus dorsalis; Dens ventralis** (Fig. 11). Arcus dorsalis is a well developed flange along the dorsal edge of the base of a proximal barbule (see Annot. 35) which serves to engage the various types of barbicels or hooklets of the Pennulum of a distal barbule in the process of uniting barb to barb and establishing the intact surface of the vane of a feather.

Dens ventralis is a strong tooth-like process on the ventral edge of the apical end of the base of a distal barbule adjacent to the stalk of its Pennulum. See Lucas and Stettenheim (1972, pp. 246–247) for illustrations and descriptions of the above structures.

(38) **Flexura.** This is a special modification of back and breast contour feathers where there is a development of curved dorsal barbicels on the *base* of both distal and proximal barbules. Since they are not homologous with other types of barbicels and are always curved, Chandler (1916, p. 271) has given them a special name, "flaxules". These occur in aquatic birds, such as grebes, albatrosses, petrels, pelicans, cranes, limpkins, rails, jacanas, sandpipers, phalaropes and alcids (Lucas and Stettenheim, 1972, p. 248).

(39) **Hypopenna** (Fig. 11). Synonymy: Afterfeather; Aftershaft. Feather-like structures attached to the underside of plumaceous or pennaceous feathers that project from the rim (Ora) of the distal umbilicus. The Hypopennae range in form from a fringe of separate barbs or downy tuft to a highly organized pennaceous structure with shaft and vanes. The Hyporachis and Hypovexillum are homologous to the rachis and vane of the main feather. See Lucas and Stettenheim (1972, pp. 252–255) for types and occurrence of Hypopennae.

(40) **Setae.** The Setae are various forms of bristle feathers. Virtually all of the setae in birds are found on the head. Setae generally consist of a stiff tapered rachis devoid of barbs except at the proximal end. Semisetae have stylet barbs along all or most of the rachis (see Lucas and Stettenheim, 1972, p. 271). Stettenheim (1974) has shown long semibristles on the front of the head in the Large Owlet—frogmouth (*Aegotheles insignis*). In most birds bristle feathers are found as short "eyelashes" or "cilia", but occasionally these may be long and without barbs; for example, in the Roadrunner (*Geococcyx californianus*) and the Papuan Hornbill (*Aceros plicatus*). Berger (1960, p. 91) noted in a musophagid (*Tauraco leucotis*) that eyelashes were absent; instead both eyelids, especially the dorsal, were covered ". . . by fleshy caruncles . . . ." In species of the so-called "glossy cuckoos" (e.g. *Chrysococcyx*), the eyelashes are feather-like rather than hair-like (Berger, 1955, p. 586).

Setae nariales are bristle feathers implanted in the loral region of birds; e.g. in many hawks, Old World vultures, and owls. These have been called "vibrissae"; however, Stettenheim (1974) does not list the term "vibrissa" in his glossary. Small contour feathers are present on the head and upper neck of the turkey poult; each generation of feathers during successive moults becomes transformed by steps into typical bristle feathers.

(41) **Folliculus.** The follicle is the outer wall of an inverted tube, the inner wall in the layers derived from epidermis and dermis that produce the feather. The epidermis of the feather therefore faces the epidermis of follicle (see Lucas and Stettenheim, 1972, Figs. 239, 306, 307). The follicle is composed basically of a neck (Collum), a cavity occupied by the calamus and a wall (Paries). The wall contains the same layers as the Cutis except that a Tela subcutanea is lacking.

(42) **Pterylae; Apteria.** Pterylae are tracts or areas of skin into which contour feathers are implanted. Nitzsch (1867) defined a tract as a group of contour feathers where the distal parts of these appeared at the surface of the plumage coat. Most birds have their feathers grouped into tracts, and these are separated by featherless spaces, **Apteria.** A few birds, mostly ratites, are feathered over their entire body: for example, the kiwi, cassowary, Emu, rhea and Ostrich. The penguins among the carinate birds have almost a complete feather coat. Down feathers may be present among contour feathers as in anseriforms, but an area containing only down feathers or no contour feathers is considered to be an apterium. An Apterium may be devoid of all types of feathers but generally includes semiplume feathers; in a large bird like the turkey, some of the large semiplumes have been included as part of a tract, usually when the pattern of feather muscles and the plumage indicates that these semiplume feathers are serving the function of contour feathers. Berger (1960) used the term "tracts" in the plural for major feather groups and the singular, "tract" for the individual components, thereby solving an awkward terminological problem. This has been followed by Lucas and Stettenheim (1972). Recently Sengel (1976, p. 220), in a detailed study of feather and skin layer inductors, uses tracts in the plural. See also Stettenheim, 1972, p. 33.

(43) **Pterylae capitales.** About 12 feather tracts on the head of a bird have been delineated. In some birds for example, the duck (White Pekin), there are no apteria; since there are few or no apteria to separate the pterylae from one another, the subdivisions are arbitrarily defined on the basis of their locations. In some cases other factors are utilized to distinguish the tracts e.g. relative abundance of feathers, different types of feathers, etc. Compton (1938) bounded the individual capital tracts on the head of osprey (*Pandion haliaetus*). His plan has been applied to the heads of the chicken, turkey, quail, duck and pigeon (Lucas and Stettenheim, 1972, p. 76). Boulton (1927) gives boundaries for the location of eleven "regions" of the capital tracts in the House Wren (*Troglodytes aedon*). Clench (1970) treats the head region of the House Sparrow (*Passer domesticus*) as a single tract without subdivisions. The use of the term "region" to designate subdivisions of a tract has been avoided because region is also used to designate a topographical unit in the deplumed bird. An anatomical region and a feather tract do not always fully duplicate one another (see **Topogr. anat.** Intro.).
Pt. rictalis; Pt. buccalis. The Rictus has been identified in **Topog. anat.** Fig. 2; it has a triangular shape when the mouth is open. With the mouth closed as in Fig. 4, the maxillary and mandibular portions of the Rictus are compressed into an elongated area. Pt. buccalis lies in the suborbital region between the lores and the ear. It does not include the jaw areas which in Lucas and Stettenheim (1972, p. 98) are caudal to the Rictus and are identified as the malar region.

(44) **Pterylae spinales.** For the term "spinal tracts" Clench (1970) recommended dorsal tract for the group of tracts on the dorsal surface of the bird from base of head to tail and divided it into three parts, anterior, saddle and posterior. Unfortunately, *Pterylae dorsales* is so similar to *Pteryla dorsalis* that confusion in use of singular and plural could occur.

(45) **Pteryla trunci lateralis.** This is a single tract surrounded by the large lateral apterium. Its dorsocranial end lies in or near the axillary fossa.

(46) **Pterylae ventrales.** These ventral tracts extend from the interramal region caudally to the junction of abdomen and tail.

(47) **Pt. interramalis.** This is a triangular tract in the angle between the two mandibular rami; its caudal extent corresponds to the caudal end of the Rhamphotheca.

(48) **Pteryla submalaris [gularis].** This tract is located on the skin of the throat (Gula). Its caudal boundary is a transverse line at the level of the quadratomandibular articulations. See **Topog. anat.** Annot. 17; and Lucas and Stettenheim, 1972, p. 18.

(49) **Pteryla cervicalis ventralis.** The ventral cervical tract extends from submalar tract to the level of the shoulder. This tract in long-necked birds, like the dorsal cervical tract, may be divided into a cranial ventral cervical tract and a caudal ventral cervical tract.

(50) **Pteryla abdominalis.** In nearly all birds this is a single tract but in the chicken, particularly the adult female, there are distinctly two tracts, a Pteryla abdominalis medialis which is a strong tract, and a weak Pteryla abdominalis lateralis. Both together in poultry literature constitute the fluff. The Ventus is located in the abdominal area and on its lips is a ring of down feathers, the Circulus venti.

(51) **Pteryla ventralis caudae.** The paired ventral caudal tracts are composed of two or more rows of tectrices (coverts) in the chicken. The number of rows of tectrices is variable in different kinds of birds.

(52) **Pteryla humeralis.** Nitzsch (1967) used the term "humeral tract" rather than "scapular" or "scapulohumeral". In all of Nitzsch's (1867) illustrations for 64 species of birds, the tract crossed the upper part of the brachium. Nitzsch noted that the humeral tract was parallel to the scapula. The humeral tract terminates caudally at the angle of wing and body. As the rows approach the margin, the feathers in them become longer and larger, thereby filling the space between wing and body.

(53) **Remiges secundarii; Diastema.** Secondary remiges are defined as flight feathers implanted along the caudal edge of forearm including the region of the elbow. Among some genera and families of birds there is a Diastema or hiatus between the 4th and 6th secondaries. Van Tyne and Berger (1976, p. 131) discussed literature on the presence and absence of the diastema. When present the wing is designated "diastataxic" and when absent, "eutaxic." Originally it was thought that the space represented the loss of the fifth secondary remex. The concept has changed; Stephan (1970) failed to find a shifting during development of the secondary remiges with a loss of the fifth. Instead he noted a shifting of the major dorsal tectrices of the secondaries resulting in an extra feather, and that the fourth and fifth secondary remiges were pushed apart during development to provide the space for the extra tectrix.

(54) **Tectrices secundariae dorsales.** Dorsal secondary covert feathers form several rows: major, median, and one or two rows of minors. The dorsal marginal coverts lie on the leading edge of the Propatagium.
Tectrices minores. Steiner (1917) lists first and second rows and a possible third row of minor coverts, but in the study of feather muscles in this area the "second row" seems to be an integral part of the more cranially placed propatagial coverts (Lucas and Stettenheim, 1972; Humphrey and Clark, 1961, p. 374).

(55) **Remex carpalis.** The carpal remex is a single feather implanted in the skin of the wrist on the dorsal surface. The dorsal and ventral major carpal tectrices are identifiable cranial to the remiges, as well as some carpal tectrices belonging to other rows.

(56) **Remiges primarii.** The number of primary remiges is somewhat variable. According to De W. Miller (1924), 16 primaries are present in *Struthio*, 12 in the Rheidae, 11 + 1 in three carinate families (the +1 represents a remicle or small remex at the end of the rows, counting from the wrist), 10 + 1 in many families and 10 only in many others. In some passerines even the 10th may be vestigial. Some rows of the dorsal and ventral tectrices may be absent in whole or in part. See Pitelka (1945) and his studies on two species of the jay (*Aphelocoma*).

(57) **Remiges alulares.** The alular remiges vary in number: two in hummingbirds, 5 to 7 in certain cuckoos, the peafowl (*Pavo*), trumpeter (*Psophia*) and a touraco (*Tauraco*) (Van Tyne and Berger, 1976, p. 132). These authors place number one at the basal end of the alula. Most authors have done the opposite, designating the large, most distal quill as number one and then continuing the numbering toward the base (Van Tyne and Berger, 1976, p. 134).

(58) **Tectrices marginales manuales.** There are several groups of marginal coverts of the hand. Dorsal and ventral marginal coverts of the Propatagium have already been listed. Similarly, there are marginal coverts of the leading edge of the Manus. **Tectrices medianae.** The middle row of tectrices is absent on the ventral surface of the Manus of the chicken.

(59) **Pteryla cruralis.** The crural tract originates below the crural apterium and extends to the ankle joint although in some birds it ends short of this joint; e.g. many long-legged wading birds. It is a continuous tract; in the literature both external and internal crural tracts are described, but this distinction seems unnecessary. The feathers of the tract are farther apart on the medial surface of the crus than on its lateral surface.

(60) **Apteria capitalia.** Few, if any, apteria occur in the head of birds. If an apterium lies within or at the edge of the feathers in a capital tract, it is designated by the same name as the tract (e.g. palpebral apterium, and the caudoauricular apterium in the caudoauricular tract (Fig. 4). See Annot. 43.

(61) **Apteria spinalia.** This heading refers generally to those featherless spaces on or near the middorsal line. Apterium dorsale caudae may be considered a spinal apterium, but in the list of terms it is included with the apteria of the tail. The spinal apteria are variable among birds. Nitzsch (1867), for example, shows apteria in the dorsal and pelvic regions of *Coracina*; in the occipital tract, the cranial end of the neck, and the dorsal and pelvic regions of the back of *Trochilus*; in the interscapular, dorsal, and pelvic regions in *Nyctibius*, and in the interscapular and dorsal regions of the turkey. This last agrees with Lucas and Stettenheim (1972, p. 122). These authors also observed that a dorsal apterium is absent in the chicken; Nitzsch noted its absence in the peafowl (*Pavo*). In the duck (White Pekin) Lucas and Stettenheim (1972, p. 141) found a dorsal cervical apterium beginning near the middle of the neck and extending without interruption into the cranial part of the pelvic tract. In the pigeon it begins in the interscapular tract, and extends to the tip of the tail.

There are two pairs of featherless spaces on the back of the bird. These are placed bilaterally rather than in the midline as are the other Apteria spinalia. (1) Apterium scapulare, when present, lies between the interscapular and humeral tracts. (2) Apterium pelvicum laterale is a narrow space separating the femoral and pelvic tracts on each side of the body; however, in the duck, the separation is present only along the cranial parts of these tracts.

(62) **Apteria ventralia.** The ventral apteria may occur both in the midventral line and bilaterally, the latter located in association with bilateral feather tracts. The interramal apterium and the submalar apterium follow the medial margins of the mandibular rami; generally these are very narrow spaces. The ventral cervical apterium, the sternal apterium, and the median abdominal apterium lie in the median axis. In the owl (*Bubo virginianus*) this apterium extends from the middle of the neck to the vent (Lucas and Stettenheim, 1972, p. 87). In the chicken it extends from the caudal third of the neck to about the caudal end of the pelvic tract. In the turkey it begins immediately behind the featherless skin of the neck, includes the beard, all of the sternal region, and terminates at about the caudal third of the abdominal tract. The region of this tract is similar in the quail except that caudally it is limited to the region of the keel. In the duck (White Pekin) it ends caudally at the vent. In the pigeon it takes its origin shortly behind the submalar tract, and

terminates at the vent. An additional bilateral tract is the medial pectoral apterium which is the space between the sternal tract and pectoral tract. Generally the medial apterium is the only pectoral apterium present, but in the owl there is an additional space, Apterium pectorale laterale. A pair of lateral abdominal apteria are present between the medial surfaces of the legs and the abdominal tract. An apterium encircles the vent, and usually carried a circlet of down feathers. Many additional examples may be found in the work of Nitzsch (1867).

(63) **Apteria caudae.** The Apt. laterale caudae is found along each of the lateral edges of the tail. Apt. dorsale caudae surrounds the uropygial eminence in the duck (White Pekin). In the chicken and turkey the dorsal apterium of the tail extends transversely, and separates the dorsal pteryla of the tail from the dorsal median coverts of the rectrices. Apt. ventrale caudae in the chicken is V-shaped.

(64) **Apteria alae.** The humeral apterium (Fig. 5) is a featherless space between the humeral tract and the feathers of the Propatagium. A subhumeral apterium is present on the ventral surface of the brachium. Cubital apteria are found on the dorsal and ventral sides of the elbow. The alular apteria are small, and the one on the dorsal side is continuous with the manual apterium. A large propatagial apterium fills the triangular space between the ventral antebrachial tract and the ventral coverts of the patagial margin. Apteria of the Manus are to be found on both dorsal and ventral surfaces.
Apterium intracrurale. A definite apterium found on the medial surface of the proximal part of the crus in the pigeon.

(65) **Apterium metatarsale craniale; Apterium metatarsale caudale.** These apteria exist only when the metatarsus is feathered. An example occurs in the pigeon whose feathers extend for a short distance along the medial and lateral surfaces of the metatarsus; the apteria are the scaled areas between these feathered extensions on the dorsal and plantar surfaces of the proximal metatarsus.

(66) **Podotheca.** The keratinized, nonfeathered integument of the avian foot (i.e. ankle region, tarsometatarsus, and digits). The Podotheca exhibits varied patterns of arrangements of scales in different birds (see Annot. 67).

(67) **Scuta.** A general term for all types of scales (singular, Scutum). Scutella (singular, Scutellum) is the diminutive indicating small scales of various shapes and patterns. See Lucas and Stettenheim (1972, pp. 595–598) for an account of scales and scutellation.

(68) **Pulvinus hypotarsalis.** The thick cornified heel pads of nestling birds that cover the plantar aspect of the Hypotarsus (see **Osteo.** Annot. 288) just distal to the intertarsal joint. The heel pads are molted about the time of fledging. Heel pads occur in woodpeckers, toucans, and barbets (Harrison, 1964, p. 489).

(69) **Calcar metatarsale.** A metatarsal spur is present in most of the Phasianinae: chickens, pheasants, grouse, francolins, and others. The spur is well developed in males and usually reduced or absent in females (Thomson, 1964; Humphrey and Clark, 1964). A spur is present in turkeys (Meleagridinae); "The peacock-pheasants, *Polyplectron* spp. have up to four full-sized spurs on each leg." (Thomson, 1964, p. 429.)

(70) **Ungues digitorum manus.** According to Fisher (1940), the claw of the alular digit is usually found in the following orders: Gaviiformes, Ciconiiformes, Anseriformes, Falconiformes, Galliformes, Gruiformes, Charadriiformes and Strigiformes. He notes that claws on the major digit are ". . . restricted, as a rule to natal Anseriformes and natal and adult Ophisthocomidae, although a second claw was found on an adult *Coragyps*, a natal *Circus*, natal and adult *Rallus* and on several natal *Apus*." See Van Tyne and Berger (1976); and **Arthr.** Annot. 137.

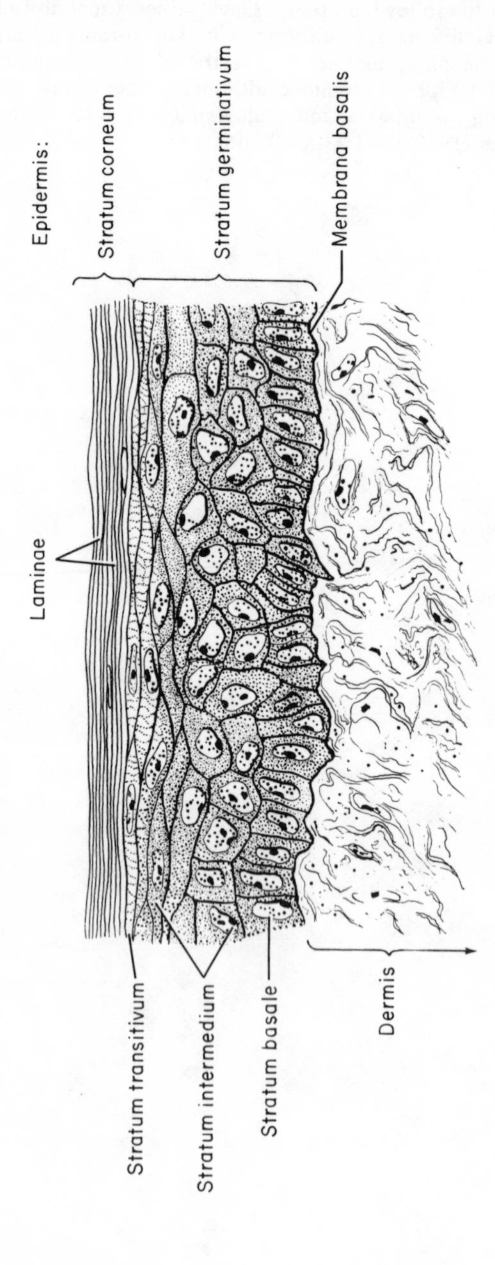

Fig. 1 Microscopic section of the Epidermis of *Gallus*. From Lucas and Stettenheim (1972). See Annot. 1–5.

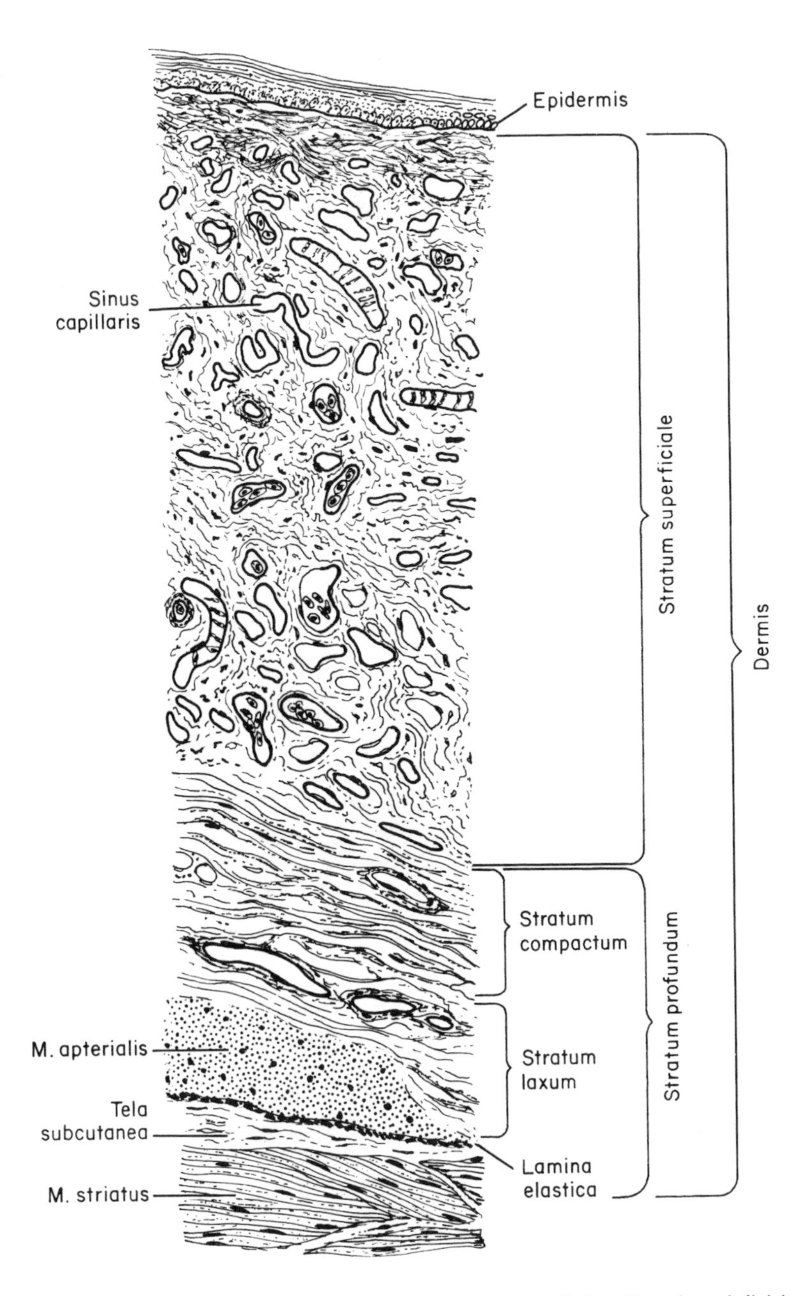

Fig. 2 Microscopic section of the skin and subjacent tissues; *Gallus*. Note the subdivisions of the Dermis. From Lucas and Stettenheim (1972).

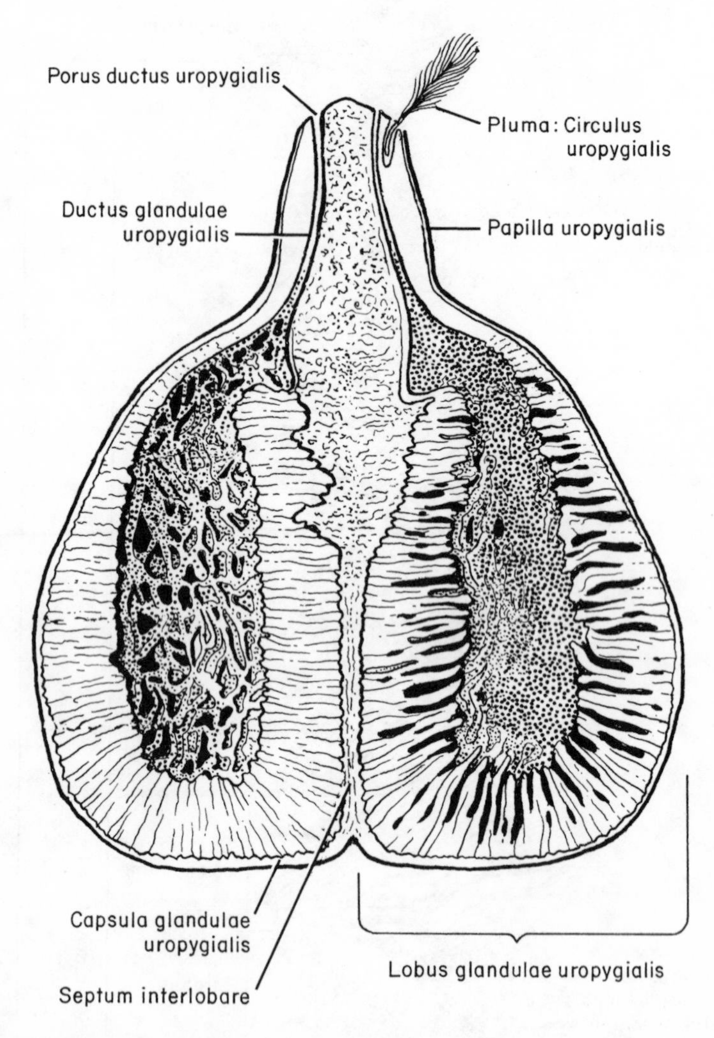

Fig. 3 Section of the uropygial gland of *Gallus*. The section is parallel to the long axis of the gland, demonstrating its right and left halves. Lucas and Stettenheim (1972) contains a very thorough description of the Gl. uropygialis.

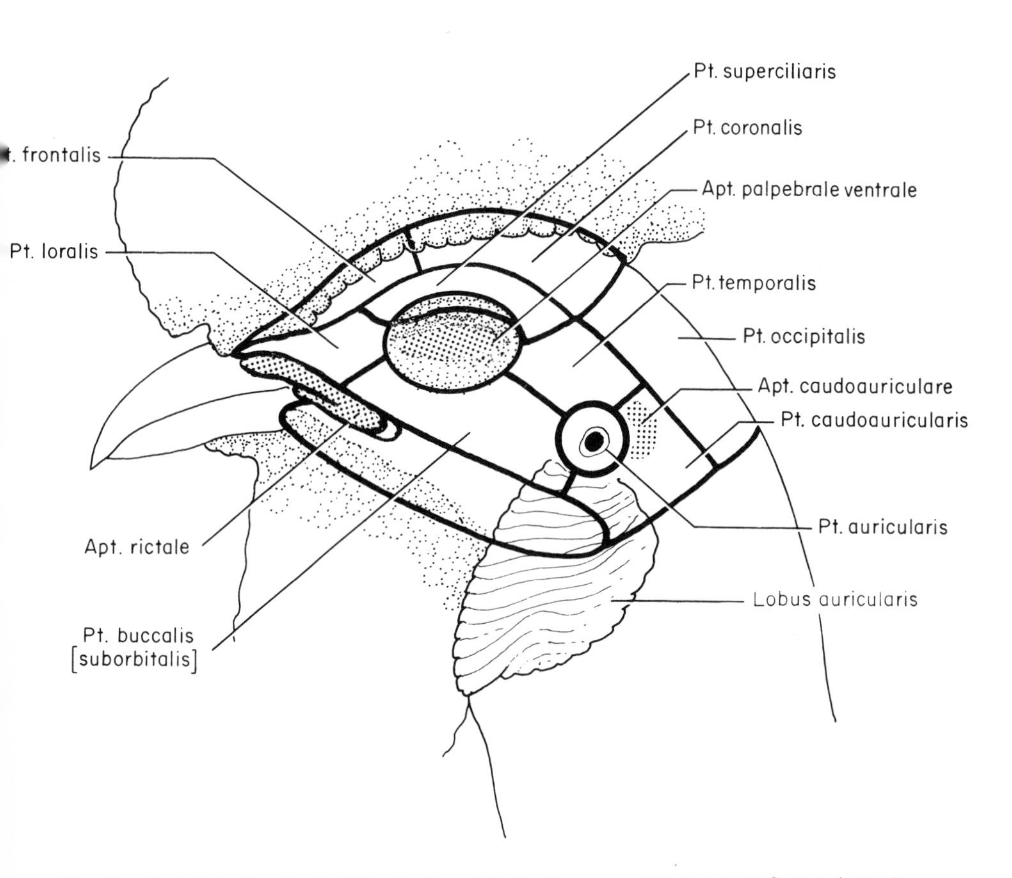

Fig. 4 Capital feather tracts (Pterylae) and featherless areas (Apteria) of *Gallus*. Lateral aspect of head, left side. See Annot. 43, 60. From Lucas and Stettenheim (1972). Apt., Apterium; Pt., Pteryla.

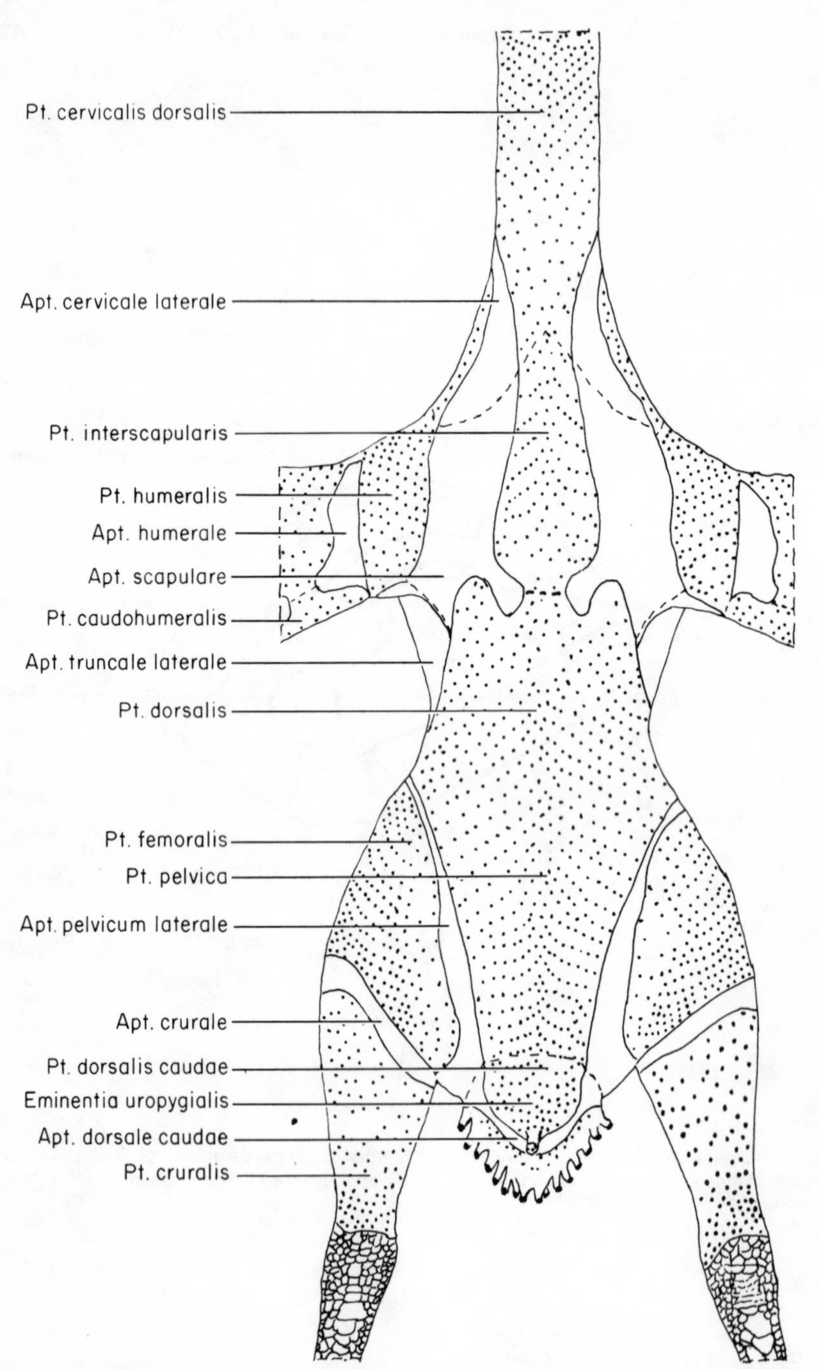

Fig. 5 Spinal (dorsal) feather tracts (Pterylae) and featherless areas (Apteria) of *Gallus*. Dorsal aspect of neck, trunk, tail, and shoulder region (including also the lateral aspect of the thigh and the caudal aspect of the crus). See Annot. 61. From Lucas and Stettenheim (1972). Apt., Apterium; Pt. Pteryla.

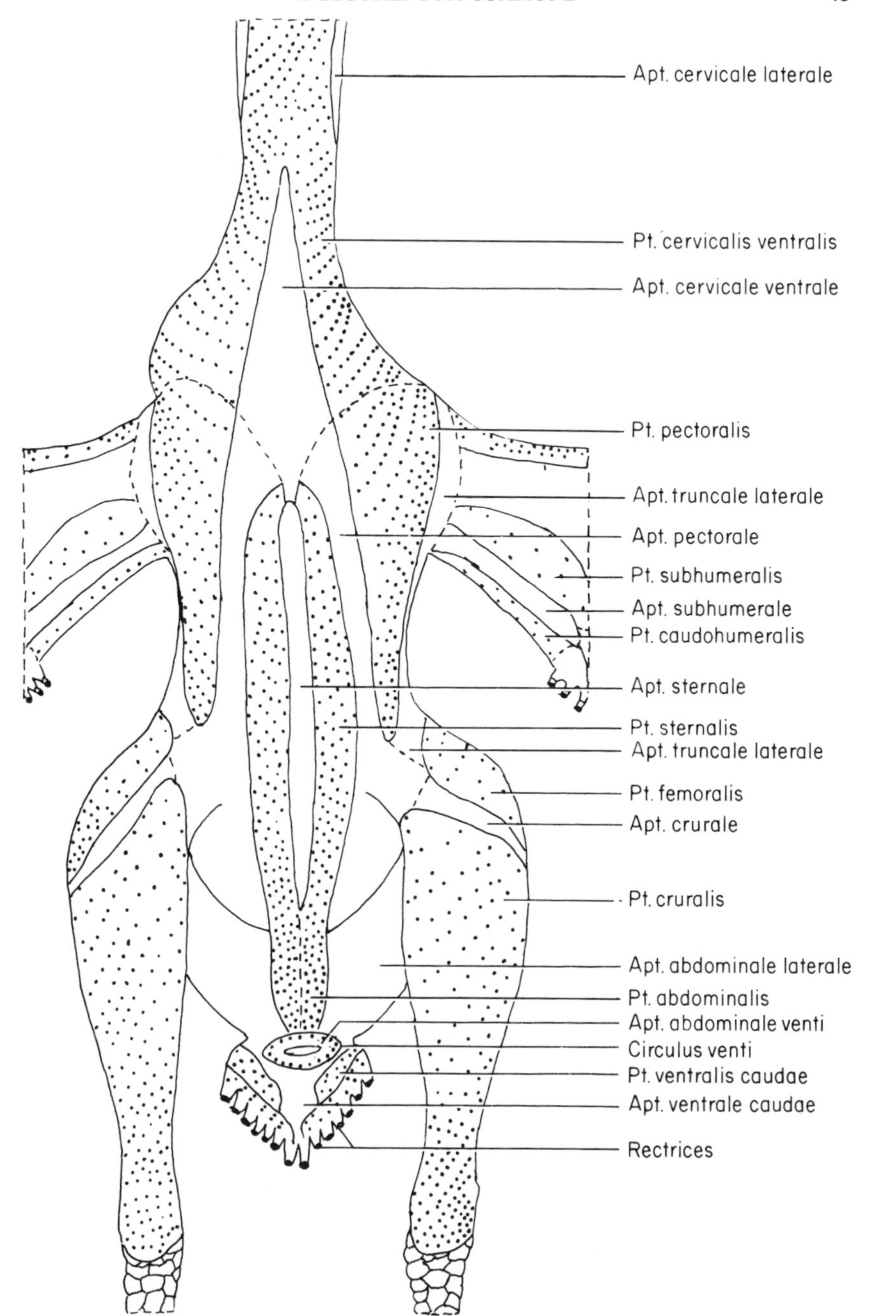

Fig. 6 Ventral feather tracts (Pterylae) and featherless areas (Apteria) of *Gallus*. Ventral aspect of neck, trunk, tail, and base of the wing (including also the cranial aspect of the knee region and the crus). See Annot. 62. From Lucas and Stettenheim (1972). Apt., Apterium; Pt., Pteryla.

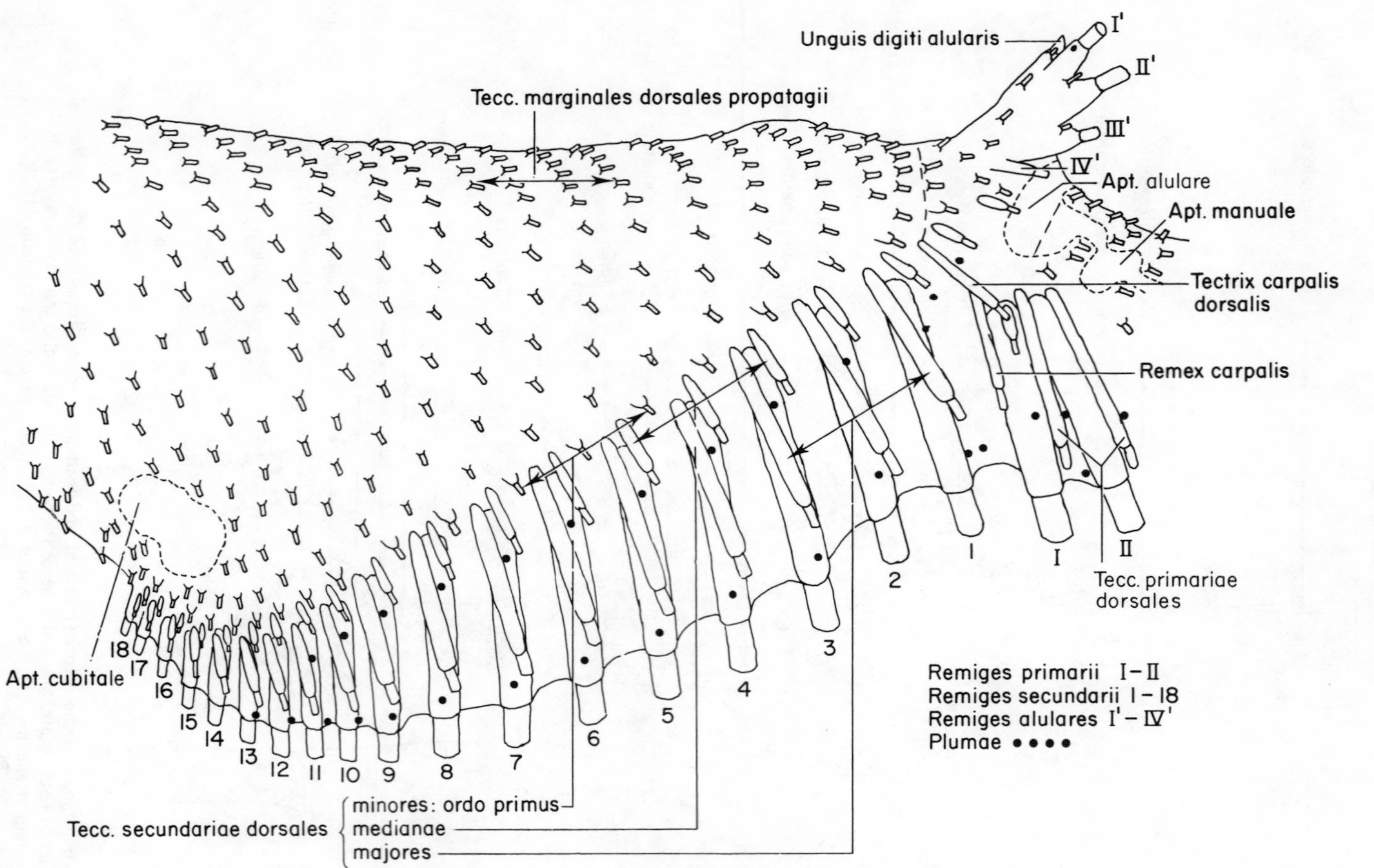

Fig. 7 Dorsal alar (antebrachial) feather tracts (Pterylae) and featherless areas (Apteria) of *Gallus*. Dorsal aspect of right forearm and wrist. From Lucas and Stettenheim (1972). Apt., Apterium; Pt., Pteryla; Tecc., Tectrices.

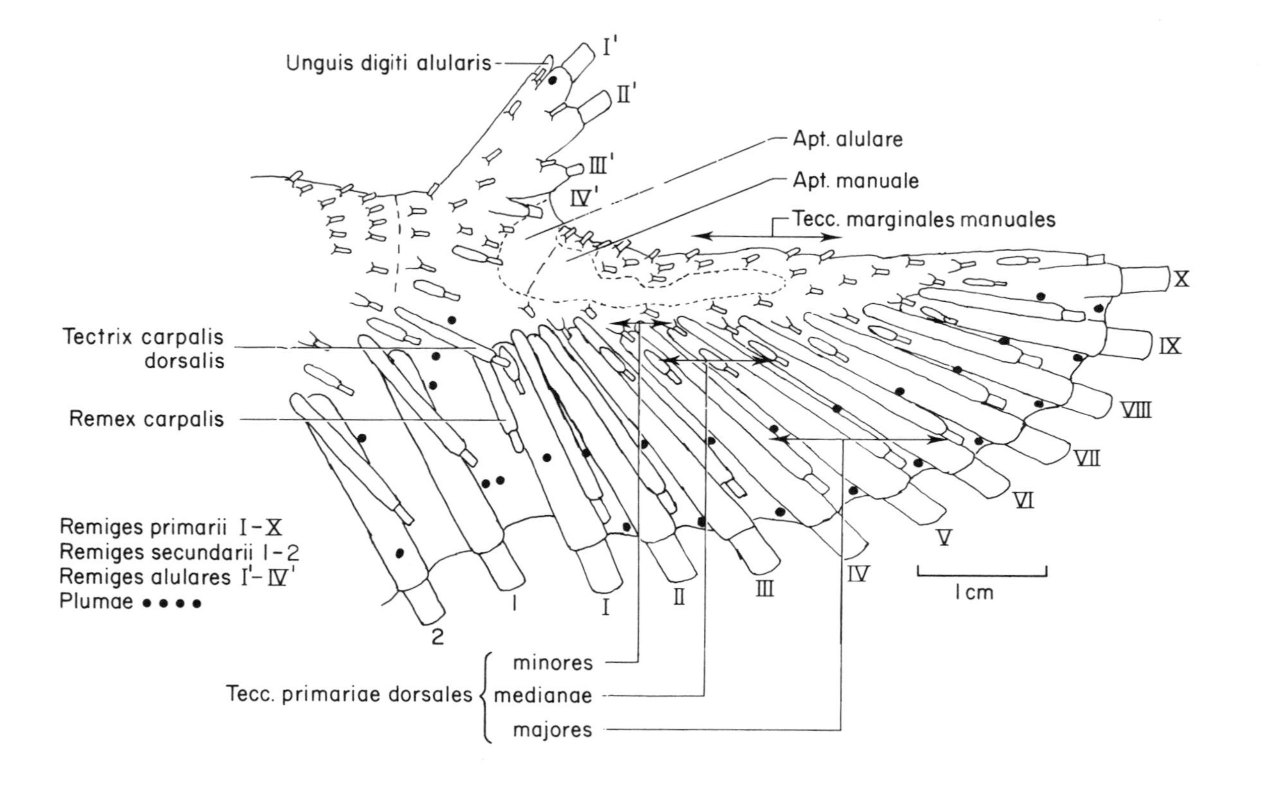

Fig. 8 Dorsal alar (manual) feather tracts (Pterylae) and featherless areas (Apteria) of *Gallus*. Dorsal aspect of right wrist and hand. From Lucas and Stettenheim (1972). Apt. Apterium; Pt., Pteryla; Tecc., Tectrices.

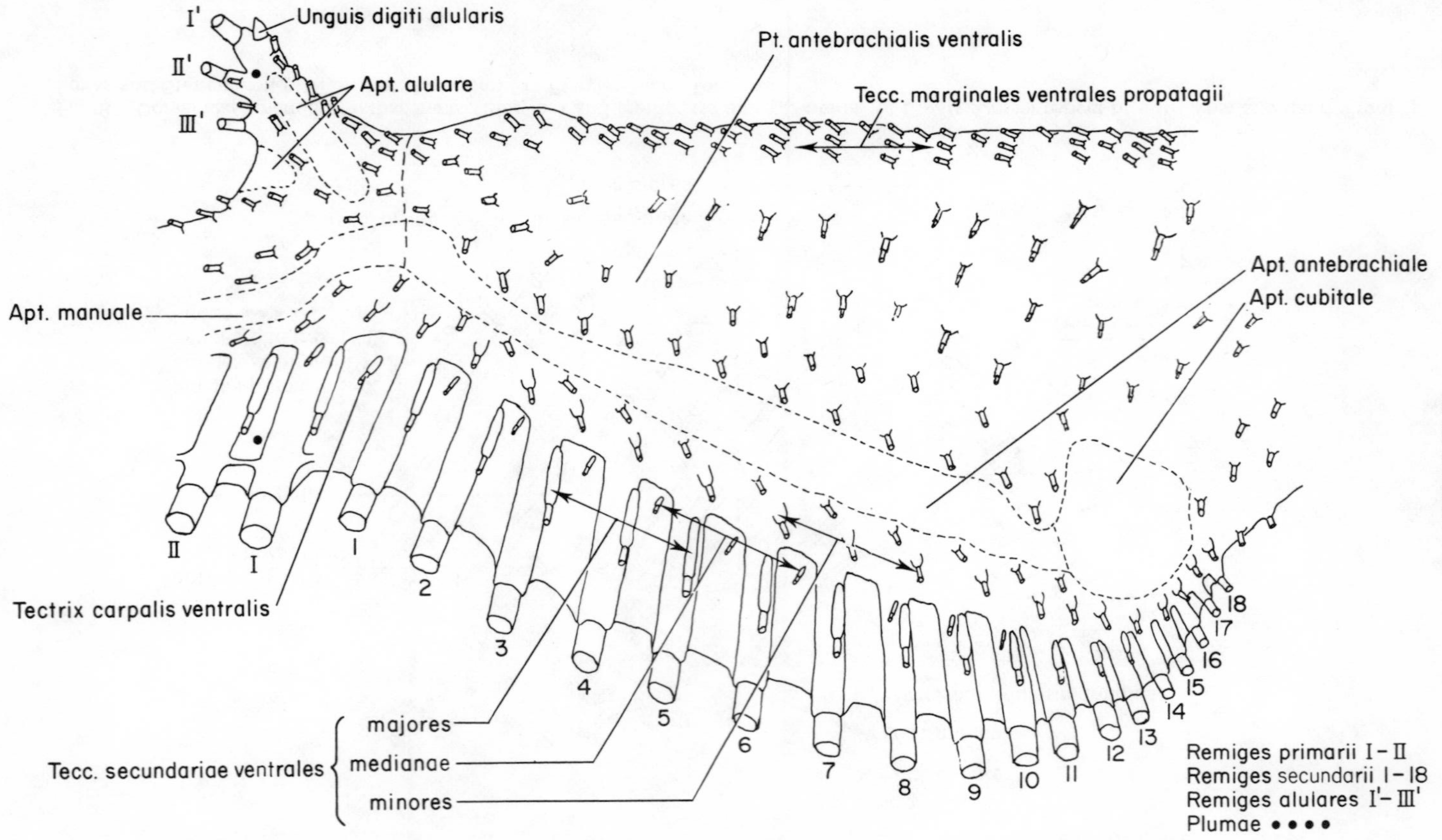

Fig. 9 Ventral alar (antebrachial) feather tracts (Pterylae) and featherless areas (Apteria) of *Gallus*. Ventral aspect of right forearm and wrist. From Lucas and Stettenheim (1972). Apt., Apterium; Pt., Pteryla; Tecc., Tectrices.

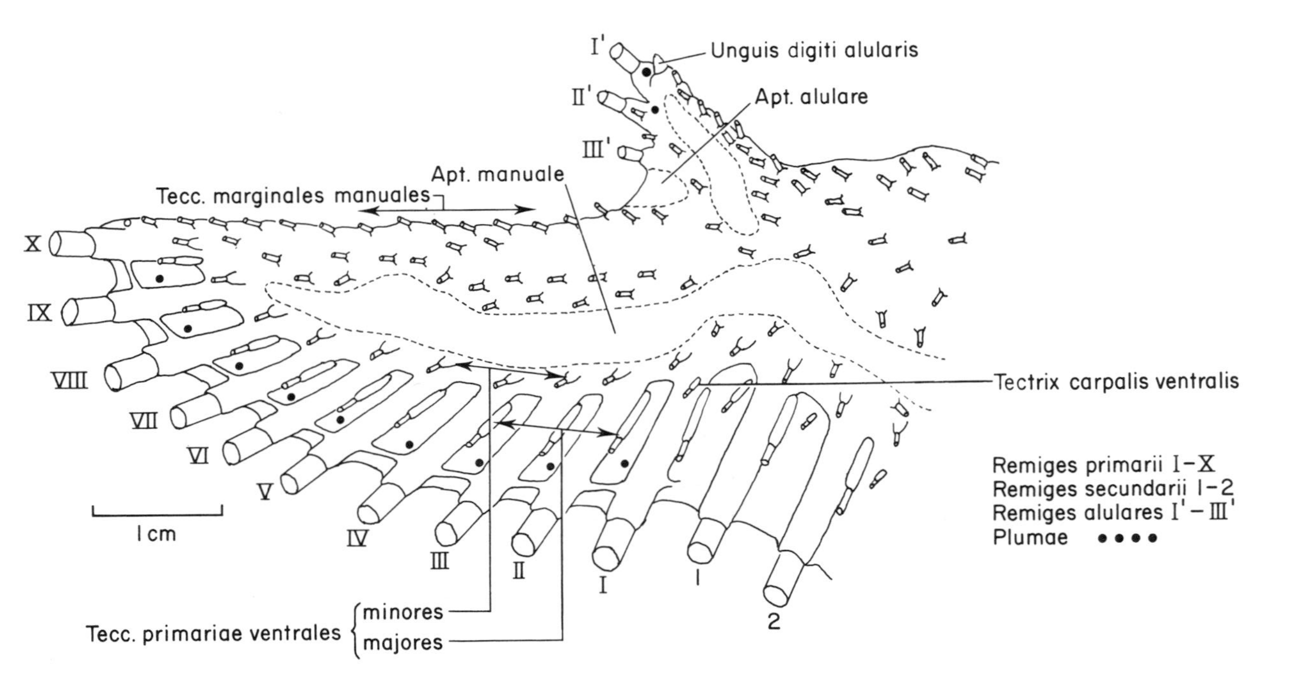

Fig. 10 Ventralalar (manual) feather tracts (Pterylae) and featherless areas (Apteria) of *Gallus*. Ventral aspect of wrist and hand. From Lucas and Stettenheim (1972). **Apt., Apterium; Pt., Pteryla; Tecc., Tectrices.**

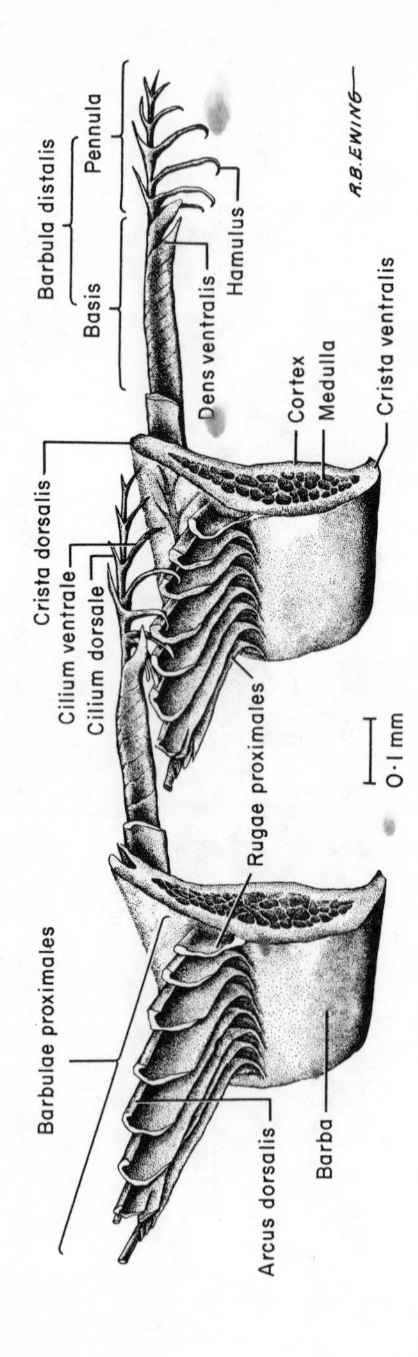

Fig. 11 Microscopic detail of the barbs and barbules of a flight feather. Short portions of two barbs from the pennaceous part of the vane of a remex of *Gallus* showing the interlocking of the proximal and distal barbules. See Annot. 32–38. From Lucas and Stettenheim (1972).

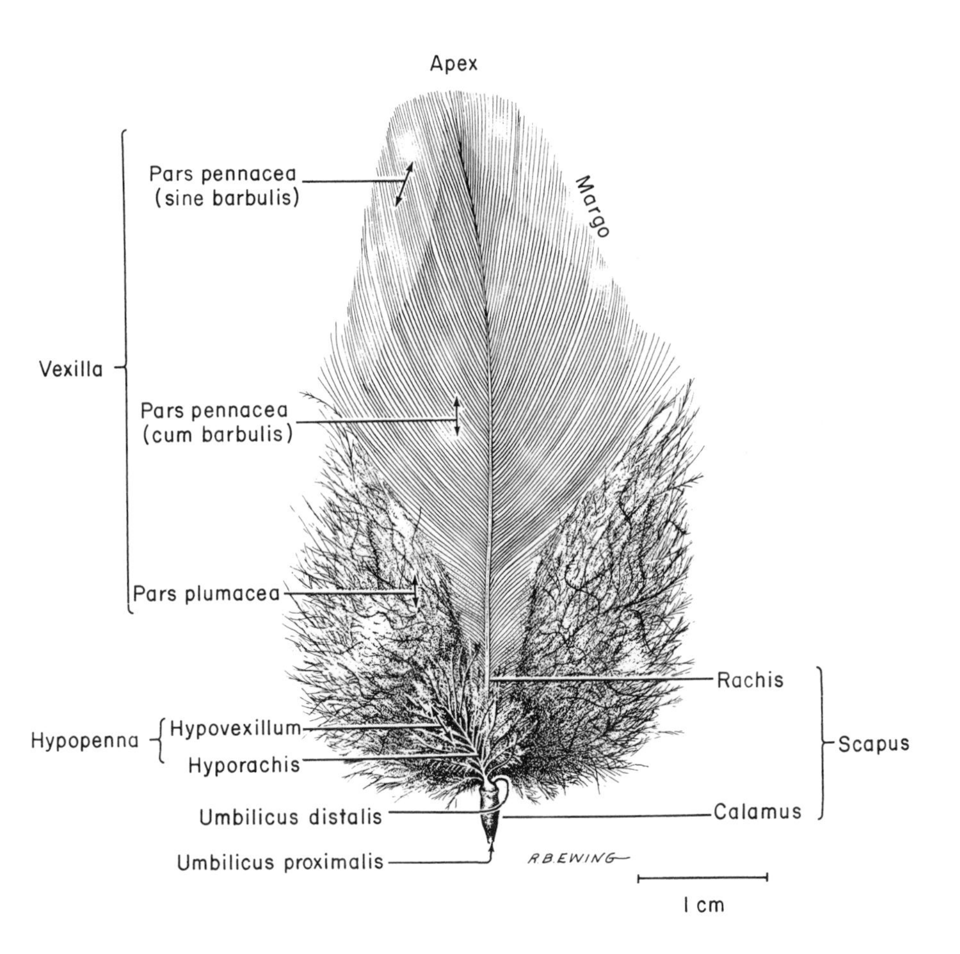

Fig. 12 The parts of a contour feather of *Gallus*. From Lucas and Stettenheim (1972).

OSTEOLOGIA

J. J. BAUMEL

Department of Anatomy, School of Medicine,
Creighton University, Omaha, Nebraska 68178, USA

With contributions from Sub-Committee Members: P. Ballmann, *Am Botanischen Garten 68, Köln*; A. J. Berger, *University of Hawaii*; W. J. Bock, *Columbia University*; P. Brodkorb, *University of Florida*; P. Bühler, *Universität Stuttgart-Hohenheim*; P. J. K. Burton, *British Museum*; H. E. Evans, *Cornell University*; M. T. Jollie, *Northern Illinois University*; Vl. Komárek, *16500 Praha 6- Lysolaje, Czechoslovakia*; E. N. Kourotchkin, *Palaeontological Museum, Moscow*; A. J. Lucas, *Michigan State University*; J. Sandoval, *Catedrático de Anatomía y Embriología Córdoba*; J. Vanden Berge, *Indiana University*; R. L. Zusi, *U.S. National Museum*

Among contemporary workers in avian osteology Dr Peter Ballmann has long been concerned with standardization of anatomical terminology. He is to be commended for his scholarly and painstaking assistance in the compilation of this present osteological terminology. Dr R. L. Zusi also contributed significantly in the development of the nomenclature of the skull and vertebral column.

The terminology of the skeleton of birds presented here is highly detailed, and will be particularly useful to avian paleontologists and myologists. In the compilation of terms the major works most heavily drawn upon were those of Fürbringer (1888), Lambrecht (1933), and Ballmann (1969a) for the limb bones; Boas (1929, 1933) and Komárek (1970) for vertebral column; and Hofer (1945, 1949, 1955) for the skull.

Nomenclature of digits of wing. The matter of homologies of the digits of the avian wing is still a disputed point among avian morphologists. There is no indisputable evidence for deciding if the elements of the avian manus represent digits I, II, III or II, III, IV. Berger (1966) summarized the controversy on the subject, citing the principal literature. More recently the paper of Seichert and Rychter (1972) has contributed additional discussion and further reviewed the literature on the topic. Rather than perpetuate a system of names of the skeleton and musculature of the manus based on the controversial numbering of the digits by having to arbitrarily choose one of the existing systems in use, an alternative system of designation has been adopted. At the suggestion of Dr. Pierce Brodkorb the digits and their skeletal elements are referred to as Digitus alularis (the so-called Pollex), Digitus major, and Digitus minor, and the bones called Os metacarpale alulare, Os metacarpale majus, and Os metacarpale minus. This terminology is partly a revival of the scheme of Milne-Edwards (1867–71). See Annot. 214; and **Myol.** Intro.

Format for listing terms. Most of the anatomical terms for skeletal features are arranged under the name of the individual bone of which they are parts. For example, Foramen n. maxillomandibularis and Proc. postorbitalis are parts of Os orbitosphenoidale. On the other hand, numerous features of the skeleton are not limited to a single bone, but are shared or extend over two or more different, adjacent bones. In fact, extensive fusion of bones of the avian skeleton often makes it difficult or impossible to distinguish adjacent bones from one another. Names of such features of the skull are listed under the headings *Facies* and *Cranium*. Subheadings under these headings are Cavitas nasalis, Orbita, Cavitas tympanica, Cavitas cranialis, and Mandibula. Features of the skull confined to individual bones of the skull are to be found under the headings: *Ossa faciei* and *Ossa cranii*. Other compound elements of the skeleton are the Notarium, Synsacrum, Os coxae, and Tarsometatarsus; these receive treatment similar to that of Facies and Cranium.

TERMINI GENERALES

Aditus
Ala
Angulus
Annulus
Antrum
Area
Apertura
Apex
Apophysis
Arcus
Area
Basis
Calvaria
Canalis
    Canalis nutricius
Canaliculus
Caput
Cartilago epiphysialis[2]
Cavitas
Cavitas medullaris[6]
Carina
Cervix
Collum
Concavitas
Condylus
Corpus
Cortex
Cornu
Cotyla[1]
Crista
Crus
Diaphysis[2]
Discus
Diverticulum
Eminentia
Epicondylus
Epiphysis[2]
Excavatio
Extremitas
Facies
    Facies articularis
Fenestra
Fissura
Fonticulus

Foramen
    Foramen nutricium
Fossa
Fossula
Fovea
Foveola
Hiatus
Incisura
Impressio
Intumescentia
Jugum
Labium
Labrum
Lacuna
Lamina
Linea
    Linea intermuscularis
Meatus
Margo
Medulla
    Medulla ossium rubra
    Medulla ossium flava
Metaphysis
Orificium
Os, ossis
    Os breve
    Os planum
    Os longum
    Os compactum
    Os spongiosum[6]
    Os medullare[6]
    Os pneumaticum
        Foramen pneumaticum[3]
        Pori pneumatici[3]
        Cavitas pneumatica ossis
    Os sesamoideum
Ostium
Phalanx
Pila[4]
Porus
Processus
Radix
Ramus
Recessus
Rostrum
Scapus[5]
Septum
Sinus
Skeleton
Spina
Squama
Stylus
Sulcus
Synostosis
Torus
Trabeculae ossium
Trochanter
Trochlea
Tuber
Tuberculum
Tuberositas[7]
Zona elastica (see **Arthr.** Annot. 45ff)
Zygapophysis

FACIES[8]

Hiatus orbitonasalis[9] (Fig. 1)
Foramen orbitonasale[31]
Hiatus craniofacialis[10]
Arcus jugalis[11]
Arcus suborbitalis[30]

Cavitas nasalis

Apertura nasalis ossea[12]

Cavitas nasalis—*continued*

Pila supranasalis[12] (see **Arthr.** Fig. 2)
Septum nasale osseum[55] (see **Arthr.** Fig. 2)
Conchae nasales[55] (see **Resp.**)

Palatum osseum

Concavitas palati (Fig. 2)
Fissura interpalatina
Fenestra palatina[13]

CRANIUM[8]

Calvaria
- Lamina externa
- Os spongiosum [Diploë][6]
- Lamina interna

Frons
- Depressio frontalis[14] (Fig. 3)

Fossa temporalis[104] (see Os squamosum)

Occiput
- Prominentia cerebellaris[16] (Fig. 2)
- Facies nuchalis
- Crista [Linea] nuchalis sagittalis[17]
- Crista [Linea] nuchalis transversa[17]
- Foramen magnum[18]

Basis cranii externa[28] (Fig. 2)

Meatus acusticus externus[19]

Cavitas tympanica[21]

Recessus antevestibularis[22] (see **Resp.** Annot. 17)
- Fenestra vestibularis
- Fenestra cochlearis

Foramen pneumaticum dorsale[23] (see **Resp.** Annot. 17)
- Antrum pneumaticum dorsale

Pila proötica[24]
- Facies articularis quadratica

Ostium tympanicum tubae pharyngotympanicae (see **Org. sens.** Fig. 3)

Foramen pneumaticum rostrale[25] (see **Resp.** Annot. 17)
- Antrum pneumaticum rostrale

Foramen pneumaticum caudale[26]
- Antrum pneumaticum caudale

Canalis ophthalmicus externus[27 86] (Fig. 2)
- Ostium rostrale canalis ophthalmici

Hiatus alae tympanicae[20 84]
Foramen m. columellae[20 84]
Columella (see **Org. Sens.** Cavitas tympanica)

Orbita[29] (Fig. 1)
Margo supraorbitalis
Proc. postorbitalis[30] (see Arcus suborbitalis)
Septum interorbitale [Paries medialis]
Paries caudalis
Paries dorsalis
Paries rostralis (see Os ectethmoidale et Os prefrontale)
Foramen orbitonasale mediale[31]
Foramen orbitonasale laterale[31]
Foramen n. olfactorii
Sulcus [Canalis] olfactorius
Foramen opticum[32] (see Os orbitosphenoidale) (Figs. 1, 3)
Fonticuli orbitales[29]

CAVITAS CRANIALIS

(see Fig. 3; also Fossa acustica interna)

Fossa cranii rostralis
Fossa bulbi olfactorii
Septum osseum fossae bulbi[33]
Foramen ethmoidale[34]
Fossa cranii media[35] (see Os basisphenoidale and Os orbitosphenoidale)
Fossa tecti mesencephali (see **CNS** Tectum mesencephali)
Fossa ganglii trigemini
Crista tentorialis
Fossa cranii caudalis
Basis cranii interna
Fossa cerebelli
Crista marginalis[36]
Crista frontalis interna (see Os frontale)
Tuberculum pineale[37]
Fossa auriculae cerebelli[38]
Eminentia arcuata canalis semicircularis rostralis
Sulcus v. semicircularis
Fovea ganglii vagoglossopharyngealis[39] (see Fossa parabasalis)

MANDIBULA

(see Ossa mandibulae) (Fig. 1; and **Arthr.** Fig. 2)

Symphysis mandibularis[40]
  Foramina neurovascularia[57]
Ramus mandibulae (see **Arthr.** Fig. 2)
  Pars symphysialis[42]
  Pars intermedia[42]
  Pars caudalis[42]
Facies medialis
Facies lateralis
  Foveae corpusculorum nervosorum[41]
Angulus mandibulae[43]
Proc. coronoideus[44]
Fenestra mandibulae rostralis[46] (Fig. 1)
Fenestra mandibulae caudalis[46] (Fig. 1)
Canalis mandibulae[47]
  Fossa aditus canalis mandibulae[48]
Proc. mandibulae lateralis (see **Arthr.** Annot. 32)
Proc. mandibulae medialis (see **Arthr.** Annot. 32)
  Foramen pneumaticum
Tuberculum pseudotemporale[45]
Fossa articularis quadratica
  Cotyla medialis
  Cotyla lateralis
  Cotyla caudalis (see Quadratum and Annot. 72)
  Tuberculum intercotylare [Crista intercotylaris]
Proc. retroarticularis[49] (see **Myol.** Annot. 24 and **Arthr.** Fig. 2)
  Recessus conicalis[50]
  Fossa caudalis[51]

OSSA FACIEI[8]

(see Intro.)

Ossa mandibulae

Os dentale[52]
  Pars symphysialis
  Proc. dorsalis
  Proc. ventralis
Os supraangulare
  Proc. mandibulae lateralis
Os angulare
  Proc. retroarticularis

Os spleniale
Os prearticulare
Proc. mandibulae medialis
Os articulare (see Fossa articularis quadratica)

Ossa maxillae et palati[53] (Figs. 1, 2)

Os nasale
Proc. frontalis
Proc. maxillaris[54]
Proc. premaxillaris[54]
Os premaxillare
Proc. frontalis
Proc. palatinus
Crista tomialis[56]
Proc. maxillaris
Foveae corpusculorum nervosorum[41]
Canalis neurovascularis[57]
Foramina neurovascularia[57]
Os maxillare[53]
Proc. palatinus [maxillopalatinus][58]
Proc. premaxillaris
Proc. jugalis[59]
Proc. nasalis
Os palatinum [pterygopalatinum][60]
Proc. premaxillaris
Lamella [Ala] caudolateralis[62]
Crista lateralis
Crista ventralis[63]
Crista dorsalis
Facies articularis sphenoidalis
Proc. pterygoideus[64]
Fossa ventralis[65]
Fossa medialis[66]
Proc. choanalis rostralis[67]
Vomer[68]
Facies articularis sphenoidalis[69]
Os pterygoideum[61]
Facies articularis palatina
Facies articularis sphenoidalis[69]
Facies articularis basipterygoidea[93]
Facies articularis quadratica
Proc. dorsalis[70]
Proc. quadraticus
Pars palatina[60]

Ossa maxillae et Palati—*continued*

Os jugale
Os quadratojugale
Condylus quadraticus
Tuberculum prefrontale (see **Arthr.** Annot. 20)
Quadratum
Corpus quadrati
Proc. oticus quadrati
Facies tympanica
Crista tympanica
Foramen pneumaticum
Sulcus pneumaticus
Condylus proöticus
Condylus squamosus
Incisura intercondylaris
Proc. orbitalis quadrati[71]
Proc. mandibularis quadrati
Condylus medialis
Condylus caudalis[72]
Condylus lateralis
Cotyla quadratojugalis[73]
Sulcus intercondylaris
Condylus pterygoideus

Dentes[82]

OSSA CRANII[8]

(see Intro. Figs. 1, 2)

Os basioccipitale
Condylus occipitalis[83]
Fossa subcondylaris
Os exoccipitale (see Cranium, Occiput)
Facies externa
Facies cerebralis
Ala tympanica[84 20]
Crista caudalis
Proc. paroticus[85] (see **Arthr.** Fig. 1)
Incisura foraminis magni
Proc. condylaris[83]
Canales n. hypoglossi
Fossa parabasalis[86]
Foramen n. vagi[39] (see Cavitas cranialis)

Foramen n. glossopharyngealis[39] (see Cavitas cranialis)
Ostium canalis carotici
Ostium canalis ophthalmici externi[86 27]
Crista fossae parabasalis[86]
Os supraoccipitale
Facies nuchalis
Facies cerebellaris (see Fossa cerebelli, Cavitas cranialis)
Fonticulus occipitalis[87]
Foramen v. occipitalis externae (see **Ven.** Annot. 32)
Os orbitosphenoidale[88]
Facies orbitalis
Area muscularis aspera[89]
Incisura n. optici
Foramen n. abducentis
Foramen n. oculomotorii
Foramen n. trochlearis
Foramen n. ophthalmici
Foramen n. maxillomandibularis[90] (Fig. 1)
Facies temporalis
Facies ventralis
Facies tecti mesencephali (Fig. 3)
Fossa ganglii trigemini (see Fossa cranii media)
Canalis n. maxillomandibularis
Sulcus n. ophthalmici
Sulcus n. trochlearis (Fig. 3)
Proc. postorbitalis[101]
Os basisphenoidale (Fig. 3)
Facies cerebralis
Canalis n. abducentis
Sella turcica[91] (Fig. 3)
Dorsum sellae
Fossa hypophysialis
Ostium canalis carotici
Foramen ophthalmicum internum[91]
Canalis craniopharyngealis[92]
Os parasphenoidale (Fig. 2)
Rostrum sphenoidale [parasphenoidale] (Fig. 1, 3)
Proc. basipterygoideus [pterygoideus][93]
Facies articularis palatina
Facies articularis vomeralis
Facies articularis pterygoidea
Tuba pharyngotympanica [auditiva] communis[94 98] (Fig. 2)
Canalis orbitalis[95]

OSSA CRANII—*continued*

- Lamina basiparasphenoidalis [basitemporalis][96]
  - Proc. lateralis (see **Arthr.** Annot. 32)
  - Proc. medialis (see **Arthr.** Annot. 32)
  - Crista fossae parabasalis[86]
  - Tuberculum basilare[97]
  - Crista basilaris transversa
- Tuba pharyngotympanica [auditiva][98]
  - Ostium tympanicum
  - Ostium pharyngeale
- Ala tympanica[84] (see Fig. 2)
  - Crista ventralis
- Canalis caroticus[99]

Os squamosum [temporale] (Fig. 1)
- Facies externa
- Facies cerebralis
- Facies articularis quadratica[100]
- Proc. postorbitalis[101]
- Proc. zygomaticus[102]
- Proc. suprameaticus[103]
- Ala tympanica[84] [20] (see Fig. 2)
  - Crista dorsalis
- Fossa temporalis[104]
  - Crista temporalis[104]

Ossa otica[105] (see Cavitas tympanica)
- Os proöticum
- Os opisthoticum
- Os epioticum
- Vestibulum
- Canales semicirculares ossei (see **Org. sens.**)
- Canalis v. semicircularis
- Cochlea (see **Org. sens.** Annot. 55)
- Fossa acustica interna[105]
  - Foramen n. facialis
  - Foramen n. cochlearis
  - Foramen n. ampullaris

Os parietale
- Facies externa
- Facies interna

Os frontale[14]
- Facies dorsalis
  - Sulcus glandulae nasalis[15]
    - Foramina neurovascularia
    - Sulci orbitonasales[31]

Facies cerebralis
Impressio eminentiae sagittalis (see **CNS** Annot. 78)
Crista vallecularis[106]
Sulcus sinus sagittalis dorsalis[31] (see **Ven.** Fig. 11)
Margo supraorbitalis
Proc. prefrontalis[107]
Proc. postorbitalis[108] (see Annot. 101)
Os mesethmoidale
Septum interorbitale
Lamina dorsalis[109] (see **Arthr.** Sut. frontoethmoidalis)
Sulcus [canalis] olfactorius
Os ectethmoidale [lateroethmoidale][110]
Facies orbitalis
Facies nasalis
Os prefrontale [lacrimale]
Proc. supraorbitalis
Proc. orbitalis
Incisura ductus nasolacrimalis (see **Org. sens.** Apparatus lacrimalis)
Facies articularis frontonasalis[111]

OSSA ACCESSORIA CRANII[75]

Ossa supraorbitalia[74]
Os prefrontopalatinum
Os suprajugale
Os nuchale[76]
Os siphonium
Ossa sclerae (see **Org. sens.** Annot. 16, 18, 19)
Annulus tympanicus[77]
Ossa suturarum[78]

LARYNX

Skeleton laryngis (see **Resp.** Larynx; and **Arthr.** Juncturae laryngis)

APPARATUS HYOBRANCHIALIS [HYOIDEUS]

(see **Myol.** Figs. 2a, b; 3)

Os entoglossum[80]
Cornua
Os basibranchiale rostrale[81]
Proc. parahyalis[81]
Arcus parahyalis[81]

APPARATUS HYOBRANCHIALIS—*continued*

Os basibranchiale caudale[81]
    Os sesamoideum[81]
Cornu branchiale
    Os ceratobranchiale
    Os epibranchiale

COLUMNA VERTEBRALIS[112, 113, 129, 139–141]
(Fig. 4)

Corpus vertebrae[113]
    Facies articularis cranialis
        Fovea cranioventralis[114]
    Facies articularis caudalis
    Facies lateralis
        Sulcus lateralis[115]
        Concavitas lateralis
        Tuberositas lig. collateralis (see **Arthr.** Annot. 60)
        Tuberculum costolaterale[116]
            Fovea costalis[117]
        Proc. costalis[118]
        Foramina pneumatica
    Facies ventralis
    Proc. [Crista] ventralis[119]
            Alae[120]
    Proc. ventrolateralis[122]
    Proc. caroticus[121]
            Sulcus caroticus
    Facies dorsalis
Arcus vertebrae (Fig. 4)
    Proc. transversus[123]
        Facies dorsalis
        Facies ventralis
            Pila costalis
        Fovea costalis[117]
        Foramen transversarium[134] [148]
    Proc. articularis [Zygapophysis] cranialis
        Facies articularis
    Proc. articularis [Zygapophysis] caudalis
        Facies articularis
        Crista transverso-obliqua[124]
        Proc. dorsalis[125]
    Lamina dorsalis arcus

Area lig. elastici[126]
Proc. dorsalis [spinosus]
Lamina lateralis arcus[127]
Incisura arcus cranialis[128]
Incisura arcus caudalis[128]
Foramen vertebrale

Canalis vertebralis
Foramen intervertebrale[128 144] (Fig. 11)

VERTEBRAE CERVICALES[129]
(Fig. 4)

Atlas
Corpus atlantis
Fossa condyloidea[130]
Incisura fossae[130]
Foramen fossae[130]
Facies articularis dentalis
Tuberositas lig. transversi
Facies articularis axialis
Proc. ventralis[131 119]
Sulcus caudoventralis
Arcus atlantis
Proc. articularis [Zygapophysis] caudalis[132]
Facies articularis
Proc. dorsalis
Proc. costalis[133]
Foramen transversarium[134]
Incisura arcus caudalis[135]

Axis
Corpus axis
Facies articularis atlantica
Dens[136]
Fovea lig. collateralis atlantoaxialis
Facies articularis atlantica
Proc. ventralis[119]
Arcus axis
Lamina dorsalis
Proc. dorsalis [spinosus][137]
Proc. articularis [Zygapophysis] cranialis[132]
Facies articularis
Proc. articularis [Zygapophysis] caudalis
Facies articularis

VERTEBRAE CERVICALES—*continued*

Proc. costalis[138]
Foramen transversarium[134 138]
Lamina lateralis[127]
Incisura arcus cranialis[128]
Incisura arcus caudalis[128]

VERTEBRAE THORACICAE[139]

Notarium [Os dorsale][140] (Fig. 5)
Crista dorsalis[143]
Crista ventralis[119]
Foramina intertransversaria
Foramina intervertebralia
Canalis notarii [vertebralis][145]

Synsacrum[141] (Figs. 5, 11)
Extremitas cranialis synsacri
Extremitas caudalis synsacri
Margo lateralis
Facies lateralis
Proc. costalis
Foramina intervertebralia[144 128]
Facies ventralis[142]
Crista ventralis[119]
Sulcus ventralis synsacri
Facies dorsalis
Crista dorsalis[143 233]
Proc. transversus
Foramina intertransversaria
Canalis synsacri [vertebralis][145]

VERTEBRAE CAUDALES
(Figs. 5, 11)

Pygostylus [Coccyx][146] (see **Arthr.** Annot. 78)
Basis pygostyli
Facies articularis cranialis
Canalis vascularis
Lamina pygostyli
Margo cranialis
Margo caudalis
Margo dorsalis

Canalis pygostyli [vertebralis]
Apex pygostyli

CAVITAS THORACICA
(see **Topog. anat.** Annot. 31)

Apertura thoracica cranialis
Apertura thoracica caudalis
Sulcus pulmonalis
Facies visceralis sterni (see Sternum)
Spatium intercostale

COSTAE[147]
(Fig. 4, 5)

Costa vertebralis
  Extremitas proximalis
    Capitulum costae
      Facies articularis vertebralis (see **Arthr.** Annot. 79)
    Collum costae
    Tuberculum costae
      Facies articularis vertebralis
    Incisura capitulotubercularis[148] (Fig. 4)
    Angulus costae
    Corpus costae
      Margo cranialis
      Margo caudalis
      Proc. uncinatus
  Extremitas distalis
Costa sternalis
  Facies articularis sternalis[156] (see **Arthr.** Annot. 81, 83)

STERNUM[149]
(Fig. 7)

Corpus sterni[149]
  Proc. craniolateralis[150]
    Impressio m. sternocoracoidei
  Trabecula lateralis [Proc. thoracicus]
  Trabecula intermedia [Proc. caudolateralis]
  Trabecula mediana
  Incisura lateralis[151]
  Incisura medialis[151]

STERNUM—*continued*

Fenestra lateralis[151]
Fenestra medialis[151]
Facies muscularis sterni[152]
  Linea intermuscularis[153]
  Planum postcarinale[154]
Facies visceralis sterni[152]
  Pars cardiaca
  Pars hepatica
  Foramen pneumaticum
  Pori pneumatici
  Sulcus medianus sterni
Margo costalis [lateralis]
  Pila costalis[155]
  Procc. costales
    Facies articularis costalis[156] (see **Arthr.** Annot. 83)
  Incisurae intercostales
Margo caudalis[157]
Margo cranialis
  Pila coracoidea[158]
  Sulcus articularis coracoideus (see **Arthr.** Annot. 89–90)
    Labrum dorsale
    Labrum ventrale
      Tuberculum labri ventralis

Rostrum sterni[148] (see Fig. 7; and **Arthr.** Annot. 86)
  Spina interna[159]
  Spina externa[159]
  Foramen rostri[160]

Carina sterni[161] (see **Integ.** Annot. 16)
  Margo cranialis
    Pila carinae
    Crista mediana
    Crista lateralis[161]
    Sulcus carinae[161]
  Margo ventralis
  Apex carinae
    Facies articularis clavicularis (see **Arthr.** Annot. 85, 88)
    Tuberositas lig. sternoclavicularis (see **Arthr.** Annot. 85, 88)
  Facies lateralis carinae
    Linea intermuscularis[153]

OSSA CINGULI MEMBRI THORACICI

Clavicula [Furcula][162] (Fig. 6)

Extremitas sternalis claviculae
- Apophysis furculae [Hypocleideum][163]

Scapus [Corpus] claviculae

Extremitas omalis claviculae [Epicleidium][164]
- Proc. acromialis[165]
  - Facies articularis acromialis (see **Arthr.** Annot. 96, 98)
- Proc. acrocoracoideus[165]
  - Facies articularis acrocoracoidea (see **Arthr.** Annot. 96, 98)

Scapula[166] (Fig. 6)

Facies lateralis[166]

Facies costalis[166]

Margo dorsalis [vertebralis]

Margo ventralis

Extremitas cranialis scapulae
- Caput scapulae
- Acromion
  - Facies articularis clavicularis
- Facies articularis humeralis[167]
- Tuberculum coracoideum[168, 173]

Collum scapulae
- Tuberculum m. scapulotricipitis[169]
- Sulcus m. supracoracoidei[172] (see Canalis triosseus)

Corpus scapulae[170]

Extremitas caudalis scapulae

Coracoideum (see Fig. 6; and **Arthr.** Fig. 3)

Extremitas omalis coracoidei
- Proc. acrocoracoideus
  - Impressio lig. acrocoracohumeralis
  - Facies articularis clavicularis
- Proc. procoracoideus
  - Sulcus m. supracoracoidei[172] (see Canalis triosseus)
- Facies articularis humeralis[171]
- Cotyla scapularis[168, 173]

Corpus coracoidei
- Facies ventralis
- Facies dorsalis
- Margo medialis
- Margo lateralis

Coracoideum—*continued*
Foramen [Incisura] n. supracoracoidei
Linea intermuscularis ventralis[174]
Extremitas sternalis coracoidei
Facies articularis sternalis
Crista ventralis
Crista dorsalis
Facies articularis intercoracoidea[177] (see **Arthr.** Annot. 90)
Angulus medialis
Proc. lateralis[176]
Angulus lateralis
Impressio m. sternocoracoidei

Canalis triosseus [supracoracoideus][177] (see **Arthr.** Fig. 3)

OSSA ALAE [MEMBRI THORACICI][178]

Ossa brachii

Humerus[178] (Fig. 8)
Extremitas proximalis humeri
Caput humeri[178 179]
Incisura capitis[180]
Facies bicipitalis[181]
Tuberculum dorsale [minus][182]
Impressio m. supracoracoidei[183]
Crista pectoralis [tuberculi dorsalis][184]
Angulus cristae
Impressio m. pectoralis
Sulcus ligamentosus transversus[185]
Impressio m. coracobrachialis cranialis[186]
Tuberculum ventrale [majus][187]
Crista bicipitalis [tuberculi ventralis]
Fossa pneumotricipitalis[188]
Crus ventrale fossae[188]
Crus dorsale forssae[188]
Foramen pneumaticum[189]
Margo caudalis[188] (Fig. 8)
Intumescentia[190]
Corpus humeri
Facies cranialis[178]
Facies caudalis[178]
Margo ventralis
Margo dorsalis

Linea [Eminentia] m. latissimi dorsi
Sulcus m. radialis[191]
Extremitas distalis humeri
Condylus dorsalis[192]
Condylus ventralis[192]
Incisura intercondylaris[193]
Fossa m. brachialis
Epicondylus dorsalis[194] [Ectepicondylus][194]
(see **Myol.** Annot. 91, 92)
Epicondylus ventralis[194] [Entepicondylus][194]
(see **Myol.** Annot. 91, 92)
Proc. flexorius[195]
Tuberculum supracondylare dorsale[196]
Proc. supracondylaris dorsalis[196]
Tuberculum supracondylare ventrale[197]
Fossa olecrani
Sulcus m. humerotricipitis
Sulcus m. scapulotricipitis

Os sesamoideum m. scapulotricipitis[202]

Ossa antebrachii

Ulna (Fig. 9)

Extremitas proximalis ulnae
Proc. cotylaris dorsalis[198] (Fig. 9)
Cotyla dorsalis[198]
Impressio m. scapulotricipitis[198]
Incisura radialis[199]
Cotyla ventralis[200]
Crista intercotylaris
Olecranon[201]
Tuberculum bicipitale[210] (Fig. 9)
Depressio m. brachialis
Tuberculum lig. collateralis ventralis[203]
Sulcus tendineus[203]
Corpus ulnae
Margo interosseus [cranialis]
Margo caudalis
Margo dorsalis
Facies cranialis[204]
Facies caudodorsalis[204]
Facies caudoventralis[204]

Ulna—*continued*

Lineae intermusculares
Papillae remigiales caudales[205]
Papillae remigiales ventrales[205]
Extremitas distalis ulnae
Trochlea carpalis
Condylus dorsalis[206]
Labrum condyli
Condylus ventralis[206]
Sulcus intercondylaris
Tuberculum carpale[207]
Sulcus radialis[208]
Sulcus tendineus[209]
Tuberculum retinaculi (see Fig. 9)

Radius (Fig. 9)

Extremitas proximalis radii
Caput radii
Cotyla humeralis
Facies articularis ulnaris
Tuberculum bicipitale[210] (see Fig. 9)
Corpus radii
Margo dorsalis
Margo ventralis
Margo interosseus [caudalis]
Lineae intermusculares
Extremitas distalis radii
Facies articularis radiocarpalis[211]
Sulcus ligamentosus[212] (see **Arthr.** Annot. 116)
Facies articularis ulnaris
Tuberculum aponeurosis[213]
Sulcus tendineus

Ossa manus

Ossa carpi

(Ossa carpi proximalia)[214]
Os carpi radiale[214] (see **Arthr.** Annot. 112, 122)
Facies articularis radialis
Facies articularis ulnaris
Facies articularis metacarpalis
Os carpi ulnare[214] (see **Arthr.** Annot. 112, 122)
Facies articularis ulnaris

Incisura metacarpalis
Facies articularis metacarpalis
(Ossa carpi centralia)[214]
(Ossa carpi distalia)[214]

Carpometacarpus (see Intro. and Fig. 10; and **Arthr.** Annot. 128)

Extremitas proximalis carpometacarpi
Trochlea carpalis
Facies articularis ulnocarpalis[215] (see **Arthr.** 124)
Facies articularis radiocarpalis[215] (see **Arthr.** 124)
Os metacarpale alulare[216] (see Intro.)
Proc. extensorius
Proc. alularis
Facies articularis alularis
Fovea carpalis cranialis[217]
Fovea carpalis caudalis[217]
Fossa infratrochlearis[218]
Fossa supratrochlearis[219]
Proc. pisiformis[220]
Corpus carpometacarpi
Facies dorsalis
Facies ventralis
Margo cranialis
Margo caudalis
Os metacarpale majus (see Intro.)
Sulcus tendineus
Proc. intermetacarpalis[221]
Os metacarpale minus (see Intro.)
Spatium intermetacarpale
Extremitas distalis carpometacarpi
Synostosis metacarpalis distalis[222]
Sulcus interosseus[223]
Facies articularis digitalis major
Facies articularis digitalis minor

Ossa digitorum manus[224] (see Intro.)

Phalanx digiti alulae
Phalanx proximalis digiti majoris
Pila cranialis[225]
Fossa dorsalis
Fossa ventralis
Facies articularis metacarpalis
Facies articularis phalangealis

Ossa digitorum manus—*continued*

Phalanx distalis digiti majoris
Phalanx digit minoris

OSSA CINGULI MEMBRI PELVICI

Os coxae (Figs. 5, 11)
Acetabulum
Fossa acetabuli
Foramen acetabuli[226]
Foramen obturatum[227 252]
Sulcus m. obturatorii[228]
Fenestra ischiopubica[229 227]
Foramen ilioischiadicum[230]
Incisura marginis caudalis[231]
Proc. marginis caudalis[231]
Antitrochanter[232]
Crista iliosynsacralis[233]
Sulcus iliosynsacralis[234]
Canalis iliosynsacralis[234]
Concavitas infracristalis[235]
Pila ilioischiadica[236]
Fossa renalis[250]
Pars ischiadica[237]
Pars pudenda[237]
Recessus iliacus[250]
Incisura caudalis pelvis[238]

Ilium (Figs. 5, 11)
Corpus ilii[239]
Incisura acetabularis[240]
(Proc. ischiadicus)[241]
(Proc. pubicus)[241]
Proc. antitrochantericus[232]
Facies articularis femoralis
Sulcus antitrochantericus
Crista iliaca obliqua[242] (Fig. 5)
Crista iliaca intermedia[243]
Tuberculum preacetabulare [Proc. pectinealis][244] (see Pubis)
Facies renalis ilii
Ala [Pars] preacetabularis ilii
Margo cranialis
Margo lateralis

Margo medialis [vertebralis]
Facies dorsalis
Fossa iliaca dorsalis
Crista iliaca dorsalis[245]
Facies ventralis
Facies costalis[246]
Ala [Pars] postacetabularis ilii
Margo lateralis[247]
Crista iliaca dorsolateralis[248]
Proc. [Spina] dorsolateralis[248] (Fig. 5)
Margo medialis [vertebralis]
Margo caudalis[238]
Facies lateralis[235]
Facies dorsalis
Fossa iliocaudalis[249] (Fig. 11)
Facies ventralis
Facies renalis ilii
Crista iliaca caudalis
Margo foraminis ilioischiadici
Proc. terminalis ilii
Lamina ischiadica[251] [235]
Proc. [Spina] dorsolateralis[248] (Fig. 5)
Recessus iliacus[250]

Ischium (Figs. 5, 11)
Corpus ischii
Incisura acetabularis[240]
Margo foraminis ilioischiadici
Margo obturatoria
(Proc. pubicus)[241]
(Proc. iliacus)[241]
Proc. antitrochantericus[232]
Proc. obturatorius[252]
Ala ischii
Facies lateralis
Facies medialis (see **Arthr.** Annot. 150)
Margo ventralis [pubica]
Crista pilae ilioischiadicae[236]
Proc. terminalis ischii[253]

Pubis[254] (see Figs. 5, 11; and Os coxae)
Corpus pubis
Incisura acetabularis

Pubis—*continued*

Incisura obturatoria[240]

(Proc. iliacus)

Tuberculum preacetabulare [Proc. pectinealis][244] (see Ilium)

Scapus pubis[5 254] (Fig. 5)

Apex pubis (see **Arthr.** Annot. 149)

OSSA MEMBRI PELVICI

Femur [Os femoris] (Fig. 12)

Extremitas proximalis femoris

Caput femoris

Fovea lig. capitis

Facies articularis acetabularis

Collum femoris

Facies articularis antitrochanterica[255]

Trochanter femoris[256]

Crista trochanteris

Fossa trochanteris

Impressiones iliotrochantericae[257]

Impressiones obturatoriae[257]

Corpus femoris[258]

Facies cranialis

Linea intermuscularis cranialis

Facies caudalis

Linea intermuscularis caudalis

Facies medialis

Facies lateralis

Extremitas distalis femoris

Sulcus patellaris[259]

Facies articularis patellaris

Fossa poplitea

Condylus medialis[260]

Condylus lateralis[260 261]

Fovea tendinis m. tibialis cranialis

Trochlea fibularis[261]

Crista tibifibularis[262]

Impressio lig. cruciati caudalis[263]

Sulcus intercondylaris

Impressio lig. cruciati cranialis

Epicondylus medialis

Impressio lig. collateralis medialis

Epicondylus lateralis

Impressio lig. collateralis lateralis
Crista supracondylaris medialis[264]
Tuberculum m. gastrocnemialis medialis
Crista supracondylaris lateralis
Tuberculum m. gastrocnemialis lateralis
Impressio ansae m. iliofibularis (see **Arthr.** Annot. 186)

Patella[265]

Tibiotarsus (see Fig. 13; and **Arthr.** Annot. 167)
Extremitas proximalis tibiotarsi[266]
Facies articularis medialis[267]
Facies articularis lateralis[267]
Area interarticularis
Fossae retrocristales[268]
Crista patellaris[269]
Crista cnemialis [tibialis] lateralis
Crista cnemialis [tibialis] cranialis[269, 270]
Facies gastrocnemialis[271]
Impressio lig. collateralis medialis
Sulcus intercristalis[272]
Incisura tibialis[273] (Fig. 13)
Tuberositas poplitea
Fossa flexoria[274]
Corpus tibiotarsi[275]
Facies cranialis
Facies medialis
Facies caudalis
Margo cranialis
Margo caudalis
Margo lateralis [fibularis]
Crista fibularis (see **Arthr.** Annot. 162)
Lineae m. fibularis [peronei]
Linea extensoria[276]
Tuberositas retinaculi mm. extensorum (see **Arthr.** Annot. 187)
Sulcus extensorius
Pons supratendineus[277]
Canalis extensorius[278]
Extremitas distalis tibiotarsi
Tuberositas retinaculi m. fibularis [peronei][282]
Sulcus m. fibularis [peronei]
Trochlea cartilaginis tibialis (see **Arthr.** Annot. 164)

Tibiotarsus—*continued*
Sulcus cartilaginis tibialis[279]
Cristae sulci
Condylus medialis[280]
Depressio epicondylaris medialis[281]
Condylus lateralis[280]
Depressio epicondylaris lateralis[281]
Epicondylus medialis
Epicondylus lateralis
Incisura intercondylaris
Area intercondylaris
Impressio lig. intercondylaris (see **Arthr.** Annot. 171)

Fibula (Fig. 13)
Caput fibulae[261] [262]
Facies articularis femoralis
Facies articularis tibialis (see **Arthr.** Annot. 157, 159)
Tuberositas lig. collateralis lateralis
Fovea m. poplitei
Corpus fibulae
Crista articularis tibialis (see **Arthr.** Annot. 161, 162)
Tuberculum m. iliofibularis
Spina fibulae

Os sesamoideum intertarsale (see **Arthr.** Annot. 164)

Ossa pedis

Ossa tarsi[283]
(Ossa tarsi proximalia)
(Os tarsi tritibiale)
(Os tarsi fibulare)
(Ossa tarsi centralia)
(Os tarsi intermedium)
(Ossa tarsi distalia)

Ossa metatarsalia (see **Arthr.** Annot. 173, 174)
Os metatarsale I [hallucis] (see **Arthr.** Fig. 9)
Proc. articularis tarsometatarsalis
Trochlea metatarsi I [primi]
Tuberositas lateralis
Ossa metatarsalia II–IV

Tarsometatarsus[284] (see Fig. 14; and **Arthr.** Annot. 173, 174)

Extremitas proximalis tarsometatarsi

Cotyla medialis

Cotyla lateralis

Eminentia intercondylaris (see **Arthr.** Annot. 171)

Area intercotylaris[285]

Fovea menisci lateralis

Sulcus ligamentosus[286]

Tuberositas m. fibularis [peronei] brevis

Impressio lig. collateralis lateralis

Sulcus m. fibularis [peronei] longi

Hypotarsus[288] (Fig. 13)

Crista medialis hypotarsi[289]

Crista lateralis hypotarsi[289]

Cristae intermediae hypotarsi[289]

Sulci hypotarsi

Canales hypotarsi

Crista plantaris mediana[290]

Fossa parahypotarsalis medialis

Fossa parahypotarsalis lateralis

Fossa infracotylaris dorsalis[291]

Tuberositas m. tibialis cranialis

Foramina vascularia proximalia (see **Art.** Annot. 79)

Corpus tarsometatarsi

Facies dorsalis[292]

Facies plantaris[292]

Facies subcutanea medialis[293]

Facies subcutanea lateralis[293]

Sulcus flexorius[294]

Crista plantaris medialis[294]

Crista plantaris lateralis[294]

Sulcus extensorius[295]

Arcus extensorius[287]

Impressiones retinaculi extensorii

Fossa metatarsi I (see **Arthr.** Annot. 174, 175, 180)

Proc. calcaris[296]

Extremitas distalis tarsometatarsi[297]

Foramen vasculare distale

Canalis interosseus tendineus[298]

Trochlea metatarsi II [secundi]

Foveae ligg. collateralium

Trochlea metatarsi III [tertii]

Foveae ligg. collateralium

Tarsometatarsus—*continued*

Trochlea metatarsi IV [quarti]

Foveae ligg. collateralium

Trochlea accessoria[297]

Incisura intertrochlearis medialis

Incisura intertrochlearis lateralis

Fossa supratrochlearis plantaris

Ossa digitorum pedis[299]

Phalanges proximales et intermediae

Basis phalangis

Cotyla articularis

Tuberculum extensorium (see **Arthr.** Annot. 183)

Tuberculum flexorium (see **Arthr.** Annot. 182)

Corpus phalangis

Facies dorsalis

Facies plantaris

Capitulum phalangis

Trochlea articularis

Phalanx distalis [ungularis][300]

Basis phalangis

Cotyla articularis

Tuberculum flexorium (see **Arthr.** Annot. 182)

Tuberculum extensorium (see **Arthr.** Annot. 183)

Apex phalangis

ANNOTATIONS

(1) **Cotyla** -ae; L. small vessel from Gr. referring to a cup or cup-like cavity. In this work "cotyla" is used for a shallow cup-shaped articular surface (Howard, 1929; Lambrecht, 1933).

(2) **Cartilago epiphysialis; Epiphysis; Diaphysis; Metaphysis.** During development and growth of long bones the ossification begins in the middle of the shaft (Diaphysis), and extends proximally and distally by growth of ossifying zones (Metaphysis) into the cartilaginous ends (Epiphysis). Unlike the mammal the epiphyses of birds do not ossify endochondrally from separate centers of ossification, but only by extension from the metaphysial centers.

(3) **Os pneumaticum.** Bone that has been invaded by air sacs. Air sacs of bones can be outgrowths of air sacs of lungs, tympanic cavity, or nasal cavity.
Foramen pneumaticum; Pori pneumatici. The skull, vertebrae, bones of limb girdles are usually pneumatized; the limb bones are variably pneumatized in different species. The pneumatization may extend to the extremities of the limb, or only the proximal elements may be involved. Inasmuch as the pneumatization is so widespread in the

skeleton, the foramina and the smaller pores are listed only for the bones in which they form especially distinctive features. In palaeontological literature the pneumatic foramina are commonly known as "pneumatopores". See **Resp.** Annot. 70, 75–77.

(4) **Pila.** From L.: pillar or column. Pila refers to a strengthening of reinforcing element of a bone that may be a distinct, prominent part, or may be a thickening that blends almost imperceptibly into the bone of which it is a part.

(5) **Scapus.** From L.: shaft or stem. Used in this work to refer to slender attenuated bones or parts of bones where "corpus" is not applicable (e.g. clavicle, pubis). See **Integ.** Partes pennae.

(6) **Os medullare.** Birds are unique in that females possess a special system of highly labile, secondary bone (medullary bone) within the marrow cavities of much of the skeleton during the reproductive period. This bone grows as spicules into the medullary cavity from the endosteal surface. Medullary bone is the main peculiarity of the avian skeleton. The significance of medullary bone is that it serves as a labile reserve of mineral that can be mobilized to provide calcium for egg shell formation. Taylor *et al.* (1971) present an extensive review of the structure and function of medullary bone; the above annotation was extracted from their account. See also Hodges (1974). **Os spongiosum; Cavitas pneumatica ossis.** Spongy bone (also known as cancellous or trabecular bone) is widespread throughout the avian skeleton. In early post-natal life the spongy bone of the Calvaria (diploë), vertebrae, limb bones, etc., is filled with red marrow (see Hodges, 1974). Later the marrow is replaced by fatty marrow (Medulla ossium flava) or by pneumatic spaces that invade the bones from the nasal cavity, tympanic cavity (see Stork, 1972), lungs, and air sacs forming the pneumatic cavities of the bones. See Annot. 189; and **Resp.** Annot. 21, 70, 72, 75–77.

(7) **Tuberositas.** L. tuberosus, full of lumps; from tuber, a lump. By conventional anatomical usage Tuberositas usually refers to a roughened or knobby area or projection of bone for attachment of tendons or ligaments (Donath, 1969).

(8) **Facies; Cranium.** Following the human *Nomina Anatomica* (IANC, 1966) and the *Nomina Anatomica Veterinaria* (ICVAN, 1973) "Cranium" refers to neurocranium and "Facies" refers to splanchnocranium.

(9) **Hiatus orbitonasalis.** Synonymy: antorbital vacuity (Shufeldt, 1909). The pronounced gap, often triangular, between the maxillary process of Os nasale and the prefronto-ectethmoid complex. The space within the Hiatus comprises the caudal part of the nasal cavity. See Fig. 1.

(10) **Hiatus craniofacialis** (Hofer, 1955). Interval between rostral edge of interorbital septum and the caudal border of nasal septum.

(11) **Arcus jugalis.** Synonymy: Arcus zygomaticus. The jugal arch consists of three ankylosed elements: Proc. jugalis of Os maxillare, Os jugale, and Os quadratojugale (see Fig. 2; and **Arthr.** Intro.).

(12) **Apertura nasalis ossea.** Some pelecaniform, sphenisciform, and other birds have minute osseous external nares. The shape of the osseous naris (schizorhinal, holorhinal) is related to kinesis of the upper jaw (see Hofer, 1955; Bock, 1964; Yudin, 1965). See also **Arthr.** Zonae elasticae.

(12)—*cont.* **Pila supranasalis.** The median column of bone forming the dorsal border of the external nares; formed by processes of nasal and premaxillary bones. See **Arthr.** Fig. 2.

(13) **Fenestra palatina.** Oval or elongate opening in rostral part of osseous palate as in certain anseriform birds; set off distinctly from Fissure interpalatina; confluent with the Fissura as in *Cathartes*. Fenestra palatina also known as Fonticulus palatinus (Hofer, 1949).

(14) **Depressio frontalis.** The frontal region of the skull in some birds (e.g. *Ardea*, *Anser*) is indented by a shallow longitudinal concavity. In other birds the frontal region is flat or dorsally convex.

(15) **Sulcus gl. nasalis.** In certain birds the nasal gland occupies a pronounced sulcus on the dorsal aspect of the supraorbital margin of the orbit, involving mostly the frontal bone (e.g. penguins, loons, albatrosses, gulls). In others the gland is intra-orbital, ventral to the supraorbital margin of the orbit.

(16) **Prominentia cerebellaris** (Shufeldt, 1909). External median convexity of Os supraoccipitale and Os parietale that overlies the dorsum of cerebellum and reflects its contour externally (e.g. *Columba*, *Buteo*, *Corvus*).

(17) **Crista [Linea] nuchalis sagittalis.** Median line or crest dorsal to Foramen magnum for attachment of median lamina of deep fascias of epaxial muscles. The Crista may surmount Prominentia cerebellaris in some forms (e.g. *Gavia immer*, *Morus bassanus*).
Crista [Linea] nuchalis transversa. Synonymy: Crista temporalis (Hofer, 1945); Crista occipitalis (Davids, 1952).

(18) **Foramen magnum.** Synonymy: Foramen occipitale magnum. Bounded by supra-, ex-, and basioccipital bones. Duijm (1951) reviews the topic of the position of the plane of the Foramen magnum in the major skull types of birds.

(19) **Meatus acusticus externus.** Synonymy: Fossa auricularis cutanea (Freund, 1926). Wall of meatus mostly formed by cutaneous, fibrous, and cartilaginous tissues; the osseous wall of meatus is formed by alae of parasphenoid, squamosal, and occipital bones. See Annot. 21.

(20) **Hiatus alae tympanicae.** Hiatus or deficiency in the bone of the junctional area between Ala tympanica of Os parasphenoidale and Ala tympanica of Os exoccipitale.

(21) **Cavitas tympanica.** Consists of a shallow open concavity in the dried skull. On account of the placement of Membrana tympanica, only the rostodorsal part of the osseous concavity is tympanic cavity; the caudoventral part is Meatus acusticus externus (Freund, 1926). See above, Annot. 19.

(22) **Recessus antevestibularis.** Synonymy: Antivestibulum (Magnus, 1870); Recessus cavi tympani (Hasse, 1871); Recessus stapedialis (Stresemann, 1934). Recess in complex of otic bones in which the basal part of Columella is located. Fenestra vestibularis, Fenestra cochlearis, and pneumatic pores open into the Recessus antevestibularis (see Annot. 26).

(23) **Foramen pneumaticum dorsale.** Consistently located between squamosal and prootic articular facets for quadrate bone.

(24) **Pila prootica.** Synonymy: Pila paroccipitalis. Short or elongate column of bone that bears the articular facet for the prootic condyle of Quadratum; Pila intervenes between dorsal pneumatic foramen and antevestibular recess.

(25) **Foramen pneumaticum rostrale.** In rostral part of Cavitas tympanica; located dorsal to ostium of auditory tube and dorsolateral to osseous Canalis caroticus. Air sacs from this foramen invade the parabasisphenoid bone.

(26) **Foramen pneumaticum caudale.** An opening in the caudal wall of Recessus antevestibularis that leads caudally into a chamber ventrolateral to the rostral semicircular canal. Air cells in the exoccipital bone communicate with this chamber, Antrum pneumaticum caudale.

(27) **Canalis ophthalmicus externus.** Synonymy: Canalis facialis; Canalis stapedialis. Canal conducts the A. et V. ophthalmica externa. The canal arches caudal and dorsal to Foramen antevestibulum; its lateral wall may project in relief into the tympanic cavity, or may be incomplete so that the interior of the canal may be observed in the dry skull. The caudal ostium of the canal is located in Fossa parabasilis; the rostral ostium is located medial to the otic process of the Quadratum, lateral to Foramen n. maxillomandibularis.

(28) **Basis cranii externa.** The usual definition of Basis cranii includes the entire ventral aspect of the skull, including facial skeleton, but excluding mandible. In a strict sense, the term Basis cranii externa should be limited to the exterior aspect of the cranial bones that form the floor of the cranial cavity proper. This would make Basis cranii interna and Basis cranii externa opposites that correspond closely to one another.

(29) **Orbita.** The osseous orbit of birds is bounded principally by cranial bones. The orbit is usually open ventrally; exceptions are the snipes and woodcocks in which the orbit is almost completely enclosed by bone. (Hofer, 1955).
Fonticuli orbitales (Barkow, 1856). Synonymy: Foramina obturata orbitalia. Deficiencies in the bone of the interorbital septum, caudal wall, or dorsal wall of orbit; in life closed by fibrous membranes.

(30) **Proc. postorbitalis.** Synonymy: Proc. orbitalis posterior; Proc. postfrontalis. The postorbital process commonly forms the caudoventral border of the orbit; however in certain birds (e.g. anseriforms) it projects rostrally and contributes to the ventral margin of the orbit (see below, Arcus suborbitalis). The tips of Proc. postorbitalis and Proc. zygomaticus are ankylosed in some birds (e.g. psittaciforms and galliforms). See Annot. 101.
Arcus suborbitalis (Portmann, 1950). A complete osseous arch bounds the orbit ventrally in psittaciforms, in certain snipes (Gadow and Selenka, 1891), in *Dendrocygna* (Shufeldt, 1909), and in tinamous according to Newton and Gadow (1896). This arch is formed by junction of a caudal extension of Os prefrontale and rostral extension of the Proc. postorbitalis or the postorbital–zygomatic process complex; these structures are united in other birds by Lig. suborbitale (see **Arthr.** Annot. 30).

(31) **Foramen orbitonasale.** Located between the orbital surface of Os frontale and the dorsal part of Os ectethmoidale. A medial foramen only may be present in some birds (e.g. *Ardea herodias*, *Aythya collaris*); both medial and lateral foramina are found in other birds (e.g. *Columba*, *Gallus*, *Coragyps atratus*, *Corvus brachyrhynchos*).
(Annot. 31 cont. on p. 84)

Foramen orbitonasale mediale transmits N. olfactorius and the medial ramus of N. ophthalmicus; the lateral foramen conducts the lateral ramus of N. ophthalmicus and the duct(s) of Gl. nasalis.

(32) **Foramen opticum.** Located in boundary zone between caudal edge of interorbital septum and caudal wall of orbit. Foramen opticum is usually a single opening inside the cranial cavity, and is bilateral externally.

(33) **Septum osseum fossae bulbi.** The fossa for the olfactory bulb is divided by an osseous septum in *Apteryx* (Starck, 1955) and certain albatrosses (pers. obs.).

(34) **Foramen ethmoidale.** Transmits A. et V. ethmoidalis into orbit from cranial cavity.

(35) **Fossa cranii media.** Synonymy: Fossa mesencephalica. The fossa houses the Diencephalon and Chiasma opticum medially and Tectum mesencephali on each side. The avian middle cranial fossa is not homologous with that of the mammals.

(36) **Crista marginalis.** Separates the cerebellar fossa from the general chamber of the vault of the Calvaria that contains the telencephalic hemispheres.

(37) **Tuberculum pineale.** On the inner aspect of calvaria the Tuberculum is a triangular eminence at the junction of Crista frontalis interna with marginal cristae of cerebellar fossa. The dorsal, expanded end of Gl. pinealis (Corpus gl. pinealis) is closely related to Tuberculum pineale.

(38) **Fossa auriculae cerebelli.** Synonymy: Fovea hemispherii cerebelli; Fossa subarcuata. The fossa in birds contains the cerebellar auricle; whereas the subarcuate fossa of mammals lodges the endolymphatic sac, therefore is not homologous to the avian fossa. See **Ven.** Annot. 35.

(39) **Fovea ganglii vagoglossopharyngealis.** The fovea is located in the suture between exoccipital and opisthotic bones. The fovea contains the combined proximal [root] ganglia of cranial nerves IX and X, and communicates with the exterior by separate foramina for N. glossopharyngeus and N. vagus.

(40) **Symphysis mandibularis.** This is the region of the junction of the two mandibular rami. See Annot. 42; and **Arthr.** Annot. 21.

(41) **Foveae corpusculorum nervosorum.** Especially numerous and conspicuous pits that house sensory corpuscles in anseriforms, sandpipers and snipes (Scolopacidae), and *Apteryx*; present in both jaws.

(42) **Pars symphysialis, intermedia, et caudalis.** Symphysial part is the rostral segment of the mandibular ramus that articulates with the contralateral ramus at Sutura intermandibularis. The intermediate part extends caudally to the oblique Zona elastica intramandibularis proximalis, often marked by the Fenestra mandibulae rostralis. Pars caudalis extends from the line of the proximal elastic zone to the retroarticular process, and includes the area of attachment of the jaw muscles and articulation with Quadratum. Segments named after Lebedinsky (1920). See **Arthr.** Annot. 21.

(43) **Angulus mandibulae.** Point on dorsal border of ramus where ramus becomes angulated or curved ventrally (e.g. caprimulgiforms, columbiforms, falconiforms, and passerines).

(44) **Proc. coronoideus.** Synonymy: Proc. pseudocoronoideus; Proc. m. adductoris mandibulae. Any process of Pars caudalis of the mandible to which is attached the strong "aponeurosis" of M. adductor mandibulae externus, pars temporalis. Commonly located on dorsal margin of mandible; may be located on lateral aspect as in anseriforms (Zweers, 1974). See Fig. 1.

(45) **Tuberculum pseudotemporal.** Synonymy: Proc. pseudotemporalis. The tendon of insertion of M. pseudotemporalis superficialis is inserted into the Tuberculum pseudotemporale. The tuberculum can be quite large in heavy-billed finches (Fringillidae); it is located slightly rostral to quadratomandibular articulation near the base of Proc. mandibulae medialis (see Fig. 1).

(46) **Fenestrae mandibulae.** Synonymy: Foramen mandibulare anterior, Foramen mandibulare posterior (Lebedinsky, 1920); Foramen ovale; interangular vacuity or fenestra (Shufeldt, 1909). Fenestrae may be completely lacking in some species. Two fenestrae occur in certain birds (e.g. some charadriiforms, gruiforms, psittaciforms, and strigiforms). Fenestra mandibulae rostralis is found in the region of the proximal intramandibular elastic zone (**Arthr.** Fig. 2); Fenestra mandibulae caudalis occurs within Pars caudalis of the mandibular ramus (see Annot. 33). Other birds may possess only one of the two fenestrae. Consult Lebedinsky (1920) for a detailed summary of the fenestrae.

(47) **Canalis mandibulae.** Conducts vessels and R. intramandibularis of N. mandibularis from region of coronoid process to symphysial region of Ramus mandibulae.

(48) **Fossa aditus canalis mandibulae.** Distinct depression on internal aspect of Pars caudalis of mandibular ramus; forms a trough leading to the aditis or opening of the mandibular canal. Floor of fossa often consists of very thin bone, sometimes fenestrated. Fossa is conspicuous in many birds (e.g. *Pygoscelis adeliae*, *Gavia immer*, *Cathartes aura*, and *Anser*).

(49) **Proc. retroarticularis.** Synonymy: Proc. mandibularis posterior (Lebedinsky, 1920); Proc. angularis posterior (Hofer, 1945). The process is formed mostly by Os angulare.

(50) **Recessus conicalis.** Synonymy: Conical fossa (Shufeldt, 1909); Recessus posterior (Lebedinsky, 1920); Cavum mandibulare (Zweers, 1974). In anseriform birds a deep recess ventral to the quadrate articular facies of the mandible; its opening is located between the blade-like Proc. retroarticularis and Proc. mandibulae medialis.

(51) **Fossa caudalis.** Synonymy; Fossa posterior (Lebedinsky, 1920); postarticular surface (Milne-Edwards, 1867–71). In many birds this term refers to the shallow concavity on the caudal surface of Proc. mandibulae medialis located medial to the retroarticular process; area for insertion of M. depressor mandibulae. See Lebedinsky (1920) for a detailed comparative account of Fossa caudalis in numerous avian species.

(52) **Os dentale.** Principal element of each mandibular ramus. Articulates with supraangular and splenial by squamous sutures at the junction of intermediate and caudal parts of the ramus. See Lebedinsky (1920) and Jollie (1957).

(53) **Os maxillare; Maxilla.** As an individual bone Os maxillare constitutes one of the components of the avian upper jaw. As a general term "Maxilla" refers to the entire complex of structures that make up the upper jaw; i.e. the opposite of "Mandibula", the lower jaw (see **Topog. anat.** Annot. 8).

(54) **Proc. maxillaris.** Synonymy: Proc. postnarialis.
Proc. premaxillaris. Synonymy: Proc. dorsonarialis.

(55) **Septum nasale osseum; Conchae nasales.** May be formed in part by ectethmoid, maxillae, and vomer. The rostral part of the nasal septum and nasal conchae in the caudal part of the nasal cavity vary in the extent that they ossify in different birds, usually remaining more or less cartilaginous. These structures characteristically ossify extensively in some birds of prey (e.g. *Buteo lineatus*). See **Resp.** Annot. 16.

(56) **Crista tomialis.** The sharp lateral edges of the upper jaw; formed mostly by Os premaxillare (see **Topog. anat.** Annot. 12).

(57) **Canalis neurovascularis.** Longitudinal canal (paired) conducting terminal branch of N. ophthalmicus and vessels from the rostral wall of the nasal cavity into the tip of the upper jaw.
Foramina neurovascularis. Rami of nerves and vessels leave Canalis neurovascularis via canaliculi that lead to the surface via the foramina.

(58) **Proc. palatinus.** Synonymy: Proc. maxillopalatinus. This process has a variety of configurations in different birds; may be inflated, pneumatized, trabeculated, excavated, etc. The process contributes to formation of the nasal cavity in most birds and to the palate in the so-called desmognathous birds (e.g. Anseriformes, Ciconiiformes). See Hofer (1949).

(59) **Proc. jugalis.** Synonymy: Proc. labialis.

(60) **Os palatinum.** Synonymy: Os pterygopalatinum. The definitive Os palatinum of neognathine birds is usually a compound bone since a segment of Os pterygoideum becomes fused to its caudal end during development (Jollie, 1957, 1958). The caudal end of Proc. pterygoideus of Os palatinum is in fact the detached rostral end of Os pterygoideum. Hofer (1945) discusses the variety of shapes and relationships of palatine bones in different groups of birds.

(61) **Os pterygoideum.** Pars palatina of Os pterygoideum that joins Os palatinum is also known as the Mesopterygoid or Hemipterygoid. In paleognathine birds and some neognaths the Pars palatina remains as a part of the definitive Os pterygoideum (Jollie, 1957). According to Hofer (1945) the Os pterygoideum may be arched (e.g. *Mergus*), bent (e.g. *Vanellus*), or extended (e.g. *Ardea*) (see Annot. 60 and 93).

(62) **Lamella [Ala] caudolateralis.** Synonymy: Proc. posterolateralis (Beddard, 1898; Hofer, 1945; Bühler, 1970). The flared, wing-like Lamella is pronounced as in *Ceryle alcyon*, *Picoides borealis*, *Caprimulgus carolinensis* and *Corvus ossifragus*.

(63) **Crista ventralis.** Crest that bounds choanal fissure. In birds whose palatines are apposed or fused in the median plane the Cristae ventrales form an unpaired, median ventral palatine crest.

(64) **Proc. pterygoideus.** Synonymy: Proc. caudalis. Articulates with Os pterygoideum (see **Arthr.** Annot. 28).

(65) **Fossa ventralis.** Synonymy: Fossa muscularis. Shallow depression between ventral and lateral cristae of Os palatinum.

(66) **Fossa medialis.** On medial aspect of Os palatinum; located between dorsal crista and ventral crista. Fossae mediales bound the nasal choanae on each side.

(67) **Proc. choanalis rostralis** (Bühler, 1970). Splint-like process (paired) that articulates with the side of the Vomer as in *Anser anser*, *Ardea herodias*, *Caprimulgus carolinensis*, *Corvus ossifragus*.

(68) **Vomer.** Synonymy: Prevomer. Not present in all birds; weakly developed in galliforms. The vomer varies in shape, and ranges from a horizontally flattened plate, strongly V-shaped in cross section, to a vertical plate. In passerine birds the vomer is extended into the "ethmoid tissue", a unique condition limited to this group. See Hofer (1949).

(69) **Facies articularis sphenoidalis.** Surface for articulation with Rostrum sphenoidale (see **Arthr.** Annot. 27).

(70) **Proc. dorsalis.** Prominent dorsal muscular process of Os pterygoideum characteristic of woodpeckers (Piciformes) (Hofer, 1945).

(71) **Proc. orbitalis quadrati.** Synonymy: Proc. pterygoideus (Hofer, 1945). Absent in Psittacidae, some Alcedinidae, etc.

(72) **Condylus caudalis.** Synonymy: Proc. postmandibularis. Quadrate of most birds has three condyles (see Bock, 1960; and **Arthr.** Annot. 35).

(73) **Cotyla quadratojugalis.** Cotyla for articulation with Os quadratojugale is located on a distinct Proc. quadratojugalis on the superficial surface of the lateral condyle of the Quadratum of some birds.

(74) **Ossa supraorbitalia** (Beddard, 1898). Synonymy: Os supraciliare.

(75) **Ossa accessoria cranii.** See Jollie (1957) for discussion of accessory bones of the head.

(76) **Os nuchale.** Synonymy: Stylus postoccipitalis. Occurs in *Phalacrocorax* (Dullemeijer, 1951).

(77) **Annulus tympanicus.** Occurs in strigiforms and *Gallus* (Stellbogen, 1930),

(78) **Ossa suturarum.** Supernumerary bones developed in cranial sutures; seen in young turkeys and ducks.

(79) **Apparatus hyobranchialis** (Goodrich, 1958). Synonymy: Apparatus hyoideus. The avian "tongue skeleton" is made up principally of branchial elements with little or no contribution from the hyoid arches. The terminology adopted is that of McLelland (1968). See Zweers (1974) for detailed description of the hyobranchial apparatus in *Anas*.

(80) **Os entoglossum.** Synonymy: Os paraglossum. Os entoglossum of most birds has the shape of an arrow head, and bears short, caudolaterally directed Cornua. Parrots are characterized by the presence of a wide, flat Os entoglossum with a large central foramen or, more commonly, paired entoglossals united rostrally by cartilage. In *Melopsittacus* (Evans, 1969) the entoglossal is paired; each is bifurcate rostrally, and each has an Os sesamoideum embedded in the fibrous tissue surrounding the ventrorostral tip. In most other parrots the two entoglossals are united by a bony isthmus.

(81) **Os basibranchiale rostrale.** Synonymy: Basihyoideum; Os basihyale; Pars basihyalis copulae.
Proc. parahyalis; Arcus parahyalis. In parrots (Mivart, 1895) Os basibranchiale rostrale possesses a caudal enlargement from which a dorsolateral process arises on each side. These processes fuse mid-dorsally to form the paraglossal arch in several Australian and Indopacific genera (*Melopsittacus*, *Eos*, *Vini*, *Lorius*, *Nestor*).
Os basibranchiale caudale. Synonymy: Urohyoideum; Os urohyale; Pars urohyalis copulae; Basibranchiale I. The rostral and caudal basibranchial bones are separate in young birds, but fused to one another in adults.
Os sesamoideum. A close-fitting sesamoid plate on the ventral surface of Os basibranchiale caudale in *Melopsittacus* (Evans, 1969).

(82) **Dentes.** True teeth with enamel are known only from fossil birds, *Archaeopteryx* and *Hesperornis* (Brodkorb, 1971) and Ichthyornis (Gingerich, 1972). Teeth of two kinds: Dentes veri and Osteodentes.

(83) **Condylus occipitalis.** Main part of condyle formed by Os basioccipitale; lateral contributions from Ossae exoccipitalia. See Goedbloed (1958) for comprehensive comparative study of the avian occipital condyle.

(84) **Ala tympanica.** Term proposed by Coues (1927) for lateral flared part of Os exoccipitale that forms part of the boundary of Cavitas tympanica. Also used for flared part of Os parasphenoidale that bounds tympanic cavity ventrally.

(85) **Proc. paroticus** (Hofer, 1945). Synonymy: Proc. paraoccipitalis (Stellbogen, 1930); Ala posttympanica; Proc. occipitalis lateralis (Davids, 1952); Proc. opisthoticus (Zusi, 1962); Proc. exoccipitalis. "Paroticus" from Gr. meaning "situated near the ear". This structure is prominent as in *Anser anser, Gavia immer, Pelecanus occidentalis*, and *Caprimulgus carolinensis*, and serves for attachment of Lig. occipitomandibularis and M. depressor mandibulae.

(86) **Fossa parabasalis** (Kesteven, 1925). Synonymy: Fossa jugularis. The fossa is usually just medial to the ventral margin of the tympanic cavity on the base of the skull. Several canals for cranial nerves VII, IX, X, (XI), the cerebral carotid vessels, and the external ophthalmic vessels open into the fossa.
Crista fossae parabasalis. Prominent raised medial margin of the parabasal fossa as in some species of anseriforms and phoenicopterids (flamingos).

(87) **Fonticulus occipitalis.** Synonymy: Fonticulus occipitalis lateralis (Barkow, 1856). Large paired openings lateral or dorsolateral to Foramen magnum as in anseriforms, scolopacids, alcids, gruids, and *Phoenicopterus* (Beddard, 1898) (see **Veni** Annot. 34).

(88) **Os orbitosphenoidale.** Synonymy: Os pleurosphenoidale. This bone forms the ventral portion of the caudal wall of the orbit, and extends from the interorbital septum laterally to the temporal fossa and the postorbital process. Jollie (1957, p. 393) treats all this as a single bone. Goodrich (1958) on the other hand considers the bone to consist of two elements: (1) a smaller medial element that fuses with its counterpart and the mesethmoid bone (the true orbitosphenoid) and (2) the larger lateral part, (the pleurosphenoid). D. A. Hogg (pers. comm.) in a recent study follows Goodrich; for purposes of simplification of the terminology, and descriptive aptness, Jollie's viewpoint has been adopted.

(89) **Area muscularis aspera.** The orbital surface of Os orbitosphenoidale of many large birds shows a roughened area for attachment of the jaw muscles (see **Myol.** Mm. mandibulae; and Figs. 2, 3).

(90) **Foramen n. maxillomandibularis.** Some birds possess separate foramina for the maxillary and mandibular nerves.

(91) **Sella turcica.** Consult Jollie (1957) for development of floor of skull and Sella turcica in *Gallus*. For anatomy of sella in different birds see Wingstrand (1951), Starck (1955) and Baumel (1968). See Cavitas cranialis.
Foramen ophthalmicum internum. Opening that conducts the internal ophthalmic vessels; connects the Sella turcica with the orbit. See **Art.** Annot. 19.

(92) **Canalis craniopharyngealis.** Persistent vestiges of this embryonic structure are commonly seen in sections of avian skulls.

(93) **Proc. basipterygoideus.** Synonymy: Proc. pterygoideus. Process on each side of Rostrum sphenoidale for articulation with pterygoid bone. Occurs in anseriforms and ratites; some caprimulgiforms, procellariiforms, trogoniforms; many charadriiforms; and cathartid vultures (see Beddard, 1898).

(94) **Tuba pharyngotympanica [auditiva] communis.** The common tube or chamber for the pharyngeal ostia of the right and left pharyngotympanic tubes. The common tube is located on the ventral aspect of the base of Rostrum sphenoidale; the tube opens into Infundibulum pharyngotympanicum which communicates with the pharynx (see **Digest.** Annot. 19).

(95) **Canalis orbitalis.** Short canal opens on side of base of Rostrum sphenoidale; an offshoot of Canalis caroticus.

(96) **Lamina basiparasphenoidalis.** Previously known as Lamina basitemporalis. The lamina serves mostly for attachment of ventral neck muscles to Basis cranii; assumes markedly different configurations in various birds.

(97) **Tuberculum basilare.** Synonymy: Bulla basitemporalis (Davids, 1952). Unpaired, median eminence (*Anser*) or bilateral swellings (*Columba*, *Corvus brachyrhynchos* on Lamina basiparasphenoidalis.

(98) **Tuba pharyngotympanica [auditiva].** Synonymy: Eustachian tube. Osseous tube located between Rostrum sphenoidale and Lamina parabasisphenoidalis. In some species there is incomplete osseous closure of the tube on its rostral aspect that is completed by connective tissue. See above, Annot. 94.

(99) **Canalis caroticus.** Osseous canal in pneumatized bone of Basis cranii; extends from parabasal fossa to Sella turcica (see Wingstrand, 1951; Jollie, 1957; Baumel, (1968).

(100) **Facies articularis quadratica.** Articulates with squamosal condyle of Proc. oticus of the Quadratum.

(101) **Proc. postorbitalis.** This small process of Os squamosale forms part of the base of the large Proc. postorbitalis of Os orbitosphenoidale.

(102) **Proc. zygomaticus:** Synonymy: Proc. lateralis. Strongly developed in gaviiform, galliform, piciform, and passeriform birds. See Annot. 30.

(103) **Proc. suprameaticus** (Hofer, 1945). Synonymy: Proc. quadratus. Proc. suprameaticus projects rostroventrally on the lateral aspect of the quadrate-squamosal articulation. (see **Arthr.** Annot. 25).

(104) **Fossa temporalis.** Floored mostly by Os squamosale; may extend onto adjacent bones. Fossa is bounded by Crista temporalis, Proc. postorbitalis and Proc. zygomaticus. The aponeurosis covering jaw muscle(s) that occupy Fossa temporalis is attached to Crista temporalis. See Annot. 30, 101–102.

(105) **Ossa otica.** In early postnatal development the otic elements become incorporated with one another and the apposing squamosal, exoccipital, and basiparasphenoid bones. The osseous labyrinth of the inner ear is located within this complex of bones (see Sandoval, 1963).
Fossa acustica interna. Consult Ghetie *et al.* (1976) for detailed close-up illustrations of the nerve foramina of the internal acoustic fossa in the goose and turkey.

(106) **Crista vallecularis.** Crest of bone on the inner aspect of Calvaria that marks the Vallecula telencephali, the sulcus that bounds the Eminentia sagittalis of the telencephalon.

(107) **Proc. prefrontalis.** Some birds (e.g. *Cathartes aura, Morus bassanus*) possess a lateral flared projection of Os frontale immediately caudal to its articulation with Os prefrontale.

(108) **Proc. postorbitalis.** A variable lateral process of the supraorbital margin of Os frontale directly rostral to, and distinct from, Proc. postorbitalis of Os orbitosphenoidale (e.g. *Ardea herodias*). See Annot. 101.

(109) **Lamina dorsalis.** The dorsal, horizontal plate of the rostral part of Os mesethmoidale that articulates with the ventral aspect of the frontal bones; noted by Shufeldt (1909).

(110) **Os ectethmoidale.** Synonymy: Os lateroethmoidale; Pars plana; Proc. antorbitalis; Aliethmoid (Shufeldt, 1909).

(111) **Facies articularis frontonasalis.** Os prefrontale articulates with Os frontale and Os nasale; often articulates medially with ectethmoid; occasionally with Os jugale.

(112) **Columna vertebralis.** The number of vertebrae as well as the number of regional vertebrae varies in different avian taxa. The total numbers ranges from 39–64 (pygostyle counted as one vertebra). Fewest vertebrae occur in passerine birds; most occur in the swans and ratites. See the review paper of Komárek (1970) for a treatment of the nomenclature of the detailed features of avian vertebrae.

(113) **Corpus vertebrae.** Birds are the only vertebrate animals in which the articular facets of the corpora are mostly heterocoelous or saddle-shaped. Of infrequent occurrence (e.g. penguins, auks, gulls) in thoracic region are opisthocoelous (concave caudal articular facet) vertebrae (Beddard, 1898). See **Arthr.** Annot. 60.

Generally the Corpus vertebrae of typical cervical and thoracic vertebrae has expanded cranial and caudal ends, with a constricted midsection.

(114) **Fovea cranioventralis.** Synonymy: Fovea anteroventralis (Boas, 1929). Depression that accommodates the ventral lip of the articular facet of the body of the preceding vertebra during ventral flexion of the neck (Fig. 4).

(115) **Sulcus lateralis.** Groove on side of the body of a cervical vertebra (Facies lateralis) for ascending vertebral artery and vein.

(116) **Tuberculum costolaterale.** Synonymy: Proc. costolateralis (Boas, 1929); parapophysis. Prominence on lateral surface of Corpus vertebrae that bears Fovea costalis for articulation with Capitulum costae. See **Arthr.** Annot. 79.

(117) **Fovea costalis.** Articular surface on the lateral end of transverse process for articulation with Tuberculum costae. Fovea costalis is also present on Tuberculum costolaterale (see Annot. 116).

(118) **Proc. costalis.** Synonymy: Spina laminae ventralis (Komárek, 1970); Pleurapophysis. Attenuated vestigial rib having its proximal end ankylosed to the Corpus and Proc. transversus of cervical and synsacral vertebrae. See Annot. 134 and Fig. 4.

(119) **Proc. [Crista] ventralis.** Synonymy: Hypapophysis. Median, unpaired, laterally compressed process or crest on ventral side of Corpus vertebrae; most highly developed on the cranial thoracic vertebrae of spheniscids, *Gavia*, alcids, and some anseriforms (Beddard, 1898). Cristae ventrales on the cranial thoracic vertebrae are for attachment of M. longus colli ventralis; these strong crests are considered as adaptations for powerful underwater use of the neck (Kuroda, 1954). See **Arthr.** Annot. 72.

(120) **Alae.** Bilateral flared wings that extend horizontally from the ventral border of Crista ventralis (e.g. *Gavia immer*, *Anser*, *Alca* (Kuroda, 1954), and *Gallus*.

(121) **Proc. caroticus.** Has been known as Proc. sublateralis (Boas, 1929); Proc. hemalis; Catapophysis, (Beddard, 1898). Paired, incurved processes on ventral aspect of the intermediate series of cervical vertebrae; processes form the lateral walls of Sulcus caroticus (Fig. 4). The ventral ends of a pair of carotid processes

ankylose producing a complete osseous canal in some birds (e.g. *Pelecanus occidentalis*, *Ardea herodias*). The homologies of the avian Proc. caroticus with Proc. hemalis of other vertebrates is not established.

(122) **Proc. ventrolateralis.** Synonymy: Proc. inferolateralis (Boas, 1929). Ventrolateral paired projections at base of Crista ventralis. See Fig. 4.

(123) **Proc. transversus.** Synonymy: Diapophysis.

(124) **Crista transverso-obliqua** (Boas, 1929). Caudolaterally directed paired crests on caudal part of dorsal aspect of vertebrae extending out onto caudal articular processes; for attachment of Mm. intercristales.

(125) **Proc. dorsalis** (Boas, 1929). Synonymy: Hyperapophysis (Beddard, 1898). Process on dorsum of caudal articular process at the root of Facies articularis for attachment of Mm. ascendentes.

(126) **Area lig. elastici.** Elastic ligaments are usually attached cranially and caudally at the base of Proc. dorsalis (or spinosus). Bony markings of the ligaments are variously developed as roughened areas, foveae, or tuberosities.

(127) **Lamina lateralis arcus.** Substitute term for Pediculus of vertebrae of mammals. In birds the lateral part of the vertebral arch is plate-like rather than a constricted stem as in mammals; this is especially pronounced in the "long vertebrae" of birds (see Komárek, 1970).

(128) **Incisurae arcus.** Synonymy: Incisura vertebralis (cranialis et caudalis). Notches at cranial and caudal ends of Lamina lateralis that form margins of Foramina intervertebralia for passage of spinal nerves and vessels.
Foramen intervertebrale. See directly above and Annot. 144.

(129) **Vertebrae cervicales.** Fewest (14–15) in passerines and coraciiforms; greatest number in ratites (ca. 20) and swans (23–25).

(130) **Fossa condyloidea.** Synonymy: ventral semi-ring (Boas, 1929). Cup-shaped or semicircular concave surface for articulation with the occipital condyle.
Incisura fossae; Foramen fossae. The condyloid fossa on the cranial aspect of the atlas may have a central perforation (foramen) or an open dorsal notch (incisura). See **Arthr.** Fibrocartilago atlantis.

(131) **Proc. ventralis.** Synonymy: Hypapophysis; Proc. latus (Boas, 1929). Caudally directed process located ventral to Facies articularis axialis. See Annot. 119.

(132) **Proc. articularis caudalis [Zygapophysis caudalis].** Occurs on Atlas of most birds studied by Boas (1929); articulates with Proc. articularis cranialis of the Axis, consequently paired atlantoaxial zygapophysial articulations are present (see **Arthr.** Annot. 68).

(133) **Proc. costalis.** A vestigial rib forms the "handle" or ventrolateral segment of the margins of the Foramen transversarium. Boas (1929) refers to this segment as Ansa costotransversaria indicating that the rib ankyloses with Proc. transversus (and

Corpus vertebrae). Komárek (1970) calls this part the Lamina ventralis. The costal processes of some of the cervical certebrae possess a caudally directed spine (see Fig. 4).

(134) **Foramen transversarium.** The transverse foramen is characteristic of the cervical vertebrae of birds. The Atlas ordinarily lacks the foramen; however, the foramen does occur in the Atlas of alcids and anseriforms (Boas, 1929). See Annot. 148.

(135) **Incisura arcus caudalis.** The caudal incisure of the arch of the atlas forms the rostral boundary of the atlanto-axial intervertebral foramen for the second cervical spinal nerve (see Annot. 128).

(136) **Dens.** Synonymy: Proc. odontoideus. The articulation of the avian axis with the atlas differs markedly from that of mammals. In addition to articulation of Dens with Atlas, an Artc. intercorporea, and paired zygapophysial articulations are present. The atlas and axis are fused together in adults of *Bucorvus* (Boas, 1929).

(137) **Proc. dorsalis.** Lacking from Axis of some species (e.g. *Gallinago delicata*).

(138) **Proc. costalis.** A vestigial rib on the Axis is present in many, but not all birds; occasionally weak projecting tips of the costal processes are found (Boas, 1929). When present, Proc. costalis forms an arch and completes the Foramen transversarium.

(139) **Vertebrae thoracicae.** The first thoracic vertebra is defined as the cranialmost vertebra with a complete rib (having vertebral and sternal parts) that articulates directly or indirectly with the sternum (see Annot. 147). The vertebrae at the root of the neck that bear freely movable ribs (so-called Vertebrae cervicodorsales of Gadow, 1890) are transitional in configuration between "typical" cervical and thoracic vertebrae. See **Arthr.** Fig. 10.

(140) **Notarium.** Synonymy: Os dorsale. In many avian species the vertebrae of the thoracic region tend to form a rigid system by ossification of ligaments and tendons. The Notarium is formed in this zone by ankylosis of 2–5 vertebrae (Portmann, 1950, Barkow, 1856). The Notarium is characteristically present in Threskiornithidae (ibis and spoonbills), galliform, and columbiform birds. In most birds the Notarium is separated from the synsacrum by one or more freely movable intervertebral articulations (exception: *Pelecanus*, Barkow, 1856). See **Arthr.** Annot. 71.

(141) **Synsacrum.** Synonymy: Os pelvicum; Vertebrae pelvicae (Gadow, 1890). A rigid unit of ankylosed vertebrae of variable number in different birds. Contains one or more thoracic vertebrae, the lumbar and sacral series, and part of the caudal vertebrae. The synsacrum forms the median dorsal wall of the pelvis. See Barkow (1856), Boas (1933), van Oort (1905) for detailed comparative studies in different species; see also **Arthr.** Artcc. synsacri.

(142) **Facies ventralis.** Synonymy: Facies abdominalis. See Barkow (1856) for details of the features of the ventral surface of the Synsacrum.

(143) **Crista dorsalis.** The ankylosed dorsal (or spinous) processes of the cranial series of synsacral vertebrae.

(144) **Foramina intervertebralia.** Dual foramina are located on each side over most of the length of the synsacrum; these are separate foramina for dorsal and ventral radices of the synsacral spinal nerves. See Annot. 128.

(145) **Canalis synsacri [vertebralis].** Vertebral canal of the fused synsacral vertebrae. Canal is enlarged along the middle of length of Synsacrum and contains the lumbosacral intumescence of Medulla spinalis. The enlarged chamber is called Cranium inferior or ischiadicus by older authors (Barkow, 1856).

(146) **Pygostylus.** Synonymy: Urostylus; Coccyx. Compound bone formed by ankylosis of 3–6, commonly 5–6, of the terminal caudal vertebrae. The development of the pygostyle is reviewed by Steiner (1938); see also van Oort (1905). According to Holmgren (1955) the pygostyle of the Ostrich (*Struthio*) is not homologous with that of carinate birds.

(147) **Costae.** Free ribs of birds vary in number. Ribs of the cervico-thoracic transitional region of the vertebral column are short "floating" ribs that fail to reach the sternum (Costae incompletae). The "true" ribs (Costae completae verae) are jointed between vertebral and sternal segments; the sternal segment articulates with Margo costalis sterni. In some instances the sternal part of one or more of the ribs does not articulate directly with the sternum (Costae completae spuriae), but instead articulate with the true complete ribs cranial to them. Caudal to the true ribs a variable number of floating ribs may occur; these and the last of the series of true ribs may articulate with the preacetabular ilium (Beddard, 1898).

(148) **Incisura capitulotubercularis.** The interval between head and tuberculum of rib that partly encloses space between incisure and vertebra; space corresponds to Foramen transversarium of cervical vertebrae. See Fig. 4.

(149) **Sternum.** Fürbringer (1888) distinguished a cranial part, the "Costosternum" and a caudal part, the "Xiphosternum."
Corpus sterni. Synonymy: Tabula sterni.

(150) **Proc. craniolateralis.** Synonymy: Proc. sternocoracoideus; Proc. precostalis.

(151) **Incisurae et fenestrae.** The caudal part of the sternum is notched or perforated in a variety of ways in different avian taxa. See Fürbringer (1888) for illustrations of patterns of incisurae and fenestrae. The openings in the sternum are completed by fibrous membranes (see **Arthr.**).

(152) **Facies muscularis sterni.** Synonymy: Facies ventralis or externa.
Facies visceralis sterni. Synonymy: Facies dorsalis or interna.

(153) **Linea intermuscularis.** Attachments of the dense deep fascia of M. supracoracoideus on the ventral surface of Corpus sterni and lateral aspect of Carina sterni (Fig. 7).

(154) **Planum postcarinale.** Synonymy: Planum postpectorale (Fürbringer, 1902). Flat surface of Facies muscularis of sternum caudal to carina; carina does not reach caudal margin of sternum in some forms (e.g. pelecaniforms).

(155) **Pila costalis.** Column of bone that reinforces Margo costalis of sternum; prolonged onto Trabecula lateralis in some birds.

(156) **Facies articularis costalis.** Articular facets are doubled on costal process of sternum in some avian species, i.e. two separate facets for articulation with corresponding facets on the ventral end of each sternal rib, single in other species.

(157) **Margo caudalis.** Caudal margin of sternum is highly variable in shape; may be squared, rounded, intact, or notched. See Fürbringer (1888) for figures of characteristic sterna of numerous species.

(158) **Pila coracoidea.** Transversely disposed, curved column of bone at cranial margin of sternum; strengthens articular sulcus for Os coracoideum.

(159) **Rostrum sterni.** Synonymy: Manubrium sterni; Rostrum sterni commune. For attachment of parts of Membrana sternocoracoclavicularis. Spina externa usually present; spina interna of less frequent occurrence (e.g. *Columba*). When both spines are present they may be fused with one another (e.g. *Gallus*). The internal spine may be represented by a tubercle(s) on Labrum dorsale of Sulcus articularis coracoideus. See Fürbringer (1888) for detailed treatment of Spinae rostri sterni.

(160) **Foramen rostri.** Synonymy: Foramen interspinale. Opening at base of ankylosed external and internal spines of rostrum (e.g. galliform and coraciiform birds).

(161) **Carina sterni.** Synonymy: Crista sterni (Fürbringer, 1888). The ratite birds are characterized by a raft-like sternum lacking a distinct carina (*Struthio*). The sternum of *Apteryx* and *Casuarius* possesses an indistinct, low crest; *Rhea* shows a slight crest and a median ventral torus (Beddard, 1898). In the Kakapo, *Strigops* (a carinate bird), the keel is lacking.
Crista lateralis. Bilateral crests on dorsal, thick part of cranial margin of carina (e.g. *Gallus*, *Cathartes*).
Sulcus carinae. Shallow sulcus between Cristae laterales.

(162) **Clavicula [Furcula].** Furcula refers to the ankylosed right and left clavicles. When not fused at their ventral ends the clavicles may be joined by cartilage or fibrous tissue (many parrots, owls, *Buceros*, *Alcedo* (Newton, 1896). Glenny and Friedmann (1954) discuss the reduction or suppression of the clavicle in various birds (e.g. the Australian parrots). See **Arthr.** Annot. 85.

(163) **Apophysis furculae.** Synonymy: Hypocleideum. In most birds the ventral part of the Furcula is drawn out into a median projecting blade, knob, or rod that is attached to Apex carinae (see **Arthr.** Annot. 85).

(164) **Extremitas omalis claviculae.** Synonymy: Epicleideum. This is the dorsal, expanded end of each clavicle at its shoulder end.

(165) **Proc. acromialis; Proc. acrocoracoideus.** Clavicles of certain avian species possess distinct processes for articulation with scapula and coracoid. In diomedeids, ciconiiforms, and falconiforms only the caudally directed Proc. acromialis that articulates with the scapula is well developed. Both processes are present in *Alcedo*, *Merops*, *Ramphastos* and *Sturnus* (Fürbringer, 1888).

(166) **Scapula.** See Fürbringer (1888) for additional terms on other detailed features of scapula not listed here.
Facies lateralis. Synonymy: Facies dorsolateralis.
Facies costalis. Synonymy: Facies ventromedialis.

(167) **Facies articularis humeralis.** The Scapula contributes only part of Cavitas glenoidalis. The humeral articular facies of the scapula adjoins (Symphysis coracoscapularis) the humeral articular facet of the coracoid; the two form Cavitas glenoidalis. The nearly plane osseous articular facets of both bones are invested with the thick fibrocartilaginous Lig. coracoscapulare interosseum; the elevated margins of this ligament produce the shallow glenoid cavity (see Annot. 171; and **Arthr.** Annot. 93).

(168) **Tuberculum coracoideum.** In many species the convex spherical or ellipsoidal torus of the cranial extremity of the scapula for articulation (symphysis) with Os coracoideum. See Annot. 173; and **Arthr.** Annot. 93.

(169) **Tuberculum m. scapulotricipitis.** In some birds a distinct tubercle for muscle attachment on the ventral border of the Scapula directly caudal to its Facies articularis humeralis.

(170) **Corpus scapulae.** The cranial half of the shaft of Scapula is usually a rounded cylinder in cross section; its caudal half is flattened and usually blade-like. In penguins the caudal half is paddle-shaped.

(171) **Facies articularis humeralis**. Synonymy: Labrum [Tuberculum] glenoidale coracoideum (Fürbringer, 1888); Apophysis glenoidale; Tuberositas [Tuberculum] humeralis or brachialis. This is the coracoidal portion of Cavitas glenoidalis for articulation with the humerus (see **Arthr.** Annot. 93). The term Tuberositas brachialis has also been used (erroneously) for the small pointed elevation of the Proc. acrocoracoideus that overhangs the Sulcus m. supracoracoidei (see Howard, 1929; and Annot. 167).

(172) **Sulcus m. supracoracoidei.** Term of Fürbringer (1902).

(173) **Cotyla scapularis.** Synonymy: Facies scapularis (Ballmann, 1969a). Spherical or ellipsoidal concavity on Os coracoideum adjacent to its Facies articularis humeralis Cotyla receives the Tuberculum coracoideum of Scapula and forms the Symphysis coracoscapulare (see Annot. 168; and **Arthr.** Annot. 93).

(174) **Linea intermuscularis ventralis.** Synonymy: Linea intermuscularis externa (Lambrecht, 1933); anterior intermuscular line (Fisher, 1945).

(175) **Facies articularis intercoracoidea.** Occurs in birds whose coracoids articulate with one another in the median plane (see **Arthr.** Annot. 90),

(176) **Proc. lateralis.** Synonymy: Proc. lateralis posterior; Proc. sternocoracoideus; Proc. externus.

(177) **Canalis triosseus [supracoracoideus].** Synonymy: Foramen triosseum; Canalis supracoracoideus (Fürbringer, 1888). Canal that transmits the tendon of M. supracoracoideus. In many birds the canal is formed between only two bones, the coracoid

and scapula. The canal may be formed completely by the coracoid alone when there is an ossified bridge connecting acrocoracoid and procoracoid processes (e.g. Musophagidae, Meropidae, Upupidae, Bucerotidae; Fürbringer, 1888). See **Arthr.** Annot. 87, 95.

(178) **Ossa alae [membra thoracici]; Humerus.** Terms of direction are based on the wing in the defined anatomical position (completely extended and abducted) (see Gen. Intro.). In this anatomical position the embryonic extensor (dorsal) aspect of the humerus faces caudally and the flexor (ventral) aspect faces cranially. The long axis of the ellipsoidal Facies articularis of Caput humeri is nearly vertical with the wing outstretched (Fürbringer, 1888).

Consult Fürbringer (1888, 1902) for complete synonomy of terms on parts of the humerus. Consult also Ballmann (1969a) for comprehensive terminology of skeletal elements of wing including attachments of muscles and ligaments; this work is of special interest to paleontologists.

(179) **Caput humeri.** Synonymy: Caput articulare humeri (Fürbringer, 1888).

(180) **Incisura capitis.** Synonymy: Incisura collaris. Notch between Caput humeri and Tuberculum ventrale; accomodates the scapular labrum of Cavitas glenoidalis (see Annot. 167).

(181) **Facies bicipitalis.** Synonymy: Planum intertuberculare.

(182) **Tuberculum dorsale [minus].** Synonymy: Tuberculum m. supracoracoidei. Point of attachment of discrete tendon of M. supracoracoideus. See Impressio m. supracoracoidei.

(183) **Impressio m. supracoracoidei.** Rough area located on dorsal aspect of Crista pectoralis immediately distal to Tuberculum dorsale for attachment of expanded part of the tendon of M. supracoracoideus present in certain species.

(184) **Crista pectoralis.** Synonymy: Crista tuberculi minoris; Crista deltoidea.

(185) **Sulcus ligamentosus transversus.** For attachment of strong Lig. acrocoraco-humeralis. Located on cranial surface just distal to Caput humeri.

(186) **Impressio m. coracobrachialis cranialis.** Distinct triangular fossa in some birds (e.g. charadriiforms) on cranial aspect of Crista pectoralis.

(187) **Tuberculum ventrale [majus].** Synonymy: Tuberculum mediale. Mostly for insertion of tendons of short muscles originating on scapula and coracoid.

(188) **Fossa pneumotricipitalis.** Synonymy: Fossa pneumatica; Fossa pneumo-anconaea (Fürbringer, 1888). Fossa pneumotricipitalis is an excavation in the proximal extremity of the Humerus, varying in its form and development in different groups of birds. When strongly developed the fossa extends into the Caput humeri and Tuberculum ventrale. In most birds only a single fossa occurs; it is bounded ventrally and dorsally by the crura of the fossae, both of which converge on the apex of the Tuberculum ventrale. The single fossa is occupied by both heads of M. humero-triceps and the insertion of M. scapulohumeralis cranialis. In other birds (see Fig. 8)

a second or additional fossa is formed between Crus dorsale fossae and the Margo caudalis dorsally. The second fossa is occupied mainly by the dorsal head of M. humerotriceps. Consult Bock (1962) for a comprehensive treatment of this topic; see also Annot. 189 concerning the Foramen pneumaticum of the Humerus.

(189) **Foramen pneumaticum.** Located in Fossa pneumotricipitalis (Annot. 188) when present. The pneumatic foramen does not occur in the humeri of all birds; humeri of birds with dual fossae are generally not pneumatized. The pneumatic foramen is lacking in the following: penguins, Procellariidae (except albatrosses), loons, grebes, cormorants and anhingas, several tribes of ducks (mainly the diving ones), most charadriiforms, rails (Rallidae), and many oscine passerines (pers. comm. of Storrs Olson and Peter Ballmann).

(190) **Intumescentia.** Term of Fürbringer (1888) and Buri (1900) for the convex swelling on the proximal end of the humerus on the cranial aspect of Crista bicipitalis directly opposite Fossa pneumotricipitalis on the caudal aspect of humerus. See Fig. 8.

(191) **Sulcus n. radialis.** A distinct sulcus on the shaft of the humerus caused by N. radialis occurs rarely (e.g. *Casuarius*; humming birds and swifts (Apodiformes) (Buri, 1900).

(192) **Condylus dorsalis.** Synonymy: Condylus radialis, lateralis or externus. This is the condyle on the radial side of the abducted and extended humerus. Condylus dorsalis articulates with both Radius and Ulna.
Condylus ventralis. Synonymy: Condylus ulnaris, medialis or internus.

(193) **Incisura intercondylaris.** Synonymy: Vallis intercondylica (Ballmann, 1969a).

(194) **Epicondylus dorsalis.** Synonymy: Epicondylus radialis, lateralis, or externus; Ectepicondylus.
Epicondylus ventralis. Synonymy: Epicondylus ulnaris, medialis, or internus; Entepicondylus.

(195) **Proc. flexorius.** Process at distal end of humerus, ventral to Condylus ventralis; for attachment of tendinous head of M. flexor carpi ulnaris (Ballmann, 1969a). See Fig. 8.

(196) **Tuberculum supracondylare dorsale.** Tubercle for attachment of M. extensor metacarpi radialis. In some birds (e.g. Charadriiformes, Passeriformes, Fregatidae) a strong pointed **Proc. supracondylaris dorsalis** serves as the point of attachment of this muscle. In some species the Tuberculum supracondylare dorsale is closely related to the dorsal epicondyle, therefore difficult to distinguish from it. See Fig. 8.

(197) **Tuberculum supracondylare ventrale.** For attachment of Lig. collaterale ventrale. See Fig. 8; and **Arthr.** Annot. 105.

(198) **Proc. cotylaris dorsalis.** Prominent dorsal extension of the proximal end of the Ulna that bears the Cotyla dorsalis on its cranial aspect and Impressio m. scapulotricipitis on its caudal aspect (see Fig. 9).
Cotyla dorsalis. Synonymy. Apophysis glenoidalis externa (Lambrecht, 1933); Facies glenoidalis (Ballmann, 1969a). Cup-shaped articular surface for Condylus dorsalis humeri.

(199) **Incisura radialis.** Synonymy: Depressio radialis proximalis (Howard, 1929; Ballmann, 1969a). Concave articular facet for Caput radii; located at distal margin of Cotyla dorsalis of Ulna (Fig. 9).

(200) **Cotyla ventralis.** Synonymy: See above, Cotyla dorsalis. Ventral cotyla is the larger of the two cotylae; it articulates with Condylus ventralis of the humerus. Ventral cotyla is located at the base of the Olecranon. (Fig. 9).

(201) **Olecranon.** Synonymy: Proc. coronoideus ulnaris (Lambrecht, 1933). Attachment of tendon of M. humerotriceps.

(202) **Os sesamoideum m. scapulotricipitis.** Synonymy: Patella ulnaris (Fürbringer, 1888). Sesamoid bone in tendon of M. scapulotriceps of some species. Unusually well-developed in the hummingbird (*Eugenes*) (Berger, 1966) and in penguins.

(203) **Tuberculum lig. collateralis ventralis** (Fig. 9). Synonymy: Facies lig. interni (Ballmann, 1969a).
Sulcus tendineus (Fig. 9). In many birds a distinct sulcus is present on the ventral aspect of the proximal end of the Ulna located in the interval between the Olecranon and Tuberculum lig. collateralis ventralis; the tendon of M. flexor carpi ulnaris glides in this sulcus (see **Arthr.** Fig. 4).

(204) **Facies corporis ulnae.** The caudodorsal facies is separated from the caudoventral facies by the row of Papillae remigiales caudales. Facies caudodorsalis is subcutaneous with no muscles intervening between ulna and integument. Proximal half of ulnar shaft often shows distinct intermuscular lines or crests in some species.

(205) **Papillae remigiales caudales.** Synonymy: Papillae ulnares anconeales (Lambrecht, 1933); quill knobs (Edington and Miller, 1941): See Fig. 9; and **Arthr.** Annot. 199.
Papillae remigiales ventrales. Synonymy: Papillae ulnares inferior (Lambrecht, 1933); quill knobs (Edington and Miller, 1941). See Fig. 9; and **Arthr.** Annot. 199.

(206) **Condylus dorsalis.** Synonymy: Condylus externus or caudalis.
Condylus ventralis. Synonymy: Condylus internus, cranialis, or metacarpalis.
The Os carpi radiale and Meniscus intercarpalis articulate with both condyles of Trochlea carpalis of the ulna. Os carpi ulnare articulates mainly with the Condylus dorsalis. On account of the torsion of the ulnar shaft the dorsal condyle of the ulna is located somewhat more caudally than the ventral condyle. See **Arthr.** Annot. 117–119.

(207) **Tuberculum carpale.** Synonymy: Tuberositas carpalis (Lambrecht, 1933).

(208) **Sulcus radialis.** Synonymy: Depressio radialis distalis (Ballmann, 1969a). For ulnar attachment of Lig. interosseum radioulnare. Radius flides in this sulcus during flexion-extension movements of elbow and wrist joints.

(209) **Sulcus tendineus.** Synonymy: Incisura tendinea (Lambrecht, 1933).

(210) **Tuberculum bicipitale.** Synonymy: Tuberculum externum (Lambrecht, 1933).

(211) **Facies articularis radiocarpalis.** Synonymy: Artc. scapholunaris (Lambrecht, 1933).

(212) **Sulcus ligamentosus.** Synonymy: Depressio ligamentalis. On distal end of caudal surface of Radius; place occupied by Lig. interosseum radioulnare distale.

(213) **Tuberculum aponeurosis.** Synonymy: Ligamental process (Howard, 1929). The Tuberculum is located on the distal end of the radius ventral to the articular surface for Os carpi radiale. The Tuberculum serves as the attachment of the Aponeurosis ventralis (see **Arthr.** Annot. 113) that fans out to attach to carpals, Carpometacarpus, and remiges.

(214) **Os carpi radiale.** Synonymy: Os scapholunare.
Os carpi ulnare. Synonymy: Os cuneiforme.

Os carpi radiale and Os carpi ulnare are derived from elements of the proximal row of fetal carpal bones (Ossa carpi proximalia). Certain of the fetal Ossa carpi centralia and Ossa carpi distalia ankylose with the proximal ends of the metacarpals in early postnatal life to produce the compound bone, the Carpometacarpus. See Steiner, Montagna, 1945; Holmgren, 1955; Romanoff, 1960; and Berger, 1966 for treatments of the developments of the avian forearm and hand as well as homologies of the digits. See Intro.; and **Arthr.** Annot. 112, 122, 128.

(215) **Facies articularis ulnocarpalis.** Os carpi ulnare articulates with caudal part of Trochlea carpalis.
Facies articularis radiocarpalis. Os carpi radiale articulates with cranial part of Trochlea carpalis.

(216) **Os metacarpale alulare.** Synonymy: Metacarpus pollicis; Proc. metacarpalis pollicis or digiti I or digiti II.

(217) **Foveae carpales.** Synonymy: Fossa carpalis anterior; Fossa carpalis posterior (Ballmann, 1969a). Located at cranial and caudal ends of the articular surfaces of Trochlea carpalis. Fovea cranialis accommodates Os carpi radiale when the carpometacarpal joint is extended; Fovea caudalis accommodates the distal end of Os carpi ulnare when the joint is flexed.

(218) **Fossa infratrochlearis.** Synonymy: Fossa carpalis interna (Ballmann, 1969a).

(219) **Fossa supratrochlearis.** Synonymy: Facies ligamentalis externa (Ballmann, 1969a).

(220) **Proc. pisiformis.** Prominent process for attachment of ventral carpometacarpal ligaments.

(221) **Proc. intermetacarpalis** (Milne-Edwards, 1867–71). Synonymy: Tuberositas muscularis (Ballmann, 1969a). Process (or tuberosity) serves as the insertion of M. extensor metacarpi ulnaris. See Fig. 10.

(222) **Synostosis metacarpalis distalis.** Synonymy: Symphysis metacarpalis distalis (Lambrecht, 1933; Ballmann, 1969a). See **Arthr.** Annot. 129.

(223) **Sulcus interosseus.** Groove on dorsal aspect of region of distal metacarpal synostosis for tendons of Mm. interossei.

(224) **Ossa digitorum manus.** The more common phalangeal formula of birds: one alular phalanx, two phalanges of Digitus major, and one phalanx of Digitus minor. In a number of avian orders the alular digit possesses two phalanges, the terminal phalanx often bearing a claw. See **Integ.** Annot. 70 for comment on supernumerary digital claws (phalanges); see also **Topog. anat.** Annot 40; and **Arthr.** Annot. 137.

(225) **Pila cranialis.** Leading edge of the large proximal phalanx of Digitus major possesses a strong reinforcing bar of bone. The caudal border of the phalanx is thin and fenestrated in some avian species.

(226) **Foramen acetabuli.** Floor of acetabulum is incomplete in birds (see **Arthr.** Annot. 151).

(227) **Foramen obturatum.** The oval-shaped opening caudoventral to acetabulum that transmits the tendon of M. obturatorius medialis. Boas (1933) contends that the Foramen obturatum is merely the detached cranial part of Fenestra ischiopubica.

(228) **Sulcus m. obturatorii.** Long, wide, shallow groove on medial surface of Ala ischii. M. obturatorius medialis lies in Sulcus m. obturatorii and also on the medial surface of pubis and Membrana ischiopubica.

(229) **Fenestra ischiopubica.** Synonymy: Foramen obturatorium, pars caudalis; Foramen oblongum. See Annot. 227.

(230) **Foramen ilioischiadicum.** Bounded caudodorsally by ilium; bounded cranioventrally by ischium. Foramen transmits the ischiadic nerves and vessels (see **Arthr.** Membrana ilioischiadica). The foramen is open caudally in *Apteryx* and tinamous.

(231) **Incisura marginis caudalis; Proc. marginis caudalis.** Boas (1933) calls the process Spina iliocaudalis. In lateral view the caudal border of Os coxae of many birds is indented between Proc. dorsolateralis and the tip of Proc. terminalis ischii. This notch is in the region of the ilioischiadic synostosis (see **Arthr;** and Annot. 230). In some species a caudally directed process occurs along the caudal margin. This is the Proc. marginis caudalis as seen in *Gallus, Ardea herodias, Coccyzus americanus.*

(232) **Antichrochanter.** Formed partly from ilium, partly from ischium.

(233) **Crista iliosynsacralis.** Formed by ankylosis of Crista dorsalis of Synsacrum with both right and left Cristae iliacae dorsales. See Annot. 143.

(234) **Sulcus iliosynsacralis.** Furrow between Crista dorsalis and Crista iliaca dorsalis on each side that contains epaxial muscles.
Canalis iliosynsacralis (Fig. 11). Synonymy: Canalis iliosacralis (Nauck, 1938); subiliac space (Howard, 1929); canalis ilioneuralis. Formed by fusion of Cristae iliacae dorsales with the median Crista dorsalis (see above: Annot 233). Canal is found on each side of the dorsal crest of the synsacral spines. The ventral wall of the canal is formed by transverse processes of the synsacral vertebrae; the dorsal wall is the ventral surface of Ala preacetabularis ilii. As with the Sulcus iliosynsacralis, the canal contains epaxial muscles. Canalis iliosynsacralis is present in *Gallus*, *Cathartes aura*, and other birds.

(235) **Concavitas infracristalis.** The moderately concave lateral surface of the pelvis caudal to the Foramen ilioischiadicum of many birds; formed largely by Lamina ischiadica of the Ilium. The depth of the concavity is accentuated by the overhanging Crista iliaca dorsolateralis. See Boas (1933).

(236) **Pila ilioischiadica.** Reinforcing column of bone that extends along the ventral border of pelvis (bilateral) from level of the cranial border of Fossa renalis to Proc. terminalis ischii. The cranial part of the Pila forms Crista iliaca obliqua; the Pila also contributes to the ventral wall of the acetabulum and the ilioischiadic foramen (see Crista pilae ilioischiadicae, Os ischii).

(237) **Pars ischiadica; Pars pudenda.** Pars ischiadica refers to the smaller cranial part of the Fossa renalis that contains the cranial division of the kidney and the ischiadic (lumbosacral) nerve plexus; Pars pudenda is the caudal part of the fossa that also houses part of kidney and the pudendal nerve plexus. See Radu (1975) for a comparative study of Fossa renalis in galliforms and anseriforms.

(238) **Incisura caudalis pelvis.** Viewed from its ventral or dorsal aspects, the intact osseous pelvis of most birds demonstrates at its caudal end a wide, semilunar or rectangular space. The Incisura is bounded on each side by Proc. dorsolateralis. The basal part of the free caudal vertebral column occupies the incisure. The incisure is notably deep in falconiform, ciconiiform, and strigiform pelves.

(239) **Corpus ilii.** Strongly developed part of ilium in the region of the acetabulum; cranial and dorsal to acetablum.

(240) **Incisura acetabularis.** The acetabular notch of the body of the pubis forms a segment of the margin of the Acetabulum. Ilium ischium, and pubis all contribute to formation of the Acetabulum as in mammals and some reptiles. See Annot. 241.

(241) **Proc. ischiadicus; Proc. pubicus.** During postnatal development these processes of the ilium ankylose with ischium and pubis forming a complete acetabular ring of bone.

(242) **Crista iliaca obliqua** (Boas, 1933). Heavy oblique bar of bone that forms the ventrolateral border of Fossa renalis; extends between ventral surface of Ala preacetabularis ilii and the ventral wall of the acetabulum (see Annot. 236).

(243) **Crista iliaca intermedia** (Boas, 1933). Slightly developed in most birds. In some birds this transverse crest is formed on the ventral aspect of Ala postacetabularis ilii, within the renal fossa, at the level of the caudal margin of the acetabular foramen. The costal process(es) of the so-called true sacral vertebrae articulate with the medial end of the crest. The crest is well developed in *Cathartes aura*, *Morus bassanus*, some Laridae, *Columba*, and other birds (see **Arthr.** Annot. 76).

(244) **Tuberculum preacetabulare** [Proc. pectinealis]. Synonymy: Proc. preacetabularis; Proc. prepubica. Tuberculum preacetabulare is the term of Boas (1933). According to Beddard (1898) the tuberculum in most birds is formed by the ilium; in other birds it is formed partly by the pubis. In most birds the Tuberculum preacetabulare is a rather slightly developed torus of bone; in *Geococcyx*, a cuculiform bird (Larson, 1930); *Struthio*, tinamous, and galliforms a prominent process is present (Beddard, 1898). See Fig. 11; and **Arthr.** Annot. 184.

(245) **Crista iliaca dorsalis.** Synonymy: Linea iliodorsalis (Lambrecht, 1933); Crista iliaca superior (Milne-Edwards, 1867–71).

(246) **Facies costalis.** Synonymy: Fossa costalis (Lambrecht, 1933). Ventral surface of preacetabular ilium that articulates with the transverse processes of the cranial series of synsacral vertebrae.

(247) **Margo lateralis.** Synonymy: Crista iliaca externa (Milne-Edwards, 1867–71).

(248) **Crista iliaca dorsolateralis.** Synonymy: Linea iliolateralis (Lambrecht, 1933). **Proc. [Spina] dorsolateralis.** Synonymy: Proc. iliolateralis (Boas, 1933). This is the caudal prolongation of Crista iliaca dorsolateralis (see Figs. 5, 11).

(249) **Fossa iliocaudalis.** Depression at base of Proc. terminalis ilii in many birds; for origin of dorsal extrinsic tail muscles.

(250) **Recessus iliacus.** (Fig. 5) Synonymy: obturator depression (Harvey *et al.*, 1968). The recess of Fossa renalis that invaginates the base of Proc. terminalis ilii. The recess is deep in *Gallus*, some owls, and in the gruiform birds (Boas, 1933). In *Gallus* the recess contains part of the origin of M. obturatorius medialis.

(251) **Lamina ischiadica.** Vertical lamina of postacetabular ilium that ankyloses with Ala ischii caudal to Foramen ilioischiadicum.

(252) **Proc. obturatorius.** Synonymy: Proc. ventralis. Ventrally directed process at the caudal border of Foramen obturatum; attachment of Lig. ischiopubicum. The ligament ossifies in some birds, producing an osseous caudal margin of Foramen obturatum.

(253) **Proc. terminalis ischii.** Synonymy: Angulus ischiadicus (Lambrecht, 1933); Proc. terminalis ischiadicus (Boas, 1933).

(254) **Pubis**. The shaft of the Pubis (Scapus pubis) in most birds closely parallels the ventral border of Ala ischii. The two are separated by the obturator foramen and the narrow ischiopubic fenestra. The fenestra is very wide in *Gavia immer*.

(255) **Facies articularis antitrochanterica.** Synonymy: Artc. iliacalis (Lambrecht, 1933); Facies glenoidea proximalis (Ballmann, 1969b). Located on proximal aspect of Collum and medial aspect of Trochanter of femur. See **Arthr.** Annot. 152.

(256) **Trochanter femoris.** Synonymy: Trochanter major.

(257) **Impressiones iliotrochantericae.** Markings for attachment of muscles and ligaments on the lateral aspect of the Trochanter are detailed by Ballmann (1969b).

(258) **Corpus femoris.** The shaft of the avian femur is commonly circular in cross section. No sharply defined borders are present except in atypical femora, e.g. *Gavia* in which the femur is strongly flattened and laterally compressed.

(259) **Sulcus patellaris.** Synynomy: Fossa patellaris (Lambrecht, 1933; rotular groove (Howard, 1929).

(260) **Condylus medialis; Condylus lateralis.** Synonymy: Condylus internus; Condylus externus.

(261) **Trochlea fibularis.** Synonymy: Sulcus fibularis. Trochlea on Condylus lateralis of femur for articulation with Caput fibulae.

(262) **Crista tibiofibularis** (Howard, 1929). Synonymy: Crista peroneo-tibialis (Ballmann, 1969b). Crest on Condylus lateralis of femur that separates its tibial and fibular articular surfaces; Crista tibiofibularis forms the medial wall of Trochlea fibularis.

(263) **Impressio lig. cruciati caudalis.** Located at proximal end of the caudal aspect of Condylus lateralis (see Fig. 12).

(264) **Crista supracondylaris medialis.** Synonymy: Adductor crest.

(265) **Patella.** Sesamoid bone of the tendon of Mm. femorotibiales; Patella is grooved, sometimes perforated, by tendon of M. ambiens in different species. See Annot. 269.

(266) **Extremitas proximalis tibiotarsi.** Synonymy: Caput tibiae (Lambrecht, 1933).

(267) **Facies articularis medialis et lateralis.** Neither of the facets on Tibiotarsus for articulation with the femoral condyles is cup-shaped; therefore, "cotyla" or "glenoid fossa" is inappropriate here. Well-developed intra-articular menisci intervene between femoral and tibial articular surfaces (see **Arthr.** Artcc. genus). Facies lateralis faces laterodorsally and articulates with the medial aspect of the lateral condyle of the femur. The area of Facies articularis lateralis is much smaller than that of Facies medialis.

(268) **Fossae retrocristales.** Synonymy: Fossae synoviales (Ballmann, 1969a). Fossae are situated between Crista patellaris and the articular facets of the proximal end of Tibiotarsus; fossae are separated from one another by a low ridge. In the intact joint the fossae are occupied by the articular fat pad (Ballmann, 1969a).

(269) **Crista patellaris** (Lambrecht, 1933). Synonymy: Crista rotularis (Milne-Edwards, 1867–71). Transverse crest on the cranial aspect of the proximal end of Tibiotarsus for attachment of Lig. patellae. Patellar crest connects the proximal ends of the two cnemial crests. See Fig. 13.

(270) **Crista cnemialis [tibialis] cranialis.** Synonymy: Crista cnemialis interna or medialis; Crista cnemialis anterior (Ballmann, 1969a). The cranial cnemial crest is enormously developed in the loon (*Gavia immer*).

(271) **Facies gastrocnemialis** (Ballmann, 1969a). The medial surface of Crista cnemialis cranialis and the area of Tibiotarsus caudal to crest; for attachment of M. gastrocnemius medialis.

(272) **Sulcus intercristalis.** Synonymy: Sulcus intercnemialis (Kolda and Komarek, 1958). Wide sulcus between cranial and lateral cnemial crests; for origin of M. extensor digitorum longus.

(273) **Incisura tibialis.** Between Crista cnemialis lateralis and Facies articularis lateralis: for Caput femorale of M. tibialis cranialis (Ballmann, 1969a).

(274) **Fossa flexoria** (Ballmann, 1969a). Depression on plantar (caudal) aspect of the proximal end of Tibiotarsus between Facies articularis lateralis and fibular crest; for origin of M. flexor digitorum longus.

(275) **Corpus tibiotarsi.** The proximal two-thirds of the shaft is for the most part three-sided, with cranial, medial, and caudal surfaces (Ballmann, 1969a).

(276) **Linea extensoria** (Ballmann, 1969a). Intermuscular line on cranial surface of tibiotarsus prolonged from Crista cnemialis cranialis along length of shaft; continuous distally with medial margin of Sulcus extensorius.

(277) **Pons supratendineus.** Synonymy: Lig. transversum ossificatum (Lambrecht, 1933); supratendinal bridge. Located at distal end of tibiotarsus proximal to condyles.

(278) **Canalis extensorius** (Fig. 13). The canal deep to Pons supratendineus; located at the distal end of the Tibiotarsus on its cranial aspect, slightly proximal to the intercondylar incisure. See **Arthr.** Annot. 187; and **Myol.** Annot. 115.

(279) **Sulcus cartilaginis tibialis.** On plantar aspect distal end of tibiotarsus; forms articular surface for Cartilago tibialis. Crests of sulcus are continuous with those of the Condyli tibiotarsi (see **Arthr.** Annot. 164 and Fig. 8).

(280) **Condyli tibiotarsi.** Articular surfaces of condyles for articulation with Tarsometatarsus face distally and cranially.

(281) **Depressio epicondylaris.** Shallow concavity proximal to tibiotarsal condyle; area between Crista sulci and the epicondyle on each side. (Fig. 13).

(282) **Tuberositas retinaculi m. fibularis [peronei].** Located at the proximal end of Condylus lateralis near Epicondylus lateralis of the Tibiotarsus; for attachment of the retinaculum which restrains the tendon of M. fibularis brevis in Sulcus m. fibularis [peronei].

(283) **Ossa tarsi.** Consult Romanoff (1960) for a summary of the development of the tarsal bones and their definitive ankylosis with the distal tibia and proximal metatarsals during early post-natal life in several avian species. See **Arthr.** Annot. 167, 173, 174.

(284) **Tarsometatarsus.** Metatarsal bones II, III, and IV of most birds ankylose extensively with one another, forming a "cannon" bone which is fused on its proximal end with the distal tarsal bones. This constitutes the definitive avian Tarsometatarsus. Metatarsal I has a ligamentous junction with the medial aspect of the Tarsometatarsus. See **Arthr.** Annot. 167, 173, 174; and **Topog. anat.** Annot. 43.

(285) **Area intercotylaris** (Fig. 14). Relatively flat area between the plantar parts of the cotylae of Tarsometatarsus; between the Eminentia intercondylaris and the Hypotarsus. See **Arthr.** Annot. 171.

(286) **Sulcus ligamentosus.** Transverse sulcus in some birds (e.g. *Pelecanus occidentalis*, *Cathartes aura*) at junction of proximal Hypotarsus and Area intercotylaris for attachment of the ligament from the distal end of Cartilago tibialis. See Fig. 14; and **Arthr.** Annot. 166.

(287) **Arcus extensorius.** An osseous arch on the dorsal aspect of the proximal end of Tarsometatarsus, the arch restrains and acts as a pulley for the tendon of M. extensor digitorum longus (Strigidae, Picidae, Rallidae, et al.; see Berger, 1966). The arch is the ossified Retinaculum extensorium tarsometatarsi. See Annot. 277, 295; and **Arthr.** Annot. 188.

(288) **Hypotarsus.** Synonymy: Calcaneus. Process on plantar aspect of proximal Tarsometatarsus formed mostly by tarsal elements capping the proximal end of Metatarsal III. The Hypotarsus is simple in some birds, consisting of a wide sulcus between low crests (e.g. most falconiforms). In most birds it is complex having sulci and high crests, and perforated by one or more canals (Newton, 1896). Sulci and canals conduct flexor tendons of the pedal digits. See Fig. 14; **Arthr.** Fig. 8; and **Integ.** Annot. 68.

(289) **Cristae hypotarsi.** Synonymy: Crista ectogastrocnemialis, Crista entogastrocnemialis (Lambrecht, 1933); calcaneal ridges (Howard, 1929).

(290) **Crista plantaris mediana.** Synonymy: Crista plantaris (Neugebauer, 1845). Median, curved crest that forms a buttress from the middle of the Hypotarsus distally onto the plantar aspect of the shaft of Tarsometatarsus.

(291) **Fossa infracotylaris dorsalis.** An excavation on the dorsal aspect of the proximal end of the Tarsometatarsus immediately distal to its cotylae. See Annot. 287 and Fig. 14.

(292) **Facies corporis tarsometatarsi.** The shaft of the Tarsometatarsus demonstrates much variation: In cross section of shaft may be: (1) rectangular, laterally compressed (e.g. *Gavia immer*); (2) rectangular, compressed in its dorsoplantar dimension (e.g. *Cathartes aura*); (3) triangular, plantar aspect flat (e.g. *Buteo borealis*); (4) triangular, dorsal aspect flat (e.g. *Pelecanus occidentalis*); U-shaped, concave on plantar aspect (*Strix varia*), etc.

(293) **Facies subcutanea.** Generally the medial and lateral aspects (see Annot. 295). of the Tarsometatarsus are covered only with integument (Podotheca). By contrast the plantar and dorsal aspects of the Tarsometatarsus have bundles of tendons interposed between skin and bone (Ballmann, 1969a).

(294) **Sulcus flexorius.** Synonymy: Sulcus longitudinalis plantaris. In certain species (e.g. *Strix varia*, *Buteo borealis*) the plantar (flexor) surface of Tarsometatarsus is grooved longitudinally with a deep sulcus bounded by prominent, sharp Cristae plantares.

(295) **Sulcus extensorius.** Synonymy: Sulcus longitudinalis dorsalis. In some species a shallow, longitudinal sulcus indents the dorsal (extensor) surface of Tarsometatarsus; the sulcus contains the intrinsic extensor muscles of the digits. See Annot. 287.

(296) **Proc. calcaris.** The osseous core of the metatarsal spur (Calcar). Proc. calcaris is ankylosed to the medial border of Tarsometatarsus in males of some galliform birds.

(297) **Trochlea accessoria.** In piciform, cuculiform and psittaciform birds the trochlea of the fourth metatarsal possesses an accessory trochlea (Milne-Edwards, 1867–71; Steinbacher, 1935). See Ballmann (1969a) for diagrams of atypical forms of the tarsometatarsal trochleae in several major taxa of birds (e.g. cuculiforms, psittaciforms, coliiforms, piciforms).

(298) **Canalis interosseus tendineus.** Synonymy: Canalis interosseus distalis. May consist only of a sulcus in some birds. Canal conducts the tendon of M. extensor brevis digiti IV and vessels into the lateral intertrochlear incisure.

(299) **Ossa digitorum pedis.** General avian phalangeal formula: Hallux, two phalanges; Digitus secundus, three phalanges; Digitus tertius, four phalanges; Digitus quartus, five phalanges. See Coues (1927) for complete discussion of toes of birds.

(300) **Phalanx distalis.** Synonymy: Phalanx ungularis or terminalis. Forms the osseous core of the heavily keratinized claw.

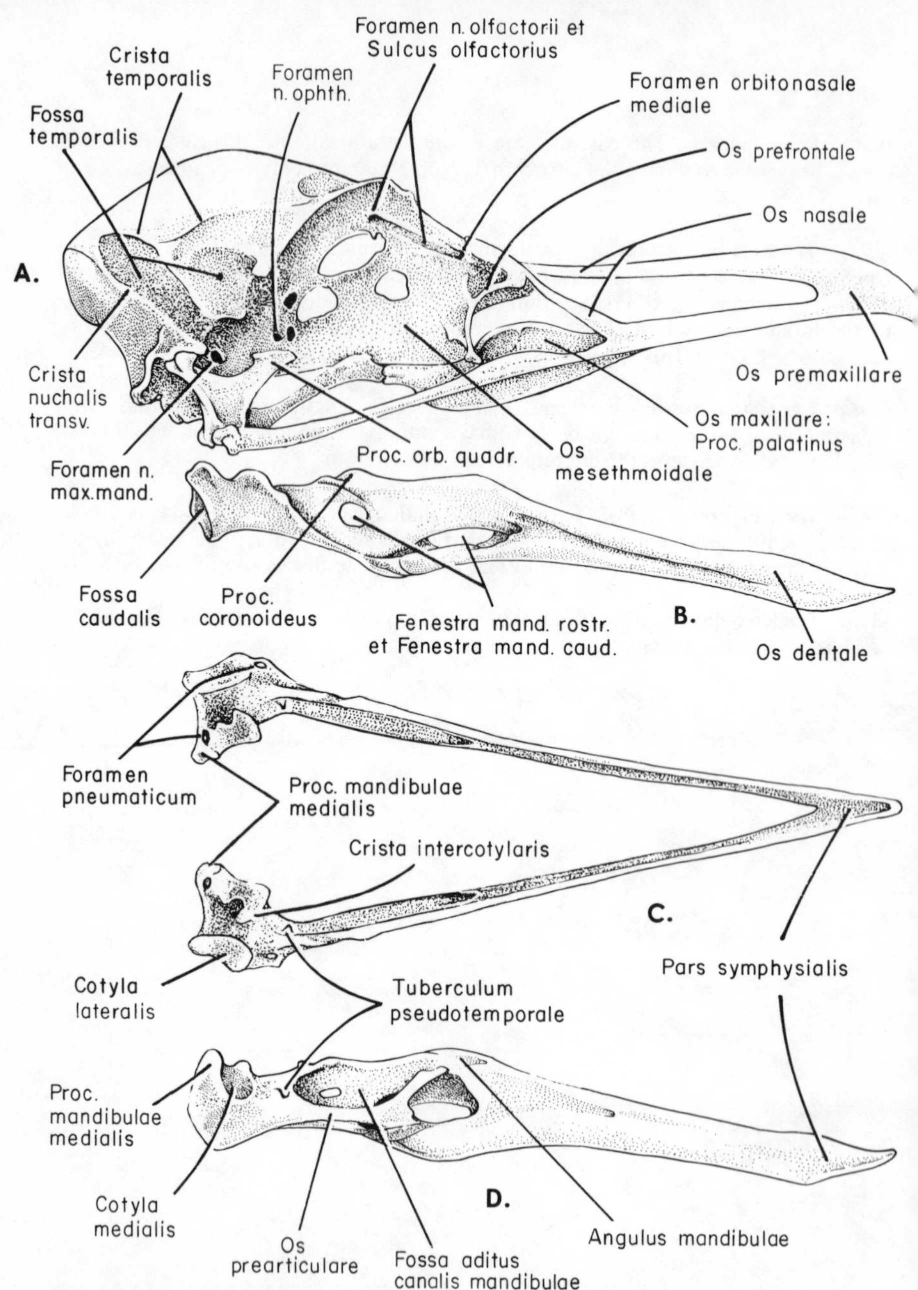

Fig. 1 Cranium and mandible of the gull, *Larus argentatus*. Original drawing, William P. Hamilton. A, cranium, right lateral aspect; B, ramus of mandible, right lateral aspect; C, mandible, dorsal aspect; D, medial aspect of left ramus of mandible. Observe that: (1) The Fossa temporalis of this species is subdivided by a pronounced crest (see Annot. 104); (2) The Sulcus of gl. nasalis (not labeled) is situated immediately caudal to the leader for Foramen n. olfactorii; (3) the proximal intramandibular elastic zone that crosses the rostral fenestra of the mandible demonstrates well several of the separate bony elements that make up the mandible (see Annot. 46 and **Arthr.** Annot. 48); (4) The retroarticular process is not a prominent feature in this species; however the caudal end of the mandible does possess a distinct Fossa caudalis (Annot. 51).
Abbreviations: mand., mandibulae; max.mand., maxillomandibularis; ophth., ophthalmici; orb., orbitalis; quad., quadrati; transv., transverse.

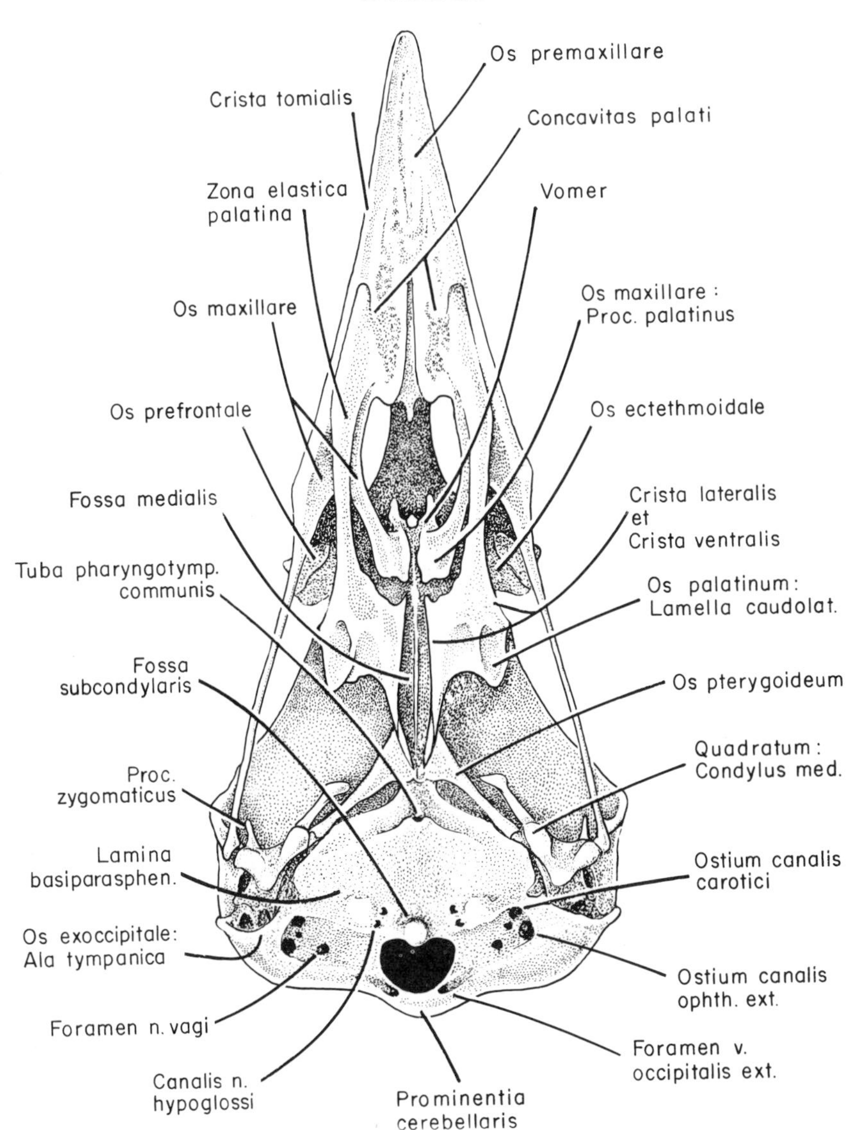

Fig. 2 Base of the skull, palate, and maxillary jaw of the crow, *Corvus brachyrhynchos*. Ventral aspect. Redrawn from Bock (1964). The leaders for Ostium canalis carotici and Lamina basiparasphenoidalis cross the Proc. oticus quadrati of each side. The depression in which the openings for the A. carotis cerebralis and A. ophthalmica externa are located is the Fossa parabasalis (see Annot. 86). The Foramen n. glossopharyngealis is seen just caudomedial to the Ostium canalis carotici (not labeled).
Abbreviations: basiparasphen., basiparasphenoidalis; ophth., ophthalmici; pharyngotymp., pharyngotympanica.

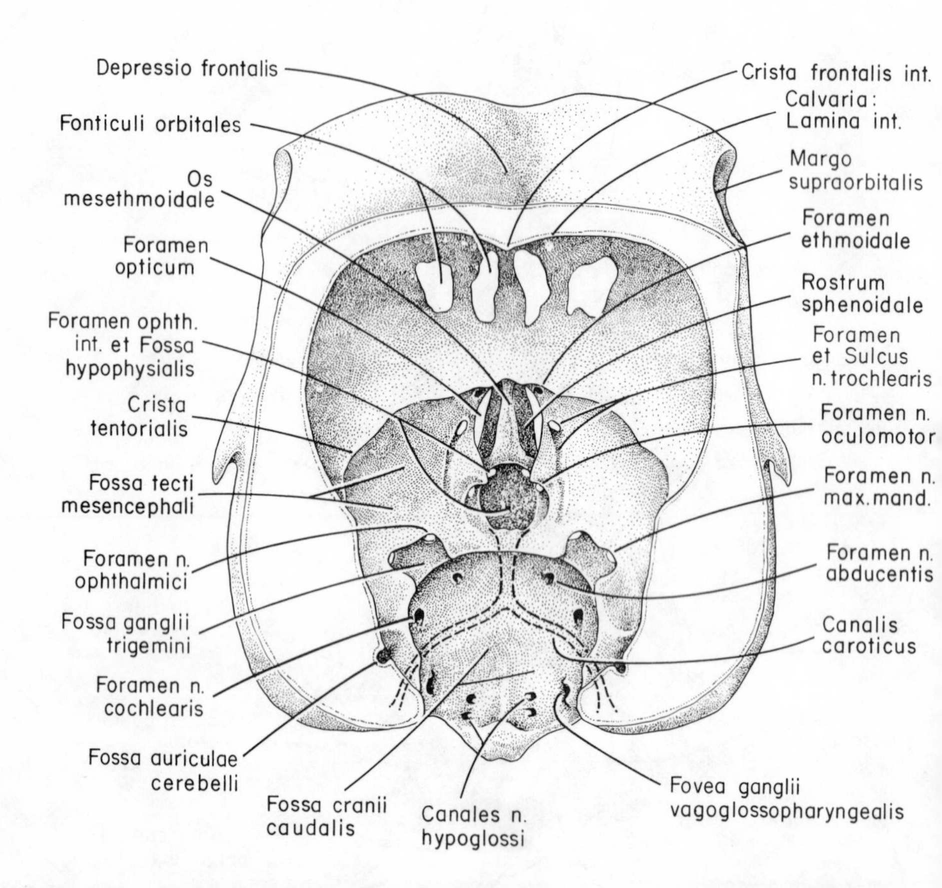

Fig. 3 Cranial cavity of the pigeon, *Columba livia.* Caudal view; transverse section through the Foramen magnum, section inclined dorsorostrally. Redrawn from Baumel (1968). (1) The Fossa cranii rostralis (perforated by the Fonticuli orbitales) houses the telencephalic hemispheres; (2) At its ventral end the carotid canal opens into the Fossa parabasalis on the Basis cranii externa and at its rostral end into Fossa hypophysialis (Annot. 99); (3) the proximal ganglia of both N. glossopharyngeus and N. vagus occupy a common depression in the floor of Fossa cranii caudalis, the Fovea ganglii vagoglossopharyngelis (Annot. 39). Abbreviations: Max.mand., maxillomandibularis; ophth., ophthalmicum.

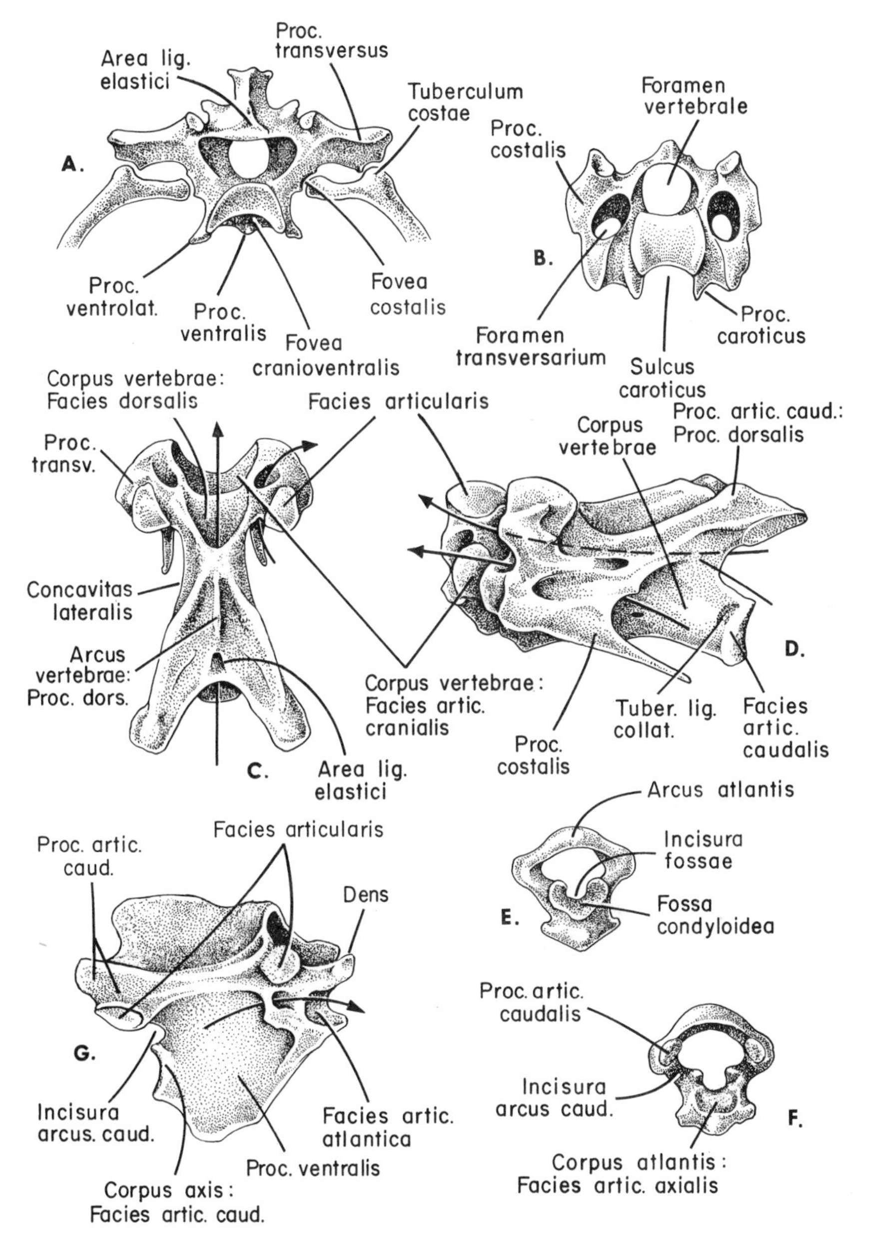

Fig. 4 Features of the cervical and thoracic vertebrae. See Figs. 5, 11 for illustrations of Notarium, Synsacrum, and caudal vertebrae.
A, Thoracic vertebra 2, *Larus*, cranial aspect; redrawn from Boas (1929).
B, Cervical vertebra 9, *Gavia*, cranial aspect; redrawn from Boas (1929).
C, Cervical vertebra 9, *Meleagris*, dorsal aspect; redrawn from Ghetie, et al. (1976).
D, Generalized cervical vertebra, craniolateral oblique view of left side; redrawn from Ghetie et al. (1976).
E, Atlas, *Meleagris*; cranial aspect, redrawn from Harvey, et al. (1968).
F, Atlas, *Meleagris*, caudal aspect; redrawn from Harvey, et al. (1968).
G, Axis, *Meleagris,* right lateral aspect; redrawn from Ghetie, et al. (1976).
Arrows in Figs. C, D, and G traverse the vertebral foramina and the transverse foramina.
Abbreviations: artic., articularis; collat., collateralis; transv., transversus; Tuber., Tuberositas.

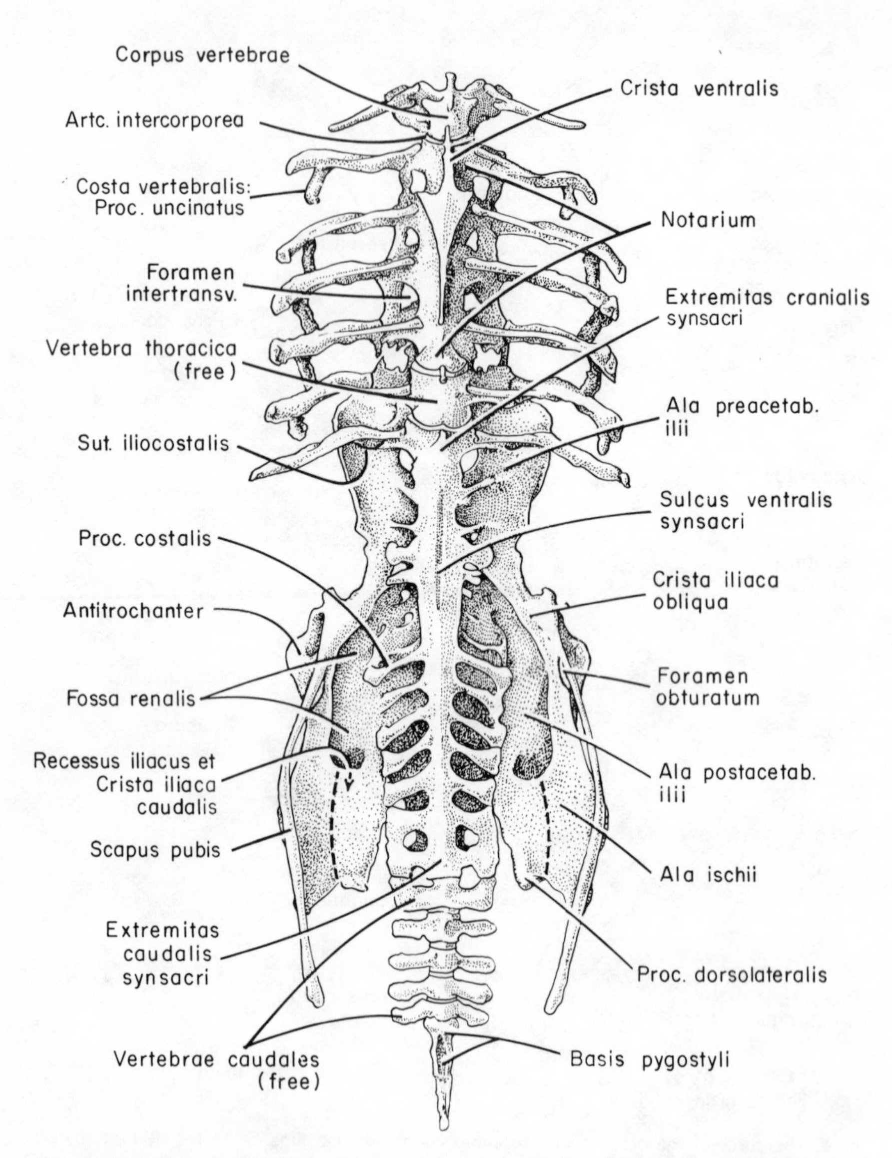

Fig. 5. Notarium, Synsacrum, and Os coxae of the turkey, *Meleagris gallapavo*; ventral view. Redrawn from Harvey, et al. (1968). The dashed lines represent the Sync. ilioischiadica (see Fig. 11). On the left side of the drawing the arrow inserted into Recessus iliacus passes dorsal to a sharp ledge of bone, the Crista iliaca caudalis.
Abbreviations: intertransv., intertransversarium; preacetabl., preacetabularis; postacetab., postacetabularis.

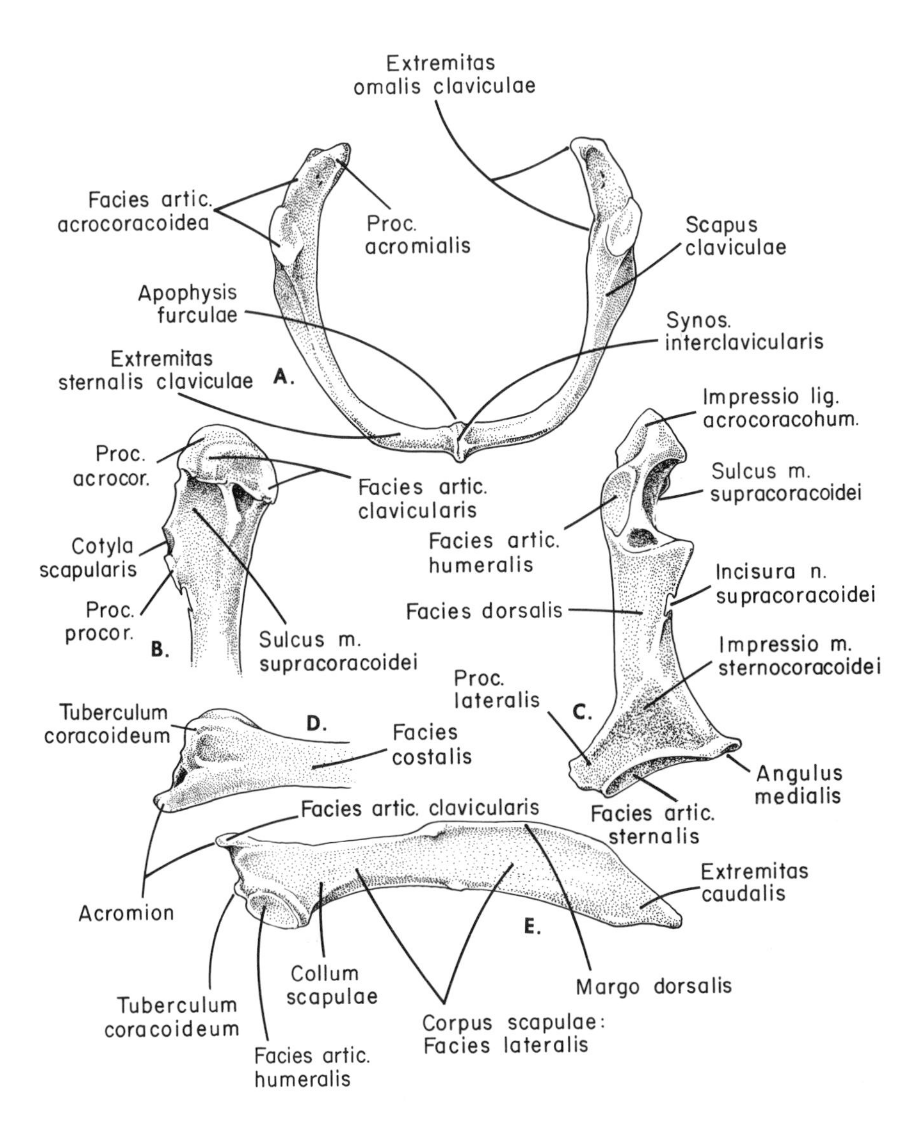

Fig. 6 Bones of the girdle of the thoracic limb of the eagle, *Aquila chrysaetos.* Redrawn from Howard (1929). A. Clavicula, caudal aspect; B, Coracoideum, left shoulder extremity, medial aspect; C, Coracoideum, left, dorsal surface; D, Scapula, left, cranial extremity, costal surfaces; E, Scapula, left, lateral surface. The ankylosed left and right clavicles form the Furcula (see Annot. 162).
Abbreviations: acrocor., acrocoracoideus; artic., articularis; procor., procoracoideus; acrocoracohum., acrocoracohumeralis.

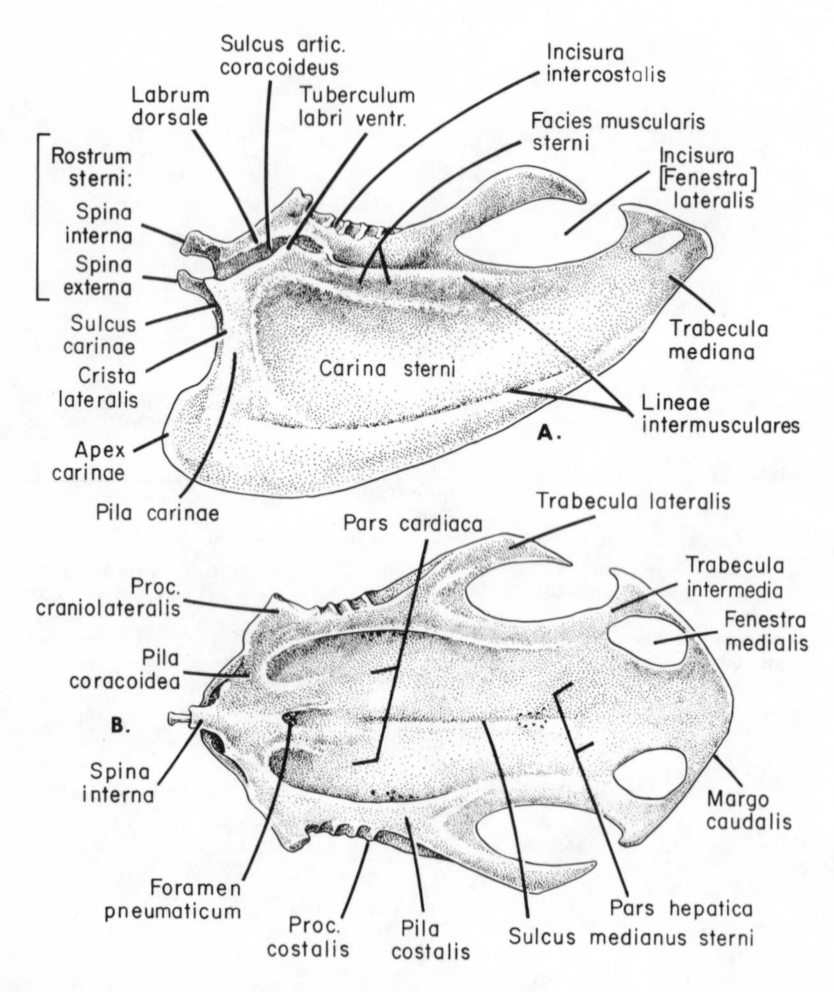

Fig. 7 Sternum of the pigeon, *Columba livia.* Adapted from original drawing of J. J. Baumel. A, left lateral aspect; B, visceral (dorsal) aspect.
Abbreviation: artic., articularis.

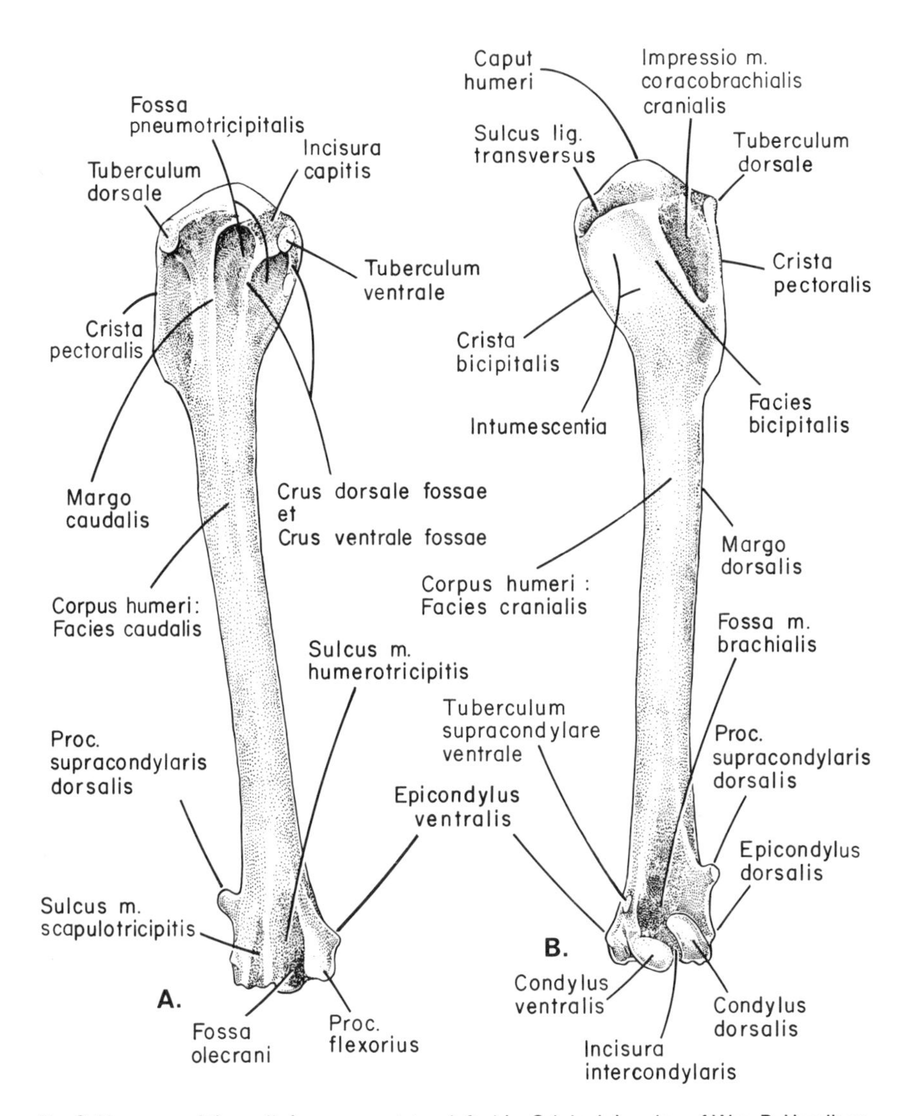

Fig. 8. Humerus of the gull, *Larus argentatus*; left side. Original drawing of Wm. P. Hamilton. A, caudal surface; B, cranial surface. In this species the pneumotriciptial fossa is the dual type, not pneumatized (see Annot. 188, 189), and the impression for M. coracobrachialis cranialis is very strongly developed.

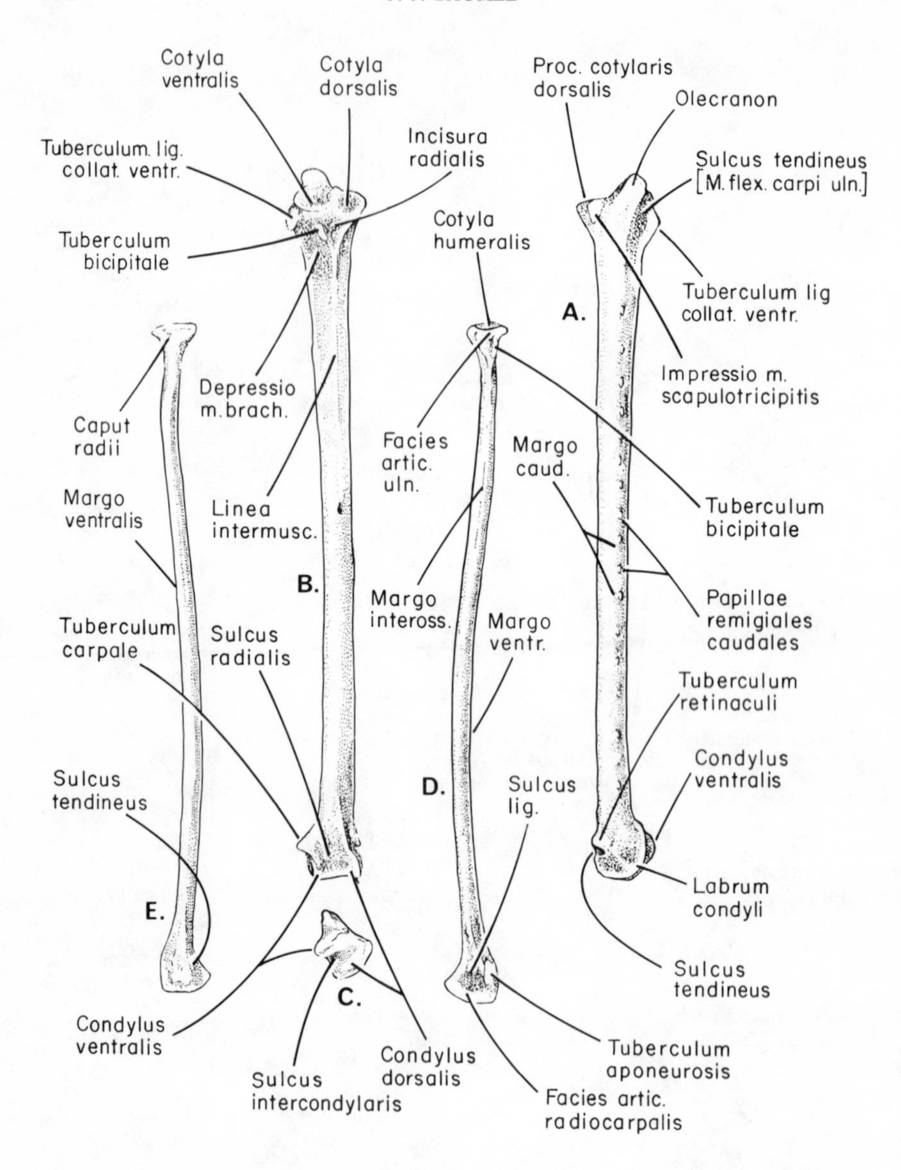

Fig. 9. Radius and Ulna of the gull, *Larus argentatus*; from the left wing. Original drawing, Wm. P. Hamilton. A, Ulna, caudal aspect; B, Ulna, cranial aspect; C, Ulna, distal end; D, Radius, caudal (interosseous) aspect; E, Radius, cranial (propatagial) aspect. Note that the terminology of these bones is based on the wing in the anatomical position (see Annot. 178 and General Intro.). In Fig. C, the Trochlea carpalis and its parts are identified; the pointed Tuberculum carpale is not labeled.
Abbreviations: artic., articularis; collat., collateralis; flex., flexor; intermusc., intermuscularis; inteross., interosseus; uln., ulnaris.

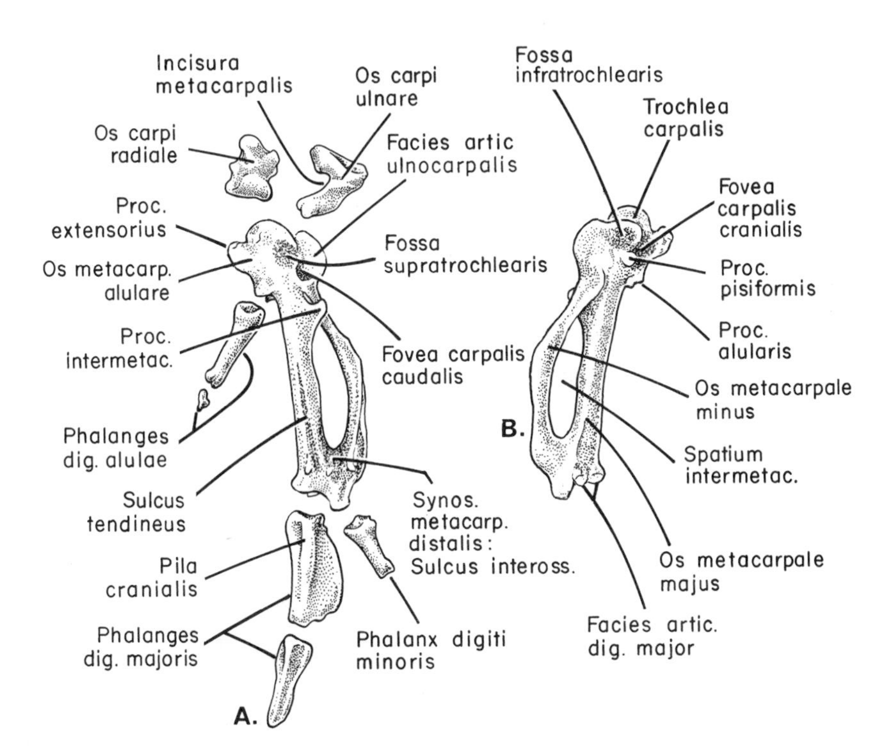

Fig. 10. Carpal and metacarpal bones, phalanges of the turkey, *Meleagris gallapavo*; bones from left wing. Redrawn from Ghetie et al. (1976). A, dorsal aspect; B, ventral aspect. Note that the alular digit of the turkey has two phalanges (see Annot. 224 and **Integ.** Annot. 70).
Abbreviations: artic., articularis; dig., digitalis or digiti; intermetac., intermetacarpale or intermetacarpalis; inteross., interosseus; metacarp., metacarpalis.

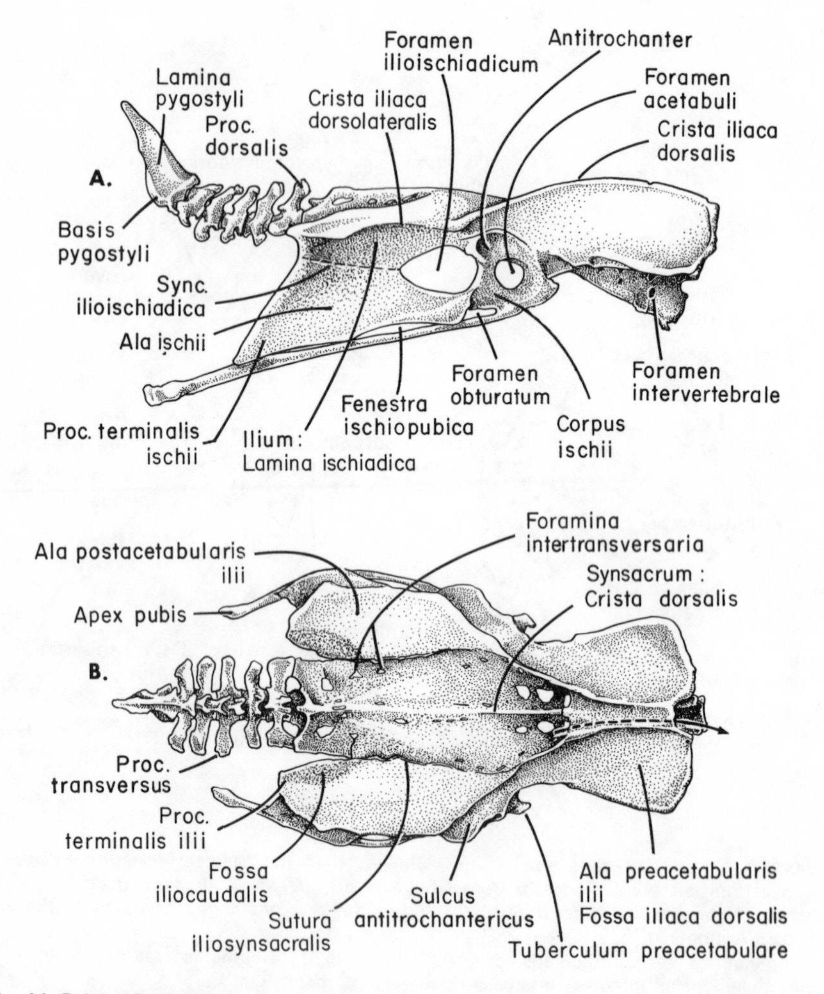

Fig. 11. Pelvis (Os coxae; Synsacrum) and tail skeleton of the turkey, *Meleagris gallapavo*. Redrawn from Harvey, et al. (1968). A, lateral aspect, right side; B, dorsal aspect. In B the arrow traverses Canalis iliosynsacralis. In A notice the dual intervertebral foramina in the synsacral part of the vertebral column (see Annot. 144).

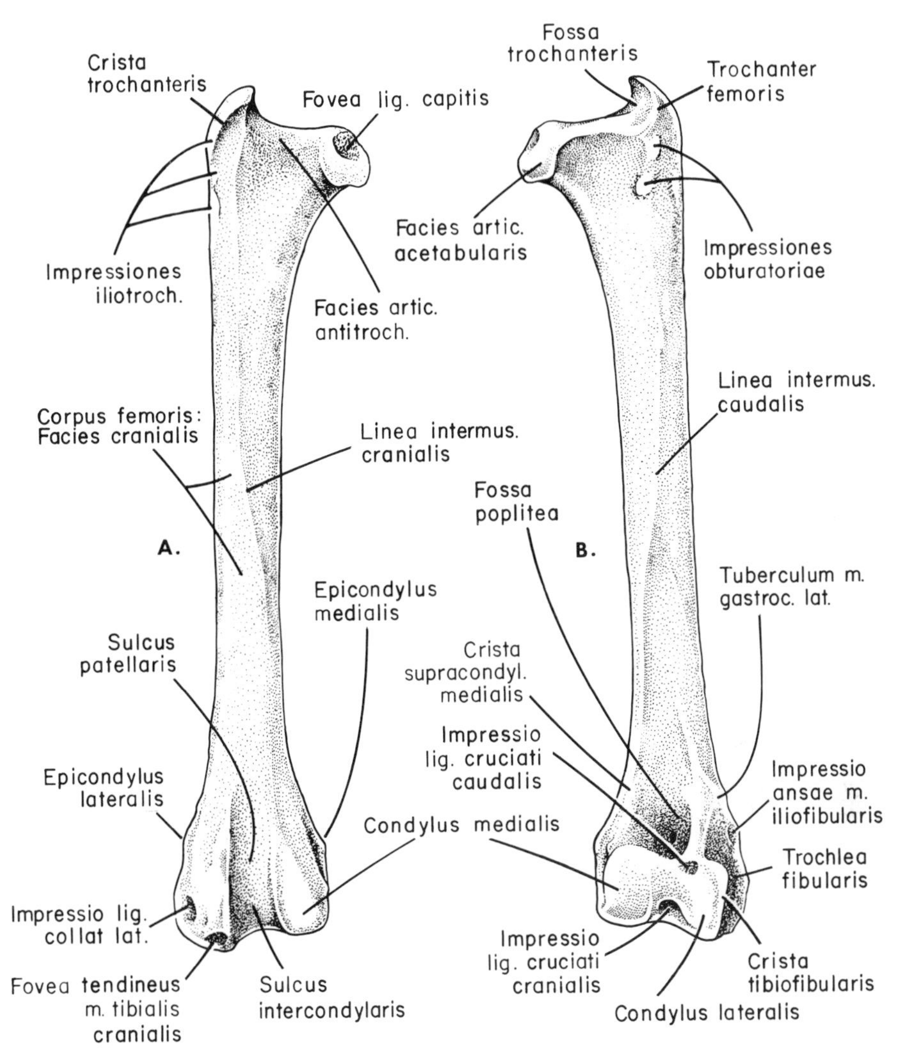

Fig. 12 Femur of the gull, *Larus argentatus*; right side. Original drawing Wm. P. Hamilton. A, cranial aspect; B, caudal aspect.
Abbreviations: antitroch., antitrochanterica; artic. articularis; gastroc., gastrocnemialis; iliotroch, iliotrochantericae; intermus., intermuscularis; supracondyl., supracondylaris.

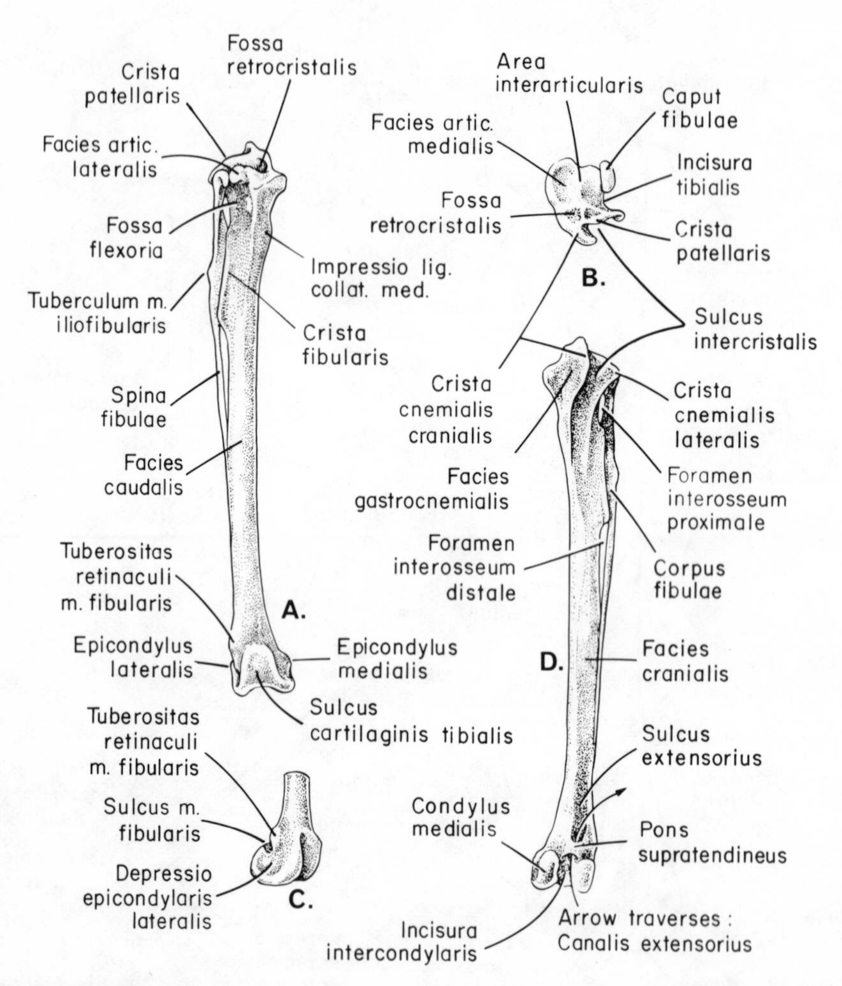

Fig. 13. Tibiotarsus and Fibula of the goose, *Branta canadensis*; left side. Original drawing Wm. P. Hamilton. A, caudal view; B, proximal articular surfaces of both bones; C, distal extremity of Tibiotarsus, lateral aspect; D, cranial view. In A Tuberositas poplitea (not labeled) is located just medial to the tip of leader on the Fossa flexoria.
Abbreviations: artic., articularis; collat., collateralis.

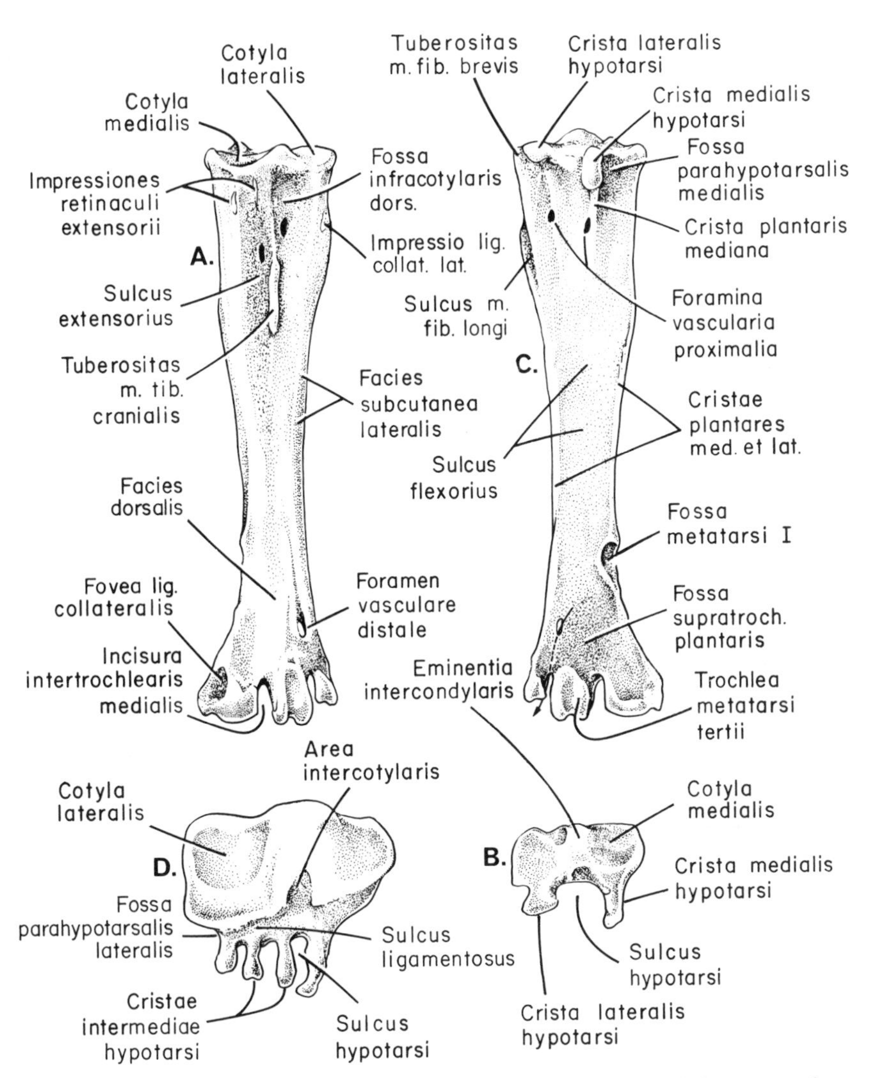

Fig. 14 Tarsometatarsus of the eagle, *Aquila chrysaetos* (A-C) and the goose, *Anser caerulescens* (*Chen hyperborea*); all bones from the left limb. Redrawn from Howard (1929). A, dorsal aspect; B, proximal end; C, plantar aspect; D, proximal end. In C the arrow traverses Canalis interosseus tendineus which conducts the tendon of M. extensor digiti IV. The Eminentia intercondylaris is a process of the proximal end of the bone that, in the articulated joint, projects between the tibiotarsal condyles and receives the distal attachment of Lig. intercondylare tibiometatarsale (see **Arthr**. Annot. 171).
Abbreviations: fib., fibularis; supratroch., supratrochlearis; tib., tibialis.

ARTHROLOGIA

J. J. BAUMEL

Department of Anatomy, School of Medicine, Creighton University, Omaha, Nebraska 68178, USA

With contributions from Sub-Committee Members: P. Ballmann, *Am Botanischen Garten 68 Köln*; A. J. Berger, *University of Hawaii*; P. Bühler, *Universität Stuttgart-Hohenheim*; J. Kaman, *University of Veterinary Medicine, Brno*; A. M. Lucas, *Michigan State University*; R. J. Raikow, *University of Pittsburgh*; P. Stettenheim, *Lebanon, New Hampshire, USA*; J. Vanden Berge, *Indiana University*.

This section is entitled Arthrologia rather than Syndesmologia, the heading used in both the Third Edition of the human *Nomina Anatomica* (IANC, 1966) and *Nomina Anatomica Veterinaria* (ICVAN, 1973). The recently published Fourth Edition of *Nomina Anatomica* (IANC, 1977) has also adopted the designation, Arthrologia. Trotter and Peterson (1966) pointed out that "Syndesmology" means the study of ligaments, and in the literal sense is a far less comprehensive term than "Arthrology" which encompasses the consideration of all joint structures and functions. These authors further commented that anatomical terminology includes numerous structures in the body called ligaments (e.g. atrophied fetal blood vessels, thickenings of mesenteries, meninges, etc.) that are excluded from the category of syndesmology.

Feather ligaments. Avian arthrological terminology differs in a major respect from that of mammals on account of the presence of feathers. The large flight feathers of wing and tail constitute an exoskeleton of sorts. An elaborate system of ligaments connect these feathers to one another and to the bony skeleton. Not only do these ligaments anchor the feathers, but some are so disposed as to produce (passively) movements of opening and closing of the flight feathers (see Annot. 114, 189–200 and terms under Ligamenta pennarum).

Elastic zones of face. The terminology of the elastic or bending zones (Zonae elasticae) of the upper and lower jaw has been included in this section rather than with Osteologia for sake of convenience. Functionally the elastic zones operate similarly to slightly movable joints such as the cartilaginous joints between certain vertebral bodies, between sternal and vertebral ribs, or ligamentous ones such as the coracoscapular joint (Annot. 93). Moreover, certain of the elastic zones in some species of birds are actually developed into freely movable synovial joints, syndesmoses, or junctional regions where different bones have undergone ankylosis (synostosis) during fetal or neonatal development. Examples of these are the craniofacial and proximal intramandibular zones (see Annot. 45–49).

Naming of ligaments and joints; Hyphenation of names. Ligaments are named generally by indicating their osseous attachments (e.g. Lig. iliofemorale) or some aspect of their topography to the joint (e.g. Lig. collaterale ventrale).

Joints are often named by the bones or parts of bones that are united at the joint (e.g. Sut. vomeropalatina). In order to make quite clear what osseous elements are involved in the joint, some anatomical names must be hyphenated (e.g. Artc. quadrato-squamoso-otica, a joint in which three bones are involved, or Artc. quadrato-quadratojugalis, a term in which the hyphen clarifies that the joint connects Quadratum with Os quadratojugale.

Juncturae ossium; *Articulationes*. Junctura(e) ossium (literally: junction of bones) is adopted in this work as the general term for all joints, i.e. fibrous, cartilaginous, and synovial. This term was used in the BNA (1895) and JNA (1936) terminologies (see **Gen. Intro.**). This usage, however, disagrees with the contemporary *Nomina Anatomica* (IANC, 1977) and the *Nomina Anatomica Veterinaria* (ICVAN, 1973); in both of these nomenclatures the former terms, "Diarthrosis", "Articulus", and "Junctura synovialis" are set aside, and "Articulationes" is adopted as the general term for all sorts of joints. In practice the IANC and ICVAN terminologies use "Articulationes" primarily to refer to synovial joints; in the present avian nomenclature the latter practice has been continued. See Annot. 1.

Simplification of terms. In a number of instances the combining form of the names of various skeletal elements has been abbreviated in order to simplify and make more manageable the anatomical names of articulations and ligaments, and to clearly differentiate terms having a degree of similarity. For example, terms involving articulations between Radius and Ulna with Os carpi radiale and Os carpi ulnare (see Annot. 111) must be distinguished.

In the limbs on account of the incorporation of carpal and tarsal bones with the bones of the crus, the foot and hand, compound names have been given to the skeletal elements. In order to avoid at least some of the cumbersome arthrological terms, abbreviated combining forms are used. For example, "tibio-" for "tibiotarso-" (see Annot. 155); "radiocarpo-" and "ulnocarpo-" for Os carpi radialis and Os carpi ulnaris (see Annot. 112), etc.

In the literature the existing names for the various wrist joints and their ligaments are confusing, difficult to learn and remember. The nomenclatural precepts of making anatomical names descriptive, meaningful, and an aid to memory were applied in developing the terminology of the wrist joints in the present list. This terminology may be employed for birds generally notwithstanding interspecific variation in the occurrence, attachments, and configuration of the ligaments of the wrist region. See Annot. 112.

In the head the combining form "jugo-" refers to any part of jugal arch, not necessarily Os jugale proper (see Annot. 38, 40). "Rostro-" refers to the Rostrum sphenoidale [parasphenoidale] (see Annot. 22, 23); "otico-" has been used to refer to the optic complex of bones (see Annot. 9).

TERMINI GENERALES

Aponeurosis
Bursa synovialis
Cartilago
Discus
Fibrocartilago
Junctura fibrosa[1]
  Syndesmosis
  Sutura
    Sutura serrata
    Sutura squamosa
    Sutura foliata
    Sutura plana
  Gomphosis[2]
Junctura cartilaginea[1]
  Synchondrosis
  Symphysis
Junctura ossea[1]
  Synostosis
Junctura synovialis
    (see Intro.)
  Articulatio simplex
  Articulatio composita
  Articulatio plana
  Articulatio spheroidea
  Articulatio ellipsoidea
  Articulatio condylaris
  Articulatio trochoidea
  Articulatio sellaris
  Ginglymus
Cartilago articularis
Cavitas articularis
Discus articularis
Meniscus articularis
Capsula articularis
  Membrana fibrosa
  Membrana synovialis
    Plica synovialis
    Villi synoviales
    Synovia
  Vasa articularia
  Nervi articulares
Ligamenta
  Ligamentum collaterale
  Ligamentum interosseum
  Ligamentum extracapsulare
  Ligamentum intracapsulare
  Ligamentum elasticum[3]
Membrana
Retinaculum

JUNCTURAE OSSIUM[1]

(see Intro.)

JUNCTURAE CAPITIS[4]

(Fig. 1; and **Osteo.** Figs. 1, 2)

Suturae cranii[4]

Sut. supraoccipitoparietalis[5]
Sut. interparietalis[6]
Sut. interfrontalis[7]
Sut. frontoparietalis[8]
Sut. exoccipitosquamosa
Sut. parietosquamosa
Sut. frontosquamosa
Sut. orbitosphenosquamosa
Sut. orbitosphenofrontalis
Sut. oticosquamosa
Sut. oticoparietalis
Sut. supraoccipitosquamosa

Synchondroses cranii[4]

Syncc. interoticae[9]
Sync. oticoexoccipitalis
Sync. supraoccipito-exoccipitalis
Sync. otico-orbitosphenoidalis
Sync. oticosupraoccipitalis
Sync. oticobasioccipitalis
Sync. basioccipito-basiparasphenoidalis
Sync. basioccipito-exoccipitalis
Sync. exoccipito-basiparasphenoidalis
Sync. otico-basiparasphenoidalis
Sync. otico-alaparasphenoidalis
Sync. basiparaspheno-alaparasphenoidalis
Sync. orbitospheno-alaparasphenoidalis
Sync. basiparaspheno-rostroparasphenoidalis
Sync. alaparaspheno-rostroparasphenoidalis
Sync. basispheno-basiparasphenoidalis
Sync. basispheno-alaparasphenoidalis
Sync. basispheno-rostroparasphenoidalis
Sync. mesethmo-orbitosphenoidalis (see **Osteo.** Annot. 88)
Sync. interorbitosphenoidalis[10] (see **Osteo.** Annot. 88)

Suturae faciei

Sutt. faciei maxillaris[11, 4]

- Sut. frontopremaxillaris
- Sut. frontonasalis
- Sut. frontomesethmoidalis[12]
- Sut. mesethmonasalis
- Sut. mesethmopremaxillaris
- Sut. prefronto-ectethmoidalis[13]
- Sut. prefrontofrontalis
- Sut. prefrontonasalis
- Sut. fronto-ectethmoidalis
- Sut. nasopremaxillaris
- Sut. nasomaxillaris
- Sut. internasalis[14]
- Sut. intermaxillaris
- Sut. interpremaxillaris
- Sut. maxillopremaxillaris
- Sut. vomeropalatina[15 23]
- Sut. vomeromaxillaris[16]
- Sut. palatomaxillaris
- Sut. rostromaxillaris (see Intro.)
- Sut. palatopremaxillaris
- Sut. interpalatina[17]
- Sut. pterygopalatina[17]
- Sut. intervomeralis[18]
- Sut. jugomaxillaris
- Sut. jugoprefrontalis[19]
- Sut. jugo-quadratojugalis

Sutt. faciei mandibularis (see **Osteo.** Annot. 53)

- Symphysis mandibularis [Sut. intermandibularis][21] (Fig. 1)
- Sut. dentosplenialis
- Sut. dentosupraangularis
- Sut. supraangulosplenialis
- Sut. angulosplenialis
- Sut. angulosupraangularis
- Sut. articulosupraangularis
- Sut. articuloprearticularis
- Sut. articuloangularis

Synchondroses faciei

Sync. rostromesethmoidalis[22] (see Intro.)

- Sync. mesethmo-ectethmoidalis

Articulationes faciei (Fig. 1; and **Osteo.** Figs. 1, 2)

Artcc. maxillares[11,4]

Artc. jugoprefrontalis[20]

Lig. jugoprefrontale[20]

Artc. rostrovomeralis[23]

Lig. mesethmovomerale [vomerale]

Artc. quadrato-quadratojugalis[24]

Lig. interosseum[24]

Artc. quadrato-squamoso-otica[25]

Capsula articularis[25]

Artc. quadratopterygoidea[26]

Capsula articularis

Artc. pterygorostralis[27]

Artc. pterygobasipterygoidea[27]

Capsula articularis

Artc. pterygopalatina [intrapterygoidea][28]

Capsula articularis

Artc. rostropalatina[29] (see Intro.)

Capsula articularis

Lig. mesethmopalatinum [orbitale]

Lig. palatoprefrontale

Lig. suborbitale[30] (Fig. 1)

Membrana circumorbitalis[31]

Septum orbitale[31]

Membrana temporalis[31]

Artcc. mandibulares (see Suturae faciei mandibularis and Fig. 1)

Artc. mandibulosphenoidalis[32]

Lig. mandibulosphenoidale[33]

Lig. intramandibulare[34] (see Zonae elasticae mandibulae)

Artc. quadratomandibularis[35]

Capsula articularis

Lig. quadratomandibulare[36]

Lig. occipitomandibulare[37]

Lig. jugomandibulare mediale[38]

Meniscus articularis[39]

Lig. jugomandibulare laterale[40]

Lig. prefrontomandibulare[41]

Lig. postorbitale[42]

Lig. zygomanticomandibulare[43]

Artc. ectethmomandibularis[44]

Ligamenta columellae (see **Org. sens.** Auris media)

ZONAE ELASTICAE OSSIUM FACIEI[45]

(see Fig. 2)

Zona elastica [Ginglymus] craniofacialis [frontonasalis][46]

Zonae elasticae maxillares[47, 11]

Zona elastica premaxillonasalis proximalis[47]
Zona elastica premaxillonasalis distalis[47]
Zona elastica premaxillomaxillaris[47]
Zona elastica nasalis dorsalis[47]
Zona elastica nasalis ventralis[47]
Zona elastica arcus jugalis[47]
Zona elastica palatina[47] (see **Osteo.** Fig. 2)

Zonae elasticae mandibulares[48]

Zona elastica intramandibularis distalis[48]
Zona elastica intramandibularis proximalis[48]
Synd. intramandibularis proximalis[49]
    Lig. intramandibulare

JUNCTURAE APPARATUS HYOBRANCHIALIS[50]

(see **Myol.** Fig. 2b)

Sut. interentoglossalis
Artc. entoglosso-basibranchialis[51]
    Lig. collaterale ventrale
Sync. [Artc.] intrabasibranchialis[52]
Artc. ceratobasibranchialis[53]
Sync. intracornualis[54]

JUNCTURAE LARYNGIS

(see **Resp.** Larynx and Figs. 1, 2)

Artc. procricoarytenoidea[55]
Artc. procricocricoidea[55]
Synd. interarytenoidea[56]
Synd. intercricoidea[56]
    Lig. intercricoideum
Synd. intra-arytenoidea[55 56]
Artc. intracricoidea[55 56]
Artc. cricoarytenoidea
Synd. cricobasibranchialis[57]
Lig. arytenoglossale[58]

JUNCTURAE COLUMNAE VERTEBRALIS[59]

(Fig. 10)

Artcc. intervertebrales

Artc. intercorporea[60] (see **Osteo.** Fig. 5)
Capsula articularis
Lig collaterale[60]
Meniscus intervertebralis[61]
Fenestra centralis
Annulus fibrosus[61]
Discus intervertebralis[62]
(Lig. suspensorium corporum vertebralium)[62]
Lig. interspinosum
Lig. elasticum interlaminare[63] (Fig. 10)
Lig. elasticum obliquum (Fig. 10)
Lig. elasticum transversum (Fig. 10)
Membrana interlaminaris (Fig. 10)
Artc. zygapophysalis[65]
Capsula articularis
Meniscus articularis[66]

Artcc. regionis cervicalis

Artc. atlantoöccipitalis[64]
Capsula articularis
Membrana atlantoöccipitalis ventralis[67]
Membrana atlantoöccipitalis dorsalis[67]
Fibrocartilago atlantis[64]
Artcc. atlantoaxiales[68]
Capsulae articulares[68]
Artc. atlantodentalis[68]
Artc. intercorporea[68]
Artc. zygapophysalis[65]
Membrana atlantoaxialis
Lig. apicis dentis
Lig. transversum atlantis[69]
Lig. collaterale atlantoaxiale[69]
Synos. costotransversaria[70]
Artc. costotransversaria[70]

Artcc. regionis thoracicae

Capsulae articulares
Artcc. notarii[71] (see **Osteo.** Fig. 5)
Synos. intercorporea

Synos. intertransversaria
Synos. [Sut.] interspinalis
Lig. intercristale ventrale[72] (see **Osteo.** Annot. 119)

Artcc. synsacri[73] (see **Osteo.** Figs. 5, 11)
Synos. intercorporea[74]
Synos. intertransversaria
Synos. interspinalis[75]
Sut. [Synos.] iliosynsacralis[76]
Ligg. iliosynsacralia[76]
Synos. interiliospinalis[77]

Artcc. caudae
Lig. intertransversarium
Artc. vertebropygostyloidea
Synoss. pygostyli[78] (see **Osteo.** Annot. 146)

JUNCTURAE COSTARUM[79]
(Fig. 10)

Sync. capitis costae[79]
Lig. collaterale[79]
Artc. costotransversaria[79]
Capsula articularis
Sut. iliocostalis[80]
Sync. intercostalis[81]
Sut. costouncinata[82]
Lig. triangulare[82]
Artc. sternocostalis[83]
Capsula articularis

JUNCTURAE CINGULI MEMBRI THORACICI

Membranae incisurarum [fenestrarum] sterni[84]
Lig. corpus claviculae[85]
Synd. sternoclavicularis[85]
Synos. interclavicularis (see **Osteo.** Fig. 6 and Annot. 162–163)
Membrana sternocoracoclavicularis[86] (Fig. 3)
Lig. sternoprocoracoideum[86]
Lig. sternocoracoideum mediale[86]
Lig. intercoracoideum[86]
Lig. sternocoracoideum longum[86]
Lig. sternoacrocoracoideum[86]

JUNCTURAE CINGULI MEMBRI THORACICI—*continued*

Lig. sternoclaviculare longum[86]
Lig. sternoacromiale[86]
Lig. coracoideum[86]
Lig. acrocoraco-procoracoideum[87] (Fig. 3)
Membrana cristoclavicularis[88]

Artc. sternocoracoidea[89]
Capsula articularis
Artc. intercoracoidea[90]
Ligg. collateralia sternocoracoidea[91]
Lig. sternocoracoideum superficiale
Lig. sternocoracoideum profundum[91] (see Memb. sternocoracoclav.)
Lig. sternocoracoideum laterale[92] (see Memb. sternocoracoclav.)

Symphysis coracoscapularis[93]
Cavitas glenoidalis[93] (see **Osteo.** Annot. 167)
Lig. interosseum coracoscapulare[93] (Fig. 3)
Labrum cavitatis glenoidalis[93] (Fig. 3)
Labrum coracoideum
Labrum scapulare
Lig. coracoscapulare dorsale[94]
Lig. coracoscapulare ventrale
Lig. coracoscapulare internum[94]
Lig. acrocoracoacromiale[95]

Synd. [Artc.] acrocoracoclavicularis[96] (see **Osteo.** Annot. 165)
Lig. acrocoracoclaviculare superficiale
Lig. acrocoracoclaviculare profundum

Synd. procoracoclavicularis[97]
Lig. procoracoclaviculare[97]

Synd. acromioclavicularis[98] (see **Osteo.** Annot. 165)
Lig. acromioclaviculare
Lig. scapuloclaviculare dorsale[99]

JUNCTURAE ALAE
(Fig. 1)

Articulatio humeralis [Artc. coraco-scapulo-humeralis][93]
(see Symphysis coracoscapularis and **Osteo.** Annot. 167–171)

Capsula articularis
Fibrocartilago humerocapsularis[100]

Lig. acrocoracohumerale[101] (Fig. 3)
Bursa acrocoracoidea[101]

Lig. coracohumerale dorsale[102]
Lig. scapulohumerale dorsale[102]
Lig. scapulohumerale caudale
Lig. scapulohumerale laterale
Lig. humerocapsulare transversum
Bursa supracoracoidea[101]
Plicae synoviales et Ligg. intracapsularia[103] (Fig. 3)
Plica synovialis coracoidea
Plica synovialis scapularis
Plica synovialis transversa
Lig. intracapsulare coracoideum craniale
Lig. intracapsulare coracoideum caudale

Junctura cubiti[104] (Fig. 4)

Artc. humeroulnaris[104]
Artc. humeroradialis[104]
Artc. radioulnaris proximalis[104]
Capsula articularis
Lig. collaterale ventrale[105]
Lig. collaterale dorsale[106]
Lig. cubiti craniale[107]
Meniscus radioulnaris [Lig. annulare radii][108]
Lig. transversum radioulnare[109]
Trochlea humeroulnaris[110]
Pars tendinea[110]
Pars pennata[110]
Lig. tricipitale[111]
Membrana interossea antebrachii

Juncturae carpi et manus[112] (Fig. 5)

Aponeurosis ventralis[113] (see Annot. 114, 115)
Retinaculum flexorum[114]
Aponeurosis ulnocarporemigialis[115]
Digitationes remigiales
Capsula articularis
Synd. radioulnaris distalis
Lig. interosseum radioulnare[116]
Artcc. ulnocarpalis et radiocarpalis[112]
Artc. ulnocarpalis
Lig. ulno-ulnocarpale proximale[117]
Lig. ulno-ulnocarpale distale[117]
Lig. ulno-radiocarpale ventrale[118]
Lig. interosseum ulno-radiocarpale[119]
Lig. ulno-metacarpale ventrale[120]

JUNCTURAE ALAE—*continued*

Artc. radiocarpalis[112]
Lig. radio-radiocarpale craniale
Lig. radio-radiocarpale ventrale
Lig. radio-radiocarpale dorsale[121]
Artcc. intercarpales (see **Osteo.** Annot. 214)
Meniscus intercarpalis[122]
Syncc. intercarpales[123]
Lig. menisco-metacarpale
Artcc. carpo-carpometacarpales[124,112]
Lig. radiocarpo-metacarpale craniale
Lig. radiocarpo-metacarpale dorsale
Lig. radiocarpo-metacarpale ventrale[125]
Lig. ulnocarpo-metacarpale ventrale[126]
Lig. ulnocarpo-metacarpale dorsale[127]
Juncturae carpometacarpi
Synoss. carpometacarpales[128]
Synos. intermetacarpalis proximalis[129] (see **Osteo.** Fig. 10)
Synos. intermetacarpalis distalis[129] (see **Osteo.** Fig. 10)
Artcc. metacarpophalangeales[130]
Artc. metacarpophalangealis alulae[131] (Fig. 5)
Capsula articularis
Lig. obliquum alulae[131]
Lig. collaterale caudale
Artc. metacarpophalangealis digiti majoris
Capsula articularis
Lig. collaterale ventrale[132]
Lig. collaterale caudale[133]
Lig. obliquum intraarticulare[134]
Meniscus articularis[135]
Artc. metacarpophalangealis digiti minoris
Capsula articularis
Lig. collaterale ventrale
Lig. collaterale dorsale
Artcc. interphalangeales manus
Juncturae interphalangeales laterales[136]
Lig. interosseum
Artc. interphalangealis digiti majoris[137]
Capsula articularis
Lig collaterale ventrale
Lig. collaterale craniale
Lig. collaterale caudale[138]
Meniscus articularis
Lig. interphalango-remigiale[139, 113]

Ligamenta accessoria alae[140]

Lig elasticum propatagiale[141] (see **Myol.** Figs. 5, 6)
Retinaculum m. scapulotricipitis[142, 111]
Lig. humerocarpale[143 195]
Retinaculum m. extensoris metacarpi ulnaris[144]
Lig. m. extensoris longi digiti majoris[145]
Lig. m. ulnometacarpalis dorsalis[145] (Fig. 11)
Lig. m. extensoris metacarpi ulnaris[145]

JUNCTURAE CINGULI MEMBRI PELVICI[146]

(see Artcc. synsacri)

Sync. ischiopubica
Sut. ischiopubica[147]
    Membrana ischiopubica[147]
    Lig. ischiopubicum[147]
Sync. ilioischiadica[148] (see **Osteo.** Fig. 11)
    Membrana ilioischiadica[148]
Sync. iliopubica
(Symphysis pubica)[149]
(Symphysis ischiadica)[150]
Membrana acetabuli[151]

JUNCTURAE MEMBRI PELVICI

(Fig. 6)

Articulatio coxae[152]

Capsula articularis
Labrum acetabulare
Lig. iliofemorale[153]
Lig. pubofemorale[153]
Lig. ischiofemorale[153]
Lig. capitis femoris

Junctura genus[154] (Fig. 7)

Artc. femorotibialis[155]
Artc. femorofibularis[156]
Artc. femoropatellaris
Artc. tibiofibularis
    Lig. tibiofibulare craniale[157] (Fig. 7)
    Lig. tibiofibulare caudale[157]
Capsula articularis
Lig. patellae[158]

JUNCTURAE MEMBRI PELVICI—*continued*

Retinacula patellae
Meniscus medialis
  Cornu craniale
  Cornu caudale
  Lig. meniscotibiale caudale
  Lig. meniscofemorale
  Lig. transversum genus
Meniscus lateralis[159]
  Lig. meniscotibiale craniale
  Lig. meniscofibulare caudale
  Lig. meniscocollaterale[160]
  Lig. meniscofemorale
Lig. cruciatum craniale
Lig. cruciatum caudale
Lig. collaterale mediale
Lig. collaterale laterale

Ligamenta crusis (see Artc. tibiofibularis)

Synd. tibiofibularis
  Lig. obliquum tibiofibulare
  Lig. interosseum tibiofibulare[161]
    Foramen interosseum proximale[162] (Fig. 7)
    Foramen interosseum distale[162]
  Membrana interossea cruris

Juncturae tarsi et pedis

Artc. cartilago-tibiotarsalis[163] (Fig. 8)
  Capsula articularis
  Cartilago tibialis[164] (Fig. 8)
    Os sesamoideum intertarsale[164]
    Lig. cartilago-sesamoideum
    Retinaculum mediale[165]
    Retinaculum laterale[165]
    Retinaculum flexorum[165]
    Lig. metatarso-sesamoideum[166, 169]
    Lig. cartilago-metatarsale[166, 169]

Articulatio intertarsalis[167] (Fig. 8) (see Annot. 163, 164)

Capsula articularis[165]
Meniscus medialis[168]
  Cornu craniale
  Cornu caudale

Meniscus lateralis[168]
Cornu craniale
Cornu caudale
Lig. meniscosesamoideum[169, 166]
Lig. meniscotibiale[170, 168] (see Artcc. genus)
Crus mediale
Crus laterale
Lig. intercondylare tibiometatarsale[171]
Lig. intercondylare transversum
Lig. collaterale mediale[172]
Lig. collaterale laterale[172]
Lig. accessorium[172]

Juncturae tarsometatarsales et intermetatarsales[173]

(see **Osteo.** Annot. 284, 288)
Synos. tarsometatarsalis[173]
Synos. intermetatarsalis[173]
Synd. intermetatarsalis hallucis[174] (Fig. 9)
Lig. interosseum[175]
Lig. transversum metatarsale[176]
Canalis flexorius plantaris[178] (Fig. 9)
Lig. elasticum metatarsi I

Artcc. metatarsophalangeales[180, 179] (see Fig. 9)

Aponeurosis plantaris[179]
Corpus adiposum plantare profundum[179]
Capsulae articulares
Ligg. collateralia[181]
Ligg. plantaria[182] (Fig. 9)
Vaginae fibrosae
Lig. obliquum hallucis[177] (Fig. 9)
Lig. rectum hallucis

Artcc. interphalangeales

Capsulae articulares
Ligg. collateralia
Ligg. plantaria[182]
Ligg. elastica extensoria digitorum[183]

Ligamenta accessoria membri pelvici

Lig. inguinale[184]
Membrana iliocaudalis[185]
Ansa m. iliofibularis[186] (see **Myol.** Figs. 8, 9)

Ligamenta accessoria membri pelvici—*continued*

- Retinaculum m. fibularis [peronei]
- Retinaculum extensorium tibiotarsi [Lig. transversum][187]
- Retinaculum extensorium tarsometatarsi[188]

LIGAMENTA PENNARUM[189]

(Fig. 11)

Ligg. remigum primariorum

- Lig. elasticum interremigiale[190]
- Aponeurosis ventralis[113]
  - Retinaculum flexorum[114]
  - Aponeurosis ulnocarporemigialis[115, 113]
    - Digitationes remigiales
  - Digitationes remigiales
- Retinacula ulnocarpo-remigialia[191]
- Aponeurosis interphalango-remigialis[192, 139]
  - Digitationes remigiales
- Lig. interphalango-remigiale[139, 192]
- Ligg. phalangoremigialia distalia[193]

Ligg. remigum secundariorum

- Aponeurosis antebrachialis ventralis[194]
  - Digitationes remigiales
- Aponeurosis antebrachialis dorsalis[194]
  - Digitationes remigiales
- Trochlea humeroulnaris (see Artc. cubiti)
  - Pars pennta
    - Digitationes remigiales
- Septum humerocarpale[195, 143]
  - Digitationes remigiales[196]
- Lig. elasticum m. flexoris carpi ulnaris[197]
  - Digitationes remigiales[196]
- Lig. elasticum interremigiale[190]

Lig. elasticum intertectricale

- Lig. elasticum intertectricale carpale[198]

Lig. cubiti[199]

Ligg. rectricium

- Lig. elasticum interrectricale[200]

ANNOTATIONS

(1) **Juncturae ossium; Junctura(e); Articulatio(nes).** Juncturae ossium is the general term for all types of joints between bones; as a part of a heading for a group of Juncturae ossium the simple, shortened form "Juncturae" is used (e.g. Juncturae columnae vertebralis). "Articulatio" (plural: Articulationes) is used to refer to synovial joints. See Intro. for a discussion of these terms.

(2) **Gomphosis.** Examples of the peg and socket joint in birds are Artc. quadrato-quadratojugalis and Artc. ectethmo-mandibularis. See Annot. 24, 43.

(3) **Lig. elasticum.** Elastic ligaments are found in many parts of the avian body; e.g. propatagial skin fold, vertebral column, flight feathers of wing and tail; digits, and phallus (see **Genit. masc.** Annot. 46). See Jollie (1957) and Hofer (1945, 1949, 1957).

(4) **Juncturae capitis; Suturae et Synchondroses cranii.** The avian skull is characterized by a high degree of fusion (synostosis) of its separate elements that occurs in early life; most often the fibrous joints (suturae) and chondral joints (synchondroses) become ankylosed to the extent that traces of the joints are obliterated in mature individuals.

(5) **Sut. supraoccipitoparietalis.** Synonymy: Sutura lambdoidea. A pronounced Crista nuchalis transversa occurs along the suture in certain birds (e.g. *Gavia immer*, *Morus bassanus*, *Ardea herodias*, *Anser*, and *Ceryle alcyon*).

(6) **Sut. interparietalis.** Synonymy: Sutura sagittalis.

(7) **Sut. interfrontalis.** Synonymy: Sutura metopica.

(8) **Sut. frontoparietalis.** Synonymy: Sutura coronalis.

(9) **Syncc. interoticae.** The complex of otic bones (Os prootica, Os epiotica, and Os opisthotica) coalesce into a unit that contains most of the inner ear structures. See Jollie (1957); Sandoval (1963); **Osteo.** Annot. 105.

(10) **Sync. interorbitosphenoidalis.** Certain of the other "sphenoid" elements of the base of the skull are paired in earlier development, later becoming ankylosed; therefore there are several "Syncc. intersphenoidales" (see Jollie, 1957; **Osteo.** Annot. 88).

(11) **Sutt. faciei maxillaris; Artcc. maxillares; Zonae elasticae maxillae.** "Maxillary" in this sense refers to the entire upper jaw skeleton (see **Osteo.** Annot. 35).

(12) **Sut. frontomesethmoidalis.** In some avian forms (e.g. *Cathartes*, *Gallus*, *Anser*, *Ardea*) the rostral end of the mesethmoid plate possesses a horizontal Lamina dorsalis that articulates broadly with the ventral surface of the frontal bones near the craniofacial hinge.

(13) **Sut. prefronto-ectethmoidalis.** The prefrontal and ectethmoid bones are articulated along much of their dorsoventral extent in some birds (e.g. *Columba*,

Corvus brachyrhynchos, Crotophaga ani); whereas only their ventral ends are joined in such forms as *Larus delawarensis* and *Buteo lineatus*. See Newton and Gadow (1896, p. 876).

(14) **Sut. internasalis.** The side to side apposition of the nasal bones in the median plane is apparently of rare occurrence. An internasal suture does occur in *Falco* near the craniofacial hinge (Suschkin, 1899) and in the spoonbill, *Platalea* (Hofer, 1954).

(15) **Sut. vomeropalatina.** The vomer is commonly fused to the caudal part of Os palatina that rides along the ventrum of Rostrum sphenoidale; e.g. anseriforms (see Hofer, 1949). Actually the part of the palatine with which the vomer articulates is the so-called antepterygoid (Jollie, 1957; Bock, 1964) that is incorporated into the palatine in the adult neognathine birds. See Annot. 28; and **Osteo.** Os palatinum.

(16) **Sut. vomeromaxillaris.** Located between the palatine processes of the maxillae and the vomer as in passerine and ciconiiform birds. See **Osteo.** Fig. 2.

(17) **Sut. interpalatina.** Located between the caudal parts of right and left palatine bones that articulate with the sphenoid rostrum; these parts become ankylosed in some birds (e.g. *Eudocimus albus, Morus bassanus, Caprimulgus carolinensis*). See Hofer (1945, 1954) for different palatal configurations. See **Osteo.** Fig. 2.
Sut. pterygopalatina. See Artc. pterygopalatina and Annot. 28.

(18) **Sut. intervomeralis.** In most birds in which the vomer exists as a distinct osseous element, the definitive vomer is a paired structure. Evidence of the ultimate fusion between the two vomers is obliterated in some forms; in adult dried skulls of forms such as *Gallinago delicata, Larus delawarensis, Corvus brachyrhynchos* some evidence of the original paired elements can be distinguished.

(19) **Artc. jugoprefrontalis; Sut. jugoprefrontalis.** Os prefrontale is firmly anchored (occasionally synostosed) to the jugal arch in some forms (*Balaeniceps* and penguins, Zusi, 1974); however, this does not hinder kinesis since the prefrontal bone moves with the arch as an integral part of the upper jaw (see Bühler, 1970). In other birds (e.g. *Cathartes, Pelecanus, Corvus*) the prefrontal articulates with the jugal arch by way of a ligament of appreciable length that allows movement between the two bones, hence a syndesmosis rather than a suture. In yet another species the joint between prefrontal and jugal arch appears to be a synovial articulation (e.g. *Sula, Gavia*, tinamous, Beddard, 1898). See Davids (1952).

(20) **Lig. jugoprefrontale** (Fig. 1). Synonymy: Lig. jugolacrimale (Zweers, 1974). Connects Tuberculum prefrontale of Arcus jugalis with the ventral tip of Os prefrontale (e.g. anseriform birds). Ghetie *et al.* (1976) describe Membrana lacrimomaxillaris rostral to Lig. jugoprefrontale in Anser; according to the terminology in this list membrane should be called Membrana jugoprefrontalis.

(21) **Symphysis mandibularis.** Synonymy: Sut. intermandibularis; Sutura interdentalis. This is actually a synchondrosis between the right and left Ossa dentalia that have overgrown and covered the cartilaginous joint. Ultimately the suture is ossified, producing a synostosis in adults of most avian species. See **Osteo.** Fig. 1.

(22) **Sync. rostromesethmoidalis.** Synonymy: Sync. mesethmo-rostroparasphenoidalis.

(23) **Artc. rostrovomeralis.** Synonymy: Artc. vomero-rostroparasphenoidalis. According to Zusi (pers. comm.) all birds that have been studied so far show a movable articulation between vomer and the ventral edge of Rostrum sphenoidale. See Hofer (1949) for discussion of articulations of the vomer in different taxa.

(24) **Artc. quadrato-quadratojugalis.** Often a synovial joint in which a peg or condyle of Os quadratojugale fits into a cotyla in the lateral aspect of the quadrate bone. In *Sturnus* (De Kock, 1955) and in *Caprimulgus* (Bühler, 1970) this joint is a syndesmosis, not a synovial joint. See **Osteo.** Fig. 1.
Lig. interosseum. This is a strong ligament connecting the condyle of the quadratojugal into the bottom of its socket in the quadrate bone (*Anser*, *Cairina moschata*).

(25) **Artc. quadrato-squamoso-otica.** In some avian species each of the two condyles of Proc. oticus of the Quadratum articulates independently of the other; therefore, two separate joint cavities are present: Artc. quadratosquamosa and Artc. quadratoötica (e.g. *Columba*). In other species both condyles of the quadrate and their receiving surfaces on the proötic and squamosal bones are enclosed within one common articular capsule (e.g. *Gallus, Corvus*).

(26) **Artc. quadratopterygoidea.** Os pterygoideum articulates with the quadrate bone via two facets in the grouse, *Pedeocetes phasinellus* and *Phasianus colchicus*; between the two facets is located the pneumatic foramen of the Quadratum (*Pedeocetes*). See Hofer (1945) for differences in this joint related to degree of kinesis of the upper jaw.

(27) **Artc. pterygorostralis; Artc. pterygobasipterygoidea.** Synonymy; Artc. pterygorostroparasphenoidalis. In some avian forms (e.g. *Ardea, Coccyzus, Corvus, Fulica, Thalasseus*) the rostral end of Os pterygoideum articulates with the palatine bone (see Annot. 28) and Rostrum sphenoidale. In other birds (e.g. *Cathartes aura, Caprimulgus carolinensis, Gallinago delicata, Columba livia*) the middle of the length of Os pterygoideum also articulates with Proc. basipterygoideus of Rostrum sphenoidale. In *Rhea* the basipterygoid process articulates with the caudal end of Os pterygoideum (Bock, 1963). In yet other species (e.g. *Anser, Cairina moschata*) the basipterygoid process is located rostrally near the pterygopalatine articulation, and is the only articulation of Os pterygoideum with the Rostrum. See **Osteo.** Annot. 60 and Fig. 2.

(28) **Artc. pterygopalatina [Artc. intrapterygoidea].** Actually this joint is the Sutura intrapterygoidea in many neognathous birds in which part of the developing Os pterygoideum ("antepterygoid") fuses with the caudal part of the Palatinum; the "postpterygoid" is the definitive Os pterygoideum (Jollie, 1957). See Annot. 15 regarding Sutura vomeropalatina; see also **Osteo.** Annot. 60 and Fig. 2.

(29) **Artc. rostropalatina.** Synonymy: Articulatio rostroparasphenopalatina.

(30) **Lig. suborbitale.** Synonymy: Lig. suboculare (Bock, 1964). A thin ligamentous band, fascial in nature that extends from the ventral tip of Os prefrontale to Proc. postorbitalis and the upper end of Lig. postorbitale; assists in walling off the orbit ventrolaterally. See Fig. 1; and **Osteo.** Annot. 30.

(31) **Membrana circumorbitalis.** The thickened connective tissue margin of the orbit attached to the supraorbital margin, Os prefrontale, jugal arch, and suborbital ligament (Fuchs, 1954–55).

Septum orbitale. The attenuated, thin extension of Membrana circumorbitalis that forms the connective tissue framework of the palpebrae.
Membrana temporalis. Synonymy: Lig. temporale (Bas, 1955). Spans the temporal fossa, and serves as partial origin of the underlying muscles that fill the fossa (Bas, 1955).

(32) **Artc. mandibulosphenoidalis.** Synonymy: Artc. mandibulo-basiparasphenoidalis; Artc. articulo-basiparasphenoidalis; Artc. articulo-basitemporalis. Proc. mandibularis medialis abuts against the basiparasphenoid bone, and forms the "medial brace" of the mandible (Bock, 1960). In birds having the "medial brace" (e.g. *Rynchops nigra*, *Pygoscelis adeliae*) the medial process of the mandible articulates with either the medial or lateral basiparasphenoid (basitemporal) processes (Bock, 1960). Bock contends that the brace supports the mandible and prevents its caudal disarticulation in birds having a quadratomandibular joint wherein the condyles of the quadrate bone and the receiving surfaces of the mandible are not interlocked strongly. Zusi (1967) disagrees with the latter viewpoint. See Bock and Morioka (1971) for more recent treatment of the functional properties of Artc. mandibulosphenoidalis. See Annot. 33.

(33) **Lig. mandibulosphenoidale.** Synonymy: Lig. mandibulo-basiparasphenoidale. The mandibulosphenoid joint is a syndesmosis in some species (*Charadrius*, Bock, 1960; a synovial joint in others (*Rynchops*).

(34) **Lig. intramandibulare.** Located at junction of intermediate and caudal parts (Fig. 8) of ramus of mandible in *Caprimulgus* (Bühler, 1970). See Fig. 2; and **Osteo.** Ramus mandibulae.

(35) **Artc. quadratomandibularis.** Synonymy: Artc. quadratoarticularis. See Zusi (1967) for a thorough discussion of the avian quadratomandibular joint. The quadrate bone articulates mostly with Os articulare of the mandible usually by means of three articular condyles (Bock, 1960).

(36) **Lig. quadratomandibulare.** Present in most passerines, especially well-developed in tyrannid flycatchers. This ligament extends from Proc. pterygoideus of the Quadratum to the mandible near the base of its medial process. It is stretched during opening of the jaws; relaxation of M. depressor mandibulae causes jaws to snap shut by elastic recoil of the ligament (Bock and Morony, 1972).

(37) **Lig. occipitomandibulare**(Bock, 1964). Synonymy: Lig. exoccipitomandibulare; Lig. depressor mandibulae (Rooth, 1953); Lig. neurocranio-mandibulare (Fuchs, 1955; Davids, 1952). The medial strongest part of this wide ligamentous sheet is attached along the caudal border of the medial process of the mandible. Part of the cranial attachment of the ligament may be on the lateral part to Ala tympanica of Os exoccipitale that forms the caudal wall of the tympanic cavity. The superficial surface of the ligament serves as origin of the deep part of M. depressor mandibulae; the ligament completes the floor of the Meatus acusticus externus. See Fig. 1; and **Osteo.** Annot. 19, 21.

(38) **Lig. jugomandibulare mediale.** Synonymy: Lig. jugale (Fuchs, 1955); Lig. jugo-mandibulare caudale (Bas, 1955); Lig. quadratojugomandibulare mediale. Lig. jugomandibulare mediale actually connects the quadratojugal part of the jugal arch with the caudal end of the mandible. The ligament is attached near the

apex of the medial process of the mandible (see Lebedinsky, 1921, for a discussion of the medial point of attachment of this ligament). Two medial jugomandibular ligaments are present in the hornbill, *Tockus* (Rawal and Bhatt, 1973). See Intro., Annot. 39, 40 and Fig. 1.

(39) **Meniscus articularis.** A distinct wedge-shaped meniscus is located in the caudal part of the quadratomandibular joint, and is attached to the internal aspect of the transverse part of the medial jugomandibular ligament (Hofer, 1945) (Zweers, 1974: *Anas*); (*Anser*, *Cairina*).

(40) **Lig. jugomandibulare laterale.** Synonymy: Lig. jugo-mandibulare rostrale (Bas, 1955); Lig. quadratojugo-mandibulare laterale. This short ligament is attached to caudolateral end of the jugal arch just opposite the quadratojugo-quadrate articulation, and extends to the lateral process of the mandible. See Intro., Annot. 38 and Fig. 1.

(41) **Lig. prefrontomandibulare.** Synonymy: Lig. lacrymo-mandibulare (Davids, 1952); Lig. lacrimomandibulare (Goodman and Fisher, 1962). This ligament extends caudoventrad from the ventral tip of Os prefrontale to the lateral mandibular process. The ligament is subcutaneous; in *Anser* it forms a distinct wide band that contributes to the formation of the rostroventral wall of the orbit. See Fig. 1; and illustration in Zweers (1974).

(42) **Lig. postorbitale** (Fig. 1). The term, Lig. postorbitale, is almost universally employed; however, based on its usual attachment it could be called "Lig. orbitosphenomandibulare", The postorbital ligament usually extends from the apex of Proc. postorbitalis of Os orbitosphenoidale across the lateral aspect of the caudal end of the jugal arch where it produces a notch in certain avian species. The ligament is attached ventrally to the lateral process of the mandible. In other instances Lig. postorbitale may be attached on different bones at both ends: in *Balaeniceps* it is firmly attached to the jugal arch; in *Podilymbus* it has an accessory attachment to the zygomatic process of the squamosal bone (Zusi and Storer, 1969). Lig. postorbitale assists in supporting the quadratojugo-quadrate articulation, and plays an important role in kinesis of the upper jaw (Bock, 1964; Zusi, 1967). See **Osteo.** Annot. 30.

(43) **Lig. zygomaticomandibulare.** Synonymy: Lig. squamosomandibulare (Lebedinsky, 1921). This ligament extends from the zygomatic process of Os squamosum to the coronoid process of the mandible.

(44) **Artc. ectethmo-mandibularis.** Bock and Morioka (1971) described this unusual peg and socket synovial joint between the dorsal border of the mandible and the ventral tip of Os ectethmoidale in the family Meliphagidae.

(45) **Zonae elasticae ossium faciei.** Specialized flexible zones of the facial bones of birds are referred to as synostotic or syndesmotic kinetic zones (Bühler, 1970) or "pseudoarthroses". These elastic zones act in concert with movable syndesmoses and synovial articulations to bring about kinesis of the upper jaw and expansion of the width of the lower jaw. See Fig. 2.

(46) **Zona elastica craniofacialis.** Synonymy: Ginglymus craniofacialis or frontonasalis (see Bühler, (1970) for comments on the usage of these terms). In birds the junction between neurocranium and the maxillary facial skeleton is commonly an

elastic zone of flexible parts of the Proc. premaxillaris of Os nasale and Proc. frontalis of the Premaxilla near their junction with the rostral end of Os frontale and Lamina dorsalis of Os mesethmoidale. The cranio-facial connection occurs only rarely as a synovial joint (large parrots, e.g. *Ara*) or a syndesmosis.

Zona elastica craniofacialis has been adopted as the primary term for sake of consistency with the terms of the other elastic zones of the facial skeleton. Ginglymus craniofacialis has been designated an official alternative term. Even though "ginglymus" (Ginglymos, Gr., a hinge) is an apt descriptive term, in widespread use, and easily understood, there has been objection to its use. In its general sense a hinge may be jointed or flexible; however, in its usual anatomical sense the term, ginglymus, pertains strictly to a hinge-type synovial articulation.

Zona elastica craniofacialis permits rotatory movements of elevation and depression of the upper jaw (kinesis). In some birds the elastic zone is a long, indefinite region; in others it is marked dorsally by a distinct, transverse sulcus floored by thin bone (*Cacatua*, *Ceryle*, *Morus*, *Rynchops*). In other forms the elastic zone is a restricted, narrow, transverse seam flush with the dorsal surfaces of bones rostral and caudal to it (*Gavia*, *Cathartes*, *Coccyzus*, *Ardea*, *Aix*). See Hofer (1954) for a discussion of the different types of craniofacial zones.

Zona elastica craniofacialis is short (in its rostro-caudal dimension) in birds with holorhinal nostrils; the entire upper jaw rotates (elevation-depression) as a unit about the elastic zone at the base of the upper jaw. This is the prokinetic condition common to most birds. See Bock (1964) and Zusi (1967) for review treatments on kinesis of the avian skull and Annot. 47.

(47) **Zonae elasticae maxillares.** (see Fig. 2 and Annot. 11). In some species with schizorhinal nostrials the elastic zone has shifted rostrad away from the craniofacial zone so that kinetic movements involve dorsal bending of the upper jaw at various points along its length. This is the rhynchokinetic condition of some charadriiform and ratite birds. See Bock (1964) and Zusi (1967); and **Osteo.** Annot. 53.
Zona elastica arcus jugalis. The elastic zone at the rostral end of the jugal arch has become specialized into a syndesmosis in cardueline finches.
Zona elastica palatina (see **Osteo.** Fig. 2). The elastic zone of the premaxillary process of Os palatinum is represented by a syndesmosis in parrots.

(48) **Zonae elasticae mandibulares.** Certain sphenisciform, procellariiform, pelecaniform, ciconiiform, charadriiform, and caprimulgiform birds possess two zones of bending in each ramus of the mandible that permit widening of the interramal distance. Zona distalis is near Symphysis mandibularis. In general, Zona proximalis is situated at the junction of the middle and caudal thirds of the ramus of the mandible; the proximal zone is elongated in *Pelecanus*. See Fig. 2; and **Osteo.** Annot. 42. Consult Yudin (1961), Zusi (1962, 1974), and Bühler (1970).

(49) **Synd. intramandibularis proximalis.** In *Caprimulgus* (and probably in Nyctibiidae) the proximal bending zone of the mandible is developed as a definite, mobile, syndesmotic joint between intermediate and caudal segments of the mandibular ramus (Bühler, 1970). See Fig. 2; and **Osteo.** Ramus mandibulae, Fig. 1.

(50) **Juncturae apparatus hyobranchialis.** Names for these joints are based mostly on the terminology that McLelland (1968) adopted for the elements of the hyobranchial apparatus. Zweers (1974) designates several ligaments associated with joints between the various elements that are probably capsular ligaments rather than collateral ligaments. See **Osteo.** Annot. 79.

(51) **Artc. entoglosso-basibranchialis.** Synonymy: Artc. basihyoentoglossalis; Artc. basihyoparaglossalis. This is a saddle-type joint (artc. sellaris) that permits mostly side to side movements (e.g. *Columba, Anser, Fulica americana*).

(52) **Sync. [Artc.] intrabasibranchialis.** Synonymy: Sync. basihyobasibranchialis. Usually a cartilaginous joint between the rostral and caudal basibranchial elements (McLelland, 1968). According to Jollie (1962) the rostral element represents the basihyal and the caudal element represents basibranchiale 1. See **Osteo.** Annot. 81.

(53) **Artc. ceratobasibranchialis.** Synonymy: Artc. basihyoceratobranchialis. Os ceratobranchiale is the proximal element of the branchial cornu; in certain birds the ceratobranchial articulates with both the rostral basibranchial (basihyal) and the caudal basibranchial (urohyal).

(54) **Sync. intracornualis.** Synonymy: Artc. ceratohyo-epihyalis. According to Jollie (1962) this is the joint between Os ceratobranchiale II and Os epibranchiale II.

(55) **Artc. procricoarytenoidea; Artc. procricocricoidea.** These paired articulations and Artc. intracricoidea are the juncturae synoviales of the larynx. In *Gallus* (White, 1975) and in *Columba* and *Eudocimus albus* the procricoid is a median element that intervenes between the two arytenoids and the two cricoid alae. See **Resp.** Figs. 1, 2.

(56) **Synd. interarytenoidea; Synd. intercricoidea.** In *Strix varia* (and doubtless other avian species) the caudal ends of the right and left Corpora arytenoidei and the right and left Alae cricoideae articulate directly with one another by fibrous joints. Each artytenoid also articulates with the cricoid ala of its side. A small rudimentary semi-globular procricoid cartilage articulates via synovial joints with sockets on the rostroventral angles of the apposing surfaces of the caudal parts of the bodies of the arytenoid cartilages.
Synd. intra-arytenoidea. Fibrous joint between Corpus and Proc. caudalis of each Cartilago arytenoidea in *Corvus brachyrhynchos* (Bock, 1978) (see **Resp.** Annot. 25, 27, 29).
Artc. intracricoidea. Synovial joint between Corpus and Ala of each Cartilago cricoidea in *Corvus brachyrhynchos* (Bock, 1978) (see **Resp.** Annot. 25, 27, 29).

(57) **Synd. cricobasibranchialis.** Synonymy: Synd. urohyocricoidea. In some birds (*Columba, Strix varia*) the caudal basibranchial element is connected to the ventral surface of the Corpus of Cartilago cricoidea by fibrous tissue.

(58) **Lig. arytenoglossale.** Paired cords of elastic tissue in *Gallus* that connect the rostral processes of the arytenoids with the cornua of the entoglossal bones. The ligaments are longer and thicker in males than in females (White, 1975).

(59) **Juncturae columnae vertebralis.** Consult Barkow (1856) for an extensive comparative review of interspecific variation in the arthrology of the avian vertebral column.

(60) **Artc. intercorporea** (Barkow, 1856). The joints between adjacent vertebral bodies in the cervical region and the cranial part of the thoracic region are synovial joints, some with intraarticular menisci (see Annot. 61–64, 71, 74, 78). In the caudal thoracic region and synsacral region these joints are ankylosed, forming synostoses; the joints between the bodies of the free caudal vertebrae are symphyses. In birds

generally, the intercorporal joints throughout the length of the vertebral column are mostly Artcc. sellares (heterocoelous; saddle-shaped) (Beddard, 1898). The vertebral bodies of the thoracic vertebrae are opisthocoelous in penguins, in suborder Charadrii, and in some parrots (Parker, 1888). See **Osteo.** Figs. 4, 5.
Lig. collaterale. The lateral part of the articular capsule that connects vertebral bodies is strongly developed as a collateral ligament; the caudal end of each vertebral body exhibits a distinct tuberosity for attachment of Lig. collaterale (see Barkow, 1856).

(61) **Meniscus intervertebralis.** Synonymy: Fibrocartilago intercalaris corpum vertebralium (Jäger, 1858). The synovial joints between bodies of cervical vertebrae are incompletely divided by intraarticular menisci, attached at their periphery to the inner aspect of the articular capsule (e.g. *Eudocimus albus* and *Strix varia*); the menisci are thin toward their centers, having openings of the variable shape (Jäger, 1858). According to Jäger in *Anas* the menisci in the thoracic region are attached to the margins of adjacent vertebral bodies as well as to articular capsules.
Annulus fibrosus. In certain birds (e.g. *Columbia, Gallus*) instead of a proper meniscus the intervertebral connective tissue consists of a thickened ring attached to the inner surface of the articular capsule. According to Jäger (1858) in *Anas* the Annulus fibrosus is variably attached to the circumference of adjacent vertebral bodies and to the articular capsule.

(62) **Discus intervertebralis.** Between the free caudal vertebrae a complete articular disc connects the vertebral bodies. These joints lack a synovial cavity, and are classed as symphyses (Barkow, 1856).
(Lig. suspensorium corporum vertebralium). Vestige of the notochord that persists in some adult birds (Jäger, 1858).

(63) **Lig elasticum interlaminare.** Synonymy: Lig. flavum. Present mostly in cervical and thoracic regions; the ligaments are unpaired narrow bundles that connect laminae of adjacent vertebrae (Boas, 1929). Boas notes that the ligaments are strongest in the caudal cervical region. He describes an additional system of elastic ligaments spanning the dorsal concavity of the root of the neck in *Rhea*; these are the Ligg. elastica interspinalis profunda et superficialia of Barkow (1856). See Fig. 10; and **Osteo.** Fig. 4.

(64) **Artc. atlantooccipitalis; Fibrocartilago atlantis.** The occipital condyle fits into Fossa condyloidea of the atlas. In some species the fossa is a complete osseous cup; in others it is an osseous semiring ventrally, completed dorsally by Fibrocartilago atlantis (intercartilago atlantis, Boas, 1929). In certain species the floor of the fossa is perforated by a foramen that transmits the Lig. apicis dentis (see Goedbloed, 1958).

(65) **Artc. zygapophysialis.** See **Osteo.** Columna vertebralis. In order to remain consistent with the human and veterinary nomenclatures, "Proc. articularis" is adopted as the primary term; "Zygapophysis' is recognized as an official alternative term. For purposes of simplicity of the arthrological terminology the inconsistency of the human and veterinary nomenclatures is followed in employing the secondary term, "Zygapophysis", in construction of the name of the joints between articular processes of adjacent vertebrae.

(66) **Meniscus articularis.** According to Barnett (1954b) the zygopophysial joints of the cervical vertebral column of *Columba palumbus* possess thin, fibrous menisci containing scattered cartilage cells.

(67) **Membrana atlantooccipitales.** Synonymy: Lig. capsularis atlantooccipitalis (Goedbloed, 1958). Laterally the continuity between the ventral and dorsal membranes is interrupted by the exit of the huge V. occipitalis interna.

(68) **Artcc. atlantoaxiales.** In certain avian species the atlas and axis articulate by means of two separate synovial joints; one between the Dens and the Corpus atlantis and the second between Corpus atlantis and Corpus axis. In most birds two additional zygapophysial synovial joints connect the vertebral arches of atlas and axis. See Boas (1929) and Goedbloed (1958) for details of Artcc. atlantoaxiales; and **Osteo.** Annot. 130–138.

(69) **Lig. transversum atlantis.** Ossified in some birds (e.g. *Phalacrocorax*, *Numenius* and *Corvus*, Goedbloed, 1958); holds Dens axis against Facies articularis dentalis of the Atlas.
Lig. medianum atlantoaxiale. Boas (1929) describes this ligament that extends from the ventral surface of the root of the Dens to the dorsal aspect of Corpus atlantis caudal to the Artc. atlantodentalis.
Lig. collaterale atlantoaxiale. Strong paired ligaments corresponding to the alar ligaments of mammals in that they limit rotatory movements between atlas and axis. Ligg. collateralia extend longitudinally from foveae on each side of the root of the dens to impressions on the inner surface of the incisure (or foramen) of the condyloid fossa of the atlas lateral to Facies articularis dentalis. The Ligg. collateralia are fused with the ventral surface of Lig. transversum atlantis.

(70) **Synos. costotransversaria; Artc. costotransversaria.** With the exception of the atlas, most cervical vertebrae have costal processes ankylosed to transverse processes and to Corpus; Foramen transversarium (for the vertebral artery) is partly bounded by the costal process (Boas, 1929) (see **Osteo.** Annot. 133 and Fig. 4). In the root of the neck a variable number of short, freely movable cervical ribs are found in most birds.

(71) **Artcc. notarii** (see Barkow, 1856). The Notarium is formed by ankylosis of the intermediate group of thoracic vertebrae that form a rigid unit. Usually the joints between Notarium and the first vertebra of the Synsacrum are freely movable. In species in which the Notarium is not well developed, the vertebral spines are frequently mortised, producing rigidity in this unit of thoracic vertebrae. Thoracic vertebrae are also united by ossification of the tendons of epaxial muscles that ankylose with the transverse processes and spinous processes of the vertebrae (e.g. pelecaniform and charadriiform birds). See **Osteo.** Annot. 140 and Fig. 5.

(72) **Lig. intercristale ventrale.** The thoracic and cervical vertebrae at the root of the neck are characterized by prominent ventral crests that are connected by these unpaired longitudinal ligaments. In some adult birds Ligg. intercristalia are partially ossified. See **Osteo.** Annot. 119.

(73) **Artcc. synsacri.** The Synsacrum is formed by a variable number of the caudalmost thoracic vertebrae, the lumbar, sacral, and the cranialmost series of caudal vertebrae that are ankylosed in varying degree in different birds. The Synsacrum articulates with the pelvic girdles on each side (see Annot. 76). This topic is covered for large comparative series of birds by Barkow (1856) and Boas (1933). See **Osteo.** Figs. 5, 11.

(74) **Synos. intercorporea.** Corpora of the synsacral vertebrae are ankylosed. The fusiform lumbosacral intumescence of the spinal cord occupies the vertebral canal of the cranial half of the synsacrum. On its ventral aspect the ankylosed unit of vertebral bodies is also fusiform and corresponds to the intumescence internally; the unit is furrowed by the elongate Sulcus ventralis. See **Osteo.** Fig. 11.

(75) **Synos. interspinalis.** The ankylosed spinous processes of the synsacral vertebrae form a continuous crest that extends either partially or the full length of the synsacrum (Crista dorsalis). See **Osteo.** Annot. 143, 133 and Fig. 11.

(76) **Sut. [Synos.] iliosynsacralis;** Ligg. iliosynsacralia. The lateral ends of the transverse processes of the synsacral vertebrae articulate with the medial border of the acetabular and postacetabular ilium by an elongated linear suture. Cranial to Fossa renalis of the pelvis the transverse processes of the synsacral vertebrae ankylose extensively with Facies ventralis of the preacetabular ilium. In some species (e.g. *Thalasseus*, *Morus*, *Eudocimus*, *Buteo*) the costal processes of the "sacral" vertebrae opposite the acetabulum form prominent lateral bracing struts that articulate with the side wall of the pelvis. See **Osteo.** Annot. 243 and Figs. 5, 11.

(77) **Synos. interiliospinalis.** In certain birds (e.g. *Morus*, *Buteo*, *Aythya affinis*, *Gallus*, *Eudocimus albus*) the cranial end of the Crista dorsalis is fused with the dorsal crests of the adjacent preacetabular ilia forming the Crista iliosynsacralis; this produces sulci or canals on each side of the crest of the spines, ventral to the preacetabular ilia, that contain epaxial muscles. See **Osteo.** Annot. 233–234 and Fig. 11.

(78) **Synoss. pygostyli.** The pygostyle is formed by ankylosis of several terminal vertebrae: *Columba* (4); *Fulica*, *Cygnus*, *Strix* (5) (Steiner, 1938). See **Osteo.** Annot. 146.

(79) **Juncturae costarum.** The junction of Capitulum costae with the Corpus vertebrae (Sync. capitis costae) is a cartilaginous joint, i.e. a persistent synchrondrosis. On the ventrocranial aspect of this joint is a thickening of the perichondrium which forms a collateral ligament. The joint between Proc. transversus and the Tuberculum of the rib is a synovial joint (Artc. costotransversaria). See Fig. 10 and **Osteo.** Fig. 4.

(80) **Sut. iliocostalis.** In some birds ribs of the thoracic synsacral vertebrae distal to the Tuberculum often have fibrous articulation with the ventral aspect of the preacetabular ilium (e.g. *Morus*, *Aythya affinis*, *Rynchops*, *Corvus*. In other birds these ribs ankylose with the ilium (*Gavia*, *Ardea*). See **Osteo.** Fig. 5.

(81) **Sync. intercostalis.** The junction between each vertebral and sternal rib is a persistent cartilaginous joint (e.g. *Columba*, *Gallus*).

(82) **Sut. costouncinata; Lig. triangulare.** In some birds the joint between the uncinate process and the vertebral rib is a synostosis. Ghetie *et al.* (1976) illustrate a triangular ligament in the angle between the upper border of uncinate process and adjacent caudal border of its rib.

(83) **Artc. sternocostalis.** A typical synovial joint of the sternal rib with the articular facets of the costal margin of the sternum. The articular facets are single in some birds (e.g. *Gallus*) and double in others (e.g. *Gavia immer*, *Aythya affinis*, *Buteo lineatus*, and *Corvus ossifragus*).

(84) **Membranae incisurarum [fenestrarum] sterni.** Synonymy: Membrane intertrabeculares (Fürbringer, 1902). The notches and fenestrae of the sternum are separated by trabeculae; these openings are spanned by strong fibrous membranes.

(85) **Lig. corpus claviculae.** Ligamentous vestige of clavicle. May be thread-like or strong ligamentous cord representing the shaft or body of the clavicle. In certain birds in which the osseous clavicle is reduced the clavicle consists of only the Extremitas omalis [Epicleidium] (see Glenny and Friedmann, 1954).
Synd. sternoclavicularis. The apex of Carina sterni is usually situated near the Apophysis furculae, joined to it by ligamentous connection. In some species the union is transformed into a synovial joint (several procellariiforms and most pelecaniforms); in extreme cases the joint is synostosed (e.g. *Pelecanus*, *Sagittarius* (Fürbringer, 1888; *Fregata*).

(86) **Membrana sternocoracoclavicularis.** Synonymy: Lamina lateralis; membrana coracoclavicularis (Fürbringer, 1888). Stretches between the inner border of the clavicle, the medioventral border of the coracoid, and the rostral border of sternum dorsal to Carina sterni. Many variations in the form and development of its parts occur throughout Aves. See Fig. 3.

A number of specialized thickenings of Membrana sternocoracoclavicularis have been described as ligaments; these extend from the rostral border (especially Rostrum sterni) of the sternum to various parts of the coracoid, to the clavicle, to the scapula; some connect right and left coracoids, etc. The most important of these (Fürbringer, 1888) are listed as subordinate items under Membrana sternocoracoclavicularis in the list of terms.

(87) **Lig. acrocoracoprocoracoideum.** In birds that lack a distinct Proc. procoracoideus Membrana sternocoracoclavicularis continues uninterruptedly to an attachment on Proc. acrocoracoideus of the coracoid bone.

(88) **Membrana cristoclavicularis.** Synonymy: Lig. cristoclaviculare; Lig. sternoclaviculare; Lamina mediana of Membrana sternocoracoclavicularis; Crista membranacea, (Fürbringer, 1888). This is a median bilaminar membrane formed by the side to side fusion of the ventral ends of the paired (Laminae laterales) Membranae sternocoracoclaviculares. The median membrane connects the ventral end of the furcula (Apophysis furculae) to the Apex and Crista mediana of the cranial margin of Carina sterni; it extends dorsally to the level of Rostrum sterni. M. supracoracoideus and M. pectoralis arise from each side of Lig. cristoclaviculare. The most ventral (or cranial) part of the membrane is developed as a ligament rather than a membrane. Sy (1936) illustrates Membrana cristoclavicularis in *Bucephala*.

(89) **Artc. sternocoracoidea.** A synovial joint that permits mostly hinge type movements between the ventral end of the coracoid and the generally obliquely disposed horizontal Sulcus articularis coracoideum of the sternum. See Fürbringer (1888) for discussion of the variation in configuration of this joint.

(90) **Artc. intercoracoidea.** In most birds the medial borders of the sternal ends of the coracoids lie closely adjacent to one another. In fact, the two sternocoracoid joints of certain birds abut or overlap one another; in Ardeidae and Musophagidae the right coracoid in part lies ventral to the left one (Fürbringer, 1888). The overlapping also occurs in some procellariiforms (Kuroda, 1954), and in *Buteo lineatus* and *Bubo africanus*.

(91) **Ligg. collateralia sternocoracoidea.** Synonymy: Ligg. accessoria sternocoracoidea (Fürbringer, 1888). Fürbringer describes collateral ligaments of varying configuration and position in different birds that connect the base and shaft of coracoid with the dorsal and ventral labra of Sulcus articularis coracoideus of the sternum.

(92) **Lig. sternocoracoideum laterale.** A fascia-like ligament in *Casuarius*, *Galliformes*, *Tinamus* (Fürbringer, 1888) that extends from the lateral edge of Proc. craniolateralis of the sternum to the lateral border of Proc. lateralis of the sternal end of the coracoid.

(93) **Symphysis coracoscapularis; Lig. interosseum coracoscapulare.** The Coracoid and Scapula are united mainly by the fibrocartilaginous Lig. interosseum coracoscapularis. The ligament itself forms **Cavitas glenoidalis** that receives Caput humeri; this includes the raised margin of the cavity, **Labrum cavitatis glenoidalis.** According to Sy (1936) the floor of the cavity is highly elastic, and its shape changes in response to contact with different aspects of Caput humeri with which it articulates. See Fig. 3.

The articulating surfaces of scapula and coracoid often have the configuration of spheroidal or ellipsoidal joints. In these cases the coracoid bears the concave receiving surface (Cotyla) and the scapula bears the convex surface (Tubercuium) (see **Osteo.** Annot. 173. The above surfaces are joined by fibrocartilage; however, movement between the two bones occurs at this joint. In the ratite birds, *Struthio*, *Rhea*, and *Apteryx*, the coracoid and scapula are fused. See **Osteo.** Annot. 167.

(94) **Lig. coracoscapulare dorsale.** Synonymy: Lig. coracoscapulare accessorium dorsale, Fürbringer (1888). This ligament occurs in certain large birds in which the Lig. acrocoraco-acromiale is not strongly developed (e.g. *Ciconia*, *Egretta*, *Cathartes*, *Heliaeetus* (Fürbringer, 1888).

Lig. coracoscapulare internum. Synonymy: Lig. coracoscapulare accessorium internum, Fürbringer, 1888). This ligament is conspicuous in *Egretta*, *Haliaeetus*, and others; however, it is not as well developed as Lig. coracoscapulare dorsale.

(95) **Lig. acrocoracoacromiale.** Forms the fibrous roof over the Canalis triosseus. See Fürbringer (1888) for discussion of the relative development of this ligament in different birds. See **Osteo.** Annot. 177.

(96) **Synd. (Artc.) acrocoracoclavicularis.** Extremitas omalis of the clavicle is attached principally to the medial surface of Proc. acrocoracoideus of the coracoid bone. In some species Proc. acromialis of the clavicle is prolonged caudally, and articulates with the Acromion of the scapula (see Annot. 98; also **Osteo.** Annot. 165).

The acrocoracoclavicular joint is a syndesmosis, modified as a symphysis in some forms of birds. In other avian groups the junction is a typical synovial joint (*Spheniscus*, *Alca*, *Pelecanus*, *Ciconia*, *Haliaeetus*, *Buceros*). In *Fregata* this joint is a synostosis (see Fürbringer, 1888).

(97) **Synd. procoracoclavicularis** occurs in species whose coracoid bones possess well-developed procoracoid processes (e.g. *Buteo*, *Columba*).

(98) **Synd. acromioclavicularis.** In species in which the dorsal end of the clavicle is not apposed to Acromion of the scapula, the two are connected by a rather long Lig. acromioclaviculare. A synovial joint connects the two elements in *Picus* and *Ramphastos*; a symphysis may be present in some species (Fürbringer, 1888).

(99) **Lig. scapuloclaviculare dorsale.** This distinct collateral ligament of the scapuloclavicular joint occurs in Podicipedidae and Anseriformes; it is more or less distinct from Lig. acromioclaviculare.

(100) **Fibrocartilago humerocapsularis.** Synonymy: Os humeroscapulare (Jäger, 1857). A fibrocartilaginous or osseous mass developed in the dorsal part of the articular capsule in many avian species, deep to the origin of M. deltoideus major (see Fürbringer, 1888; Jäger, 1857).

(101) **Lig. acrocoracohumerale.** This is the principal collateral ligament of the shoulder joint which is usually quite independent of the articular capsule (Sy, 1936). Sy discusses the function of this ligament. See Fig. 3.
Bursa acrocoracoidea; Bursa supracoracoidea. Both these bursae are usually in open communication with the general articular cavity.

(102) **Lig. coracohumerale dorsale; Lig. scapulohumerale dorsale.** Fürbringer (1888) notes that the dorsal coracohumeral ligament is lacking in many birds and is very strongly developed in others. In columbiforms particularly, the ligament is independent of the capsule, and is grooved where the overlying tendon of M. supracoracoideus crosses it.

The dorsal scapulohumeral ligament varies in its development in different avian groups, but is only rarely lacking.

(103) **Plicae synoviales et ligg. intracapsularia.** Fürbringer (1888) describes the interspecific variation in occurrence and strength of these intraarticular structures. In general the plicae and ligaments are attached proximally to the scapular and coracoid labra of the articular fossa and distally to humerus, articular capsule, or other ligaments. See Fig. 3.

(104) **Junctura cubiti.** The parts of Junctura cubiti are named with the wing in the defined anatomical position: abducted and fully extended laterally (see General Intro. and **Osteo.** Annot. 178. The surface of the distal end of the humerus on which M. triceps brachii lies is its morphological dorsal or extensor surface. In the older literature descriptive terms were based on the wing in the folded position.

Junctura cubiti is a compound joint of radius and ulna with the humeral condyles and the proximal radioulnar joint; the ulna of birds articulates with both humeral condyles. All three articulations share a common synovial cavity.

(105) **Lig. collaterale ventrale.** A distinct, strongly developed, triangular ligament that connects the Tuberculum supracondylare ventrale of the humerus with the ventral aspect of ulna near the margin of its ventral cotyla. See Fig. 4; and **Osteo.** Fig. 9.

(106) **Lig. collaterale dorsale.** Varies in its strength and extent in different avian taxa. May be attached directly onto the humerus or to tendons near Epicondylus dorsalis of the humerus; attached distally to caudal border of the ulna near the remigeal papillae (see Stettenheim 1959). See Fig. 4.

(107) **Lig. cubiti craniale** (Fig. 4). This strong, poorly defined cranial part of Capsula articularis is attached to the intercondylar region of the humerus. Lig. cubiti craniale appears to consist partly of elastic tissue (*Gallus*, *Columba*) that may assist in the first stage of flexion of the fully extended elbow joint.

(108) **Meniscus radioulnaris.** The common synonym, Lig. anulare radii, is not an apt descriptive term; Alix (1874) and Sy (1936) both refer to this structure as a meniscus. The thick dorsal edge of the meniscus is not attached to Capsula articularis; the thin deep portion of the meniscus partly separates the dorsal humeral condyle from direct contact with the ulna. The meniscus unites the radius and ulna (see Fig. 4).

(109) **Lig. transversum radioulnare.** The common synonym, Lig. teres cubiti, is not appropriate for this flat, band-like ligament located on the dorsal aspect of the elbow (see Fig. 4).

(110) **Trochlea humeroulnaris** (Shufeldt, 1890) (Fig. 4). A strong ligament having the form of a sling, attached to the proximal end of the ulna; guides the proximal end of the tendon of M. flexor carpi ulnaris. **Pars tendinea** of Trochlea humeroulnaris is arranged spirally around its tendon, and blends with the muscle attachment of M. flexor carpi ulnaris on the distal humerus. **Parts pennata** of the Trochlea fans out and is fastened to follicles of calami of secondary remiges at the elbow where it is related to slips of M. expansor secundariorum (Berger, 1966).

(111) **Lig. tricipitale** (Stettenheim, 1959). Extends from ventral edge of tendon of M. scapulotriceps to the ventral margin of Sulcus m. humerotricipitis on the distal end of Humerus. The ligament is intracapsular; at its bony attachment it lies deep to tendon of M. humerotricips.

(112) **Artcc. carpi et manus.** Most of the terms on carpal articulations are based on dissections of *Gallus* and *Columba* that closely agree with the accounts of Stettenheim (1959) in several charadriiform birds and that of Sy (1936) in *Corvus* and other species. See Ghetie *et al.* (1976) for illustrations.

The complicated arrangements and configurations of the articulating elements at the carpus add to the difficulties of applying meaningful descriptive names for the various ligaments that provide topographical connotations and that indicate the bones connected by the ligaments. In order to avoid confusion of the names the following usage has been adopted: (1) "metacarpo-" is the combining term used for Carpometacarpus; (2) "radiocarpo-" and "ulnocarpo-" refer to the carpal bones, Os carpi radiale and Os carpi ulnare; (3) the conventional combining forms, "ulno-" and "radio-" are used in referring to the long bones of the antebrachium, Ulna and Radius; (4) hyphenation is used in certain names to avoid confusion concerning the bones involved in formulation of the names of the ligaments; (5) names are applied so that terms of position and direction refer to the limb in the anatomical position (see Annot. 104, 155). See also **Arthr.** Intro.

(113) **Aponeurosis ventralis** (Pelissier, 1923) (Fig. 11). Synonymy: Lig. radiale metacarpi (Stettenheim, 1959); Lig. ulni metacarpale mediale (Kolda and Komárek, 1958). Dense fibrous sheet that shows definite orientation of the collagenous fasciculi in its several parts. These parts radiate from its attachment on the distal end of the radius to the carpals, metacarpals, and flight feathers. See Annot. 114, 115; **Osteo.** Annot. 213 and Fig. 9, Tuberculum aponeurosis of the Radius.

(114) **Retinaculum flexorum.** This is the short, transverse proximal segments of Aponeurosis ventralis that extends from the end of the radius (see **Osteo.** Fig. 9) to

the distal end of Os carpi ulnare and Proc. pisiformis of the proximal end of the Carpometacarpus. It is anchored to the underlying ligaments and fibrous flexor sheaths of the flexor muscle tendons that pass deep to it as well as to Os carpi ulnare. See **Osteo.** Annot. 213.

(115) **Aponeurosis ulnocarporemigialis.** This is the distal segment of Aponeurosis ventralis that extends from the ventral surface of Os carpi ulnare and fans out in the area distal and caudal to the Carpus along the caudal border of Os metacarpale minus. The ulnocarporemigial aponeurosis sends digitations to the follicles of the proximal series of remiges (primary flight feathers). The deep part of the ulno-carporemegial aponeurosis is strongly attached to Os carpi ulnare (*Columba* and *Gallus*); the superficial collagenous bundles cross Os carpi ulnare and are continuous with Retinaculum flexorum (Annot. 114). See also Annot. 191 and Fig. 11.

(116) **Lig. interosseum radioulnare.** Only ligament that directly connects the distal ends of radius and ulna; separates the radius from the ulna and prevents their direct contact. Lig. interosseum radioulnare is situated so that it limits distal movement of the radius relative to the ulna. See Fig. 5.

(117) **Lig. ulno-ulnocarpale proximale.** Synonymy: Lig. posticum ulnare carpi ulnaris.
Lig. ulno-ulnocarpale distale. Synonymy: Lig. obliquum carpi ulnaris. Both these ligaments are distinct in *Gallus* and *Columba*; in *Corvus* (Sy, 1936); and in several charadriiforms (Stettenheim, 1959); however, in the piciform, *Pteroglossus*, the two appear to be represented by one ligament.

(118) **Lig. ulno-radiocarpale ventrale.** Thickened part of Capsula articularis of the ulno-radiocarpal joint.

(119) **Lig. interosseum ulno-radiocarpale.** Synonymy: Lig. ulnare carpi radialis. This intraarticular ligament extends ventrally from the Sulcus intercondylaris of the ulna to the Os carpi radiale. It is the major ligament that connects ulna to the carpals; it limits distal movement of the radius.

(120) **Lig. ulno-metacarpale ventrale.** Synonymy: Lig. ulnare internum metacarpi. This ligament is independent of Lig. radiocarpo-metacarpale ventrale in certain charadriiforms (Stettenheim, 1959) and *Columba*; in *Gallus* the two ligaments are combined distally (Fig. 5).

(121) Lig. **radio-radiocarpale dorsale.** Synonymy: Lig. radiale externum carpi radialis. In *Gallus* two digitations from the radius merge to form this ligament.

(122) **Meniscus intercarpalis.** Synonymy: Lig. carpi interni. The meniscus connects Os carpi ulnare with Os carpi radiale; it represents the embryonic carpal, centrale III (Romanoff, 1960). Attachments of the meniscus to Os carpi ulnare are only at dorsal and ventral points; the intervening thin caudal border of the meniscus next to Os carpi ulnare is free. The slit-like opening between the free border and the ulnar carpal permits communication between the joint cavities proximal and distal to the meniscus. The dorsal edge of the meniscus is thick; its cranial attachment to the caudal, sharp border of Os carpi radiale forms the thick part of the wedge-like meniscus. See Fig. 5; and **Osteo.** Annot. 214.

(123) **Syncc. intercarpales.** Each of the definitive ossa carpi is formed during embryonic development by coalescence of several carpal anlagen (Romanoff, 1960). See also **Osteo.** Annot. 214.

(124) **Artcc. carpo-carpometacarpales.** The ulna and radius do not articulate directly with the Carpometacarpus; carpal bones and Meniscus intercarpalis intervene.

(125) **Lig. radiocarpo-metacarpale ventrale.** Synonymy: Lig. internum ossis carpi radialis et metacarpi (see Ahnot. 120).

(126) **Lig. ulnocarpo-metacarpale ventrale.** Synonymy: Lig. internum ossis carpi ulnaris et metacarpi. Consists of two separate parts in *Gallus.*

(127) **Lig. ulnocarpo-metacarpale dorsale.** Synonymy: Lig. externum ossis carpi ulnaris et metacarpi.

(128) **Synoss. carpometacarpales.** Midway during embryonic development centralia II and IV and the distal row of carpal elements coalesce with the proximal end of the metacarpal elements (Romanoff, 1960) forming the definitive carpometacarpus; therefore, Artc. carpo-carpometacarpalis is an intercarpal articulation. See **Osteo.** Annot. 214.

(129) **Synoss. intermetacarpales.** The metacarpal elements ankylose with one another and with the carpals at their proximal end; distally the major and minor metacarpals fuse. Romanoff (1960) points out how the development of the manus in ratites and penguins differs from the general avian pattern. See **Osteo.** Fig. 10.

(130) **Artcc. metacarpophalangeales.** Ligaments of the digital articulations play an important role in guiding and limiting movements and strengthening the distal part of the wing, since the major primary flight feathers (remiges) are attached to Carpometacarpus and phalanges. In general the ligaments of the ventral side of the joints are stronger than those of the dorsal side; the ventral ligaments resist forces against the ventral surface of the wing during flight.

(131) **Lig. obliquum alulae.** Synonymy: Lig. pollicare. Extends from distal edge of Proc. extensorius of the alular metacarpal to the base of the ventral surface of the alular phalanx. See Fig. 5.

(132) **Lig. collaterale ventrale.** Consists of a broad, flat part and an elongate, cord-like part; the latter is attached along the caudal crest of the proximal phalanx of Digitus major.

(133) **Lig. collaterale caudale.** The caudal collateral ligament of the metacarpophalangeal joint of Digitus major is present in *Columba*, absent in *Gallus.*

(134) **Lig. obliquum intraarticulare.** Present in *Columba* and *Gallus*; apparently limits rotation around long axis of Digitus major caused by elevation of the caudal border of the wing.

(135) **Meniscus articularis.** Loosely organized fatty-fibrous intraarticular structure located in the dorsal part of the joint cavity, attached to inner aspect of articular capsule of the metacarpophalangeal joint of Digitus major.

(136) **Juncturae interphalangeales laterales.** The phalanx of Digitus minor is attached along most of its length to the caudal edge of the proximal phalanx of Digitus major by an interosseous ligament (Pelissier, 1923); therefore, movements of the minor digit follow those of the major digit. Each of the two digits has an individual synovial cavity at its articulation with the Carpometacarpus.

In penguins (e.g. *Pygoscelis adeliae*) the alular phalanx becomes fused with the cranial margin of Os metacarpale major and at its base with Os metacarpale alulare (Synos. metacarpophalangealis alulae). See **Integ.** Annot. 70.

(137) **Artc. interphalangealis.** In most birds Digitus major usually possesses two phalanges therefore, only one interphalangeal joint is present. In some groups of birds supernumerary phalanges in the form of inconstantly or regularly occurring wing claws are found on the alular and major digits (Fisher, 1940); Digitus major has three phalanges in certain birds, e.g. *Gavia* and Anatidae (Berger, 1966); members of several avian orders possess two alular phalanges (see **Osteo.** Annot. 224; **Integ.** Annot. 70; and **Topog. anat.** Annot. 40).

(138) **Lig. collaterale caudale.** The caudal collateral ligament of the interphalangeal articulation of Digitus major is a strong, poorly defined part of Capsula articularis in *Gallus* and *Columba*.

(139) **Lig. interphalango-remigiale.** Part of Aponeurosis interphalango-remigialis that consists of a tough ligamentous band; strengthens the ventral side of Artc. interphalangealis of Digitus major. Distally the ligament is attached to the follicle of a remex rather than to bone. See Annot. 192.

(140) **Ligg. accessoria alae.** Ligamentous structures not associated with articulations of bones with one another *per se* (cf. retinacula of Cartilago tibialis).

(141) **Lig. elasticum propatagiale.** Located in the cranial free edge of the propatagial skin fold. M. tensor propatagialis inserts into the medial end of this ligament; Berger (1966) considers Lig. elasticum propatagiale as the elastic tendon of M. tensor propatagialis.

(142) **Retinaculum m. scapulotricipitis.** A flat, wide ligamentous band that anchors the cranial edge of M. scapulotriceps to the proximal shaft of the humerus. This condition occurs widely among birds (Berger, 1966).

(143) **Lig. humerocarpale.** Following Pelissier (1923) and Alix (1874) this ligament ("Lig. du petit palmaire") is considered as a distinct band that extends from the Epicondylus ventralis of the Humerus to Os carpi ulnare, generally independent of the more superficial Aponeurosis antebrachialis ventralis. Berger (1966) equates Lig. humerocarpale with the entire ventral antebrachial aponeurosis.

(144) **Retinaculum m. extensoris metacarpi ulnaris.** Berger (1966) discusses the occurrence and function of this retinaculum ("ulnar anchor"' that binds the proximal end of the muscle against the ulna.

(145) **Ligg. accessoria musculi.** On the dorsal side of the wrist these accessory ligaments are connected proximally to the distal end of the ulna, and join or envelop certain tendons that cross Artcc. carpi. The accessory ligaments are arranged mechanically so that they passively contribute to the actions produced by the muscles with which they are associated (*Columba*: Berger, 1966).

(146) **Juncturae cinguli membri pelvici** (see Artcc. synsacri). Os coxae is formed by coalescence of ilium, ischium, pubis. In many birds these elements develop embryologically as separate cartilaginous anlagen in the acetabular region, and subsequently undergo ossification and ultimate synostosis in postnatal life. All three elements contribute to the formation of the acetabulum. See Annot. 148.

(147) **Sut. ischiopubica.** The caudal end of the ischium (Proc. terminalis ischii) articulates by sutural ligaments with the dorsal border of the shaft of the pubis. In some passeriforms and piciforms this joint is ankylosed (Boas, 1933). In all ratites the end of the ischium is united to the pubis by cartilage that eventually ossifies (Boas, 1933).
Membrana ischiopubica. This membrane closes off the elongate Fenestra ischiopubica. See **Osteo.** Annot. 227, 229.
Lig. ischiopubicum. Extends from Proc. obturatorius of the ischium to the pubis, and forms the caudal border of Foramen obturatorium; in many avian species this ligament becomes ossified (Boas, 1933). See **Osteo.** Annot. 227.

(148) **Sync. ilioischiadica.** Ilium and ischium articulate at the acetabulum in all birds. In young carinate birds the postacetabular ilium and ischium are also joined by cartilage, they become synostosed caudal to Fenestra ilioischiadica in older birds. In some ratites they are separated or incompletely joined (Nauck, 1938). See **Osteo.** Annot. 231, 251 and Fig. 11.
Membrana ilioischiadica. Stretches across the caudal two-thirds of Fenestra ilioischiadica; muscles are attached to its superficial and deep surfaces, Nerves and vessels traverse only the cranial part of the fenestra.

(149) **(Symphysis pubica).** The two pubes articulate and partially complete the bony pelvis ventrally only in *Struthio* and *Archaeopteryx* (Heilmann, 1926).

(150) **(Symphysis ischiadica).** Occurs only in *Rhea*; the two ischia meet in a long median symphysis ventral to the synsacrum and kidneys, and separate these structures from the abdominal viscera (Newton, 1896; Barkow, 1856).

(151) **Membrana acetabuli.** Closes Foramen acetabuli; its ventral part blends with the acetabular attachment of Lig. capitis femoris.

(152) **Artc. coxae.** Actually two distinct articulations are included within the single joint cavity: (a) Caput femoris with the Acetabulum, and (b) Trochanter and dorsal surface of neck of femur with the Antitrochanter of Os coxae. See Fig. 6; and **Osteo.** Annot. 255.

(153) **Lig. iliofemorale.** Located on the cranial aspect of the hip joint; its thick dorsal edge is distinct from the thin dorsal part of Capsula articularis.
Lig. ischiofemorale. Proximally this ligament is attached to the cranial border of the ilioischiadic foramen and the Labrum acetabulare of the caudal part of the Antitrochanter. Distally the ligament is attached to the caudal part of the trochanteric crest of the femur (see Fig. 6).

(154) **Junctura genus.** A compound joint formed of four separate Synovial joints whose synovial cavities all intercommunicate; Femur articulates with Fibula as well as with Tibiotarsus and Patella. The proximal tibio-fibular joint is a synovial articulation. See Annot. 156, 157 and Fig. 7.

(155) **Artc. femorotibialis.** The major bone of the crus is the Tibiotarsus. For purposes of simplification of the terminology of the arthrology of the knee region a form of "tibia" instead of tibiotarsus will be used.

(156) **Artc. femorofibularis.** Caput fibulare articulates with Trochlea fibularis of the lateral condyle of the Femur in birds. See Annot. 159 and Fig. 7.

(157) **Ligg. tibiofibulares.** Both of these are discrete intraarticular ligaments. Near its fibular attachment to the medial aspect of Caput fibulare, Lig. tibiofibulare caudale partly intervenes between the tibial and fibular articular facies; Caput fibulare articulates directly with the Tibiotarsus just distal to the attachment of Meniscus lateralis to Caput fibulae. See Annot 154.

(158) **Lig. patellae.** This so-called ligament is actually the tendon of Mm. femorotibiales that extends from the distal border of the Patella to Crista patellaris of the Tibiotarsus. Lig. patellae forms much of the cranial wall of the articular cavity of Artc. femorotibialis.

(159) **Meniscus lateralis.** Synonymy: Femoro-fibular disc (Haines, 1942). Usually a solid, oblong disc that separates the lateral condyle of the femur from the tibiotarsus; it only partially intervenes between the lateral condyle of the femur and Caput fibulae. The lateral meniscus is perforated or grooved by Caput femorale of M. tibialis cranialis; the lateral meniscus is attached to the inner aspect of Caput fibulae. See Fig. 7.

(160) **Lig. meniscocollaterale.** Connects the cranial edge of Meniscus lateralis with the cranial border of Lig. collaterale laterale (Haines, 1942; Cracraft, 1971).

(161) **Lig. interosseum tibiofibulare.** Connects Crista fibularis of the Tibiotarsus with Corpus fibulae and Spina fibulae.

(162) **Foramen interosseum proximale; Foramen interosseum distale.** These are openings between Tibiotarsus and Fibula proximal and distal to Crista fibularis. The foramina transmit nerve and vessels that pass between the flexor and extensor compartments of the crus. See Fig. 7; **Art.** Annot. 76; and **Ven.** Annot. 71.

(163) **Artc. cartilago-tibiotarsea.** Cartilago tibialis articulates with Trochlea cartilaginis tibialis on the caudal surface of the distal end of the Tibiotarsus; this articulation is continuous with the intertarsal joint. Cartilago tibialis forms the caudal wall of Artc. intertarsalis (see Fig. 8).

(164) **Cartilago tibialis** (Fig. 8). Synonymy: Cartilago semilunaris; Sustentaculum (Fujioka, 1962). Shufeldt (1890) described Cartilago tibialis as the fibrocartilaginous block that lies on the caudal aspect of the Trochlea cartilaginis tibialis on the distal end of the Tibiotarsus. The tendons of M. gastrocnemius and superficial flexors pass over Cartilago tibialis, and the deep flexor tendons of the pedal digits glide through

canals in the Cartilago tibialis. Hudson (1937) notes the presence of Cartilago tibialis in many different avian species.
Os sesamoideum intertarsale. In certain birds the mediodistal angle of Cartilago tibialis forms this distinct ossified process (Cracraft, 1971); this sesamoid articulates extensively with the Tarsometatarsus (see Annot. 166).

(165) **Retinaculum mediale; Retinaculum laterale** (Barnett, 1954a). These tough, transverse ligamentous sheets connect the medial and lateral edges of Cartilago tibialis to the tibiotarsal epicondyles caudal to the collateral ligaments of the intertarsal joint; the retinacula form parts of Capsula articularis (see Fig. 8).
Retinaculum flexorum. Arches over the tendons that lie in the sulci on the caudal surface of Cartilago tibialis; attached to the curved lateral and medial margins of Cartilago tibialis.

(166) **Lig. metatarso-sesamoideum.** The strong Lig. metatarso-sesamoideum connects Os sesamoideum intertarsale with the Tarsometatarsus distal to the rim of the medial cotyla, caudal to the attachment of Lig. collaterale mediale; at times it is continuous with the medial collateral ligament (e.g. *Columba*: Cracraft, 1971; *Gallus*). See Annot. 169 and Fig. 8.
Lig. cartilago-metatarsale. Connects the middle of the distal end of Cartilago tibialis to Sulcus ligamentosus on the proximal aspect of the Tarsometatarsus. See Annot. 164, 169; and **Osteo.** Annot. 286 and Fig. 14.

(167) **Artc. intertarsalis.** Synonymy: Artc. tibiotarso-tarsometatarsalis. This joint is analogous to the mammalian ankle joint (Artc. tarsocruralis). The Fibula does not take part in the formation of the avian intertarsal joint. No separate tarsal elements are found in adult birds. During development the proximal row of tarsal bones (Fibulare and Tritibiale) becomes incorporated with the distal end of the tibia (Romanoff, 1960). See **Osteo.** Annot. 284, 286.

(168) **Meniscus medialis; Meniscus lateralis.** According to the comparative study of Stolpe (1932) only the lateral meniscus of the intertarsal joint is well developed. The medial meniscus is absent or poorly developed in some species: *Gallus*, *Acryllium* (Stolpe, 1932); *Columba*, *Cyanocitta*, *Asio*; well-developed in *Ara* (Stolpe, 1932) and *Meleagris* (Barnett, 1954a). Meniscus lateralis is strongly attached to Lig. collaterale laterale in many birds. Both menisci are absent in flamingos (Phoenicopteridae); two complete menisci exist in psittaciforms (Stolpe, 1932). See Fig. 8 and Annot. 168).

(169) **Lig. meniscosesamoideum.** Connects Cornu caudalis of Meniscus lateralis with the intertarsal sesamoid **(Columba, Gallus).** See Annot. 166.

(170) **Lig. meniscotibiale.** Synonymy: Kreuzband (Stolpe, 1932). Attached to intercondylar incisure of Tibiotarsus proximocranial to attachment of Lig. intercondylare tibiometatarsale. Distally the ligament bifurcates; the medial crus is usually better developed than the lateral one. The crura become continuous with the cranial edges of the menisci (see Stolpe, 1932, p. 171). See Fig. 8 and Annot. 168).

(171) **Lig. intercondylare tibiometatarsale.** Synonymy: Lig. anticum (Gadow and Selenka, 1890; Stolpe, 1932); Lig. tibiometatarseum mediale (Ghetie *et al.*, 1976). Strong intracapsular ligament that extends from Eminentia intercondylaris of Tarsometatarsus to a distinct impression on Incisura intercondylaris of the distal end of the Tibiotarsus. This ligament appears to be twisted on itself; its proximal part may be blended with ligamentous bands from the menisci of the joint (see Annot. 170).

(172) **Ligg. collateralis.** Stolpe (1932) indicates that one collateral ligament on each side of the intertarsal articulation is the usual avian condition. According to Stolpe the flamingo has two clear-cut collateral ligaments on each side of the joint: one short, one long. Barnett (1954a) describes an accessory band of Lig. collaterale in *Meleagris*.

(173) **Juncturae tarsometatarsales et intermetatarsales.** Metatarsal bones II, III, and IV ossify separately perichondrally, then ankylose near their middles with each other. The common metatarsal bone fuses with the distal tarsal elements, forming the definitive Tarsometatarsus of the adult (Hamilton, 1952). The evidence of the fusion of the separate metatarsal bones persists in the adults of different species in varying degrees. See Annot. 167; and **Osteo.** Annot. 288.

(174) **Synd. intermetatarsalis hallucis.** Joint between the movable metatarsal I and the shallow, elongate depression on metatarsal II this joint shows no evidence a synovial cavity in *Gallus* or *Columba*. See Fig. 9; and **Osteo.** Fig. 14.

(175) **Lig. interosseum.** The articular surface of Os metatarsale I (at its proximal end) is held in the Fossa metatarsi I on metatarsal bone II by the Lig. interosseum. In the macerated Tarsometatarsus conspicuous roughened areas at each end of the fossa indicate the attachment of the proximal and distal thickened parts of this ligament (*Columba, Gallus*). See below, next paragraph.
Lig. elasticum metatarsi I. This ligament extends from the distal part of Os metatarsale I to the medial aspect of the distal end of Os metatarsale II near its trochlea. This elastic ligament permits the distal end of Os metatarsale I to be drawn away from the plantar surface of the Tarsometatarsus and assists in its return movement. See Fig. 9.

(176) **Lig. transversum metatarsale.** Synonymy: Lig. a (Cracraft, 1971). This ligament unites Tuberositas lateralis of the distal end of Os metatarsale I and the proximal phalanx of the Hallux with the lateral epicondyle of Os metatarsale IV (*Columba, Gallus*) (see Fig. 9). The ligament acts as a retinaculum which restrains the flexor tendons of the digits against the plantar surface of the Tarsometatarsus. See Fig. 9.

(177) **Lig. obliquum hallucis.** Synonymy: Lig. c (Cracraft, 1971). Lig. obliquum hallucis extends from the plantar surface of the base of the proximal phalanx of the Hallux distad to the medial side of the base of the proximal phalanx of Digitus II (*Columba, Gallus*), and apparently limits hyperextension of the Hallux. See Fig. 9.

(178) **Canalis flexorius plantaris** (Fig. 9). Tendons of the flexor muscles of the toes pass through this canal into the distal foot. The plantar aspect of the canal is formed by Lig. transversum metatarsale and Tuberositas lateralis of Os metatarsale I; the dorsal aspect of the canal is formed by the surface of the Tarsometatarsus proximal to its trochleae (Fossa supratrochlearis plantaris). See **Osteo.** Fig. 14.

(179) **Aponeurosis plantaris.** Synonymy: Lig. b. (Cracraft, 1971). Consists of a distal, tough connective tissue sheet that covers the plantar aspect of metatarso-phalangeal joints II, III, and IV. The strongest attachments of this part of the aponeurosis are to the bases of the proximal phalanges of digit II (medial), digit III (lateral), and digit IV (lateral). The proximal thin part of the Aponeurosis plantaris is attached mainly to Lig. transversum metatarsale and to the lateral border of Os metatarsale I. See Fig. 9.
Corpus adiposum plantare profundum. An organized fat pad that occupies the compartment deep to Aponeurosis plantaris and cushions the more deeply situated tendons and joints. Corpus adiposum plantare superficiale forms the mass of Pulvinus metatarsalis superficial to Aponeurosis plantaris (see **Topog. anat.** Annot. 46).

(180) **Artcc. metatarsophalangeales.** In most birds the hallux is directed caudally, and opposes the three cranially oriented digits; however in piciforms, cuculiforms, and psittaciforms the lateral digit is also pointed caudally; modifications of the metatarsophalangeal joint of digit IV in these groups of birds are not included in this terminology (see Steinbacher, 1935). The metatarsophalangeal joint of the hallux of non-perching birds is usually located more proximal ("elevated") than the metatarsophalangeal joints of the other digits. In some species, the hallux is vestigial. See Coues (1927) for a thorough account of the feet of birds. See **Osteo.** Annot. 297.

(181) **Ligg. collateralia.** (Proximally the collateral ligaments of the metatarso-phalangeal joints are attached to Foveae ligg. collateralium of the metatarsal trochleae and distally to the bases of the proximal phalanges and to Ligg. plantaria.

(182) **Ligg. plantaria.** Synonymy: Ligg. subarticularia (Cracraft, 1971). Well-developed fibrocartilaginous plantar ligaments form the plantar wall of the metatarsophalangeal articulations, and are firmly attached to the bases of the proximal phalanges and the collateral ligaments. Deep sulci in the plantar surfaces of the ligaments transmit flexor tendons that are restrained by fibrous flexor sheaths (Vaginae fibrosae) that stretch across the sulci. The plantar ligament of the metatarsophalangeal joint of the hallux is poorly developed by comparison with the others (*Columba, Gallus*). Ligg. plantaria also occur at all interphalangeal joints except those that involve the distal phalanges. See Annot. 181 and Fig. 9 and **Osteo.** Ossa digitorum pedis.

(183) **Ligg. elastica extensoria digitorum.** Judson (1937) and Berger (1952) have described in *Corvus* and in cuculids elastic extensor ligaments on the dorsal aspects of the pedal digits that are attached to the distal ends of penultimate phalanges and the bases of the distal phalanges; these elastic ligaments extend the ungual phalanges automatically when not resisted by contraction of M. flexor digitorum longus. See **Osteo.** Ossa digitorum pedis.

(184) **Lig. inguinale.** In larger birds this distinct fibrous band stretches craniad from the ventral margin of the acetabulum (Tuberculum preacetabulare) to Margo lateralis of the preacetabular ilium (see **Osteo.** Fig. 11). Lig. inguinale forms the ventral boundary of the neurovascular hiatus for the femoral nerve and vessels; abdominal muscles arise from the ventral convex edge of the inguinal ligament.

(185) **Membrana iliocaudalis.** Fibrous sheet connecting the transverse processes of free caudal vertebrae with the dorsolateral process of the Ilium and the caudal margin of Os coxae (see **Osteo.** Annot. 231 and Fig. 11).

(186) **Ansa m. iliofibularis.** Synonymy: Ansa bicipitalis. The Ansa has two femoral bands and one fibular band; caudal to the knee the ansa acts as a fibrous trochlea for the tendon of M. iliofibularis (M. biceps femoris). See **Myol.** Figs. 8, 9.

(187) **Rectinaculum extensorium tibiotarsi.** Synonymy: Lig. transversum. The tendon of M. tibialis cranialis is restrained by this tough fibrous arch that is located on the distal end of cranial surface of Tibiotarsus just proximal to the osseous Pons supratendineus. The retinaculum is obliquely disposed, not transversely as its older name indicates. On a deeper level the tendon of M. extensor digitorum longus also passes deep to the extensor retinaculum. See Fig. 8 and **Myol.** Fig. 10.

(188) **Retinaculum extensorium tarsometatarsi.** Fibrous arch that restrains and acts as a pulley for the tendon of M. extensor digitorum longus; located on the dorsal (cranial) aspect of the proximal end of the Tarsometatarsus. The Retinaculum is an osseous arch in certain avian groups (see Fig. 8 and **Osteo.** Annot. 287).

(189) **Ligg. pennarum.** Specialized ligamentous connections of wing and tail flight feathers other than the general fibrous attachment of follicles of remiges and retrices to bones, integument, and fascias. Certain of the ligaments interconnect adjacent flight feathers, others pass from parts of the skeleton to the feathers. See Pelissier (1923) for a comparative study of the feather ligaments of the wing; see also Robin and Chabry (1884) and Stettenheim (1959).

(190) **Lig. elasticum interremigiale.** Synonymy: Grand ligament palmaire superieur (Alix, 1874); grand ligament marginal (Pelissier, 1923); interremigeal lig. (Stettenheim, 1959). This elastic ligament is found in the free caudal edge of the skin fold between adjacent remiges. The part of the ligament between feathers splits and passes around each feather follicle on its dorsale and ventral aspects (Pelissier, 1923). Sy (1936) illustrates differences in form and relationship of the ligament to the follicles of the remiges in birds of five different avian orders. The elastic ligament of the primary remiges is continuous with that of the secondary remiges. See Fig. 11.

(191) **Retinacula ulnocarpo-remigialia.** Variable bands from Os carpi ulnare or from Aponeurosis ventralis that extend to follicles of the remiges of the carpal region (see Pelissier, 1923). See Annot. 113, 115.

(192) **Aponeurosis interphalango-remigialis.** Located on ventral side of Digitus major. The Aponeurosis is attached to the skeleton in the region of Artc. interphalangealis; it radiates to the follicles of several adjacent remiges.

(193) **Ligg. phalangoremigialia distalia.** Individual ligamentous slips from the ventral surface of the terminal phalanx of Digitus major to several of the most distal primary remiges.

(194) **Aponeurosis antebrachialis ventralis** (see Annot. 143); **Aponeurosis antebrachialis dorsalis.** (Both terms introduced by Pelissier, 1923.) Dense fascial sheets enveloping the flexor and extensor muscles of the forearm. The expanded tendon of

M. tensor propatagialis, Pars brevis blends with the proximal end of the dorsal aponeurosis to a variable degree in different birds. Digitations pass from the caudal margin of Aponeurosis antebrachialis dorsalis to the follicles of the secondary remiges.

(195) **Septum humerocarpale.** Well defined intermuscular fascial sheet that passes dorsally from the humerocarpal ligament to the ventral aspects of the secondary remiges. The septum is described in detail by Alix (1884) and Pelissier (1923). See Annot. 143, Lig. humerocarpale and Figs. 5, 11.

(196) **Digitationes remigiales.** Synonymy: Ligg. sous-remigien (Pelissier, 1923). These ligamentous slips from Septum humerocarpale and from the elastic ligament of M. flexor carpi ulnaris are attached to the follicles of the secondary remiges. See Annot. 197 and Fig. 11.

(197) **Lig. elasticum m. flexoris carpi ulnaris.** Continuous elastic ligament stretching the length of the antebrachium; receives the insertion of Pars caudalis of M. flexor carpi ulnaris. Alix (1874) noted the elastic structure of this ligament, and referred to the part of the muscle inserting on it as "rotateur des remiges" inasmuch as it sends digitations to each secondary feather. Pelissier (1923) called it the "lig. du cubital anterieur"; he noted that in the coraciiform bird, *Lophoceros*, the ligament is not continuous but consists of a series of individual digitations to the secondary remiges.

(198) **Lig. elasticum intertectricale.** Synonymy: intercovertal lig. (Stettenheim, 1959). Connects adjacent major ventral tectrices (covert feathers) of the antebrachium, carpus, and manus. Lig. elasticum intertectricale is strongest and best-developed in the carpal region (*Lig. elasticum intertectricale carpale*) (e.g. *Columba*, charadriiforms, Stettenheim, 1959) and tapers proximally and distally in the wing.

(199) **Ligg. cubiti** (Pelissier, 1923). Short ligaments attached to follicles of the calami of the secondary remiges that insert on the series of Papillae remigiales caudales of the shaft of the ulna; not attached to the apices of the follicles, but slightly distal to the apices.

(200) **Lig. elasticum interrectricale.** Synonymy: "lig. elastique souscaudal ou inferior" (Robin and Chabry, 1884). The form of this ligament varies markedly between different species, e.g.: In *Columba* it consists of a thin, wide transverse band stretching between the follicles of the rectrices on their ventral aspects. In *Gallus* strong, thick parts of Lig. elasticum are found between adjacent retrices and a thin band ventral to the rectrices. In both species the ligament is placed just deep to the fold of integument that connects rectrices. See Fig. 11.

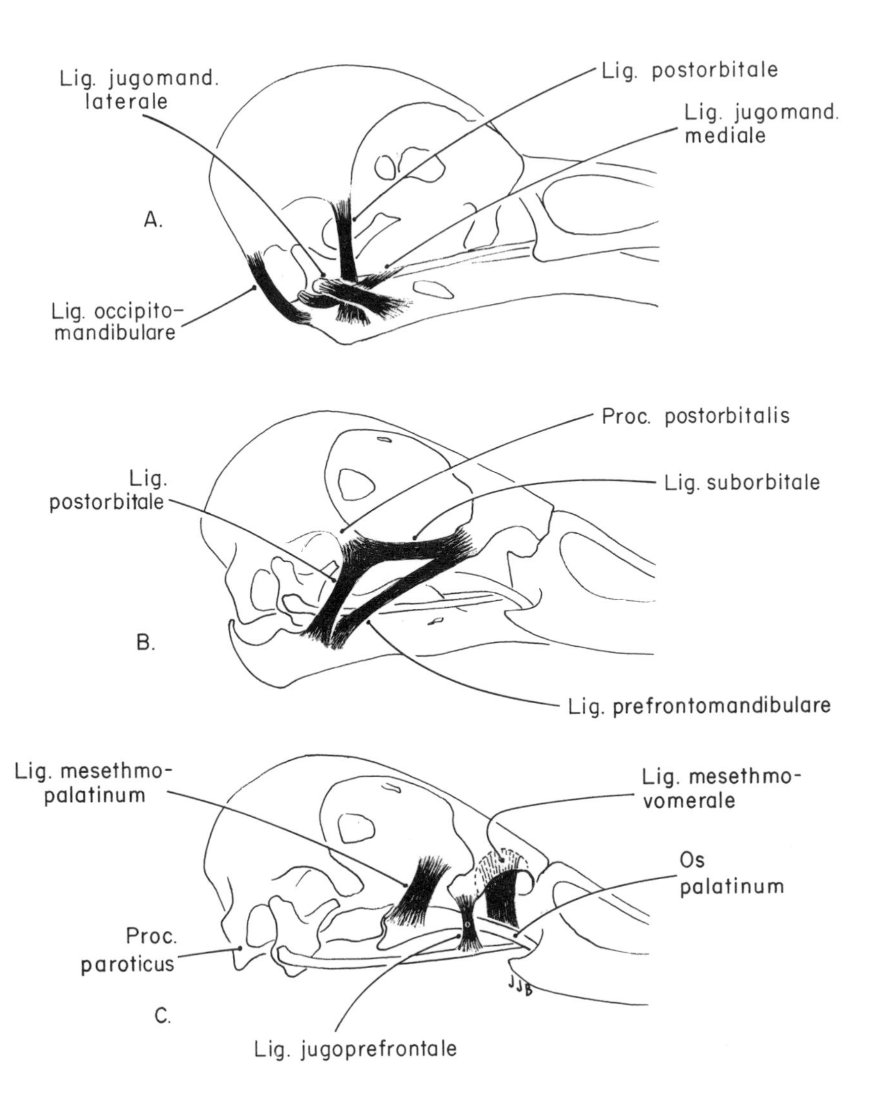

Fig. 1 Ligaments of the face. A, *Corvus*; lateral view, right side of the skull showing the superficial ligaments; B and C, *Anser*, lateral view, right side of the skull. In B the superficial ligaments are shown; the more deeply disposed ligaments are shown in C. Redrawn from Bock (1964).
Abbreviations: Lig. jugomand., Lig. jugomandibulare.

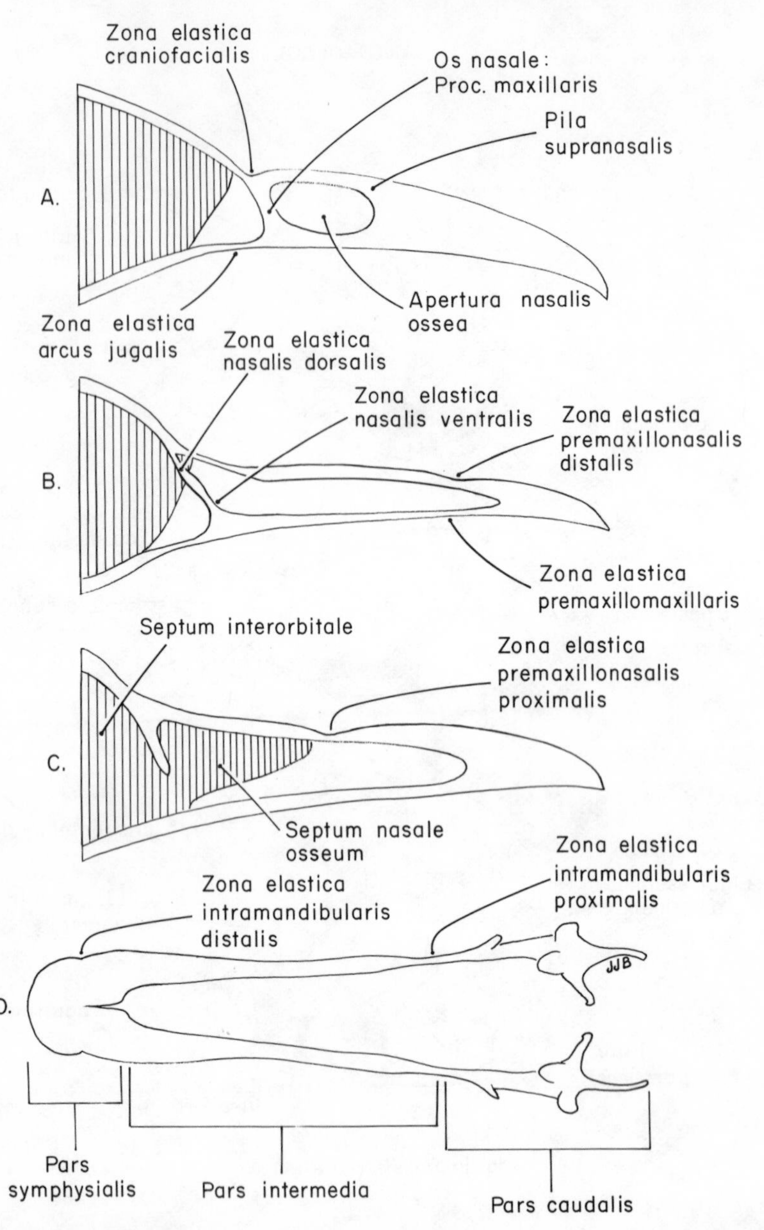

Fig. 2 Elastic zones of the skull. A–C, lateral views of the craniofacial region; redrawn from Bock (1964); vertically-lined areas = Septum interorbitale and Septum nasale. D, dorsal view of the mandible; redrawn from Lebidinsky (1920).

A, Prokinetic skull, holorhinal nostril (*Corvus*).

B, Rhynchokinetic skull, schizorhinal nostril (charadriiiform type). Note: not all of the different elastic zones shown would appear in the skull of any one species.

C, Rhynchokinetic type (ratite type).

D, Mandible of the swan (*Cygnus*). See **Osteo.** Fig. 1 for a lateral view of the mandible of a gull (*Larus*) demonstrating its structure at the level of the proximal intramandibular elastic zone.

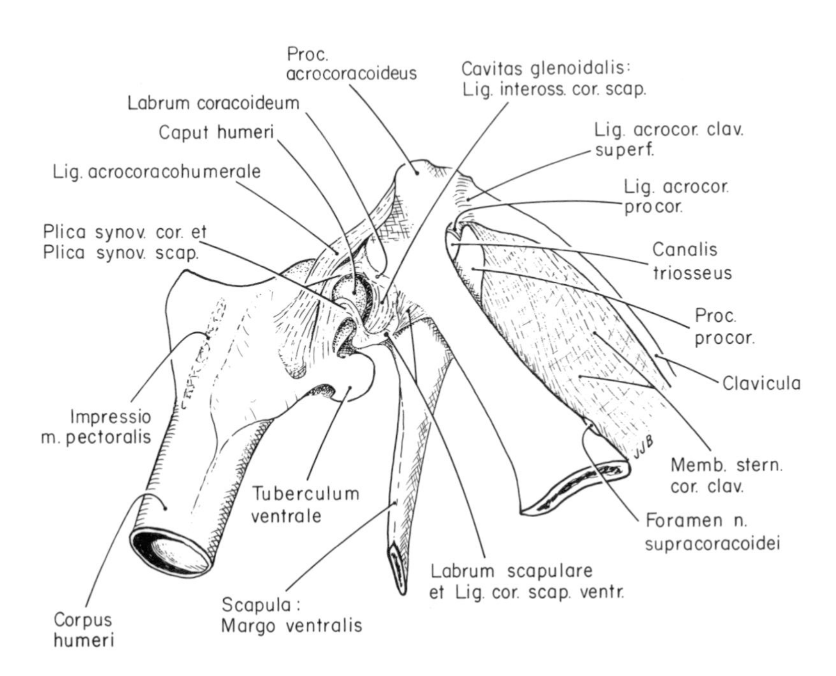

Fig. 3 The shoulder joint (Artc. humeralis) of the pigeon, *Columba livia*. Right shoulder ventral aspect; the general articular capsule is not depicted. Original drawing, J. J. Baumel. Abbreviations: Lig. acrocor. clav. superf., Lig. acrocoracoclaviculare superficiale; Lig. acrocor. procor., Lig. acrocoraco-procoracoideum; Lig. cor. scap. ventr., Lig. coracoscapulare ventrale; Lig. inteross. cor. scap., Lig. interosseum coracoscapulare; Proc. procor., Proc. procoracoideus.

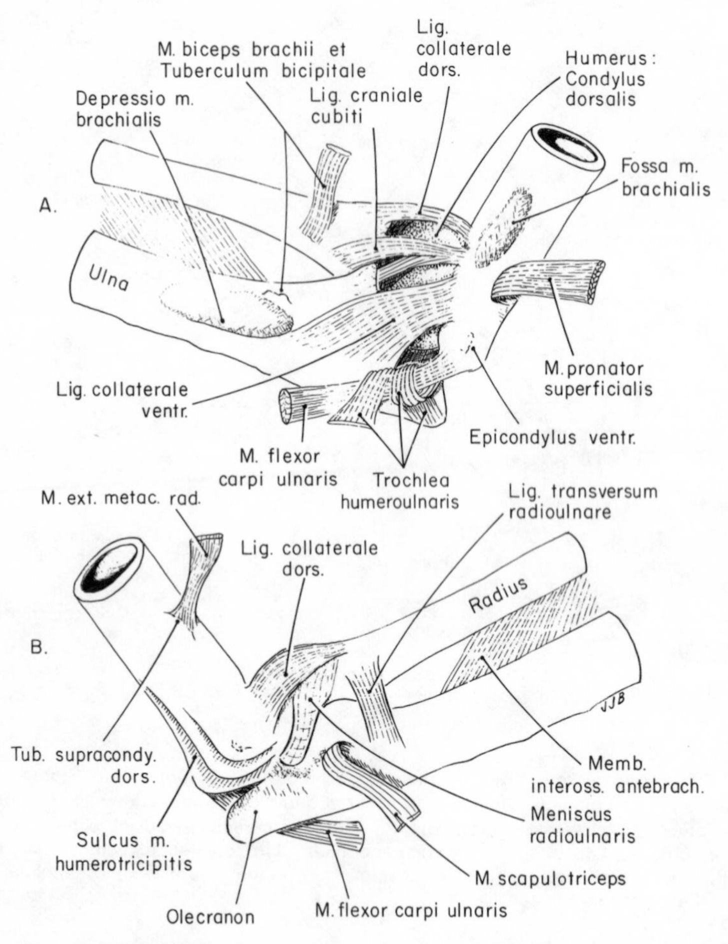

Fig. 4 The elbow joint (Junctura cubiti) of the pigeon, *Columba livia*.
Right side. Original drawing, J. J. Baumel.
A, ventral aspect; B, dorsal aspect.
Note: Lig. craniale cubiti splits into two slips, one attached to the Radius, the other attached to the Ulna; the latter intervenes between the two bones; 2) in A the well-developed Capsula articularis has been removed between the three ligaments; 3) in B the articular capsule is quite delicate (not shown); 4) the Meniscus radioulnaris has strong attachments to Radius and Ulna; 5) the dorsal and caudal aspects of the humeroulnar joint are strengthened by the tendons of the triceps muscles and possess no collateral ligaments.
Abbreviations: M. ext. metac. rad., M. extensor metacarpi radialis; Memb. inteross. antebrach., Membrana interossea antebrachialis; Tub. supracondy. dors., Tuberculum supracondylare dorsale.

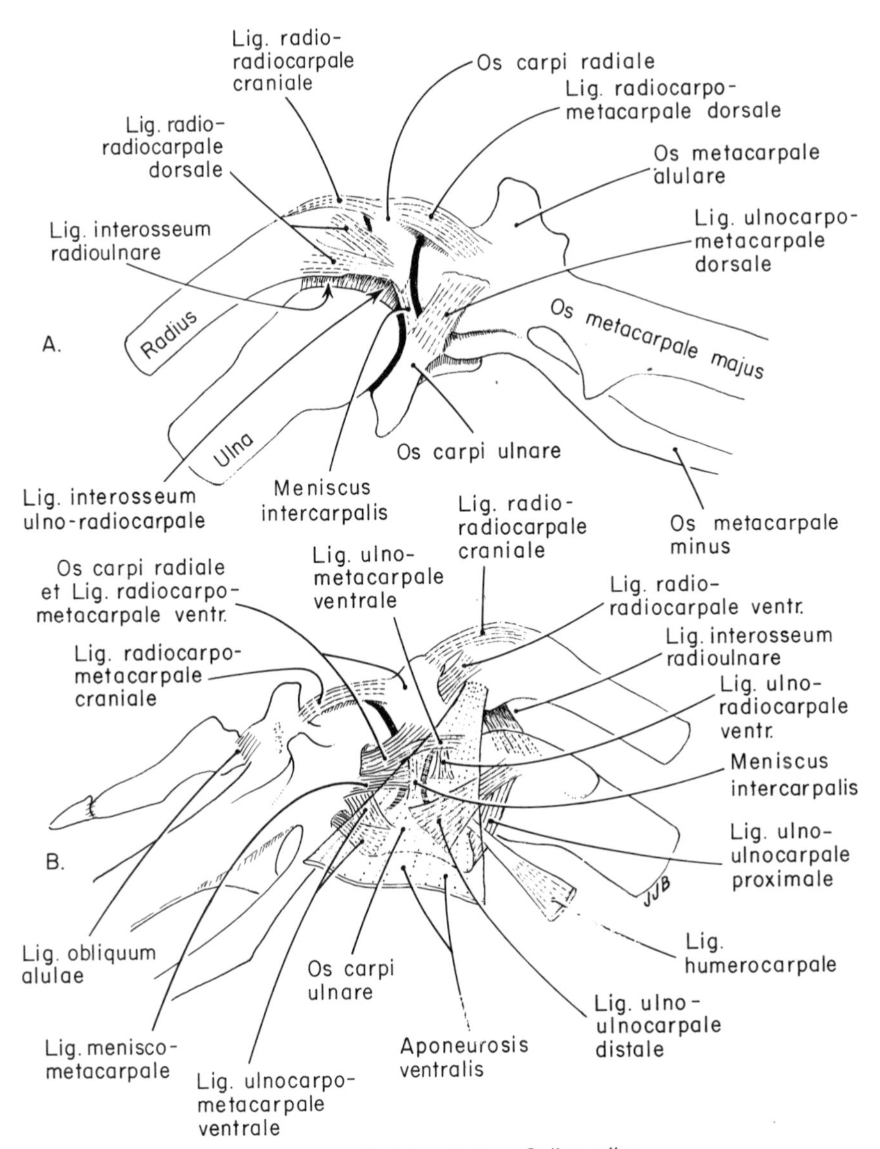

Fig. 5 The wrist joint (Junctura carpi) of the chicken, *Gallus gallus*.
Original drawing, J. J. Baumel.
A, dorsal aspect; B, ventral aspect.
Note: 1) The bones are drawn apart slightly to better demonstrate the interosseous ligaments and meniscus; the joint spaces are depicted in solid black; 2) the Lig. interosseum radioulnare and Lig. interosseum ulno-radiocarpale are continuous with one another; 3) the ventral aponeurosis is superficial to many of the ventral ligaments of the joint; the main part only of the aponeurosis is drawn (see Annot. 114, 115, and Fig. 11); 4) the Lig. interosseum radioulnare intervenes between the distal extremities of the Radius and Ulna, so that the two bones articulate directly with one another only slightly, if at all.

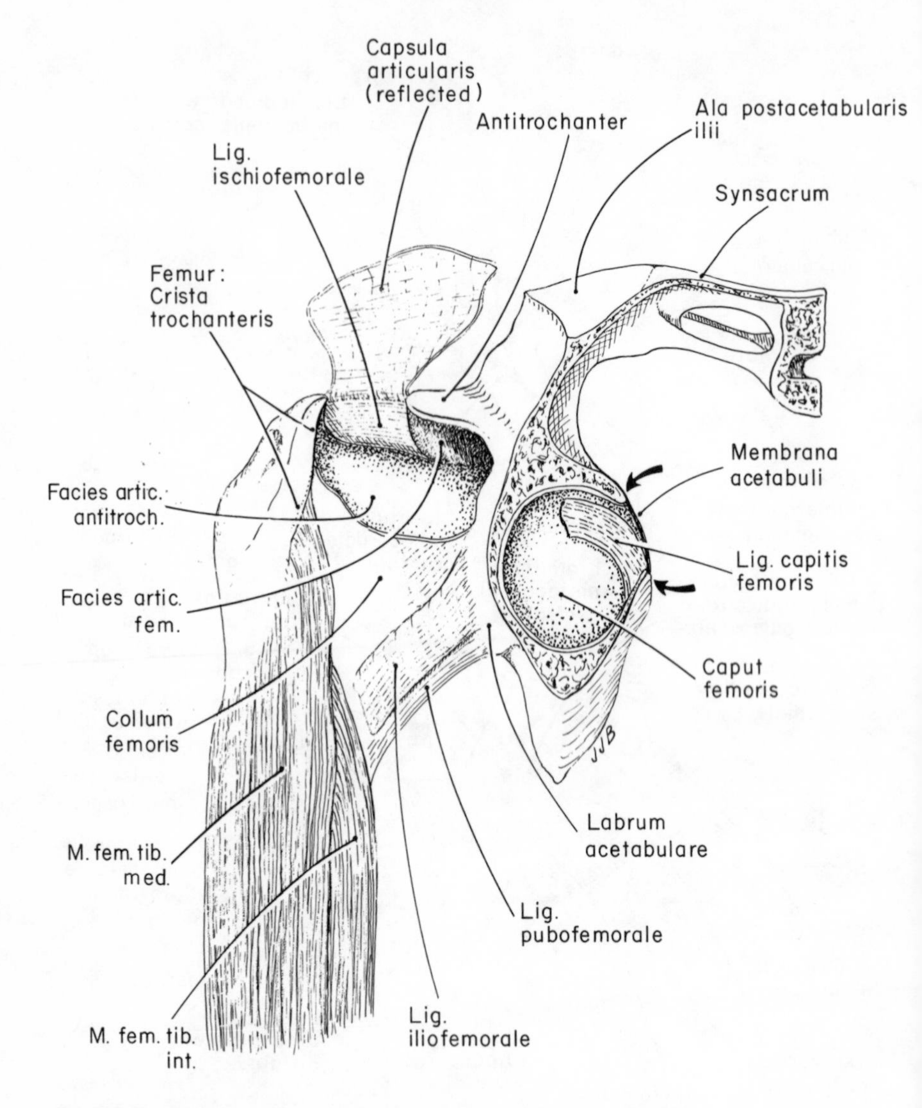

Fig. 6 The hip joint (Artc. coxae) of the pigeon, *Columba livia*.
Right side, cranial aspect. Original drawing, J. J. Baumel.
The pelvis is sectioned approximately transversely, opening a window into the Acetabulum and exposing the head of the Femur and its ligament. The Femur is rotated laterally and withdrawn slightly from the Acetabulum. The two arrows indicate the upper and lower margins of Foramen acetabuli.
Note: 1) the cranial and superior parts of Capsula articularis are thin and delicate; the pubofemoral and ischiofemoral ligaments are thickened parts of the capsule, whereas the iliofemoral ligament is strongly distinct from the capsule; (2) the articulation between the neck of the Femur and the Antitrochanter.
Abbreviations: Facies artic. antitroch., Facies articularis antitrochanterica; Facies artic. fem., Facies articularis femoralis; M. fem. tib., M. femorotibialis.

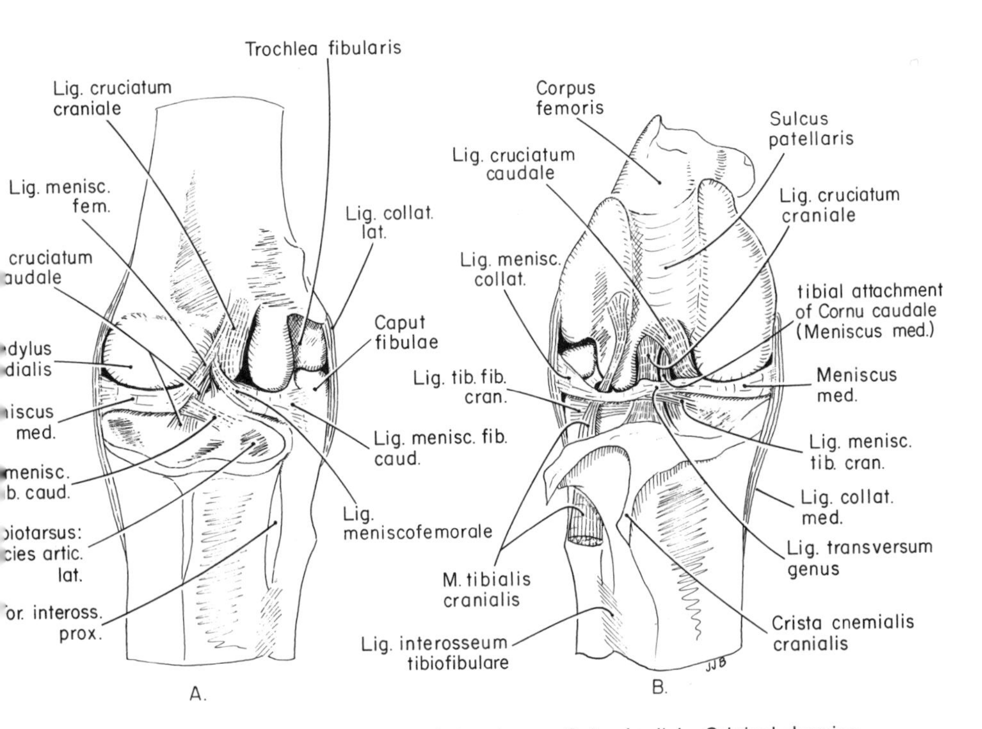

Fig. 7 The knee joint (Junctura genus) of the pigeon, *Columba livia*. Original drawing, J. J. Baumel.
A, caudal aspect of right knee; joint hyperextended.
B, cranial aspect of right knee; joint fully flexed.
Note: 1) the tendon of M. tibialis cranialis perforates the Meniscus lateralis; 2) in B, Meniscus lateralis sends a dorsal extension between Caput fibulae and the lateral surface of the lateral condyle of the femur; 3) the strong attachment of both menisci to collateral ligaments of the joint.
Abbreviations: For. inteross. prox., Foramen interosseum proximale; Lig. collat., Lig. collaterale; Lig. menisc. collat., Lig. meniscocollaterale; Lig. menisc. fem., Lig. menisco-femorale; Lig. menisc. fib. caud., Lig. meniscofibulare caudale; Lig. menisc. tib., Lig. meniscotibiale.

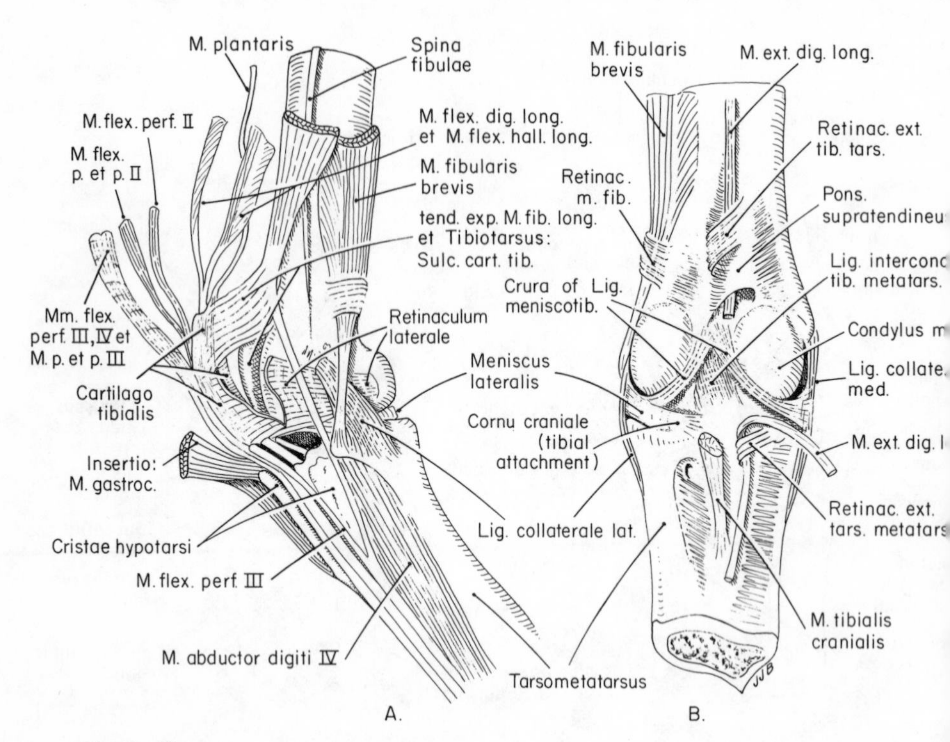

Fig. 8 The ankle joint (Artc. intertarsalis) of the pigeon, *Columba livia*. Original drawing, J. J. Baumel.
A, lateral aspect, right limb; B, cranial aspect, right limb.
Note: 1) In A, the attachments of M. plantaris and M. fibularis longus to Cartilago tibialis; 2) in A, the tibial cartilage is drawn caudally out of its articular sulcus on the distal extremity of the Tibiotarsus; 3) the Artc. cartilago-tibiotarsalis is in open communication with the Artc. intertarsalis; 4) in A, the distal part of the Retinaculum laterale serves as the articular capsule of the intertarsal joint; 5) the tendon of M. tibialis cranialis (cut) traverses the Retinaculum extensorium tibiotarsi; 6) the medial crus of Lig. meniscotibiale is attached to the margin of the medial cotyla of the Tarsometatarsus since no medial meniscus is present in the pigeon and other birds (see Annot. 168).

Abbreviations: Lig. intercondy. tib. metatars., Lig. intercondylare tibiometatarsale; Lig. meniscotib., Lig. meniscotibialis; M. ext. dig. long., M. extensor digitorum longus; M. fib. long., M. fibularis longus; M. flex dig. long., M. flexor digitorum longus; M. flex, hall. long., M. flexor hallucis longus; M. flex. perf. II, III, IV; M. flexor perforatus digiti II, III, or IV; M. flex. p. et p. dig. II, III, M. flexor perforans et perforatus digiti II or III; Retinac. m. fib., Retinaculum m. fibularis; Retinac. ext. tars. metatars., Retinaculum extensorium tarsometatarsi; Retinac. ext. tib. tars., Retinaculum extensorium tibiotarsi; Sulc. cart. tib., Sulcus cartilaginis tibialis; tend. exp., tendinous expansion.

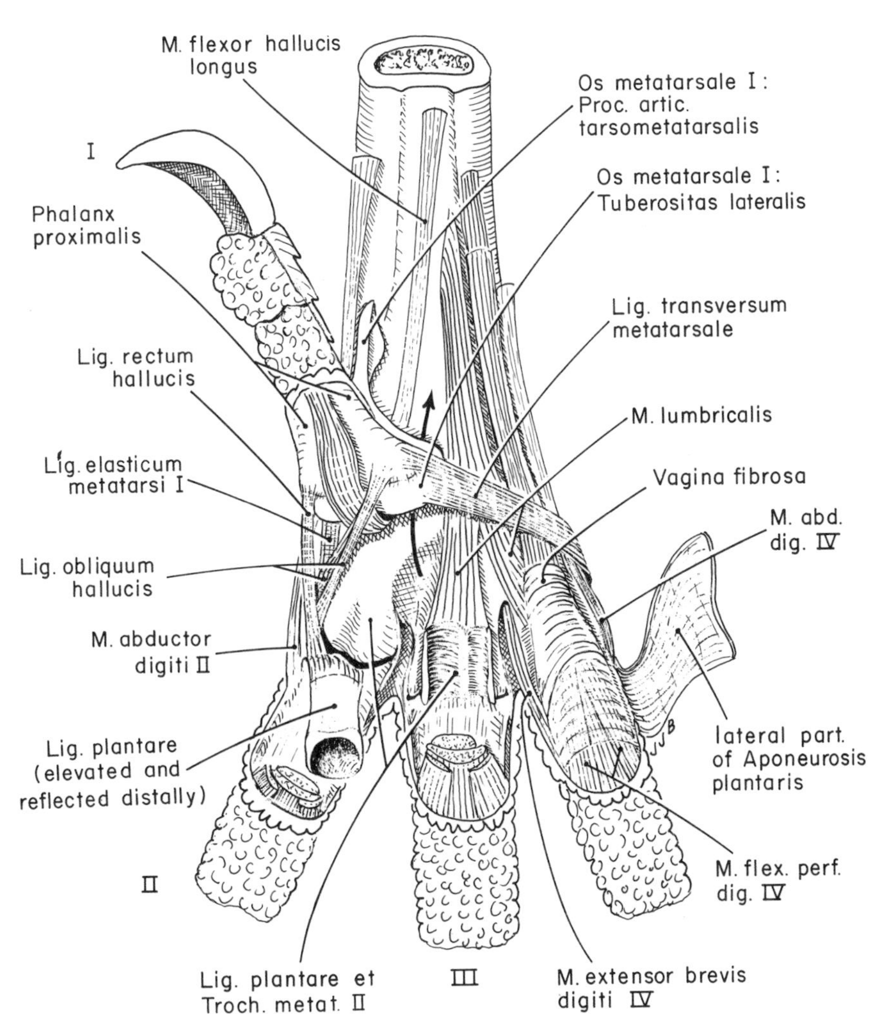

Fig. 9 Joints of the foot (Juncturae pedis) of the pigeon, *Columba livia*. Plantar aspect; right foot; Digitus I [Hallux] hyperextended and drawn medially; the arrow is situated in Canalis flexorius plantaris through which pass the long flexor tendons to the digits.
Note: 1) The deep part of the Lig. obliquum hallucis is attached to the medial aspect of the trochlea of metatarsal II; its superficial part is attached to the proximal phalanx of Digitus II. 2) On the fourth digit the Vagina fibrosa is intact, attached to the margins of the planatar ligament and to the proximal phalanx. On the third digit the flexor sheath and tendons are removed showing the plantar ligament and its attachment to the base of the proximal phalanx. On the second digit the plantar ligament is elevated from its trochlea and reflected distally showing its dorsal surface. 3) A strong part of Aponeurosis plantaris (only a portion depicted) extends transversely across the metatarsophalangeal joints; a thin portion (not shown) extends proximally to become continuous with the transverse metatarsal ligament and the lateral tuberosity of the first metatarsal bone.
Abbreviations: M. abd. dig. IV, M. abductor digiti IV; M. flex. perf. dig. IV, M. flexor perforatus digiti IV; Proc. artic., Proc. articularis.

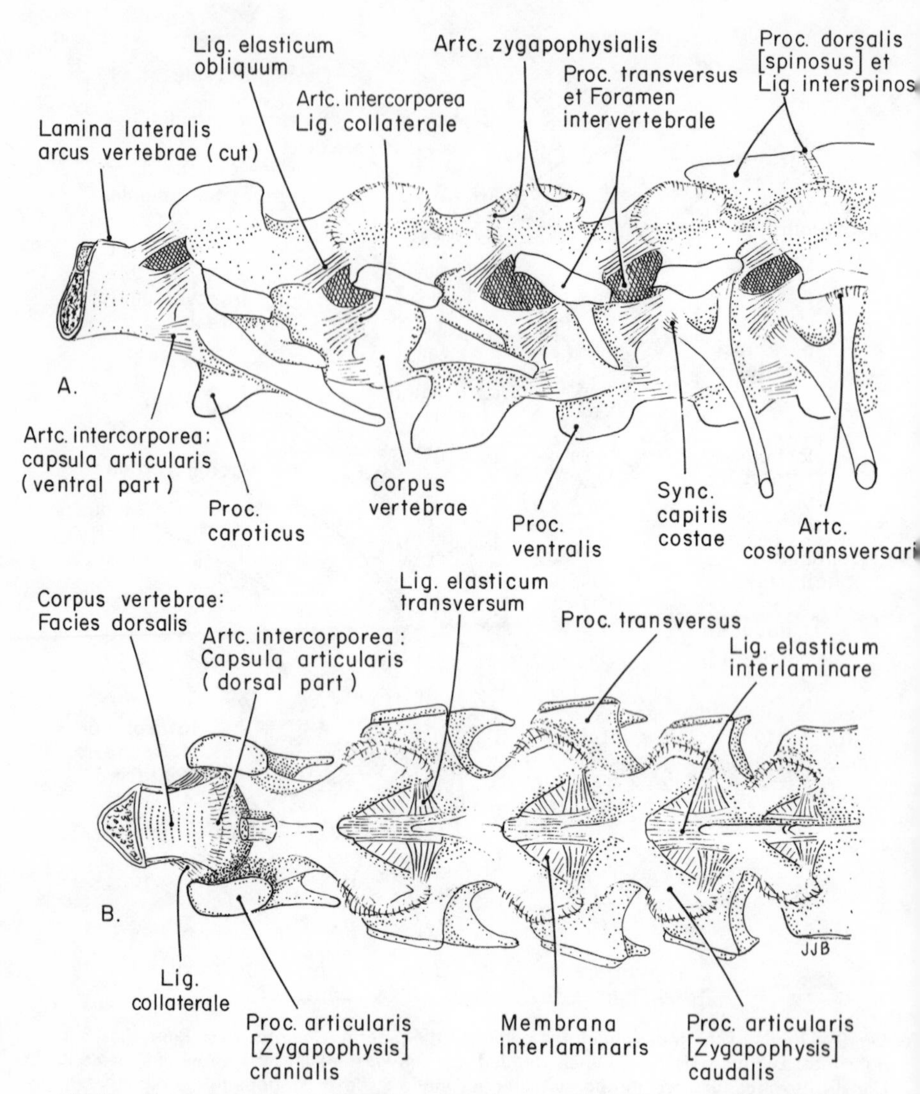

Fig. 10 Ligaments of the joints of the vertebral column (Juncturae columnae vertebralis) of the pigeon, *Columbia livia.* Vertebral segment is from the root of the neck. A, lateral view, left side; B, dorsal view. Original drawing, J. J. Baumel.

Note: (1) the arch of the cranialmost vertebra in both A and B is cut away; (2) only the lateral margin of Lig. elasticum obliquum is depicted, the deeper part of the ligament being situated on the ventral surface of Proc. articularis caudalis of the vertebra that forms the cranial attachment of the ligament; (3) Lig. elasticum obliquum assists Lig. elasticum interlaminare in maintaining the dorsal concavity of this part of the neck; (4) both the transverse and the oblique elastic ligaments of one side are tensed when the vertebral column is flexed to the contralateral side; (5) most of the joints between vertebral bodies are synovial joints having a meniscus; (6) the intervertebral foramina of this vertebral segment are extraordinarily large since they transmit the roots of the brachial plexus.

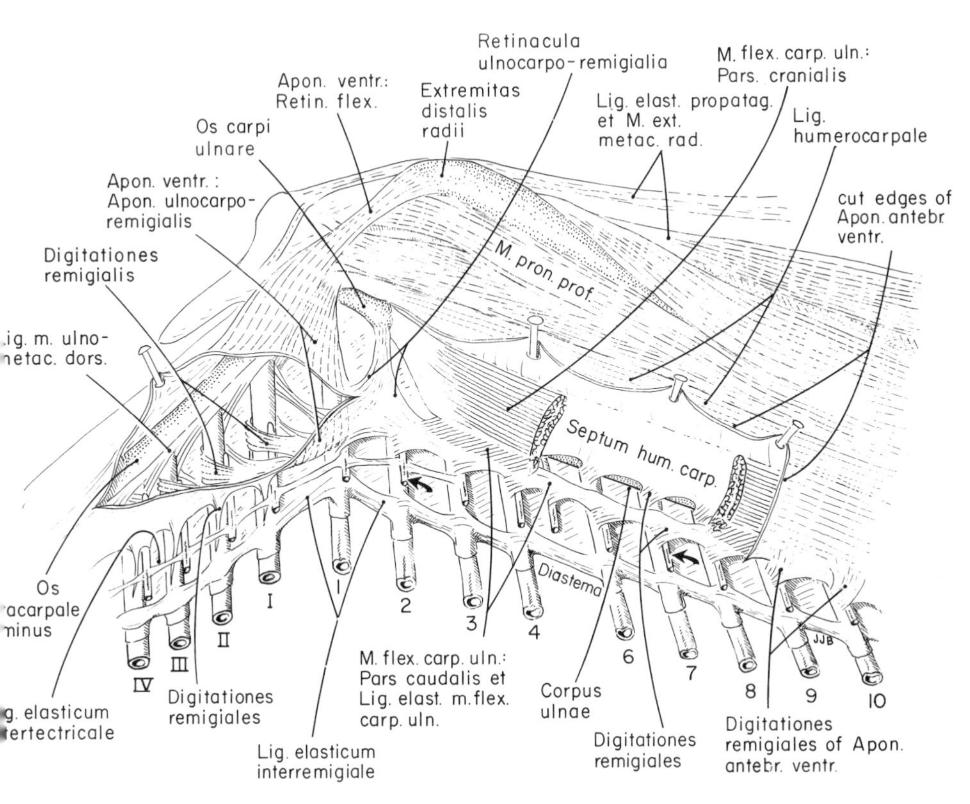

Fig. 11 Ligaments of the feathers (Ligg. pennarum) of the wing of the pigeon, *Columba livia*. Ventral aspect, right wing. Original drawing, J. J. Baumel. Arrows indicate lower major coverts of the secondary remiges (Tectrices secundariae ventrales majores); Roman numerals indicate Remiges primarii, Arabic numerals indicate Remiges secundariae.
Note: 1) the Diastema or hiatus in the series of secondary remiges (see **Integ.** Annot. 53); 2) the digitations to the remiges from the ventral antebrachial aponeurosis, from the elastic ligament of M. flexor carpi ulnaris, and from the humerocarpal septum; 3) the Lig. humerocarpale splits to enclose M. flexor digitorum superficialis, the cranial and caudal laminae blending on the dorsal side of the muscle and becoming continuous with the Septum humerocarpale. (4) the space deep to the reflected parts of Aponeurosis ventralis contains the belly of M. ulnometacarpalis dorsalis which has been removed to display the calami of the primary feathers.
Abbreviations: Apon. antebr. ventr., Aponeurosis antebrachialis ventralis; Apon. ventr., Aponeurosis ventralis; Lig. m. ulnometac. dors., Lig. m. ulnometacarpalis dorsalis; Lig. elast. m. flex. carp. uln., Lig. elasticum m. flexoris carpi ulnaris; Lig. elast. propatag., Lig. elasticum propatagiale; M. ext. metac. rad., M. extensor metacarpi radialis; M. flex. carp. uln., M. flexor carpi ulnaris; M. pron. prof., M. pronator profundus; Retin. flex., Retinaculum flexorum; Septum hum. carp., Septum humerocarpale.

MYOLOGIA

J. C. VANDEN BERGE

Northwest Center for Medical Education and Department of Anatomy, Indiana University School of Medicine, Gary, Indiana 46408, USA

With Contributions from Sub-Committee Members: J. J. Baumel, *Creighton University*; A. J. Berger, *University of Hawaii*; W. J. Bock, *Columbia University*; P. Bühler, *Universität Stuttgart-Hohenheim*; P. J. K. Burton, *British Museum*; J. Cracraft, *University of Illinois*; T. Fujioka, *Nagoya University*; H. Hoerschelmann, *Universität Hamburg*; R. Raikow, *University of Pittsburgh*; R. L. Zusi, *U.S. National Museum.*

Nomenclature and homology. For many years there have been repeated attempts to establish homology between the muscle in birds and their counterparts in mammals. To date, such attempts have been largely unsuccessful and, as a result, have been the source of a great deal of confusion in nomenclature. Various investigators have adopted the same name for different muscles in the same species, or in different species (see synonymy in Berger (1966) and in Vanden Berge (1975)). Nevertheless, certain monumental works in avian anatomy continue to be invaluable references, among these Lakjer (1926) on the jaw muscles; Boas (1929) on the cervical muscles; Fürbringer (1888, 1902) on the muscles of the shoulder and brachium, and the comprehensive volume by Gadow and Selenka (1891).

Two groups of muscles, and their associated nomenclature, still constitute sources of considerable confusion in the avian literature. The first group, and most difficult to resolve, are the muscles associated with the metacarpals and digits of the wing. As mentioned elsewhere (see **Osteo.** Intro.) there is no indisputable evidence for deciding if the digits of the avian manus represent I, II, III or II, III, IV. For this reason, a new set of highly descriptive names has been adopted, in which the muscles are named in reference to three principal digits: Digitus alularis, Digitus major, Digitus minor. Each new name is annotated with synonymy based on some presumed digital enumeration although the subcommittee strongly recommends that such synonyms no longer be used.

The second group of muscles is that associated with the hip and thigh. Most avian anatomists have chosen to follow Hudson (1937), based on Gadow and Selenka (1891), or Fisher (1946), based on Howell (1938). The confusion in nomenclature for the group of muscles is based in part on different interpretations of the development and evolution of the pelvic girdle of birds as compared to mammals, a corollary of which is the presumed homology of the corresponding musculature (see Howell (1938) and Romer (1927)). In this case, the Sub-Committee chose either to adopt a name which had been proposed by earlier authors but had not been widely used, e.g. M. iliofemoralis externus (Annot. 105) or to adopt a name which was itself descriptive and which did not convey a presumed homology, e.g. M. pubo-ischiofemoralis (Annot. 113).

In general the Sub-Committee chose to adopt a nomenclature for the muscles based on the following criteria:

(1) The particular name has a long history of use in avian anatomy and has been consistently used for only one muscle, e.g. M. ambiens.
(2) The particular name may suggest homology with a mammalian counterpart, but even though this homology is not necessarily substantiated, other criteria justify continued use of the name, e.g. M. pectoralis.
(3) The particular name may suggest a functional role which is not always established by reference to electromyography but which may be assumed by development or other morphological criteria, e.g. M. extensor metacarpi ulnaris (Annot. 88).
(4) A particular collective name that designates a muscle complex is more appropriate than individual names for the separate muscles of the complex, e.g. M. longus colli dorsalis (Annot. 47).

Musculi alae [*membri thoracici*]. It has become standard practice among avian anatomists to describe all muscles attaching to the pectoral girdle and wing skeleton as Musculi alae, i.e. muscles of the wing, rather than Musculi membri thoracici, i.e. muscles of the pectoral or thoracicic limb. Although such categorization of muscles differs from the *Nomina Anatomica Veterinaria* (ICVAN, 1973) which lists certain mammalian counterparts as Musculi dorsi (e.g. M. latissimus dorsi) or Musculi thoracici (e.g. M. pectoralis), the Sub-Committee chose to follow the precedent of long standing ing, namely, to continue use of the collective term, Musculi alae.

Musculi membri pelvici and the Garrod formula. The original Garrod formula was first expanded by Hudson (1937) to include 11 symbols to represent the variation in the leg muscles in birds; recently, an additional pair of symbols (M and N) were suggested by Kurochkin (1968) and adopted by Van Tyne and Berger (1976). Although the formula has become associated with the technical diagnoses of avian taxa, it is more properly a somewhat artificial measure of anatomical variation within which there may be individual as well as species differences (see e.g. Vanden Berge, 1970; and Raikow, 1975). The designation of these muscles or structural features by their formula letters is summarized as follows:

A = M. caudofemoralis (Annot. 110)
B = M. iliofemoralis (Annot. 110)
C = M. iliotrochantericus medius (Annot. 104)
D = M. iliofemoralis externus (Annot. 105)
E = M. iliofemoralis internus (Annot. 106)
F = M. plantaris (Annot. 119)
G = M. popliteus (Annot. 120)
M = M. fibularis [peroneus] longus (Annot. 116)
N = M. fibularis [peroneus] brevis (Annot. 117)
X = M. flexor cruris lateralis, pars pelvica (Annot. 108)
Y = M. flexor cruris lateralis, pars accessoria (Annot. 108)
Am = M. ambiens (Annot. 103)
V = Vinculum tendinum flexorum, M. flexor perforatus digiti III (Annot. 122)

TERMINI GENERALES

Musculus striatus [skeleti] (see **Integ.** Fig. 2)
  Origo (see *Nomina Anatomica*, IANC, 1977, footnote 78, p. A37)
  Caput
  Venter
    Fascia
      Epimysium
      Perimysium
    Fasciculus muscularis
  Cauda
  Insertio (see *Nomina Anatomica*, IANC, 1977, footnote 78, p. A37)
Musculus nonstriatus (see **Integ.**)
Musculus subcutaneus
Musculus unipennatus
Musculus bipennatus
Musculus multipennatus
Musculus fusiformis
Musculus planus
Musculus abductor
Musculus adductor
Musculus articularis
Musculus constrictor
Musculus depressor
Musculus dilator
Musculus expansor
Musculus extensor
Musculus flexor
Musculus levator
Musculus orbicularis
Musculus protractor
Musculus retractor
Musculus rotator
  Musculus pronator
  Musculus supinator
Musculus sphincter
Musculus tensor
Tendo
  Peritendineum
  Intersectio tendinea[2]
  Aponeurosis
  Arcus tendineus
  Vagina fibrosa tendinis[3]
  Vagina synovialis tendinis[3]
    Mesotendineum
  Bursa synovialis[3]
Tendo ossificans[1]

MUSCULI SUBCUTANEI[4]

M. constrictor colli[5]
  Pars intermandibularis
M. cucullaris
  M. cucullaris capitis[6]
    Pars interscapularis[7]
    Pars propatagialis[8]
    Pars clavicularis[9]
  M. cucullaris cervicis[10]
    Pars nuchalis (Fig. 1)
    Pars clavicularis
  M. latissimus dorsi[77, 4]
    Pars interscapularis[7]
    Pars metapatagialis
  M. serratus superficialis[70, 4]
    Pars metapatagialis
  M. pectoralis[75, 4]
    Pars subcutanea thoracica[11]
    Pars subcutanea abdominalis[11]
    Pars propatagialis[75]
  M. biceps brachii
    Pars propagatagialis[11]
  M. expansor secundariorum[12]

MUSCULI CAPITIS

(Figs. 2, 3)

Musculi bulbi oculi[13] (see **Org. sens.**)

M. rectus dorsalis
M. rectus ventralis
M. rectus medialis
M. rectus lateralis
M. obliquus dorsalis[14]
M. obliquus ventralis
M. quadratus membranae nictitantis[15]
  Vagina fibrosa tendinis
M. pyramidalis membranae nictitantis[15]
  Tendo m. pyramidalis[15]
M. levator palpebrae dorsalis[16]
M. depressor palpebrae ventralis[16]
M. tensor periorbitae[16]

M. columellae (see **Org. sens.** Annot. 64)

Musculi mandibulae[17] (Figs. 2, 3)

M. adductor mandibulae externus[18]
M. pseudotemporalis superficialis[19]
M. pseudotemporalis profundus[20]
M. adductor mandibulae caudalis[20]
M. pterygoideus[21]
M. ethmomandibularis[22]
M. protractor pterygoidei et quadrati[23]
M. depressor mandibulae[24]

MUSCULI APPARATUS HYOBRANCHIALIS[25]

(Figs. 2, 3)

M. intermandibularis
  M. intermandibularis ventralis[26]
  M. intermandibularis dorsalis[26]
M. serpihyoideus[27]
M. stylohyoideus[28]
M. branchiomandibularis[29]
M. interceratobranchialis [ceratohyoideus][30]
M. ceratoglossus[31]
M. hypoglossus rostralis [medialis][32]
M. hypoglossus obliquus [lateralis][33]
M. mesoglossus[34]
M. genioglossus[35]

MUSCULI TRACHEALES

(see Fig. 2b; **Resp.** Annot. 45 and Figs. 6a–g)

M. sternohyoideus
M. sternotrachealis
M. cleidotrachealis
M. tracheolateralis

MUSCULI LARYNGEALES

(see Fig. 2b; **Resp.** Annot. 30)

M. dilator glottidis[38]
M. constrictor glottidis[39]
M. cricohyoideus[36]

MUSCULI SYRINGEALES[40]

(see **Resp.** Annot. 46 and Figs. 6a–g)

M. tracheobronchialis dorsalis
M. tracheobronchialis brevis
M. tracheobronchialis ventralis
M. syringealis dorsalis
M. syringealis ventralis
M. vocalis dorsalis
M. vocalis ventralis
M. obliquus ventralis
M. obliquus lateralis

MUSCULI COLLI[41]

M. biventer cervicis (Fig. 1)
M. complexus[42]
M. splenius capitis[43]
M. rectus capitis dorsalis[44]
M. rectus capitis lateralis
M. rectus capitis ventralis
  Pars lateralis
  Pars medialis
Mm. iliocostalis et longissimus dorsi[45]
  M. cervicalis ascendens[46]
  M. thoracicus ascendens[46]
M. longus colli dorsalis[47]
  Pars cranialis[48]
  Pars caudalis[49]
  Pars profunda[50]
  Pars thoracica[51]
Mm. intercristales[52]
Mm. intertransversarii[53]
Mm. inclusi[54]
M. longus colli ventralis[56]
M. flexor colli lateralis [brevis][55]
M. flexor colli medialis [profundus][55]

MUSCULI TRUNCI[57]

(Fig. 4)

Mm. levatores costarum
M. scalenus[58]
Mm. intercostales externi[59]

Mm. intercostales interni
M. costosternalis[60]
    Pars major
    Pars minor
M. costoseptalis[61]
M. sternocoracoideus[62]
M. rectus abdominis[63]
M. obliquus externus abdominis[63]
M. obliquus internus abdominis[63]
M. transversus abdominis[63]

MUSCULI CAUDAE[64]
(Fig. 4)

M. levator caudae[64]
M. lateralis caudae[64]
M. depressor caudae[65]
M. pubocaudalis externus[66]
M. pubocaudalis internus[66]
M. caudofemoralis[110]
M. bulbi rectricium[67]
M. adductor rectricium[67]
M. transversus cloacae[68]
M. sphincter cloacae[68]
M. levator cloacae [M. retractor phalli][68] (see **Genit. masc.** Annot. 47)

MUSCULI ALAE [MEMBRI THORACICI]
(see **Myol.** Intro and Figs. 5, 6; and **PNS** Annot. 34)

M. rhomboideus superficialis[69]
M. rhomboideus profundus
Mm. serrati
    M. serratus superficialis[70] (Fig. 5)
        Pars cranialis
        Pars caudalis
    M. serratus profundus[70]
M. scapulohumeralis cranialis[71]
M. scapulohumeralis caudalis (Fig. 5)
Mm. subcoracoscapulares[72]
    M. subscapularis[73]
        Caput laterale
        Caput mediale
    M. subcoracoideus

MUSCULI ALAE—*continued*

M. coracobrachialis cranialis[74]
M. coracobrachialis caudalis
M. pectoralis[75]
  Pars propatagialis
M. supracoracoideus[76]
M. latissimus dorsi[77]
  Pars cranialis
  Pars caudalis
M. tensor propatagialis[78] (Figs. 5, 6)
  Pars longa
    Tendo longa
  Pars brevis
    Tendo brevis
M. deltoideus major[79]
M. deltoideus minor
  Caput dorsale
  Caput ventrale[80]
M. triceps brachii
  M. scapulotriceps[81] (Fig. 5)
  M. humerotriceps[82]
  M. coracotriceps[83]
M. biceps brachii
  Caput coracoideum
  Caput humerale
  Pars propatagialis[11]
M. expansor secundariorum[12] (Fig. 6)
M. brachialis
M. pronator superficialis[84]
M. pronator profundus[84]
M. flexor carpi ulnaris[85]
  Pars cranialis
  Pars caudalis
M. flexor digitorum superficialis[86]
M. flexor digitorum profundus
M. extensor metacarpi radialis[87] (Fig. 5)
  Caput dorsale
  Caput ventrale
M. extensor digitorum communis
M. extensor metacarpi ulnaris[88]
M. extensor longus alulae[89]
M. extensor longus digiti majoris[90]
  Pars proximalis
  Pars distalis

M. supinator (Fig. 5)
M. ectepicondylo-ulnaris[91]
M. entepicondylo-ulnaris[92]
M. ulnometacarpalis dorsalis
M. ulnometacarpalis ventralis
M. interosseus dorsalis
M. interosseus ventralis[93] (Fig. 6)
M. extensor brevis alulae[94]
M. abductor alulae[95]
M. flexor alulae[96]
M. adductor alulae[97]
M. abductor digiti majoris[98]
M. flexor digiti minoris[99]

MUSCULI MEMBRI PELVICI

(see Intro. and Figs. 7–11; and **PNS** Annot. 39–46)

Mm. iliotibiales[100]
    M. iliotibialis cranialis[101]
    M. iliotibialis lateralis
    M. iliotibialis medialis[101]
M. iliofibularis[102] (Figs. 7, 8)
    Ansa m. iliofibularis (Figs. 8, 9) (see **Arthr.** Annot. 186)
M. ambiens[103]
Mm. iliotrochanterici[104]
    M. iliotrochantericus caudalis
    M. iliotrochantericus cranialis
    M. iliotrochantericus medius
M. iliofemoralis externus[105]
M. iliofemoralis internus[106]
Mm. femorotibiales (**Osteo.** Annot. 265, and **Arthr.** Annot. 158)
    M. femorotibialis externus[107]
    M. femorotibialis medius
    M. femorotibialis internus
M. flexor cruris lateralis[108]
    Pars pelvica
    Pars accessoria
M. flexor cruris medialis[109]
M. caudo-ilio-femoralis[110]
    M. caudofemoralis (see Mm. caudae) (Fig. 9)
    M. iliofemoralis
M. ischiofemoralis (Fig. 9)

MUSCULI MEMBRI PELVICI—*continued*

M. obturatorius lateralis[111] [112]
  Pars dorsalis
  Pars ventralis
M. obturatorius medialis[111] (Fig. 10)
M. pubo-ischio-femoralis[113] (Figs. 8, 9)
  Pars lateralis [cranialis]
  Pars medialis [caudalis]
M. tibialis cranialis[114]
  Caput femorale
  Caput tibiale
M. extensor digitorum longus[115]
M. fibularis [peroneus] longus[116] (Figs. 7–10)
M. fibularis [peroneus] brevis[117] (Figs. 7–10)
  Caput tibiale
  Caput fibulare
M. gastrocnemius[118]
  Pars lateralis [externa]
  Pars intermedia
  Pars medialis [interna]
M. plantaris[119]
M. popliteus[120]
M. flexor perforans et perforatus digiti II[121] (Fig. 9)
M. flexor perforans et perforatus digiti III[121] (Fig. 9)
M. flexor perforatus digiti II
M. flexor perforatus digiti III
  Vinculum tendinum flexorum[122] (Fig. 9)
M. flexor perforatus digiti IV
M. flexor hallucis longus[123] (see **Osteo.** Ossa metatarsalia)
M. flexor digitorum longus
  Vinculum tendinum flexorum[124]
M. extensor hallucis longus[125] (see **Osteo.** Ossa metatarsalia)
  Pars proximalis
  Pars distalis
M. flexor hallucis brevis[126]
M. abductor digiti II[127]
M. adductor digiti II
M. extensor proprius digiti III[128]
M. extensor brevis digiti III (Fig. 11)
M. extensor brevis digiti IV[129]
M. abductor digiti IV[130]
M. lumbricalis

ANNOTATIONS

(1) **Tendo ossificans.** General term referring to discrete, nonpathologic, initially mineralized, and later ossified, segments of many tendons in the wing and especially in the leg of *Phasianus* and other Galliformes (Hudson *et al.*, 1965, 1966) and in other avian groups (Hoff, 1966; Hudson, 1937).

(2) **Intersectio tendinea.** Replaces Inscriptio tendinea in agreement with *Nomina Anatomica Veterinaria* (ICVAN, 1973). See Annot. 7, 63 and Fig. 4.

(3) **Vagina fibrosa tendinis; Vagina synovialis tendinis; Bursa synovialis.** Gross anatomical observations concerning the tendon sheath around certain extensor and flexor tendons of the foot in *Gallus* have been described by Frewein (1967). More recently, the development of the chick flexor digital tendon, the synovial sheath, elastic vincula, and certain fibrocartilaginous adaptations on the surface of the tendon have also been described (Greenlee *et al.*, 1975). Following *Nomina Anatomica Anatomica Veterinaria* (ICVAN, 1973). See Annot. 7, 63 and Fig. 4.

(4) **Musculi subcutanei** (see Annot. 70, 75, 77). Thin sheets or narrow bands of striated muscle lying between a well-defined innermost Lamina elastica of the dermis and loose connective tissue and fat of the superficial fascia (Tela subcutanea). Although often called "dermal muscles", Mm. subcutanei typically originate from the skeleton or are formed as muscular slips given off from somatic muscle. They traverse beneath featherless areas (apteria) of the skin, to which they may adhere, and some (e.g. the metapatagial part of M. latissimus dorsi) subtend certain feather tracts (e.g. Pteryla humeralis). These muscles have no direct continuity with true dermal muscles which are smooth (Mm. nonstriati), confined to the dermis (Stratum laxum), and subtend feathers (Mm. pennales) or interconnect feather tracts (Mm. apteriales). See Fig. 294, p. 483, in Lucas and Stettenheim (1972); see also **Integ.** The special anatomical relationships of the non-striated M. expansor secondariorum are dealt with in Annot. 12.

Some of the so-called "true dermal muscles" which have been described under several different names are subcutaneous slips from otherwise well-known somatic muscles (see Annot. 11). Those listed here are the most frequently described and the best documented. Certain exceptions should be noted, however: "dermoglossus" (Fisher and Goodman, 1955) is a special muscular slip from M. ceratoglossus (Annot. 31); "cleidotrachealis" (Burt, 1930), "sternomaxillaris" (Owen, 1842) and similar names are associated with the Mm. tracheales complex (see Annot. 37). The actual existence of muscles known as "occipitalis" (Helm, 1884), "dermofrontalis" (Burt, 1930; Shufeldt, 1890) and "circumconcha" (Shufeldt, 1890) remains to be confirmed.

(5) **M. constrictor colli.** Synonymy: M. cucullaris, superficial layer (Gadow and Selenka, 1891); M. sphincter colli (Fürbringer, 1888); M. cutaneus colli transversus (Fujioka, 1963). Typically a thin and rather loosely organized sheet of muscle fasciculi, oriented in the transverse (circumferential) plane opposite the cranial half or less of the neck. M. constrictor colli is superficial to M. cucullaris and most often is closely adherent to the skin. Separate muscle bundles are most organized ventrally where they attach on either side of a median ventral raphe. These bundles are more widely spaced dorsally and caudally where they terminate in superficial fascia, or in a dorsal median raphe beneath the integument. The most rostral bundles may also attach to the cranium, especially on the caudal and dorsal rim of the external acoustic

meatus where they blend with the attachment (origin) of M. cucullaris capitis. The constrictor colli receives its major innervation from the cervical ramus of the facial nerve near this cranial attachment.

M. constrictor colli, pars intermandibularis (Fig. 2). Synonymy: M. mylohyoideus posterior (Ghetie and Atanasiu, 1962); M. intermandibularis, pars caudalis (Zweers, 1974); M. articulomylohyoideus (Harvey *et al.*, 1968); M. basibranchialis mandibularis, pars superficialis (Vanden Berge, 1975, based on McLelland, 1968); M. platysma myoides (Burt, 1930; Fujioka, 1963; Owen, 1842). If separate and distinct, this is a fleshy band of muscle with origin on the lateral aspect of Proc. retroarticularis of the mandible (see **Osteo.** Annot. 49), adjacent (caudal) to the origin of M. serpihyoideus if the latter is attached to the lower jaw (see Annot. 27). It lies superficial (ventral) to M. serpihyoideus, and meets its counterpart in a midventral raphe. Caudally this typically flattened sheet is more or less continuous with the main (cervical) portion of M. constrictor colli; rostrally, it is united by fascia with the caudal edge of M. intermandibularis ventralis and/or with the deeper hyobranchial musculature (M. serpihyoideus; M intermandibularis dorsalis).

Pars intermandibularis is well-developed in *Gallus*, *Phasianus*, *Chrysolophus*, *Pavo*, and *Meleagris* (Ghetie and Atanasiu,1962; Harvey *et al.*, 1968; Osborne, 1968; Vanden Berge, 1975). It is the muscle identified as "M. gularis" in *Gallus*(Edgeworth, 1935, Fig. 619). It is illustrated and described in *Anser* (Ghetie and Atanasiu, 1962, Fig. 4) and at least suggested in *Anas* (Zweers, 1974); in other anseriforms, Pars intermandibularis may not be separable from M. constrictor colli, but is at least suggested by the mandibular attachment of the latter muscle in *Oxyura* (see Goodman and Fisher, 1962). In the Kakapo, *Strigops*, this muscle was identified as a posterior portion of "M. mylohyoideus anterior" (Mudge, 1903, Plate XXVI, Fig. 2.).

(6) **M. cucullaris capitis.** Synonymy: M. craniocervicalis, pars cranialis (Edgeworth, 1935); M. cutaneus colli lateralis (Fujioka, 1963; Harvey *et al.*, 1968); M. dermotemporalis (Goodman and Fisher, 1962; Shufeldt, 1890; Weymouth *et al.*, 1964). An extensive sheet of muscle fibers oriented longitudinally in the neck between M. constrictor colli and M. cucullaris cervicis, and deep to the lateral cervical apterium. Its origin is typically on the skull; its insertion (i.e. caudal extent of the belly) typically consists of three slips: Pars interscapularis, Pars propatagialis, and Pars clavicularis (see Annot. 7, 8, 9, and Figs. 1, 3). The motor innervation is principally from the accessory nerve via the external ramus of the vagus nerve.

(7) **M. cucullaris capitis, pars interscapularis.** Synonymy: M. cucullaris dorsocutaneus (Fürbringer, 1888); M. cutaneus spinalis dorsalis (Harvey *et al.*, 1968); M. dermodorsalis (Shufeldt, 1890); M. dermospinalis (Owen, 1842). Dorsal terminal part of M. cucullaris capitis inserting into the interscapular dorsal tract (see **Integ.** Pteryla interscapularis) where it may unite by means of a tendinous intersection with M. latissimus dorsi, pars interscapularis.

(8) **M. cucullaris capitis, pars propatagialis.** Synonymy: M. dermotensor patagii (Burt, 1930; Shufeldt, 1890). Lateral terminal part of M. cucullaris capitis inserting into humeral feather tract (Pteryla humeralis) or adjacent under marginal coverts of the propatagium (Tectrices marginales ventrales propatagii). In psittacine birds, woodpeckers, and passerines, the principal insertion is on Tendo longa, M. tensor propatagialis (Berger, 1966, 1968; Burt, 1930; Evans, 1969; Hudson and Lanzillotti, 1955; Raikow, 1977; see Annot. 78).

(9) **M. cucullaris capitis, pars clavicularis.** A ventral terminal part of M. cucullaris capitis inserting principally on the clavicle, although the insertion may be somewhat more diffuse and include attachments on the integument about the shoulder (M. cucullaris omocutaneus Fürbringer, 1888) and on fascia of M. pectoralis. In those birds which have a pendulant crop (Ingluvies), Pars clavicularis may form superficial (ventral) and deep (dorsal) sheets which ensheath this organ and serve to support it in a sling-like fashion. An additional function of facilitating periodic emptying of the crop has also been suggested (Helm, 1884); although this has not been experimentally verified.

(10) **M. cucullaris cervicis.** Synonymy: M. craniocervicalis, pars cervicalis (Edgeworth, 1935); M. cucullaris, hals part (Fisher and Goodman, 1955); M. dermodorsalis (Burt, 1930; Shufeldt, 1890); M. dermotransversalis (Owen, 1842).
Pars nuchalis. Synonymy: M. cutaneus colli nuchalis (Harvey *et al.*, 1968). Typically a loose arrangement of three to six widely spaced, obliquely oriented, ribbonlike fasciculi which insert into the interscapular pteryla (see Harvey *et al.*, 1968, Plate 7; also Osborne, 1968). See Fig. 1.
Pars clavicularis. Synonymy: M. cutaneus cleidodorsalis (Harvey *et al.*, 1968); M. dermocleidodorsalis (Burt, 1930). A more or less separate sheet of muscle arises on the clavicle and inserts on the interscapular pteryla, intermingling with the insertions of M. cucullaris, pars interscapularis (Annot. 7) and the corresponding part of M. latissimus dorsi (Annot. 11, 77). See Fig. 1.

(11) **M. pectoralis, pars propatagialis; M. biceps brachii, pars propatagialis.** May be fleshy slips (e.g. "biceps slip", Hudson *et al.*, 1964); partly or entirely aponeurotic, or absent. When present, both generally have some relationship to M. tensor propatagialis (Annot. 78). Pars propatagialis of M. biceps brachii has become part of the technical diagnosis of major avian groups (see Berger, 1966).
M. pectoralis, pars subcutanea abdominalis (Osborne, 1968). M. gastrolumbalis, described in *Gallus* by Helm (1884) is a variant or modification of Pars subcutanea abdominalis.

(12) **M. expansor secundariorum.** This nonstriated muscle lies beneath the subhumeral apterium and is attached to skin of the caudohumeral and humeral pterylae. At the latter attachment it may be confluent with the insertions of the metapatagial parts of Mm. serratus superficialis and latissimus dorsi. However, it is the only subcutaneous muscle which also inserts directly on the follicles of certain feathers (two to six of the most proximal secondaries opposite the elbow). For this reason, Lucas (1970) proposed that it be considered an apterial muscle (Mm. apterii), but since it is ordinarily described with striated muscles of the wing, it is listed here and again under Musculi alae for convenience. See Fig. 6.

M. expansor secundariorum may have a single belly, or it may consist of a proximal tendon and a distal fan-shaped belly associated with a distal tendon. Tendo proximalis is variously attached to bones of the pectoral girdle and to deep fascia of associated muscles (see Annot. 83). Tendo distalis (Tendo humeralis) is attached to the humerus (Epicondylus ventralis, **Osteo.** Annot. 194) or to the humeroulnar pulley (Trochlea humeroulnaris, pars pennata, **Arthr.** Annot. 110). See descriptions of Berger (1956, 1966, 1968), Hudson *et al.* (1964, 1969, 1972) and Vanden Berge (1970).

(13) **Musculi bulbi oculi.** See Abraham and Stammer (1966) for a review of the structure and innervation of the eight extrinsic muscles of the avian eye.

(14) **M. obliquus dorsalis.** There is no trochlea associated with the distal end of this muscle of the avian eye, although M. obliquus dorsalis is in all probability a homologue of the same muscle of the mammalian eye.

(15) **M. quadratus membranae nictitantis; M. pyramidalis membranae nictitantis.** Derived from the homologous primordium and innervated by N. abducens, similar to M. retractor bulbi and M. rectus lateralis in mammals. The aponeurosis of M. quadratus membranae nictitantis is reflected on itself to form a fibrous sheath (Vagina fibrosa tendinis) which facilitates passage of the Tendo m. pyramidalis as the latter arches over N. opticus (Arcus tendineus nervi optici). See Slonaker (1918) for a most complete description and also Fisher and Goodman (1955), Fujioka (1963) and Simić and Jablan-Pantič (1959). See **Org. sens.** Annot. 36.

(16) **M. levator palpebrae dorsalis; M. depressor palpebrae ventralis; M. tensor periorbitae.** These three striated muscles do not insert on the eyeball, but are functionally related to the eye since they move the eyelids and/or place tension on the orbital fascia (periorbita). They are listed here for convenience.

M. levator palpebrae dorsalis inserts well down toward the marginal folds (Plicae marginales dorsales, **Org. sens.** Annot. 38) of the upper eyelid; it is innervated by the dorsal ramus of the oculomotor nerve (Baumel, 1975; Slonaker, 1918). M. depressor palpebrae ventralis inserts on the proximal edge of the fibrous tarsus (**Org. sens.** Annot. 37). of the lower eyelid; according to most investigators, it is innervated by a branch from the mandibular nerve (Baumel, 1975; Hofer, 1950; Lakjer, 1926; Starck and Barnikol, 1954; Watanabe and Yasuda, 1970). According to Slonaker (1918), M. depressor palpebrae ventralis is innervated by the oculomotor nerve.

According to Lakjer (1926), M. tensor periorbitae and M. depressor palpebrae ventralis are phylogenetically derived from M. levator bulbi and from the muscle primordium in the mandibular arch. M. tensor periorbitae originates on the interorbital septum, principally on the orbitosphenoid bone, approximately at the same level as the Musculi bulbi. Insertion is on the laterocaudal aspect of the periorbita on a level with the insertion of M. depressor palpebrae ventralis into the lower eyelid. Both M. tensor periorbitae and M. depressor palpebrae ventralis contribute to the formation of a muscular sling on the ventral aspect of the eyeball and thereby support the eye (Berger, 1966). In addition, these thin sheets of muscle also separate the orbit and its contents from jaw muscle (principally M. adductor mandibulae externus, M. pseudotemporalis superficialis, and M. protractor pterygoidei et quadrati). See Baumel (1975), Fujioka (1963), Hofer (1950), Starck and Barnikol (1954), and Watanabe and Yasuda (1970) for additional information on anatomical relationships and innervation in birds generally.

According to Slonaker (1918, especially Fig. 4), M. levator palpebrae dorsalis and M. depressor palpebrae ventralis are well developed in *Passer*; no mention is made of M. tensor periorbitae. In addition, however, he has described a few nonstriated muscle fibers which are arranged approximately parallel to the free margins of the eyelids. These nonstriated muscle fibers represent M. orbicularis palpebrarum. From his illustration, the muscle fibers are fewest in the upper eye-lid and seem to be within the dorsal marginal folds. In the lower eyelid, the muscle fibers are most numerous and occupy a considerable part of the area in front of the fibrous tarsus, separated from it by lymphatic spaces. The innervation of M. orbicularis palpebrarum is not mentioned by Slonaker although he presumes that the muscle is a functional sphincter of the eyelids. If the muscle fibers are nonstriated, then M. orbicularis palpebrarum is most likely a dermal muscle. For this reason, it is best listed as an accessory structure of the eye; see **Org. sens.** Organa accessoria oculi.

In *Crypturellus*, *Buteo*, *Cepphus*, *Aprosmictus*, and *Corvus*, Lakjer (1926) illustrated a fleshy slip which he identified as M. pseudotemporalis bulbi. Although he did not further describe this slip, he did indicate that some fibers of M. pseudotemporalis superficialis (Annot. 19) attain a secondary attachment to the bulb, presumably to the outer aspect of the periorbita near the insertion of M. tensor periorbitae. Lakjer may have misinterpreted some superficial muscle fibers of M. pseudotemporalis superficialis which were adherent to the fascial sheath surrounding N. maxillo-mandibularis and Rete mirabile ophthalmicum (A. ophthalmica externa) (Baumel, 1975) which lie between jaw muscles (M. adductor mandibulae externus and M. pseudotemporalis superficialis) and the contents of the orbit proper. Hofer (1950) also doubted the existence of M. pseudotemporalis bulbi and suggested a similar explanation.

(17) **Musculi mandibulae** (Fig. 3). A collective term for those muscles which are involved in closing and opening the jaws. They are most conveniently described on the basis of the arrangement of their fasciculi to their constituent aponeuroses (Gans and Bock, 1965), of their position relative to the facial and craniofacial articulations (Bock, 1968; **Arthr.** Annot. 45, 46), and of their consequential role in kinetics of the avian skull (Bock, 1964; Richards and Bock, 1973; Zweers, 1974). With the exception of M. depressor mandibulae (Annot. 24), they are derived from the muscle primordium of the mandibular arch and are innervated by N. mandibularis. The terminology is based on Lakjer (1926).

(18) **M. adductor mandibulae externus.** Synonymy: M. masseter superficialis, medius and profundus (Fujioka, 1963). A muscle complex rather than a single muscle, consisting of three general parts: pars rostralis (=superficialis, Lakjer, 1926); pars ventralis (=media, Lakjer, 1926), and pars caudalis (= profunda, Lakjer, 1926). Pars rostralis is also subdivided in most species (Bock, 1964; Bock and Morioka, Burton, 1974a, b; Richards and Bock, 1973; see also Hofer, 1950). At least three principal aponeuroses (Starck and Barnikol, 1954), or several more (Zweers, 1974), have also been described as a basis for organization of the muscle; see **Osteo.** Annot. 44.

(19) **M. pseudotemporalis superficialis.** Synonymy: M. adductor mandibulae medius (Dubale, 1969; Rawal, 1971); M. temporalis (Fjuioka, 1963; Watanabe and Yasuda, 1970). M. pseudotemporalis superficialis has been considered the most lateral (at its origin) of the so-called "adductor mandibulae internus" complex of muscles (including Mm. pseudotemporalis profundus, adductor mandibulae caudalis, pterygoideus and ethmomandibularis). In some birds, M. pseudotemporalis superficialis is said to be separated from M. adductor mandibulae externus, pars rostralis, by N. maxillaris and branches of N. mandibularis. See Hofer (1950), Starck and Barnikol (1954), Annot. 20, 21, 22; and **Osteo.** Annot. 45 and Fig. 1.

(20) **M. pseudotemporalis profundus.** Synonymy: M. quadratomandibularis (Hofer, 1950).
M. adductor mandibulae caudalis. Synonymy: M. adductor mandibulae "posterior" of many authors; not to be confused with a more rostrolateral slip of M. adductor mandibulae externus, i.e. pars caudalis (see Annot. 18).

M. pseudotemporalis profundus + M. adductor mandibulae caudalis. Synonymy: quadratomandibularis (Fujioka, 9163); M. pseudotemporalis (Dubale, 1969; Rawal, 1971). (Annot. 20 cont. on p. 190).

M. pseudotemporalis profundus is functionally independent from M. adductor mandibulae caudalis (Bock, 1964). Both are equivalent to a "quadrate" portion of a so-called "adductor mandibulae internus" muscle complex (Annot. 19) since they have a principle attachment on the Quadratum. M. pseudotemporalis profundus is sometimes separated from M. adductor mandibulae caudalis by a fibrous gap through which passes a branch of the mandibular nerve (N. pterygoideus, Lakjer, 1926); M. pseudotemporalis profundus is absent in the parrots and reduced or absent in some other birds (Burton, 1974c).

(21) **M. pterygoideus.** Synonymy: M. adductor mandibulae internus (Dubale, 1969; Rawal, 1971). M. pterygoideus is the "palatine (= pterygoid)" portion of an "adductor mandibulae internus" complex (see Annot. 19, 20), and is a muscle complex itself rather than a single muscle. It has been subdivided into: (1) pars dorsalis and pars ventralis, each of which has a lateralis and medialis component (Lakjer, 1926); (2) three principal aponeuroses and pars lateralis, pars ventralis medialis, pars dorsalis medialis (Burton, 1974a, c; Zusi, 1962), or (3) pars ventralis, pars dorsalis lateralis, pars dorsalis medialis, and pars retractor, the latter a separate slip which attaches on the brain case (Lamina basiparasphenoidalis, **Osteo.** Annot. 96; Richards and Bock, 1973).

(22) **M. ethmomandibularis.** Apparently derived from M. pterygoideus but characteristic of only the Psittaciformes, in which it is an important part of the distinctive system of feeding structures of this avian group (Burton, 1974c).

(23) **M. protractor pterygoidei et quadrati.** Synonymy: M. craniopterygoquadratus (Fujioka, 1963); M. sphenopterygoquadratus (Dubale, 1969; Rawal, 1971). M. protractor pterygoidei et quadrati is best considered as one muscle having two parts, defined primarily on the basis of their respective insertions, namely on Os pterygoideum and Quadratum.

(24) **M. depressor mandibulae** (Figs. 2, 3). Synonymy: M. occipitomandibularis (Fujioka, 1963); M. diagastricus (Berger, 1966). This muscle is innervated by N. facialis (see Annot. 17), and may consist of more than one part, each distinct and separable, with different functional roles (Bock, 1964; Burton, 1974a, b; Zusi, 195a, 1962; Zweers, 1974). Bühler (1970; and pers. comm.) suggests that in the Caprimulgidae an internal head, which is fused with Lig. occipitomandibulare (**Arthr.** Annot. 37; **Osteo.** Annot. 85), may not be part of this muscle complex. The structure of M. depressor mandibulae appears to be correlated with the size and shape of Proc. retroarticularis and Proc. mandibulae medialis of the mandible on which it inserts. See **Osteo.** Annot. 49, 51; and **Arthr.** Annot. 37.

(25) **Musculi apparatus hyobranchialis.** Collective term for a group of muscles which are functionally related in that their action effects movement of the hyobranchial apparatus (see **Osteo.** Annot. 79) although the muscles may be of diverse developmental origin, namely, from the mandibular, hyomandibular, and/or branchial muscle plates, and from occipital myotomes (Engels, 1938; Kallius, 1905). They are also described as "tongue muscles" (Richards and Bock, 1973).

(26) **M. intermandibularis ventralis** (Figs. 2, 3). Synonymy: M. mylohyoideus (anterior) (Bock, 1972; Bock *et al.*, 1973; Burton, 1974a, b; Ghetie and Atanasiu, 1962; Mudge, 1903; Zusi and Storer, 1969); M. intermandibularis, pars rostralis

(Dubale and Rawal, 1965; Goodman and Fisher, 1962; Zweers, 1974). M. intermandibularis ventralis is the most superficial and rostral of the two muscles and the most frequent in occurrence. The fleshy fibers from either side typically unite in a median ventral raphe and its fascial sheath may be continuous with that of M. serpihyoideus (see Annot. 27) or with M. constrictor colli, pars intermandibularis (see Annot. 5), or both. Apparently, M. intermandibularis ventralis may also be secondarily divided by other topographical relationships; see "M. mylohyoideus anterior" and "posterior" in Bock *et al.* (1973).

M. intermandibularis dorsalis. Synonymy: M. mylohyoideus anterior, "anterior" belly (Mudge, 1903; see also Annot. 5); M. hyomandibularis transversus (Kallius, 1905); M. suspensor hyoideus (Ghetie and Atanasiu, 1962). M. intermandibularis dorsalis is derived from a horizontal splitting of the muscle primordium of the mandibular arch (Kallius, 1905) and lies dorsal to M. intermandibularis ventralis, in part or entirely covered by the latter. This dorsal component may also insert with its counterpart on a median ventral raphe, but separate from that of M. intermandibularis ventralis (Zweers, 1974); it may insert directly on Os basibranchiale caudale (see **Osteo.** Annot. 81) or on a cartilaginous or bony nodule ("sesamoid plate, Evans, 1969, Fig. 5–9) which articulates with the hyobranchial skeleton (Mudge, 1903). See Fig. 3. Both muscles are innervated by the intermandibularis ramus of N. mandibularis (Baumel, 1975) which also innervates M. interceratobranchialis (Annot. 30).

(27) **M. serpihyoideus** (Figs. 2, 3). Synonymy: M. articulohyoideus, caudal slip (Harvey *et al.*, 1968); M. basibranchialis mandibularis, pars medialis (McLelland, 1968); M. digastricus (Burt, 1930); M. gularis posterior (Dubale and Rawal, 1965); M. hyomandibularis medialis (Fujioka, 1963; Kallius, 1905); M. intermandibularis dorsalis (Rawal, 1970, 1971); M. mylohyoideus posterior, medial part (Gadow and Selenka, 1891; Mudge, 1903); M. stylohyoideus, pars ventralis (Ghetie and Atanasiu, 1962).

The origin of the prefix "serpi" is unknown but undoubtedly is derived from the Latin verb "serpo", i.e. "to creep", from which is derived "serpigo", any creeping or serpiginous eruption that extends with an arciform border. In this sense, the prefix may refer to: (1) the topographic position of this muscle as it develops by a medial invagination between M. intermandibularis (see Annot. 26), M. interceratobranchialis (see Annot. 30), and the hypoglossal musculature, or (2) its variable proximal attachments (origin; Bock, pers. comm.), or (3) its arciform rostral border in some species. In any case, the name is in common use in many recent papers (Bock, 1972, 1973; Burton, 1974a, b; Zusi and Storer, 1969; Zweers, 1974) and elsewhere in the literature (Fisher and Goodman, 1955; Goodman and Fisher, 1962; Weymouth *et al.*, 1964).

M. serpihyoideus and M. stylohyoideus (Annot. 28) are innervated by the hyomandibular ramus of N. facialis (Baumel, 1975).

(28) **M. stylohyoideus.** Synonymy: M. articulohyoideus, cranial slip (Harvey *et al.*, 1968); M. basibranchialis mandibularis, pars lateralis (McLelland, 1968); M. gularis anterior (Dubale and Rawal, 1965); M. hyomandibularis lateralis (Fujioka, 1963; Kallius, 1905); M. mylohyoideus posterior, lateral part (Gadow and Selenka, 1891; Mudge, 1903); M. stylohyoideus, pars dorsalis (Ghetie and Atanasiu, 1962). In terms of its embryonic development from the muscle primordium of the hyoid arch, Kallius (1905) indicated that it was a homologue of the same muscle in mammals, and that M. stylohyoideus and M. serpihyoideus are innervated by the hyomandibular ramus of N. facialis. M. stylohyoideus is the name in general use in many recent papers

(Bock, 1972, 1973; Burton, 1974a, b; George, 1962; Zusi and Storer, 1969). See Figs. 2, 3.

(29) **M. branchiomandibularis.** Synonymy: M. ceratomandibularis (Fujioka, 1963; Kallius, 1905); M. geniohyoideus (Berger, 1966; Ghetie and Atanasiu, 1962; Zweers, 1974); M. mandibularis epibranchialis (McLelland, 1968). A complex muscle in some avian groups (Psittaciformes, Mudge, 1903), and derived from the muscle primordium of the first branchial arch (third visceral), innervated by N. glossopharyngeus. See Figs. 2, 3.

(30) **M. interceratobranchialis.** Synonymy: M. ceratohyoideus (Bock, 1972; Zweers, 1974); M. interceratoideus (Fujioka, 1963) or interkeratoideus (Kallius, 1905). Considered to be derived from the mandibular muscle primordium (Kallius, 1905) although topographically it is distinctly separated from M. intermandibularis and is associated primarily with the hypoglossal musculature. It is innervated by the intermandibularis ramus of N. mandibularis (Baumel, 1975). See Figs. 2, 3.

(31) **M. ceratoglossus.** Synonymy: M. ceratoentoglossus (Harvey *et al.*, 1968); M. ceratoglossus posterior (Engels, 1938); M. ceratohyoideus (Kallius, 1905); M. hyoglossus (Dubale and Rawal, 1965; Rawal, 1970); M. paraglossoceratobranchialis (McLelland, 1968; Vanden Berge, 1975). There may be one or more fleshy muscular slips (M. ceratoglossus anterior, Burton, 1974a, b, c; Zusi and Storer, 1969) arising from the insertion (tendon) on Os entoglossum. M. dermoglossus (Fisher and Goodman, 1955) also represents one of the muscular slips. See Figs. 2, 3.

(32) **M. hypoglossus rostralis [medialis].** Synonymy: M. ceratoglossus anterior (Engels, 1938); M. depressor glossus (Harvey *et al.*, 1968); M. hypoglossus (or hyoglossus) anterior, medialis, rectus or rostralis (Bock *et al.*, 1973; Burton, 1974a, c; Fisher and Goodman, 1955; Fujioka, 1963; Kallius, 1905; Mudge, 1903; Zweers, 1974); M. paraglossobasibranchialis medialis (McLelland, 1968; Vanden Berge, 1975). See Figs. 2, 3.

(33) **M. hypoglossus obliquus [lateralis].** Synonymy: M. basientoglossus (Harvey *et al.*, 1968); M. depressor glossus (Ghetie and Atanasiu, 1962); M. hyoglossus lateralis (Fujioka, 1963) or obliquus (Zweers, 1974); M. paraglossobasibranchialis lateralis (McLelland, 1968; Vanden Berge, 1975). The adjective, "lateralis", refers to the anatomical position of the muscle relative to M. hypoglossus rostralis [medialis] (Annot, 32); the transverse or oblique-transverse orientation of the muscle fasciculi is the basis for the term "obliquus" (see "hypoglossus posterior" in George, 1962, for some types of morphological variation). See Figs. 2, 3.

(34) **M. mesoglossus.** A muscle considered to be peculiar to Psittaciformes (Mudge, 1903) although Burton (1974c) suggests that it may be homologous with M. hypoglossus rostralis in other birds.

(35) **M. genioglossus.** Synonymy: M. geniohyoideus- genioglossus (Dubale and Rawal 1965; Engels 1938; Kallius 1905); M. geniothyreoideus (Leiber 1907); M. basioglossus (Ghetie and Atanasiu 1962).

M. genioglossus has its origin on Pars symphysialis of the mandible (**Osteo.** Annot. 42) but the insertion is somewhat variable. There is typically (1) an attachment on the hyobranchial skeleton or on the fascia of associated muscles and (2) on the connective tissue underlying the oral mucosa lateral to the tongue. Occasionally, the insertion extends as far caudally as the cricoid cartilage. See Bock, 1972; Bock *et al.*, 1973; Burton, 1974a, b, and Mudge, 1903 for descriptions of this muscle in various species. M. geniohyoideus, as given in synonymy, is not equivalent to a muscle of the same name described by others (see synonymy in Annot. 29).

(36) **M. cricohyoideus.** Synonymy: M. basibranchialis laryngeus (McLelland, 1968); M. laryngohyoideus (Fujioka, 1963). Many authors use the term M. thyrohyoideus (=thyreohyoideus) despite the fact that a thyroid cartilage is not present in the avian larynx. M. cricohyoideus takes origin from the cricoid cartilage and inserts on Os basibranchiale rostrale (**Osteo.** Annot. 81). At both attachments, but especially at its origin, it may interdigitate with the Mm. tracheales complex (see Annot. 37) and Figs. 2, 3.

In some species of birds, M. cricohyoideus is subdivided into multiple fleshy slips, at least one of which is M. thyroglossus of Mudge (1903). According to Burton (1974c), M. cricohyoideus is the more frequent in occurrence among birds generally; other ventral and ventrolateral fleshy slips, including "M. thyroglossus", probably have arisen independently in several groups of birds. His recommendation to consider all such slips as subdivisions of M. cricohyoideus is adopted here.

(37) **Musculi tracheales.** A collective term for a superficial group of ventral cervical muscles which have attachment on the sternum, clavicle, skin, trachea, larynx, hyobranchial apparatus, and associated membranes and/or ligaments. Burton (1974a) noted variation in the attachments of one such muscle ("M. cleidohyoideus") within a single species (*Calidris alpina*) and other evidence in the literature suggests that there is apparently no consistent pattern within major avian groups (see "M. sternohyoideus" in Mudge, 1903). One or more of the several muscles of this complex may interdigitate with M. cricohyoideus, Mm. laryngeales, or Mm. syringeales, and occasionally with certain hyobranchial muscles. See Figs. 2, 3, and **Resp.** Annot. 45.

According to Edgeworth (1935), these muscles are said to be derived from a primitive, reptilian M. rectus cervicis system (= sternohyoideus system, Gadow and Selenka, 1891; Mudge, 1903; =posthyoidean hypobranchial spinal muscles, (Engels, 1938) which is comprised of an "external fasciculus" (="M. sternohyoideus") and an "internal fasciculus" (="M. cleidohyoideus", consisting of "M. sternotrachealis" and "M. tracheohyoideus"). The entire complex is said to be innervated by N. hypoglossocervicalis;; however, this is an unresolved point. See **PNS** Annot. 29, 59; Lockner and Youngren, 1976; Watanabe, 1964).

Gadow and Selenka (1891) and Mudge (1903) described "M. sternohyoideus" as an independent muscle of this complex in at least some birds. The so-called "internal fasciculus" however, seems to have been retained in all birds which have been investigated, and apparently gave rise to the major primordium of the tracheal muscles. However, there appears to be considerable overlap in the use of descriptive names for these muscles and interpretation of homology is sometimes difficult.

As mentioned elsewhere (**Resp.** Annot. 45), there is one general characteristic of the attachment of these tracheal muscles, namely, on the upper airway tract and specifically on the trachea. M. sternotrachealis is typically attached to the Process craniolateralis of the sternum (**Osteo.** Annot. 150). The essential morphological characteristic of M. cleidotrachealis, and the basis for the name, is an origin on the clavicle. M. tracheolateralis is discussed as a tracheal muscle (**Resp.** Annot. 45) and as a syringeal muscle (**Resp.** Annot. 46). M. sternohyoideus is attached to the larynx and to the first few millimeters of the trachea, at least in *Gallus* (McLelland, 1965).
(Annot. 37 cont. on p. 194).

In Table I, a list of selected "tracheal" muscles is presented with mention of their general (principal) attachments and appropriate references in which such muscles are described. It seems reasonable to assume that any of these other named muscles are homologs of one of the four muscles listed officially.

(38) **M. dilator glottidis.** Synonymy: M. laryngeus superficialis (White, 1975); M. thyreoarytenoideus (Bock, 1972). M. dilator glottidis is an intrinsic muscle of the larynx superficial in relation to M. constrictor glottidis (Annot. 39). Functionally the muscle is a dilator of the glottis as demonstrated in *Gallus* by electrical stimulation (White, 1975). See **Resp.** Annot. 30.

(39) **M. constrictor glottidis.** Synonymy: M. laryngeus profundus (White, 1975). M. constrictor glottidis is a deep intrinsic muscle of the larynx, to a large extent concealed by the fibers of M. dilator glottidis (Annot. 38). It is subdivided into separate slips in *Gallus* (White, 1975) and in *Corvus* (Bock, 1978); in both genera, and in other birds, it is functionally a constrictor of the glottis. See **Resp.** Annot. 30.

(40) **Musculi syringeales.** See **Resp.** Annot. 46 for an extensive review of the variation in the syringeal muscles in passerine and non-passerine birds. See also **PNS** Annot. 30.

(41) **Musculi colli.** Most recent studies of the cervical muscles (Fisher and Goodman, 1955; Palmgren, 1949; Zusi and Storer, 1969) adopt the terminology of Boas (1929). Kuroda (1962) used a slightly different set of names. In general, the Sub-Committee recommends that authors follow the suggestion of Zusi and Storer (1969, p. 27), namely, "in complex serially arranged muscles, we regard the functional units to be those fibers inserting on a given vertebra rather than those originating from a vertebra".

(42) **M. complexus.** Commonly known as the "hatching muscle" (Fisher, 1966) even though its specific contractile properties and other internal changes may not be directly related to the actual "pipping" of the shell (Bock and Hikida, 1968, 1969; Brooks and Garrett, 1970; Hays and Hikida, 1976).

(43) **M. splenius capitis.** Descriptive aspects, variability, and general occurrence of this muscle in many groups of birds are summarized by Burton (1971). It may be equivalent to either M. obliquus capitis cranialis or M. rectus capitis dorsalis major, or both, of Fujioka (1963).

(44) **M. rectus capitis dorsalis.** Synonymy: M. rectus capitis superior (Boas, 1929; Burton, 1974a, b; Palmgren, 1949; Zusi, 1962); M. trachelomastoideus (Fujioka, 1963).

(45) **Mm. iliocostalis et longissimus dorsi.** In mammals these two sets of epaxial muscle are typically separable from each other; in birds such a separation is not so readily apparent.

(46) **M. cervicalis ascendens; M. thoracicus ascendens.** Synonymy: Mm. ascendentes. These are derived from the iliocostalis–longissimus system and have an anatomical position somewhat lateral to M. longus colli dorsalis. The oblique orientation of these muscles is the basis for the synonym, M. obliquus colli (Gadow and Selenka, 1891;

Kuroda, 1962). Mm. obliquotransversales are said to be present in some birds (Harvey *et al.*, 1968; Shufeldt, 1890); most authors since Boas (1929) consider them to be a deeper component of Mm. ascendentes.

(47) **M. longus colli dorsalis.** At least four separate muscles (Annot. 48–51) form this muscle complex which apparently constitutes an avian spinalis muscle system. Kuroda (1962) described it as one muscle complex, M. longus colli posticus. The several parts are variously joined together and/or interdigitate with other cervical muscles, particularly M. cervicalis ascendens and Mm. intertransversarii.

(48) **M. longus colli dorsalis, pars cranialis.** Synonymy: pars anterior (Kuroda, 1962); M. splenius colli (Boas, 1929; Zusi, 1962; Zusi and Storer, 1969).

(49) **M. longus colli dorsalis, pars caudalis.** Synonymy: pars posterior (Kuroda, 1962); M. spinalis cervicis (Boas, 1929; Zusi, 1962; Zusi and Storer, 1969).

(50) **M. longus colli dorsalis, pars profunda.** Synonymy: Mm. dorsales pygmaei (Boas, 1929; Zusi, 1962); M. profundus colli posticus (Kuroda, 1962). May be of inconstant occurrence in avian species generally (see Zusi and Storer, 1969, for example).

(51) **M. longus colli dorsalis, pars thoracica.** Synonymy: M. spinalis thoracis (Boas, 1929; Zusi, 1962; Zusi and Storer, 1969).

(52) **Mm. intercristales.** Intersegmental muscles which typically extend from the Crista transverso-obliqua and/or Proc. dorsalis of one cervical vertebra (**Osteo.** Annot. 124, 125) to those of the next cervical vertebra in the series. What appear to be more or less discrete medial slips connecting adjacent spinous processes of the vertebrae are considered to be Mm. interspinales. The latter are not as clearly developed (see Zusi and Storer, 1969) as Mm. intercristales.

(53) **Mm. intertransversarii.** Synonymy: M. colli lateralis (Kuroda, 1962). Intersegmental multipennate muscles connecting transverse processes of adjacent vertebrae to form the principal lateral musculature of the neck (Zusi and Storer, 1969).

(54) **Mm. inclusi.** Essentially intersegmental muscles often described as medial slips from Mm. intertransversarii. Dorsal slips (from the transverse process to the vertebral arch of consecutive vertebrae) and ventral slips (from the transverse process to the body of consecutive vertebrae) constitute one muscle system.

(55) **M. flexor colli lateralis; M. flexor colli medialis.** Lateralis (=brevis) and medialis (=profundus) are somewhat more descriptive than the alternate terms although the latter are frequently used (Boas, 1929; Burton, 1974b; Zusi, 1962; Zusi and Storer, 1969).

M. flexor colli lateralis lies ventral to and in part concealed by M. rectus capitis dorsalis. These two muscles may be difficult to separate in dissection because of their close anatomical relationship (Burton, pers. comm.)

M. flexor colli medialis appears to be a continuation of certain muscular components of M. longus colli ventralis (see Zusi and Storer, 1969). Nevertheless, it is usually described as an independent muscle.

(56) **M. longus colli ventralis.** The major muscle lying on the ventral surface of the cervical vertebrae, essentially a muscle complex consisting of a series of muscular slips and aponeuroses of origin and insertion. It lies between the two sets of Mm. intertransversarii and the attachments on the first few cervical vertebrae interdigitate with similar attachments of M. flexor colli lateralis and M. flexor colli medialis; see also **Osteo.** Annot. 119.

(57) **Musculi trunci.** Collective term for the most frequently described muscles of the thoracic and abdominal walls (see **Topog. anat.** Annot. 31). See also **PNS** Annot. 43 and de Wet (1967) for information on innervation of the trunk muscles and George and Berger (1966) for illustrations of abdominal musculature.

M. quadratus lumborum was described in *Rhea americana* by Gadow and Selenka (1891), but this muscle is apparently greatly reduced (vestigial) or absent in birds generally.

(58) **M. scalenus.** de Wet *et al.* (1967) recognize a pars cranialis and pars caudalis in *Gallus*.

(59) **Mm. intercostales externi.** Harvey *et al.* (1968) refer to special muscular slips (Mm. intercostales superficiales) which are attached to the uncinate processes of the ribs. These appear to be additional slips of the usual external intercostal muscles.

(60) **M. costosternalis.** Preferred over other terms which apparently refer to the same muscle, namely, "subcostalis", "triangularis sterni", "transversus thoracis", and other names. According to de Wet *et al.* (1967), Pars major is functionally distinct from Pars minor (inspiratory vs. expiratory, respectively), in *Gallus*.

(61) **M. costoseptalis.** Synonymy: M. costopulmonaris (de Wet *et al.*, 1967) Striated muscle associated with the Septum horizontale. See King (1966, 1975); and **Pericar.** Annot. 2, 3.

(62) **M. sternocoracoideus.** Usually described in relationship with muscles of the pectoral girdle (and wing) although it probably has no significant functional role in terms of wing movement.

(63) **M. rectus abdominis.** The aponeuroses of the oblique and transverse abdominal muscles do not form a fibrous sheath over M. rectus abdominis such as that of mammals. What has been called a tendinous intersection opposite the last short ribs is actually a vestige of yolk sac resorption (Table 23b, Fig. 8, Gadow and Selenka, 1891). See George and Berger (1966), pp. 184–185, for illustrations of the abdominal muscles of the pigeon and chicken.

(64) **Musculi caudae.** Terminology for the muscles of the tail is based on Baumel (1971): certain descriptive aspects are given in Vanden Berge (1975). See **PNS** Annot. 47, 48 and Baumel (1975) for innervation of tail muscles.

M. levator caudae. Synonymy: M. levator coccygis + M. levator caudae (Berger, 1966, and many others). See Fig. 4.

M. lateralis caudae. Synonymy: M. levator rectricum (Harvey *et al.*, 1968). See Fig. 4.

M. levator caudae and M. lateralis caudae comprise the dorsal, intrinsic (axial) musculature of the tail. M. levator caudae is a single muscle mass having two slightly distinct parts, "levator coccygis" and "levator caudae" (see Berger, 1966). Both parts are enclosed in a common sheath of dense deep fascia and fasciculi of both parts blend with one another.

(65) **M. depressor caudae.** Synonymy: M. depressor coccygis (Gadow and Selenka, 1891; Porta, 1908; Liebe, 1914); M. lateralis coccygis (Berger, 1966).

M. depressor caudae is the ventral counterpart of M. levator caudae in general structure, position and function, and is named accordingly. A cruciate aponeurosis of insertion (see Baumel, 1971) serves as an attachment for the ventral, extrinsic musculature of the tail, in particular for M. pubocaudalis internus and M. caudofemoralis. Certain extrinsic muscles of the tail have been named "depressor caudae (=coccygis)"; these, however, originate on the caudal margin of the pelvic girdle. See synonymy in Annot. 66.

(66) **M. pubocaudalis externus.** Synonymy: M. pubococcygeus externus (Gadow and Selenka, 1891; Liebe, 1914; Porta, 1908); M. depressor caudae (Berger, 1966); M. depressor rectricum (Harvey *et al.*, 1968). See Fig. 4.

M. pubocaudalis internus. Synonymy: M. pubococcygeus internus (Gadow and Selenka, 1891; Liebe, 1914; Porta, 1908); M. depressor coccygis (Berger, 1966); M. caudalis lateralis (Harvey *et al.*, 1968). See Fig. 4.

M. pubocaudalis externus, M. pubocaudalis internus, and M. caudofemoralis (Annot. 110) comprise the extrinsic ventral musculature of the tail. The name "pubocaudalis externus" or "internus" is particularly appropriate because it conveys information concerning (1) the principal origin (pubis; caudal margin of pelvis), (2) insertion (on the tail) and (3) topographic position of either of these two muscles. Both participate in depression (ventral flexion) of the tail; M. pubocaudalis externus, in particular, is a functional counterpart to M. lateralis caudae (Annot. 64) in this regard.

M. pubocaudalis internus sometimes consists of a ventral and dorsal belly separated by a tendinous intersection. This tendinous intersection on one side is continuous with its counterpart on the opposite side by way of a strong, horizontal, membranous structure which Baumel (1971) has called the "supracloacal septum". The septum intervenes between the ventral surface of M. depressor caudae and the dorsal surface of the cloaca, cloacal bursa, and distal ends of the ureters and genital ducts. The septum is continuous with the deep fascia of M. sphincter cloacae (Annot. 68). For these anatomical reasons Baumel (1971) postulates an important functional component of M. pubocaudalis internus directly influencing the shape and the functional activity of the cloaca.

(67) **M. bulbi rectricium; M. adductor rectricium.** M. bulbi rectricium is a striated muscle which ensheaths a well-organized fibro-adipose mass known as the rectrical bulb. M. adductor rectricium, on the other hand, is a straited muscle associated primarily with the inner aspect of the ventral elastic ligament of the rectrices; it supplies striated slips of insertion to rectrical follicles. See Baumel (1971) for a detailed description.

(68) **M. transversus cloacae.** Synonymy: M. transversus analis (=transversoanalis) (Berger, 1966; Gadow and Selenka, 1891); M. transversus perinei (Harvey *et al.*, 1968); M. sphincter cloacae (Vanden Berge, 1975). See Fig. 4.

M. sphincter cloaceae. Synonymy: M. suspensor ani (Porta, 1908); M. eversor urodeum (Harvey *et al.*, 1968); M. retractor penis caudalis (=posterior) (King, 1975; Liebe, 1914). See Fig. 4. (Annot. 68 cont. on p. 198).

M. levator cloacae [M. retractor phalli] (see **Genit. masc.** Annot. 47). M. transversus cloacae and M. levator cloacae are extrinsic striated muscles associated with the wall of the proctodeum of the cloaca. M. sphincter cloacae is an intrinsic striated muscle of the cloacal wall, extending into the dorsal and ventral lips bordering the external opening (see **(Cloaca** Annot. 21, 22).

M. transversus cloacae may consist of one or two separate heads of origin and/or separate bellies in certain groups of birds, e.g. Apodiformes (Zusi, pers. comm.; Vanden Berge, pers. obs.). The muscle originates on the caudal margin of the pelvis lateral (superficial) to the origin of M. pubocaudalis externus and M. pubocaudalis internus. The muscle fibers of insertion interdigitate with M. sphincter cloacae to form a nearly continuous muscle mass (see Gadow and Selenka, 1891; Leibe, 1914; Klemm *et al.*, 1973).

M. levator cloacae originates on the ventral aspect of the tail somewhat between the insertions of M. pubocaudalis externus and the dorsal belly of M. pubocaudalis internus (Annot. 66). The belly (or tendon) crosses the tendinous intersection of M. pubocaudalis internus and enters the ventral wall of the proctodeum, passing deep to M. sphincter cloacae.

According to Komárek (1969), M. levator cloacae [M. retractor phalli] inserts near the deeply invaginated "root" (Pars extrema phalli, **Genit. masc.** Annot. 42) of the copulatory organ in the duck; see also Liebe, 1914. In *Gallus*, the muscle inserts on the median phallic body (Corpus phallicum medianum, **Genit. masc.** Annot. 29). For these reasons, M. levator cloacae has also been called "M. retractor penis caudalis (=posterior)". A specific functional role in retracting the everted phallus, following detumescence, has not been conclusively demonstrated; see details in King (1975).

In addition to M. transversus cloacae (including M. sphincter cloacae) and M. levator cloacae, a "M. constrictor cloacae" and "M. dilatator cloacae" have also been described. M. dilatator cloacae, in particular, has been described more often as "M. retractor penis cranialis (=anterior)". The latter is said to insert more laterally in the cloacal wall (Liebe, 1914), or near the lymphatic folds (Plicae lymphaticae, **Cloaca** Annot. 18) in *Gallus* and *Meleagris* (King, 1975). Both M. constrictor cloacae and M. dilatator cloacae are most probably fleshy slips which are derived from M. pubocaudalis internus, especially from the ventral belly (see Anot. 66). They are of infrequent occurrence and presumably have only a minor functional role when present.

(69) **M. rhomboideus superficialis.** Sometimes given the name M. trapezius (e.g. Fujioka, 1959), although Fürbringer (1886, 1902) suggested that M. rhomboideus superficialis and M. rhomboideus profundus are derived from the same muscle primordium, and that M. serratus profundus also may be a derivative. There is no firm evidence that M. trapezius occurs in birds. M. rhomboideus superficialis may consist of a Pars clavicularis and Pars scapularis. See Fig. 5.

(70) **M. serratus superficialis** (Fig. 5). Pars cranialis and Pars caudalis are typically present, often nearly continuous proximally, but each part has a separate insertion on the scapula. Since M. serratus profundus exists, "superficialis" should be used. A separate, striated, subcutaneous slip, Pars metapatagialis of M. serratus superficialis, is typically present (Annot. 11) and has been studied in terms of its microstructure (Hikida, 1972; Hikida and Bock, 1974, 1976).

M. serratus profundus. Usually consists of one to several fasciculi which insert along a continuous line on the scapula.

(71) **M. scapulohumeralis cranialis.** Also described as M. proscapulohumeralis (see Berger, 1966, 1968). Both Mm. scapulohumeralis cranialis and caudalis have been considered homologs of the teres muscles in mammals (Fukioka, 1959; Sullivan, 1962).

(72) **Mm. subcoracoscapulares.** A collective name for a muscle complex which consists of four heads in many birds (Fürbringer, 1902). Most recent authors describe two separate muscles, M. subscapularis, typically having two heads, and M. subcoracoideus which may or may not have two heads (see Berger, 1966).

(73) **M. subscapularis.** Typically has two well defined heads, often named Pars externa and Pars interna (Hudson and Lanzillotti, 1964). Substitution of the terms Caput laterale and Caput mediale, for the Pars externa and Pars interna, respectively, is somewhat more appropriate since there are two heads of one muscle which unite to form a common tendon of insertion. See **Osteo.** Annot. 187.

(74) **M. coracobrachialis cranialis.** According to Sullivan (1962), this muscle should be designated "M. coracobrachialis" whereas M. deltoideus minor (Annot. 80), represents his "M. coracobrachialis anterior". Since the names M. coracobrachialis cranialis (or "anterior") and M. deltoideus minor have long been in use for well-known muscles a substitution of the same name for very different muscles would be particularly inappropriate. In some birds, the muscle occupies a distinct triangular fossa on the cranial aspect of Crista pectoralis of the humerus (**Osteo.** Annot. 184, 186).

(75) **M. pectoralis** (Figs. 5, 6). Simić and Andrejevic (1963, 1964) recognize Pars sternobrachialis and Pars thoracobrachialis, separated by an extension of the aponeurosis of insertion ("membrana intermuscularis") in such a way that the muscle appears to have a modified bipennate form. Both parts are probably characteristic of this muscle in birds generally, but are not necessarily the same as other subdivisions which have been described (Fisher, 1946; Kuroda, 1960, 1961a, b; Vanden Berge, 1970).

Similarly, the fleshy and/or aponeurotic connections of M. pectoralis with M. tensor propatagialis (Annot. 78) are often described as Pars propatagialis of M. pectoralis (Annot. 11). There is considerable variation in the relative development of these connections (see Berger, 1966).

M. pectoralis abdominis of Berger (1966), Hudson *et al.* (1964, 1969, 1972) and others may consist of a single, striated, subcutaneous muscle (M. pectoralis, pars subcutanea thoracica and pars subcutanea abdominalis) (see Annot. 11). Considerable variation may exist, as among the Ciconiiformes. Pars subcutanea thoracica is present in the Ardeidae; Pars subcutanea abdominalis is present or absent. Both subcutaneous slips are absent in *Balaeniceps*, Ciconiidae, and Threskiornithidae. Both are present, however, in the Phoenicopteridae (Vanden Berge, 1970).

(76) **M. supracoracoideus.** For synonymy see Vanden Berge (1975). The so-called "deep head" of M. supracoracoideus (Berger, 1966; Hudson and Lanzillotti, 1964) is most probably a derivative of M. deltoideus minor as originally suggested by Fürbringer (M. deltoideus minor, pars ventralis, 1902) and substantiated by Sullivan (M. coracobrachialis anterior, pars ventralis, 1962); see Fig. 5 and Annot. 80. For relationships to the Canalis triosseus and the attachments of the muscle, see **Osteo.** Annot 153, 174, 177, 182, 183; and **Arthr.** Annot. 88.

(77) **M. latissimus dorsi** (Fig. 5). Pars cranialis and Pars caudalis arise by cleavage in a single muscle primordium (Sullivan, 1962), although each part typically has a separate insertion on the humerus (see **Osteo.** Linea m. latissimi dorsi). Presumbably the striated subcutaneous slips, Pars interscapularis and Pars metapatagialis arise from the same primordium. Considerable variation exists with respect to the presence or absence of one or the other of the skeletal slips in avian species generally, and in either or both of the subcutaneous slips themselves (Berger, 1966, 1969). Certain structural characteristics of the skeletal slips have been described by Hikida and Bock (1971).

(78) **M. tensor propatagialis.** Synonymy: M. propatagialis longus et brevis, of numerous authors. M. tensor propatagialis is a structural complex in which both Pars longa and Pars brevis may be separately developed, or in which only a single belly is present; for examples, see Berger, 1966, 1969; Raikow, 1977; Vanden Berge, 1970, and others. In either case, there is present a long tendon to the wrist and a short tendon to the proximal antebrachium. Tendo longa inserts on the "elastic tendon" (Lig. elasticum propatagiale, **Arthr.** Annot. 141; see Oakes and Bialkower 1977), situated on the cranial edge of the propatagial skin fold; it may enclose a sesamoid near the insertion (Bock and McEvey, 1969). Tendo brevis is often structurally subdivided into multiple tendons proximal to its insertion (see Annot. 87; Figs. 5, 6; also **Arthr.** Annot. 194). See Fürbringer (1888); Hudson *et al.* (1969); Raikow (1977); and Vanden Berge (1970) for examples of morphological variation.

(79) **M. deltoideus major** (Fig. 5). A cranial head of origin, associated with Fibrocartilago humerocapsularis (**Arthr.** Annot. 100), and a caudal head, associated with bones of the shoulder joint, have been described in certain birds (Berger, 1966, 1969; Raikow, 1977). In other birds, the muscle has a single belly with only a suggestion of internal division. As Berger (1966) has mentioned, there is considerable variation in the number of heads present, relative development of these heads, presence or absence of a "scapular anchor" (retinaculum), and the extent of insertion on the humerus (see **Osteo.** Annot. 184).

(80) **M. deltoideus minor, caput ventrale.** Synonymy: M. supracoracoideus (Fujioka, 1959, and others; see synonymy in Vanden Berge, 1975). Caput ventrale is a separate, ventral, fleshy slip arising from Membrana sternocoracoclavicularis (**Arthr.** Annot. 86) within Canalis triosseus (**Osteo.** Annot. 177), and inserting more or less in common with Caput dorsale. Caput ventrale is distinct in several avian groups (Hudson *et al.*, 1969; Vanden Berge, 1970) (see also Berger, 1966, Hudson and Lanzillotti, 1964; Annot. 76; and Fig. 5).

(81) **M. scapulotriceps.** This muscle is anchored to the humerus (Retinaculum m. scapulotricipitis, **Arthr.** Annot. 142; and Lig. tricipitale, **Arthr.** Annot. 111) and in some birds to the scapula (**Osteo.** Annot. 169) (see Berger, 1966, Hudson *et al.*, 1969, 1972; Vanden Berge, 1970). Os sesamoideum m. scapulotricipitis (**Osteo.** Annot. 102) is present in the tendon of insertion of some species. See Figs. 5, 6.

(82) **M. humerotriceps.** The proximal origin on the humerus (see **Osteo.** Annot. 188) may be partially subdivided into "caput mediale", "caput posticum" and "caput breve" by the insertion of other muscles (M. scapulohumeralis cranialis; M. scapulohumeralis caudalis; M. latissimus dorsi) (Buri, 1900), but these are not usually considered important functional subdivisions. See Fig. 6.

(83) **M. coracotriceps.** Synonymy: Caput coracoideum m. anconei (Fürbringer, 1886). Presence or absence of this vestigial muscle in birds generally is reviewed by Berger (1966, M. anconeus coracoideus). Although it is present in some genera of ciconiiforms (Vanden Berge, 1970), it is apparently of infrequent occurrence, presumably on account of its vestigial nature. When present this muscle is typically associated with the proximal tendon of M. expansor secundariorum.

(84) **M. pronator superficialis; M. pronator profundus.** Alternate terms, "brevis" = superficialis and "longus" = profundus, do not necessarily apply since the so-called "pronator longus" is not always the longer of the two muscles. See Fig. 6.

(85) **M. flexor carpi ulnaris.** Tendon of origin is typically associated with a bony process on humerus **(Osteo.** Annot. 195) and "humero-ulnar pulley", Trochlea humeroulnaris, pars tendinea **(Arthr.** Annot. 110), and may enclose a sesamoid.
Pars cranialis is a fusiform belly, closely invested by the intermuscular Septum humerocarpale **(Arthr.** Annot. 195) which is derived from the "humerocarpal band", Lig. humerocarpale **(Arthr.** Annot. 143).
Pars caudalis is represented by a series of fleshy slips obliquely oriented to an insertion on Lig. elasticum m. flexoris carpi ulnaris (**Arthr**. Annot. 197).
Both pars cranialis and pars caudalis insert on Os carpi ulnare. See Fig. 6; also Berger (1966) and Buri (1900).

(86) **M. flexor digitorum superficialis.** Formerly known as M. flexor digitorum sublimis; it appears to be a superficial derivative of a common muscle primordium from which is also derived M. flexor digitorum profundus and M. flexor carpi ulnaris (Sullivan, 1962). This muscle demonstrates considerably anatomical variation in birds generally; at its origin, it is typically associated with Lig. humerocarpale (see Fig. 6; **Arthr.** Annot. 143; and review in Berger, 1966).

(87) **M. extensor metacarpi radialis** (Figs. 5, 6). Synonymy: M. extensor carpi radialis. According to Berger (1966) there is typically a dorsal and a ventral head of origin from the Humerus (see **Osteo.** Annot. 196), and one or two tendons of insertion on Proc. extensorius of Os metacarpale alulare (see **Osteo.** Annot. 216). Tendo brevis of M. tensor propatagialis generally inserts on the tendon of origin and/or fascial sheath of Caput dorsale as well as on the Aponeurosis antebrachialis dorsalis (see **Arthr.** Annot. 194).

(88) **M. extensor metacarpi ulnaris.** A part of the extensor group of muscles on the basis of its origin, position and nerve supply, even though it has secondarily assumed a function of flexion of the wrist and flexion of the elbow joint. It has often been named M. extensor carpi ulnaris although "metacarpi" is more descriptive of its true insertion (see **Osteo.** Annot. 220). See Berger (1966) for review of synonymy and also the occurrence and function of the "ulnar anchor" or retinaculum **(Arthr.** Annot. 144, 145). See Fig. 5.

(89) **M. extensor longus alulae.** Synonymy: M. extensor pollicis longus; M. extensor longus digiti II. See Fig. 5.

(90) **M. extensor longus digiti majoris** (Figs. 5, 6). Synonymy: M. extensor indicis longus; M. extensor longus digiti III. Pars distalis is topographically located in the manus proper and when present it unites with the tendon of the proximal or main

head. This distal head has been designated as M. flexor metacarpi brevis (Fisher and Goodman, 1955; see also Berger, 1966) but, according to Sullivan (1962) Pars distalis ("M. extensor medius brevis") is derived from M. extensor longus digiti major ("M. extensor medius longus"). See also **Arthr.** Annot. 145.

(91) **M. ectepicondylo-ulnaris** (Fig. 5). Synonymy: M. anconeus (Berger, 1966, and many others). The muscle has its origin on the Epicondylus dorsalis ("ectepicondyle") of the humerus (see **Osteo.** Annot. 194); its anatomical position is more or less opposite that of M. entepicondylo-ulnaris (Annot. 92). Although "M. anconeus" has been more widely used for this muscle, the name "anconeus" also has been used for M. triceps brachii; this precludes the continued use of M. anconeus for M. ectepicondylo-ulnaris.

(92) **M. entepicondylo-ulnaris.** This muscle is also known as M. anconeus medialis (Fujioka, 1959) and, to many avian anatomists, as the "gallinaceous" muscle. It is present only in the kiwi (*Apteryx*), in the Tinamiformes (Hudson *et al.*, 1972), in the Galliformes (Hudson and Lanzillotti, 1964), and in the Anatidae (Sy, 1936). Beddard (1884) described an "anconeus internus" in *Scopus* which suggests M. entepicondylo-ulnaris, but this has never been substantiated. The origin of this muscle is on Epicondylus ventralis ("entepicondyle") of the Humerus. See Annot. 91; and **Osteo.** Annot. 193.

(93) **M. interosseous ventralis** (Fig. 6). Synonymy: M. interosseus palmaris or volaris.

(94) **M. extensor brevis alulae.** Synonymy: M. extensor pollicis brevis; M. extensor brevis digiti II. See Fig. 5.

(95) **M. abductor alulae** (Fig. 6). Synonymy: M. abductor pollicis; M. abductor digiti II.

(96) **M. flexor alulae** (Fig. 6). Synonymy: M. flexor pollicis; M. flexor digiti II.

(97) **M. adductor alulae** (Figs. 5, 6). Synonymy: M. adductor pollicis; M. adductor digiti II.

(98) **M. abductor digiti majoris.** Synonymy: M. abductor indicis; M. abductor digiti III. See Fig. 6.

(99) **M. flexor digiti minoris.** Synonymy: M. flexor digiti III or IV. See Figs. 5, 6.

(100) **Mm. iliotibiales.** According to Romer (1927), the superficial portion of the dorsal muscle mass in the developing thigh is differentiated into a cranial, preacetabular, or preaxial mass from which is derived Mm. iliotibialis cranialis and lateralis, M. ambiens, and **Mm. femorotibiales.** A caudal, postacetabular, or postaxial mass gives rise to M. iliofibularis.

(101) **M. iliotibialis cranialis.** Synonymy: M. sartorius (Hudson, 1937); M. extensor iliotibialis anterior (Fisher, 1946). See Figs. 7, 10.

M. iliotibialis medialis. A medial muscle, described only in flamingos, originally as "an extra muscle in the thigh, associated with the sartorius" (Vanden Berge, 1970) and later by this formal name (Vanden Berge, 1976).

(102) **M. iliofibularis** (Figs. b, g, 9). Synonymy: M. biceps femoralis (Hudson, 1937, and later co-workers); M. extensor iliofibularis (Fisher, 1946; Klemm, 1969). Together with Mm. flexor cruris lateralis (see Annot. 108) and medialis (see Annot. 109) these three muscles have been considered homologues of the mammalian "hamstring" muscles. However, which of these three is the counterpart of which mammalian muscle is anything but clear (see synonymy for each in Vanden Berge, 1975). The tendon of M. iliofibularis typically traverses a fibrous loop (Ansa m. iliofibularis) and inserts on Tuberculum m. iliofibularis of Corpus fibulae. See **Arthr.** Annot. 186; and **Osteo.** Fig. 13.

(103) **M. ambiens** (Fig. 10). This muscle is designated "Am" in the Garrod formula (see Intro.).

(104) **Mm. iliotrochanterici** (Figs. 8, 9). Gadow and Selenka (1891) considered this muscle group to be peculiar to birds and without homology in mammals. Others, following Fisher (1946), refer to M. iliotrochantericus caudalis as "gluteus profundus", to M. iliotrochantericus cranialis as "iliacus", and retain the name M. iliotrochantericus medius. M. iliotrochantericus medius is designated "C" in the Garrod formula. Attachments on the lateral aspect of Trochanter femoris are detailed by Ballmann (1969b) (**Osteo.** Annot. 256, 257). See **Myol.** Intro. regarding the Garrod formula muscles.

(105) **M. iliofemoralis externus.** Synonymy: M. gluteus medius et minimus (Hudson, 1937); M. piriformis (Fisher, 1946). The original term, M. iliofemoralis externus, was the name given by Gadow and Selenka (1891) and, although it is infrequently used today, is highly descriptive of topographic relationships and implies no homology. It is designated by the letter "D" in the Garrod formula. See Intro. and Fig. 8.

(106) **M. iliofemoralis internus.** Synonymy: M. iliacus (Hudson, 1937); M. psoas (Fisher, 1946). M. iliofemoralis internus is the name originally proposed by Gadow and Selenka (1891). Garrod formula: "E". See Intro. and Fig. 10.

(107) **M. femorotibialis externus.** According to Berger (1966), the chief variations pertain to the origin of Pars proximalis and the presence or absence of a discrete Pars distalis. See Fig. 9.

(108) **M. flexor cruris lateralis.** Synonymy: M. caudilioflexorius (Gadow and Selenka, 1891; M. semitendinosus (Hudson, 1937). M. flexor cruris lateralis is the name proposed by Fisher (1946). The recognition of two parts has been known for many years; the distal part is often known as the "accessory semitendinosus". The designation, Pars pelvica, for the main belly suggests the fact that this arises primarily from the pelvis (ilium) although it may also arise from proximal caudal vertebrae. Pars accessoria might also be called pars femoralis; however, the femoral attachment is usually considered to be an insertion, not a second head of origin. It joins with the pelvic part in a distal raphe. See Figs. 7, 8, 9.

The two parts of this muscle have been designated as "X" and "Y" (see Intro.) in the Garrod formula. Pars pelvica (X) may be present without pars accessoria

(Y), but the latter never is present alone. In some major avian groups, both parts are absent (see Berger, 1966).

(109) **M. flexor cruris medialis.** Synonymy: M. ischioflexorius (Gadow and Selenka, 1891); M. semimembranosus (Hudson, 1937). The principal name is used by Fisher (1946). See Figs. 7, 8, 9.

(110) **M. caudo-ilio-femoralis.** Synonymy: M. piriformis (Hudson, 1937); the principal name, or slight variation of it, is common to Fisher (1946), Gadow and Selenka (1891), and Howell (1938). According to Romer (1927), this muscle appears to be a homolog of the coccygeofemorales muscle system in reptiles which was subject to considerable variation in the course of avian evolution during which the tail became reduced. M. caudofemoralis, then, represents an ancestral muscle whereas M. iliofemoralis is a secondarily derived portion whereby the muscle also gained access to the ilium (and ischium in some species). This is seen in the differentiation of both parts during development. See Figs. 7, 8, 9.

One or the other part, or both, may be absent and this presence or absence has long been considered in the technical diagnoses of avian taxa. Pars caudofemoralis is "A", Pars iliofemoralis is "B" in the Garrod formula. See **Myol.** Intro.

(111) **Mm. obturatorius lateralis and medialis** (Figs. 9, 10). Lateralis is sometimes known as "externus"; medialis, as "internus". Such alternate names suggest homology, but, according to Romer (1927), the obturator muscles differentiate from a common primordium which would appear to be the equivalent of M. obturatorius externus in mammals, not the internus. Substitution of the terms lateralis and medialis are consistent with the anatomical relationship between them (see **Osteo.** Annot. 227).

(112) **M. obturatorius lateralis.** Pars dorsalis and Pars ventralis are clearly defined parts of this muscle in the nine-primaried New World oscines (Passeriformes: see Raikow, 1976), and in some other avian groups (Berger, 1966, 1969).

(113) **M. pubo-ischio-femoralis** (Figs. 8, 9, 10). The synonym M. adductor, with several qualifying terms (longus et brevis, superficialis et profundus, longus et magnus) has been in common use. However, as Cracraft (1971) has demonstrated, the muscle probably has no significant functional role in adduction of the femur: it is more likely a postural muscle (see also Helmi and Cracraft, 1977). This would preclude the use of the term "adductor".

The name M. pubo-ischio-femoralis (Gadow and Selenka, 1891) is highly descriptive in terms of the origin (pubis and/or ischium) and insertion (femur) of this muscle. In nonpasserine species the two subdivisions are Pars lateralis and Pars medialis; in passerines, lateralis becomes cranialis and medialis, caudalis (see Raikow, 1976).

(114) **M. tibialis cranialis.** Caput femorale takes origin (Fovea tendinis) on Condylus lateralis of Femur (**Osteo.** Fig. 11); perforates or merely grooves Meniscus lateralis (**Arthr.** Annot. 159), and passes distally across Incisura tibialis (**Osteo.** Annot. 273). The tendon of insertion is restrained by a fibrous arch (Retinaculum extensorium tibiotarsi) (**Arthr.** Annot. 187) and inserts on the Tarsometatarsus (**Osteo.** Fig. 14). See Figs. 8, 10.

(115) **M. extensor digitorum longus** (Fig. 10). Origin in Sulcus intercristalis of Tibiotarsus (**Osteo.** Annot. 272); tendon restrained, in common with that of M.

tibialis cranialis (Annot. 114), by a fibrous arch (Retinaculum extensorium tibiotarsi, **Arthr.** Annot. 187). Distally, the tendon passes beneath the ossified "supratendinal bridge" (Pons supratendineus, **Osteo.** Annot. 227) across the intertarsal joint and then beneath the fibrous Retinaculum extensorium tarsometatarsi (**Arthr.** Annot. 188) which may sometimes be ossified (see **Osteo.** Annot. 287).

(116) **M. fibularis [peroneus] longus** (Fig. 7). According to Kurochkin (1968), it should be included in an expanded Garrod formula and designated by the symbol "M". Variation in the Mm. peroneus longus and brevis in a large number of birds was discussed by Mitchell (1913), although Hudson (1937) and Berger (1966) did not consider the wide variation in relative development of the two muscles significant in terms of the "muscle formulas".

The tendon of insertion grooves the proximal end of the Tarsometatarsus (see **Osteo.** Fig. 14) and attaches on the tibial cartilage by an aponeurosis (see **Arthr.** Fig. 8) and on the tendon of M. flexor perforatus digiti III.

(117) **M. fibularis [peroneus] brevis** (Fig. 7). Variably developed in different species and taxonomic groups of birds. In the New World nine-primaried oscines (Passeriformes), a Caput tibiale distinct from a Caput fibulare is said to be an important taxonomic character (Raikow, 1976). According to Kurochkin (1968), this muscle is designated "N" in an expanded Garrod formula. The tendon lies in a sulcus on the distal end of the tibiotarsus where it is restrained by a retinaculum (see **Osteo.** Annot. 282). Insertion is on the proximal end of the Tarsometatarsus.

(118) **M. gastrocnemius.** Typically consists of three parts identified as Pars lateralis (=externa), Pars intermedia and Pars medialis (=interna). These are best recognized by the term "Pars" rather than "Caput" since they are not merely separate heads of one muscle (see **Osteo.** Annot. 271), but may be entirely separate bellies, sharing in the formation of a common tendon of insertion (see **Arthr.** Annot. 164), without continuity of their respective contractile tissue. The presence or absence of accessory heads associated with Pars lateralis (Vanden Berge, 1970) or Pars *medialis* Raikow, 1970) in different species or groups of birds may be useful as taxonomic characters in future studies. See Figs. 7, 10.

(119) **M. plantaris.** No definitive homology with the mammalian muscle of this name has been established, but the presence or absence of this muscle (symbol "F" in the Garrod formula) has been a part of the technical taxonomic diagnosis of avian groups since the formula was first proposed (see **Myol.** Intro.).

(120) **M. popliteus.** Symbol "G" in the Garrod formula (see **Myol.** Intro.).

(121) **Mm. flexores perforantes et perforati digiti II et III.** Considered "intermediate flexors" in the sense that the tendons may pass through those of the "superficial flexors" (Mm. flexores perforati II et III) but are, in turn, perforated by the corresponding digital tendons of M. flexor digitorum longus. All of these long flexors, and M. flexor hallucis longus, pass through the tibial cartilage (**Arthr.** Annot. 164, Fig. 8) and Hypotarsus (**Osteo.** Annot. 288 and Fig. 14) to enter Sulcus flexorius (**Osteo.** Annot. 294), the exact method being very definite for different species (see Hudson, 1937; Vanden Berge, 1970). The tendons of the flexor muscles to the long toes also traverse Canalis flexorius plantaris (**Arthr.** Annot. 178 and Fig. 9). See Figs. 7, 8, 9.

(122) **Vinculum tendinum flexorum** (Figs. 9, 12). The fibrous band connecting the tendons of M. flexor perforans et perforatus digiti III and M. flexor perforatus digiti III in the foot is highly variable among birds generally (Hudson, 1937). It is designated "V" in the Garrod formula (see Intro.).

(123) **M. flexor hallucis longus.** Distinctly separate from M. flexor digitorum longus at its origin. The relationship between these two muscles and their tendons, distally, has been classified into eight different types (Berger, 1966). Such detail should be considered in descriptive studies. See Fig. 12.

(124) **Vinculum tendineum flexorum.** The fibrous band which unites the tendons of Mm. flexor hallucis longus and flexor digitorum longus. The tendons of these in muscles may be totally independent in some birds, partially or completely fused in other birds. See Fig. 12.

(125) **M. extensor hallucis longus.** Pars proximalis and Pars distalis are well defined in some birds (Hudson, *et al.*, 1959, for example), but the entire muscle is one of the more variable of the group of short muscles of the toes (see Berger (1966) and Raikow (1976)). Since there is but one short extensor of the hallux, the use of the qualifier, "longus", may seem unnecessary. However, Pars proximalis may represent a "longus" and Pars distalis, a "brevis"; the name as given here has been in common use. See Fig. 11.

(126) **M. flexor hallucis brevis.** The tendon of insertion may be perforated by that of M. flexor hallucis longus, wrapping sheath-like around the latter, or both tendons pass toward their respective insertions independent of each other.

(127) **M. abductor digiti II.** Functionally, this muscle probably extends the second digit (Cracraft, 1971). See Fig. 11.

(128) **M. extensor proprius digiti III** (Fig. 11). In most taxa of birds, this muscle is of rare and irregular occurrence. Hudson *et al.* (1972) have seen it in five living genera of ratites and in the tinamous (Tinamiformes). See Holmes (1962) for further information.

(129) **M. extensor brevis digiti IV.** Tendon of insertion passes through Canalis interosseous tendineus (**Osteo.** Annot. 298 and Fig. 14). See Fig. 11.

(130) **M. abductor digiti IV.** Present in most groups of birds (Berger, 1966), but often minute and identified only by staining techniques (Bock and Shear, 1972).

M. adductor digiti IV has also been described by Gadow and Selenka (1891) and by Hudson (1937), but it is apparently insignificant and extremely rare in occurrence.

TABLE 1 Musculi Tracheales

Muscle	Origin	Insertion	Selected References
M. cleidohyoideus (= M. cleidotrachealis?)	Clavicle	(1) Hyobranchial apparatus (2) Cricoid cartilage (3) Tracheal cartilages	Burton, 1974a; M. ypsilotrachealis, Lockner and Youngren, 1976
M. sternohyoideus	Sternum (Proc. craniolateralis)	(1) Hyobranchial apparatus (2) Larynx (3) Cranial tracheal cartilages	Gadow and Selenka, 1891; Mudge, 1903; McLelland, 1965
M. sternotrachealis	(1) Sternum (Proc. craniolateralis) (2) Coracoid and sternal ribs	(1) Cricoid cartilage (2) Trachea (3) Syrinx	Ames, 1971; Burton, 1974a; Lockner and Youngren, 1976
Mm. sternotracheolaryngeales M. sternotracheolaryngeus lateralis M. sternotracheolaryngeus medialis M. sternolaryngeus M. tracheolaryngeus dorsalis M. tracheolaryngeus ventralis	(1) Sternum (Proc. craniolateralis) (2) Trachea	(1) Cricoid cartilage (2) Trachea	McLelland, 1965, in *Gallus*
M. tracheohyoideus	(1) Trachea (2) Cervical fascia (3) Sternum? Clavicle?	(1) Hyobranchial apparatus (2) Cricoid cartilage	Gadow and Selenka, 1891; Bock and Morioka, 1971; Burton, 1974b
M. trachealis lateralis (= M. tracheolateralis)	(1) Tracheal cartilages (2) Syrinx	(1) Cricoid cartilage (2) Syrinx	Ames, 1971; Bock, 1972; Burton, 1974b; Lockner and Youngren, 1976

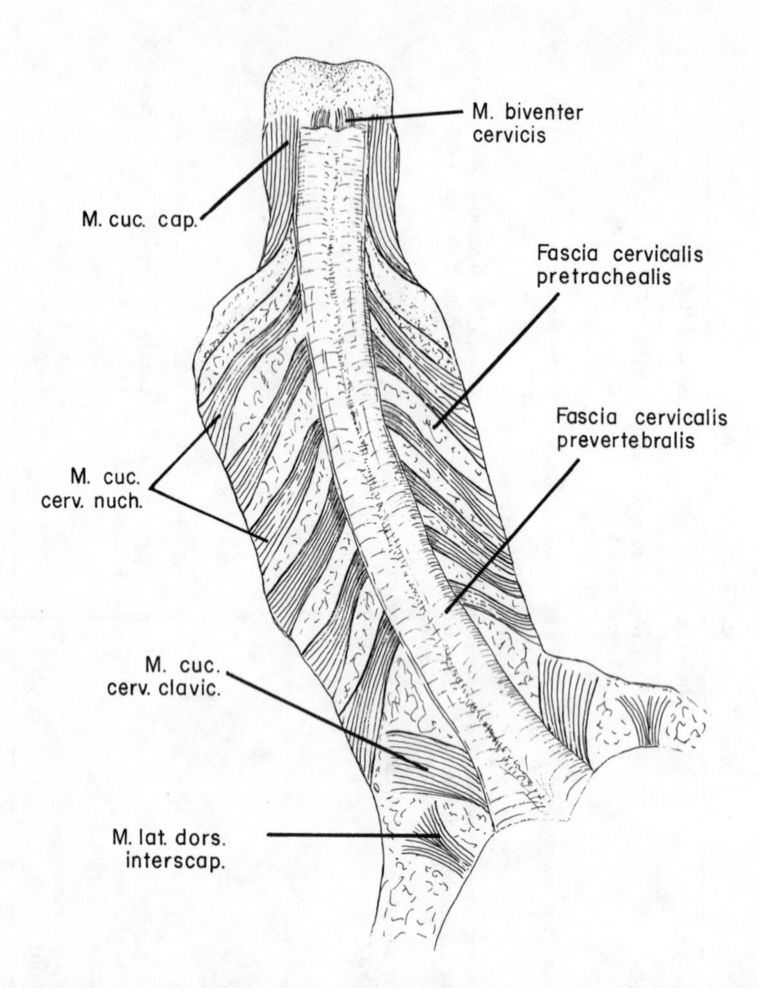

Fig. 1 Dorsal view of subcutaneous muscles of the neck; *Meleagris gallopavo*. Redrawn with modification from Ghetie, *et al.* (1976). M. cuc. cap., M. cucullaris capitis; M. cuc. cerv. nuch. or clavic., M. cucullaris cervicis, Pars nuchalis or Pars clavicularis; M. lat. dors. interscap., M. latissimus dorsi, Pars interscapularis.

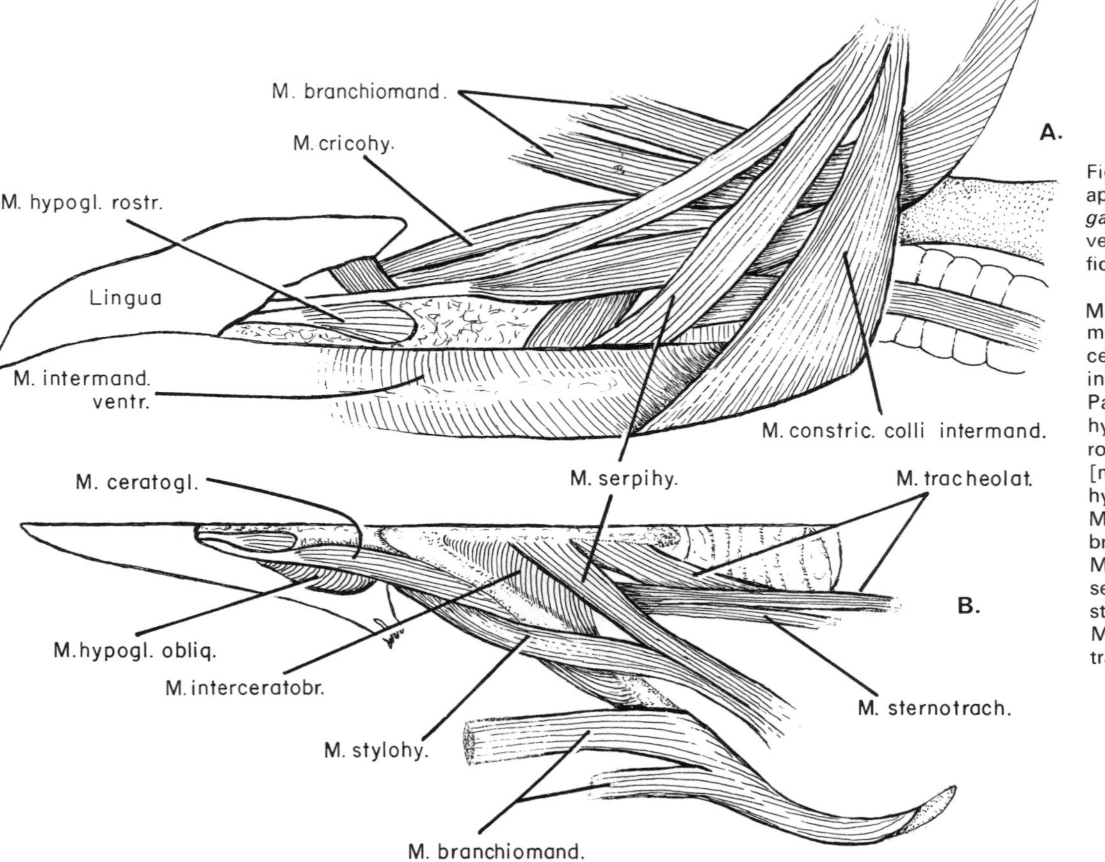

Fig. 2 Muscles of the hyoid apparatus and tongue; *Meleagris gallopavo*. A, lateral view; B, ventral view. Redrawn with modification from Ghetie, *et al.* (1976).

M. branchiomand., M. branchiomandibularis; M. ceratogl., M. ceratoglossus; M. constric. colli intermand., M. constrictor colli, Pars intermandibularis; M. cricohy., M. cricohyoideus; M. hypogl. rostr., M. hypoglossus rostralis [medialis]. N. hypogl. obliq., M. hypoglossus obliquus [lateralis]; M. interceratobr., M. interceratobranchialis; M. intermand. ventr., M. intermandibularis ventralis; M. serpihy., M. serpihyoideus; M. sternotrach., M. sternotrachealis; M. stylohy, M. stylohyoideus; M. tracheolat., M. tracheolateralis.

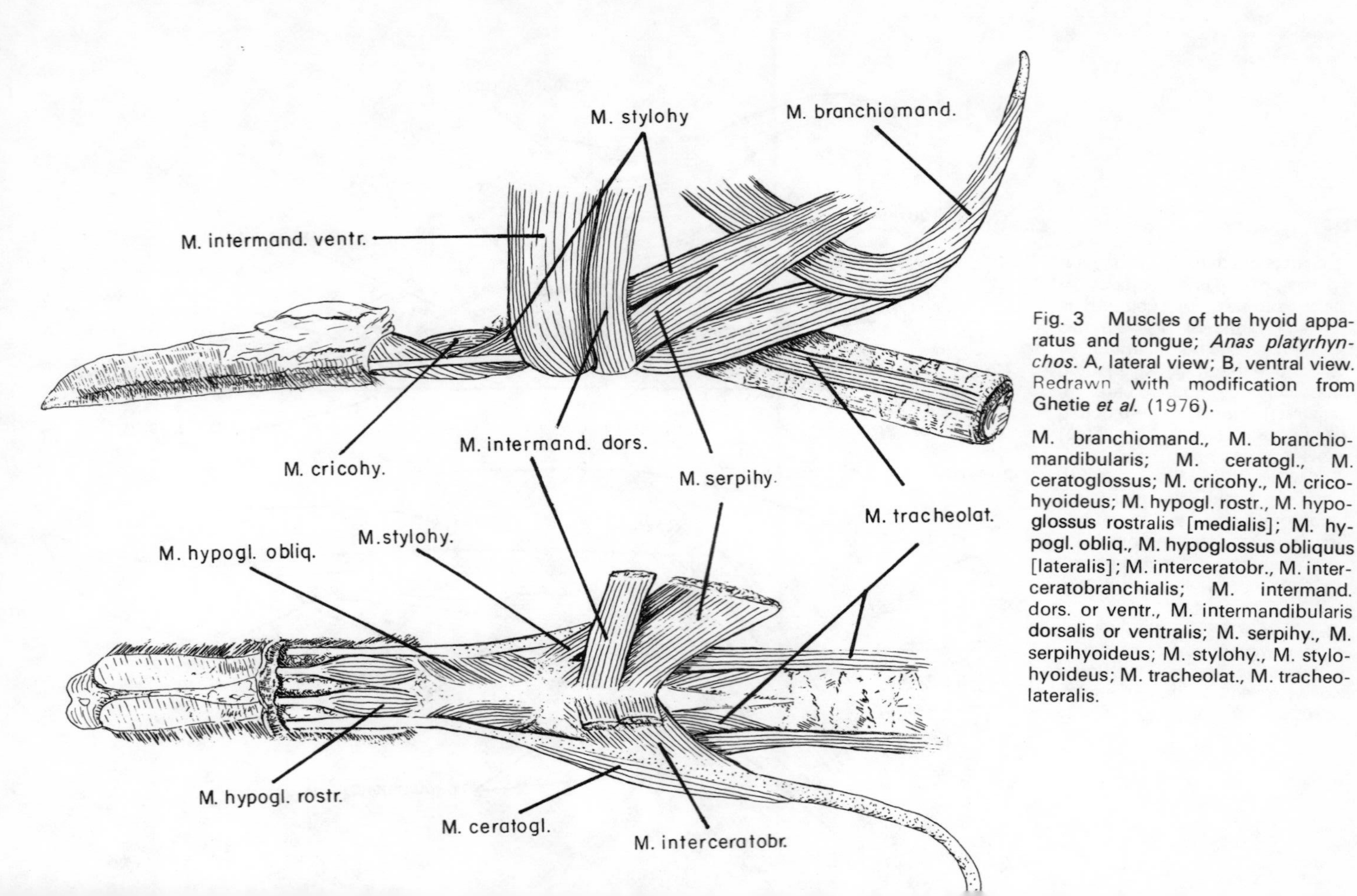

Fig. 3 Muscles of the hyoid apparatus and tongue; *Anas platyrhynchos*. A, lateral view; B, ventral view. Redrawn with modification from Ghetie *et al.* (1976).

M. branchiomand., M. branchiomandibularis; M. ceratogl., M. ceratoglossus; M. cricohy., M. cricohyoideus; M. hypogl. rostr., M. hypoglossus rostralis [medialis]; M. hypogl. obliq., M. hypoglossus obliquus [lateralis]; M. interceratobr., M. interceratobranchialis; M. intermand. dors. or ventr., M. intermandibularis dorsalis or ventralis; M. serpihy., M. serpihyoideus; M. stylohy., M. stylohyoideus; M. tracheolat., M. tracheolateralis.

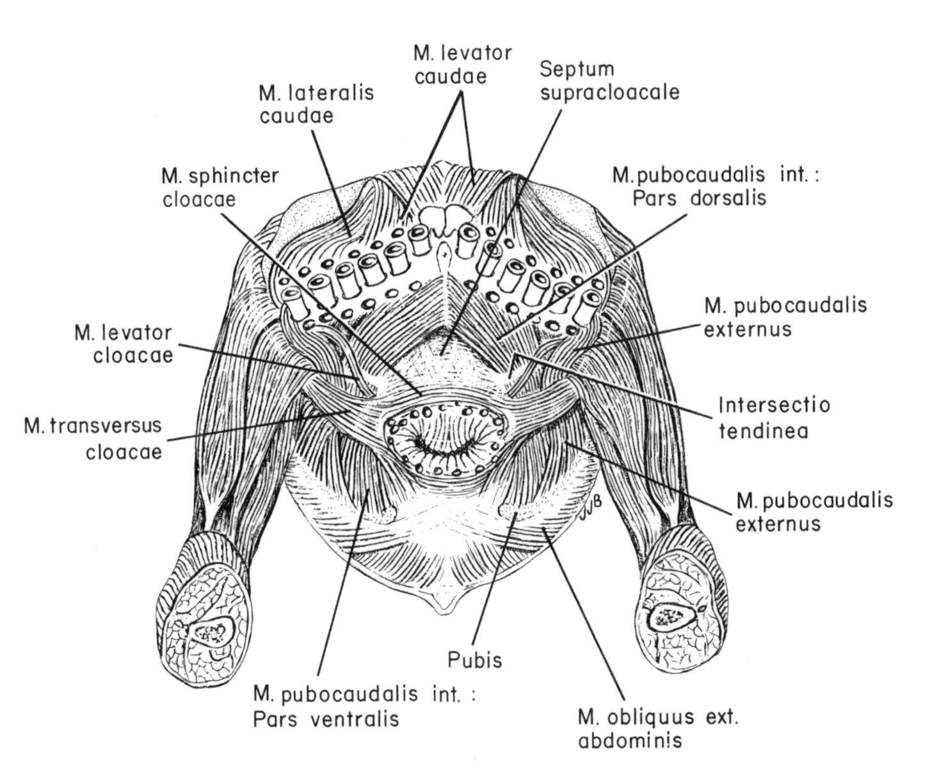

Fig. 4 Muscles of the tail of the pigeon, *Columba livia*. Caudal view. Original drawing, J. J. Baumel.

Tail partially elevated from the supracloacal septum (see Annot. 66) which is situated above the dorsal wall of the cloaca and the cloacal bursa. In this view the dorsal part of M. pubocaudalis internus covers the insertions of M. depressor caudae and M. caudofemoralis (not shown). Note the tendinous intersection that subdivides M. pubocaudalis internus in this species.

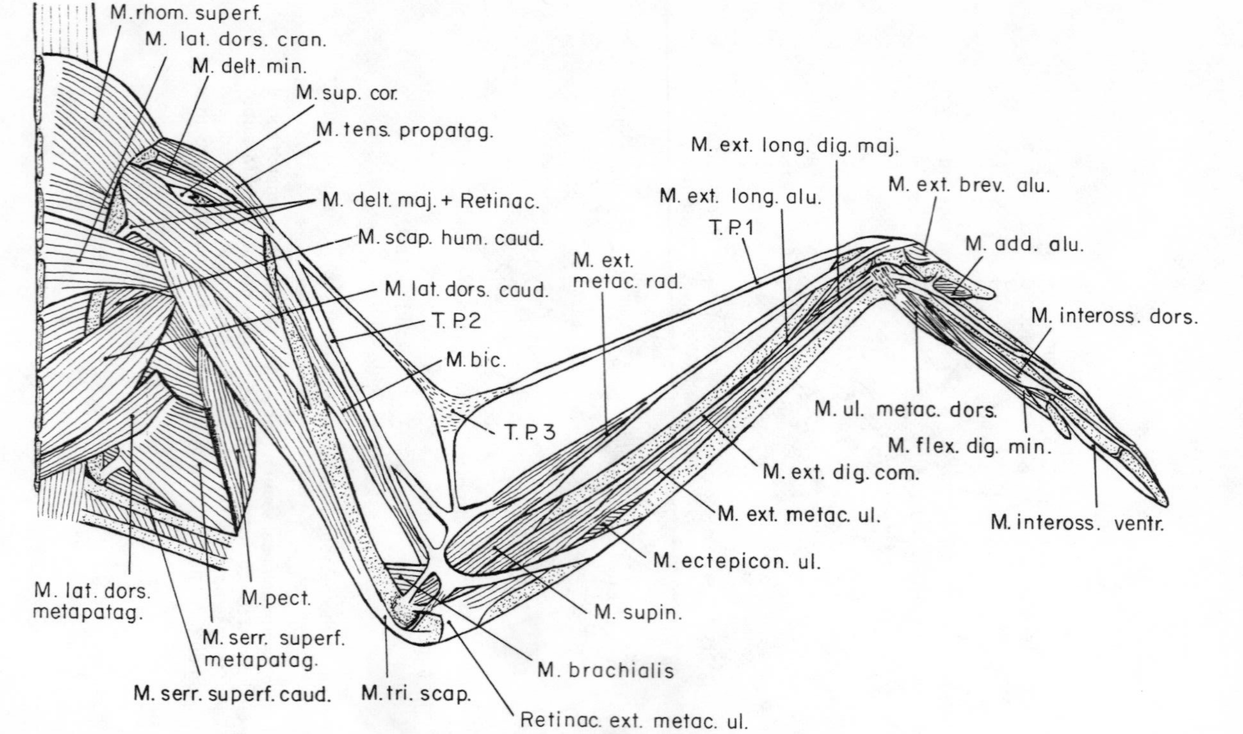

Fig. 5 Muscles of the wing of the night-heron, *Nycticorax nycticorax*. Dorsal view. From Vanden Berge (1970).

M. add. alu., M. adductor alulae; M. bic., M. biceps brachii; M. delt. min., M. deltoideus minor; M. delt. maj., M. deltoideus major and retinaculum; M. ectepicon. ul., M. ectepicondyloulnaris; M. ext. brev. alu., M. extensor brevis alulae; M. ext. dig. com., M. extensor digitorum communis; M. ext. long. alu., M. extensor longus alulae; M. ext. long. dig. maj., M. extensor longus digiti majoris (Pars proximalis); M. ext. metac. rad., M. extensor metacarpi radialis; M. ext. metac. ul., M. extensor metacarpi ulnaris; M. flex. dig. min., M. flexor digiti minoris; M. inteross. dors. or ventr., M. interosseus dorsalis or ventralis; M. lat. dors. cran. or caud. or metapatag., M. latissimus dorsi, Pars cranialis, Pars caudalis, Pars metapatagialis; M. pect. M. pectoralis; Retinac. ext. metac. ul., Retinaculum m. extensoris metacarpi ulnaris; M. rhom. superf., M. rhomboideus superficialis; M. scap. hum. caud., M. scapulohumeralis caudalis; M. serr. superf. caud. or metapatag., M. serratus superficialis, Pars caudalis or Pars metapatagialis and retinaculum; M. sup. cor., M. supracoracoideus; M. supin., M. supinator; M. tens. propatag

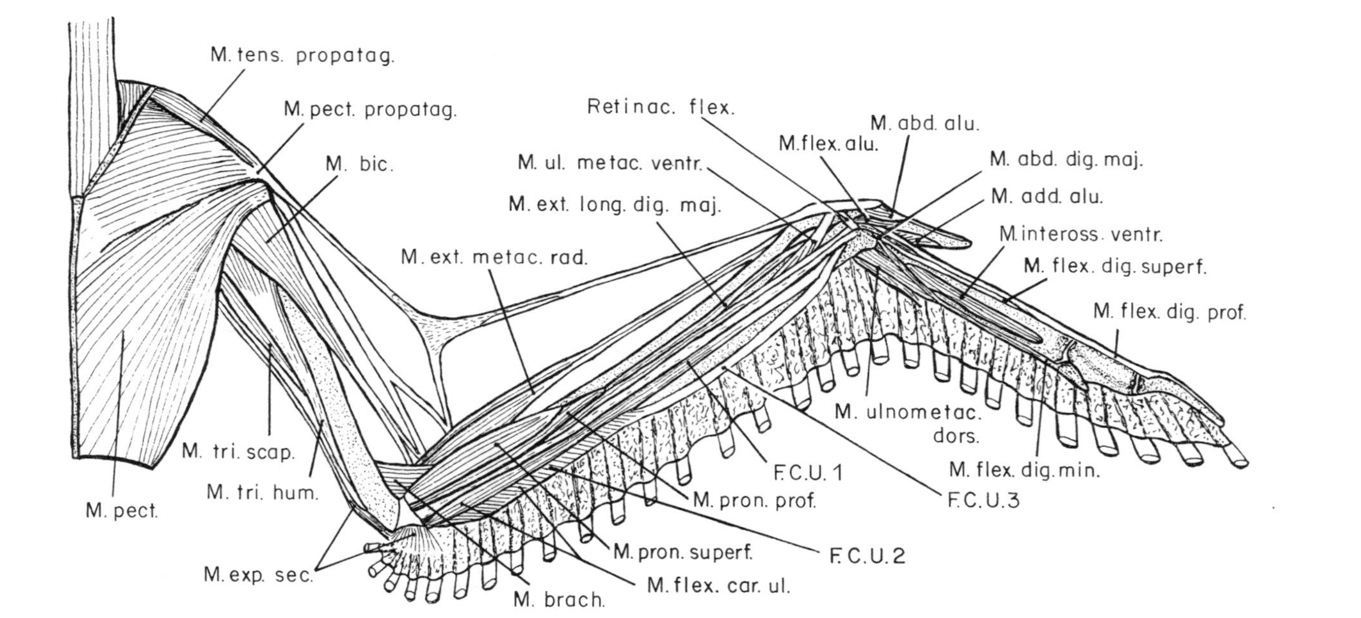

Fig. 6 Muscles of the wing of the night-heron, *Nycticorax nycticorax*. Ventral view. From Vanden Berge (1970).

M. abd. alu., M. abductor alulae; M. abd. dig. maj., M. abductor digiti majoris; M. add. alu., M. adductor alulae; M. bic., M. biceps brachii; M. brach., M. brachialis; M. exp. sec., M. expansor secundariorum, Tendo proximalis and Tendo distalis; M. ext. long. dig. maj., M. extensor longus digiti majoris; M. ext. metac. rad., M. extensor metacarpi radialis, Caput ventrale; M. flex. alu., M. flexor alulae M. flex. car. ul, M. flexor carpi ulnaris, F.c.u.1, Pars cranialis, F.c.u.2, Pars caudalis, F.c.u.3, Ligamentum elasticum m. flexoris carpi ulnaris; M. flex. dig. min., M. flexor digiti minoris; M. flex. dig. prof., M. flexor digitorum profundus; M. flex. dig. superf., M. flexor digitorum superficialis; M. inteross. ventr., M. interosseus ventralis; M. pect., M. pectoralis; M. pect. propatag., M. pectoralis, Pars propatagialis., M. pron. prof. or superf., M. pronator profundus or superficialis; Retinac. flex., Rectinaculum flexorum; M. tens. propatag; M. tensor propatagialis, M. tri, hum. or scap., M. humerotriceps or M. scapulotriceps; M. ulnometac. dors. M. ulnometacarpalis dorsalis.

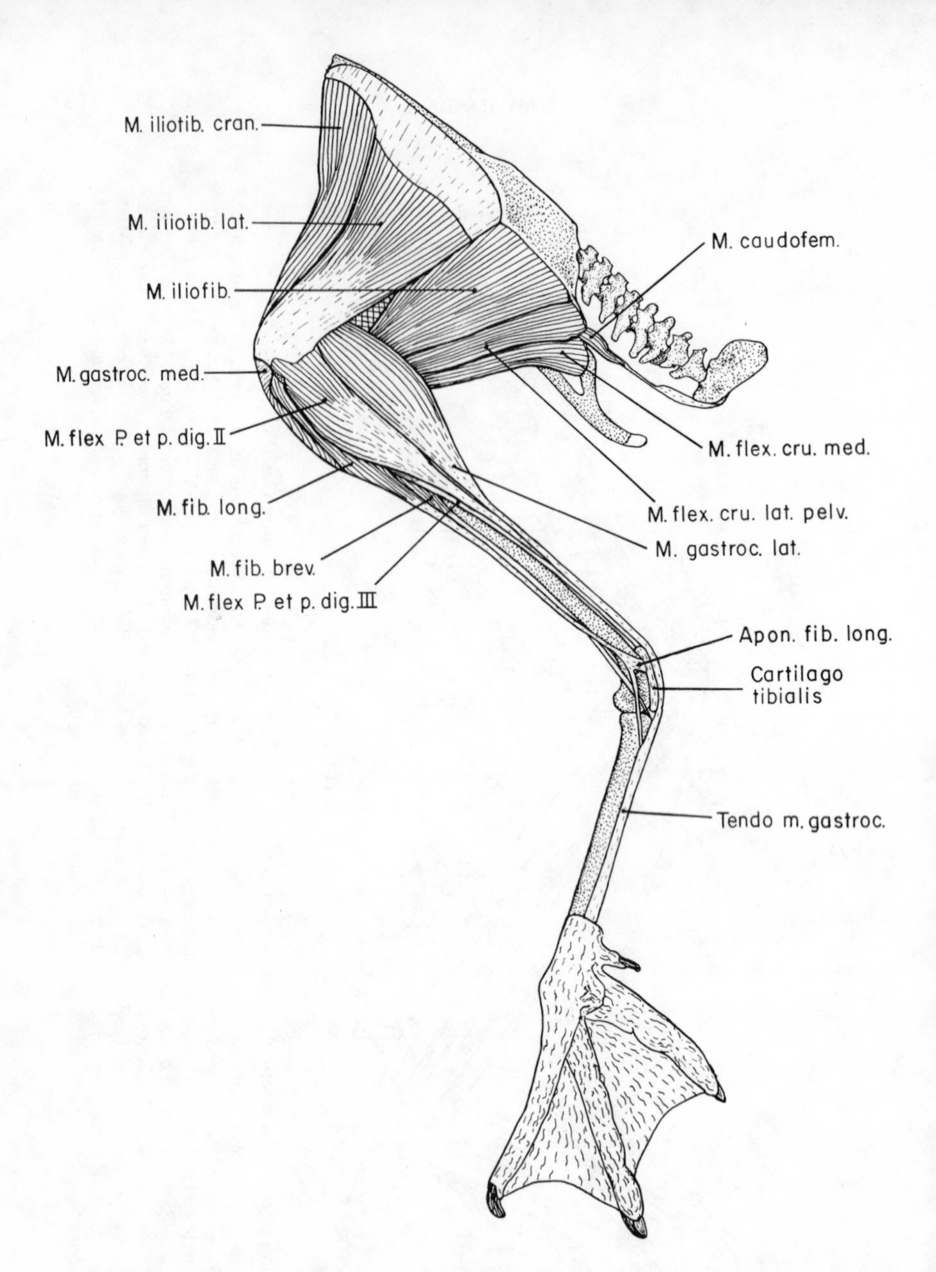

Fig. 7 Muscles of the pelvic limb of the gull, *Larus delawarensis*. Lateral view, superficial layer. From Hudson, *et al.* (1969).

Apon. fib. long., Aponeurosis m. fibularis longi; M. caudofem., M. caudofermoralis; M. fib. brev. or long., M. fibularis brevis or longus; M. flex. cru. lat. pelv., M. flexor cruris lateralis, Pars pelvica; M. flex. cru. med., M. flexor cruris medialis; M. flex. p. et p. dig. II or III, M. flexor perforans et perforatus digiti II or III; M. gastroc. lat. or med. M. gastocnemius, Pars lateralis or Pars medialis; M. iliofib., M. iliofibularis; M. iliotib. cran. or lat., M. iliotibialis cranialis or lateralis; Tendo m. gastroc., Tendo m. gastrocnemialis. See **Osteo.** Annot. 271 concerning the tibial attachment of M. gastrocnemius medialis and **Arthr.** Annot. 164 and **Osteo**. Annot. 279 regarding Cartilago tibialis.

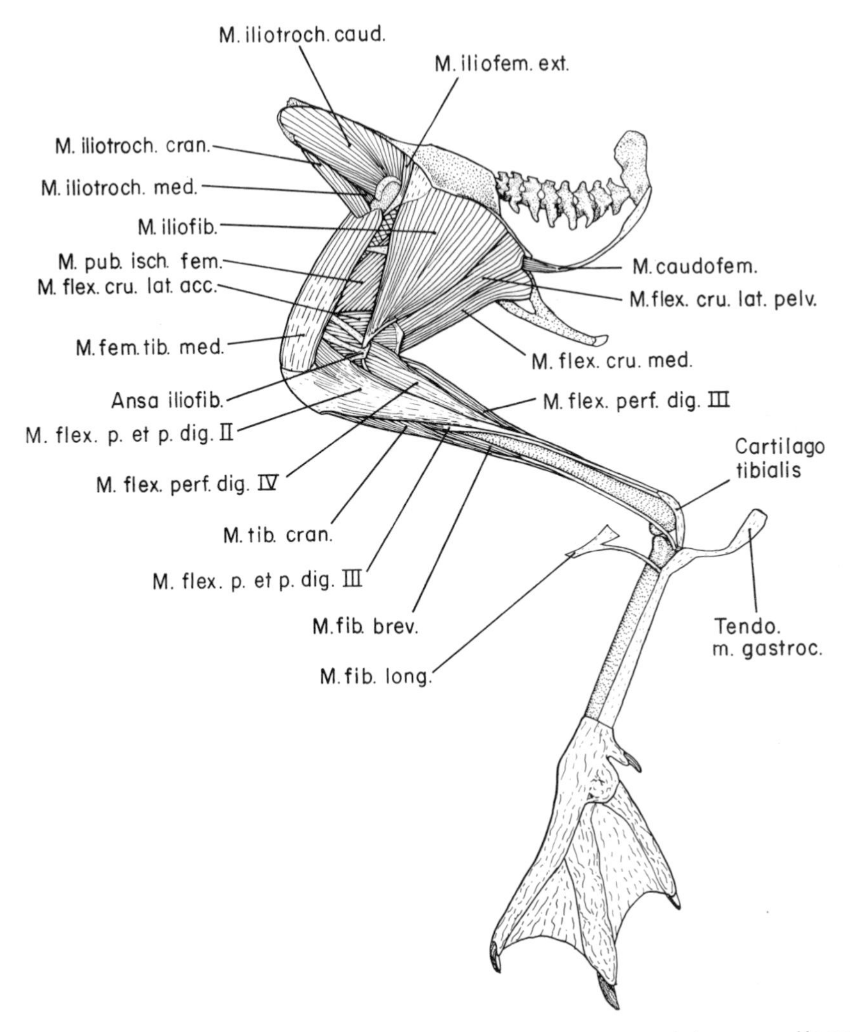

Fig. 8 Muscles of the pelvic limb of the gull, *Larus delawarensis*. Lateral view, second layer. From Hudson, *et al.* (1969).

Ansa iliofib., Ansa m. iliofibularis; M. caudofem., M. caudofemoralis; M. fem. tib. med., M. femorotibialis medius; M. fib. brev. or long., M. fibularis brevis or longus; M. flex. cru. lat. acc. or pelv., M. flexor cruris lateralis, Pars accessoria or Pars pelvica; M. flex. cru. med., M. flexor cruris medialis; M. flex. p. et p. d. II or III, M. flexor perforans et perforatus digiti II or III; M. flex. perf. dig. II or IV, M. flexor perforatus digiti II or IV; M. iliofem. ext., M. iliofemoralis externus; M. iliofib., M. iliofibularis; M. iliotroch. caud. or cran. or med., M. iliotrochantericus caudalis or cranialis or medius; M. pub. isch. fem., M. pubo-ischio-femoralis; M. tib. cran., M. tibialis cranialis.

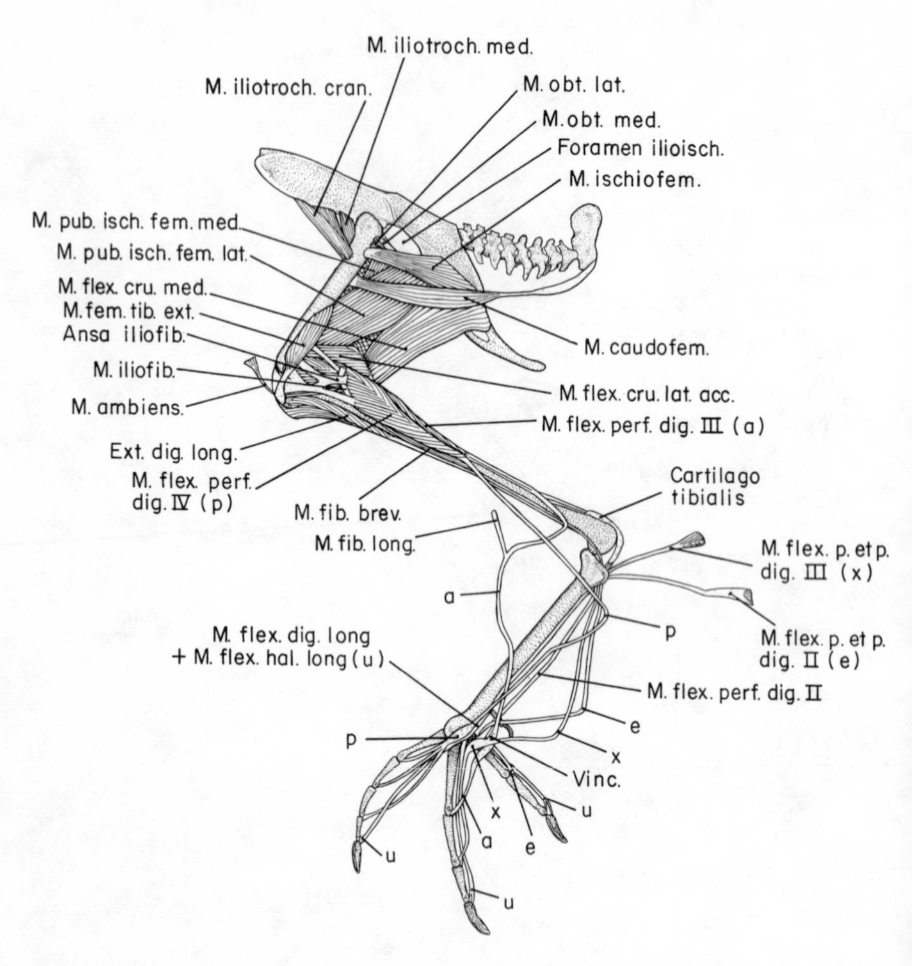

Fig. 9 Muscles of the pelvic limb of the gull, *Larus delawarensis*. Lateral view, third layer. From Hudson, *et al.* (1969).

Ansa iliofib., Ansa m. iliofibularis; M. caudofem., M. caudofemoralis; M. ext. dig. long., M. extensor digitorum longus; M. fem. tib. ext., M. femorotibialis externus; M. fib. brev. or long., M. fibularis brevis or longus; M. flex. cru. lat. acc., M. flexor cruris lateralis, Pars accessoria; M. flex. cru. med., M. flexor cruris medialis; M. flex. dig. long. +M. flex. hal. long., M. flexor digitorum longus +M. flexor hallucis longus; M. flex. p. et p. dig. II or III, M. flexor perforans et perforatus digiti II or III; M. flex. perf. dig. II or III or IV, M. flexor perforatus digiti II or III or IV; Foramen ilioisch., Foramen ilioischiadicum; M. iliofib., M. iliofibularis; M. iliotroch. cran. or med., M. iliotrochantericus cranialis or medius; M. obt. lat. or med., M. obturatorius lateralis or medialis; M. pub. isch. fem. lat. or med., M. puboischio-femoralis, Pars lateralis or Pars medialis; Vinc., Vinculum tendinum flexorum.

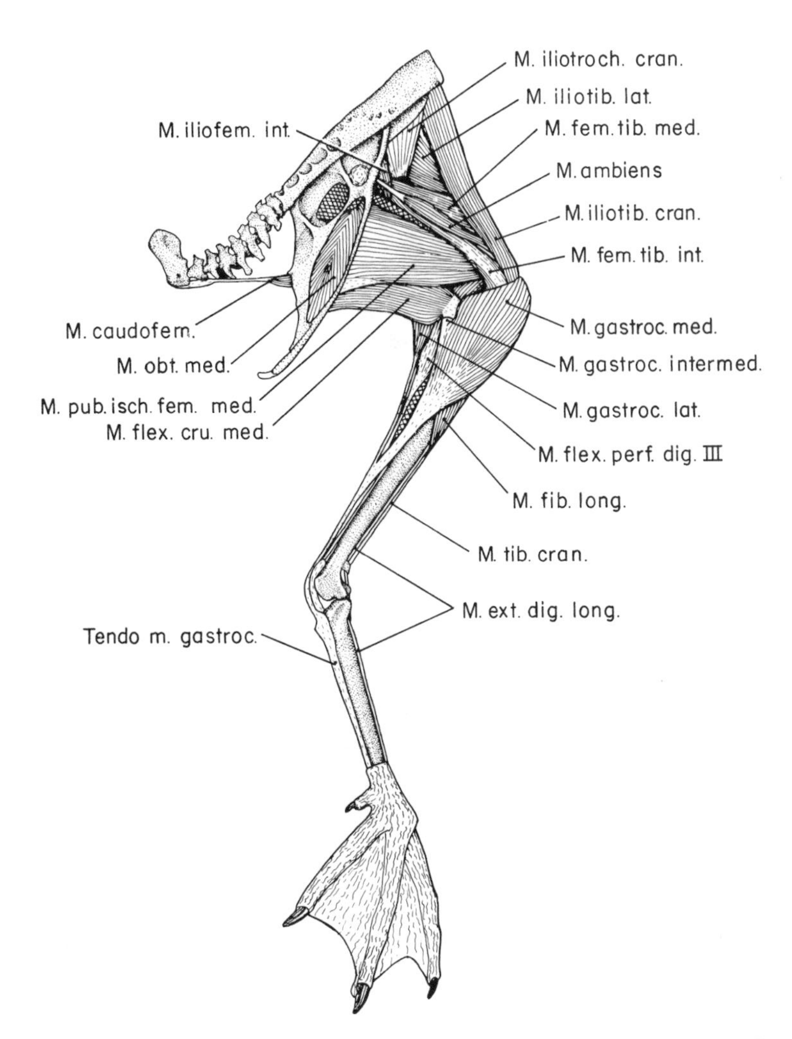

Fig. 10 Muscles of the pelvic limb of the gull, *Larus delawarensis*. Medial view. From Hudson, *et al.* (1969).

M. caudofem., M. caudofemoralis; M. ext. dig. long., M. extensor digitorum longus; M. fem. tib. int. or med., M. femorotibialis internus or medius; M. fib. long., M. fibularis longus; M. flex. cru. med., M. flexor cruris medialis; M. flex. perf. dig. III, M. flexor perforatus digiti III; M. gastroc. intermed. or lat. or med., M. gastrocnemius, Pars intermedia or Pars lateralis or Pars medialis; M. iliofem. int., M. iliofemoralis internus; M. iliotib. cran. or lat., M. iliotibialis cranialis or lateralis; M. iliotroch. cran., M. iliotrochantericus cranialis; M. obt. med., M. obturatorius medalis; M. pub. isch. fem. med., M. pubo-ischio-femoralis, Pars medialis; M. tib. cran., M. tibialis cranialis; Tendo m. gastroc., Tendo m. gastrocnemialis.

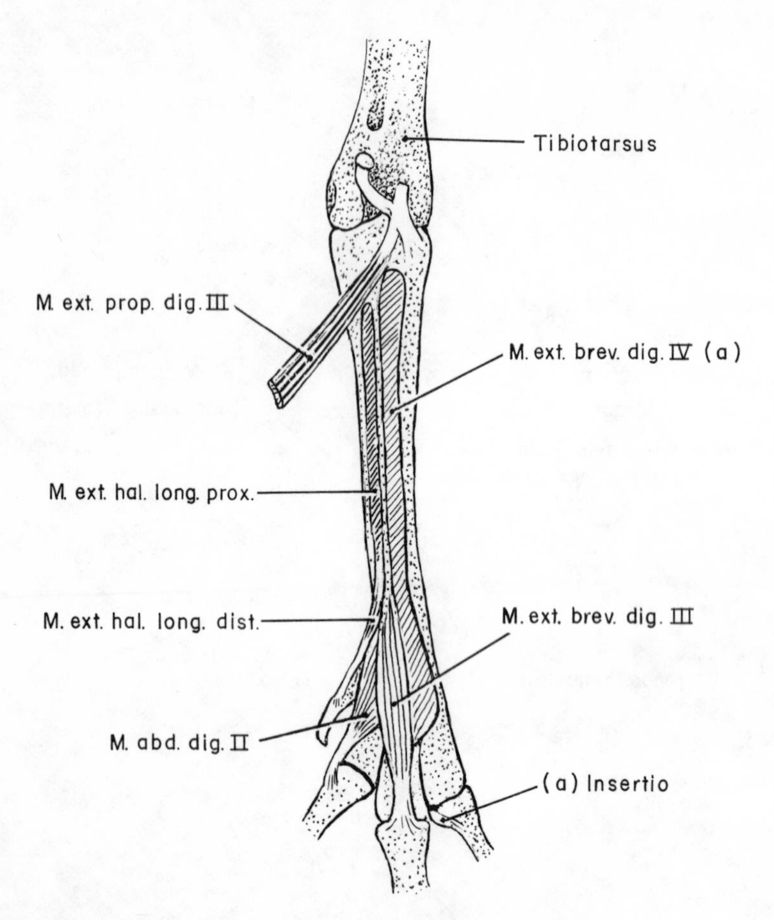

Fig. 11 Muscles of the foot of the tinamou, *Crypturellus tataupa*. Dorsal view of the distal Tibiotarsus and Tarsometatarsus. Hudson, *et al.* (1972).

M. abd. dig. II, M. abductor digiti II; M. ext. brev. dig. III or IV, M. extensor brevis digiti III or IV; M. ext. hal. long. prox. or dist., M. extensor hallucis longus, Pars proximalis or Pars distalis; M. ext. prop. dig. III, M. extensor proprius digiti III.
Note: The tendon of M. extensor brevis digiti IV passes through Canalis interosseus tendineus from the dorsum of the Tarsometatarsus into the lateral intertrochlear notch [see Insertio (a)].

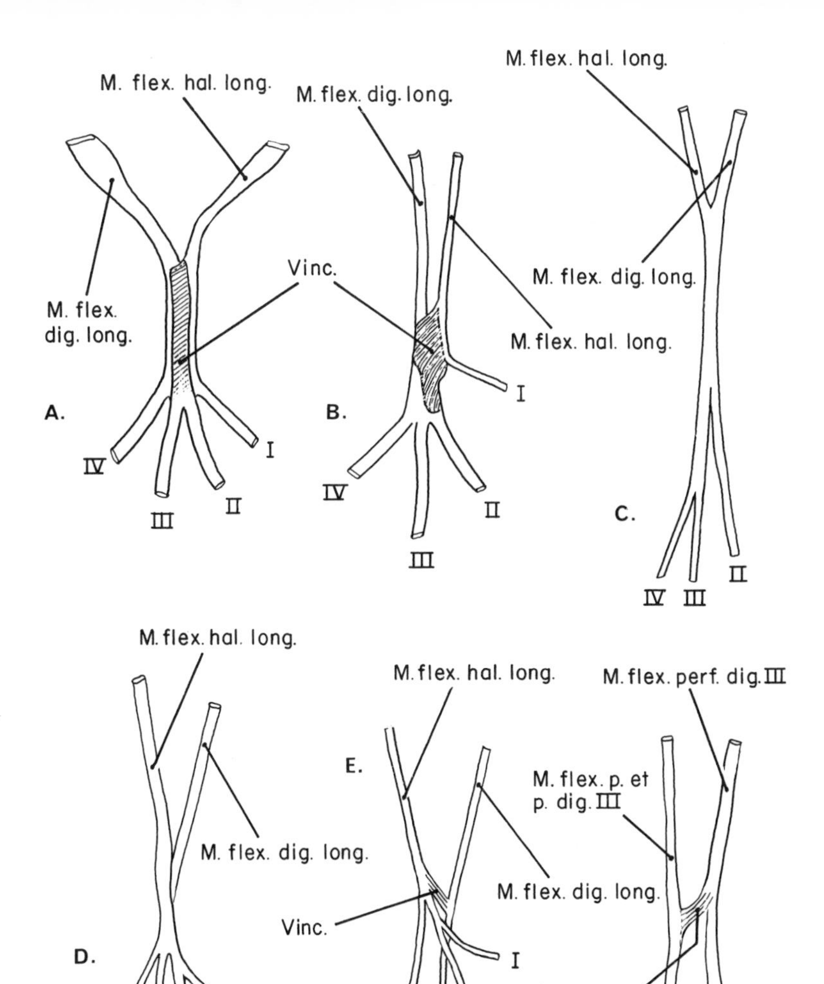

Fig. 12 Arrangement of the long flexor tendons in the foot of various birds. From Hudson (1937).
M. flex. dig. long., M. flexor digitorum longus; M. flex. hal. long., M. flexor hallucis longus; M. flex p. et p. dig. III, M. flexor perforans et perforatus digiti III; M. flex., perf. dig. III, M. flexor perforatus digiti III; Vinc., Vinculum tendinum flexorum.
A, the owl, *Bubo virginianus,* shows lengthy vinculum connecting the two deep flexor tendons.
B, the grouse, *Pedioecetes phasianellus,* shows a broad vinculum connecting the two deep flexor tendons.
C, the loon, *Gavia immer* demonstrates a side-to-side fusion of the two deep flexor tendons with no branch to the Hallux.
D, the frigatebird (*Fregata magnificens*) displays an overlapping fusion of the deep flexor tendons.
E, the flicker, *Colaptes auratus*, exhibits an undivided tendon of M. flex. dig. long. that serves the third digit and the tendon of M. flex. hal. long. is trifurcated, sending branches to digits I, II, and IV.
F, the crane, *Grus canadensis,* possesses a vinculum that connects the tendons of the two superficial flexors of digit III.

PERICARDIUM, PLEURA ET PERITONEUM

J. McLELLAND

Department of Anatomy, Royal (Dick) School of Veterinary Studies, Edinburgh EH9 1QH, Scotland

With Contributions from Sub-Committee Members: H.-R. Duncker, *Justus Liebig-Universität*; A. S. King, *University of Liverpool*

In contrast to the format in *Nomina Anatomica Veterinaria* (ICVAN, 1973) the nomenclature for the serosal membranes and celomic cavities is placed under one heading. This arrangement will provide ease of reference as well as assist in the viewing of this notoriously complex aspect of avian anatomy as a continuous whole. The pericardial terminology was compiled by Dr J. J. Baumel. To him and to the members of the Sub-Committee I should like to express my sincere thanks.

PERICARDIUM

Basis pericardii[1]
Pericardium fibrosum
    Lig. hepatopericardiacum[1]
Pericardium serosum
    Pericardium serosum parietale
    Pericardium serosum viscerale [Epicardium]
Cavitas pericardialis
    Sinus transversus pericardialis

PLEURA
(Fig. 1)

Pleura parietalis (Fig. 1)
    Septum horizontale[2] (Fig. 1)
        Mm. costoseptales[3] (Fig. 1)
Pleura visceralis [Pleura pulmonalis] (Fig. 1)
Cavitas pleuralis[4] (Fig. 1)

PERITONEUM

(Figs. 1, 2)

- Tunica serosa
- Tela subserosa
- Peritoneum parietale (Figs. 1, 2)
    - Septum obliquum[2] (Fig. 1)
        - M. septi obliqui[5] (Fig. 1)
- Peritoneum viscerale (Figs. 1, 2)
- Cavitas peritonealis[6] (Figs. 1, 2)
    - Cavitas peritonealis heptatica[6] (Figs. 1, 2)
        - Cavitas peritonealis hepatica ventralis[6] (Figs. 1, 2)
        - Cavitas peritonealis hepatica dorsalis[6] (Fig. 1)
    - Cavitas peritonealis intestinalis[6] (Fig. 2)
- Lig. hepaticum[7] (Fig. 1)
- Septum posthepaticum[8] (Fig. 2)
- Mesenterium dorsale (Fig. 2)
    - Radix mesenterii
    - Mesoduodenum
        - Lig. gastroduodenale[9]
    - Mesojejunum (see **Digest.** Annot. 63)
    - Mesoileum (see **Digest.** Annot. 63)
        - Lig. ileocecale
- Mesorectum
- Mesenterium ventrale (Figs. 1, 2)
    - Lig. falciforme hepatis
    - Lig. hepatoduodenale[10]
    - Lig. ileodiverticulare[11]
- Mesorchium
    - Mesovarium (Fig. 2)
    - Mesoviductus
        - Lig. dorsale oviductus (see **Genit. fem.** Annot. 34)
        - Lig. ventrale oviductus (see **Genit. fem.** Annot. 34)
            - Funiculus musculosus

ANNOTATIONS

(1) **Lig. heptatopericardiacum.** The caudal part of the fibrous pericardium is drawn out into a pointed bilaminar sheet that becomes continuous with the part of the ventral mesentery between the hepatic lobes (*Gallus*, *Columba*). Wolf (1967) describes

lateral pericardial ligaments that attach to the abdominal wall at the level of the caudal border of the liver.
Basis pericardii. The surface of the pericardial sac that rests dorsally against the bifurcation of the trachea, the esophagus, and the horizontal septum on each side.

(2) **Septum horizontale; Septum obliquum.** Two gross partitions, one dorsal and the other ventral, are formed on each side of the body by penetration of the developing cranial and caudal thoracic air sacs into the pulmonary fold. See Figs. 1, 2.

Septum horizontale, the dorsal partition, is composed of parietal pleura. Among the many different names by which it has been referred are: diaphragme pulmonaire (Sappey, 1847, pp. 21–26); diaphragmite antérieur (Milne-Edwards, 1865); pulmonary aponeurosis (Huxley, 1882; Butler, 1889; Goodrich, 1930, p. 633); horizontal septum (Beddard, 1896, 1898, p. 37; Poole, 1909; Duncker, 1971); horizontal diaphragm (Juillet, 1912); septum pulmonale (Kern, 1963, p. 50); bronchopleural membrane (McLelland and King, 1970); saccopleural membrane (McLelland and King, 1975).

Septum obliquum, the ventral partition, which consists of parietal peritoneum, has also been given various names: diaphragme thoracoabdominal (Sappey, 1847, pp. 21–26; Juillet, 1912); diaphragmite thoracoabdominal (Milne-Edwards, 1865); oblique septum (Huxley, 1882; Beddard, 1885, 1888, 1896, 1898, p. 38; Butler, 1889; Goodrich, 1930, p. 633; Duncker, 1971); septum thoraco-abdominale (Kern, 1963, p. 50; abdominal diaphragm (Salt and Zeuthen, 1960); bronchoperitoneal membrane (McLelland and King, 1970); saccoperitoneal membrane (McLelland and King 1975).

Much ambiguity surrounds the terms applied to these dorsal and ventral partitions since only a few authors defined their terms precisely, e.g. Duncker (1971) restricted his terms to the serosal component of each partition; Juillet (1912) and McLelland and King (1970, 1975) explicitly included both the serosal component and the air sac component in their terms. The terms Septum horizontale and Septum obliquum are now restricted to the serosal derivatives in each of the two partitions, the horizontal septum being derived from the parietal pleura of the dorsal partition, and the oblique septum being derived from the parietal peritoneum of the ventral partition.

(3) **Mm. costoseptales.** The striated muscles which insert into the lateral part of the horizontal septum (Fedde *et al.*, 1964). See Fig. 1.

(4) **Cavitas pleuralis.** There is considerable disagreement in the literature as to whether the pleural cavity exists in birds and it seems likely that in all species a certain amount of obliteration, sometimes total, of the cavity takes place during embryonic development. Nevertheless, extensive areas of cavity are known to persist in the adult in a number of species including *Gallus* (Groebbels, 1932, p. 45; Kern, 1963, pp. 49–50; McLelland and King, 1975). In *Gallus* the pleural cavity is best developed on the dorsolateral aspect of the lung where the filaments uniting the Pleura parietalis to the Pleura visceralis are relatively delicate; in contrast the parietal pleura and visceral pleura on the ventromedial surface of the lung are extensively fused together. See Fig. 1.

(5) **M. septi obliqui.** The smooth muscle, absent in some species, e.g. *Apteryx* (Huxley 1882), which lies in the medial part of the Septum obliquum (Fig. 5).

(6) **Cavitas peritonealis hepatica.** The terminology for the subdivisions of the peritoneal cavity is based on Grau (1943). The four hepatic peritoneal cavities lie

cranial and lateral to the Septum posthepaticum. Synonymy of the right and left dorsal hepatic cavities; pulmohepatic recesses (Butler, 1889, p. 43; Beddard, 1898; Poole, 1909); dorsalen Bauchfellhöhlen (Bittner, 1925); pulmonary recesses (Goodrich, 1930, p. 636); dorsalen Leberbauchfellsäcken (Kern, 1963, p. 29). Synonymy of the right and left ventral hepatic cavities: ventral liver sacs (Butler, 1889; Poole, 1909); ventralen Bauchfellhöhlen (Bittner, 1925); liver sacs (Goodrich, 1930, p. 636); compartiments abdominaux inférieurs (Petit, 1933); ventralen Leberbauchfellsäcken (Kern, 1963, p. 29). See Figs. 1, 2.
Cavitas peritonealis intestinalis. Synonymy: posthepatic intestinal cavity (Poole, 1909); intestinal coelomic chamber (Goodrich, 1930, p. 636; compartiment abdominal supérieur (Petit, 1933); Eingeweidebauchfellsack (Kern, 1963, p. 50). The intestinal peritoneal cavity is a midline space lying between the left and right partitions of the posthepatic septum (see Annot. 8). The left dorsal hepatic peritoneal cavity and intestinal peritoneal cavity connect with each other, but otherwise the peritoneal cavities are blind ones. See Figs. 1, 2.

(7) **Lig. hepaticum.** Synonymy: pulmohepatic ligament (Butler, 1889); horizontal hepatic ligament (Goodrich 1930, p. 636; McLelland and King, 1975). Right and left hepatic ligaments extend between the Septum obliquum and the Peritoneum viscerale of the liver and separate the dorsal and ventral hepatic peritoneal cavities (Fig. 1).

(8) **Septum posthepaticum.** The term used by Butler (1889) for the right and left double-layered partitions separating the intestinal peritoneal cavity from the hepatic peritoneal cavities. The left partition extends between the dorsolateral parietal peritoneum and the left surface of the Ventriculus. The right partition extends between the dorsolateral parietal peritoneum and the right surface of the Ventriculus. See Fig. 2.

(9) **Lig. gastroduodenale.** Described by Bittner (1924), Pilz (1937), Grau (1943) and Kern (1963, p. 17).

(10) **Lig. hepatoduodenale.** Described by Bittner (1924), Pilz (1937) and Kern (1963, p. 18).

(11) **Lig. ileodiverticulare.** The short peritoneal ligament which extends between the Diverticulum vitelli and the Ileum.

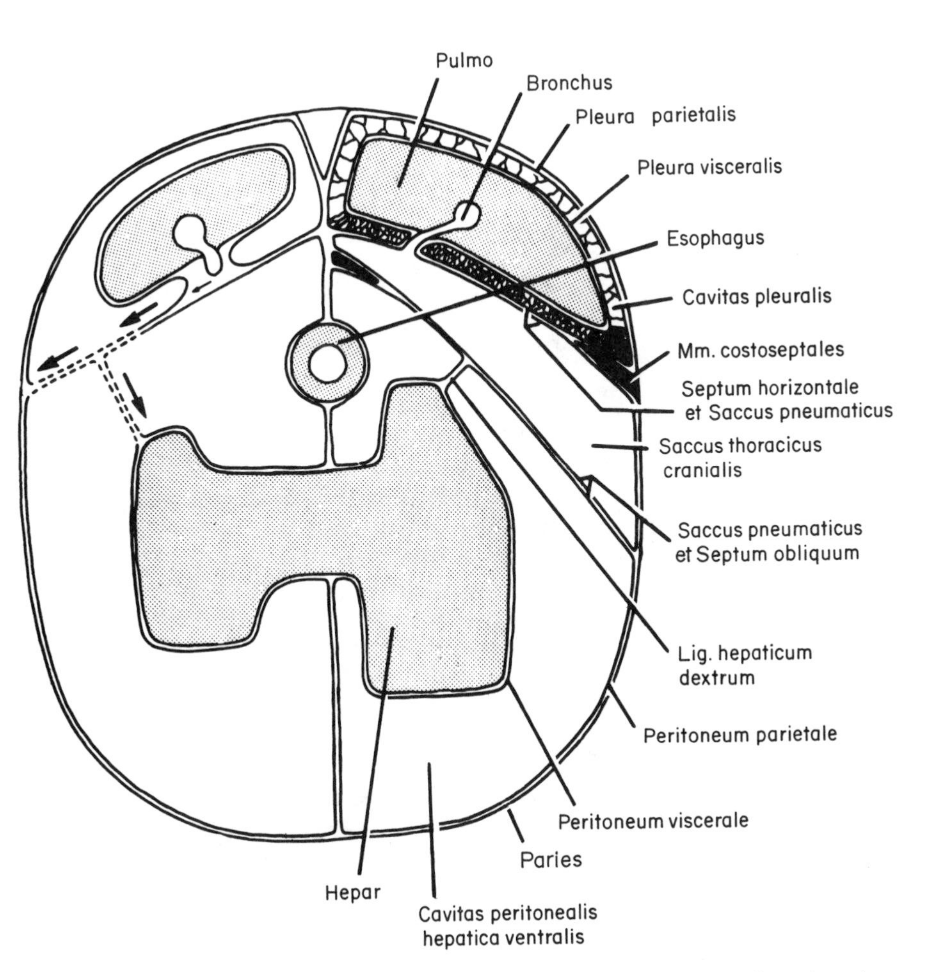

Fig. 1 Pleural and peritoneal cavities of *Gallus*. Schematic transverse sections through the trunk at the level of the lungs and liver. From McLelland and King (1970). The left side of the figure (based on Goodrich, 1930) shows how the pleural cavity is separated from the peritoneal cavity in the embryo by extension of the pulmonary fold ventrolaterally to the lateral body wall and to the liver (dashed lines); the three large arrows indicate the direction in which these extensions grow. The single small arrow indicates the subsequent penetration by the air sacs.

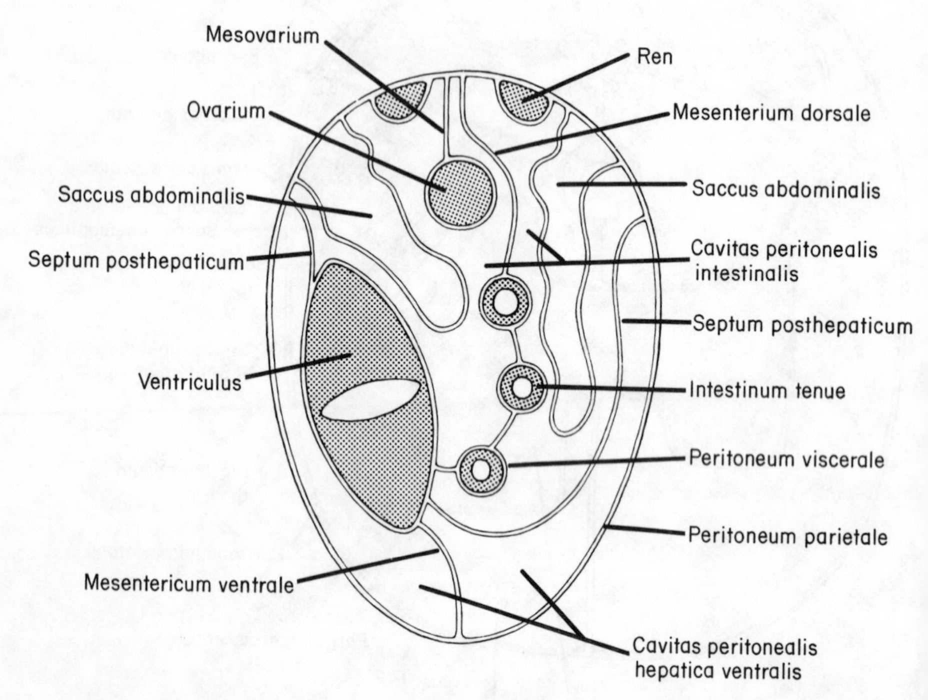

Fig. 2 Subdivisions of the peritoneal cavity of *Gallus*. Schematic transverse section through the trunk at the level of the ovary and muscular stomach. From McLelland and King (1970).

SYSTEMA RESPIRATORIUM

A. S. KING

Department of Veterinary Anatomy, University of Liverpool, P.O. Box 147, Liverpool L69 3BX, England

With contributions from Sub-Committee Members: P. L. Ames, *Chicago, USA*; Betsy G. Bang, *Johns Hopkins University*; H.-R. Duncker, *Justus Liebig Universität*; M. R. Fedde, *Kansas State University*; G. V. Morejohn, *California State University*; W. S. Tyler, *National Center for Primate Biology, Davis, California*; R. W. Warner, *London, England*

Especially valuable advice was received from Dr H.-R. Duncker on the list as a whole and on the lung in particular, from Dr Betsy G. Bang on the Cavitas nasalis, and from Drs Ames and Warner on the Syrinx. Dr M. R. Fedde played a crucial role in establishing the new terminology for the lung.

A number of difficulties were encountered in the preparation of this list. The terms for the larynx had to be based on anatomical observations on a rather small number of species. The syrinx has been much more extensively studied than the larynx, and in many more groups of birds, but no comprehensive recent analysis of syringeal anatomy is available. The facts about its skeletal elements seem fairly well established, but their names are controversial. The identities and attachments of some of the syringeal muscles are still uncertain, and there is disagreement on how to name them. Generally a conservative approach has been used to resolve these difficulties for the syrinx, the best possible terms being drawn eclectically from the rival terminologies.

The nomenclature for the bronchi, particularly the Bronchi secundarii, was extremely controversial. The cause here was not a lack of facts but a proliferation of differing terminologies, a process which had been going on actively among anatomists and physiologists for more than a hundred years. Contemporary researchers and teachers were naturally not eager to relinquish familiar terms, but in the end a new nomenclature has been evolved which is both descriptive and not complicated by previous usage. The Sacci pneumatici were also a problem because of their considerable variability among the different groups of birds, an unusual difficulty in avian anatomy. It was tempting to list names for the many minor diverticula, notably those of the Saccus clavicularis. But this would have added greatly to the length of the list, and in the end the precept was followed that "no attempt should be made to name every minor structure ever discovered".

CAVITAS NASALIS

Naris[1]
 Operculum nasale[2]
 Lamella verticalis naris[3]
Choana
 Pars rostralis[4]
 Pars caudalis[5]
Septum nasale[16] (see **Osteo.** Annot. 55)
Concha nasalis rostralis[6]
Concha nasalis media[7]
Concha nasalis caudalis[8]
Concha septalis[9]
Meatus nasalis[10]
Valvula nasalis[11]
Crista nasalis[12]
Tunica mucosa nasalis
 Regio vestibularis[13]
 Regio respiratoria[14]
  Glandulae mucosae nasi
 Regio olfactoria[15]
Sinus septalis[16]
Sinus infraorbitalis[17]
 Sinus conchoinfraorbitalis[18]
 Apertura sinus infraorbitalis[19]
Ostium ductus nasolacrimalis
Glandula nasalis[20]
 Lobus medialis[20]
  Ductus lobi medialis
 Lobus lateralis[20]
  Ductus lobi lateralis
Diverticula cervicocephalica[21]

LARYNX

Mons laryngealis[22]
Cavitas laryngealis

Glottis[23]
Sulcus laryngealis[24]
Tunica mucosa laryngis
Cartilagines laryngeales (Figs. 1, 2)
Cartilago cricoidea (Fig. 2)
Corpus[25] (Fig. 2)
Crista ventralis[26]
Ala [Cartilago cricoidea dorsalis][27] (Fig. 2)
Cartilago procricoidea[28] (Figs. 1, 2)
Corpus (Fig. 1)
Cauda (Fig. 1)
Cartilago arytenoidea (Figs. 1, 2)
Corpus (Figs. 1, 2)
Processus rostralis (Figs. 1, 2)
Processus caudalis[29] (Figs. 1, 2)
Musculi laryngeales (see **Myol.** Fig. 2b)
M. dilator glottidis[30]
M. constrictor glottidis[30]

TRACHEA

Cartilagines tracheales[31]
Tunica mucosa tracheae
Ansa trachealis[31]
Saccus trachealis[31]
Bulbus trachealis[31]
Musculi tracheales[45] (see **Myol.** Annot. 37)
M. sternohyoideus[45]
M. sternotrachealis[45] (Figs. 6a–g)
M. cleidotrachealis[45] (Fig. 6a)
M. tracheolateralis[45] (Figs. 6a–g)

SYRINX[32]

Cavitas syringealis
Tunica mucosa syringis
Tympanum[33] (Figs. 3, 4, 5a, 5b, 6a, 6c, 6d, 6g)
Bulla syringealis[34] (Figs. 5b, 6a)
Cartilagines syringeales[35]
Cartilagines tracheales syringis[36] (Figs. 3, 4, 5a, 5b)
Cartilagines bronchiales syringis[37] (Figs. 3, 4, 5a, 5b)
Pessulus[38] (Figs. 3, 5a, 5b)
Membrana semilunaris[39] (Fig. 5a)

SYRINX—*continued*

Cartilagines accessoriae[40]
Cartilago membranosa dorsalis[40] (Fig. 4)
Cartilago membranosa ventralis[40]
Processus vocalis[40] (Figs. 6c, 6d)
Ligamentum interbronchiale[41] (Figs. 4, 5a, 5b, 6a)
Foramen interbronchiale[41] (Figs. 5a, 5b, 6a)
Membrana tympaniformis lateralis[42] (Figs. 4, 5a)
Membrana trachealis[42] (Figs. 6c, 6d)
Membrana tympaniformis medialis[43] (Figs. 5a, 5b)
Labium laterale[44] (Figs. 5a, 5b)
Labium mediale[44] (Fig. 5a)

Musculi syringeales[46]
M. tracheobronchialis dorsalis (Figs. 6f, 6g)
M. tracheobronchialis brevis
M. tracheobronchialis ventralis (Figs. 6f, 6g)
M. syringealis dorsalis (Figs. 6f, 6g)
M. syringealis ventralis (Fig. 6g)
M. vocalis dorsalis (Figs. 6c, 6d)
M. vocalis ventralis (Fig. 6d)
M. obliquus ventralis (Fig. 6e)
M. obliquus lateralis (Fig. 6e)

Bronchus primarius, Pars extrapulmonalis[47]
Cartilagines bronchiales[47]
Tunica mucosa bronchi

PULMO

Facies costalis[48]
Facies vertebralis[48] (Fig. 7)
Sulci costales[49] (Fig. 8)
Tori pulmonales[49]
Tori intercostales[49] (Fig. 8)
Tori marginales[49] (Fig. 8)
Facies septalis[48] (Fig. 7)
Margo costovertebralis[50] (Fig. 7)
Margo costoseptalis[50] (Figs. 7, 8)
Margo vertebroseptalis[50] (Fig. 7)
Margo cranialis[50] (Fig. 7)
Margo caudalis[50] (Fig. 7)

Angulus craniodorsalis[50] (Fig. 7)
Angulus caudodorsalis[50] (Fig. 7)
Angulus cranioventralis[50] (Fig. 7)
Angulus caudoventralis[50] (Fig. 7)
Hilus pulmonalis
Planum anastomoticum[51] (Fig. 9)
Linea anastomotica[51] (Fig. 8)
Paleopulmo [Pulmo arcuiformis][52] (Fig. 9)
Neopulmo [Pulmo reteformis][52] (Fig. 9)

Bronchus primarius, Pars intrapulmonalis[47 53] (Figs. 8, 9)
Bronchi secundarii[54]
Bronchi medioventrales[55] (Figs. 8, 9)
Bronchi mediodorsales[56] (Figs. 8, 9)
Bronchi lateroventrales[57] (Figs. 8, 9)
Bronchi laterodorsales[58] (Fig. 9)
Parabronchus[59] (Fig. 8)
Atria[60] (Fig. 10)
Musculi atriales[61] (Fig. 10)
Septa interatrialia[62] (Fig. 10)
Infundibula[63] (Fig. 10)
Cellula granularis[64]
Pneumocapillares[66] (Fig. 10)
Cellula squamosa[67]
Septa interparabronchialia[65]
Ostium[68] (Fig. 8)
Saccobronchus[69]

Vasa sanguinea intrapulmonalia
Aa. interparabronchiales (see Fig. 10; and **Art.** Truncus pulmonalis)
Arteriolae intraparabronchiales (Fig. 10)
Vv. interparabronchiales (see Fig. 10; and **Ven.** Vv. pulmonales)
Vv. intraparabronchiales (Fig. 10)
Vv. atriales (Fig. 10)
Venulae septales (Fig. 10)
Venulae intraparabronchiales (Fig. 10)

SACCI PNEUMATICI

Saccus cervicalis
Diverticula vertebralia[70]
Diverticula intermuscularia[71]
Diverticula subcutanea[72]

SACCI PNEUMATICI—*continued*

Saccus clavicularis[73]
  Pars medialis[74]
  Pars lateralis[74]
  Diverticula intrathoracica[75] (see **Topog. anat.** Annot. 31)
  Diverticula extrathoracica[76] (see **Topog. anat.** Annot. 31)
Saccus thoracicus cranialis
Saccus thoracicus caudalis
Saccus abdominalis
  Diverticula perirenalia[77]
  Diverticula femoralia[77]

ANNOTATIONS

(1) **Naris.** The external aperture of the Cavitas nasalis. The many variations in form are categorized by various terms: Nares gymnorhinales, exposed nares; Nares perviae, open (most birds); Nares imperviae, secondarily closed (some Pelecaniformes). Other terms relate to the shape of the *bony* aperture, which has taxonomic significance: holorhinal, caudal margin rounded; schizorhinal, caudal margin a slit; amphirhinal, two bony openings on each side. In Sulidae even the bony aperture is closed (Thompson, 1964, p. 505).

(2) **Operculum nasale.** A horny flap above the nares, which is present in *Gallus* and some other species (Thomson, 1964, p. 554; Bang, 1971), and below the nares in wrynecks (*Jynx*) (van Tyne and Berger, 1976, p. 548). See **Topog. anat**. Annot. 20.

(3) **Lamella verticalis naris.** Synonymy: Atrial concha (Bang, 1971). A cartilaginous sheet arising from the ventral border of the Naris in *Gallus* (Jungherr, 1943; Bang, 1971; King, 1975, p. 1886), and a few other species as in *Turnix* and members of Apodiformes (Bang, 1971).

(4) **Choana, Pars rostralis.** A narrow rostral slit, possibly homologous to the median palatine suture of mammals (Heidrich, 1908).

(5) **Choana, Pars caudalis.** A triangular caudal opening, caudal to the palatine processes of the maxillae, between the palatine bones. It is divided dorsally in the midline by the Vomer and Septum nasale, and is possibly homologous to the mammalian Choanae. This is the interpalatine cleft, or Choana I, of Lucas and Stettenheim (1972, Fig. 362), the left and right channels which connect it dorsally to the Cavitas nasalis being their internal nares or Choana II.

(6) **Concha nasalis rostralis.** Sometimes called the Concha ventralis, but rostralis is preferred because in birds with long nasal chambers e.g. *Hydrophasianus*, see Bang, 1971), there is no room for a ventrodorsal relationship between the conchae, but only a rostrocaudal one. In some groups, e.g. Sulidae, the Concha rostralis is absent (Bang, 1971).

(7) **Concha nasalis media.** The maxillary concha of Bang (1971) and maxilloturbinal of Bellairs and Jenkin, (1960, p. 285). This concha is rarely absent in birds (Bang, 1971). The term maxillary implies homology to the maxillary concha of mammals and other vertebrates. Until the homology is established Concha nasalis media is preferred, since this is topographically accurate for most birds and could not be misleading in homology.

(8) **Concha nasalis caudalis.** A fairly constant structure in birds (absent, however, in *Sula*), which seems to be regularly lined by olfactory epithelium (Bang, 1971), and is probably homologous with the single Concha of most reptiles (Bellairs and Jenkin, 1960, p. 285).

(9) **Concha septalis.** Unique to petrels (e.g. *Pagodroma*), this concha arises from the Septum nasale and interdigitates with the Concha caudalis, its epithelial lining being olfactory (Bang, 1971).

(10) **Meatus nasalis.** Since there are three Conchae nasales located in a rostrocaudal sequence, the mammalian terms, Meatus nasi dorsalis, medius, and ventralis are inapplicable in birds.

(11) **Valvula nasalis.** A paired crescentic mucosal fold which is attached to the Septum nasale or to the adjoining roof of the Cavitas nasalis, at the level of the caudal end of the Concha media. It occurs mainly in wading and aquatic species (e.g. *Gavia*) and passively deflects water from the Regio olfactoria (Bang, 1971).

(12) **Crista nasalis.** Synonymy: Schwelle. A ridge separating the Regio vestibularis from the Regio respiratoria in birds generally (Bang, 1971).

(13) **Regio vestibularis.** Synonymy: Nasal vestibule (Bang, 1971); vestibular zone or anterior zone (Sandoval, 1964). The vestibular region is separated from the Regio respiratoria in birds generally by the Crista nasalis. The vestibular region contains the Concha nasalis rostralis when this concha is present; it is lined by stratified squamous epithelium, and receives the secretion of the Glandula nasalis. Structural diversity is greater in the area of the Naris and Regio vestibularis than in any other part of the nasal cavity (Bang, 1971).

(14) **Regio respiratoria.** The middle or respiratory area of Bang (1971) and Sandoval (1964), containing the Concha nasalis media and lined by a mucociliary epithelium (Bang, 1971).

(15) **Regio olfactoria.** The olfactory chamber of Bang (1971) and inner zone of Sandoval (1964), containing the Concha nasalis caudalis and lined by olfactory epithelium.

(16) **Sinus septalis.** In a few species (Bang, 1971) there are bony spaces in the Septum nasale which are continuous with the nasal cavity, some of these spaces being lined by nasal mucociliary epithelium (*Tephrodornis*, *Lanius*) and others by a non-ciliated epithelium (*Coracina*). See **Osteo.** Annot. 55.

(17) **Sinus infraorbitalis.** Synonymy: Orbital sac, subocular sac (Romanoff, 1960, p. 537; Bellairs and Jenkin, 1960, p. 290); maxillary sinus (Bang, 1971). The term "infraorbital sinus" is in fairly wide use in veterinary literature (e.g. McLeod and

Wagers, 1939; Cover, 1953). Homology to the maxillary sinus of mammals was proposed by Bleicher and Legait (1932) because: (1) A similar cavity is always present in amphibia and reptiles; (2) it always connects to the Cavitas nasalis; (3) it is partitioned by a fold carrying the nasopalatine branch of the maxillary nerve. These grounds appear too weak to establish this homology.

(18) **Sinus conchoinfraorbitalis.** A compound term for the Sinus infraorbitalis and the interior of the Concha nasalis caudalis, which in many species are continuous with one another (Bang, 1971).

(19) **Apertura sinus infraorbitalis.** The opening of the Sinus infraorbitalis into the Cavitas nasalis.

(20) **Glandula nasalis; Lobus medialis; Lobus lateralis.** Sometimes known as the supraorbital gland (Beddard, 1898), and also referred to as the salt gland, but it only secretes salt in marine birds and in some desert species and raptors. In the large majority of terrestrial species, including the very large order Passeriformes, the nasal gland does not secrete salt (Peaker and Linzell, 1975, p. 220). Therefore the term Glandula salsa is not appropriate for this gland in birds generally. Most birds possess a Lobus lateralis and a Lobus medialis each with its own Ductus (Technau, 1936). In *Gallus* and closely related forms only the Lobus medialis, with its Ductus, is present (Marples, 1932). See **Osteo.** Annot. 15.

(21) **Diverticula cervicocephalica.** A system of diverticula pneumatizing particularly the skull, and sometimes other parts of the head and also the neck (Bignon, 1889; Groebbels, 1932, pp. 59–62; Coe, 1960; Bellairs and Jenkin, 1960, p. 290; Romanoff, 1960, p. 537; King, 1966, p. 217). The diverticula which invade the skull come from both the tympanic cavity and the nasal cavity (Baumel, pers. comm.). In some birds, e.g. *Leptoptilos*, there are apparently major connexions between the Cavitas nasalis and the air pouches of the neck (Coe, 1960; Akester *et al.* 1973). See **Osteo.** Cavitas tympanica and Annot. 23, 25.

(22) **Mons laryngealis.** The conspicuous mound carrying the opening into the larynx. The mound is covered with caudally pointing papillae, and its caudal border is prolonged in many species by a fringe of particularly well developed, caudally projecting papillae. The edges of the glottis are also generally guarded by similar well developed papillae.

(23) **Glottis.** The slit-like opening into Cavitas laryngealis, for which White (1970, p. 7) found seven different names in the literature for birds since 1883. The term, Glottis, seems to be in general use for the entrance into the larynx in amphibians, reptiles, and birds (e.g. Marshall, 1961, pp. 403, 478, 585), and has found acceptance in current ornithological literature (e.g. Bock, 1978).

(24) **Sulcus laryngealis.** A prominent furrow, continuing the glottis caudally, present in *Gallus* and *Anser* (White, 1970, p. 8; 1975), and the crow (*Corvus brachrhynchos*) (Bock, 1978).

(25) **Corpus.** The large median gutter-like ventral part of the Cartilago cricoidea (White, 1970, 1975). There is a tendency for incomplete fused tracheal cartilages to be attached to its ventrocaudal end, as in species of *Corvus* (Bock, 1978; White, pers. comm.). See **Arthr.** Annot. 56.

(26) **Crista ventralis.** A median ossified ridge on the Corpus of the Cartilago cricoidea in some species, e.g. *Anas*, (White, 1975, p. 1892).

(27) **Ala** [Cartilago cricoidea dorsalis]. In *Gallus* each paired Ala, or cricoid wing, is a thin flattened cartilaginous plate attached dorsolaterally to the Corpus of the Cartilago cricoidea. In *Corvus brachyrhynchos* (Bock, 1978), and in *Corvus corax* and *Corvus orru* (White, pers. comm.), this structure is represented by a separate rod-like skeletal element, articulating by true joints with the main component (the Corpus) of the cricoid cartilage and with the procricoid cartilage. The rod-like nature of this element in *Corvus corax* had already been shown in Fig. 10 of Shufeldt (1890). See **Arthr.** Annot. 56.

(28) **Cartilago procricoidea.** A small dorsal median comma-shaped element, intervening between the two arytenoids and the two cricoid alae in *Gallus* (White, 1970; 1975, p. 1892), and in *Columba* and *Eudocimus* (Baumel, pers. comm.). In species of *Corvus* (Bock, 1978; White, pers. comm.) it is a more compact and somewhat cuboidal structure. According to White (1970, p. 12) the term Cartilago procricoidea was adopted by most of the earlier authors, but others regarded this structure as simply a part of the cricoid. See **Arthr.** Annot. 55.

(29) **Processus caudalis.** Synonymy: dorsal arytenoid (Bock, 1978). The Proc. caudalis of the arytenoid cartilage was observed by Bock (1978) to be a separate skeletal element, making a synovial joint with the Corpus of the arytenoid, in the crow *Corvus branchyrhynchos*; White (pers. comm.) found no joint in *Corvus orru* and *Corvus corax*, and concluded that the Processus caudalis in these species is fused to the corpus of the arytenoid cartilage essentially as in *Gallus*. See **Arthr.** Annot. 56.

(30) **M. dilator glottidis; M. constrictor glottidis.** Synonymy: superficial and deep intrinsic laryngeal muscles (White, 1975). These are the intrinsic muscles of the larynx, i.e. they attach exclusively to laryngeal elements. They have been studied fully in *Gallus* by White (1975) and shown by electrical stimulation to dilate and constrict the glottis respectively in this species (White and Chubb, 1967). The dilator is superficial, immediately beneath the mucosa. It runs essentially from the wing and body of the cricoid, to the arytenoid cartilage. The constrictor runs from the caudal midline of the larynx (especially from the procricoid cartilage) to the arytenoid and cricoid cartilages, embracing the glottis in a somewhat horseshoe-shaped arrangement. These attachments are basically similar in *Corvus brachyrynchos* (Bock, 1978), and in *Corvus orru* and *Corvus corax* as well as in a wide variety of birds from other orders (White, pers. comm.). See **Myol.** Annot. 38, 39.

(31) **Cartilagines tracheales; Ansa trachealis; Saccus trachealis; Bulbus trachealis.** The cartilage rings are closely interlocked and leave minimal intervals between them (McLelland, 1965), and therefore the Ligamenta anularia of mammals are lacking. In many species the caudal rings ossify, and in some large birds (e.g. *Cygnus*, Gruidae) all the rings become bony (Duncker, 1971). The rings are generally ossified in Passeriformes (Warner, 1972b). Since the rings are complete in birds, there is no counterpart of the Musculus trachealis of mammals in which the C-shaped cartilages are open dorsally.

In Sphenisciformes the trachea is divided into left and right channels by a median septum containing cartilaginous bars which are continuous with the tracheal rings, the craniocaudal length of the septum varying with the species (Watson, 1883; Zeek, 1951). A septum also occurs in certain Procellariiformes (Beddard, 1898, pp. 439, 449). A longitudinal dorsal ridge, resembling an incomplete septum, occurs in *Casuarius* species (Forbes, 1881).

Tracheal coils, the **Ansae tracheales,** occupy an excavation in the sternum in *Cygnus* and Gruidae (see Portmann, 1950, p. 265 for review, and Johnsgard, 1961). In some other species the coils lie between the skin and pectoral muscles, as in *Crax*, *Platalea*, and *Aramus* (see Portmann, 1950, p. 266) and in *Anseranas* (Johnsgard, 1961). Among passerine birds, *Manucodia*, is known to have a coiled trachea (Ames, 1971, p. 137), as is *Phonygammus* (Rüppell, 1933) both of these species being members of the Paradisaeidae.

In a few species such as *Dromaius* (Murie, 1867) and *Oxyura jamaicensis* (Wetmore, 1918) a sac-like diverticulum, the **Saccus trachealis,** opens from the trachea.

A bulbous expansion of the trachea, the **Bulbus trachealis,** occurs a short distance rostral to the syrinx in the males of many Anseriformes including *Melanitta fusca* (Yarrell, 1833; Beddard, 1898, p. 463).

(32) **Syrinx.** Huxley (1877) introduced this term to avoid the confusion of "upper" and "lower" larynx. The anatomy of the syrinx is exceedingly variable among the avian species. However, the structures listed in the *Nomina* appear to include the basic components which are present in many birds.

Three types of syrinx are recognised: (1) the tracheobronchial syrinx, occurring in the great majority of birds (Fürbringer, 1888, p. 1088; Gadow, 1896, p. 941; Beddard, 1898, p. 161), based on "tracheal" and "bronchial" elements (Fig. 6b); (2) the tracheal syrinx, occurring in the suboscine Passeriformes of the superfamily Furnarioidea (Figs. 6c, 6d) these groups being known collectively as the "tracheophonae" (Fürbringer, 1888, p. 1088; Gadow, 1896, p. 940), and in certain Ciconiidae (Beddard, 1898, p. 69), characterized by the modification of many "tracheal" rings; and (3) the bronchial syrinx, occurring in certain Caprimulgiformes (Garrod, 1973), in some Cuculiformes (Beddard, 1885), and to some extent in Strigidae (Beddard, 1896, p. 251), and founded on modified "bronchial" elements. This classification is useful anatomically and is adopted in the subsequent Annotations, with the reservation that its functional implications must be treated with caution.

(33) **Tympanum**. Synonymy: Trommel (Wunderlich, 1884); tracheal box (Gadow, 1896, p. 941; Forbes, 1881; Beddard, 1898, p. 289); drum (Ames, 1971, p. 15); tympanic box or tympanic chamber (Warner, 1972b). This structure is the rigid cylinder formed by the very close apposition, or more often the fusion of tracheal syringeal cartilages. The Tympanum is often present in the tracheobronchial type of syrinx, being found in oscines (Figs. 4, 5a) and some suboscine Passeriformes, as well as in most of the nonpasseriform groups (Beddard, 1898, p. 61) (e.g. Figs. 3, 5b, 6a). The cartilages frequently become ossified as in *Larus argentatus* (Rüppell, 1933), *Gallus* (Gross, 1964a), and oscines generally (Haecker, 1900).

The Tympanum is also present in some forms possessing the tracheal type of syrinx, as in certain suboscine Passeriformes, including the Dendrocolaptidae (Fig. 6d), most Furnariidae, and *Melanopareia* (Fig. 6c) among the Rhinocryptidae (Ames, 1971, pp. 21, 24, 31).

In the bronchial type of syrinx the Tympanum seems generally to be absent as in *Steatornis caripensis* (Garrod, 1873).

(34) **Bulla syringealis.** Synonymy: Bulla tympanica. Typically an asymmetrical dilation of the left side of the syrinx (Fig. 6a) in males of the subfamily Anatinae, except the Oxyurini (Johnsgard, 1961), possibly arising embryologically from the left bronchial syringeal cartilages (Broman, 1942; see also King, 1975, p. 1910); in

Tadorna tadorna the asymmetrical dilation is greater on the right side (Gadow and Selenka, 1891, p. 727; A. S. King, pers. obs.). In the males of most Anatini the bulla is largely or entirely osseous, but in all Aythyini and most Mergini the Bulla is partly or extensively membranous (Johnsgard, 1961) as in *Aythya fuligula* (Fig. 5b) (Warner, 1971). Among the subfamily Anserinae there is a bilaterally symmetrical Bulla in the Dendrocygnini, but the Bulla is absent in the true geese and swans (Anserinae) and in the Anseranatinae (Johnsgard, 1961).

(35) **Cartilagines syringeales.** Numerous descriptions of the syrinx since the second half of the nineteenth century have been based on its "tracheal" and "bronchial" skeletal elements (e.g. Fürbringer, 1888, p. 1087; Haecker, 1900; Köditz, 1925; Rüppell, 1933). Usually the tracheal elements are numbered in a caudocranial sequence and the bronchial elements in a craniocaudal sequence. Some authors (e.g. Miskimen, 1951) called the first three bronchial elements "intermediary bars". Unfortunately there is no agreement on the anatomical boundary between the tracheal cartilages proper and the tracheal elements of the syrinx. One of the few to attempt to establish this boundary was Haecker (1900) who defined it as the cranial limit of the Tympanum; this criterion could be used in a great many species, but it fails in those in which the Tympanum is indistinct or absent. The extent of the specialized syringeal musculature could be used to determine the extent of the skeletal elements; this could be effective in oscines, but would fail in the many species devoid of such muscles.

On functional grounds, Ames (1971, p. 131) questioned the validity of the terms "tracheal" and "bronchial", and therefore proposed a new nomenclature (Ames, 1971, p. 14) based solely on the form of the individual cartilages: his type A cartilages are flat in transverse section, stiff in texture, and have a concave curvature of their caudal border; his type B cartilages are rounded in transverse section, flexible, and have a concave curvature of their cranial border. His type A cartilages are always the more cranial, the two series being continuous; the Tympanum is formed of A elements.

The functional arguments against "tracheal" and "bronchial" are compelling, but the terms "tracheal" and "bronchial", being based on the gross arrangement of the airway, have the advantage of being descriptively informative and therefore easy to remember. Retaining them also meets the requirement of not interfering with long-established and widely accepted terms. For these reasons the skeletal elements of the syrinx are classified in the *Nomina* into four basic components, namely the Cartilagines tracheales syringis and the Cartilagines bronchiales syringis, together with the Pessulus and the Cartilagines accessoriae. Nevertheless the system devised by Ames (1971) could be combined with the classical terminology adopted here.

The syringeal "cartilages" are commonly ossified, but the extent of ossification is very variable (Fürbringer, 1888, p. 1086). The terms "cartilage", "cartilaginous", etc., have been applied to the syrinx by various authors (e.g. Garrod, 1879; Miskimen, 1963; Ames, 1971, 1975; Lockner and Youngren, 1976). In mammals the laryngeal and tracheal skeletal elements are quite often ossified but are nevertheless known as "cartilages".

(36) **Cartilagines tracheales syringis.** Essentially these cartilages constitute a direct continuation of the trachea: thus they lie in the midline cranial to the bifurcation of airway, and typically are complete rings. They may be modified in various ways, but especially by being more or less closely attached to each other to form the Tympanum (Annot. 33) as in the majority of birds. In *Gallus* there are about eight tracheal syringeal cartilages (Fig. 3); the most cranial three or four are strong and well curved

forming the Tympanum, but the most caudal four are very thin and flattened (constituting the syringeal bars of Morejohn, 1966).

In the tracheal type of syrinx many tracheal syringeal cartilages may be included in the syrinx. For example as many as 78 are said to occur in some Ciconiiformes (Beddard, 1898, p. 69), and 10 to 12 in members of the Furnarioidea (e.g. *Chamaeza*) and some Rhinocryptidae (e.g. *Melanopareia*) as in Fig. 6c (Ames, 1971, Plate 4).

(37) **Cartilagines bronchiales syringis.** These elements typically lie caudal to the bifurcation of the airway and appear to constitute the skeleton of the most cranial part of the left and right primary bronchi; thus they are usually paired, incomplete, and C-shaped (Halbringe in the German literature). In addition, they tend to be more or less extensively differentiated from the true bronchial cartilages which follow them caudally, being relatively broad, irregular in shape with expanded ends, and of greater diameter. A further major characteristic is that their free ends typically support the Membrana tympaniformis medialis. There are three bronchial syringeal cartilages (Figs. 4 and 5a) in oscines (Haecker, 1900, Köditz, 1925; Warner, 1972b), five or six in *Larus argentatus* (Rüppell, 1933), and typically three in *Gallus* (Myers, 1917) (Fig. 3). In some species the first cartilages caudal to the trachea are complete rings rather than C-shaped elements. These occur especially in the bronchial type of syrinx, where 10 or more complete bronchial rings may follow the bifurcation of the trachea, as in *Crotophaga* (Beddard, 1898, p. 276) and *Steatornis caripensis* (Garrod, 1873); the complete rings are followed by several C-shaped cartilages bearing a disc-like Membrana tympaniformis medialis between their free ends.

(38) **Pessulus.** Synonymy: Steg (Haecker, 1900). This median cartilage (Figs. 3, 5a) splits the airway of the syrinx in most birds. It is absent in certain ratites (Forbes, 1881; Rüppell, 1933), in the oscine Alaudidae (Mayr, 1931; Ames, 1971, p. 104), and in suboscine Passeriformes such as the Furnarioidea and many of the Pittidae and Tyrannidae (Köditz, 1925; Ames, 1971, p. 104). It is represented only by connective tissue in Columbidae (Warner, 1972a). In most Passeriformes the Pessulus is well developed and ossified, though in *Corvus corone* it is small and wholly cartilaginous or only partly ossified (Warner, 1972b).

(39) **Membrana semilunaris.** A projection of the mucosa (Fig. 5a) which sometimes extends the Pessulus cranially. It is not present in Strigidae (Miller, 1934). Rüppell (1933) claimed it to be absent in *Larus argentatus* but present in many oscines. Warner (1972b) confirmed its presence in many oscines though not as a vibratile structure, but noted its absence in *Delichon urbica*.

(40) **Cartilagines accessoriae.** This term was used by Ames (1971, p. 143) to include two types of small accessory cartilages in the syrinx: (1) the two pairs of cartilages which are sometimes embedded in the medial tympaniform membrane, i.e. the Cartilago membranosa dorsalis and ventralis); and (2) the paired Processus vocalis. **Cartilago membranosa dorsalis; Cartilago membranosa ventralis.** Synonymy: Cartilagines arytenoideae (Müller, 1847; Owen, 1866, p. 223); Stellknorpel (Haecker, 1900); Cartilagines tensores (Wunderlich, 1884); internal cartilages (Ames, 1971, pp. 104, 144). These thin paired cartilages of irregular shape are nearly always blended into the left and right medial tympaniform membranes, being found in suboscine Passeriformes, especially in Tyrannidae, where they are amongst the most uniform of all the cartilaginous elements of the syrinx at the intrageneric level (Ames, 1971, pp. 104, 144). Haecker (1900) illustrated a dorsal cartilage in the oscine *Pica*

(Fig. 4). Köditz (1925) believed that the ventral cartilage occurs in all oscines, a dorsal cartilage being present in only some members of this group.

A dorsal cartilage was illustrated by Müller (1847, Plate V, Fig. 3) in *Anthracothorax dominicus* (Trochilidae). Beddard (1898, p. 191) described a separate rounded piece of cartilage at the ventral aspect of the first and second bronchial cartilages of the syrinx of Ramphastidae. These reports appear to be the only mention of a Cartilago membranosa outside the Passeriformes.

Processus vocalis. A paired cartilage (Figs. 6c, 6d) which apparently occurs almost exclusively in all members of the superfamily Furnarioidea (Garrod, 1877; Ames, 1971a, p. 143). It consists of a cartilaginous or bony rod, or a thin curved plate, intimately attached to the lateral aspect of the trachael and bronchial syringeal cartilages. It gives attachment to M. sternotrachealis, and in some genera also to M. tracheolateralis and to the M. vocalis ventralis and/or dorsalis. Beddard, 1898, pp. 69, 423) described a rudimentary Processus vocalis in two species of Ciconiidae.

(41) **Ligamentum interbronchiale.** Synonymy: Bronchidesmus (Garrod, 1879; Myers, 1917). The term used by Wunderlich (1884) for the connective tissue bridge (Figs. 4, 5a, 5b) joining the left and right bronchi to each other (and often to the esophagus also P. L. Ames, pers. comm.) in many birds including most Passeriformes, although it is sometimes absent as in *Delichon urbica* (Warner, 1972b).

Foramen interbronchiale. Synonymy: Subpessular air-space (Warner, 1971, 1972a). The space between the bifurcation of the trachea and the Ligamentum interbronchiale (Figs. 5a, 5b), present in most passerines and in *Anas*, *Aythya*, and Columbidae (Warner, 1971, 1972a, 1972b).

(42) **Membrana tympaniformis lateralis.** Synonymy: Membrana tympaniformis externa (Gadow, 1896, p. 937; Greenewalt, 1968, p. 27). The lateral tympaniform membrane is an annular membrane or series of membranes between the lateral aspects of certain of the Cartilagines bronchiales syringis (Figs. 4, 5a). Ames (1971, p. 16) and Warner (1972b) doubted the presence of true membranous areas in the lateral walls of the syrinx in Passeriformes, and no evidence for them was found by Warner (1971) in *Anas*. On the other hand Stein (1968) believed them to be present in oscines and a source of frequency modulation. In Columbidae the lateral wall of the tracheobronchial junction consists of thickened connective tissue which could be called a lateral tympaniform "membrane", but is not really membranous and would be incapable of vibration (Warner, 1972a).

According to Gadow (1896, pp. 937, 941), in the majority of birds a number of lateral tympaniform membranes are present which generally extend between the second and third, and also the third and fourth bronchial cartilages of the syrinx; but sometimes the main lateral membrane lies between the last tracheal and first bronchial cartilages, as in *Rhea*, *Spheniscus*, *Perdix* and others. In *Gallus* the membrane is thin and extensive, being stretched across the wide space between the most caudal tracheal syringeal cartilage and the first bronchial syringeal cartilage; it is the main source of sound in this species (Gross, 1964b; Gaunt *et al.*, 1976).

Membrana trachealis. This is a modified form of Membrana tympaniformis lateralis, occurring only in suboscine Passeriformes possessing the tracheal type of syrinx, i.e. the Furnarioidea. Ames (1971a, p. 20) described it as a membranous window on the dorsal and/or ventral surface of the trachea (Figs. 6c, 6d), spanning several of the tracheal synringeal cartilages which are reduced to delicate rods. Thus the Membrana trachealis is really an almost continuous series of smaller Membranae tympaniformes laterales (Ames, 1971, p. 20).

(43) **Membrana tympaniformis medialis.** Synonymy: Membrana tympaniformis interna (Gadow, 1896, p. 937; Ames, 1971, p. 16; Warner, 1972b). The paired medial tympaniform membrane (Figs. 5a, 5b) seems to be present in virtually all species, being without question the primary vibrating membrane in many groups of birds (Greenewalt, 1968, p. 28). In some instances, however, such as the male of *Anas* it is too thick to be readily vibratile (Warner, 1971). In *Larus argentatus* the membrane is held between the free ends of five bronchial cartilages of the syrinx, numbers 2 to 6 (Rüppell, 1933), whereas in oscines it is supported by the three bronchial cartilages of the syrinx. Usually the Pessulus separates the left and right membranes. Caudal to the syrinx, the medial tympaniform membranes is continuous with the sheet of soft tissue which completes medially the C-shaped cartilages of the extrapulmonary part of the Bronchus primarius (Fig. 4).

(44) **Labium laterale; Labium mediale.** Synonymy: Labium externum and Labium internum (Haecker, 1900); Stimmpolster or Stimlippen (Rüppell, 1933). Paired band-like pads of elastic tissue, projecting respectively from the lateral and medial wall into the airway of the syrinx (Figs. 5a, 5b). Their occurrence is variable, neither labium being present for example in *Larus argentatus*, but both being present in oscines (Haecker, 1900; Rüppell, 1933), although the medial pair is small (Haecker, 1900) amounting to little more than a thickening of the medial tympaniform membrane (Greenewalt, 1968, p. 27). The Labium laterale appears to be one of the few constant features of the oscine syrinx (Greenewalt, 1968, p. 180).

(45) **Musculi tracheales.** The muscles of the upper airway are usually divided into tracheal and syringeal muscles, but this distinction is not really clear. A muscle which has extensive contact with the trachea could reasonably be called a tracheal muscle; yet by pulling on the trachea such a muscle will amost certainly act indirectly on the syrinx, and could therefore be regarded as a syringeal muscle. Here, the M. sternohyoideus, M. sternotrachealis, M. cleidotrachealis, and M. tracheolateralis are classified as tracheal muscles, but doubtless they all act indirectly on the syrinx.
M. sternohyoideus. At least in *Gallus* (McLelland, 1965) this muscle is attached to the larynx and to the most cranial part of the trachea.
M. sternotrachealis (Figs. 6a–g). Owen (1886, p. 224) regarded the M. sternotrachealis as "the most constant of all the muscles affecting the lower larynx", and George and Berger (1966, p. 263) believed that it has been found in all birds which have been investigated. It is, however, relatively feeble in oscines (Ames, 1975; Warner, 1972b). Moreover it has been reported as absent in some Psittaciformes, i.e. *Barnardius barnardi* (Beddard, 1898, p. 258) and *Brotogeris jugularis* (P. L. Ames, pers. comm.).
M. cleidotrachealis (Fig. 6a). Synonymy: M. ypsilotrachealis (Gadow and Selenka, 1891, p. 730; Rüppell, 1933; George and Berger, 1966, p. 263. The synonym, M. ypsilotrachealis, has been very widely employed, but the term M. cleidotrachealis has also been used for a long time (e.g. see Coues, 1890; Beddard, 1898). Unfortunately, the meaning of the term M. ypsilotrachealis has been confused because German veterinary anatomists have applied it to the M. sternotrachealis; this term has therefore been used for two entirely different muscles, and for this reason M. cleidotrachealis has been preferred. The M. cleidotrachealis has been described in many members of the Anseriformes (Gadow, 1896, p. 938; Beddard, 1998; p. 464; Rüppell, 1933; Lockner and Youngren, 1976). It has also been reported in representatives from four other orders, i.e. in *Crypturellus tataupa*, *Tockus*, and *Crax daubentoni* by

Beddard (1898, pp. 487, 222, 292) and in Sphenisciformes by Watson (1883). The essential characteristic of this muscle is its origin from the clavicle.

M. tracheolateralis (Figs. 6a–g). Synonymy: M. contractor tracheae (Watson, 1883); M. tracheobronchialis (Haecker, 1900); M. bronchotrachealis (Rüppell, 1933). Beddard's (1898) review indicates that this muscle is present throughout most of the subpasseriform orders, and Ames (1971, p. 105) implied its presence in all Passeriformes. It occurs in the great majority of avian orders (Fürbringer, 1888, p. 1089; Beddard, 1898; p. 61; Gadow, 1896, p. 938; Ames, 1971, p. 133), being absent only in some of the ratites, most Ciconiidae and Cathartidae, and some Galliformes (Gadow, 1896, p. 939; Beddard, 1898).

For further details of possible tracheal muscles see **Myol.** Annot. 37 and its accompanying Table 1.

(46) **Musculi syringeales.** These are very small muscles, not always sharply distinguished from each other anatomically. In species with several syringeal muscles it could be a matter of opinion whether some of these are fascicles of a single muscle or two or more genuine entities. This no doubt accounts for the different numbers of syringeal muscles claimed by various authors in oscines, even though this group is notable for the anatomical uniformity of these muscles (Ames, 1971, p. 94; 1975); for example the maximum total number of paired muscles (excluding M. sternotrachealis and M. tracheolateralis) is four according to Ames (1971, p. 89), five according to Warner (1972b), seven according to Fürbringer (1888, p. 1091), and eight according to Köditz (1925).

Ames (1971, p. 89) noted that two types of terminology have been devised for the syringeal muscles, one based on function (e.g. that of Wunderlich, 1884) and the other on topographical location (e.g. that of Owen, 1866, p. 223). He pointed out the danger of naming these muscles on a functional basis when so little is known about the operation of the syrinx. In any event, there is now a myriad of names. Moreover, the same name has sometimes been applied to non-homologus muscles throughout the avian orders (Berger, 1960, p. 306). As noted by George and Berger (1966, p. 268), Ames (1971, p. 10), and Warner (1972b), the nomenclature for the syringeal muscles which has most commonly been used in major works in English has originated from Owen (1866, p. 223). Ames himself employed this terminology. Fürbringer's (1888) nomenclature, however, was adopted by Gadow (1896, p. 939) and Haecker (1900), and has been closely followed by Warner (1972b) also. Among other major terminologies are those of Wunderlich (1884), Setterwall (1901), and Köditz (1925). The synonyms for the oscines were worked out by Köditz (1925), Ames (1971, p. 90), and (1972b), essentially as in Table 2. The terminology adopted here is based on that of Fürbringer (1888).

The syringeal muscles are often classified as extrinsic and intrinsic muscles. An extrinsic syringeal should have an attachment beyond the syrinx, whereas an intrinsic syringeal muscle should have all its attachments on the syrinx. Unequivocal intrinsic syringeal muscles appear to be quite rare, even in Passeriformes where only the very short muscles of the suboscine passeriforms (e.g. the M. vocalis dorsalis in Fig. 6c) are undeniably attached solely to the skeletal elements of the syrinx. In this work, all of these muscles are classified simply as Musculi syringeales.

The anatomy of the syringeal muscles is less well-known in subpasseriform birds than in the passeriform groups. Nevertheless the literature of the 19th century contains a wealth of information on the subpasseriform syrinx, which was fully reviewed by Beddard (1898), and summarized by Fürbringer (1888, p. 1087) and Gadow (1896, p. 937). The principal sources on the passeriform syringeal muscles are the wide-ranging investigation by Ames (1971, 1975) and the studies by Miskimen (1951) and Warner (1972b).

The first five muscles in the list (M. tracheobronchialis dorsalis, brevis, and ventralis, and the M. syringealis dorsalis and ventralis) all of which are paired, are present in oscines as in Fig. 6g (Owen, 1866, p. 223; Fürbringer, 1888, p. 1091; Shufeldt, 1890, p. 48; Miskimen, 1951; Warner, 1972b), their synonyms being listed in Table 2. Ames (1971, pp. 89–94) confirmed the presence of these muscles in oscines, but interpreted the M. tracheobronchialis dorsalis and brevis as two parts of the same muscle, and considered the M. syringealis ventralis to consist of medial and lateral parts (Fig. 6g). The M. tracheobronchialis dorsalis and ventralis, and the M. syringealis dorsalis also occur in the suborder Menurae as in Fig. 6f (Ames, 1971, pp. 85, 87). The four remaining muscles in the list (M. vocalis dorsalis and ventralis, and M. obliquus ventralis and lateralis) occur in suboscine passeriforms as in Figs. 6c, d, and e (Ames, 1971, pp. 20–79).

In the non-passeriform groups there is usually only a single (clearly extrinsic) syringeal muscle, the M. tracheolateralis, which typically inserts on the tracheal or bronchial elements of the syrinx. This is the basic tracheobronchial syringeal pattern, which is generally agreed to be present in the majority of avian orders (Fürbringer, 1888, p. 1089; Gadow, 1896; p. 938; Beddard, 1898; p. 61; Ames, 1971, p. 133). It is also present in several families of suboscine Passeriformes as in Fig. 6b (Ames, 1971, p. 133). The presence of one additional pair of short specialized syringeal muscles in *Gallinago* and *Falco* was mentioned by Wunderlich (1884), Fürbringer (1888, p. 1089), and Gadow (1896, p. 939), but without reliable descriptions. There are certainly two pairs of short syringeal muscles in some Psittaciformes, e.g. *Brotogeris jugularis* (Ames, pers. comm.), *Psittacula* (Gadow and Selenka, 1891, Plate 50, Figs. 8–9), *Ara ararauna* (Yarrell, 1833; A. S. King, pers. obs.) and *Melopsittacus undulatus* (Evans, 1969, p. 70); furthermore Müller (1847, Plate V, Fig. 1) clearly illustrated two such pairs of muscles in the trochilid *Anthracothorax dominicus*. In a small number of groups, i.e. some ratites, most Ciconiidae and Cathartidae, and some Galliformes (Gadow, 1896, p. 939; Beddard, 1898), there are said to be no syringeal muscles whatsoever, not even the extrinsic syringeal M. tracheolateralis.

(47) **Bronchus primarius, Pars extrapulmonalis** et **Pars intrapulmonalis; Cartilagines bronchiales.** The term Bronchus primarius was introduced by Juillet (1912). The cartilages of the extrapulmonary part of the primary bronchus are incomplete C-shaped structures, except in the Hirundinidae in which they are complete rings (Warner, 1972b). See Annot. 53.

(48) **Facies costalis; Facies vertebralis; Facies septalis.** The Facies costalis adjoins the thoracic wall, the Facies vertebralis (Fig. 7) is in contact with the vertebrae, and the Facies septalis faces the Septum horizontale (see Cavitas pleuralis). Other terms in fairly common use depend on the orientation of the surface, e.g. Facies dorsolateralis, medialis, ventralis, respectively, but these terms are not always meaningful. For instance, as Juillet (1912) pointed out, the so-called ventral surface is particularly complicated; it varies in its orientation, some parts of it having a strong medial component. The terms Facies costalis, vertebralis, septalis, are preferred as being less ambiguous.

(49) **Sulci costales; Tori pulmonales; Tori intercostales; Tori marginales.** The ribs make impressions, Sulci costales, on the Facies costalis and Facies vertebralis (Fig. 8). Between two Sulci costales (Fig. 8) is a Torus intercostalis (Quitzow, 1970). At the

Angulus craniodorsalis and at the Angulus caudodorsalis there is a Torus marginalis cranialis and caudalis respectively (Fig. 8). Together, the Tori intercostales and Tori marginales constitute the Tori pulmonales. The term Torus pulmonis was used by Schulze (1908) and adopted by Groebbels (1932, p. 44).

(50) **Margo costovertebralis; Margo costoseptalis; Margo vertebroseptalis; Margo cranialis; Margo caudalis; Angulus craniodorsalis; Angulus caudodorsalis; Angulus cranioventralis; Angulus caudoventralis.** The following usage is based in principle on suggestions by Quitzow (1970). The Margo costovertebralis (Fig. 7) separates the Facies costalis from the Facies vertebralis, while the Margo costoseptalis (Fig. 7) separates the Facies costalis from the Facies septalis; the Margo vertebroseptalis (Fig. 7) forms the boundary between the Facies vertebralis and the Facies septalis. The Margo cranialis and the Margo caudalis form the cranial and caudal borders of the lung respectively (Fig. 7). The Margo costovertebralis and the Margo vertebroseptalis join cranially with the Margo cranialis at the Angulus craniodorsalis; caudally they join with the Margo caudalis at the Angulus caudodorsalis (Fig. 7). The Margo costoseptalis meets the Margo cranialis at the Angulus cranioventralis, and meets the Margo caudalis (Fig. 7). This description applies to the essentially quadrilaterally-shaped lung of most birds. In primitive species with a poorly developed Neopulmo the Angulus cranioventralis and the Angulus caudoventralis are indistinct; in these birds the lung is triangular with the apex pointing ventrally (Duncker, 1972).

(51) **Planum anastomoticum; Linea anastomotica.** On the surface of the lung the Planum anastomoticum is visible as the Linea anastomotica of Locy and Larsell (1916a) (the Linea serpta of Quitzow, 1970); the Linea anastomotica is a continuous line extending dorsoventrally on the cranial end of the Facies costalis and craniocaudally on the Facies vertebralis (Fig. 7). The Planum anastomoticum and its superficial Linea are caused mainly by (a) the terminal anastomoses of the Parabronchi of the medioventral secondary bronchi with those of the mediodorsal secondary bronchi (Locy and Larsell, 1916a); also by (b) the terminal anastomoses of the Parabronchi of the medioventral secondary bronchi with those of the Bronchi lateroventrales (Fig. 9).

(52) **Paleopulmo; Neopulmo.** These terms were proposed by Duncker (1971), to indicate a phylogenetic relationship between the two main components of the lung. The Paleopulmo is present in all birds. It consists (Fig. 9) of (a) Bronchi medioventrales and mediodorsales, and their Parabronchi; (b) the large Bronchus lateroventralis which connects directly to the Saccus thoracicus caudalis; and (c) two or three intermediate-sized Bronchi lateroventrales, which medially form Parabronchi joining Parabronchi from the fourth Bronchus medioventralis at the Planum anastomoticum. In supposedly primitive birds (e.g. sphenisciform species) the Paleopulmo forms the whole of the lung.

In most birds (including the domestic species) the Neopulmo also is present, typically forming about one tenth of the lung. It is a network (Fig. 9) of anastomosing bronchi consisting of: (a) the small Bronchi laterodorsales and their Parabronchi; (b) laterally directed Parabronchi of the Bronchi lateroventrales; and (c) the connexions of this network to the caudal air sacs. When well developed, as in *Gallus*, the Neopulmo displaces the primary bronchus and the Bronchi mediodorsales medially, so that they no longer lie on the surface of the lung. It appears to correspond with the réseau anastomotique of Campana (1875) and Locy and Larsell (1916a). (Annot. 52 cont. on p. 244).

In the present state of knowledge of the anatomy of the avian lung the phylogenetic validity of the terms Paleo- and Neopulmo may, however, be somewhat uncertain. As possible alternative terms, Pulmo arcuiformis and Pulmo reteformis are anatomically descriptive. Pulmo arcuiformis, as an alternative to Paleopulmo, indicates the arc-like character of the Parabronchi in this part of the lung. Pulmo reteformis, as an alternative to Neopulmo, indicates the net-like Parabronchi which typify the Neopulmo.

(53) **Bronchus primarius, Pars intrapulmonalis.** That part of the Bronchus primarius which lies within the lung is the Pars intrapulmonalis (Juillet, 1912). The possible presence of a dilated region, called the Vestibulum, was described by Huxley (1882) and mentioned by many other authors (e.g. Duncker, 1971). Its presence in *Gallus* has been denied (Juillet, 1912; Payne and King, 1959; Akester, 1960; Quitzow, 1970). Akester (1960) found no convincing evidence for it in *Columba* and *Anas*. The term Vestibulum has also been applied to that region of the Bronchus primarius which carries the openings of the Bronchi medioventrales (e.g. by Juillet (1912) and by Groebbels (1932, p. 47); or, on the contrary, it has been applied to the region of the Bronchus primarius which is devoid of branchings (Duncker, 1971). The term Mesobronchus was used by Huxley (1882) and a number of later authors for the region of the Bronchus primarius which lies caudal to the supposed Vestibulum.

(54) **Bronchi secundarii.** These are the bronchi of the second order, and therefore include all those bronchi which arise from the primary bronchus (Campana, 1875, p. 31; Locy and Larsell, 1916a). The earlier authors (e.g. Campana, 1875; Schulze, 1908) named them according to either the orientation of their origin from the primary bronchus, or their subsequent direction, or both their origin and their direction. At that time, unfortunately, both their orientation and their direction were erroneously described (see Annot. 57), and consequently the suggested terminology was misleading. A further complication has been the use of the same term for two or more different groups of secondary bronchi, as with "ventrobronchi", "dorsobronchi", and "laterobronchi" (see the terms in italics in Table 1). The resulting confusion has pervaded the literature on the anatomy and physiology of the avian lung. A completely new approach has now become imperative. This approach is based on the two following principles: (1) All terms for the Bronchi secundarii should be based on the territory which these Bronchi supply, as in Fig. 9 (a precept which has hitherto been relatively neglected, as pointed out by Quitzow, 1970); (2) Terms which have already been in use, but with different meanings, should be avoided. The terminology for the secondary bronchi which is adopted here is based on these principles.

(55) **Bronchi medioventrales.** These are the four secondary bronchi which supply the medial and ventral regions of the lung (Fig. 9). They arise from the dorsomedial wall of the cranial region of the Pars intrapulmonalis of the Bronchus primarius (Fig. 8), as described by Fischer (1905), King (1966, p. 181), and Quitzow (1970). Many of their main branches pass in the medial direction over the Facies septalis to end by turning dorsally onto the Facies vertebralis (Duncker, 1971). Other names used in the literature for these Bronchi are shown in Table 1.

(56) **Bronchi mediodorsales**. These bronchi supply the medial and dorsal regions of the lung (Fig. 9). They arise from the dorsal wall of the caudal part of the Bronchus primarius (Fig. 8), and then travel dorsally (Duncker, 1971). Other names used in the literature for these Bronchi are shown in Table 1.

(57) **Bronchi lateroventrales.** These supply the lateral and ventral regions of the lung (Fig. 9). Their origin is directly opposite to those of the Bronchi mediodorsales (Fig. 8), from the ventral wall of the caudal part of the Bronchus primarius, and their course is ventral or caudoventral (Payne and King, 1960; King, 1966, p. 181; Quitzow, 1970; Duncker, 1971). In birds generally, the first or second of these Bronchi lateroventrales connects with the Saccus thoracicus caudalis. The subsequent two or three intermediate-sized ones contribute Parabronchi to the Paleopulmo in all species; in species in which the Neopulmo is developed, these intermediate-sized Bronchi lateroventrales and yet other more caudal and smaller ones contribute Parabronchi to the Neopulmo (Fig. 9).

Other names used in the literature for the Bronchi lateroventrales are shown in Table 1. The use of the term "external" or "lateral", or "laterobronchi" in the earlier literature arose from faulty orientation in the observations published by Campana (1875, pp. 32–33), Schulze (1908), and Juillet (1912); these authors believed that these bronchi arose from the *lateral* wall of the Bronchus primarius, and they named the bronchi accordingly. The term laterobronchi was then adopted by Locy and Larsell (1916a) and Groebbels (1932), who were not aware of the error. Subsequently the wrong orientation was again repeated by McLeod and Wagers (1939) and King (1956). Eventually the error was noted by Payne and King (1960), and analysed by King (1966, pp. 180–182).

(58) **Bronchi laterodorsales.** These supply the lateral and dorsal regions of the lung (Fig. 9). They form a large part of the Neopulmo; since the degree of development of the Neopulmo varies so greatly in the different species, the position, diameter, and number of the Bronchi laterodorsales are also correspondingly variable. These bronchi constitute the fourth group of Bronchi secundarii, being recognized as such by Campana (1875), Locy and Larsell (1916a), Groebbels (1932), McLeod and Wagers (1939), and King (1966). Quitzow (1970) included them, with the other three series of secondary bronchi, among her "regional bronchi". They arise from the lateral wall of the caudal part of the Bronchus primarius and extend mainly laterally. For other names used in the literature see Table 1. The first two or three of these Bronchi secundarii are of large diameter (certainly in *Gallus*, see King, 1966, p. 182; and in *Cygnus*, see Fig. 30b of Duncker, 1971). The more caudal of these secondary bronchi are similar in diameter to Parabronchi; because of the small diameter of these more caudal bronchi, and their great variability between species, some authors (e.g. Akester, 1960; Duncker, 1971) excluded the whole of this group of bronchi from the category of Bronchi secundarii and classified them as Parabronchi.

(59) **Parabronchus.** Synonymy: Bronchus tertiarius (King, 1966); Lungenpfeife (Krause, 1922); Bronchi fistularii (Quitzow, 1970; Gerisch, 1917). The Parabronchi are bronchi of the third or subsequent orders, i.e. those arising from the Bronchi secundarii or from the subdivisions of the Bronchi secundarii (Fig. 8). The term Bronchi tertiarii was introduced by Campana (1875) for these bronchi, and has subsequently remained in wide use (e.g. McLeod and Wagers, 1939; Akester, 1960; King, 1966; Evans, 1969; Lasiewski, 1972). It has the advantage of logically completing the sequence of the three orders of bronchi, i.e. primary, secondary and tertiary bronchi. On the other hand the term Parabronchi is also long established, having been introduced by Huxley (1882) and used by numbers of recent authors (e.g. Duncker, 1971). It has the great advantage of being a single compact term, and is preferred for this reason.

The term Parabronchus includes both the parabronchial airway and the mantle of exchange tissue which surrounds that airway, the boundary of the Parabronchus being the Septum interparabronchiale (in those species which have it). Within a given species, the lumen of the Parabronchi is relatively small and uniform in calibre.

(60) **Atria.** The Atria of the avian lung are the polygonal chambers (Fig. 10) leading into the air capillaries from the Parabronchi and from many secondary bronchi (Duncker, 1971). They are not homologous to the atria of the mammalian lung. The term is deeply entrenched in the literature, having been in general use for at least the half century since Krause's (1922) book. There have been many alternative terms, reviewed by Gerisch (1971), which have fallen into disuse. See also Annot. 65.

(61) **Musculi atriales.** The network of smooth muscle bundles (Fig. 10) surrounding the openings into the atria (King and Cowie, 1969; Gerisch, 1971). See Annot. 65.

(62) **Septa interatrialia.** The thin septa (Fig. 10) which separate adjacent atria (Gerisch, 1971), and in fact constitute the atrial wals. See also Annot. 65.

(63) **Infundibula.** The term Infundibulum appears to have originated from Krause (1922) and remained in fairly general use (e.g. King, 1966; Duncker, 1971; Gerisch, 1971). It refers to the funnel-shaped ducts which open from the floor of the Atria (Fig. 10) and lead to the air capillaries. According to Gerisch (1971) the infundibula may give rise not only to Pneumocapillares but also to Rami respiratorii, which are intermediate in size between the Infundibulum and the Pneumocapillares.

(64) **Cellula granularis.** This cell lines the Atria and contains osmiophilic laminated bodies (see King and Molony, 1971, p. 119). It is probably homologous to the similar cell which lies in the epithelial lining of the mammalian alveolar wall and is variously known in mammals as the septal cell, great alveolar cell, type 2 epithelial cell, and granular pneumocyte.

(65) **Septa interparabronchialia.** The septa of connective tissue separating the exchange tissue of adjacent Parabronchi. According to Duncker (1971), these septa, and also the Atria and the Musculi atriales, are best developed in birds which fly poorly or not at all (e.g. *Fulica* and *Gallus*). In other species (e.g. *Columba*) all of these structures are somewhat reduced. Extreme reduction occurs in yet other species (e.g. small Passeriformes and *Melopsittacus*); in these the Atria are almost completely reduced and the Musculi atriales are thinner; furthermore the Septa interparabronchialia are entirely lost.

(66) **Pneumocapillares.** Synonymy: Tubuli respiratorii, Ductuli respiratorii. The term air capillary or Luftkapillaren has been in quite general use since the beginning of the 20th century. Gerisch (1971) suggested Ductuli respiratorii, because he believed that anastomoses are not sufficiently common to justify the term air capillary, but further observations are needed to verify this claim.

(67) **Cellula squamosa.** The squamous epithelial cell which lines the Pneumocapillares. This cell is homologus to the similar cell which lines the mammalian alveolar wall and is variously known as the respiratory cell, small alveolar cell, and type 1 epithelial cell.

(68) **Ostium.** This term has been widely used, but few authors have defined it. Muller (1908) and Schulze (1908) appeared to apply it to the general zone of attachment of the air sac to the lung, enclosing one or more actual orifices which open into the bronchi; King (1966, p. 224) defined it explicitly in this way. On the other hand, Campana (1875, p. 27) used Infundibulum for this general zone, but this usage remained unnoticed. Quitzow (1970) introduced the term Area saccopulmonalis for the same region; like Groebbels (1932) she appeared to restrict the term Ostium to the actual opening from an air sac into a secondary (or the primary) bronchus. All of these definitions are reasonable, but the employment of only one set of terms is now necessary. The older and relatively widely used meaning of Muller (1908), Schulze (1908), and King (1966) is therefore adopted here. Thus Ostium refers to the general area of attachment of an air sac to the lung (Fig. 8).

Within the Ostium there are two possible types of opening. One type connects the air sac to Parabronchi; such connexions are the Bronchi recurrentes of Schulze (1910), Juillet (1912), and other authors, and the indirect connexions of King (1966, p. 223). The other type connects the air sac to a secondary (or to the primary) bronchus; such a connexion is the "bronche directe" and "orifice direct" of Juillet (1912), the "direkt Bronchen" and "direct Ostien" of Groebbels (1932), and the "direct connexion" of King (1966, p. 223). If any of these or similar terms should indicate the direction in which the air flows in these connexions, it should be avoided, in view of the discoveries between 1968 and 1972 of the complex air flow between the lungs and air sacs (see King, 1975, pp. 1915–1916 for a summary. This certainly excludes Bronchi recurrentes, and probably also the terms Bronchi pulmosaccales and saccopulmonales of Quitzow (1970). The terms "direct" and "indirect connexions" avoid suggesting the direction of air flow, and have the advantage of being relatively long established. Consequently they are recommended for descriptive purposes, but without being listed as official terms. Other acceptable alternatives for descriptive use are "parabronchial connexions", and "secondary" or "primary bronchial" connexions, indicating the connexions between the air sac and Parabronchi, secondary bronchi, or primary bronchus, respectively.

(69) **Saccobronchus.** This is an old term adapted to a new meaning by Duncker (1971). It now refers to a single large funnel-like bronchus which collects many parabronchi and connects with an air sac. This new usage of the term Saccobronchus must be carefully distinguished from the old usage. Originally the term was applied by earlier authors (e.g. Schultze, 1910) to the many small parabronchial connexions of an air sac, the term thus being synonymous with Bronchus recurrens; this older meaning of Saccobronchus has still been retained by some more recent authors (e.g. Quitzow, 1970).

A Saccobronchus occurs only in the Saccus abdominalis and Sacchus thoracicus caudalis. The Saccus abdominalis has been found to possess one such Saccobronchus in birds generally, except Spheniscidae and Dromiceiidae (Duncker, 1917). A Saccobronchus is particularly well developed in *Pluvialis*, *Anas*, and *Ciconia*, in both the Saccus abdominalis and the Saccus thoracicus caudalis (see Duncker, 1971, Figs. 24, 25, 26, 32).

(70) **Diverticula vertebralia.** Diverticula of the Saccus cervicalis passing cranially and caudally along the vertebral column in birds generally (Groebbels, 1932, p. 56; King, 1966; Duncker, 1971), that typically invade the vertebrae (Groebbels, 1932, p. 56).

(71) **Diverticula intermuscularia.** In some birds diverticula of the Saccus cervicalis invade the cervical musculature and accompany branches of the brachial plexus (Duncker, 1971).

(72) **Diverticula subcutanea.** Extensive subcutaneous diverticula of the Saccus cervicalis have been reported in certain Pelecaniformes and a few other species (Groebbels, 1932, pp. 62–64; King, 1966, p. 211) probably including *Leptoptilos* (Akester *et al.*, 1973).

(73) **Saccus clavicularis.** The rival terms Saccus clavicularis (Schulze, 1908) and Saccus interclavicularis (Juillet, 1912) have enjoyed about the same length and frequency of usage. Clavicularis is preferred for its relative brevity.

(74) **Pars medialis; Pars lateralis.** The Pars medialis and Pars lateralis of the Saccus clavicularis are the primordial paired medial and lateral components of Locy and Larsell (1916a, b). These four primordial sacs fuse in most species, to form a single unpaired Saccus clavicularis. In *Meleagris*, however, the paired Pars medialis persists in the adult as a pair of very small separate sacs (King and Atherton, 1970). It has been claimed that among the Lari and Ciconiiformes the paired Pars medialis and the paired Pars lateralis remain perfectly separate, giving four separate Sacci claviculares in the adult (for review see King, 1966, p. 214).

(75) **Diverticula intrathoracica.** The intrathoracic diverticula of the Saccus clavicularis extend variably around the heart (Diverticula cardiaca) and along the sternum (Diverticula sternalia) (Groebbels, 1932, pp. 57, 74–76; King, 1966, p. 212; Duncker, 1971).

(76) **Diverticula extrathoracica.** The extrathoracic diverticula spread between the muscles and bones of the pectoral girdle and shoulder joint, sometimes invading certain of the bones as well as the subcutaneous tissues (Groebbels, 1932, pp. 58, 74–76; King, 1966, p. 212; Duncker, 1971). Important examples of these Diverticula in *Gallus*, *Meleagris*, and other species are the Diverticulum axillare and Diverticulum humerale. The latter often penetrates inside the humerus, but this is not a universal feature being absent, for example in *Phalacrocorax* although present in related forms (e.g. *Sula*). See **Osteo.** Annot. 188–189. Groebbels (1932, pp. 58–59) recognized four main extrathoracic diverticula: the subscapular diverticulum lies essentially between the scapula and the thoracic cage; the axillary diverticulum spreads between the muscles around the shoulder region, and in many species gives off the humeral diverticulum which invades the humerus; the subpectoral diverticulum lies under the pectoral muscles; the suprahumeral diverticulum covers the head of the humerus and communicates with the subscapular and axillary diverticula. However, caution is necessary about these generalizations, since there is no agreement in the literature about this terminology and furthermore very little is known about species variations.

(77) **Diverticula perirenalia; Diverticula femoralia.** In many birds, variable diverticula extend laterally along the kidneys invading the adjacent vertebrae and pelvic girdle (Diverticula perirenalia), and into the bones and muscles of the pelvic limb (Diverticula femoralia) (Groebbels, 1932, p. 59; King, 1966, p. 216; Duncker, 1971).

TABLE 1. Some of the synonyms for the Bronchi secundarii

	Bronchi medioventrales	Bronchi mediodorsales	Bronchi lateroventrales	Bronchi laterodorsales
Campana (1875)	Divergent	Internal	External	Posterior or dorsal
Huxley (1882)	Entobronchia	Ectobronchia		
Schulze (1908)	Ventral	Dorsal	Lateral	Intermediate
Juillet (1912)	Entobronchi	Ectobronchi		
Locy and Larsell (1916a,b)	Entobronchi	Ectobronchi	*Laterobronchi*	*Dorsobronchi*
Groebbels (1932)	Ventral	Dorsal	Lateribronchi	Dorsolateral
Vos (1934)	*Ventrobronchi*	*Dorsobronchi*	*Laterobronchi*	
McLeod and Wagers (1939)	*Ventrobronchi*	Dorsomedial	Dorsolateral	Dorsal
Grau (1943)	*Ventrobronchi*	*Dorsobronchi*		
Akester (1960)	Anterior dorsal	Posterior dorsal	Posterior ventral	
King (1966)	Craniomedial	Caudodorsal	Caudoventral	Caudolateral
Quitzow (1970)	Mediobronchi	*Dorsobronchi*	*Ventrobronchi*	*Laterobronchi*
Duncker (1971)	*Ventrobronchi*	*Dorsobronchi*	*Laterobronchi*	

The terms in italics are those which have different meanings to different authors

TABLE 2. Synonymy of syringeal muscles in the oscines

N.A.A.	Owen, 1866; Shufeldt, 1890; Miskimen, 1951	Ames, 1971	Fürbringer, 1888; Gadow, 1896; Haecker, 1900	Warner, 1972b	Köditz, 1925	Wunderlich, 1884	Setterwall, 1901
M. tracheolateralis	M. tracheo-lateralis	M. tracheolateralis	M. tracheo-bronchialis or M. trachealis	M. tracheolateralis	M. laryngosyringeus		
M. tracheobronchialis dorsalis	M. broncho-trachealis posticus	M. broncho-trachealis posticus	M. tracheo-bronchialis longus	M. tracheobronchialis dorsalis	M. laryngosyringeus dorsolateralis (longus)	M. levator longus posterior arcus secundi	M. dorsolateralis longus
M. tracheobronchialis brevis	M. broncho-trachealis brevis	part of M. broncho-trachealis posticus	M. tracheo-bronchialis dorsalis brevis	M. tracheobronchialis brevis	M. laryngosyringeus dorsalis (longus)	M. tensor membranae tympaniformis internae	M. dorsalis longus
M. tracheobronchialis ventralis	M. broncho-trachealis anticus	M. broncho-trachealis anticus	M. tracheo-bronchialis ventralis	M. tracheobronchialis ventralis	M. laryngosyringeus ventralis (longus)	M. levator longus anterior arcus secundi	M. ventralis longus
M. syringealis dorsalis	M. bronchialis posticus	M. bronchialis posticus	M. syringeus dorsalis	M. syringeus obliquus	M. syringeus dorsolateralis	M. levator brevis posterior arcus secundi	M. dorsolateralis brevis

M. syringealis ventralis	M. bronchialis anticus	M. bronchialis anticus, pars medialis and pars lateralis	M. syringeus ventralis and M. syringeus ventrolateralis	M. syringeus ventralis	M. syringeus ventralis (brevis)	M. levator brevis anterior arcus secundi	M. ventralis brevis
					M. syringeus ventrilateralis internus (brevis)	M. laxator membranae tympaniformis externae	M. ventrilateralis brevis internus
			Uncertain: M. tracheo-bronchialis obliquus		*Uncertain:* M. laryngosyringeus ventrilateralis (longus)	*Uncertain:* M. rotator arcus tertii	*Uncertain:* M. ventrilateralis brevis externus and M. ventrilateralis longus
					M. syringeus ventrilateralis externus		
					M. syringeus dorsalis		M. dorsalis brevis

Based on the text of Fürbringer (1886, p. 1091) and the Tables of Köditz (1925, p. 140), Ames (1971, p. 90), and Warner (1972b).

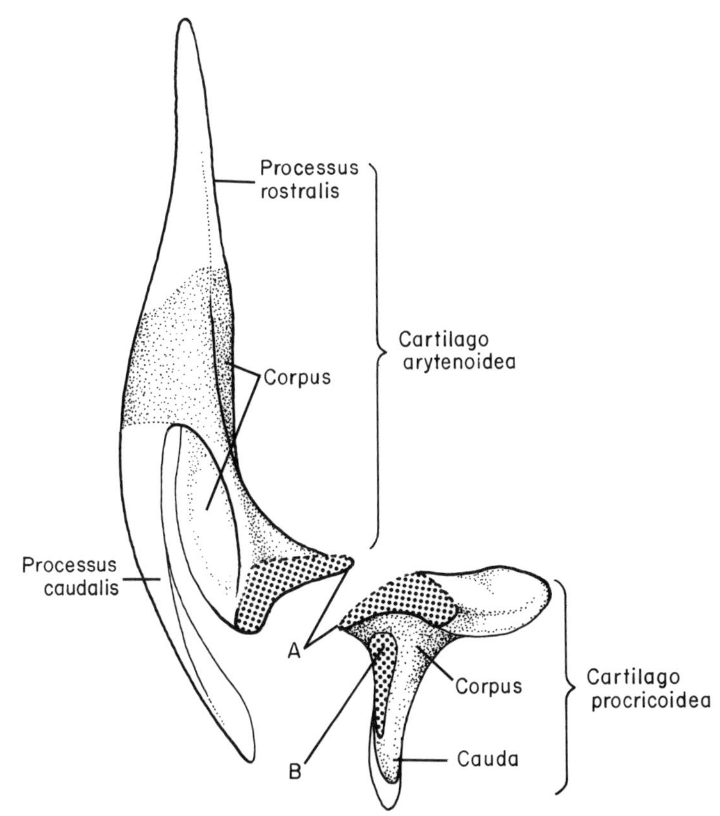

Fig. 1 Dorsal view of the left Cartilago arytenoidea and the Cartilago procricoidea, disarticulated, of *Gallus*. The dorsal surface of the left arytenoid cartilage has been rotated to the left, and the dorsal surface of the procricoid has been rotated to the right, to display the two articular surfaces (A) of the Artc. procricoarytenoidea. B is a facet of the Artc. procricocricoidea. From White (1975) with permission of the author and Saunders, Philadelphia.

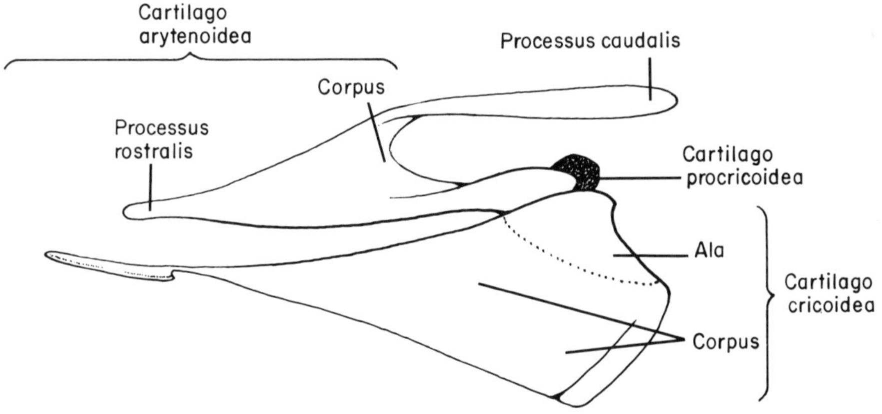

Fig. 2 Cartilagines laryngeales of *Gallus*, left lateral view. From White (1975), with permission of the author and Saunders, Philadelphia.

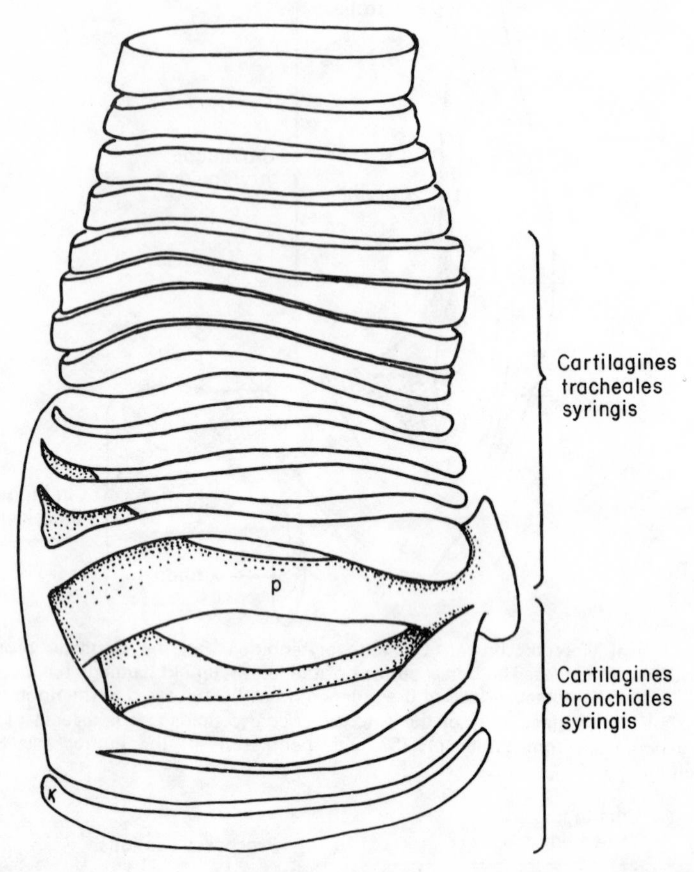

Fig. 3 Lateral view of left side of Cartilagines syringeales of *Gallus*. Of the Cartilagines tracheales syringis, the four cranial elements are rigid rings closely attached to each other to form the Tympanum; the four caudal elements are slender flexible bars. P, pessulus. From King (1975) after Myers (1917), with permission of Saunders, Philadelphia.

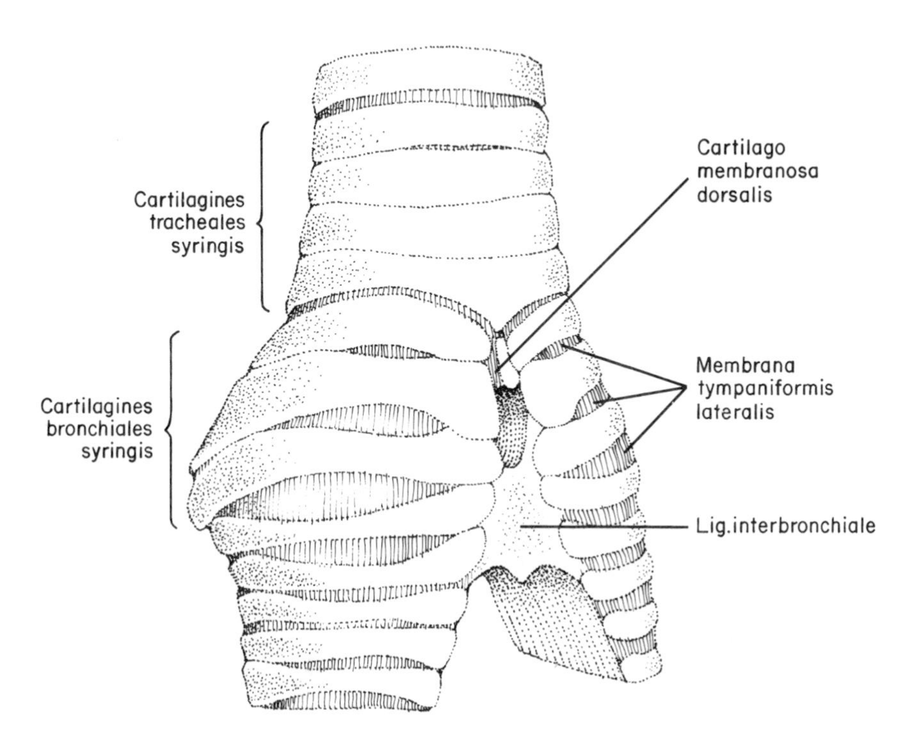

Fig. 4 Dorsal view of the syrinx of the magpie, *Pica*. The so-called Membrana tympaniformis lateralis is not a thin vibrating membrane in most oscines (see Fig. 5a). The Cartilagines tracheales syringis are fused to form the Tympanum. A sheet of soft tissue completes medially the C-shaped cartilages of the primary bronchus. From Haecker (1900).

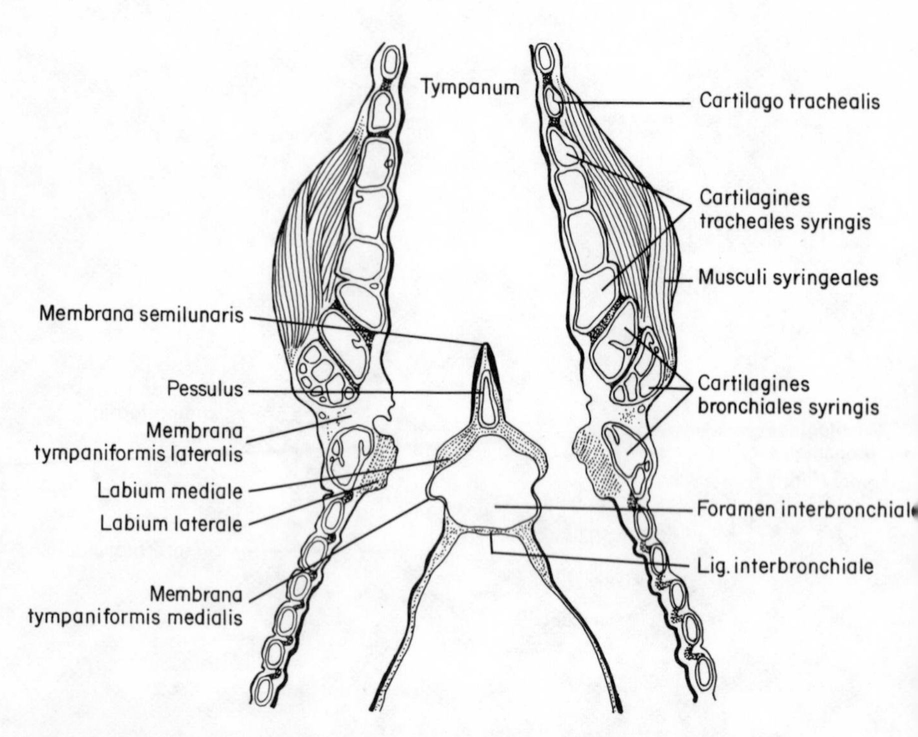

Fig. 5a Horizontal section through the syrinx of the Blackbird, *Turdus merula*. The region labelled Membrana tympaniformis lateralis was so named by Haecker and consists of soft tissue, but is too thick to be regarded as a true membrane in most oscines. The Cartilagines tracheales syringis form the Tympanum. From Haecker (1900).

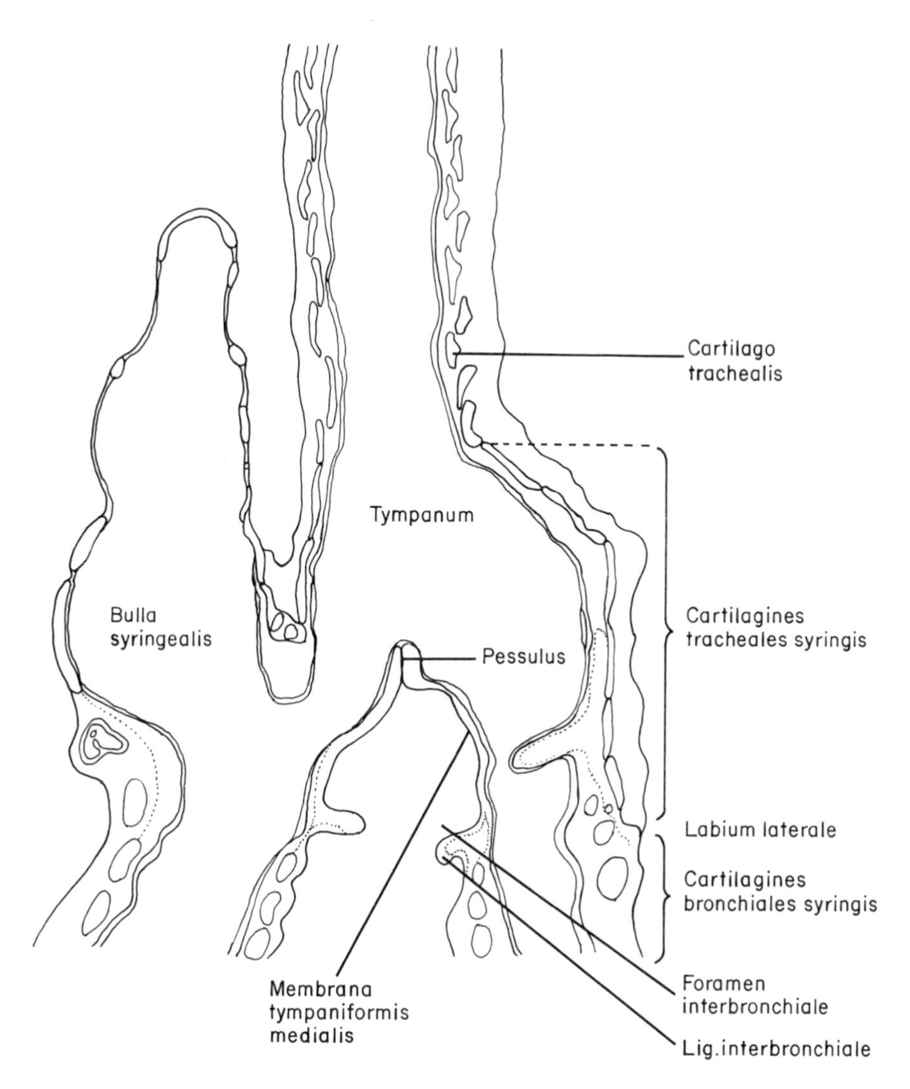

Fig. 5b Horizontal section through the syrinx of a male duck, *Aythya fuligula.* From Warner (1971), with permission of the author and the Zoological Society of London.

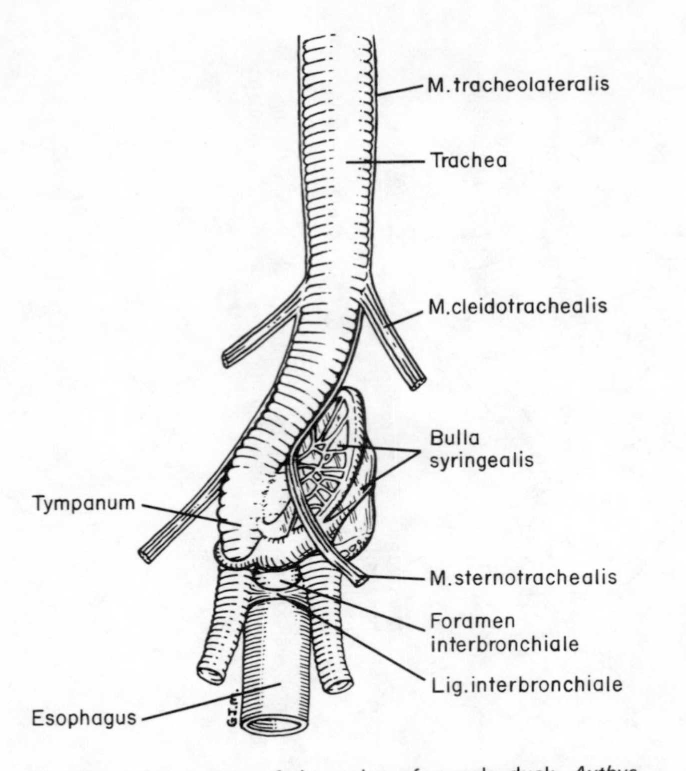

Fig. 6a Ventral view of the syrinx of a male duck, *Aythya fuligula*. In this species the M. tracheolateralis is shortened, inserting on the tracheal syringeal cartilages. The M. cleidotrachealis is present. (A.S. King, unpublished).

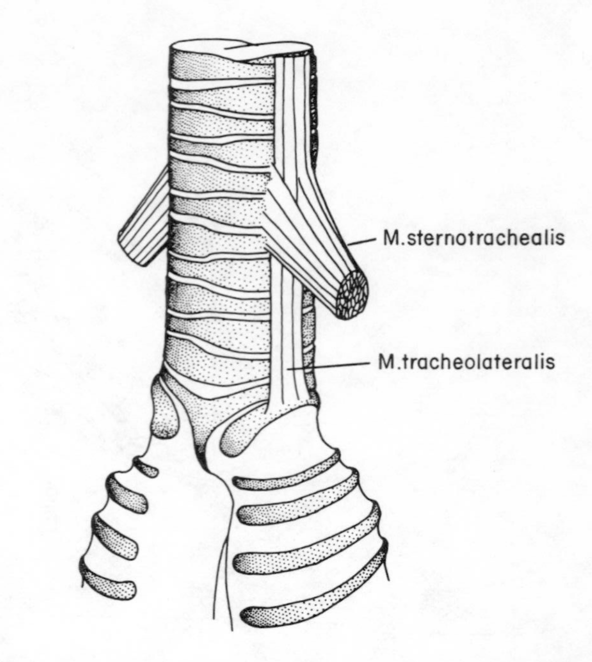

Fig. 6b Dorsolateral view of the syrinx of the broadbill, *Psarisomus dalhousiae*, showing the muscles of the typical tracheobronchial type of syrinx, the M. tracheolateralis constituting an extrinsic syringeal muscle. From Ames (1971) with permission of the author and the Peabody Museum of Natural History.

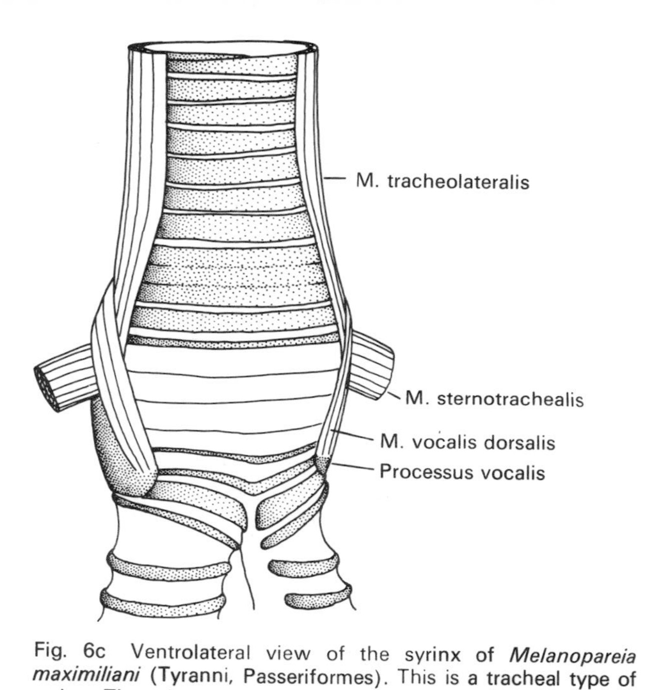

Fig. 6c Ventrolateral view of the syrinx of *Melanopareia maximiliani* (Tyranni, Passeriformes). This is a tracheal type of syrinx. There is one pair of syringeal muscles, i.e. M. vocalis dorsalis. Three tracheal syringeal cartilages are fused ventrally and five dorsally to form the Tympanum (stippled). The unstrippled region is the Membrana trachealis, the lines which cross it being tracheal syringeal cartilages reduced to delicate rods. Altogether there are about 10 tracheal syringeal cartilages in this species. From Ames (1971) with permission of the author and the Peabody Museum of Natural History.

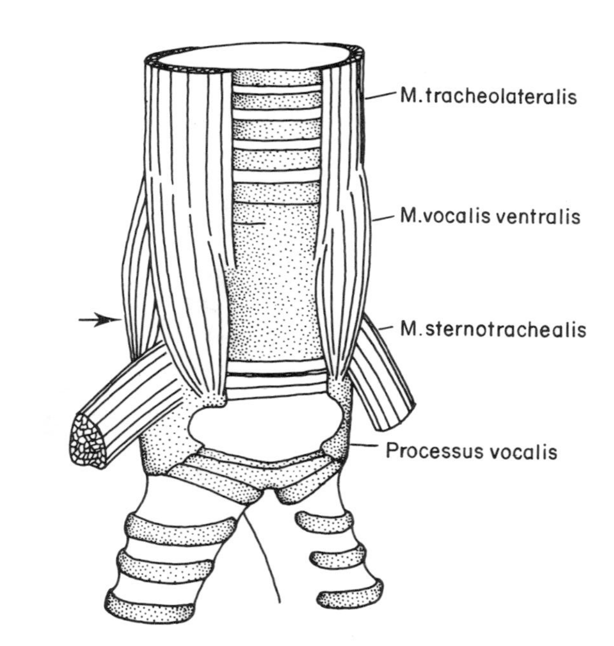

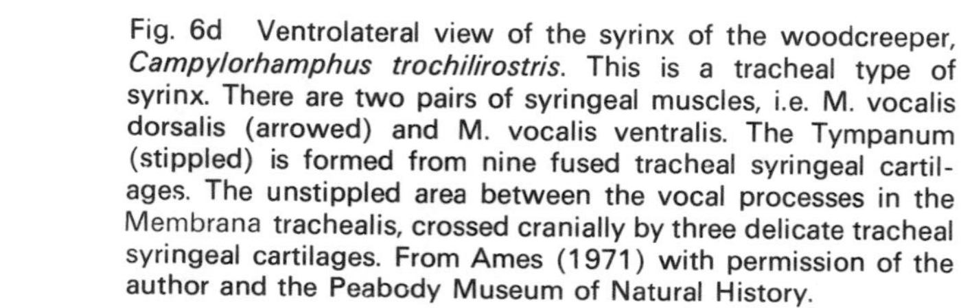

Fig. 6d Ventrolateral view of the syrinx of the woodcreeper, *Campylorhamphus trochilirostris*. This is a tracheal type of syrinx. There are two pairs of syringeal muscles, i.e. M. vocalis dorsalis (arrowed) and M. vocalis ventralis. The Tympanum (stippled) is formed from nine fused tracheal syringeal cartilages. The unstippled area between the vocal processes in the Membrana trachealis, crossed cranially by three delicate tracheal syringeal cartilages. From Ames (1971) with permission of the author and the Peabody Museum of Natural History.

Fig. 6e Ventrolateral view of the syrinx of the flycatcher, *Nuttallornis borealis*, showing a syrinx with two pairs of syringeal muscles, i.e. M. obliquus ventralis and M. obliquus lateralis. There is minimal fusion of tracheal syringeal cartilages so that there is virtually no Tympanum. From Ames (1971) with permission of the author and the Peabody Museum of Natural History.

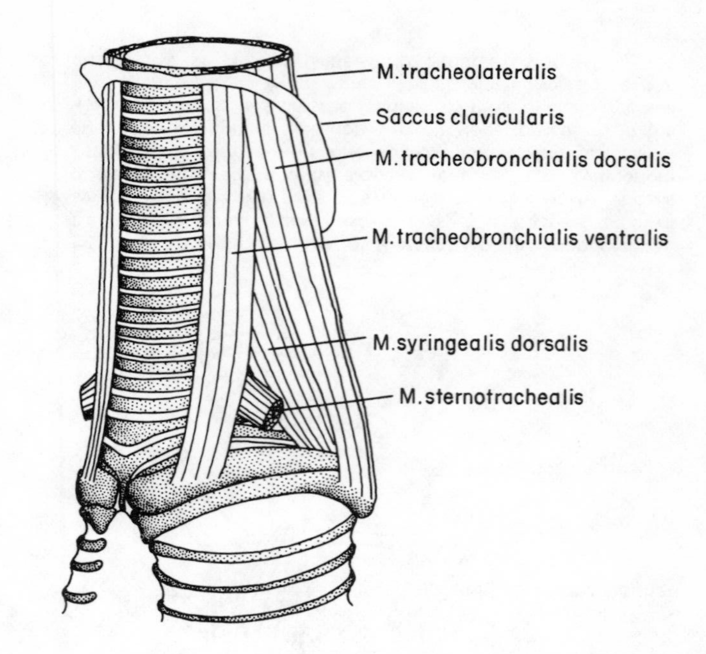

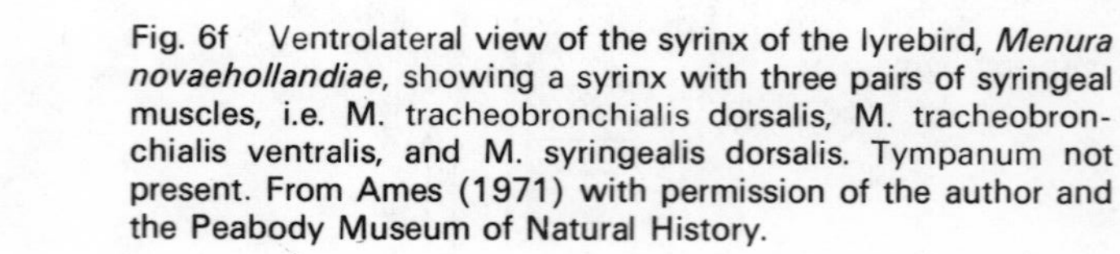

Fig. 6f Ventrolateral view of the syrinx of the lyrebird, *Menura novaehollandiae*, showing a syrinx with three pairs of syringeal muscles, i.e. M. tracheobronchialis dorsalis, M. tracheobronchialis ventralis, and M. syringealis dorsalis. Tympanum not present. From Ames (1971) with permission of the author and the Peabody Museum of Natural History.

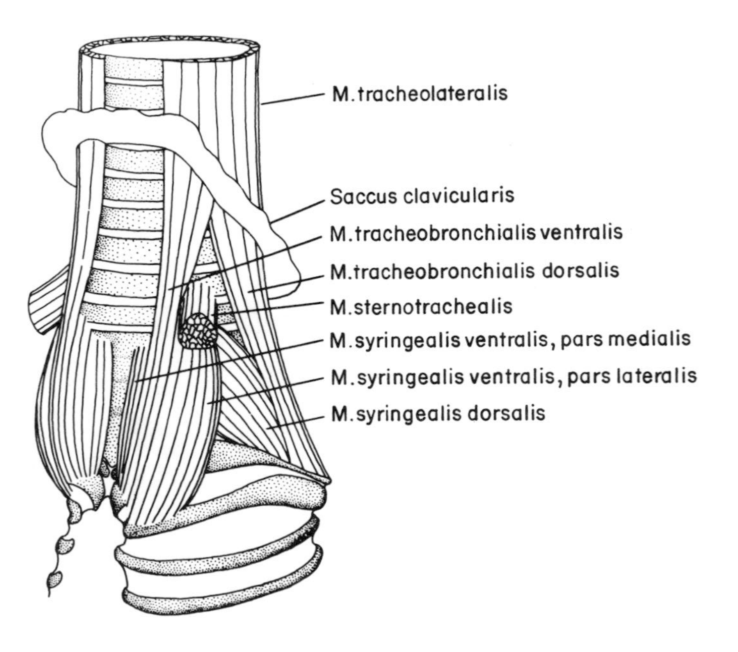

Fig. 6g Ventrolateral view of an example of the oscine syrinx; the crow, *Corvus brachyrhynchos*. Of the five pairs of syringeal muscles only four are shown: M. tracheobronchialis ventralis, M. tracheobronchialis dorsalis, M. syringealis ventralis, and M. syringealis dorsalis; M. tracheobronchialis brevis is not indicated. The Tympanum (strippled and visible in the midline) is formed from four fused tracheal syringeal cartilages. From Ames (1971) with permission of the author and the Peabody Museum of Natural History.

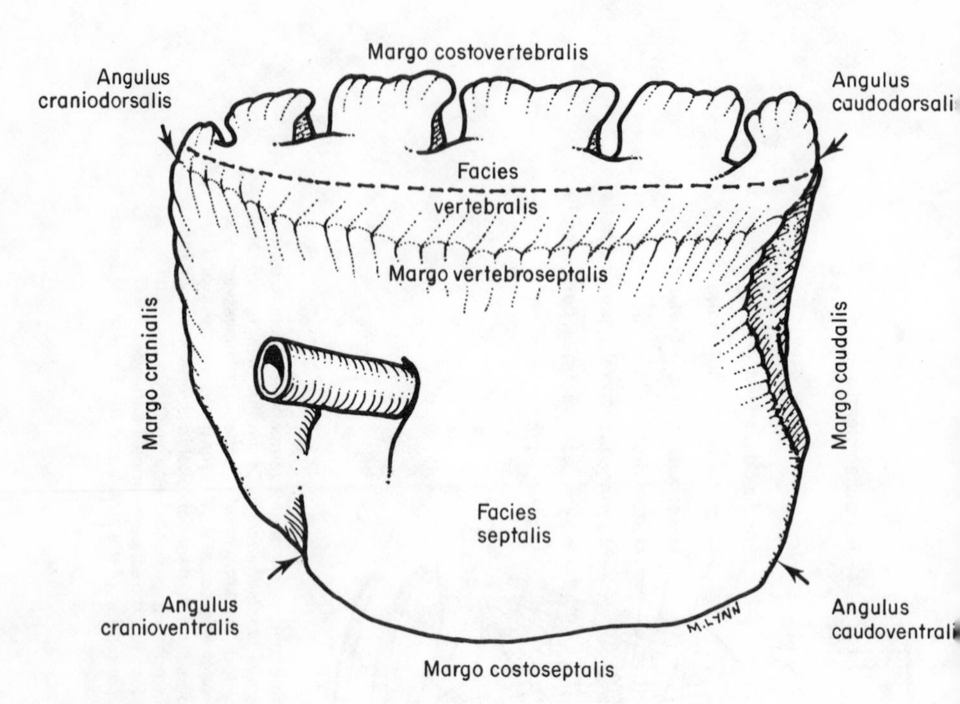

Fig. 7 Semidiagrammatic medial view of the right lung of *Gallus*. The broken line along the Facies vertebralis is the Linea anastomotica.

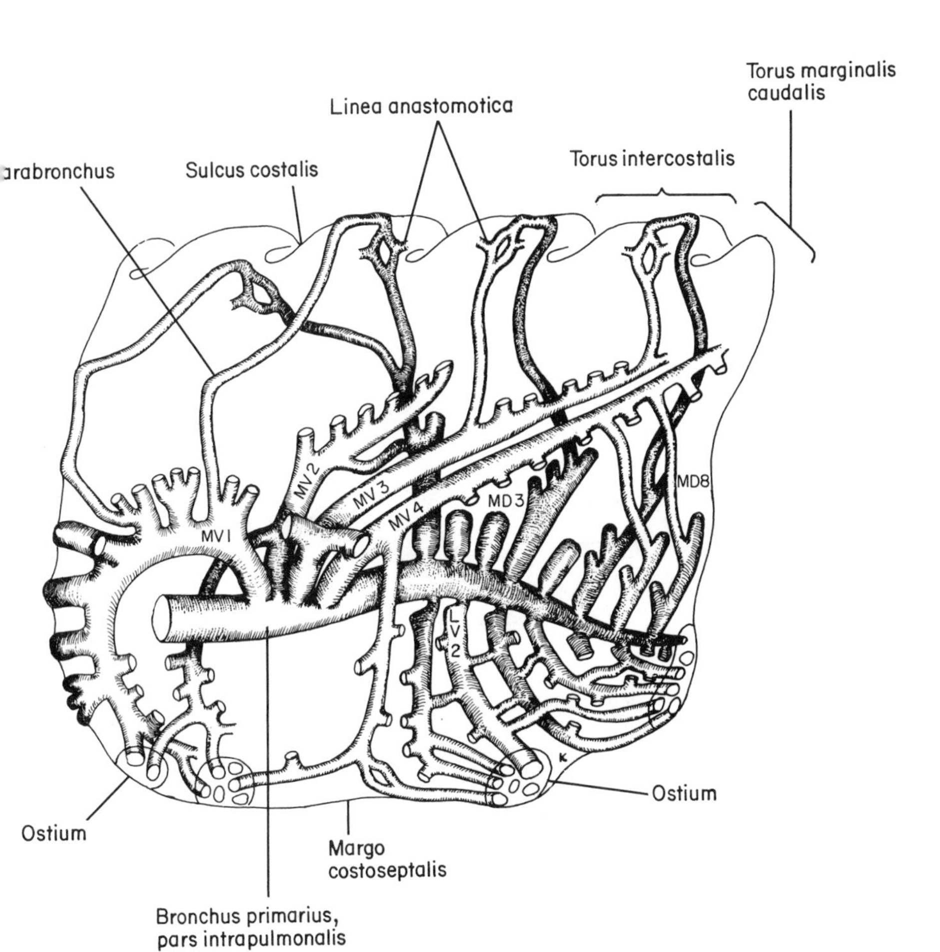

Fig. 8 Medial view of the right lung of *Gallus,* from King (1966). The lung is drawn as though transparent, to show the main bronchi. MV, Bronchi medioventrales; MD, Bronchi mediodorsales; LV, Bronchi lateroventrales. The four Ostia along the Margo costoseptalis are indicated by rings, except for the Ostium of the Saccus cervicalis from the arch of MV1, and the Ostium of the Saccus clavicularis (Pars medialis) and the Ostium of the Saccus thoracicus cranialis from the root of MV3; these are shown as short transected tubes of large diameter. The Linea anastomotica is caused by the terminal anastomoses of Parabronchi of the medioventral secondary bronchi with those of the mediodorsal secondary bronchi.

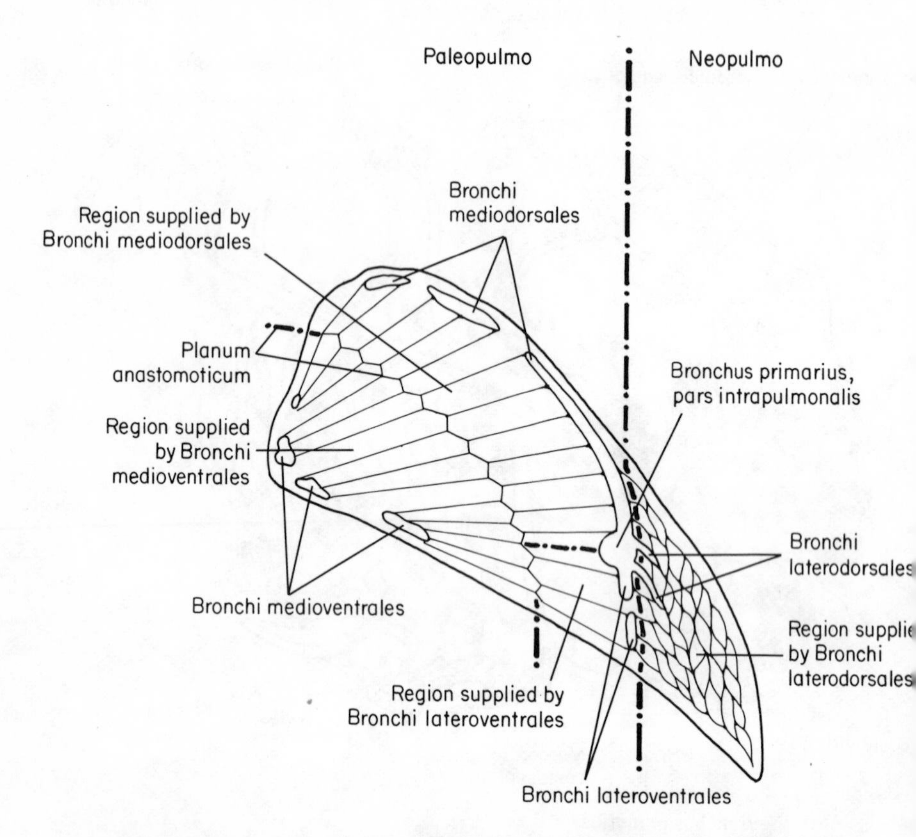

Fig. 9 Diagrammatic transverse section of the right lung, viewed from the caudal aspect, of a bird with a well developed Neopulmo, e.g. *Gallus*. The diagram shows the approximate regions supplied by the four groups of Bronchi secundarii. In birds such as *Gallus* it is difficult to distinguish the territory of the Bronchi laterodorsales from the territories of the Bronchi mediodorsales and lateroventrales with the degree of precision suggested by the diagram; this is because the latter two groups of Bronchi secundarii have, in their lateral walls, highly developed respiratory tissue which anastomoses immediately with the parabronchial network of the Bronchi laterodorsales (H.-R. Duncker, unpublished).

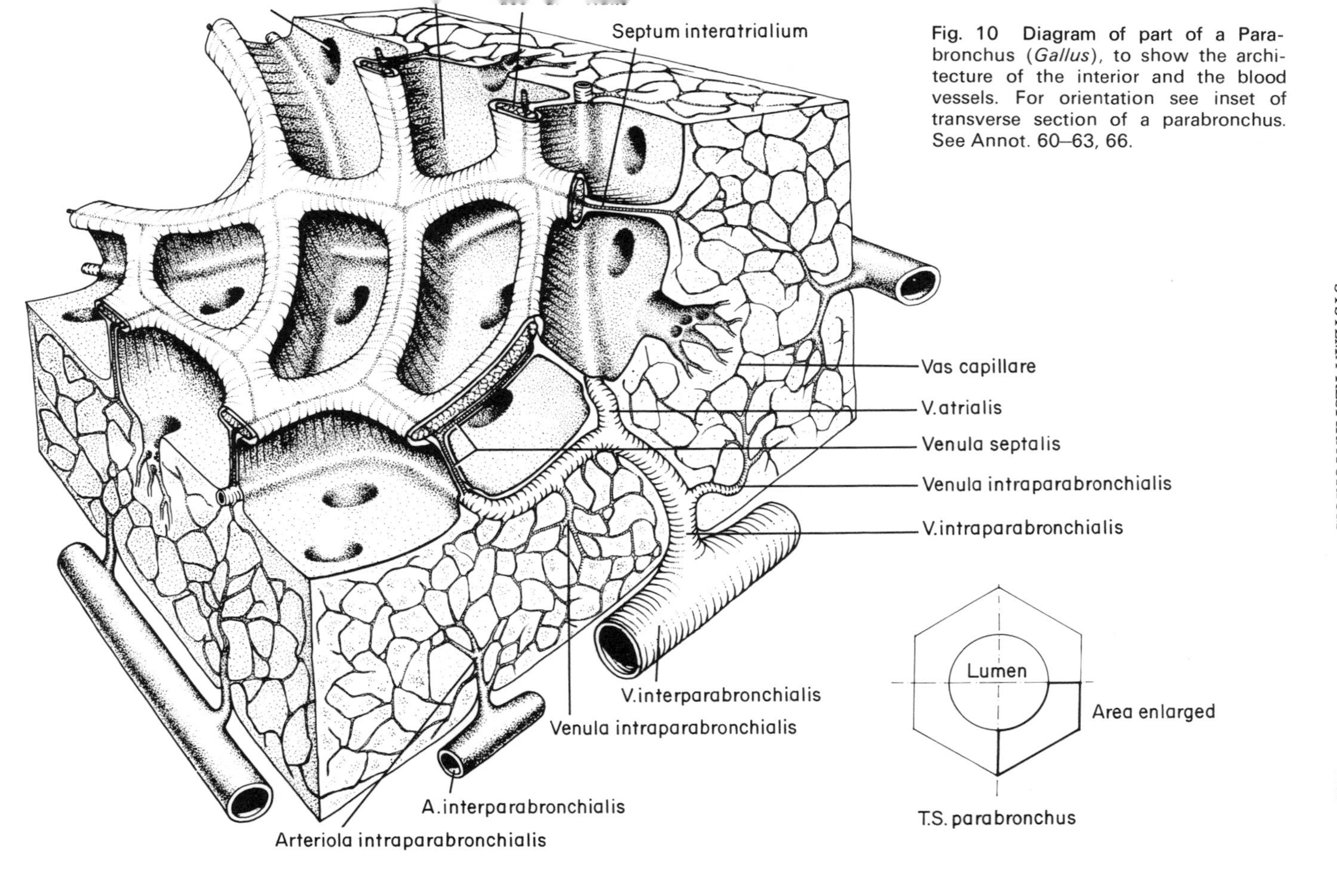

Fig. 10 Diagram of part of a Parabronchus (*Gallus*), to show the architecture of the interior and the blood vessels. For orientation see inset of transverse section of a parabronchus. See Annot. 60–63, 66.

SYSTEMA DIGESTORIUM

J. McLELLAND

Department of Anatomy, Royal (Dick) School of Veterinary Studies, Edinburgh EH9 1QH, Scotland

With contributions from Sub-Committee Members: L. N. Das, *Orissa Veterinary College, India*; D. S. Farner, *University of Washington, Seattle*; Gy. Fehér, *Veterinärmedizinischen Universität, Budapest*; R. D. Hodges, *Wye College, University of London*; G. Michel, *Karl-Marx Universität, Leipzig*; T. Nishida, *University of Tokyo*; E. Pastea, *Faculty of Veterinary Medicine, Bucharest*; U. M. Rawal, *Gujarat University, India*; V. Simić, *Universitè de Belgrade*; R. Tucker, *University of Dar Es Salaam*; V. Ziswiler, *Zoologisches Museum, Zürich*

I should like to take this opportunity of acknowledging with gratitude the forceful contributions made by the Members of the Systema Digestorium Sub-Committee. The large size of the Sub-Committee made it possible to canvas a broad spectrum of opinion, whilst at the same time to have available advice of specialists on most aspects. Inevitably, it was not possible to obtain total agreement for all terms; consequently some of the nomenclature is based on a consensus of views. For a number of structures, on the other hand, a range of terms was available, all of which appeared to have equal merit. The final choice then rested with the Sub-Committee Chairman. In several instances the selection of a term was influenced by the knowledge that the term was to be utilized in the formation of other terminologies especially those dealing with blood vessels and nerves.

The primary aim of the Sub-Committee has been to provide a terminology for the digestive system of birds generally. Structures which are restricted to a relatively small number of species are also covered when the structures are well-documented. Future editions of *Nomina Anatomica Avium* will doubtless include more terms dealing with interspecific differences. Suggestions on this are welcomed. Finally, for some terms, e.g. "crop" and "gizzard", which are deeply established in the scientific literature, it was not possible to provide closely corresponding Latin equivalents. In the conversion of the Latin form of anatomical names into modern languages, it is hoped that a certain license will be permitted so that terms like these and others may be retained.

CAVITAS ORALIS[1]

(Fig. 1)

Bucca
Rostrum (see Fig. 1; and **Topog. anat.** Annot. 14, 15)
  Rostrum maxillare (see **Osteo.** Annot. 53).
  Rostrum mandibulare
Dentes (see **Osteo.** Annot. 82)
Rima oris
Angulus oris
Palatum (Fig. 1) (see **Resp.** Annot. 4, 5)
  Torus palatinus[2]
  Choana (Fig. 1)
    Pars rostralis
    Pars caudalis
Saccus oralis[3]

Tunica mucosa oris (see below, Tunica mucosa linguae)
  Papillae palatinae
  Rugae palatinae laterales[4] (Fig. 1)
  Rugae palatinae intermediales[4]
  Ruga palatina mediana[4] (Fig. 1)
  Sulcus palatinus[5]
  Sulcus palatinus medianus[5]
  Sulci palatini laterales[5]

Glandulae oris[6] (see Fig. 2; and below, Glandulae pharyngis)

Gl. maxillaris[7]
Gll. palatinae[8]
Gl. anguli oris
Gll. mandibulares rostrales[10] (see Gll. mand. caudales)
Gll. mandibulares externae[11]
Gll. mandibulares intermediales[11]
Gll. mandibulares internae[11]
Gll. linguales[12]
Ductuli glandularum oralium[13]

Lingua[14] (Fig. 1)

Radix linguae
Corpus linguae

Apex linguae
Dorsum linguae
    Torus linguae[15]
Margo linguae
Ventrum linguae
Tunica mucosa linguae
    Fenulum lingualis
    Cuticula cornea lingualis[16]
    Papillae linguales[17] (Fig. 1)
    Sulcus lingualis[18]
    Rugae linguae
    Gemmae gustatoriae (see **Org. sens.** Annot. 71)
Musculi linguales (See **Myol.**)

Apparatus hyobranchialis (see **Osteo.; Arthr.; Myol.**)

PHARYNX[1]

(Fig. 1)

Cavitas pharyngealis (Fig. 1)
    Rima infundibuli
        Infundibulum pharyngotympanicum [auditivum][19]
            Tuba pharyngotympanica [auditiva]
                (see **Osteo.** Annot. 94, 98)
            Tuba pharyngotympanica [auditiva] communis[19]
                (see **Osteo.** Annot. 94, 98)
            Ostium pharyngeale tubae pharyngotympanicae[19]
            Sulcus infundibularis medianus (Fig. 3)
            Sulcus infundibularis lateralis (Fig. 3)
            Plica infundibularis (Fig. 3)
Mons laryngealis (see Fig. 1; and **Resp.** Annot. 22)
Glottis (Fig. 1)
Tunica mucosa pharyngis
    Papillae pharyngeales (Fig. 1)
    Plicae pharyngeales[20]
    Lymphonoduli pharyngeales [Tonsilla pharyngea]
Tela submucosa pharyngis
Tunica muscularis pharyngis

Glandulae pharyngis[6] (see Fig. 2; and above, Glandulae oris)
    Gll. sphenopterygoideae[21]
    Gll. mandibulares caudales[10] (see Gll. mand. rostrales)
    Gll. cricoarytenoideae
    Ductuli glandularum pharyngealium[22]

CANALIS ALIMENTARIUS

ESOPHAGUS

(Figs. 4, 7)

Pars cervicalis
    Saccus esophagealis[23] (Fig. 4)
Pars thoracica

Ingluvies[24] (Fig. 4)
    Ostium ingluviale
    Fundus ingluvialis
    Diverticulum dextrum ingluviale[25]
    Diverticulum sinistrum ingluviale[25]
    Diverticulum medianum ingluviale[25]

Tunica adventitia[26]
Tunica muscularis esophagi
    Stratum longitudinale[27]
    Stratum circulare
Tela submucosa esophagi
Tunica mucosa esophagi
    Lamina muscularis mucosae
    Plicae esophageales
    Plicae ingluviales
    Rugae ingluviales[28]
    Gll. esophageales
        Gll. ingluviales[29]
    Lymphonoduli esophageales [Tonsilla esophagealis][30]

GASTER[31]

(Figs. 5, 6, 7)

Proventriculus [Pars glandularis][32] (Figs. 5, 6, 7)

Regio glandularis[33]
Diverticulum proventriculare[33] (Fig. 7)
Isthmus gastris (Fig. 5)
Zona intermedia gastris[34] (Fig. 6)

Ventriculus [Pars muscularis][35] (Figs. 5, 6, 7)

Facies tendinea[36]
Facies annularis[37]
Curvatura minor[38]
    Incisura angularis[39]
Curvatura major
Corpus[40]
Saccus cranialis[40]

Saccus caudalis[40]
Sulcus cranialis[41]
Sulcus caudalis[41]
Plica angularis[42]
Ostium ventriculopyloricum (Figs. 5, 6)

Pars pylorica[43] (Fig. 7)
Canalis pyloricus
Bulbus pyloricus[44]
Torus pyloricus[45]
Pylorus
Ostium pyloricum

Tunica serosa
Tela subserosa
Tunica muscularis gastris
Stratum longitudinale[46]
Stratum circulare[47]
M. crassus cranioventralis[48] (Figs. 5, 6)
M. crassus caudodorsalis[48] (Figs. 5, 6)
M. tenuis craniodorsalis[48] (Figs. 5, 6)
M. tenuis caudoventralis[48] (Figs. 5, 6)
Centrum tendineum[49] (Fig. 5)
Tela submucosa gastris
Tunica mucosa gastris
Lamina muscularis mucosae[50]
Gll. proventriculares superficiales
Gll. proventriculares profundae[51, 33]
Plicae proventriculares[52]
Papillae proventriculares (Fig. 6)
Rugae proventriculares[53]
Sulci proventriculares[53]
Gll. ventriculares
Plicae ventriculares[54]
Rugae ventriculares[55]
Sulci ventriculares[55]
Gll. pyloricae
Plicae pyloricae[56]
Cuticula gastrica[57] (Fig. 6)
Columnae verticales
Proc. dentatus
Matrix horizontalis
Procc. conicales[58]
Papillae filiformes pyloricae[59] (Fig. 7)

INTESTINUM

Intestinum tenue[60]

Duodenum
- Ansa duodenalis[61] (Fig. 8)
  - Pars descendens
  - Pars ascendens
- Flexura duodenojejunalis
- Papilla duodenalis[62]

Jejunum[63]
- Ansae jejunales[64] (Fig. 8)

Diverticulum vitellinum[65]
- Saccus vitellinus
- Ductus vitellinus
- Papilla ductus vitellini[66]

Ansa axialis[64]

Ileum[43]
- Ansae ileales[64] (Fig. 8)
  - Ansa supraduodenalis[64]
  - Ansa supracecalis[64]
- Valva ileorectalis[67]

Intestinum crassum

Cecum[68] (Fig. 8)
- Basis ceci[69]
- Corpus ceci[69]
- Apex ceci[69]
- Ostium ceci[70]

Rectum[71] (Fig. 8)

Tunica serosa

Tela subserosa

Tunica muscularis intestini
- Stratum longitudinale
- Stratum circulare
  - M. sphincter ilealis
  - M. sphincter cecale

Tela submucosa intestini

Tunica mucosa intestini
- Lamina muscularis mucosae
- Villi intestinales[72]
- Plicae intestinales villosae[72]
- Plicae intestinales submucosae[72]
- Gll. duodenales[73]

Gll. intestinales
Lymphonoduli solitarii (see **Lym.**)
Lymphonoduli aggregati (see **Lym.**)

Cloaca (see **Cloaca, Urogen.**)

PANCREAS

(see Fig. 8; and **Endoc.** Annot. 7)

Lobus pancreatis dorsalis
Pars dorsalis
Pars ventralis
Lobus pancreatis ventralis
Pars dorsalis
Pars ventralis
Lobus pancreatis splenalis
Ductus pancreaticus dorsalis
Ductus pancreaticus ventralis
Ductus pancreaticus accessorius

HEPAR

Tunica serosa
Tela subserosa
Tunica fibrosa
Facies parietalis
Facies visceralis
Incisura interlobaris cranialis
Incisura interlobaris caudalis
Margo hepaticus[75]

Lobus hepaticus dexter (Fig. 9)
Proc. intermedius dexter[76]
Proc. papillaris
Lobus hepaticus sinister (Fig. 9)
Pars caudodorsalis[78] (Fig. 9)
Pars caudoventralis[78] (Fig. 9)
Proc. intermedius sinister[76]
Pars interlobaris
Incisura lobaris[79]
Porta hepatis
Foramen craniale venae cavae caudalis
Foramen caudale venae cavae caudalis

HEPAR—*continued*

Fossa vesicae felleae
Impressio cardiaca
Impressio proventricularis
Impressio ventricularis
Impressio duodenalis[80]
Impressio jejunalis
Impressio splenalis
Impressio testicularis[81]
Lobuli hepatici[82]
Capsula fibrosa perivascularis
Vasa sanguinea hepatis
  Arteriae interlobulares (see **Cardvas.**)
  Venae interlobulares (see **Cardvas.**)
  Venae centrales (see **Cardvas.**)
Ductuli interlobulares
Ductuli biliferi[83]
Ductus hepaticus dexter[84]
Ductus hepaticus sinister[84] (Fig. 9)
Ductus hepatocysticus[85]
Ductus hepatoentericus dexter[86]
Ductus hepatoentericus communis[87] (Fig. 9)
Ductus hepatoentericus accessorius[88]
Vesica fellea[89] (Fig. 9)
  Tunica serosa
  Tela subserosa
  Tunica muscularis
  Tunica mucosa
  Ductus cysticoentericus[90] (Fig. 9)

ANNOTATIONS

(1) **Cavitas oralis; Pharynx.** Because there is no soft palate or oropharyngeal isthmus, the oral and pharyngeal cavities of birds together form a common chamber often referred to as the "oropharynx". On the basis of the embryology of the visceral arches, Lucas and Stettenheim (1972, p. 18) proposed that the dorsal transverse boundary between the oral and pharyngeal cavities lies between the Choana and the Rima infundibuli extending laterally to the angles of the jaws. They suggested that the ventral transverse boundary between the two cavities lies between the entoglossal and rostral basibranchial bones (see **Osteo.**). (In *Gallus* this boundary coincides with the transverse row of lingual papillae.)

(2) **Torus palatinus.** This term covers the highly vascular palatal cushion or pad of

many Psittacidae (Evans, 1969) as well as the relatively broad elevations on the rostral part of the palate of many granivores (Ziswiler, 1965).

(3) **Saccus oralis.** This term covers the inflatable type of "gular pouch" of some bustards (Otididae) described by Murie (1868, 1869) and Garrod (1874a, b) which are used during courtship for display or as resonating chambers, as well as the various forms of food-carrying diverticula such as the paired pouches of the rosy finch (*Leucosticte*) (Miller, 1941).

(4) **Rugae palatinae laterales; Rugae palatinae intermediales; Ruga palatina mediana.** A lateral palatine ridge situated on each side of the Choana appears to be relatively common in birds (Göppert, 1903). In some species, e.g. *Gallus*, there is also a median ridge in the rostral portion of the palate (McLelland, 1975a). Lateral, intermediate, and median palatine ridges occur in granivorous passerines, a variable number of secondary and tertiary ridges arising from the sides of the intermediate and lateral ridges (Ziswiler, 1965).

(5) **Sulcus palatinus; Sulci palatini laterales; Sulcus palatinus medianus.** A transversely-oriented V-shaped groove occurs in the palate of the Budgerigar (*Melopsittacus*) (Feder 1969). In most seed-eating passerines a groove occurs between the lateral palatine ridge and the Tomium (see **Topog. anat.** Annot. 12) of the beak (Ziswiler, 1965).

(6) **Gll. oris; Gll. pharyngis.** The terminology for the salivary glands does not cover the subdivisions of the glands which show many interspecific variations (see Antony, 1920; Groebbels, 1932, pp. 476–468; Farenholz, 1937).

(7) **Gl. maxillaris.** Synonymy: Gll. palatinae maxillares (Antony, 1920)

(8) **Gll. palatinae.** Synonymy: Gll. palatinae posteriores (Antony, 1920). In *Gallus* are Gll. palatinae mediales and Gll. palatinae laterales (Saito, 1966).

(9) **Gll. mandibulares rostrales.** Synonymy: Gll. inframaxillares (Cholodkowski, 1892); Gll. submaxillares anteriores (Heidrich, 1908). In *Caprimulgus* the rostral and caudal mandibular glands are represented by a single group of glands, the Gll. submandibulares of Antony (1920).

(10) **Gll. mandibulares caudales.** Synonymy: Gll. submaxillares posteriores (Heidrich, 1908; Gll. mandibulares posteriores (Antony, 1920). The part of Gll. mandibulares caudales in woodpeckers (Picidae) which produces the extremely sticky fluid that coats the tongue is called Gl. picorum by Antony (1920). In *Gallus* the caudal mandibular glands consist of medial, intermediate, and lateral groups (Saito, 1966).

(11) **Gll. mandibulares externae; Gll. mandibulares intermediae; Gll. mandibulares internae.** In many granivores and insectivores the mandibular glands consist of external, intermediate and internal groups (Antony, 1920; Foelix, 1970).

(12) **Gll. linguales.** Gll. linguales rostrales and Gll. linguales caudales occur in *Gallus* (Saito, 1966).

(13) **Ductuli glandularum oralium.** The palatine, rostral mandibular and lingual salivary glands are polystomatic, whereas the maxillary gland and the Gl. anguli oris are generally monostomatic.

(14) **Lingua.** The lingual nomenclature is general and does not cover detailed interspecific variations (see Lucas, 1897; Gardner, 1926).

(15) **Torus linguae.** The raised caudal part of the tongue in Anatidae.

(16) **Cuticula cornea lingualis.** The heavily keratinized epithelium forming the nail-like plate on the ventral surface of the tongue of some species e.g. *Gallus* (Susi, 1969).

(17) **Papillae linguales.** Papillae linguales marginales, Papillae linguales radicales rostrales and Papillae linguales radicales laterales occur in *Gallus* (Marvan and Těšik, 1970).

(18) **Sulcus lingualis.** In *Gallus* there is a well-defined Sulcus lingualis radicis and a much less distinct Sulcus lingualis medianus (Marvan and Těšik, 1970).

(19) **Infundibulum pharyngotympanicum [auditivum]; Tuba pharyngotympanica [auditiva] communis; Rima infundibuli.** The paired pharyngotympanic [auditory] tubes join one another at their medial ends within Basis cranii near the base of Rostrum sphenoidale. Here they form Tuba pharyngotympanica communis; the common tube (e.g. in *Gallus* and *Columba*) continues rostrally, and opens into the dorso-caudal part of Infundibulum pharyngotympanicum, a laterally compressed, median chamber located caudal to the nasal cavity. The expanded, ventral end of the Infundibulum is connected with the pharynx via a short median cleft, Rimi infundibuli, located just caudal to Pars caudalis of the Choana. See Fig. 1 and consult Heidrich (1908) for details of the Infundibulum. See also **Osteo.** Os parasphenoidale and Annot. 94, 98.

(20) **Plicae pharyngeales.** In *Gallus* the Mons laryngealis is bounded rostrally by transverse folds of mucous membrane.

(21) **Gll. sphenopterygoideae.** Synonymy: Gll. pterygoideae (Antony, 1920).

(22) **Ductuli glandularium pharyngealium.** All the pharyngeal salivary glands are polystomatic.

(23) **Saccus esophagealis** (Fig. 4). An inflatable enlargement of the Esophagus which functions during the breeding season in "showing-off", or as a resonating chamber for the production of mating calls. Amongst the species possessing this sac are the Sage Grouse *Centrocercus urophasianus* (Clarke *et al.*, 1942; Honess and Allred, 1942), the bustard *Choriotis australis* (Garrod, 1874a, b), and the Painted Snipe *Rostratula benghalensis* (Niethammer, 1966).

(24) **Ingluvies.** The Ingluvies or crop is an expansible portion of the esophagus in which food is stored. Usually situated in the neck, it may be spindle-shaped as in *Casuarius*, *Anas* and *Otis* and the Emberizidae and Fringillidae, or pendulant and sac-like as in tinamiform, falconiform, psittaciform, galliform and columbiform species (Gadow, 1879a, b, 1891a, pp. 671–672; Swenander, 1899, 1902; Niethammer,

1933; Ziswiler, 1967). The Ingluvies of *Opisthocomus* is unusual in consisting of both cervical and thoracic parts (Gadow, 1891; Böker, 1929). In addition to its storage function the Ingluvies of pigeons and doves (Columbidae) produces the nutritive "crop milk", whilst in *Opisthocomus* it takes over the role of the very reduced Ventriculus in physical digestion.

(25) **Diverticulum dextrum ingluviale; Diverticulum sinistrum ingluviale, Diverticulum medianum ingluviale.** These terms refer to the bilobed crop of *Columba.*

(26) **Tunica adventitia.** In the adventitial tunic of the pendulant type of crop like that of galliform species, striated muscle fibres of the M. cucullaris capitis, pars clavicularis may be present (see **Myol.** Annot. 6–10).

(27) **Stratum longitudinale.** Restricted to species of only a small number of families (Swenander, 1902; Hanke, 1957).

(28) **Rugae ingluviales.** The mucous membrane of the exceptionally large crop of *Opisthocomus hoazin* is raised into approximately 20 roughly parallel ridges (Gadow, 1891; Böker, 1929).

(29) **Gll. ingluviales.** Limited in number in the more sac-like forms of crop (Swenander, 1902).

(30) **Lymphonoduli esophageales [Tonsilla esophagealis].** See **Lym.** Annot. 9.

(31) **Gaster.** Synonymy: Ventriculus (Schummer, 1973, p. 48). Two basic types of stomach can usually be distinguished depending on whether the organ is adapted primarily for storage or has an important role in the physical preparation of food. The 'type 1' stomach characteristic of fish and meat-eaters is a relatively undifferentiated, poorly muscled sac-like structure. The 'type 2' stomach characteristic of omnivores, insectivores, herbivores, and granivores is clearly divided externally into cranial and caudal chambers, the muscle tunic of the caudal chamber or Ventriculus being massively developed. The form of stomach in many species, e.g. frugivores and testacivores, is intermediate to these types.

(32) **Proventriculus [Pars glandularis].** Synonymy: Vormagen, Infundibulum, cardiac cavity, estomac glanduleux, ventricule pepsique, ventricule succenturié, jabot (Schepelmann, 1906); Drüsenmagen, Ventriculus glandularius (Schummer, 1973, p. 48).

(33) **Regio glandularis; Diverticulum proventriculare.** In some species the deep glands (Annot. 51 are not uniformly distributed throughout the Proventriculus but are restricted to certain regions as in *Grus grus*, *Buteo buteo* (Swenander, 1902), *Struthio* (Cazin, 1887) and *Chauna chavaria* (Mitchell, 1895), or to a diverticulum as in *Anhinga anhinga* (Garrod, 1876). (See Fig. 7).

(34) **Zona intermedia gastris.** Synonymy: Cardia, Schaltstuck, Zwischenschlund, Verbingungsstück (Schepelmann, 1906); Zwischenstück (Pernkopf and Lehner, 1937). The variably developed region with histological features intermediate to those of the Proventriculus and Ventriculus.

(35) **Ventriculus [Pars muscularis].** Synonymy: Kaumagen, Muskelmagen, Reibmagen, Fleischmagen, Pylorusraum, estomac proprement dit, ventricule charnu, pré gizzard (Schepelmann, 1906); Ventriculus muscularis (Schummer, 1973, p. 50).

(36) **Facies tendinea.** The term used by Groebbels (1932, p. 472) for each of the two relatively flat surfaces of the Ventriculus which contains a tendinous centre (Annot. 49). In the well-differentiated Ventriculus of the "type 2" stomach the surfaces are clearly directed to the left and right.

(37) **Facies annularis.** The two relatively narrow surfaces of the 'type 2' stomach which dorsally and ventrally unite the left and right tendinous surfaces.

(38) **Curvatura minor.** The short length of Ventriculus between the Proventriculus and the Duodenum.

(39) **Incisura angularis.** In some species with a sac-like stomach, e.g. *Phalacrocorax carbo* and *Ardea cinerea*, the lesser curvature of the Ventriculus is strongly angled (Pernkopf and Lehner, 1937).

(40) **Corpus; Saccus cranialis; Saccus caudalis.** Synonymy for Saccus cranialis and Saccus caudalis: superior sac and inferior sac (Garrod, 1872); poche superieure and cul-de-sac inferieur (Cazin, 1887); craniodorsal sac and caudoventral sac (Dziuk and Duke, 1972); kranialer Blindsack and kaudaler Blindsack (Schummer, 1973, p. 54). These sub-divisions cannot be distinguished in the less developed forms of Ventriculus like those of most fish and meat eaters.

(41) **Sulcus cranialis; Sulcus caudalis.** Synonymy: sillon antéro-supérieur and sillon postéro-inférieur (Cazin, 1887). The transverse grooves between the cranial and caudal sacs and the body of the Ventriculus.

(42) **Plica angularis.** The ventricular entrances to the Proventriculus and Pars pylorica in some species, e.g. *Podiceps cristatus* and *Apteryx* appear to be separated by this fold (Pernkopf and Lehner, 1937).

(43) **Pars pylorica.** Synonymy: pyloric stomach (Garrod, 1876); pyloric lobe (Forbes, 1882); poche pylorique (Cazin, 1887); pyloric chamber (Beddard, 1898, p. 415); Bulbus pyloricus (Gadow, 1891, p. 679); Pylorialerweiterung (Swenander, 1902); Nebenmagen, Endstuck (Pernkopf and Lehner, 1937). The variably developed portion between the Ventriculus and Duodenum. Forms a distinct chamber of the stomach in some species, e.g. *Phalacrocorax carbo*, *Ardea cinerea*, *Botauris stellaris*, *Struthio* and *Anhinga*, but in domestic birds is very reduced (Hodges, 1974, pp. 63–64; Larsson *et al.* 1974). For further details consult Cazin (1887), Swenander (1902) and Cornselius (1925).

(44) **Bulbus pyloricus.** In *Ardea cinerea* a constriction divides the pyloric part of the stomach into proximal and distal portions, the smaller distal portion being named pyloric bulb by Pernkopf and Lehner (1937).

(45) **Torus pyloricus.** This term refers to the conical protuberance in the pyloric part of the stomach of *Anhinga rufa* (Garrod, 1878b).

(46) **Stratum longitudinale.** Usually absent over the Ventriculus except the lesser curvature (Pernkopf, 1930).

(47) **Stratum circulare.** In the "type 2" stomach as well as in the better developed intermediate forms of stomach, the circular muscle layer of the Ventriculus is clearly differentiated into four semi-autonomous masses. See Annot. 48 and Fig. 5.

(48) **M. crassus cranioventralis, M. crassus caudodorsalis; M. tenuis craniodorsalis, M. tenuis caudoventralis.** Synonymy: vorderer Hauptmuskel, hinterer Hauptmuskel, oberer Zwischenmuskel, and unterer Zwischenmuskel (Groebbels, 1932, p. 472); M. lateralis ventralis, M. lateralis dorsalis, M. intermedius cranialis and M. intermedius caudalis (Schummer, 1973, p. 51). The terminology of the semiautonomous muscles of the Ventriculus is that suggested by Dziuk and Duke (1972). See Fig. 5.

(49) **Centrum tendineum.** Synonymy: Operculum (Newton and Gadow, 1896, p. 917). Two tendinous centres provide the attachments of the ventricular circular muscle. See Fig. 5.

(50) **Lamina muscularis mucosae.** In the Proventriculus the muscularis mucosae is split by the deep glands into inner and outer layers. In the Ventriculus it is generally absent.

(51) **Gll. proventriculares profundae.** These are the compound glands which are the source of gastric juice. For a description of the glands in *Gallus* see Hodges (1974, p. 51).

(52) **Plicae proventriculares.** This term can be used for the large folds which serve to increase the storage capacity of the Proventriculus in some species, e.g. fish and meat eaters (Cazin, 1887; Swenander, 1902; Magnan, 1912), as well as the folding of the proventricular wall in procellariiform species which is an arrangement to increase the number of deep glands (Matthews, 1949).

(53) **Rugae proventriculares; Sulci proventriculares.** The low anastomosing ridges and grooves between the openings of the deep proventricular glands.

(54) **Plicae ventriculares.** When the primary function of the Ventriculus is that of a storage organ as in fish and meat eaters, its inner surface is often strongly folded, the folds usually disappearing when the chamber dilates with food (Swenander, 1902).

(55) **Rugae ventriculares; Sulci ventriculares.** The muscosal ridges and grooves of the Ventriculus in the "type 2" and intermediate forms of stomach.

(56) **Plicae pyloricae.** Frequently, as in many birds of prey, the mucosa of the pyloric part is strongly folded (Pernkopf and Lehner, 1937). In some species, annular folds of the mucosa subdivide the pyloric part as in *Ardea cinerea* (Cazin, 1887), or separate it from the adjacent regions of the digesive tract as in *Pelecanus* (Pernkopf and Lehner, 1937).

(57) **Cuticula gastrica.** Synonymy: couche cornée, revétement coriace (Cazin, 1887); keratinoid layer (Hedonius, 1892); koilin (Hofmann and Pregl, 1907). The variably developed internal lining of the Ventriculus and the Pars pylorica, consisting of a

carbohydrate-protein complex secreted by the ventricular and pyloric glands. For a recent account of the cuticle in *Gallus* (see Hill, 1971).

(58) **Procc. conicales.** The ventricular cuticle in certain fruit-eating pigeons is raised into a number of hard pointed conical processes (Garrod, 1878a; Wood, 1924; Cadow, 1933).

(59) **Papillae filiformes pyloricae.** The hair-like cuticular papillae which project into the pyloric part of the stomach in darters (Anhingidae) (Garrod, 1876; Cazin, 1887).

(60) **Intestinum tenue.** Different patterns of arrangement of the convolutions of the small intestine are described by Gadow (1889) and have been used in taxonomy.

(61) **Ansa duodenalis.** The duodenal loop is a "closed" loop of intestine as described by Gadow (1889) with both of its limbs held closely together by mesentery. In a few species, e.g. *Spheniscus demersus*, the primary duodenal loop is thrown into a series of secondary folds whilst in *Morus bassanus* it is compound (Mitchell, 1901). A duodenal loop is reported to be absent in certain fruit-eating pigeons (Beddard, 1911).

(62) **Papilla duodenalis.** The pancreatic and bile ducts in domestic birds open into the duodenum on a papilla (Batojeva and Batojev, 1972; Paik, Fujioka and Yasuda, 1974).

(63) **Jejunum, Ileum.** Whilst there is no morphological justification in birds for adopting the terms Jejunum and Ileum, they have still been retained for descriptive purposes, the junction between the two regions being arbitrarily placed at the Diverticulum vitellinum. See **Art.** Annot. 60.

(64) **Ansae jejunales; Ansa axialis; Ansae ileales; Ansa supraduodenalis; Ansa supracecalis.** The Jejunum and Ileum in most species are arranged in one or more closed loops (see Annot. 61). The Ansa axialis (Mitchell, 1901) carries the Diverticulum vitellinum and therefore has both jejunal and ileal components. The Ansa supraduodenalis (Mitchell, 1901) is usually the most distal loop of the Ileum. In a relatively small number of species, one or more small supracecal loops are present distal to the supraduodenal loop (Beddard, 1911). See Figs. 7 and 8.

(65) **Diverticulum vitellinum.** Synonymy: Meckel's diverticulum. Consistently absent in some species (Mitchell, 1901).

(66) **Papilla ductus vitellini.** Described in *Gallus* by Fehér and Gyürü (1971).

(67) **Valva ileorectalis.** Amongst domestic species best developed in *Gallus* and *Meleagris*.

(68) **Cecum.** In most groups of birds right and left ceca open into the cranial part of the Rectum. Ceca are absent in a number of species, e.g. psittaciforms and some Columbidae. For a classification of ceca see Naik and Dominic (1963, 1969).

(69) **Basis ceci; Corpus ceci; Apex ceci.** These terms refer to the regions which can be identified externally in the ceca in many galliform species.

(70) **Ostium ceci.** Usually the right and left ceca open separately into the rectum, but in *Struthio* they share a common orifice.

(71) **Rectum.** This term refers to the straight portion of the intestinal tract between the Ileum and the Cloaca, part of which is probably homologous to the colon of mammals (Romanoff, 1960, p. 482).

(72) **Villi intestinales; Plicae intestinales villosae; Plicae intestinales submucosae.** The intestinal mucosa is arranged into villae or folds which show great interspecific variations (Müller, 1922; Ziswiler, 1967). The term Plicae intestinales submucosae refers to the relatively uncommon type of intestinal fold which is based on a core of submucosa (Müller, 1922; Jacobshagen, 1937; Fenna and Boag, 1974; Johnson and Skadhauge, 1975).

(73) **Gll. duodenales.** The avian equivalent of Brünner's glands of mammals. Evidence for these glands in a number of species is provided by Clara (1934) and Patzelt (1936).

(74) **Pancreas.** The terminology of the Pancreas is that suggested for *Gallus* by Paik, Fujioka and Yasuda (1974). Whilst three lobes of the pancreas can be identified in most species, the gross morphology of these lobes shows considerable interspecific variation. The number of pancreatic ducts varies from one to three depending on the species (Groebbels, 1932, p. 487). Part of Lobus pancreatis ventralis is sometimes referred to as the third main lobe (McLelland, 1975a).

(75) **Margo hepaticus.** In domestic birds right and left dorsal, ventral, cranial and caudal parts of the liver margin can be distinguished (Simić and Janković, 1959).

(76) **Proc. intermedius dexter; Proc. intermedius sinister.** In the domestic species the left process only occurs in *Gallus* and *Meleagris*. Both processes are absent in *Columba*.

(77) **Proc. papillaris.** Amongst domestic species absent in *Columba* (Simić and Janković, 1959).

(78) **Pars caudodorsalis; Pars caudoventralis.** In many species one or both of the liver lobes are partly subdivided. Amongst domestic species the division of the left lobe of the liver into two parts by a deep incision extending cranially from the caudal margin occurs only in *Gallus* and *Meleagris*.

(79) **Incisura lobaris.** In *Gallus* incisions are constant features of the cranial and caudal margins of the right lobe of the liver and the caudal margin of the left lobe.

(80) **Impressio duodenalis.** The descending and ascending parts of the duodenal loop make separate impressions on the visceral surface of the liver.

(81) **Impressio testicularis.** The impression formed on the liver by the right testis.

(82) **Lobuli heptaci.** Hepatic lobules are difficult to identify in birds except close to the Porta hepatis.

(83) **Ductuli biliferi.** For details of the lobar bile ducts in *Gallus* see Miyakı (1973).

(84) **Ductus hepaticus.** In *Gallus*, bile ductules from the right lobe of the liver are drained by the right hepatic duct, and those from the left lobe by the left hepatic duct.

(85) **Ductus hepatocysticus.** This duct branches from the right hepatic duct and enters the gall bladder.

(86) **Ductus hepatoentericus dexter.** In species like *Columba* which have no gall bladder, the branch from the right hepatic duct opens directly into the Duodenum. This duct is absent in *Struthio* (Newton and Gadow, 1896, p. 299).

(87) **Ductus hepatoentericus communis.** The right and left hepatic ducts unite on the visceral surface of the right lobe of the liver to form in the majority of species a Ductus hepatoentericus communis which opens into the Duodenum.

(88) **Ductus hepatoentericus accessorius.** Observed in *Anser* by Simić and Janković (1959).

(89) **Vesica fellea.** A gall bladder is absent in some species (Gorham and Ivy, 1938). (See above, Annot. 86.)

(90) **Ductus cysticoentericus.** The duct that connects the gallbladder with the Duodenum. In contrast to mammals, a common bile duct (Ductus choledochus) is absent in birds.

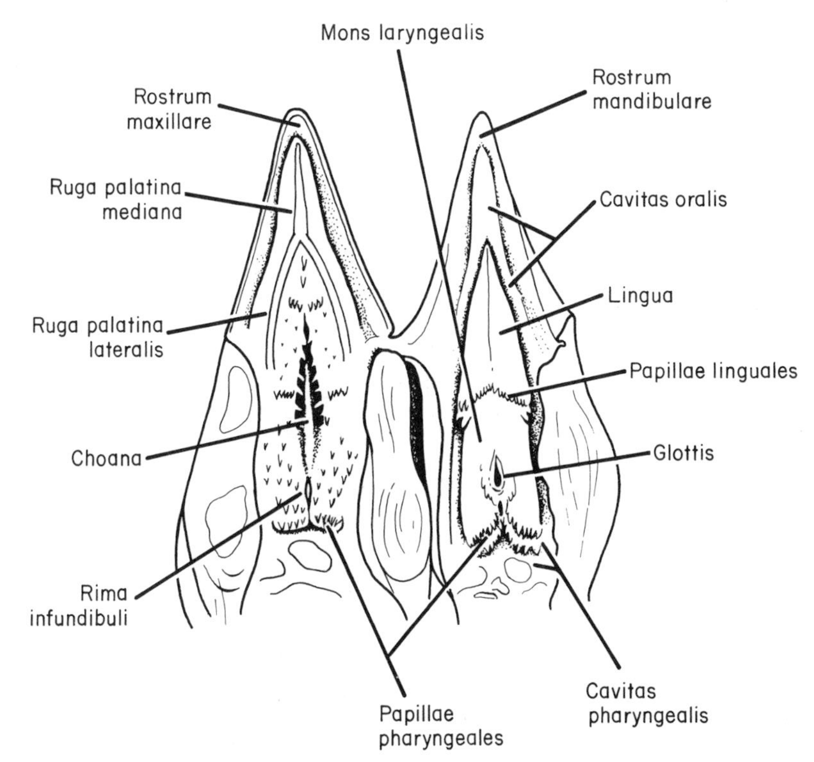

Fig. 1 Oral cavity and pharynx of the Black Grouse (*Lyrurus tetrix*). From Göppert (1903). The right cheek (Bucca) of the specimen has been sectioned in the dorsal plane and the oral cavity and pharynx spread like an open book. The left side of the figure shows the palate and dorsal wall of the pharynx; the right side of the figure shows the tongue and laryngeal eminence in the floor of the oral cavity and pharynx.

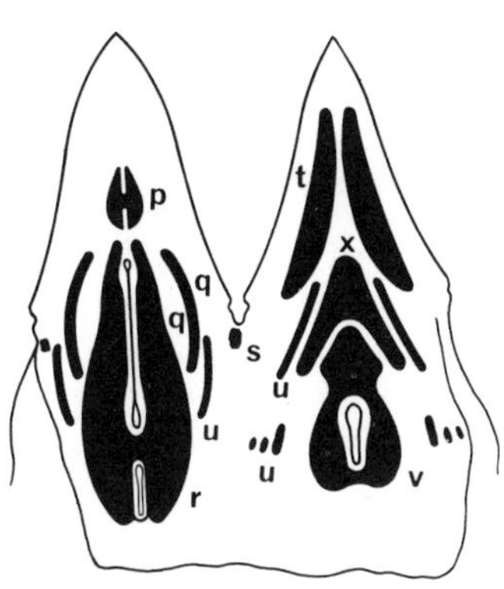

Fig. 2 Glands (salivary) of the oral cavity and pharynx of *Gallus*. Preparation of the specimen is similar to that in Fig. 1 (see legend for Fig. 1). The left side of the figure depicts the salivary glands of the palate and dorsal wall of the pharynx; the right side of the figure depicts the glands of the tongue, of the ventral wall of the oral cavity, and of the pharynx. From Saito (1966).

Abbreviations: p, Gl. maxillaris; q, Gll. palatinae; r, Gll. sphenopterygoideae; s, Gl. anguli oris; t, Gll. manidibulares rostrales; u, Gll. mandibularea caudales; v, Gll. cricoarytenoideae; x, Gll. linguales.

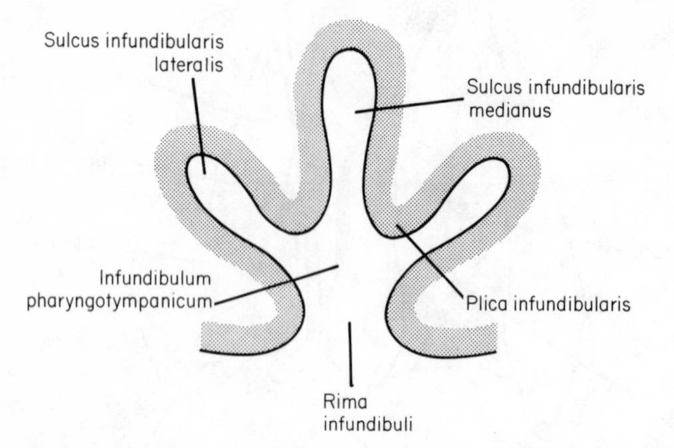

Fig. 3 Transverse section through the pharyngotympanic infundibulum of *Gallus*. From Heidrich (1908). The bottom of the figure represents the dorsal wall of the pharynx, with its opening into the chamber called the Infundibulum. In its caudodorsal part the Infundibulum communicates with the pharyngotympanic tube(s) and ventrally communicates with the Pharynx via a slit, the Rima infundibuli.

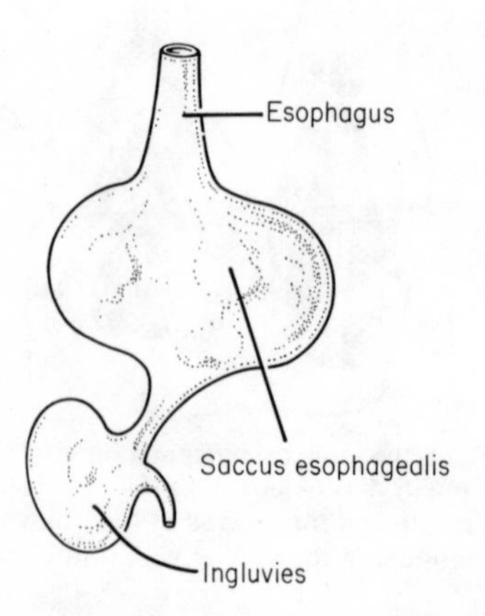

Fig. 4 Inflatable enlargement of the Esophagus of the male Sage Grouse (*Centrocercus urophasianus*). Ventral view. Based on Honess and Allred (1942). See Annot. 23 for occurrence of the esophageal sac in other birds.

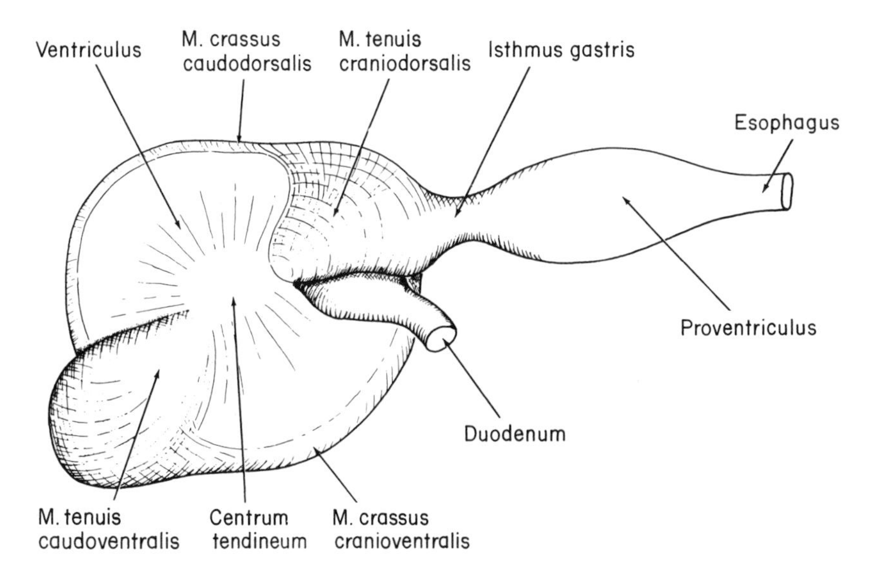

Fig. 5 Exterior of the stomach of *Gallus*. Lateral aspect, right side. From McLelland (1975).

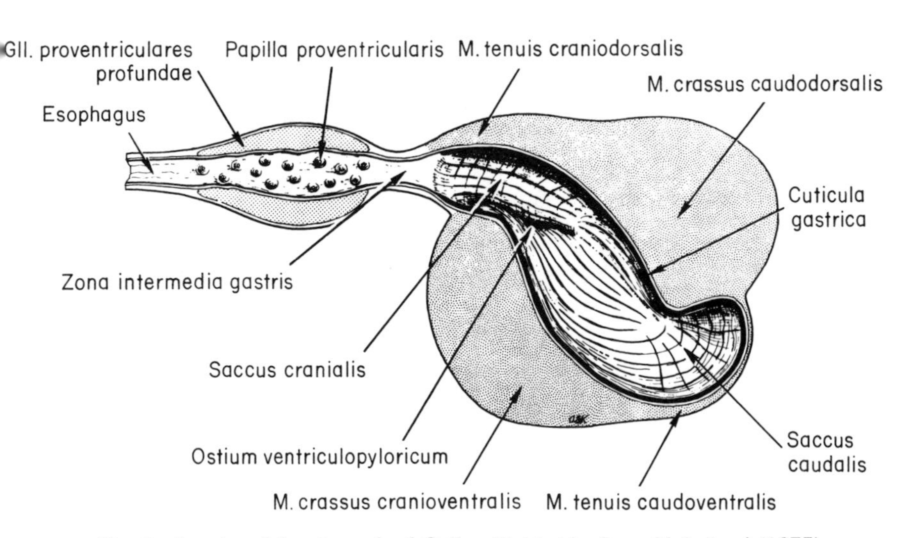

Fig. 6 Interior of the stomach of *Gallus*. Right side. From McLelland (1975).

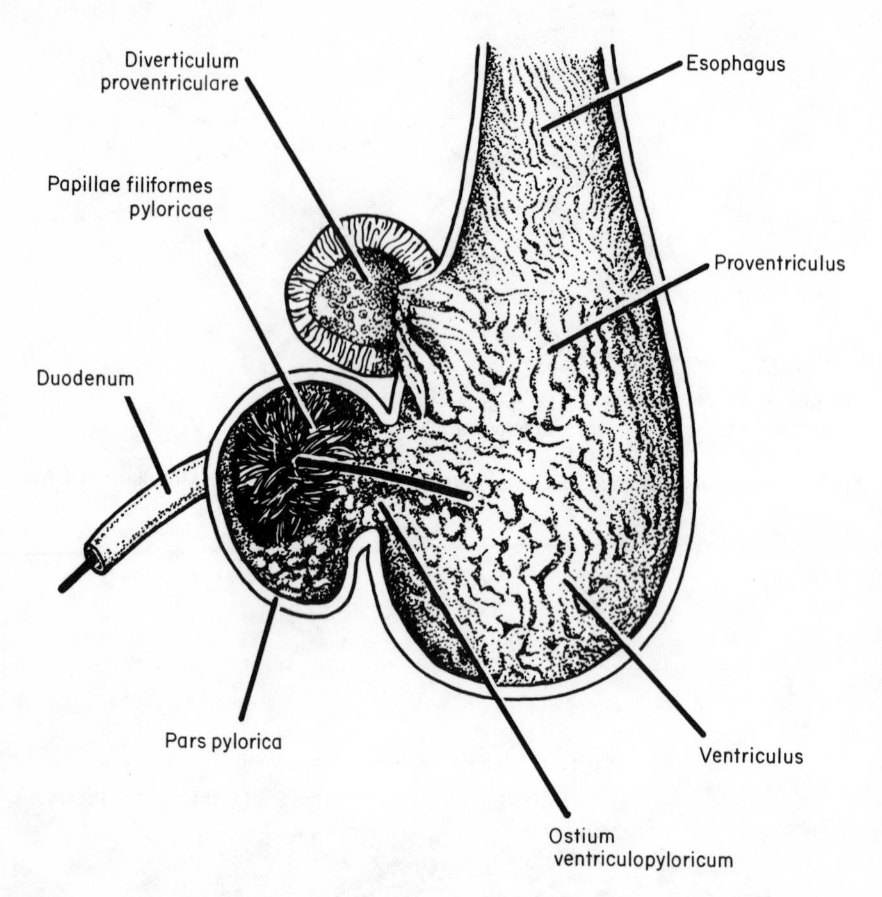

Fig. 7 Interior of the stomach (Gaster) of the Anhinga (*Anhinga anhinga*). From Garrod (1876). This species possesses the Diverticulum proventriculare (see Annot. 33 for the significance of this structure) and a well-developed pyloric part of the stomach (see Annot. 43). Note probe passing from the Ventriculus through the Pars pylorica and into the Duodenum.

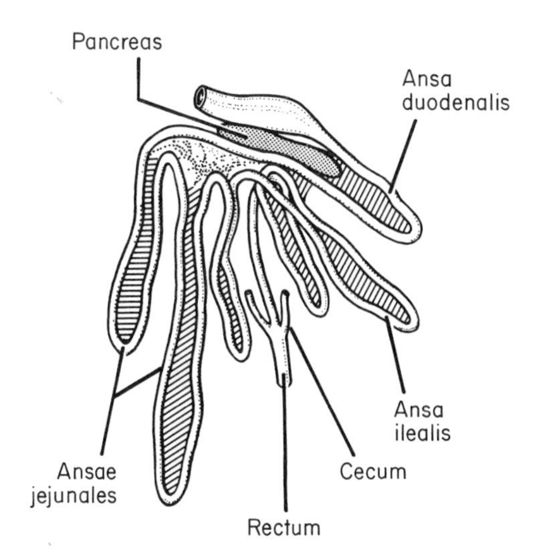

Fig. 8 Intestinal tract of the Japanese Crane (*Grus japonensis*). Ventral view. From Beddard (1911). This figure demonstrates one of the several basic configurations of intestinal loops (see Annot. 64).

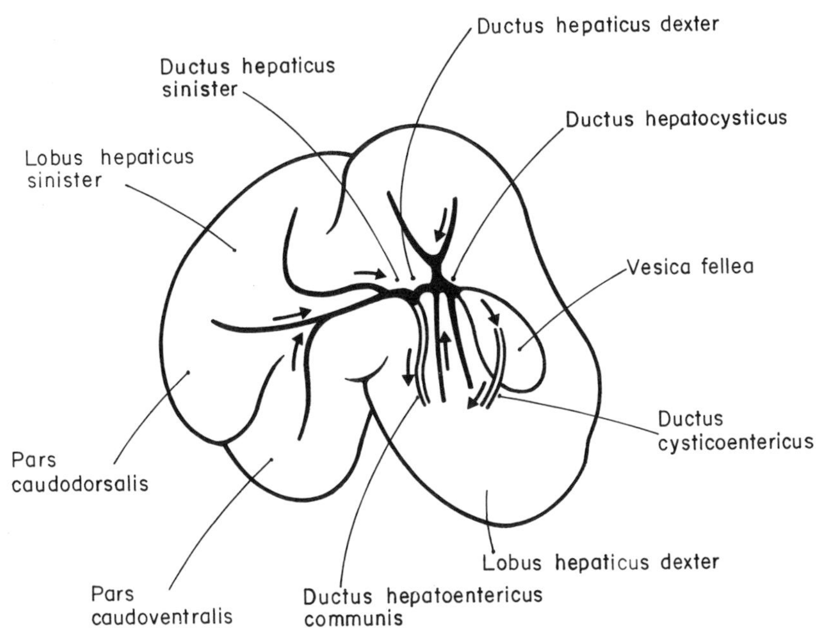

Fig. 9 Biliary system of *Gallus*. Caudal view of liver showing the major intra- and extra-hepatic bile passages. From Miyaki (1973).

SYSTEMA UROGENITALE

A. S. KING

*Department of Veterinary Anatomy,
University of Liverpool, P.O. Box 147,
Liverpool L69 3BX, England*

With contributions from Sub-Committee Members: R. N. C. Aitken, *University of Glasgow*; A. B. Gilbert, *Agricultural Research Council, Edinburgh*; O. W. Johnson, *Moorhead State College, Minnesota*; Vl. Komárek, *16500 Praha 6- Lysolaje, Czechoslovakia*; P. E. Lake, *Agricultural Research Council, Edinburgh*; T. Nishida, *University of Tokyo*; M. D. Tingari, *University of Khartoum*; E. Traciuc, *Institut de Biologie, Bucharest*; M. Yasuda, *Nagoya University*

Originally it was planned that Dr V. Komárek, one of the leading contributors to research in this field, would prepare the list of terms and the annotations for the Organa genitalia masculina. Unfortunately, for reasons beyond his control Dr Komárek found himself unable to take the leading role. Nevertheless he contributed many suggestions and much valuable advice. The list for Organa urinaria has depended greatly on the numerous contributions made by Dr O. W. Johnson. The terminology for Organa genitalia femina gained greatly from the many comments and suggestions made by Dr A. B. Gilbert. The terminology for the spermatozoon could not have been prepared without the advice of Dr P. E. Lake. Furthermore, all the other members of the Sub-Committee have participated actively in the preparation of this work. Whatever merit these lists and annotations may have must be credited to all of these contributors.

ORGANA URINARIA

REN

Margo lateralis[1]
Margo medialis[2]
Facies ventralis[3]

REN—*continued*

Facies dorsalis
Extremitas cranialis
Extremitas caudalis
Divisio renalis cranialis[4] (Fig. 1)
Divisio renalis media[4] (Fig. 1)
Divisio renalis caudalis[4] (Fig. 1)

Lobus renalis[5] (Fig. 2)
Lobulus renalis[6] (Figs. 2, 3)
Cortex renalis[7]
Medulla renalis[7]

TUBULUS RENALIS

Nephronum

Nephronum corticale[8] (Figs. 2, 3, 4)
Nephronum medullare[8] (Figs. 2, 3, 4)

Corpusculum renale
Glomerulus corpusculi renalis
Mesangium
Mesangiocytus[9]
Polus vascularis
Capsula glomerularis
Pars externa
Pars interna
Podocytus[10]
Lamina basalis[10]
Lumen capsulae
Polus tubularis
Tubulus convolutus proximalis (Fig. 4)
Pars tenuis[11]
Pars crassa[11]
Pars convoluta[11]
Pars recta[11]
Segmentum intermedium[12] (Fig. 4)
Ansa nephroni[12]
Pars descendens ansae (Fig. 4)
Pars ascendens ansae (Fig. 4)
Tubulus convolutus distalis[13]
Pars paraglomerularis[13]
Pars convoluta[13] (Fig. 4)
Pars conjungens[13] (Fig. 4)

Tubulus colligens[14]
Tubulus colligens perilobularis (Figs. 3, 4)
Tubulus colligens medullaris (Fig. 3)
Ductus colligens[14]
Complexus juxtaglomerularis[15]
Macula densa[15]
Cellula maculae densae
Cellula juxtaglomerularis[15]
Insula juxtavascularis [Mesangium extraglomerulare][16]
Cellula insulae juxtavascularis[16]

VASA SANGUINEA INTRARENALIA

Aa. intralobulares[17,6] (Fig. 3)
Arteriola glomerularis afferens (Fig. 3)
Rete capillare glomerulare (Fig. 3)
Ansa capillaris
Arteriola glomerularis efferens (Fig. 3)
Arteriolae rectae[18] (Fig. 3)
Rr. renales afferentes[19] (Figs. 3, 5)
Vv. interlobulares[19,6] (Fig. 3)
Rete capillare peritubulare corticale[20] (Fig. 3)
Rdxx. renales efferentes[19] (Fig. 5)
Vv. intralobulares[19,6] (Figs. 2, 3, 4)
Venulae rectae[21]

URETER

Pars renalis (Fig. 1)
Rr. ureterici primarii[22] (Figs. 1, 2)
Rr. ureterici secundarii[22] (Figs. 1, 2, 3)
Pars pelvica (Fig. 1)
Ostium cloacale ureteris (see **Cloaca** Urodeum)

ANNOTATIONS

(1) **Margo lateralis.** This border is deeply indented where major blood vessels (e.g. A. iliaca externa, A. ischiadica) enter and leave the kidney (Fig. 1).

(2) **Margo medialis.** In many species the left and right kidneys fuse along the Margo medialis, particularly in the Divisio caudalis, the fusion being confined to cortical tissue (Johnson, 1968).

(3) **Facies ventralis.** This surface is superficially incised by the vessels that cross it. (Annot. 3 cont. on p. 292).

Variations in the grooves so formed probably account for claims of more than three divisions in some species (see Annot. 4).

(4) **Divisio renalis cranialis; Divisio renalis media; Divisio renalis caudalis** (Fig. 1). These divisions are believed to be characteristic of birds in general, although there are marked species variations in their relative sizes. The boundaries between them are arbitrary; that between the cranial and middle divisions may be the A. iliaca externa, and that between the middle and caudal divisions the A. ischiadica. The three divisions are in no way homologous to the lobes or renculi of the mammalian kidney (Goodchild, 1956; Siller, 1971, pp. 197, 202), and should not be referred to as "lobes". It has been claimed that there are four divisions in Ciconiiformes and Charadrii and five in *Apteryx* (Francis, 1964), but Johnson (1978) examined several species from the first two of these groups and found only three divisions (see Annot. 3). There are also reports of species with only two divisions (Feinstein, 1962).

(5) **Lobus renalis.** The conical bundles of collecting tubules belonging to several adjacent lobules (see Annot. 6) converge into a cone-shaped assemblage of collecting tubules enclosed by a connective tissue sheath (Fig. 2). One such assemblage, together with the lobules which drain into it, may be regarded as Lobus renalis (Lindgren, 1868; Goodchild, 1956; Siller, 1971, p. 202; Johnson, 1974). The cone-shaped assemblage of collecting tubules in the avian kidney is probably homologous to the mammalian medullary pyramid; the avian Lobus renalis in toto is probably homologous to the lobe of the mammalian multilobar kidney (King, 1975, p. 1922).

(6) **Lobulus renalis.** In histological sections (Spanner, 1925; Feldotto, 1929; von Möllendorff, 1930, p. 232 and Fig. 201; Sperber, 1960, pp. 470–478) the lobule is a pear-shaped area surrounded by the Tubuli colligentes which drain it (Fig. 3). The terminal branches of the Rr. renales afferentes lie between these lobules and therefore become Vv. interlobulares. On the other hand the initial tributaries of the Rdxx. renales efferentes lie in the centres of the lobules, and are therefore Vv. intralobulares, being in the same position as the Aa. intralobulares. The tapering stalk of the pear is formed by its Tubuli colligentes, which converge into a conical bundle enclosed by a connective tissue capsule; this is the medullary region of the lobule (MRL 1 to 5, in Fig. 2) and contains not only Tubuli colligentes but the ansae of the medullary nephrons. The wide part of the pear is the cortical region of the lobule (CRL in Fig. 2), containing cortical and medullary nephrons, but not the ansae of the medullary nephrons.

The three-dimensional studies of Johnson and co-workers have shown that the cortical region of the lobule is not simply pear-shaped, but is really elongated like a loaf of bread (Johnson *et al.*, 1972; Johnson, 1974). This elongated cortical region typically contributes Tubuli colligentes to several independent medullary regions; moreover any one medullary region may receive Tubuli colligentes from several independent cortical regions. Thus it is clear that the avian Lobulus renalis is much more complicated than hitherto realized. Since to some extent a given Lobulus shares its drainage with its neighbours it could be argued that the term Lobulus is not strictly applicable; nevertheless the Lobulus renalis is a convenient structural concept.

This analysis of the avian renal lobule places the collecting tubules and afferent veins in an interlobular position. As pointed out by Johnson (1978), most workers have followed this approach. Goodchild (1956), following Smith (1956), argued for a reversal of this relationship, proposing that the collecting tubules and afferent veins should comprise the intralobular axis. This would have the advantage of corresponding to the mammalian pattern, wherein the medullary ray forms the axis of the renal lobule. However, it is difficult to equate the elaborately arranged collecting tubules of birds with medullary rays. Also the arterial distribution is less orderly in the avian kidney than in the mammalian kidney. Hence it is more convenient to regard the efferent venous drainage as intralobular. This sacrifice of the direct homology with the mammalian renal lobule is compensated by the greater ease with which the organization of the avian renal lobule can then be appreciated, together with the emphasis this places on the presence of a renal portal system in birds (see Annot. 19; and **Ven.** Annot. 59).

(7) **Cortex renalis; Medulla renalis.** Cortex renalis is formed by the wide cortical regions of the Lobuli renales, and Medulla renalis by the tapering stalk-like medullary regions of the Lobuli renales (see Annot. 5 and 6). However, the Lobuli (as well as the Lobi renales) are embedded at varying depths in the avian kidney. Consequently the Cortex and Medulla do not form the continuous outer and inner strata which typify the kidney of many mammalian species. The avian kidney does, however, resemble the kidney of the extreme renculus type shown by whales (Cetacea) (Sperber, 1944, p. 401). In both birds and cetaceans there are many lobes at varying depths in the kidney. In the cetacean kidney the lobes are loosely joined by connective tissue, but in the avian kidney the lobes are fused into a continuous mass of renal tissue.

(8) **Nephronum corticale; Nephronum medullare.** Nephronum corticale (Fig. 4) is reptilian in type. In many avian species the great majority of nephrons are cortical ones (Huber, 1917; Feldotto, 1929; von Möllendorff, 1930; Sperber, 1960, pp. 470–474). The essential feature of the cortical nephron is its short Segmentum intermedium with no ansa. The Nephronum medullare (Fig. 4) is mammalian in type, its Segmentum intermedium forming a distinct Ansa. In certain Passeriformes adapted to habitats where water is scarce (e.g. *Ammodramus*) the Nephrona medullaria appear to be substantially more numerous (Johnson and Ohmart, 1973). The functional characteristics of the two types of nephron were described by Braun and Dantzler (1972).

(9) **Mesangiocytus.** The mesangial cells form the core of the Glomerulus (Siller, 1971, p. 210).

(10) **Podocytus; Lamina basalis.** The cells of the inner layer of the Capsula glomerularis appear to be more cuboidal than in mammals. The essence of a Podocytus is its foot-like processes, and these are present in *Gallus* (Siller, 1971, p. 209). Therefore this cell is almost certainly homologous to the mammalian Podocytus. As in mammals the Lamina basalis has an essential role in filtration, being the only continuous layer in the Pars interna and probably the main filter of large molecules.

(11) **Pars tenuis; Pars crassa; Pars convoluta; Pars recta.** Pars tenuis and Pars crassa occur only in the Nephronum corticale; Pars convoluta and Pars recta are restricted to the Nephronum medullare. In the Nephronum corticale the first part of the proximal convoluted tubule is of relatively small diameter, hence Pars tenuis, while the remainder is relatively thick, hence Pars crassa; these two parts correspond to the Pars convoluta and Pars recta of the Nephronum medullare (Johnson, 1978).

(12) **Segmentum intermedium; Ansa nephroni.** The Ansa nephroni has been called the loop of Henle. In cortical nephrons the Segmentum intermedium is very short and and convoluted, but the segment is extended into an ansa in medullary nephrons

(Huber, 1917; Marshall, 1934) (Fig. 4). The ansa resembles exclusively the mammalian short type, since the loop turns in the thick segment in those species which have been examined (Sperber, 1960, p. 474; Johnson and Mugaas, 1970; Siller, 1971, p. 215). The thin segment of the Pars descendens differs from that of mammals in having taller and more continuous cells (Siller, 1971, p. 215). The thick segment resembles its mammalian homologue.

(13) **Tubulus convolutus distalis; Pars paraglomerularis; Pars convoluta; Pars conjungens.** The distal convoluted tubule may begin with the Pars paraglomerularis which contributes the Macula densa to the Complexus juxtaglomerularis as in mammals (Siller, 1971, p. 221). Alternatively the distal convoluted tubule may have a preglomerular portion (Siller, pers. comm.). The coils of the Pars convoluta lie near the V. intralobularis. The Pars conjungens is histologically similar to the Tubulus colligens; it is believed by Siller (1971, p. 221) to form the final link between the latter and the Nephronum (Siller, 1971, p. 221), but other workers have tended to regard the Pars conjungens as the first part of the Tubulus colligens (see Johnson, 1978).

(14) **Tubulus colligens; Ductus colligens.** The Tubulus colligens perilobularis (Fig. 5) lies superficially on the cortical region of the lobule, while the Tubulus colligens medullaris lies within the medullary region (Fig. 3) of the lobule (Siller, 1971, p. 225). At the narrow apex of the medullary region of the renal lobule there is typically only a single large collecting vessel, the Ductus colligens; this is the same as the third order of branching of the ureter.

(15) **Complexus juxtaglomerularis; Macula densa; Cellula juxtaglomerularis.** A complete juxtaglomerular complex is present in birds, comprising the Macula densa, Cellulae juxtaglomerulares (the renin-secreting, modified myocytes of the tunica media of the afferent glomerular arteriole), and the Insula juxtavascularis (Johnson, 1978).

(16) **Insula juxtavascularis** [Mesangium extraglomerulae]; **Cellula insulae juxtavascularis.** The Insula is commonly known as the extraglomerular mesangium. Its Cellulae insulae juxtavascularis are also called Polkissen, Goormaghtigh, or Lacis cells. In birds on a salt free diet, substantial numbers of granules appear in these cells (Siller, 1971, p. 210).

(17) **Aa. intralobulares.** These arteries arise from branches of the renal arteries, and are central in the Lobulus renalis (Fig. 3). Siller and Hindle (1969) believe that Aa. interlobares have not yet been identified; see Kurihara and Yasuda (1975) for the opposite view. See Annot. 6.

(18) **Arteriolae rectae.** These arteries descend in the medullary part of each lobule, arising from efferent arterioles near the medula essentially as in mammals (Siller and Hindle, 1969).

(19) **Rr. renales afferentes; Vv. interlobulares; Rdxx. renales efferentes; Vv. intralobulares.** The Rr. renales afferentes are the branches of the V. portalis renalis cranialis and caudalis (see **Ven.** Annot. 59); they end by forming the Vv. interlobulares (Fig. 3). The Rdxx. renales efferentes are formed by union of the Vv. intralobulares, and they discharge into the cranial and caudal renal veins (see Annot. 6 and Fig. 5).

(20) **Rete capillare peritubulare corticale.** The peritubular capillary network, which ramifies among the nephronal tubules within the cortex, receives blood both from the venules of the Vv. interlobulares and from the Arteriolae efferentes (Fig. 3). The Ansa nephroni is also surrounded by a capillary network within the medulla, but this medullary network should be considered as an entity separate from the cortical network (Johnson, pers. comm.).

(21) **Venulae rectae.** These venules drain the blood which reaches the medullary part of each lobule via the Arteriolae rectae. The Venulae rectae and Arteriolae rectae together constitute the Vasa recta.

(22) **Rr. ureterici primarii et secundarii.** Rr. ureterici primarii (Figs. 1, 2) are direct tributaries of the ureter. Each R. **uretericus** secundarius drains a cone-shaped assemblage of Tubuli colligentes which (see Annot. 5) forms the basis of a Lobus renalis (Fig. 2).

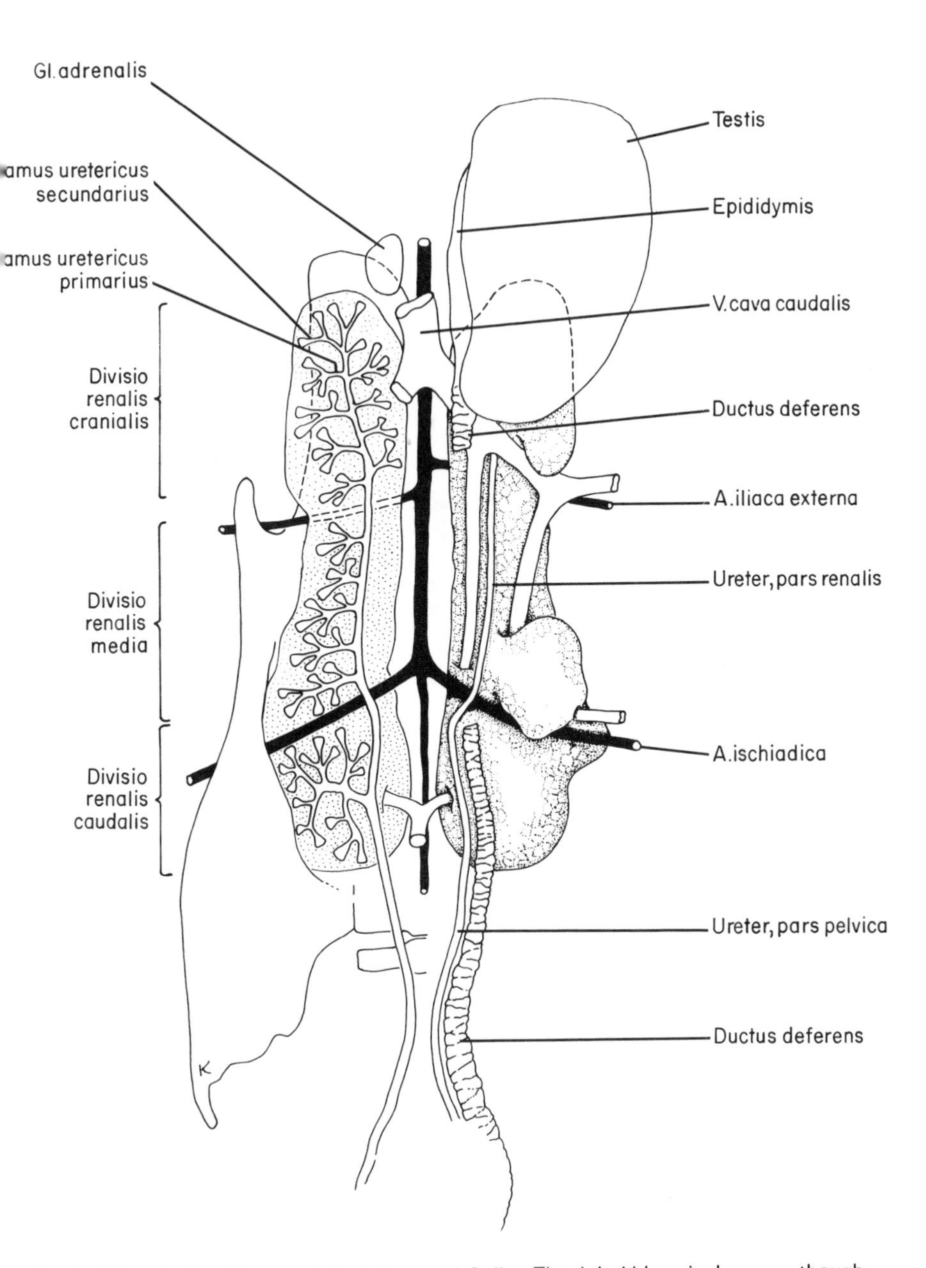

Fig. 1 Ventral view diagram of the kidneys of *Gallus*. The right kidney is drawn as though transparent to show the Rr. ureterici primarii and some of the Rr. ureterici secundarii. Each secondary branch ends in a cone-shaped assemblage of collecting tubules belonging to a Lobus renalis. Most of the cranial half of the left Ductus deferens has been removed. From King (1975), with permission from W. B. Saunders, Philadelphia.

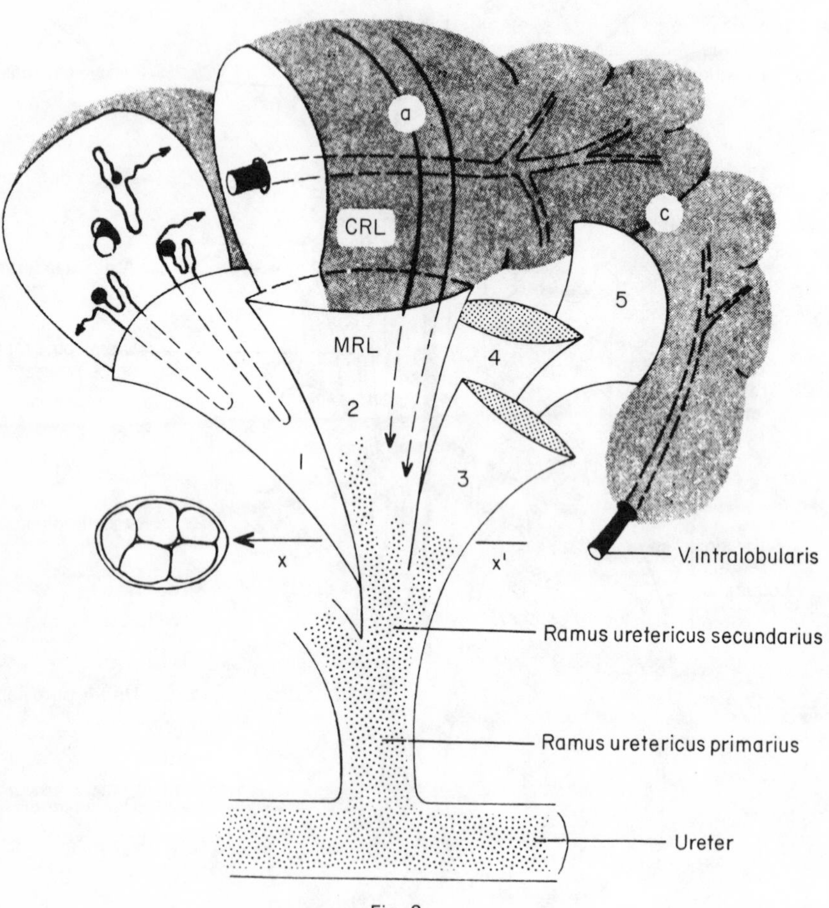
a
CRL
c
MRL
5
4
2
1
3
x
x'
V. intralobularis
Ramus uretericus secundarius
Ramus uretericus primarius
Ureter

Fig. 2

Fig. 2 Basic organizational plan of the avian kidney. Starting from a R. uretericus primarius, a R. uretericus secundarius receives five medullary regions (numbered 1 through 5). Each medullary region is cone-shaped and can be likened to a funnel bounded by connective tissue (dark lines). As the cluster of medullary regions converges upon the secondary branch, the whole group gradually acquires a common connective tissue coat (depicted in the transverse section at level X–X''). Following the development of a common investment, the connective tissue surrounding each individual medullary region largely disappears. All of the foregoing connective tissue layers are continuous with the connective tissue sheath (dark lines) of the ureter. MRL = medullary region of Lobulus renalis; CRL = cortical region of Lobulus renalis. Arrows represent Tubuli colligentes.

The Cortex is arranged concentrically around the Vv. intralobulares, three of which are shown. Tubuli colligentes (such as those illustrated at "a") encircle the Cortex peripherally. The Tubuli shown are on the surface facing the viewer; other tubules would course similarly on the other cortical surface, i.e. on the surface away from the viewer.

The Tubuli colligentes enter the funnel-like end of a medullary region and undergo gradual dendritic fusion (in the direction of the arrows) to form increasingly larger tubules. The latter are directly continuous with the lumen of the ureter. Medullary regions 3 and 4 are transected; the stippling on their cut surfaces represents collecting tubules and other medullary components.

Three nephrons (the upper one being a Nephronum corticale of the reptilian type) are shown in the cut surface of cortex at the upper left. The arrows indicate the direction by which distal convoluted tubules of these nephrona will eventually empty into peripherally located Tubuli colligentes perilobulares. The relationship between the Ansa nephroni (loop of Henle) and the medullary region is typified by the two ansae shown in medullary region 1.

A Lobulus renalis (renal *lobule*) consists of a medullary region (MRL) and a cortical region (CRL). The latter is not sharply delimited from adjacent cortical regions, being simply that region of cortex which contributes collecting ducts and loops of Henle to a single medullary region. A given medullary region also often drains the cortical regions associated with adjacent Vv. intralobulares. For example, at "c" two cortical regions lie in close approximation; medullary region 5 would receive tubular elements from both of these cortical regions. A Lobus renalis (renal *lobe*) is a group of medullary regions which drain into a secondary branch of the ureter, plus their associated cortical regions. Hence, the diagram essentially represents one Lobus renalis.

The foregoing pattern of organization is repeated many times throughout the kidney, with medullary regions often elaborately curved. From Johnson (1974).

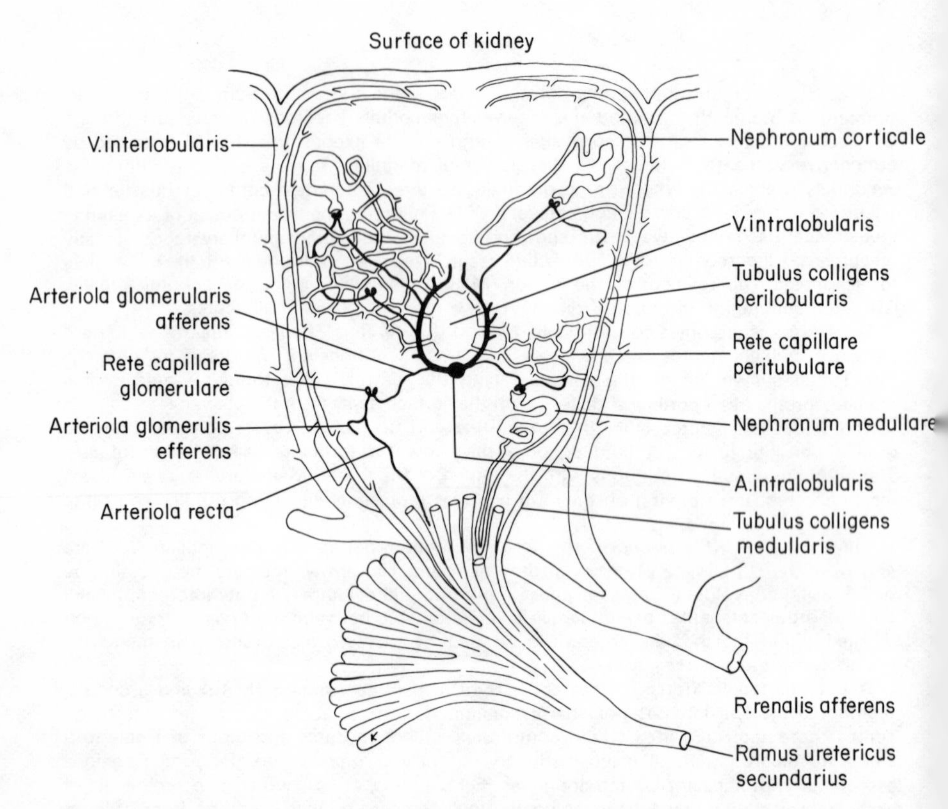

Fig. 3 Diagram of a Lobulus renalis as seen in a histological section. The Lobulus appears pear-shaped. The wide part, containing the Nephrona corticalia and Nephrona medullaria (except for the ansae of the latter), is the cortical region of the Lobulus renalis. The tapering "stalk" of the pear, consisting of the conical bundle of Tubuli colligentes and the Ansae of the Nephrona medullaria, is the medullary region of the Lobulus renalis. The lower, left side of the diagram also shows the conical bundles of Tubuli colligentes belonging to the medullary regions of two other Lobuli renales. The cone-shaped assemblage of the Tubuli colligentes belonging to the medullary regions of all three Lobuli renales, plus their associated cortical regions, constitute a Lobus renalis. The Lobus renalis in turn drains into a Ramus uretericus secundarius. From King (1975), with permission from W. B. Saunders, Philadelphia.

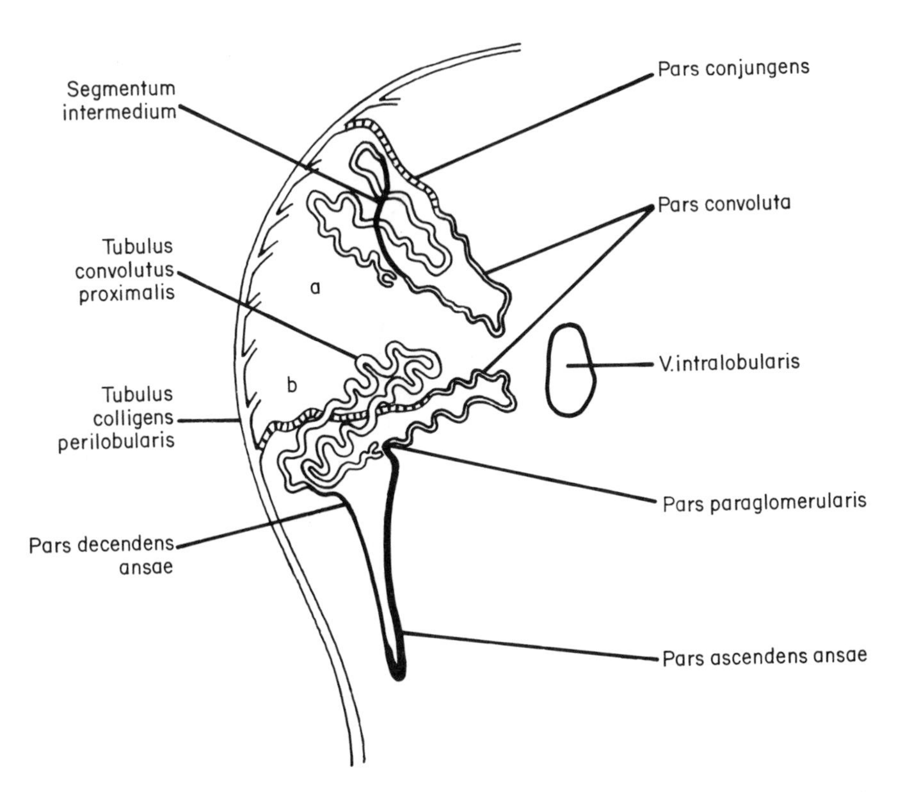

Fig. 4 Diagram showing the main parts of (a) Nephronum corticale, and (b) Nephronum medullare. From King (1975) with permission from W. B. Saunders, Philadelphia.

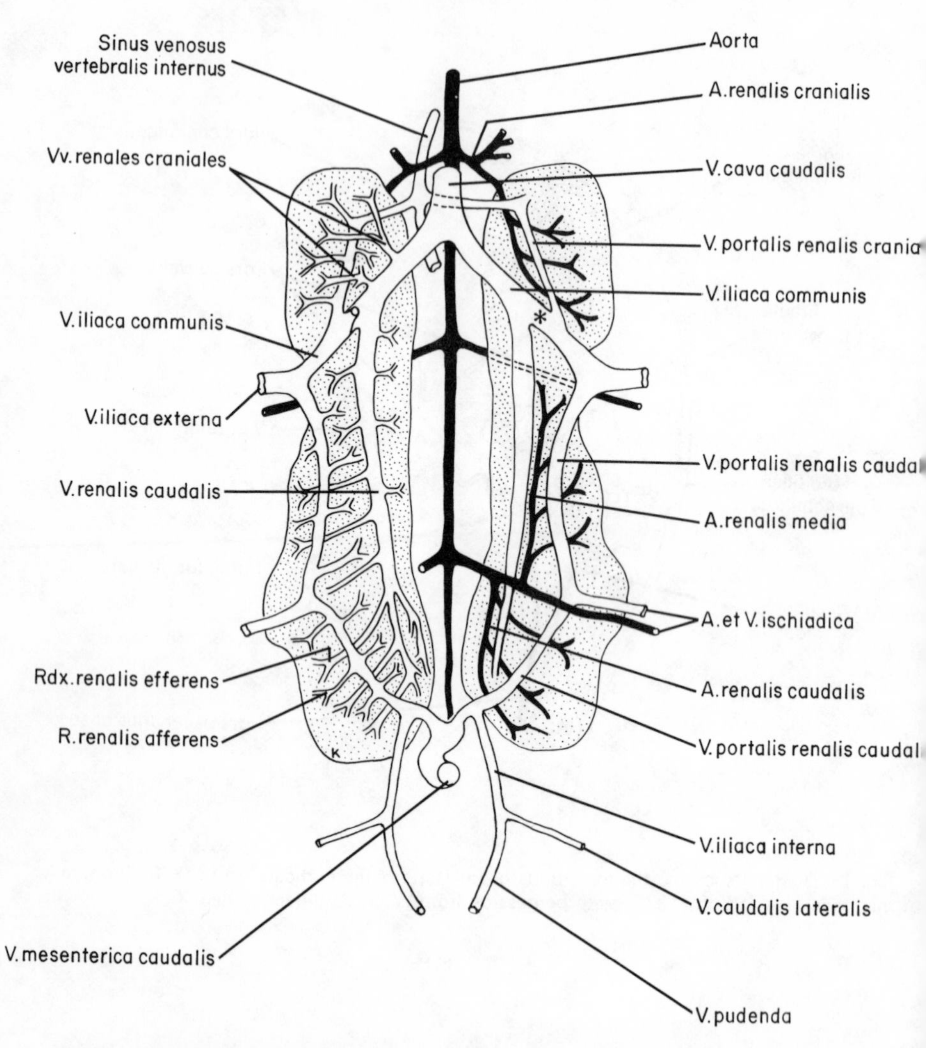

Fig. 5 Ventral view diagram of the kidneys of the domestic fowl, drawn as though transparent to show their blood vessels. *, site of Valva portalis renalis. From King (1975), with permission from W. B. Saunders, Philadelphia.

ORGANA GENITALIA MASCULINA

TESTIS

Extremitas cranialis
Extremitas caudalis
Facies lateralis
Facies medialis
Margo liber
Margo epididymalis
Mesorchium

Tunica albuginea[1] (Fig. 1)
Parenchyma testis (Fig. 1)
- Tubuli seminiferi[2] (Fig. 1)
  - Cellula sustentacularis[3]
  - Epithelium spermatogeneticum
    - Spermatogonium[4]
    - Spermatocytus primarius
    - Spermatocytus secundarius
    - Spermatidium
      - Acroblastus
        - Vesicula acrosomatica[5]
        - Granulum acrosomaticum[6]
      - Tubuli paraxiales[7]
    - Spermatozoon
      - Acrosoma
        - Galerum acrosomae[8]
        - Spina acrosomae[8]
        - Caverna subacrosomatica[9]
      - Caput spermatozoi
        - Nucleus spermatozoi
      - Cauda spermatozoi
        - Collum spermatozoi[10]
          - Fossa articularis
          - Discus basalis
          - Centriolum proximale
        - Pars media
          - Centriolum distale
          - Vagina mitochondrialis
          - Annulus
        - Pars principalis
          - Vagina amorpha[11]

TESTIS—*continued*

Complexus filamenti axialis
Subfibrilla A
Subfibrilla B
Brachium
Pars terminalis
Stratum fibrosum[12]
Interstitium testis
Cellulae interstitiales[13]
Melanocyti[14]
Tubuli recti[15] (Fig. 1)

Rete testis[16] (Fig. 1)

A. testicularis[17]
Vv. testiculares[17]

EPIDIDYMIS

(Figs. 1, 2)

Extremitas cranialis[18]
Extremitas caudalis[18]
Ductuli efferentes[19] (Fig. 1)
Ductuli conjungentes[20]
Ductuli aberrantes[21] (Fig. 1)
Tubuli paradidymales[21]
Appendix epididymalis[21]
Ductus epididymalis (Fig. 1)

DUCTUS DEFERENS

(Figs. 1, 2, 3)

Ansae ductus deferentis[22]
Pars recta ductus deferentis[23]
Receptaculum ductus deferentis[24]
Papilla ductus deferentis (Figs. 2, 3, 5, 6)
Glomus seminale[25]
Promontorium cloacale[26]
Ostium cloacale ductus deferentis

APPARATUS COPULATIONIS

Corpus vasculare paracloacale[27] (Figs. 2, 3)

Phallus nonprotrudens[28]
  Corpus phallicum medianum [mediale][29] (Fig. 2)
  Corpus phallicum laterale[30] (Fig. 2)
  Plicae lymphaticae[30] (Fig. 2)
Phallus protrudens[31]
  Fossa ejaculatoria[32] (Figs, 3, 5, 6)
    Corpus fibrocartilagineum[33]
  Cavitas phalli[34]
    Aditus ad cavitatem phalli[35] (Fig. 5)
  Basis phalli[36] (Figs. 5, 6)
  Pars cutanea phalli[37] (Figs. 5, 6)
    Sulcus phalli[38] (Figs. 3, 4, 6)
    Rugae phalli[39]
  Flexura phalli [Apex phalli][40] (Figs. 5, 6)
  Pars glandularis phalli[41] (Figs. 3, 5)
  Pars extrema phalli[42] (Fig. 5)
  Suspensorium phalli[43]
  Corpus fibrolymphaticum[44] (Figs. 3, 4)
  Cavitas lymphatica[45] (Fig. 3)
  Lig. elasticum phalli[46] (see **Arthr.** Annot. 3)
M. retractor phalli [M. levator cloacae][47] (see **Myol.** Annot. 68)

ANNOTATIONS

(1) **Tunica albuginea.** A membranous and extremely thin layer, septula and mediastinum testis being absent (King, 1975, p. 1928).

(2) **Tubuli seminiferi.** This is equivalent to the Tubulus seminifer contortus of *Nomina Histologica* (IANC, 1975) and *Nomina Anatomica Veterinaria* (ICVAN, 1973). Since the Tubulus rectus is not seminiferous, in *Gallus* at least, there is only one type of Tubulus seminiferus; therefore the adjective, "contortus" is superfluous.

(3) **Cellula sustentacularis.** Formerly known as Sertoli cells, with a mechanical or nutritive role.

(4) **Spermatogonium.** The mammalian types A and B have not been recognized in birds (see Reviers, 1971a,b).

(5) **Vesicula acrosomatica.** A vacuole in which the first sign of the Acrosoma appears.

(6) **Granulum acrosomaticum.** This is the **entrenching** phase of the development of the acrosome.

(7) **Tubuli paraxiales.** The Tubuli paraxiales or "manchette" constitute a scaffolding of microtubules around the nucleus. It is believed that elongation of the nucleus is a function of the double helix, while the manchette promotes the final curvature of the head (McIntosh and Porter, 1967).

(8) **Galerum acrosomae; Spina acrosomae.** The Granulum acrosomaticum soon elongates, and the outer Galerum acrosomae (acrosomal cap) and inner Spina acrosomae then appear. The Spina acrosomae may be homologous to the "perforatorium" of certain mammals (Lake *et al.*, 1968; Tingari, 1973). Further research on the Spina acrosomae may reveal its derivation more precisely, and hence lead to modification of its terminology. The ultrastructure of ejaculated spermatozoa of *Gallus* was described by Lake *et al.* (1968).

(9) **Caverna subacrosomatica.** The space between the Galerum and Spina acrosomae.

(10) **Collum spermatozoi.** The region joining the Caput to the Pars media; short and not elaborate in *Gallus*, but it may be better developed in other species of bird.

(11) **Vagina amorpha.** In birds the sheath is amorphous rather than fibrous (Lake *et al.*, 1968).

(12) **Stratum fibrosum.** This peritubular boundary tissue, with homogeneous electron dense material, collagen fibers, and fibroblasts, is present as in man, cat, ram, and boar (Rothwell and Tingari, 1973).

(13) **Cellulae interstitiales.** Formerly the cells of Leydig.

(14) **Melanocyti.** In some seasonal species (e.g. *Sturnus*) numerous melanocytes impart a dense black colour to the Testis in the eclipse phase (Marshall, 1961, p. 187), as in the Black Leghorn breed of chickens. When the testis enlarges with the approach of sexual activity the melanocytes disperse and the testis goes grey and finally white.

(15) **Tubuli recti.** The presence of Tubuli recti in birds is controversial. According to Gray (1937) the Tubuli seminiferi of *Gallus* lose their germ cells just before they join the Rete testis, and then become very short Tubuli recti with cuboidal epithelium. Tingari (pers. comm.) denied the presence of Tubuli recti in this species. Possibly the "efferent cones" described by Traciuc (1967, 1969) in other species are homologous to the supposed Tubuli recti of *Gallus*.

(16) **Rete testis.** The Rete testis is absent in some birds (Traciuc 1967, 1969) but present in others (Alverdes, 1924; Gray, 1937; Gailey, 1953; Stoll and Maraud, 1955; Tingari, 1971). Bailey (1953) was certain of its homology to the Rete testis of other vertebrates. However, when it is present it lies on the surface of the Testis or Epididymis, thus differing from most but not all mammals (in the rat the position of the Rete testis is similar to that of birds) (Figs. 1, 2).

(17) **A. testicularis; Vv. testiculares.** A. testicularis arises from the cranial renal artery, an inconstant accessory testicular artery arising sometimes from the aorta; the several Vv. testiculares drain into the caudal vena cava (Nishida, 1964). See **Art.** Annot. 66; and **Ven.** Annot. 56.

(18) **Extremitas cranialis; Extremitas caudalis.** The terms Caput, Corpus, and Cauda of the Epididymis are inappropriate since the Ductuli efferentes arise throughout the whole length of the Epididymis.

(19) **Ductuli efferentes.** Where they arise from the Rete testis, the Ductuli efferentes

have a wide calibre and contain relatively numerous Spermatidia and relatively few Spermatozoa. Further on, the ductuli become progressively narrower, and finally continue directly as Ductuli conjungentes. Budras and Sauer (1975) named the wide ductuli as Ductuli efferentes proximales, and the narrow ones as Ductuli efferentes distales; they claimed that the wide ductuli arise from the glomerular capsules of the mesonephros, and the narrow ductuli from the proximal and distal tubules and intermediate segments of the mesonephros.

(20) **Ductuli conjungentes.** These are initially narrower than the Ductuli efferentes. As they approach the Ductus epididymalis they begin to join each other and become gradually wider. The Ductuli conjungentes nearer to the Ductus epididymalis contain relatively few Spermatidia and relatively numerous Spermatozoa (Tingari, 1971). The Ductuli conjungentes end by joining the Ductus epididymalis. They apparently arise from the collecting tubules of the mesonephros (Budras and Sauer, 1975).

(21) **Ductuli aberrantes; Tubuli paradidymales; Appendix epididymalis.** These vestigial structures were seen in *Gallus* by Stoll and Maraud (1955) and Tingari (1971). Budras and Sauer (1975) studied them in detail in the same species, including their embryonic origin; they found that most of the Ductuli aberrantes lay in the Appendix epididymalis, connecting with the cranial end of the mesonephric duct (the Ductus aberrans of Budras and Sauer) which extends into the Appendix. Only a few Ductuli aberrantes were found within the main part of the Epididymis; of these, some connected to the Rete testis and others to the Ductus epididymalis. The Tubuli paradidymalis, which are blind at both ends and therefore connect with nothing, are rare but can be found in all areas of the Epidymis. Budras and Sauer (1975) also emphasized that the Appendix epididymalis in this species is firmly attached by connective tissue to the adrenal gland; when the Testis is removed, this connective tissue breaks and the Appendix remains on the adrenal gland.

(22) **Ansae ductus deferentis.** The sinuous curves of the main part of the Ductus deferens.

(23) **Pars recta ductus deferentis.** The short straight caudal end of the Ductus deferens (Marvan, 1969).

(24) **Receptaculum ductus deferentis.** Synonymy: Ampulla ductus deferentis (Marvan, 1969; Kudo *et al.*, 1975). The Receptaculum is the spindle-shaped terminal dilation of the Ductus deferens. The term, Ampulla ductus deferentis, is descriptively accurate, but has the disadvantage of suggesting homology to the mammalian ampulla.

(25) **Glomus seminale.** The convoluted caudal end of the Ductus deferens in passerines (Wolfson, 1952, 1954; Bailey, 1953; Middleton, 1972). The Glomus seminale is also known as the seminal vesicle or seminal sac; it enlarges in the breeding season, and may enable storage and maturation of sperm.

(26) **Promontorium cloacale.** The external projection caused by the Glomus seminale during the nuptial phase in Passeriformes, enabling the male to be sexed (Wolfson, 1952; Salt, 1954).

(27) **Corpus vasculare paracloacale.** Synonymy: Corpus cavernousm, Tannenberg's body, Lymphobulbus phalli (see Rautenfeld, 1973). This structure had various names in the earlier literature, which were reviewed by Rautenfeld (1973), who preferred

Lymphobulbus phalli. But the term (Gefasskörper is nearly 150 years old, originating from J. Muller's classic paper of 1836 (Rautenfeld, 1973). It is an arterial rete from the A. pudenda, comprising many Glomera (see **Cloaca**) which produce lymph for erection in *Anas* and *Gallus* (Liebe, 1914; Nishiyama, 1950, 1955; Knight, 1970; Kudo *et al.*, 1975; Sugimura *et al.*, 1975), and probably in birds generally (see **Cloaca** Annot. 6). The lymph is carried to Cavitas lymphatica of the Phallus in the (paired) Ductus lymphaticus in **Anas** (Guzsal, 1974) and to the Corpus phallicum and Plicae lymphaticae by several or many Vasa lymphatica in **Gallus** (Knight, 1970) (see Fig. 2).

(28) **Phallus nonprotrudens.** The non-intromittent copulatory organ mounted on the ventral lip of the vent in Galliformes (King, 1975, pp. 1930–1932), and probably in birds generally.

(29) **Corpus phallicum medianum [mediale].** Synonymy: White body (Nishiyama, 1955; Lake, 1971, p. 1420). In *Gallus* this is a single midline structure; therefore "medianum" is appropriate. In *Meleagris* it is paired, making "mediale" the correct term. Rautenfeld (1973) suggested Colliculus phalli for this structure. (See Fig. 2).

(30) **Corpus phallicum laterale; Plicae lymphaticae.** Corpus phallicum laterale is a paired structure, with various names in the literature, mostly descriptive such as "round folds" (Nisiyama, 1955; Lake, 1971, p. 1420). Rautenfeld (1973) suggested Crura phalli, but this indicates similar relationships to the Corpus phallicum medianum as those of the Crura penis to Corpus penis in mammals. So long as there is any doubt of such relationships, a neutral and topographically descriptive term is preferred. Plicae lymphaticae are the lymph folds of Nishiyama (1955) and Lake (1971) that contribute the so-called transparent fluid to the semen.

(31) **Phallus protrudens.** The true intromittent organ of the ratites, Tinamidae, Anseriformes, and possibly the curassows (*Crax*) (Gerhardt, 1933, pp. 305–311).

(32) **Fossa ejaculatoria.** Synonymy: Ejaculatory groove (Fujihara *et al.*, 1976). A depression in the proctodeal floor leading to the Sulcus phalli (Komárek, 1969). (Fig. 6). The fossa is formed by the left and right components of the Plica uroproctodealis (see **Cloaca**) which converge on the Basis phalli (Fujihara *et al.*, 1976).

(33) **Corpus fibrocartilagineum.** A gutter-shaped fibro-elastic cartilage plate, embedded in the floor of the Fossa ejaculatoria. Between the cartilage plate and the mucous membrane is the paired Ductus lymphaticus (Guzsal, 1974) leading to the caudal end of the Cavitas lymphatica (Fujihara *et al.*, 1976).

(34) **Cavitas phalli.** The cavity inside the invaginated detumescent Phallus.

(35) **Aditus ad cavitatem phalli.** Synonymy: Ostium penis (Guzsal, 1974). The orifice leading to the cavity inside the invaginated Phallus (Komárek, 1969). The Aditus is bounded by the caudal end of the Corpus fibrocartilagineum (Guzsal, 1974).

(36) **Basis phalli.** The base of the Phallus attaching to the proctodeal floor (Komárek, 1969) (Fig. 5).

(37) **Pars cutanea phalli.** Synonymy: Portio proximalis (Fehér, pers. comm.; Guzsal, 1974); Pars cavernosa (Guzsal, 1974). In the invaginated Phallus, this is the

region between the Basis phalli and the Flexura phalli, and contains an internal lumen which is lined by a stratified squamous epithelium. It is everted during erection, and carries the Sulcus phalli on its surface (Komárek and Marvan, 1969). Fehér (pers. comm.) and Guzsal (1974) have suggested naming this the Portio proximalis: this term would be useful in the invaginated phallus, since it would distinguish this segment from Pars glandularis phalli and the Pars extrema phalli, which would together form the Portio distalis in the non-erect state; but in the erect Phallus both the Portio proximalis and the Pars glandularis of the Portio distalis are included in the protruded component, so that "proximalis" and "distalis" are no longer so clearly understandable (Figs. 5, 6).

(38) **Sulcus phalli.** The Sulcus is constantly present in all avian species with a Phallus protrudens, and is spiral in Anseriformes. But is straight in the ratites and Crocodilia (Gerhardt, 1933, pp. 305–311) (Figs. 4, 6). The sulcus is commonly believed to carry semen, but this has now been questioned (Rautenfeld *et al.*, 1974).

(39) **Rugae phalli.** The more or less circular folds on the Pars cutanea phalli when erect (Komárek, 1969) (Fig. 6).

(40) **Flexura phalli** [Apex phalli]. The Flexura phalli is the region which is strongly curved where the Phallus is invaginated, but becomes the Apex phalli when it is erect (Komárek, 1969) (Figs. 5, 6). Note that the Apex penis of Guzsal (1974) is the "root" of the Phallus (see Annot. 42, and is therefore not the same structure as the Apex phalli of Komárek (1969) and Fig. 6.

(41) **Pars glandularis phalli.** The region that contains an internal lumen lined by a glandular epithelium, located between Flexura phalli and Pars extrema when the Phallus is invaginated; it forms a glandular tube inside the erect Phallus (Komárek and Marvan, 1969). Fehér (pers. comm.) and Guzsal (1974) have suggested that this should form part of their Portio distalis; this term would be meaningful in the invaginated Phallus, but would be difficult to understand in the erect Phallus (see Annot. 37). This glandular tube is absent in *Struthio*, *Apteryx*, Tinamidae, and Crocodilia (Gerhardt, 1933, pp. 305–311). See Fig. 5.

(42) **Pars extrema phalli.** Synonymy: Apex penis (Guzsal, 1974). This is the deeply invaginated "root" of the Phallus, anchored to the Corpus fibrocartilagineum (Guzsal, 1974). It is not evaginated when the Phallus is erect, but remains buried in the cloacal wall.

(43) **Suspensorium phalli.** The suspensory sheet of connective tissue that links Pars glandularis phalli to Pars extrema phalli in the non-erect condition (Komárek, 1969). It helps to return the Phallus to its invaginated state after erection (Guzsal, 1974).

(44) **Corpus fibrolymphaticum.** The pair of fibrolymphatic bodies forming the base and body of the erect Phallus. By apposition they also form the Sulcus phalli. They seem to occur in all avian species with a Phallus protrudens King (1980), the Corpus sinistrum always being much larger than the Corpus dextrum (Gerhardt, 1933, pp. 305–311). The Corpora fibrolymphatica become erect by engorgement with lymph from the Corpora vascularia paracloacalia (Liebe, 1914). See Figs. 3, 4.

(45) **Cavitas lymphatica.** Paired lymphatic cavities in the Basis phalli, surrounding Corpus vasculare paracloacale (Liebe, 1914; Guzsal, 1974) that lead into the erectile

lymphatic spaces of Corpora fibrolymphatica (Liebe, 1914) (Fig. 3). The Cavitates lymphaticae receive lymph from the Corpora vascularia paracloacalia during erection (see Annot. 27).

(46) **Lig. elasticum phalli.** A strong axial, elastic ligament present in all birds with the Phallus protrudens, except *Apteryx* and Tinamidae (Gerhardt, 1933, pp. 305–311), that aids the invagination of the Phallus in detumescence. See **Arthr.** Annot. 3.

(47) **M. retractor phalli [M. levator cloacae].** Synonymy: M. retractor penis (Nishiyama, 1955; Komárek, 1959; Guzsal, 1974). In *Gallus* Nishiyama (1955) described paired cranial and caudal retractor muscles of the penis; in *Anas* Liebe (1914, p. 636) also described them, though he regarded the cranial muscle as indistinct. Komárek (1969) and Guzsal (1974) reported only one pair of retractor muscles, presumably the caudal pair of Liebe. According to Komárek, the retractor muscle is attached to Pars extrema phalli as in Fig. 5, but Guzsal claimed that it is attached to corpus fibrocartilagineum, and this seems consistent with Liebe's report that his caudal muscle is attached to the basal part of the Phallus. See also **Myol.** Annot. 68 M. levator cloacae.

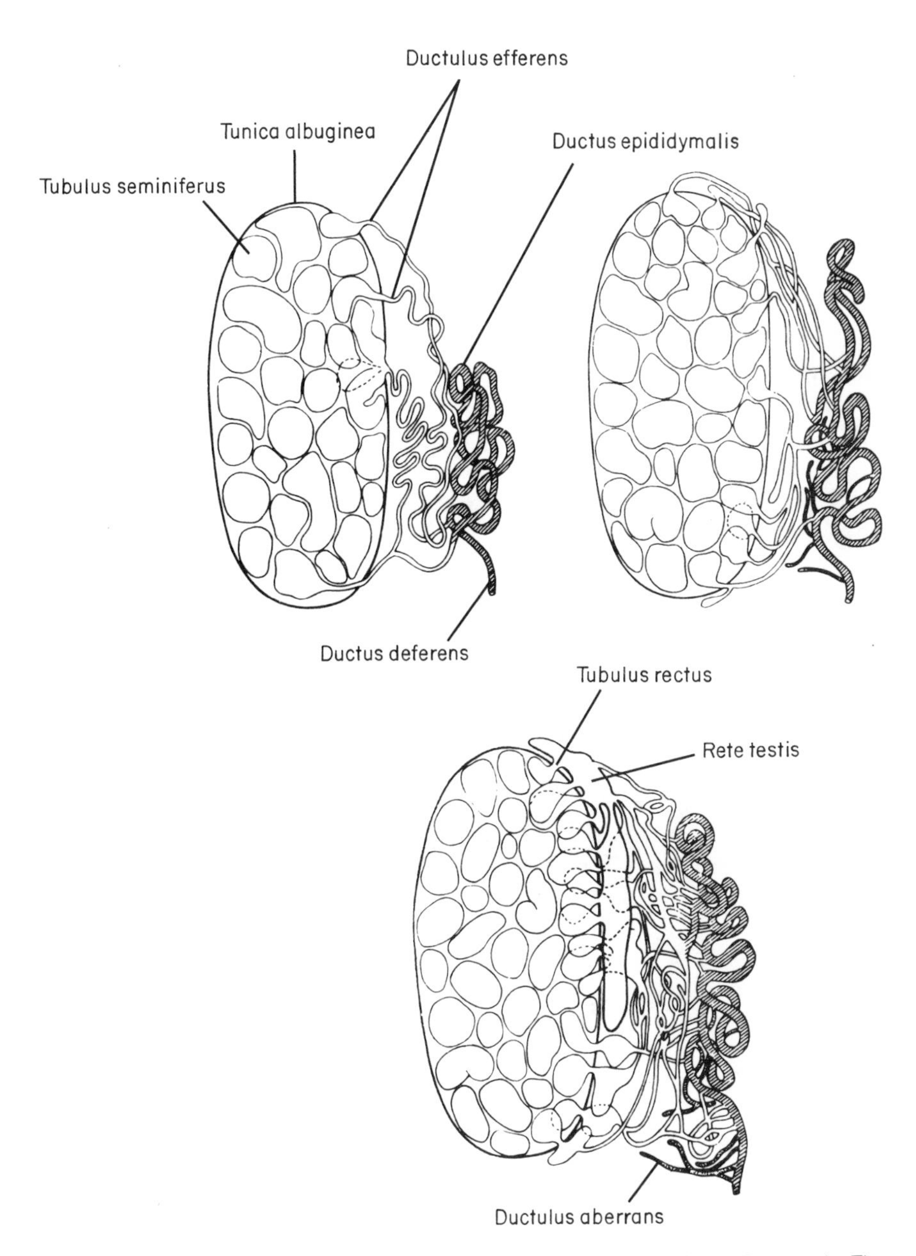

Fig. 1 Three variants of Epididymis in *Sterna*. Only the lower one has a Rete testis. The Ductus epididymalis is cross-hatched in all three diagrams. From Traciuc (1967), with permission of *Anatomischer Anzeiger*.

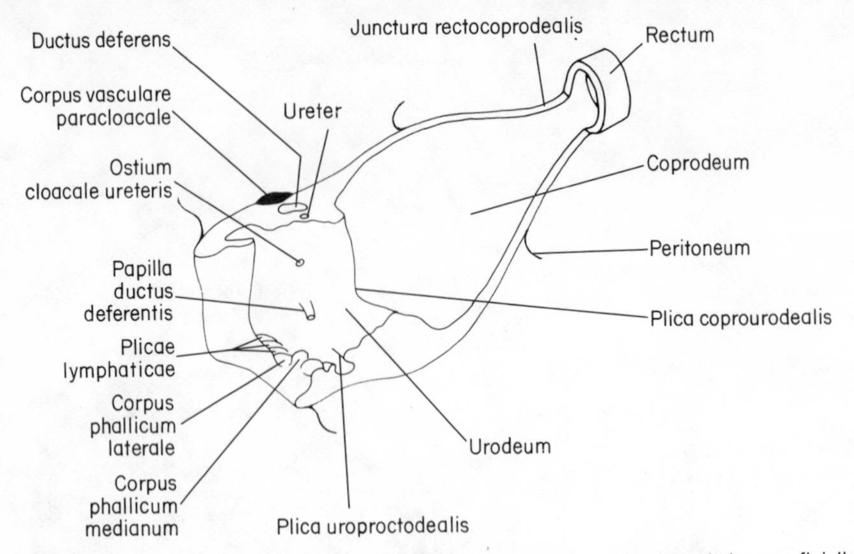

Fig. 2 Phallus nonprotrudens of *Gallus*. The Corpus vasculare paracloacale is superficially embedded in the outer surface of the Urodeum. From King (1975), with permission of W. B. Saunders, Philadelphia.

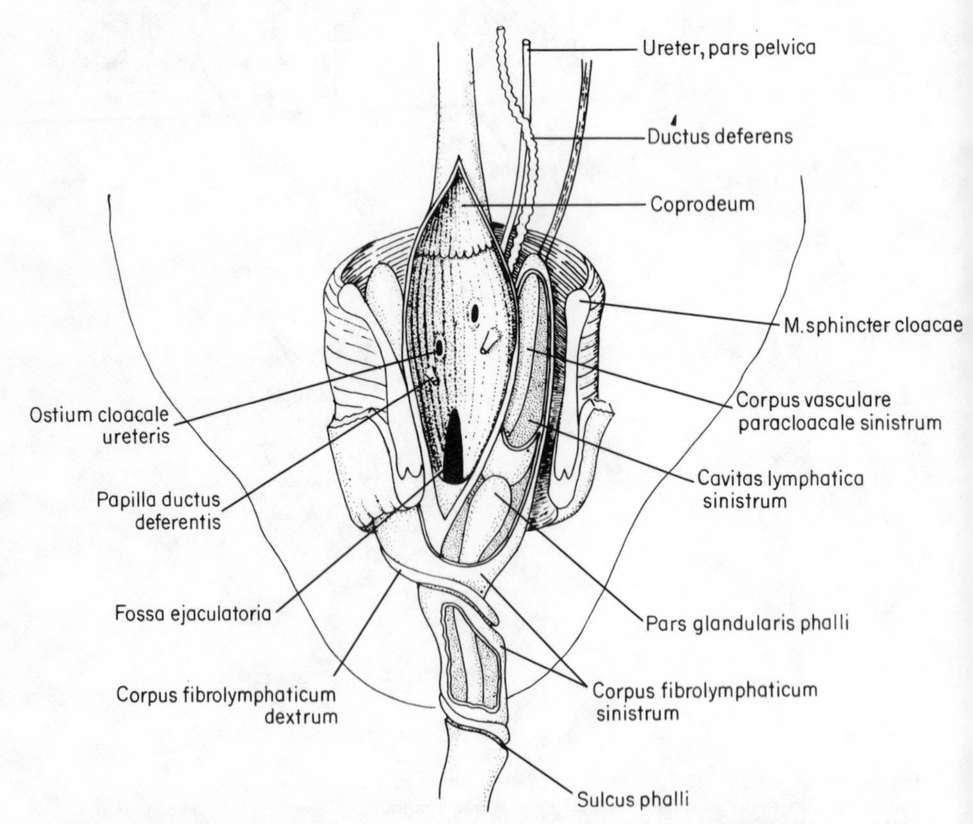

Fig. 3 Caudoventral view of the Cloaca of *Anas*, with erect Phallus. Based on Liebe (1914), from King (1975), with permission of W. B. Saunders, Philadelphia.

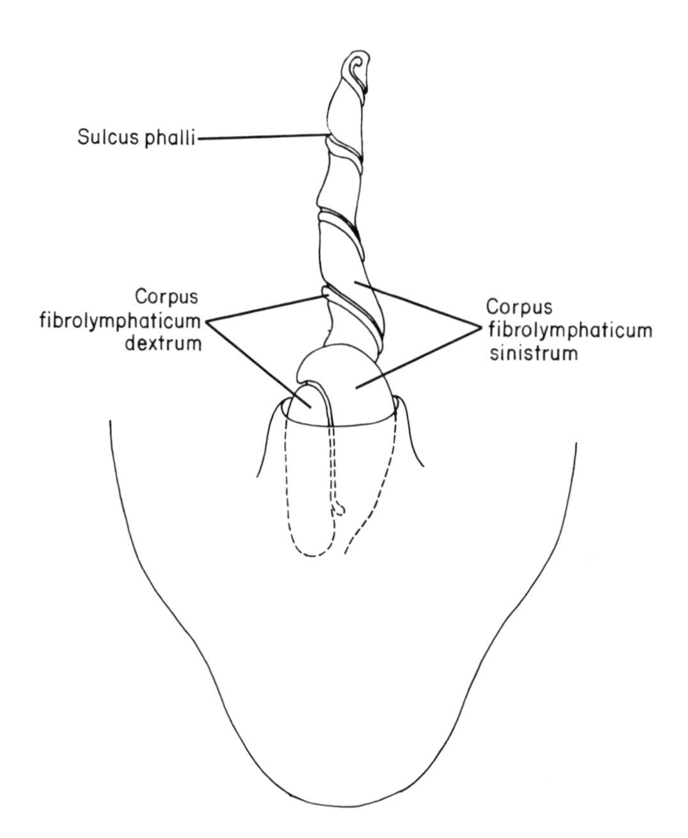

Fig. 4 Phallus protrudens of *Anas,* erect. Based on Liebe (1914), from King (1975), with permission of W. B. Saunders, Philadelphia.

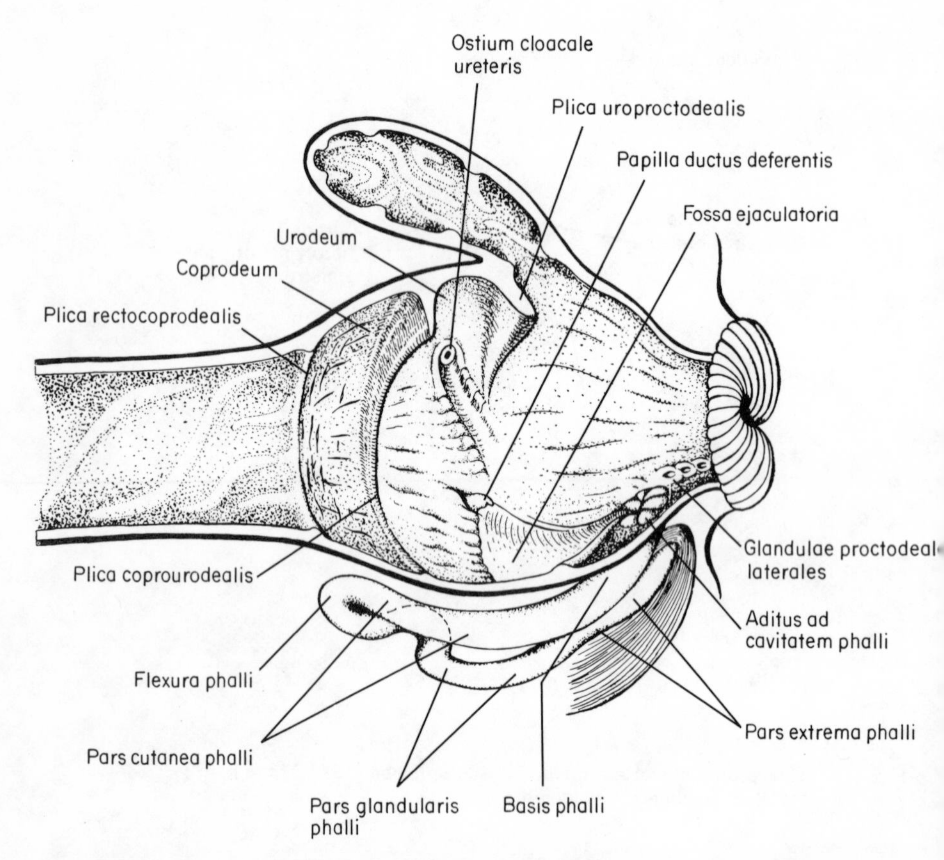

Fig. 5 Paramedian section of the Cloaca of *Anser,* with Phallus at rest. The Plica recto-coprodealis is present in some Anatidae and some ratites, but not in birds generally. From Komárek (1969), with permission of *Anatomischer Anzeiger.*

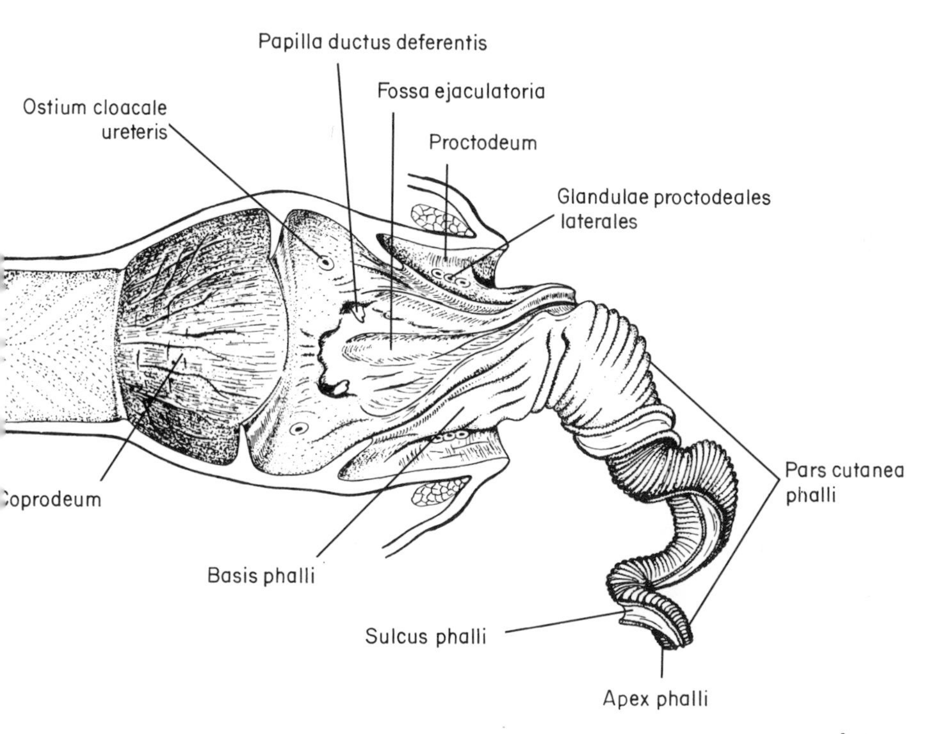

Fig. 6 Dorsal view of the interior of the Cloaca of *Anas,* with erect Phallus. From Komárek (1969), with permission of *Anatomischer Anzeiger*.

ORGANA GENITALIA FEMININA

OVARIUM SINISTRUM

Facies dorsalis[1]
  Hilus ovarii[2]
Facies ventralis[1]
Extremitas cranialis[1]
Extremitas caudalis[1]
Margo lateralis[1]
Margo medialis[1]
Mesovarium[3]
Tunica serosa
  Epithelium superficiale (Fig. 2)
Tunica albuginea[4]
Cortex ovarii[4]
Zonae parenchymatosae[4]
Medulla ovarii[4]
Zonae vasculosae[4]
Interstitiocyti ovarii[5]

FOLLICULUS OVARICUS[6]

Pedunculus folliculi
  Musculi pedunculares[7]
Cytolemma ovocyti[8,9] (Figs. 1, 2)
Zona radiata[8,10] (Fig. 1)
Lamina perivitellina[8,11] (Fig. 1)
Stratum granulosum (Fig. 2)
  Cellulae strati granulosi (Fig. 1)
Lamina basalis folliculi[12] (Figs. 1, 2)
Thecae folliculi
  Theca interna (Fig. 2)
  Theca externa (Fig. 2)
    Cellulae thecales
Tunica superficialis (Fig. 2)
  Musculi intramurales[7]
Epithelium superficiale
Nervi folliculares[13]
  Nn. pedunculares
    Nn. intramuales
Stigma folliculare
Folliculus postovulationem[14]
Folliculus atreticus[15]

Corpus aureum[16]

Epoöphoron[17]
  Tubuli epoöphori[17]
Paraoöphoron

Ovogonium
Ovocytus primarius
Ovocytus secundarius
Polocytus primarius
Polocytus secundarius
Ovum[18]

VASA SANGUINEA OVARICA[19]

Arteria ovarica (Fig. 3) (see **Art.**)
  Rr. ovarici
    Aa. contortae
      Aa. pedunculares
        Aa. intramurales (Fig. 2)
          Rete capillare terminale[20] (Fig. 2)
Venae ovaricae (see **Ven.**)
  Rdxx. ovaricae
    Vv. pedunculares
      Vv. intramurales externae (Fig. 2)
      Vv. intramurales mediae (Fig. 2)
      Vv. intramurales internae (Fig. 2)

GONADUM DEXTRUM[21]

OVIDUCTUS DEXTER[45]

OVIDUCTUS SINISTER

Infundibulum
  Ostium infundibulare
  Fimbriae infundibulares
  Tubus infundibularis[22]
    Fossae glandulares infundibuli[23]
    Glandulae tubi infundibularis[24]
Magnum[25]
  Glandulae magni
Isthmus

OVIDUCTUS SINISTER—*continued*

Pars translucens isthmi[26]
Glandulae isthmi[27]
Uterus[28]
Pars cranialis uteri[28]
Pars major uteri[28]
Recessus uterinus[29]
Glandulae uterinae
Lamellae uterinae
Vagina
M. sphincter vaginae[30]
Fossulae spermaticae [Glandulae vaginales][31]
Ostium cloacale oviductus sinistri

Tunica mucosa
Plicae primariae[32]
Plicae secundariae[32]
Epitheliocyti[33]
Lamina propria mucosae
Tunica muscularis
Tunica serosa

Lig. dorsale oviductus[34]
Lig. ventrale oviductus[34]
Funiculus musculosus[34]

VASA SANGUINEA OVIDUCTALIA

Arteriae oviductales[35] (see **Art.** Fig. 5)
A. oviductalis cranialis[36] (Fig. 3)
A. oviductalis cranialis accessoria[36] (Fig. 3)
A. anastomotica[37] (Fig. 3)
A. oviductalis marginalis ventralis (see Fig. 3; and **Art.** Annot. 72)
A. oviductalis marginalis dorsalis (see Fig. 3; and **Art.** Annot. 72)
Aa. infundibuli
Aa. magni
A. oviductalis media[38] (Fig. 3)
Aa. magni
Aa. isthmi
Aa. uterinae[39] (Fig. 3)
A. uterina medialis[40] (Fig. 3)
A. uterina lateralis[40] (Fig. 3)
A. oviductalis caudalis[41] (Fig. 3)

- Aa. uterinae
- A. vaginalis[42]

Venae oviductales[43] (see **Ven.** Fig. 5)

- V. oviductalis cranialis
  - V. oviductalis marginalis ventralis
  - V. oviductalis marginalis dorsalis
    - Vv. infundibuli
    - Vv. magni
- V. oviductalis media[44]
  - Vv. magni
  - Vv. isthmi
  - Vv. uterinae
    - V. uterina lateralis
    - V. uterina medialis
- V. oviductalis caudalis
  - Vv. uterinae
  - V. vaginalis

OVUM[46]

- Discus germinalis[47] (Fig. 4)
- Vitellus
  - Vitellus albus[48] (Fig. 4)
    - Latebra
      - Centrum latebrae (Fig. 4)
      - Collum latebrae (Fig. 4)
      - Discus latebrae[49] (Fig. 4)
  - Vitellus aureus[50] (Fig. 4)
  - Cytolemma ovocyti[51]
  - Lamina perivitellina[52]
  - Lamina continua[53]
  - Lamina extravitellina[54]
- Albumen
  - Stratum chalaziferum[55] (Fig. 4)
    - Chalaza[56] (Fig. 4)
  - Albumen rarum[57]
    - Stratum internum (Fig. 4)
    - Stratum externum (Fig. 4)
  - Albumen densum[58] (Fig. 4)
    - Albumen polare[59] (Fig. 4)
- Membranae testae[60] (Fig. 4)
  - Membrana testae interna[61] (Fig. 4)

OVUM—*continued*

Membrana testae externa[61] (Fig. 4)
Cella aeria (Fig. 4)
Testa (Fig. 4)
Stratum mamillarium[62]
Stratum spongiosum[62]
Pori testae
Canaliculi testae
Polus acutus[63]
Polus obtusus[63]
Cuticula[64] (Fig. 4)

ANNOTATIONS

(1) **Facies dorsalis; Facies ventralis; Extremitas cranialis; Extremitas caudalis; Margo lateralis; Margo medialis.** The sexually active ovary is so irregular in shape that it can scarcely be said to have extremities, surfaces or borders. However, in seasonal birds it becomes small and compact during the refractory (eclipse) period (Witschi, 1961), and these terms are then meaningful.

(2) **Hilus ovarii.** Synonymy: Ovarian stalk (Gilbert, 1969). The area of entry of the blood vessels and nerves into the Facies dorsalis is very diffuse. Nevertheless a hilus need not be perfectly discrete (e.g. the mammalian spleen).

(3) **Mesovarium.** With the approach of maturity the peritoneal Mesovarium is reinforced by the growth of the Hilus ovarii.

(4) **Tunica albuginea; Cortex ovarii; Zonae parenchymatosae; Medulla ovarii; Zonae vasculosae.** After hatching, the immature ovary has distinct Cortex and Medulla ovarii (Benoit, 1950; Bradley, 1960; Marshall, 1961). Oögonia and Ovocyti are confined to the Cortex. The Medulla consists mainly of connective tissue with blood vessels, nerves, and smooth muscle (Gilbert, 1971a). However, with the onset of sexual activity the distinction between Cortex and Medulla as two distinct strata is virtually lost. On the other hand it is still possible to distinguish: (1) irregular areas containing many immature follicles, i.e. the Zonae parenchymatosae, and (2) other irregular areas containing blood vessels, nerves and smooth muscle, i.e. the Zonae vasculosae (Prochazkova and Komárek, 1970). The primary Tunica albuginea is a thin layer of connective tissue between the germinal epithelium and the primary sex cords in the embryonic female gonad; it disappears during the second half of incubation (Romanoff, 1960, p. 833). The secondary, definitive Tunica albuginea is a connective tissue layer between the Epithelium superficiale and the Cortex; it persists in the mature ovary (Hodges, 1974, p. 330).

(5) **Interstitiocyti ovarii.** Although large vacuolated interstitial cells have often been described in the avian ovary, their origin, fate and function are obscure. "Medullary" interstitial cells were believed to arise from the primary (medullary) sex cords (Benoit, 1950, p. 391); "cortical" interstitial cells were supposed to come

from the secondary (cortical) cords (Benoit, 1950, p. 391). "Thecal" interstitial cells have been found within the Thecae of developing follicles (Marshall, 1961, p. 194) and have sometimes been termed "thecal glands" (Dahl, 1970). Narbaitz and De Robertis (1968) have suggested that all these cells are simply different stages in the life history of the same cell. The sources of these interstitial cells and their possible endocrinological functions have been reviewed by Gilbert (1971c). Since these interstitial cells may well be the same cell in different stages of development and activity, it seems at present inadvisable to name each of them individually.

(6) **Folliculus ovaricus.** Mammalian terms such as primary, growing, and vesicular folliculi do not appear to be in use for birds. However, Komárek and Prochazkova (1970) recognized four orders of Folliculus ovaricus: (I) about 6 large unstained cells surrounding one Ovocytus; (II) the Ovocytus is surrounded by not more than 26 peripheral cells which are now cuboidal; (III) the cells surrounding the Ovocytus are now columnar; (IV) at this stage the Folliculus has reached the surface, and ends ovulating. (The cells surrounding the Ovocytus are the cells of the follicular epithelium, i.e. of the Stratum granulosum). Another classification was suggested by Gilbert (1971a) on a physiological basis. This proposed three orders of Folliculus ovaricus: (1) with no true yolk; (2) with true white yolk; (3) growing, with both white and yellow yolk.

(7) **Musculi pedunculares; Musculi intramurales.** The Pedunculus folliculi contains bundles of smooth muscle, the Musculi pedunculares, which continue into the adjacent Tunica superficialis of the wall of the follicle as the Musculi intramurales, but do not reach the Stigma (Guzsal, 1966).

(8) **Cytolemma ovocyti; Zona radiata; Lamina perivitellina.** The older and very confused terminology for the innermost layers of the wall of the follicle was evolved with the light microscope, but several of its components are only distinguishable with the electron microscope. The terms listed here are based on ultrastructural observations.

It is advisable to follow Wyburn *et al.* (1965) and adopt the classification of egg "membranes" which was proposed by Boyd and Hamilton (1952). According to this classification a primary egg "membrane" is formed by the cytoplasm of the Ovocytus; a secondary egg "membrane" is formed by the cells of Stratum granulosum; a tertiary "membrane" is added by the cells of the Oviductus.

(9) **Cytolemma ovocyti.** According to the classification of Boyd and Hamilton (1952) this structure (Fig. 1) is a primary egg membrane.

(10) **Zona radiata.** This structure (Fig. 1) is caused by folding of the Cytolemma ovocyti (Wyburn *et al.*, 1965; Wyburn and Baillie, 1966). It is a transient structure, first appearing when the follicle has already undergone some enlargement and disappearing in the immediately pre-ovulatory follicle (Wyburn and Baillie, 1966). The Corona radiata of mammals is lacking in birds.

(11) **Lamina perivitellina.** This is a narrow zone between the Ovocytus and Stratum granulosum (Fig. 1), previously called the perivitelline layer. It consists of ground substance and electron dense rods. It is almost certainly secreted by the Cellulae strati granulosi (Wyburn *et al.*, 1965), and is therefore a secondary "membrane". It may well be homologous to the Zona pellucida of the mammal. See also Annot. 52.

(12) **Lamina basalis folliculi.** A distinct boundary (Fig. 1) between the Stratum granulosum and Theca interna (Wyburn *et al.*, 1965; Aitken, 1966). It has some features which raise doubts about whether it is a true Lamina basalis, including its exceptional thickness, its physical properties, and a lack of evidence as to whether it arises from the cells of the Stratum granulosum or from some other source (A. B. Gilbert, pers. comm.). Therefore this term should be kept under review.

(13) **Nn. folliculares.** The Folliculus ovaricus, including its wall and stalk, is profusely innervated (Gilbert, 1965, 1969; Bennett and Malmfors, 1970).

(14) **Folliculus postovulationem.** The postovulatory follicle is evidently a source of progesterone or progesterone-like compounds, probably from the cells of the Stratum granulosum (Gilbert, 1971c).

(15) **Folliculus atreticus.** Marshall (1961, pp. 194–198) identified two varieties of preovulatory atretic follicle, one secreting oestrogen and the other a progestin.

(16) **Corpus aureum.** The "yellow bodies", scattered throughout the ovary (Aitken, 1966), which may be late stages of regressing postovulatory follicles, or remnants of atretic follicles.

(17) **Epoöphoron; Tubuli epoöphori.** The Tubuli epoöphori, which are homologous to the Ductuli aberrantes of the Appendix epididymalis, are closely associated with the capsule of the adrenal gland, like their male homologues. The Tubuli epoöphori form secretory noduli at the onset of sexual maturity, producing a steroid sex hormone (Budras, 1972).

(18) **Ovum.** In this sense, the Ovum is the female gamete, resulting from the second maturation division. Unfortunately, exactly the same term is traditionally used for the shelled egg. See Annot. 46.

(19) **Vasa sanguinea ovarica.** The ovary and its follicles have a well organized system of blood vessels (Nalbandov and James, 1949; Oribe, 1968). See Fig. 2.

(20) **Rete capillare terminale.** The rich capillary network (Fig. 2) at the boundary between the Stratum granulosum and Theca interna.

(21) **Gonadum dextrum.** In the genetic female of birds generally, the term Ovarium dextrum is not appropriate for the right gonad, which is usually arrested at a testis-like stage of development (Romanoff, 1960, p. 835). Only in *Apteryx* is a well developed and fully functional Ovarium dextrum regularly present (Kinsky, 1971).

(22) **Tubus infundibularis.** The narrow tubular part of the infundibulum, also called the chalaziferous region (Richardson, 1935).

(23) **Fossae glandulares infundibuli.** Glandular grooves, occurring throughout the Infundibulum only (Surface, 1912; Aitken and Johnston, 1963). Some spermtozoa are stored here (van Drimmelen, 1946; Lorenz, 1964). See Annot. 31.

(24) **Glandulae tubi infundibularis.** In the Infundibulum of *Gallus*, tubular glands occur only in the narrow tubular part.

(25) **Magnum.** The longest and most coiled part of the oviduct. The region of the Magnum which immediately precedes the glandless Pars translucens isthmi possesses very numerous and very tall mucus secreting cells (Richardson, 1935, Aitken, 1971); this region has been called the mucous part of the Magnum.

(26) **Pars translucens isthmi.** The distinct but narrow (1–3 mm wide) translucent glandless zone between the Magnum and Isthmus.

(27) **Glandulae isthmi.** Tubular glands are present in all parts of the Isthmus except the Pars translucens isthmi (Aitken, 1971).

(28) **Pars cranialis uteri; Pars major uteri.** The Pars cranialis uteri is the short relatively tubular cranial portion, through which the egg passes quickly (Johnston *et al.*, 1963). The suggestion that it should be regarded as part of the Isthmus (Davidson *et al.*, 1968; Draper *et al.*, 1972) was rejected by Aitken (1971) because its glands are characteristically uterine.

The Pars major uteri is the pouch-like caudal portion of the Uterus which holds the egg during most of the period of shell formation (Johnston *et al.*, 1963).

The term "uterus" has been in use in avian anatomy for nearly 150 years (see Barkow, 1829, p. 444). The uterus is also known as the "shell gland".

(29) **Recessus uterinus.** The tapered funnel-shaped caudal region with a distinctive pink or greyish white colour and abundant cholesterin ester lipids in the ciliated cells (Fujii, 1963).

(30) **M. sphincter vaginae.** A thickening of the Tunica muscularis of the first segment of the Vagina (Bobr *et al.*, 1964), functionally analogous to the mammalian Cervix uteri.

(31) **Fossulae spermaticae [Glandulae vaginales].** Previously known as the utero-vaginal glands (Bobr *et al.*, 1964); definitely vaginal in position, being confined entirely to the region of M. sphincter vaginae (Fujii, 1963). These tubular Fossulae house the niduli spermatici (sperm nests), acting as the main storage site for spermatozoa (Bobr *et al.*, 1964). See Annot. 23.

(32) **Plicae primariae; Plicae secundariae.** Longitudinal Plicae primariae are more or less continuous throughout most of the oviduct. They carry many Plicae secundariae in the Tubus infundibuli, Isthmus, Pars cranialis uteri, and Vagina (Surface, 1912; Blom, 1973).

(33) **Epitheliocyti.** The whole oviduct is lined by alternating ciliated columnar cells and unicellular glands (Aitken and Johnston, 1963; Guzsal, 1968; Draper *et al.*, 1968). The latter are commonly called goblet cells. Since there are known to be about 40 different proteins in the egg albumen it is possible that these secretory Epitheliocyti may eventually prove to be of several, even many, different types, a possibility which is illustrated by histochemical studies (Fujii *et al.*, 1965). At present, however, only three proteins have been linked to specific cells, so it is not yet possible to classify the individual secretory Epitheliocyti which produce all of these substances. The ciliated cells vary in the different parts of the oviduct (Fujii, 1975).

(34) **Lig. dorsale oviductus; Lig. ventrale oviductus; Funiculus musculosus.** The dorsal ligament suspends the oviduct and is continued ventrally as the fan-like ventral ligament, the free edge of which is reinforced by smooth muscle; caudally the latter increases to a muscular cord, the Funiculus musculosus, attaching to the Uterus and Vagina (Curtis, 1910; Kar, 1947).

(35) **Aa. oviductales.** The blood vessels of the Oviductus sinister (Fig. 3) have been studied in detail in *Gallus*, *Meleagris*, and *Anas*, by Freedman and Sturkie (1963) and Hodges (1965), and in the goose by Gertner (1969). See also King (1975, pp. 1950 and 1958) and Baumel (1975, p. 1973) for a summary of these vessels in the common domestic species.

(36) **A. oviductalis cranialis; A. oviductalis cranialis accessoria.** The A. oviductalis cranialis arises from the left A. renalis cranialis; the A. oviductalis cranialis accessoria arises from the left A. iliaca externa. These two oviductal arteries vary between species and within the same species (see also **Art.**). In *Gallus* there is typically an A. oviductalis cranialis, but sometimes the A. oviductalis cranialis accessoria is also present (Hodges, 1965). *Meleagris* and *Anas* usually possess only the A. oviductalis cranialis accessoria (Hodges, 1965). In *Anser* both the A. oviductalis cranialis and the A. oviductalis cranialis accessoria are generally present; Gertner (1969) named these two vessels respectively the Ramus infundibularis of the A. renoovarica, and the A. oviductalis cranialis. In *Columba* (Baumel, pers. comm.) there is sometimes neither an A. oviductalis cranialis nor an A. oviductalis cranialis accessoria; the A. oviductalis media then supplies the whole of the cranial part of the oviduct, forming Aa. infundibuli as well as Aa. magni and Aa. uterinae.

(37) **A. anastomotica.** Joins the A. oviductalis cranialis to the A. oviductalis media in *Gallus*, but absent in other species such as *Meleagris* and *Anas*.

(38) **A. oviductalis media.** Synonymy: Middle oviductal artery (Westpfahl, 1961); hypogastric artery (Freedman and Sturkie, 1963; Hodges, 1965). This artery arises from the left A. ischiadica or from the left A. renalis media. See **Art.** Annot. 68.

(39) **A. uterinae.** The blood vessels ln the lateral (left) side of the uterus are much better developed than those on the medial (right) side.

All the uterine arteries are exceedingly variable, even within the same species; because of this variability it seems inadvisable, in the present state of knowledge of birds generally, to attempt to establish names for an elaborate system of A. uterinae. However, owing to the physiological importance of the uterus in the domestic birds, a summary now follows of some of the terms that have been suggested. The A. uterina cranialis (left and right) of King (1975, p. 1950) is the anterior uterine artery (lateral and medial) of Freedman and Sturkie (1963) and the anterior uterine of Hodges (1965). The A. uterina ventralis of King (1975) is the inferior uterine artery of Freedman and Sturkie (1963) and Hodges (1965); this artery prolongs the A. oviductalis marginalis ventralis. The A. uterina dorsalis of King (1975) is the superior uterine artery of Freedman and Sturkie (1963) and Hodges (1965). The right and left caudal uterine arteries of King (1975, p. 1950) are the middle uterine arteries of Freedman and Sturkie (1963) and the medial and lateral posterior uterine arteries of Hodges (1965).

(40) **A. uterina medialis; A. uterina lateralis.** The origins, courses, and branches of these two arteries are particularly variable, even within the same species. The A.

uterina medialis and lateralis were respectively named the right lateral uterine artery and left lateral uterine artery by King (1975, p. 950), and the medial lateral uterine artery and lateral lateral uterine artery by Hodges (1965).

(41) **A. oviductalis caudalis.** Synonymy: The intestinal branch of the internal pudendal (Westpfahl, 1961); the pelvic branch of the internal iliac (Freedman and Sturkie, 1963; Hodges, 1965). This artery supplies variable Aa. uterinae to both sides of the caudal part of the uterus, and may also supply rami to the vagina (Baumel, 1975, p. 1993).

(42) **A. vaginalis.** The vaginal artery may arise as a separate branch of the left A. pudenda, or may be incorporated into the vaginal rami of the A. oviductalis caudalis (Baumel, 1975, p. 1993). When present as a separate artery it supplies both sides of the vagina. Such arteries, supplying both sides of the vagina, are the vaginal arteries of Hodges (1965). They also appear to be the same arteries as the posterior uterine arteries of Freedman and Sturkie (1963).

(43) **Vv. oviductales.** The veins of the Oviductus sinister are, in general, satellites of the arteries. See **Ven.** Annot. 61.

(44) **V. oviductalis media.** This is the largest venous pathway from the Uterus. It drains the same region as that supplied by the A. oviductalis media, but the two vessels do not normally run in parallel and the vein is much more elaborate than the artery. It generally drains into the left caudal renal vein (Baumel, 1975, p. 2008). See also **Ven.** Annot. 61 and Fig. 5.

(45) **Oviductus dexter.** The presence of an Oviductus dexter as well as an Oviductus sinister has been reported in various orders and particularly in Falconiformes. But the Oviductus dexter is usually only vestigial, even in *Apteryx* where the Gonadum dextrum is a functional ovary (Kinsky, 1971). In *Gallus* small rudiments of the Oviductus dexter can nearly always be found at the Cloaca, and larger cysts are not uncommon (Webster, 1948; Winter, 1958; Williamson, 1965). Simple tubes also occur (Winter, 1958). All these most commonly appear to resemble the Magnum histologically (Winter, 1958). Occasionally a full-sized Oviductus dexter occurs in *Gallus*; none of the early evidence indicates that it is fully functional (Sell, 1959), but Bickford (1965) did report fully functional right and left oviducts in a chicken. The histological structure of the Oviductus dexter is not constant enough to warrant a list of components.

(46) **Ovum.** Unfortunately the term Ovum has long been applied to both the shelled egg and the cell which results from the second reduction division. See Annot. 18.

(47) **Discus germinalis.** In the laid egg the Discuss germinalis (Fig. 4) is a greyish white area about 3 mm in diameter, just beneath the Cytolemma ovocyti; it is referred to as the blastoderm if fertilized, and blastodisc if unfertilized (Gilbert, 1971b).

(48) **Vitellus albus.** The white yolk, containing about twice as much protein as fat (Boyd and Hamilton, 1952).

(49) **Discus latebrae.** Synonymy: The nucleus of Pander (Fig. 4).

(50) **Vitellus aureus.** This is the yellow yolk, containing about twice as much fat as protein (Boyd and Hamilton, 1952). The Vitellus aureus sometimes shows about six concentric wide dark strata alternating with the same number of narrow pale strata. This stratification is an artefact depending on the diet, the pale strata being deficient in carotinoid pigment. When the diet is well-balanced these strata disappear. True stratification of the Vitellus aureus can be visible in histological preparations, and these true strata do reflect the structure of the yolk and not simply its colour (Gilbert, 1971b); little is known about their structural basis.

(51) **Cytolemma ovocyti.** The remains of the cytolemma of the oöcyte (Fig. 1), which becomes discontinuous just before ovulation (Bellairs, 1965).

(52) **Lamina perivitellina.** Shortly before ovulation the electron dense rods of this secondary egg "membrane" become rapidly transposed into a meshwork of solid cylindrical fibres, the total thickness of the Lamina then being about 2.7 μm (Bellairs *et al.*, 1963). See Annot. 11 and Fig. 1.

(53) **Lamina continua.** This layer appears for the first time in oviductal eggs, and is therefore a tertiary egg "membrane" (see Annot. 8). It is a narrow granular layer about 50 to 100 μm thick (Bellairs *et al.*, 1963).

(54) **Lamina extravitellina.** The "outer layer of the vitelline membrane" of Bellairs *et al.* (1963). It consists of many layers of fine fibrils, with a total thickness of between 3 and 8 μm. This also is a tertiary egg "membrane" (see Annot. 8).

(55) **Stratum chalaziferum.** The thin innermost layer of dense albumen.

(56) **Chalaza.** The chalaza going to the sharp end of the egg is a double strand, but the chalaza at the blunt end is a single strand. See Fig. 4.

(57) **Albumen rarum.** Commonly called the "liquid albumen", but renamed the "thin albumen" by Gilbert (1971b). It consists of a structureless fluid, apparently free of fibres, but mucin is present.

(58) **Albumen densum.** Commonly called the "dense albumen", but renamed the "thick white" by Gilbert (1971b). It differs from the Albumen rarum only in being rich in ovomucin which may account for its gel-like quality, and in apparently having some structural organization possibly fibrous.

(59) **Albumen polare.** At each end of the egg the Chalazae are embedded in the Albumen polare, an area of Albumen densum. The Albumen polare has been known as the Lig. albuminis densi, but true ligaments are not distinct structurally in this part of the egg.

(60) **Membranae testae.** The nomenclature for the shell membranes and the shell itself has been reviewed by Simons (1971).

(61) **Membrana testae interna; Membrana testae externa.** These terms were first used by Purkinje in 1830 (Simons, 1971).

(62) **Stratum mamillarium; Stratum spongiosum.** It is generally agreed that the organic material can be classified into an inner Stratum mamillarium and an outer

Stratum spongiosum (Romanoff and Romanoff, 1949; Simons, 1971). The inorganic (crystalline) material, however, shows its own architecture, for which Simons (1971) adopted the following terms, from the inside towards the outside of the shell: layer of cones and basal caps, palisade layer, and surface crystal layer.

(63) **Polus acutus; Polus obtusus.** These terms refer to the "sharp" and "blunt" ends respectively of the egg of pyriform shape. In some species, however, the egg is spherical (e.g. certain Strigiformes), and in others (e.g. Pteroclididae and Podicipedidae) the two ends are similar in shape (Pitman, 1964, pp. 238–240).

(64) **Cuticula.** The Cuticula, which is organic, is sometimes called the true cuticle to distinguish it from a calcified layer deposited on the outer surface of the cuticle in various species of bird including specimens of the broiler type of *Gallus* (Simons, 1971); the calcified layer has been called the "cover".

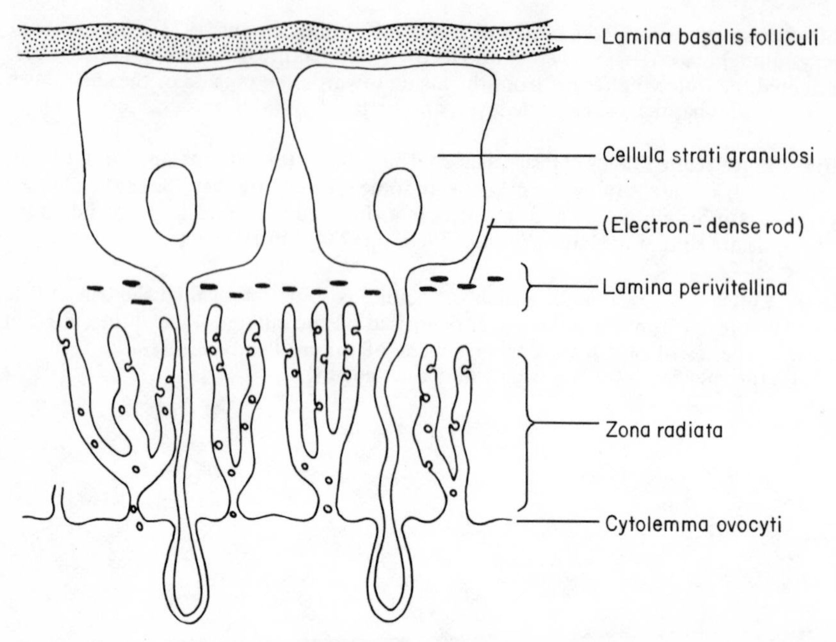

Fig. 1 Diagram of the innermost layer of the wall of the follicle of *Gallus,* in the region of the square inset in Fig. 2. From King (1975), with permission of W. B. Saunders, Philadelphia.

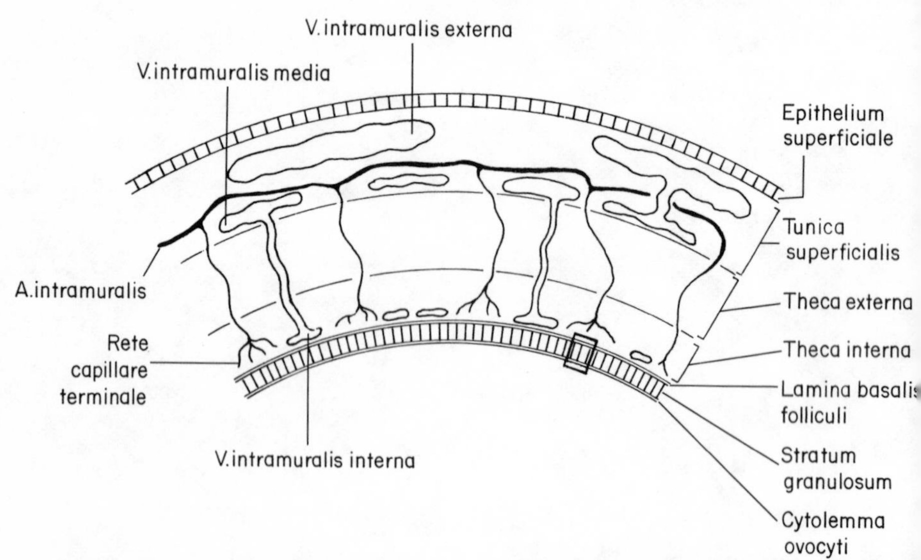

Fig. 2 Diagram of the wall of a follicle of *Gallus,* showing its layers and its blood supply, based on Nalbandov and James (1949). From King (1975), with permission of W. B. Saunders,

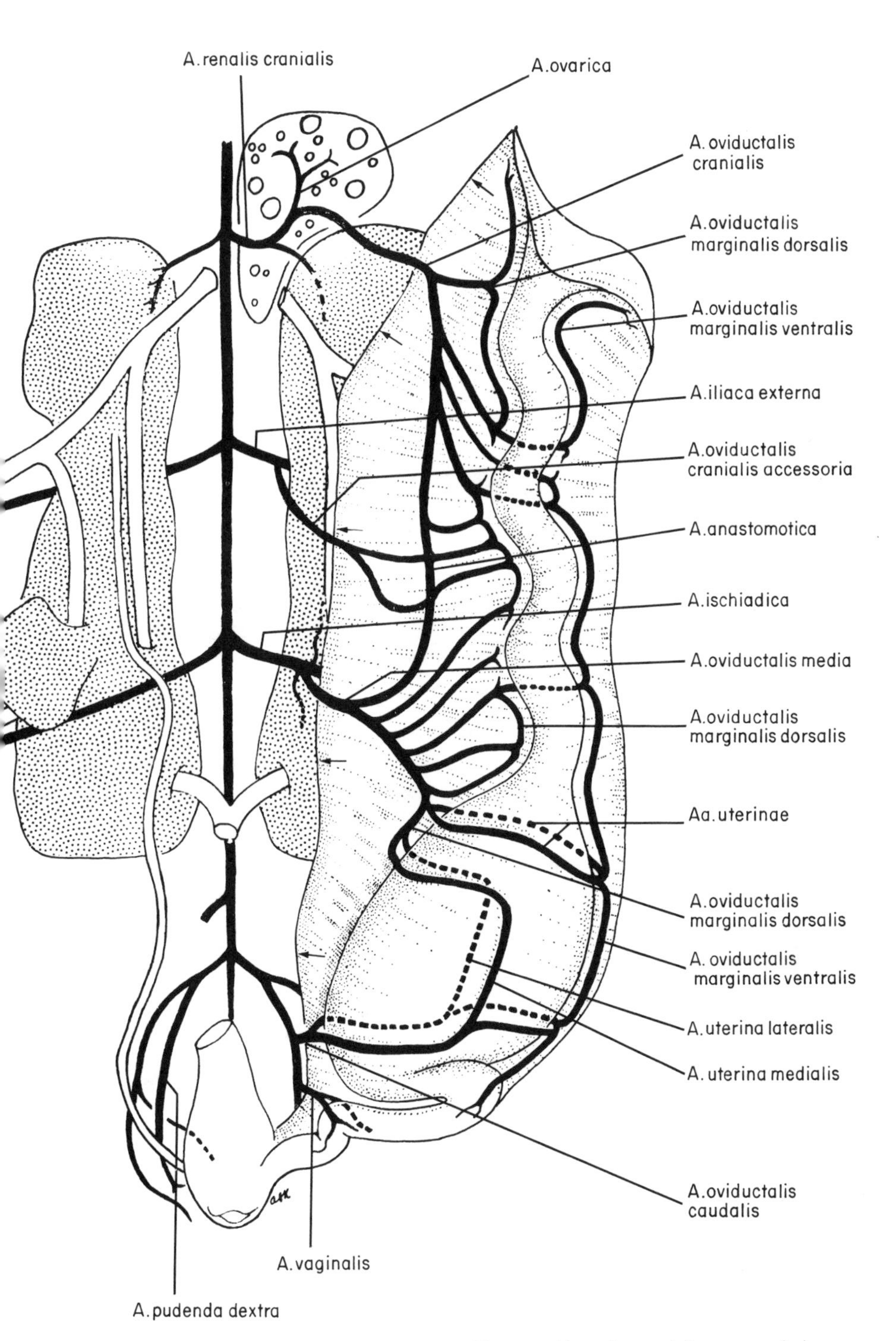

Fig. 3 Diagram of the Aa. oviductales of *Gallus*. The a. oviductalis cranialis accessoria is only occasionally present in this species (see Annot. 36). Based on King (1975), with permission of W. B. Saunders, Philadelphia.

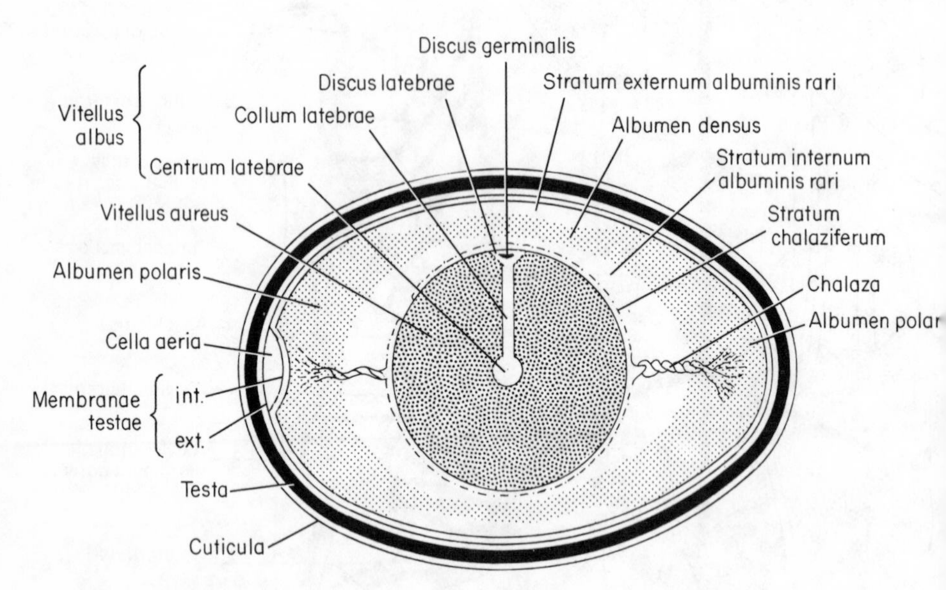

Fig. 4 Diagram of an egg. Based on King (1975), with permission of W. B. Saunders, Philadelphia.

CLOACA

COPRODEUM[1]

Junctura rectocoprodealis[2] (see **Genit. masc.** Fig. 2)
Plica rectocoprodealis[3] (see **Genit. masc.** Fig. 5)

URODEUM[1]

Plica copro-urodealis[4] (see **Genit. masc.** Figs. 2, 5)
Ostium cloacale ureteris[5] (see **Genit. masc.** Figs. 2, 3, 5, 6)
Papilla ductus deferentis (see **Genit. masc.** Figs. 2, 3, 5, 6)
  Ostium cloacale ductus deferentis (see **Genit. masc.** Fig. 2)
Corpus vasculare paracloacale[6] (see **Genit. masc.** Figs. 2, 3)
  Glomera corporis vascularis paracloacalis[6]
Ostium cloacale oviductus sinistri
  Papilla oviductus sinistri[7]
Ostium cloacale oviductus dextri[8]
Fossa oviductalis[9]
Plica uroproctodealis[10] (see **Genit. masc.** Figs. 3, 6)
Corpus para-ampullare[11]

PROCTODEUM[1]

Bursa cloacalis[12]
  Collum bursae cloacalis[13]
  Ostium bursae cloacalis[14]
Bursa cloacalis accessoria[15]
Glandula proctodealis dorsalis[16]
Glandulae proctodeales laterales[17] (see **Genit. masc.** Figs. 5, 6)
Plicae lymphaticae[18] (see **Genit. masc.** Fig. 2)
Plicae proctodeales[19]
Sinus proctodeales[19]
Phallus nonprotrudens (see **Genit. masc.**)
Phallus protrudens (see **Genit. masc.**)
Phallus femininus[20]

VENTUS[21]

Labia venti (see **Myol.** Fig. 4)
  Labium venti dorsale[22]

VENTUS—*continued*

Labium venti ventrale[22]

Pars rugosa[23]
Pars plana[23]
Glandulae externae labii venti[25]
Glandulae internae labii venti[25]

Orificium venti

MUSCULI CLOACALES

(see **Myol.** Mm. caudae and Fig. 4)

M. transversus cloacae (see **Myol.** Annot. 68)
M. sphincter cloacae (see **Myol.** Annot. 68; and **Genit. masc.** Fig. 3)
M. retractor phalli [M. levator cloacae] (see **Genit. masc.** Annot. 47)

VASA ET NERVI CLOACALES

Aa. cloacales[26] (see **Art.** A. mesenterica caudalis, A. pudenda)
Vv. cloacales[26] (see **Ven.** V. pudenda)
Nn. cloacales[27] (see **PNS** Pars visceralis, N. pudendus)
Gg. cloacalia[27]

ANNOTATIONS

(1) **Coprodeum; Urodeum; Proctodeum.** As originally defined by Gadow (1887) the Cloaca of reptiles and birds is divided by folds of mucous membrane into these three compartments. Subsequently it has proved difficult to demonstrate these folds and compartments in both reptiles and birds (Gerhardt, 1933, pp. 277–280), but Gadow himself had pointed out that especially in birds, "...the determination of the various chambers is beset with difficulties because of the extreme variability of the separating folds." Nevertheless Gadow's terminology has been universally adopted as a valuable foundation for further studies.

Gadow proposed that the interior of the Coprodeum is in principle separated from that of the Rectum (in both birds and reptiles) by a distinct annular fold, the Plica rectocoprodealis (see Annot. 3), containing a circular sphincter muscle. Although this fold has often been illustrated and described in *Gallus*, it is in fact not present in this species (Jolly, 1915; Lillie, 1952, p. 388; Komárek, 1970; King, 1975, p. 1960). It is present, however, in some ratites including *Struthio* (Gadow, 1887) and in some anseriforms including *Anas* (Komárek, 1969); therefore in these species the Coprodeum is a distinct compartment (see **Genit. masc.** Fig. 5). In *Gallus* the Coprodeum can only be identified grossly as a bell-shaped enlargement at the appropriate site. "Ampulla recti" has been suggested for this enlargement (Komárek, 1970), but the term Coprodeum is so deeply entrenched in the literature that it seems better to retain it for birds generally.

(2) **Junctura rectocoprodealis.** The beginning of the bell-shaped enlargement of the Coprodeum in *Gallus*, and birds generally (see **Genit. masc.** Fig. 2).

(3) **Plica rectoproctodealis.** The fold (see Annot. 1) between Rectum and Coprodeum, known to be present in only a few species including certain ratites (*Struthio*) and anseriforms (*Anas*) (see **Genit. masc.** Fig. 5). Reliable information about the presence or absence of this fold in other species is scarce. Few authors commit themselves one way or the other. An exception is Romanoff (1960, p. 497) who noted that "rarely the intestine and the Corprodeum are separated by a low projecting fold". In the works of others, including Salt (1954) who dissected and illustrated the passerine cloaca, the fold is notable for its absence.

(4) **Plica copro-urodealis.** The prominent annular ridge between Coprodeum and Urodeum, probably present in birds generally (see **Genit. masc.** Figs. 2, 6)

(5) **Ostium cloacale ureteris.** The popular belief that the Ureter opens on a papilla in the domestic birds evidently stems from Pilz (1937), but is certainly incorrect for *Gallus* and appears to lack evidence in other species also (King, 1975, p. 1962). See **Genit. masc.** Figs. 2, 5, 6).

(6) **Corpus vasculare paracloacale; Glomera corporis vascularis paracloacalis.** The paracloacal vascular body on each side is an oval structure embedded in the lateral wall of the Urodeum, and consists of tufts of arterial capillaries, the Glomera corporis vascularis paracloacalis, implanted in lymphatic channels (Knight, 1970; Sugimura *et al.*, 1975). See **Genit. masc.** Annot. 27 and Figs. 2, 3).

(7) **Papilla oviductus sinistri.** Synonymy: Papilla vaginalis (Preuss and Rautenfeld, 1974); Papilla feminina (Komárek, 1971). This structure, which is claimed to be the homologue of the Papilla ductus deferentis and reaches 0.5 mm in length in the 14 to 17 weeks female chicken, is said to carry the opening of the left oviduct; it converts into a slit when laying begins (Preuss and Rautenfeld, 1974), a similar papilla on the right side persisting after laying has started. Komárek (1971) reported the presence in the female of a left and right papilla, homologues of the papillae of the left and right ductus deferens; this author, however, did not claim that in the female the left papilla carried the opening of the left oviduct, but noted instead that the left papilla is close to the opening of the left oviduct.

(8) **Ostium cloacale oviductus dextri.** The Oviductus dexter occasionally has a cloacal opening in *Gallus* (Morgan and Adams, 1959).

(9) **Fossa oviductalis.** The rudimentary opening of Oviductus dexter on the interior wall of the Urodeum; usually only a pit-like depression in the adult of *Gallus*, *Meleagris*, *Anas* and *Anser* (Komárek, 1971).

(10) **Plica uroproctodealis.** A semilunar ridge, incomplete ventrally in *Gallus*, between Urodeum and Proctodeum, and present in all of the 90 species examined by Forbes (1877). See **Genit. masc.** Figs. 2, 5.

(11) **Corpusculum para-ampullare.** A globular, fluid-containing, structure about 2 mm diameter which lies about 2 mm from the base of the Papilla ductus deferentis in about 2 per cent of male chickens, usually on one side only, and is considered to be the remnant of the Müllerian duct (Marvan, 1969).

(12) **Bursa cloacalis.** Synonymy: Bursa Fabricii. Together with its dependent lymphoid tissues, the Bursa cloacalis is responsible for the synthesis of circulating antibodies. The chromolipoid body of the Black Vulture (*Coragyps atratus*) may be derived from the cloacal bursa (Correa *et al.*, 1969).

(13) **Collum bursae cloacalis.** The stalk of the Bursa cloacalis.

(14) **Ostium bursae cloacalis.** The opening of the Bursa cloacalis into the dorsal wall of the Proctodeum.

(15) **Bursa cloacalis accessoria.** Paired accessory bursae occur on each side of the caudal part of the Collum bursae cloacalis in *Gallus* (Pintea and Rizkalla, 1967).

(16) **Glandula proctodealis dorsalis.** In *Gallus* (Komárek, 1970) this median gland forms a prominent mound-like projection from the dorsal midline of the Proctodeum in the male and female, consisting of tubular glands and lymphoid tissue. Jolly (1915) named it the "lymphoglandular ridge". An encapsulated aggregation of small tubular glands occurs in the male and female *Coturnix coturnix* in the same dorsal position (Coil and Wetherbee, 1959), and has been variously named the foam gland, cloacal gland, paracloacal gland, and proctodeal gland (see Klemm *et al.*, 1973). These dorsal glands in *Coturnix* are apparently not lymphoid, but are nevertheless likely to be homologous to the gland in *Gallus*. In the sexually active male *Coturnix* the secretion, which is mucoid in nature, is released from the vent as a white froth, the gland as a whole being sufficiently enlarged to cause an externally visible bulge in the dorsal region of the vent (Coil and Wetherbee, 1959; Tamura and Fujii, 1967; McFarland *et al.*, 1968).

(17) **Glandulae proctodeales laterales.** These are comparatively isolated glands, about three on each side in Anatidae and more numerous in *Gallus* (Komárek, 1970). See **Genit. masc.** Figs. 5, 6.

(18) **Plicae lymphaticae.** The paired lymphatic folds of Nishiyama (1950, 1955) and the Plicae lymfaceae of Komárek (1970), found in *Gallus* and *Meleagris* (Komárek, 1970). See **Genit. masc.** Fig. 2.

(19) **Plicae proctodeales; Sinus proctodeales.** The Plicae proctodeales are regularly arranged folds of the mucosa in *Gallus* and *Meleagris*. The Sinus proctodeales are depressions between the Plicae (Komárek, 1970).

(20) **Phallus femininus.** A prominent structure about 4 cm long in the proctodeal floor of the female *Struthio* (Gerhardt, 1933, p. 307), and recognisable in other species in which the male has a Phallus protrudens (Marshall, 1961, p. 170). The homologous structure is also visible in females of species with Phallus nonprotrudens such as *Gallus*, especially in immature birds; indeed in the female of *Gallus* homologues of all the main phallic structures of the male, (i.e. Corpus vasculare paracloacale, Corpus phallicum medianum and laterale, and Plicae lymphaticae) have been claimed (Preuss and Rautenfeld, 1974).

(21) **Ventus.** The vent is the external opening of the cloaca (Gadow, 1896, p. 90; Berger, 1961, p. 300; Marshall, 1962, p. 585; King, 1975, p. 1963). The term anus has often been used for this opening, even in birds. Strictly, however, the term anus

should be confined to the external opening of the alimentary tract where this is separate from the urogenital openings (Berger, 1961, p. 300; Romer, 1962, p. 4), as in "Amphioxus", Teleostei, and the higher mammals.

(22) **Labium venti dorsale; Labium venti ventrale.** In birds generally (Welty, 1962, p. 93) the external opening at rest is a horizontal slit with a dorsal and ventral lip. See **Myol.** Fig. 4.

(23) **Pars rugosa.** Synonymy: Pars externa (Komárek, 1970). This is the externally visible region of each Labium, and is a cutaneous zone devoid of feathers but marked by numerous radial furrows (King, 1975, p. 1963).

(24) **Pars plana.** Synonymy: Pars interna (Komárek, 1970). The Pars plana of each Labium is the smooth shiny mucous region of each labium, about 2 to 3 mm wide in *Gallus*, which is concealed when the vent is closed (King, 1975, p. 1963).

(25) **Glandulae externae labii venti; Glandulae internae labii venti.** The mucus secreting glands of the vent have been described by Quay (1967) from observations on 72 species. The external glands open outside the canal of the vent onto areas of featherless skin; the internal glands open on the stratified squamous epithelium of the canal of the vent. The glands were found in most but not all of Quay's species; exceptions occur in Anatidae and Phasianidae.

(26) **Aa. et Vv. cloacales.** The arteries and veins of the cloaca have been investigated by Pintea *et al.* (1967) and Knight (1970), and summarized by Baumel (1975, pp. 1997, 2008). See **Art.** Annot. 80; and **Ven.** Annot. 63.

(27) **Nn. cloacales; Gg. cloacalia.** The innervation of the cloacal region of *Gallus* has been fully reviewed by Baumel (1975, pp. 2052, 2059). The N. pudendus follows the Ureter and Ductus deferens to the dorsolateral surface of the Cloaca (Watanabe, 1972), and forms there, at the caudal end of the Ureter, the G. cloacale (the bursocloacal ganglion of Pintea *et al.*, 1967, and probably the uterovaginal ganglion of Freedman and Sturkie, 1963). The cloacal ganglion communicates with the rectal ganglia (colic ganglia of Watanabe, 1972) in the mesorectum; this communication may carry "sympathetic" fibers from the pelvic plexus to the cloacal wall (Baumel, 1975, p. 2059). The connections between the cloacal ganglion (ganglia) and the N. intestinalis (Watanabe, 1972) may carry "parasympathetic" fibers to N. intestinalis for supply of the caudal end of the Rectum (Baumel, 1975, p. 2059). See **PNS**, Pars visceralis.

GLANDULAE ENDOCRINAE

A. B. GILBERT

Reproductive Physiology Section
Agricultural Research Council, Poultry Research Centre,
King's Buildings, West Mains Road, Edinburgh EH9 3JS, Scotland

With contributions from Sub-Committee Members: R. B. Chiasson, *University of Arizona*; B. K. Follett, *University of North Wales*; B. M. Freeman, *Houghton Poultry Research Station*; K. J. Hill, *Unilever Research Laboratory*; R. D. Hodges, *University of London*; A. Kjaerheim, *University of Oslo*; J. McLelland, *University of Edinburgh*; Y. K. Paik, *Chon-puk University*; A. Tixier-Vidal, *Laboratoire de Biologie Moléculaire*; K. G. Wingstrand, *University of Copenhagen*

The endocrine glands of birds have not been extensively studied throughout the class as a whole. Moreover, they do not comprise one homogeneous group, their origin, functions and structure being variable (Assenmacher, 1973). For example, not all birds may have a secretory pineal gland; the adrenal glands are, in reality, two glands in one. Sometimes anatomical divisions and nomenclature blur the known physiological divisions, as in the hypophysis and hypothalamus.

I wish to pay tribute to the members of the Sub-Committee for undertaking so well the difficult task of producing a rational system which it is hoped will be applicable to birds in general. It is not easy to single out individuals from a group, for all members have contributed much; however, without the untiring help and advice of Drs R. B. Chiasson, R. D. Hodges, and K. G. Wingstrand, my task would have been far more difficult. Notwithstanding this, the responsibility for any errors or defects rest solely on my shoulders.

GLANDULA THYROIDEA[1]

Stroma thyroidea
    Capsula thyroidea
Parenchyma thyroidea
    Folliculus thyroideus
    Colloidum thyroideum

GLANDULA PARATHYROIDEA[2]

Stroma parathyroidea
    Capsula parathyroidea
Parenchyma parathyroidea

Glandulae parathyroideae accessoriae[3]

GLANDULA ULTIMOBRANCHIALIS[4]

Stroma ultimobranchialis
Parenchyma ultimobranchialis[5]
    Vesiculae ultimobranchiales[5]

GLANDULA PINEALIS[6]

Corpus gl. pinealis
Pedunculus gl. pinealis
Parenchyma pinealis

INSULAE PANCREATICAE[7]

Parenchyma insularum
Cellulae insularum[7]

GLANDULA ADRENALIS[8]

Capsula adrenalis
Partes corticales gl. adrenalis[9]
Partes medullares gl. adrenalis[9]

Glandulae adrenales accessoriae[10]

HYPOPHYSIS [GLANDULA PITUITARIA][11]

Adenohypophysis[12]
    Pars tuberalis
    Pars distalis[13]
        Zona rostralis partis distalis
        Zona caudalis partis distalis

Neurohypophysis[14]
    Sulcus tuberoinfundibularis[15]

Eminentia mediana (see **CNS** Hypothalamus)
    Zona rostralis eminentiae medianae[16]
    Zona caudalis eminentiae medianae[16]
Infundibulum
Lobus nervosus
Recessus neurophypophysialis[17]
Tractus hypothalamohypophysialis[19]

VASA SANGUINEAE HYPOPHYSIS

A. carotis cerebralis (see **Art.** Annot. 17)
    Anastomosis intercarotica (see **Art.** Annot. 17)
        A. hypophysialis caudalis[18]
    A. infundibularis[19]
        A. eminentiae rostralis
        A. eminentiae caudalis
        A. neurohypophysialis
        A. retrochiasmatica
        Aa. hypothalamicae ventrales
    R. rostralis a. carotis cerebralis[19]
        A. cerebralis rostralis
        Aa. preopticae[20]
Systema portale hypophysiale[21]
    Vv. portales rostrales[22]
        Vasa sinusoidea portalia[23]
    Vv. portales caudales[22]
        Vasa sinusoidea portalia[23]
Sinus cavernosus[24] (see **Ven.** Annot. 18)
    Vv. adenohypophysiales[25]

ANNOTATIONS

(1) **Glandula thyroidea.** The general structure of the thyroid glands in a number of wild and domestic birds—thrush, sparrow, jay, crow, pigeon, goose, duck, partridge, turkey and chicken—has been described by Scheschin (1925). The histology of the thyroid of the chicken has been described by Fritschi (1926) and Hodges (1974). The capsule is thin, and the parenchyma consists of follicles lined by secretory cells of different heights varying according to their degree of activity (Hodges, 1974, pp. 442–446).

(2) **Glandula parathyroidea.** The anatomy and histology of the parathyroid glands in a number of birds have been described by Forsyth (1908), Benoit (1950) and Hodges (1974). The terms [III] and [IV] refer to the origin of these glands from the embryonic visceral pouches 3 and 4; the embryological evidence for the pouches

which give rise to each gland appears not to be extensive in birds generally, but it seems to be commonly assumed that the cranial and caudal glands are derived from pouches 3 and 4 respectively. In the chicken the two glands are usually attached to each other within a thick capsule (see Abdel-Magied and King, 1978, for review). The stroma is minimal and the parenchyma consists of a single cell type (Hodges, 1974, pp. 445–556). In many birds, but not the chicken (Abdel-Magied and King, 1978), part or all of the carotid body tissue may be found embedded within one of the parathyroid glands on each side (de Kock, 1958, 1959).

(3) **Glandulae parathyroideae accessoriae.** These structures are not universally present in birds or, if so, have not been identified. However, positive identification within the ultimobranchial gland has been made in several species, such as chicken (Hodges, 1974, p. 444), quail (Hodges, pers. comm.), crow, owl and pigeon (Watzka, 1933). Their occasional presence outside the ultimobranchial gland, in thymus tissue and the thyroid, has been described in the fowl by Nonidez and Goodale (1927).

(4) **Glandula ultimobranchialis.** This structure has been called both the ultimobranchial body and the ultimobranchial gland. Since it has a secretory function, the term gland is correct. The gland is not enclosed by a capsule, and is therefore not clearly delineated from the adjacent tissues (Hodges, 1974, p. 432).

(5) **Parenchyma ultimobranchialis; Vesiculae ultimobranchialis.** The histological complexity of the ultimobranchial gland in birds (Watzka, 1933; Hodges, 1970) and the uncertainty that all the tissues described occur in every species have prompted the inclusion of names of only the two most important ultimobranchial tissues; the ultimobranchial parenchyma consisting of C cells (calcitonin-producing) and the ultimobranchial vesicles.

In columbiform birds ultimobranchial C cells may be dispersed throughout the neighbouring tissues and organs, particularly the thyroid gland (Stoeckel and Porte, 1970).

(6) **Glandula pinealis.** Synonymy: Corpus pineale; Epiphysis cerebri. The pineal is included here since there is growing evidence that this structure has secretory activity in birds (Pelham *et al.*, 1972; Menaker and Oksche, 1974; Ellis, 1976). However, not all avian species may have a secretory pineal (see also Quay, 1970). The cell types in the avian Gl. pinealis are controversial (Hodges, 1974, p. 477).

(7) **Insulae pancreaticae; Cellulae insularum.** See Systema Digestorium for the general nomenclature of the pancreas. The Insulae pancreaticae comprise the only "ductless" part of the gland. The pancreatic islets can be divided into so-called "light" and "dark" islets according to their staining reactions (Nagelschmidt, 1939; Hellman and Hellerström, 1960). The dark islets (A or α islets) are composed essentially of A and D cells, while the light islets (B or β islets) contain primarily B and D cells (Hazelwood, 1973; Smith, 1974). The A_1 cells of Hellman and Hellerström (1960) apparently correspond to D cells whilst their A_2 cells correspond to the A cells of other authors. A further cell type, the type IV cell, has been described in the quail A islets by Smith (1974). A islets occur only in the ventral and splenic lobes of the pancreas while B islets are found in all three lobes (Mikami and Ono, 1962; Smith, 1974). See also **Digest.** Annot. 74.

(8) **Glandula adrenalis.** Discrete bilateral adrenal glands appear in most birds; however in a few species, e.g. a booby (*Sula variegata*) and a gull (*Larus argentatus*) the glands are fused in the mid-line.

(9) **Partes corticales; Partes medullares.** Although, as in mammals, the cells of the adrenal gland arise from two different sources (neural crest and mesoderm), these cells are not organized into distinct cortex and medulla. Instead the cortical parts of mesodermal origin are intermingled with the medullary parts of ectodermal origin. The Partes corticales consist of cylindrical cords of granular vacuolated eosinophilic cells; the Partes medullares consist of irregular clumps of polygonal basophilic cells (Hodges, 1974, pp. 469–470). The term "interrenal tissue" has been suggested for Partes corticales, and "chromaffin tissue' for the Partes medullares (see Hodges, 1974, pp. 467 for review).

(10) **Glandulae adrenales accessoriae.** In the Jackdaw (*Corvus monedula*) 1–3 accessory adrenal glands are embedded in the epididymis (Traciuc, 1969). The accessory glands should not be regarded as paraganglia; only tissue masses shown to contain mostly chromaffin material should be termed paraganglia.

(11) **Hypophysis** [Glandula pituitaria]. The nomenclature for the components of this organ are clouded by doubtful anatomical distinctions, a wealth of alternative names for the same structure, and the fact that physiological function often overlaps more than one anatomically distinct area.

(12) **Adenohypophysis.** Synonymy: Pars glandularis; anterior lobe. The latinized "Lobus anterior" may be confused with the commonly, but incorrectly, used "Pars anterior" which excludes the Pars tuberalis (Tixier-Vidal and Follett, 1973).

(13) **Pars distalis.** Synonymy: Pars anterior (see Annot. 12).

(14) **Neurohypophysis.** Synonymy: Lobus posterior. "Posterior lobe" should not be used, for it excludes the median eminence and the infundibulum.

(15) **Sulcus tuberoinfundibularis.** This furrow, particularly in species with a strongly thickened median eminence, delineates the median eminence from the surrounding hypothalamus (Wingstrand, 1966). The term was defined for mammals by Spatz *et al.* (1948), though Sulcus hypothalamo-hypophyseus was used by Kuhlenbeck and Haymaker (1949).

(16) **Zona rostralis eminentiae medianae; Zona caudalis eminentiae medianae.** The two zones of the median eminence are mainly defined by differences in innervation, affinities to neurosecretory stains, and histochemistry (Benoit and Assenmacher, 1953; Assenmacher and Benoit, 1958; Oksche, 1963; Oksche *et al.*, 1963, 1964, 1974). In *Zonotrichia leucophrys* the zones are visible macroscopically as two distinct protuberances (Oksche and Farner, 1974). The two zones appear to be drained by separate sets of portal vessels that supply the Zona rostralis and Zona caudalis of the Pars distalis of the Adenohypophysis. This is clearly shown in *Zonotrichia* (Vitums *et al.*, 1974).

(17) **Recessus neurohypophysialis.** Extension of the third ventricle into the neurohypophysis.

(18) **A. hypophysialis caudalis.** In some birds, such as the pigeon, caudal hypophysial arteries from the intercarotid anastomosis supply the neurohypophysis; in others the neurohypophysis is supplied by a branch of the infundibular artery, the A. neurohypophysialis.

(19) **A. infundibularis.** A. infundibularis is usually a branch of A. carotis cerebralis near its bifurcation into the Rami rostralis and caudalis. Vitums, *et al.* (1964) state, "Each infundibular artery (right and left) usually originates as a single artery from the anterior ramus just rostral to the origin of the posterior ramus. In a few cases the infundibular artery originates directly from the carotid artery at the point just caudal to the origin of the posterior ramus." Wingstrand (1951, p. 274) states that the blood supply to the primary plexus of the median eminence comes exclusively from the infundibular arteries. These vessels are fairly small and variable, but always start from the A. carotis cerebralis somewhere between the Diaphragma sellae and its point of division into rostral and caudal rami. Since Wingstrand discusses 25–30 different species, his description should be accepted as the usual condition.
Aa. hypothalamicae ventrales. The hypothalamohypophysial tract is supplied by these arteries (Vitums *et al.*, 1964).

(20) **Aa. preopticae.** These arteries supply the supraoptic and paraventricular (neurosecretory) hypopthalamic nuclei (Vitums *et al.*, 1964).

(21) **Systema portale hypophysiale.** The system starts from capillary beds in the median eminence and ends in the adenohypophysis. In *Zonotrichia*, *Columba*, and other birds Vv. portales form the sole afferent blood supply to the adenohypophysis (Vitums *et al.*, 1964; Wingstrand, 1951). In the chicken (Hasegawa, 1956) and duck (Assenmacher, 1953) the Pars distalis also receives a limited part of its supply from caudal and rostral hypophysial arteries (Vitums *et al.*, 1964).

(22) **Vv. portales rostrales; Vv. portales caudales.** The rostral and caudal groups of veins are derived from capillary plexuses of the rostral and caudal zones of the median eminence, respectively (Vitums *et al.*, 1964; Singh and Dominic, 1970; Sharp and Follett, 1969) (see Annot. 16).

(23) **Vasa sinusoidea portalis.** According to Mikami *et al.* (1970), these sinusoids in the portal zone of Pars distalis have a structure resembling postcapillary venules.

(24) **Sinus cavernosus.** The cavernous sinus of birds is a complicated structure subdivided into several separate components. It has not been determined that the separate components have interconnections and consequently the drainage from the neurohypophysis may be entirely separate from adenohypophysial drainage. See Annot. 14; and **Ven.** Annot. 18.

(25) **Vv. adenohypophysiales.** In *Zonotrichia* three main groups of veins drain the postal sinusoids of the adenohypophysis into different parts of the cavernous sinus (Vitums *et al.*, 1964). A similar drainage occurs in the chicken (Green 1951).

SYSTEMA CARDIOVASCULARE

J. J. BAUMEL

Department of Anatomy,
School of Medicine, Creighton University,
Omaha, Nebraska 68178 USA

With contributions from Sub-Committee Members: A. R. Akester, *University of Cambridge*; J. L. Bhaduri, *University of Calcutta*; B. Biswas, *Indian Museum, Calcutta*; P.-C. Blin, *École Nationale Veterinaire d'Alfort*; J. Kaman, *University of Veterinary Medicine, Brno*; F. E. F. Lindsay, *University of Glasgow*; L. Malinovský, J. E. Purkynev, *University Brno*; T. Nishida, *University of Tokyo*; V. Simić, *l'Universitè of Belgrade*

The efforts of Drs Jiri Kaman and Takao Nishida in formulating the cardiovascular terminology deserve special acknowledgment.

Following the lead of the *Nomina Histologica* (IANC, 1975) the more inclusive heading, Systema Cardiovasculare, has been adopted rather than Angiologia, the term used in *Nomina Anatomica* (IANC, 1966) and *Nomina Anatomica Veterinaria* (ICVAN, 1973) which literally connotes vessels only.

Intrinsic vessels of viscera. Some of the terminology of the blood vessels is also to be found in sections of *Nomina Anatomica Avium* other than this section on Systema cardiovasculare. In general the names of vessels of macroscopic dimensions that supply or drain viscera are presented in this chapter. Vessels that make up the intrinsic angioarchitectural features of several of the major viscera (e.g. lungs, kidneys, liver) are named in the appropriate chapters. Consult Apparatus respiratorius, Organa urinaria, and Apparatus digestorius for Vasa sanguinea intrapulmonaria, Vasa sanguinea intrarenalis, Vasa sanguinea intrahepatica, etc.

Heart. The treatment of the terminology of the avian heart differs from that of the mammalian nomenclatures in that the names of the various recognized parts of the atrial and ventricular musculature are presented (see Myocardium).

A departure from the conventional nomenclature of the veins of the heart has taken place. Instead of the genitive form "cordis", the adjectival form "cardiaca" has been adopted, thus: V. cardiaca sinistra instead of V. cordis sinistra. By using the term "cardiac" this puts the vessel terminology consistent with that of the heart nerves: N. cardiacus, Ganglia cardiaca, Plexus cardiacus.

Nomenclature of branches of A. celiaca and associated venous radices. The terminology for the celiac artery recommended by Malinovský (1965) has been adopted. In this scheme "proventricular" refers to the glandular stomach and "gastric" refers to the muscular stomach (ventriculus; gizzard). Malinovský defines various surfaces, margins, and parts of the muscular stomach on which the names of vessels are based in the definitive, adult condition; he also presents recommendations on the reconciliation of vessel terminology of branches of A. celiaca in various avian species of several orders.

Nishida, Paik, and Yasuda (1969) base their terms on "muscular" and "glandular" stomach (example: A. gastrica glandularis sinistra). These authors refer to the primary subdivisions of A. celiaca as A. gastrica dextra and A. gastrica sinistra. Malinovský points out the difficulties in using this nomenclature in that the stomach is only one of several major organs supplied by A. celiaca.

Splenic-lienal. Nomina Anatomica Avium follows *Nomina Histologica* (IANC, 1975) in adopting Splen for the spleen rather than Lien which has been retained as an official alternative name. This requires that the vessels be named accordingly; therefore, Aa. et Vv. splenicae.

Ascending-descending. Usage of the terms "ascendens" and "descendens" with respect to names of blood vessels requires definition. As used in this work these terms refer to the direction of blood flow in arteries. For example. blood flow in A. vertebralis ascendens is directed cranially; that in A. vertebralis descendens would flow caudally.

The companion vein(s) of an artery designated by ascendens takes the same name as its artery even though the blood flow is caudally (example: V. cutanea cervicalis ascendens parallels A. cutanea cervicalis ascendens).

TERMINI GENERALES

Arteria
Arteriola
Anastomosis arterioarteriosa
Anastomosis venovenosa
Anastomosis arteriovenosa
Anastomosis lymphovenosa
Arcus arteriosus
Arcus venosus
Capillaris
Cisterna
Circulus arteriosus
Circulus venosus
Lympha
Plexus arteriosus
Plexus vasculosus
Plexus venosus
Plexus lymphaticus
Rete arteriosum
Rete mirabile
Rete venosum
Sanguis
Sinus venosus
Tunica externa
Tunica media
Tunica intima
Valvula venosa
Valvula lymphatica
Vas anastomoticum
Vas afferens [Vas advehens]
Vas deferens [Vas revehens]
Vas capillare
Vas collaterale
Vas lymphaticum
Vasa nervorum
Vasa vasorum
Vena
Vena commitans
Vena emissaria
Venula

COR

(Figs. 1, 2)

Basis cordis [Facies pulmonalis][1]
Facies ventrocranialis [Facies sternalis][1]
Facies dorsocaudalis [Facies hepatica][1]
Apex cordis
    Fovea apicis
Sulcus interventricularis paraconalis [ventralis][2]
Sulcus interventricularis subsinuosus [dorsalis][2]
Sulcus coronarius [Sulcus atrioventricularis]
Septum interatriale[3] (see Annot. 15)
    Perforationes interatriales[3]
    Pars cavopulmonalis[3]
Septum interventriculare
Chordae tendinae
    M. papillares[4]
Trigona fibrosa[5]
Annuli fibrosi[5]
(Cartilago cardiaca)

Epicardium (see **Pericar.**)

Myocardium

Mm atriales[6]
    Arcus longitudinalis dorsalis
    Arcus transversus dexter
    Arcus transversus sinister (Fig. 2)
    M. pectinati[6]
    M. basiannularis atrii[6] (Fig. 2)

Mm. ventriculares[7]
    Lamina superficialis[7]
    M. longitudinalis ventriculi dextri
    M. sinuspiralis
    M. bulbospiralis
        Pars superficialis
        Pars profunda
    M. valvae atrioventricularis dextrae (Fig. 2)
    Trabeculae carneae
    Mm. papillares[4]

Systema conducens cardiacum[8]
Nodus sinuatrialis[9, 10]
Rr. nodi sinuatrialis[10]
Rr. subendocardiales atriales
Rr. periarteriales atriales
Nodus atrioventricularis
Fasciculus atrioventricularis[11]
Truncus[11]
R. recurrens
Crus sinistrum
Crus dextrum
Rr. fasciculi atrioventricularis[10]
Rr. subendocardiales ventriculares
Rr. periarteriales ventriculares
Annulus atrioventricularis dexter[12]

Endocardium

SINUS VENOSUS [SINUS VENARUM CAVARUM][13]
(Fig. 2)

Valva sinuatrialis
Valvula sinuatrialis dextra
Valvula sinuatrialis sinistra
M. pectinati valvae[14]
Ostium v. cavae caudalis[13]
Ostium v. cavae cranialis dextrae[13] (Fig. 2)

ATRIUM DEXTRUM

Auricula dextra
Recessus sinister atrii dextri[15] (Fig. 2)
Ostium v. cavae cranialis sinistrae (Fig. 1)
Ostium v. proventricularis cranialis[16]
Ostia venarum cardiacarum
Foramina venarum minimarum[25]
Septum sinus venosi[17] (Fig. 2)
Cavitas atrii dextri

VENTRICULUS DEXTER

(Fig. 2)

Ostium atrioventriculare dextrum
Valva atrioventricularis dextra[18]
Conus arteriosus
Ostium trunci pulmonalis
Valva trunci pulmonalis[19]
 Valvula semilunaris sinistra
 Valvula semilunaris dextra
 Valvula semilunaris dorsalis
Foramina venarum minimarum[25]
Cavitas ventriculi dextri

ATRIUM SINISTRUM

(Fig. 2)

Auricula sinistra
Ostium v. pulmonalis dextrae[20] (see **Ven.** Vv. pulmonales)
Ostium v. pulmonalis sinistrae[30]
Camera pulmonalis[21] (Fig. 2)
Valva v. pulmonalis[22]
Foramina venarum minimarum[25]
Cavitas atrii sinistri

VENTRICULUS SINISTER

(Fig. 2)

Ostium atrioventriculare sinistrum
Valva atrioventricularis sinistra[23]
 Cuspis sinistra
 Cuspis dextra[23]
 Cuspis dorsalis
Vestibulum aortae
Ostium aortae
Valva aortae[24]
 Valvula semilunaris sinistra
 Valvula semilunaris dextra ventralis
 Valvula semilunaris dextra dorsalis
Cavitas ventriculi sinistri

ANNOTATIONS

(1) **Basis cordis.** The base of the heart is defined as its dorsal or pulmonary surface (Facies pulmonalis) that consists mostly of left and right atria. Pericardium intervening, the dorsal surface of the heart is related to the trachea, bronchi, and proventriculus (near the median plane) and the ventral surface of the Septum horizontale (laterally).
Facies sternalis and **Facies hepatica** of the heart are terms of Baum (1930).

(2) **Sulci interventriculares.** Referred to in the literature as right and left longitudinal sulci. In the avian heart the indistinct sulci are obliquely disposed, and do not parallel the axis of the heart. The terms adopted here follow the *Nomina Anatomica Veterinaria* (ICVAN, 1973).

"Paraconalis" and "subsinuosis" refer to Conus arteriosus and Sinus venosus. See Fig. 1.

(3) **Septum interatriale.** Only the cranioventral part of the adult interatrial septum represents the fetal septum; this part separates the left atrium from the Recessus sinister atrii dextri, and was pierced by multiple Perforationes interatriales during fetal and neonatal life. The caudodorsal part (Pars cavopulmonalis) of the adult interatrial septum is formed by contributions of the embryonic pulmonary veins and the left cranial vena cava which become incorporated into the definitive left atrium and interatrial septum (Quiring, 1933–34). Functionally the perforations correspond to the mammalian fetal foramen ovale; there is no adult avian equivalent of the mammalian fossa ovalis (see Annot. 15).

(4) **Mm. papillares.** Found only in the left ventricle of the avian heart; in the hearts of smaller species they are poorly differentiated from the general myocardium, often distinguished only as the point of attachment of Chordae tendineae.

(5) **Annuli fibrosi; Trigona fibrosi.** These structure are the elements of the "skeleton" of the heart that are well developed about the left atrioventricular ostium and root of the aorta. Except in large birds the fibrous rings of the pulmonary trunk and the right atrioventricular ostium are weakly developed. The right fibrous trigone is the thickest, most rigid part of the heart skeleton that is located directly dorsal to the root of the aorta. The left trigone is between the left side of the aortic ring and the ventromedial part of the left atrioventricular annulus.

(6) **Mm. atriales.** Avian atrial musculature has distinctive parts; most of the terminology is that of Quiring (1933–34). The muscular arches break up into definite, internally prominent fascicles, the Mm. pectinati, that merge into **M. basiannularis,** the circular layer of muscle around the bases of the atria that bounds the coronary sulcus (see Baumel, 1975, p. 1970). See Fig. 2.

(7) **Mm. ventriculares.** Names of parts of the ventricular myocardium mostly after Shaner (1923); Lamina superficialis is a new term.

(8) **Systema conducens cardiacum.** The cardiac impulse generating and conduction system formed from modified cardiac muscle cells (Myofibra nodalis and Myofibra conducens purkinjiensis). Terminology follows that of the *Nomina Histologica* (IANC, 1975). See Baumel (1975).

(9) **Nodus sinuatrialis.** Contemporary consensus is that a discrete sinuatrial node is present in the avian heart.

(10) **Rr. nodi sinuatrialis; Rr. fasciculi atrioventricularis.** The ramifications of nodal myofibers and Purkinje conducting myofibers. In addition to the subendocardial ramifications that make contact with typical cardiac muscle cells in the atria and ventricles, parts of the conducting tissue penetrate the myocardium, and are distributed as cords within the periarterial connective tissue; these cords at times form circular "muffs" of conducting cells (Rr. periarteriales) that surround the intramyocardial rami of Aa. coronariae (Davies, 1930; Chiodi and Bortolami, 1967).

(11) **Fasciculus atrioventricularis.** A definite atrioventricular trunk is lacking in some of the avian species that have been studied; instead the trunk is represented by multiple crura which depart from the region of Nodus atrioventricularis (Chiodi and Bortolami, 1967).

(12) **Annulus atrioventricularis dexter.** A part of the conducting system that is mainly concerned with the contraction of the muscular Valva atrioventricularis dextra (Davies, 1930). See Annot. 18.

(13) **Sinus venosus.** [Sinus venarum cavarum]. The Sinus venosus in different taxa of birds is variously incorporated into the right atrium (Gasch, 1888). A distinct Sinus venosus is present in the heart of *Gallus* (Quiring, 1933–34), *Corvus*, *Struthio*, and others (Romanoff, 1960); its right boundary is usually set off from the right atrium by a prominent groove externally. Internally it is demarcated from the right atrium by the sinuatrial valve. The Sinus venosus is partially subdivided in some forms: one part contains the common opening for the right cranial vena cava and the caudal vena cava. The ostium of the left cranial vena cava usually opens into the chamber of the right atrium separated from the apical part of the Sinus venosus by the Septum sinus venosi.

(14) **Mm. pectinati valvae.** Certain of the pectinate muscles are continuous with the bases of the valvules of the sinuatrial valve (Kolda and Komárek, 1958). See Fig. 2.

(15) **Recessus sinister atrii dextri.** Typical of the avian heart is the tubular "left recess of the right atrium" that extends to the left past the median plane of the heart and dorsal to the aortic bulb (Rigdon and Frölich, 1970; Kern, 1926). This extension is separated from the left atrium by a part of the interatrial septum that represents the remnant of the fetal septum (see Annot. 3).

(16) **Ostium v. proventricularis cranialis.** V. proventricularis cranialis drains the glandular stomach; the vein may empty into the left cranial vena cava near its termination (Malinovský, 1965) or into the right atrium proper. See **Ven.** Annot. 51.)

(17) **Septum sinus venosi.** A muscular band that lies dorsal to the orifice of the left cranial vena cava. It spans the left end of the right atrioventricular ostium, and separates this ostium from the part of the right atrium into which the sinuatrial valve opens (see Fig. 2).

(18) **Valva atrioventricularis dextra.** This valve is formed of both atrial and ventricular musculature. A distinct band of M. sinuspiralis makes up the thicker external

lamina of the valve. A fibrous layer separates the external lamina from the thinner internal lamina that is derived from an invagination of right atrial musculature (Shaner, 1923). At its thicker cranial border this muscular valve contains the right fibrous annulus of the cardiac skeleton.

(19) **Valva trunci pulmonalis.** The valvules of the pulmonary valve fit into shallow sinuses of the base of the pulmonary trunk during ventricular systole (see **Art.** Annot. 1).

(20) **Ostium v. pulmonalis dextrae.** In some avian forms the right and left pulmonary veins empty into the left atrium via separate ostia (*Columba*, *Gallus*, *Anas*); in others the two pulmonary veins become confluent outside the heart and produce a common pulmonary vein (*Melopsittacus*, Szabó, 1958).

(21 **Camera pulmonalis.** On entering the left atrium the pulmonary veins coalesce into a single vessel that invaginates the left atrium, protruding into the left atrioventricular ostium (Quiring, 1933–34). The invaginated vein forms the Camera pulmonalis. The left side of the Camera has a free margin that guides blood directly into the left ventricle and separates the "pulmonary chamber" from the general cavity of the left atrium (see Annot. 22 and Fig. 2).

(22) **Valva v. pulmonalis.** The left free margin of the Camera pulmonalis appears to act as a valve to prevent regurgitation of blood into the Camera pulmonalis. In the literature this has been called the "pulmonary valve". More appropriately it should be referred to as the "valve of the pulmonary vein" in order to distinguish it from the valve of the pulmonary artery.

(23) **Valva atrioventricularis sinistra.** Certain authors have called this the "tricuspid valve" in birds. The term, tricuspid valve, should not be used in order to avoid confusion when comparisons are made with the mammalian heart where the tricuspid valve guards the right atrioventricular ostium.

Cuspis dexter has been referred to in the literature as the "septal cusp"; the latter designation is undesirable since the interventricular septum of the avian heart has both ventral and right surfaces on account of the extensive envelopment of the left ventricle by the right ventricle.

(24) **Valva aortae.** Ostia of the coronary arteries are dealt with in **Art.** Annot. 4.

(25) **Foramina venarum minimarum.** See **Ven.** Annot. 9 for comments on Vv. cardiacae minimae.

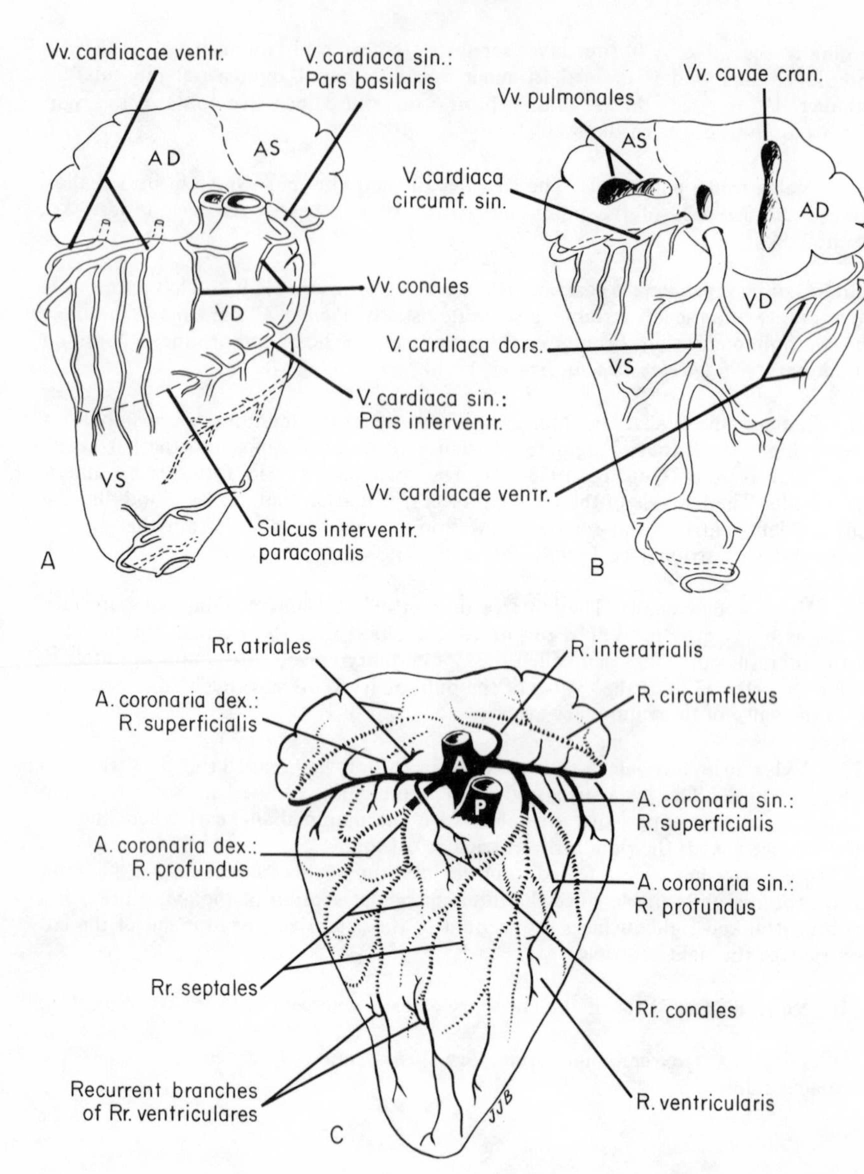

Fig. 1 Heart, coronary arteries, and cardiac veins; *Gallus*. A, B redrawn from Baumel (1975) after Lindsay (1967); C, redrawn from Baumel (1975). A, Facies ventrocranialis of heart; B, Facies dorsocaudalis; C, Facies ventrocranialis. In the ventricles the cross-hatched lines represent the parts of the deep rami of the coronary arteries that are embedded in the myocardium of the ventral and right side of the Septum interventriculare.

Abbreviations: A, Aorta ascendens; AD, Atrium dextrum; AS, Atrium sinistrum; circumf., circumflexa; dex., dextra; interventr., interventricularis; P, Truncus pulmonalis; sin., sinistra; VD, Ventriculus dexter; VS, Ventriculus sinister.

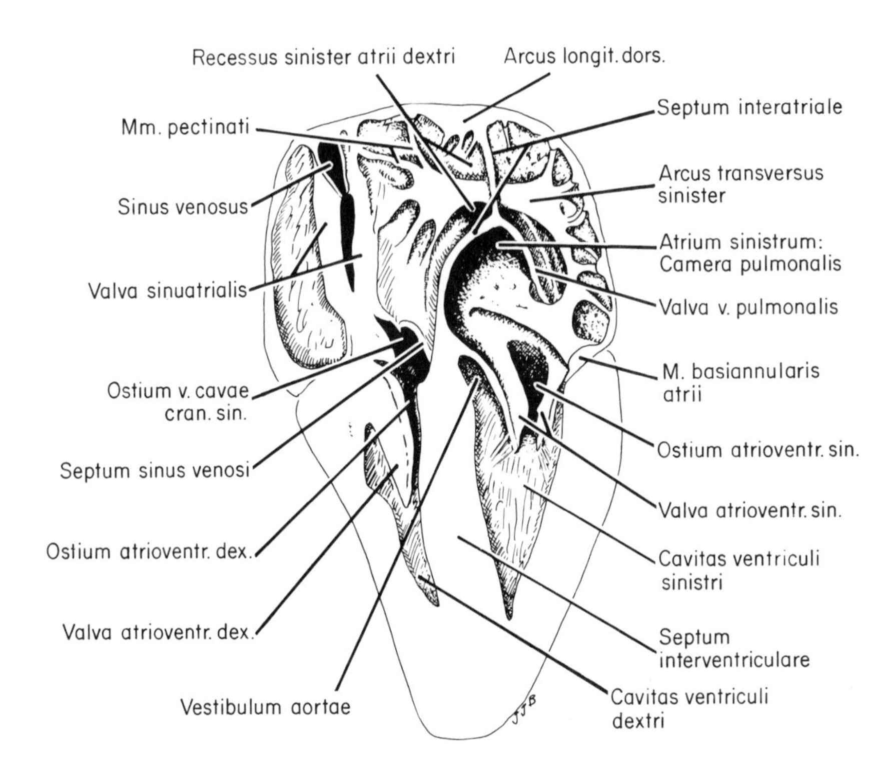

Fig. 2 Interior features of heart; *Gallus*. Sectioned in the dorsal plane; ventrocranial view. Redrawn from Baumel (1975) after Quiring (1933–34).

Abbreviations: atrioventr., atrioventricularis(arum); dex., dextra(um); longit., longitudinalis; sin., sinistra(um).

ARTERIAE

TRUNCUS PULMONALIS

(see **Cor.** Fig. 1)

Sinus trunci pulmonalis[1]
    Sinus sinister
    Sinus dexter
    Sinus dorsalis

A. pulmonalis[1]
    R. cranialis[1]
    R. caudomedialis[1]
    R. caudolateralis[1]
    R. accessorius[2]

        Aa. interparabronchiales[3]
            Arteriolae intraparabronchiales[3]

AORTA

(Fig. 7)

Aorta ascendens
    Bulbus aortae
        Sinus aortae[4] (Fig. 5)
            Sinus sinister[4]
            Sinus dexter ventralis[4]
            Sinus dexter dorsalis[4]

Glomera aortica[5]

Glomera pulmonalia[5]

Arteriae coronariae (see **Cor.** Fig. 1)
    A. coronaria sinistra[4]
        R. interatrialis
        R. superficialis
            R. circumflexus
            Rr. atriales
            Rr. ventriculares

- R. profundus
  - Rr. septales
  - Rr. ventriculares
- A. coronaria dextra[4]
  - R. superficialis
    - R. circumflexus
    - Rr. atriales
    - Rr. conales
    - Rr. ventriculares
  - R. profundus
    - Rr. ventriculares
    - Rr. septales

TRUNCUS BRACHIOCEPHALICUS[6]

(Fig. 5)

- Arteria carotis communis[7] (Fig. 3)
  - A. esophagotracheobronchialis[8, 2]
    - R. esophagealis
    - R. trachealis
    - R. bronchialis[2]
  - Aa. thyroideae[9]
  - A. esophagealis ascendens[10]
  - Truncus vertebralis (Fig. 3)
    - A. vertebralis ascendens[11]
      - Aa. intersegmentales cervicales[51, 65]
      - Rr. ventrales
      - Rr. dorsales[84]
      - Anastomoses cum a. carotica interna[11]
      - Anastomosis cum a. occipitali
    - A. vertebralis descendens[11] (Fig. 3)
      - Aa. intersegmentales truncales
      - Aa. intercostales dorsales[52]
      - Rr. dorsales[84]
    - A. comes nervi vagi[12] (Fig. 3)
      - Aa. ingluviales
      - A. suprascapularis[13]
      - A. esophagealis ascendens[10]
      - A. transversa colli
        - R. acromialis
      - Rr. cutanei colli
      - A. cutanea cervicalis ascendens
      - Rr. thymici

Glomus caroticum[14]

Arteria carotis interna[15] (Figs. 1, 3)
- Anastomoses cum a. vertebrali ascendenti
- A. occipitalis[16, 12]
  - A. occipitalis profunda[21]
  - A. occipitalis superficialis
  - Anastomosis cum a. vertebrali ascendenti[21]
- A. carotis cerebralis[17] (Figs. 1, 2)
  - A. sphenoidea[18]
    - R. palatinus
  - Anastomosis intercarotica[17] (Fig. 1)
    - A. hypophysialis caudalis (see **Endoc.** Annot. 18)
  - A. ophthalmica interna[19]
  - A. infundibularis (see **Endoc.** Annot. 19)
  - Ramus rostralis (Fig. 2)
    - A. tecti mesencephali ventralis
    - A. cerebroethmoidalis
      - A. cerebralis rostralis
        - Aa. preopticae
      - A. ethmoidalis
        - Anastomosis cum a. supraorbitali
        - Anastomosis cum a. ophthalmotemporali
        - Rr. orbitales
        - Rr. nasales
    - A. cerebralis media
      - Rr. hemispherici laterales
    - A. cerebralis caudalis[20]
      - A. cerebellaris dorsalis
      - Aa. tecti mesencephali dorsales
      - Rr. hemispherici ventrales
      - Rr. hemispherici occipitales
      - Rr. diencephali dorsales
      - A. choroidea ventriculi lateralis et tertii[20]
      - A. interhemispherica[20]
      - Rr. hemispherici dorsales
      - A. meningealis caudalis
        - R. pinealis
  - Ramus caudalis (Fig. 2)
    - A. trigeminalis
    - Aa. medullares
    - A. interpeduncularis
    - A. basilaris[21]

A. cerebellaris ventralis[22]
A. medullae lateralis
A. labyrinthi rostralis[23]
A. labyrinthi caudalis[23]
A. choroidea ventriculi quarti
A. ophthalmica externa [A. stapedia][24] (Fig. 1)
Rete mirabile ophthalmicum[24]
A. temporalis
A. intramandibularis[25]
Rr. mentales
A. supraorbitalis (Fig. 1)
Rr. glandulae nasalis
Aa. palpebrales dorsocaudales (see A. facialis)
Aa. ciliares anteriores[26]
A. infraorbitalis
Aa. palpebrales ventrales
Aa. ciliares anteriores[26]
A. ophthalmotemporalis[27] (Fig. 1)
A. ciliaris posterior longa temporalis[26, 28]
Rr. choroidei[29]
Circulus arteriosus iridicus[28]
Circulus arteriosus ciliaris[28]
Anastomosis cum circulo arterioso iridico
Aa. ciliares posteriores breves[26]
Rr. choroidei
Aa. musculares bulbi oculi
Aa. glandulae membranae nictitantes
A. pectinis oculi[29]
Anastomosis cum a. ethmoidali
Anastomosis cum a. ophthalmica interna
A. ciliaris posterior longa nasalis[26, 28]
Rr. choroidei[29]

Arteria carotis externa[30, 12] (Figs. 1, 3)
A. cutanea cervicalis descendens
A. comes nervi vagi[12]
A. auricularis rostralis[31]
A. auricularis caudalis[31]
A. mandibularis[32] (Figs. 1, 3)
A. esophagealis descendens
A. trachealis descendens
Aa. hyobranchiales [hyoideae]
A. laryngea propria[33]
Rr. pharyngeales

Arteria carotis externa—*continued*

- A. lingualis
  - A. lingualis propria
- A. sublingualis[34]
- A. submandibularis superficialis[35]
- A. submandibularis profunda

A. maxillaris[38] (Figs. 1, 3)

- A. pterygopharyngealis
  - Rr. pharyngeales
  - Plexus pterygoideus[36]
  - R. nasalis[37]
- A. facialis
  - Rr. palpebrales ventrales
  - Rr. frontales
  - A. palpebralis dorsorostralis (see A. ophthalmica externa)
  - R. nasalis[37]
- A. palatina[38] (Fig. 1)
  - R. palatinus medialis
  - R. palatinus lateralis
  - A. palatina mediana[38]
  - R. nasalis[37]
- A. pterygoidea dorsalis (Fig. 1)

Arteria subclavia (Figs. 3, 5)

A. sternoclavicularis[39]

- A. sternalis interna[40]
- A. clavicularis
- A. sternalis externa
- A. coracoidea dorsalis[41]
- A. esophagotrachealis[42]
- A. thoracica interna
  - R. ventralis
  - R. dorsalis
    - Aa. intercostales ventrales[53]
- Truncus pectoralis [A. thoracica externa][43] (Fig. 5)
  - A. pectoralis cranialis
  - A. pectoralis media[44]
    - A. infrascapularis
    - A. cutanea thoracoabdominalis[44]
  - A. pectoralis caudalis

A. axillaris (Figs. 3, 5)
- A. subscapularis
- A. supracoracoidea[45]
- A. brachialis (Fig. 4)
  - A. profunda brachii
    - A. circumflexa humeri dorsalis
    - A. antebrachialis dorsalis cranialis
  - A. bicipitalis[46]
    - Rr. propatagiales
  - A. circumflexa humeri ventralis
  - A. nutricia humeri
  - A. collateralis radialis
  - A. collateralis ulnaris
  - A. radialis[47] (Fig. 4)
    - A. recurrens radialis
    - A. cubitalis dorsalis
    - A. radialis profunda
      - Aa. interosseae dorsales[48]
      - A. antebrachialis dorsalis caudalis
        - Rr. postpatagiales
      - Rr. carpales dorsales
      - Rr. carpales ventrales
      - Rr. metacarpales dorsales
      - Rr. alulares
    - A. radialis superficialis
      - Rr. propatagiales
  - A. ulnaris (Fig. 4)
    - A. cutanea brachialis
    - A. recurrens ulnaris
    - A. cubitalis ventralis
    - A. ulnaris profunda[49]
      - Rr. metacarpales ventrales
      - A. metacarpalis interossea
      - Rr. digitales dorsales
      - Rr. digitales ventrales
  - A. ulnaris superficialis
    - Rr. postpatagiales
  - A. postpatagialis marginalis
  - A. propatagialis marginalis

Arcus aortae (Figs. 3, 5)

Aorta descendens (Figs. 3, 5)
Lig. arteriosum
Lig. aortae[50] (Figs. 3, 5)
Aa. intersegmentales truncales[51]
Aa. intercostales dorsales[52]
Rr. dorsales[84]
A. esophagealis[54]
A. musculorum colli

A. celiaca (see Figs. 5, 6; and **Cardvas.** Intro.)
A. esophagealis[54]
A. proventricularis dorsalis
Rr. esophageales
A. gastrica dorsalis[55]
Ramus sinister a. celiacae (Fig. 6)
A. proventricularis ventralis
A. gastrica sinistra
Rr. sacci cranialis
Rr. sacci caudalis
A. gastrica ventralis
A. hepatica sinistra[56, 59] (see **Digest.** Hepar)
R. cranialis
R. lateralis
R. caudalis
Ramus dexter a. celiacae (Fig. 6)
Aa. splenicae[57] (see **Cardvas.** Intro.)
A. hepatica dextra[56, 59] (see **Digest.** Hepar)
R. cranialis
R. lateralis
R. caudalis
A. vesicae felleae[58]
Anastomosis aa. hepaticarum[59]
Rr. medii[59]
A. gastroduodenalis[60, 56]
Aa. duodenales[60]
A. duodenojejunalis[56]
A. jejunalis[56, 60]
A. gastrica dextra
Rr. sacci caudalis
Aa. ileocecales[61, 63]
Aa. ileae[61]
A. pancreaticoduodenalis[62]
Rr. duodenales
Rr. pancreatici

A. mesenterica cranialis (see Fig. 5; and **PNS** Fig. 8)
- A. duodenojejunalis
- Aa. jejunales[60]
- Aa. ileae[60, 63]
- A. ileocecalis[63, 61]
- A. marginalis intestini tenuis[64]

Aa. intersegmentales synsacrales[65, 51] (Fig. 5)

A. renalis cranialis (see Fig. 5; and A. ischiadica)
- Aa. intralobulares (see **Urin.** Vasa sanguinea intrarenalia)
- A. testicularis[66]
  - Rr. epididymales
- A. ovarica[66] (see Fig. 5; and **Genit. fem.** Vasa sanguinea ovarica)
- A. adrenalis
- Rr. ureterodeferentiales craniales (see **Ven.** Annot. 57)
  - Rr. ureterici craniales
- A. oviductalis cranialis[67, 72] (see Fig. 5; and **Genit. fem.**)
  - Rr. ovarii
  - Aa. infundibuli
  - Aa. magni

A. iliaca externa[68] (Fig. 5)
- A. oviductalis cranialis accessoria[67] (see **Genit. fem.** Annot. 36 and Fig. 3)
- A. pubica[69]
  - A. umbilicalis
- A. femoralis
  - A. coxae cranialis[70]
  - A. femoralis medialis
    - A. circumflexa femoris medialis[71]
  - A. femoralis cranialis
    - A. cutanea femoris cranialis

A. ischiadica (Fig. 5)
- A. renalis media
  - Aa. intralobulares (see **Urin.** Vasa sanguinea intrarenalia)
- A. renalis caudalis
  - Aa. intralobulares (see **Urin.** Vasa sanguinea intrarenalia)
  - Rr. ureterodeferentiales medii
  - Rr. ureterici medii
- A. oviductalis media[67, 72] (see Fig. 5; and **Genit. fem.**)
  - Aa. magni
  - Aa. isthmi
  - Aa. uterinae (see **Genit. fem.** Annot. 39)

A. ischiadica—*continued*

A. oviductalis marginalis ventralis[72] (see **Genit. fem.** Fig. 3)
A. oviductalis marginalis doraslis[72] (see **Genit. fem.** Fig. 3)
A. coxae caudalis[70]
A. obturatoria
A. trochanterica
A. circumflexa femoris lateralis[71] (Fig. 7)
A. femoralis proximocaudalis[73]
A. femoralis distocaudalis
A. nutricia femoris proximalis
A. cutanea femoralis caudalis
A. suralis[74] (Fig. 7)
  A. cutanea femoralis lateralis
  A. cutanea cruralis caudalis
  A. suralis medialis
  A. suralis lateralis
A. poplitea[75] (Fig. 7)
  A. genicularis lateralis
  A. nutricia femoris distalis
  A. tibialis medialis (Fig. 7)
    A. genicularis medialis
    A. cruralis medialis
  A. tibialis caudalis
  A. fibularis [A. peronea]
    A. genicularis proximalis
A. tibialis cranialis[76]
  A. nutricia tibiae
  A. recurrens tibialis cranialis
  A. interossea
  Rete tibiotarsale[77] (Fig. 7)
  R. fibularis [peroneus] superficialis[78]
  R. fibularis [peroneus] profundus[78]
  A. metatarsalis dorsalis communis[79]
    Aa. tarsales plantares[79]
    Aa. metatarsales plantares
      Arcus plantaris
      Rr. pulvinares
    Aa. metatarsales dorsales
      Aa. digitales

A. sacralis mediana (Fig. 5)

Aa. intersegmentales synsacrales[51]

A. mesenterica caudalis (see Fig. 5; and **Digest.** Fig. 8)
R. cranialis
Rr. ilei[61,63]
R. caudalis
Rr. rectales
A. bursalis[80]
Rr. cloacales

A. iliaca interna[81,68] (Fig. 5)
A. pudenda[82]
Rr. ureterodeferentiales caudales
Rr. ureterici caudales
Aa. oviductales caudales[67,72] (see **Genit. fem.** Fig. 3)
Aa. uterinae (see **Genit. fem.** Annot. 39 and Fig. 3)
A. vaginalis (see **Genit. fem.** Fig. 3)
Aa. cloacales (see **Cloaca** Annot. 24 and A. mesenterica caudalis)
Rr. bursocloacales[80]
Rr. bursales
Rr. cloacales
A. caudae lateralis[83] (Fig. 5)
A. cutanea abdominalis[44]
Rr. ureterodeferentiales caudales
Rr. ureterici caudales

A. caudae medianae (Fig. 5)
Aa. intersegmentales caudales[51]
Rr. glandulares uropygiales

Arteriae medullae spinalis
Aa. intersegmentales[51]
Rr. ventrales
Rr. dorsales
Aa. vertebromedullares[84]
Aa. radiculares ventrales
A. spinalis ventralis[85]
Rr. marginales[86]
Rr. sulci
Rr. sulcocommissurales
As. radiculares dorsales
Aa. dorsolaterales[85]
Rr. spinales dorsales
Rr. marginales[86]
Rr. fissurae

ANNOTATIONS

(1) **Sinus trunci pulmonalis; A. pulmonalis.** Pulmonary sinuses are less well developed and thinner walled than the aortic sinuses. Names of sinuses are based on their positions *in situ*. The branching patterns of right and left pulmonary arteries are similar. Rami of A. pulmonalis closely correspond to tributaries of V. pulmonalis; neither arteries or veins correspond in their intrapulmonary branching pattern to that of the bronchial tree. Rr. caudomedialis and caudolateralis of the artery are generally dorsal to the two caudal venous radices. The R. cranialis of the artery is lateral to the cranial radix of V. pulmonalis; this is based on observations of Radu and Radu (1971) and Abdalla and King (1975)in *Gallus*, *Meleagris*, *Anser*, *Anas*, and in *Columba*.

(2) **R. accessorius.** In *Gallus* the medial aspect of each undivided pulmonary artery releases a distinct R. accessorius that supplies lung tissue cranial and medial to the hilus. Each R. accessorius gives off a bronchial ramus that anastomoses directly with the bronchial ramus of A. esophagotracheobronchialis (Abdalla and King, 1976).

(3) **Aa. interparabronchiales; Arteriolae intraparabronchiales.** For details of intrapulmonary angio-architecture see Abdalla and King (1975) and **Resp.**, Vasa sanguinea intrapulmonalia. Interparabronchial arteries are branches of each of the main rami of A. pulmonalis.

(4) **Sinus aortae.** Names of aortic sinuses are based on their *in situ* positions. Of the three sinuses, Sinus sinister and Sinus dexter ventralis are "coronary"; i.e. they contain ostia of the coronary arteries. Occasionally the superficial ramus of A. coronaria sinistra or other aberrant coronary artery may arise from the right dorsal sinus (Petren, 1926; Lindsay and Smith, 1965).

(5) **Glomera aortica; Glomera pulmonalia.** Tcheng *et al.* (1963a, b) described encapsulated receptors in the adventitia of the roots of the ascending aorta that are believed to be chemoreceptors. Jones (1969) has verified that the intramural receptors at the bifurcation of the pulmonary trunk and in the wall of the aorta and pulmonary trunk near their valves are baroreceptors.

(6) **Truncus brachiocephalicus.** Both right and left brachiocephalic trunks arise from the left side of Aorta ascendens (see Fig. 5).

(7) **A. carotis communis.** Short vessel that breaks up in the root of the neck into A. carotis interna, Truncus vertebralis, and A. comes nervi vagi. Variation of the arteries in the region of the heart in most of the orders of birds is thoroughly treated in the series of papers of Glenny cited in the bibliography of his monograph of 1955.

(8) **A. esophagotracheobronchialis.** Synonymy: A. syringotracheobronchialis (Bhaduri *et al.*, 1957); ductus shawi (Glenny, 1955 and earlier papers). An artery or complex of arteries usually arising from the medial side of the common carotid artery (*Gavia*, *Larus*, *Phoenicopterus*, *Gallus*, *Branta*, *Alcedo*, *Trogon*). The artery may also spring from the vertebral trunk (*Perisoreus*), or from the common carotid of one side and the vertebral trunk contralaterally (*Spheniscus*, *Corvus*). The general distribution of A. esophagotracheobronchialis is to the caudal trachea and its bi-

furcation (syrinx), main bronchi, pericardium, Septum horizontale dorsal to the heart, and esophagus from the origin of the artery caudally to the esophago-proventricular junction. In some species the artery supplies the thyroids, parathyroids, ultimobranchial and carotid bodies (see Annot. 42).

(9) **Aa. thyroideae.** The arterial supply of the thyroid gland varies both in number of arteries to the gland and the site of origin of the thyroid arteries. From a review of the numerous papers of Glenny (cited in Glenny, 1955) in which he systematically surveyed the main arteries in the region of the heart, it appears that the majority of species have but a single artery to each gland (*Rhea*, *Spheniscus*, *Gavia*, *Podiceps*, *Cygnus*, *Fulica*, *Apus*, *Trogon*, *Corvus*).

Two thyroid arteries to the gland are found in several species of Columbidae, several galliform species, *Larus*, and *Phoenicopterus*; three thyroid arteries supply the thyroid gland of ducks (Anatidae) according to Assenmacher (1953). Usually the origin of single thyroid arteries and the caudal thyroid artery of the dual-artery condition is directly from the common carotid artery.

Other sites of origin are: A. vertebralis, proximal part of A. esophagotracheobronchialis, A. comes n. vagi, terminal part of brachiocephalic artery, or subclavian artery. Cranial thyroid arteries arise from A. comes n. vagi, common carotid, or ascending esophageal artery. Aa. thyroideae also supply the carotid body, parathyroid glands, ultimobranchial gland, and distal ganglion of the vagus nerve (see Abdel-Magied and King, 1978).

(10) **A. esophagealis ascendens.** In certain species of unicarotid birds (see below, Annot. 15), the artery that corresponds to the single persistent carotid of the other side remains as a superficial artery, and appears to be functionally modified to serve as the A. esophagealis ascendens (*Rhea*, *Aperyx*, *Casuarius*, some coraciiform species, *Trogon*, piciform and passeriform birds) (Glenny, 1955). In *Phoenicopterus*, an unicarotid form having the single internal carotid fed by both common carotids, the A. esophagealis ascendens springs from the proximal part of A. comes nervi vagi (Bhaduri *et al.*, 1965). In bicarotid forms the ascending esophageal artery arises from the common carotids (*Grus*) or from the vertebral trunk (*Alcedo*) (Glenny, 1955), or from A. comes n. vagi (see Fig. 3).

(11) **Aa. vertebrales ascendens et descendens.** Supply the vertebral column, axial muscles, and cervical spinal cord. Ascending vertebral arteries anastomose with the cervical parts of A. carotis interna and A. occipitalis profunda (Baumel, 1964). (See Aa. intercostales dorsales.)

(12) **A. comes nervi vagi.** Synonymy: A. cervicalis superficialis (Glenny, 1955). May be a direct branch of A. carotis communis or may share a common stem with the vertebral trunk (A. cervicovertebralis, Glenny, 1955). A comes nervi vagi is the adult vessel derived from the fetal A. cartotis externa; the definitive external carotid appears as a branch of the internal carotid at the base of the skull. A. comes nervi vagi arises as a terminal branch of the common carotid, courses in a common bundle with N. vagus and V. jugularis, and anastomoses with a branch of the occipital artery or the external carotid at the base of the skull.

(13) **A. suprascapularis.** Synonymy: A. cephalica humeri (Neugbauer, 1945) and Bodrossy, 1938).

(14) **Glomus caroticum.** De Kock (1958) has surveyed the location and distribution of the carotid body proper or disseminated carotid body tissue in examples of passeriform, charadriiform, anseriform, and podicipediform birds. This tissue is found in a zone from a point proximal to the bifurcation of the common carotid up to the position of the thyroid gland; carotid body tissue is often closely associated with parathyroid, ultimobranchial body, and the vagus nerve and is supplied by local arteries. See also Kose (1904), Muratori (1934), Murillo-Ferrol (1967), and Abdel-Magied and King (1978).

(15) **A. carotis interna.** Most birds are bicarotid; the two internal carotids ascend the neck side-by-side in an osseomuscular canal on the ventral aspect of the cervical vertebral column. In most unicarotid forms the left carotid persists as the principal artery to the head. In certain unicarotid forms the single vessel represents the fused or conjugate right and left carotids; in some of these birds the basal portion of both carotids remains, while in others only the basal part of one of the conjugate vessels is present (Glenny, 1955). No matter how it be formed, the single carotid divides into right and left internal carotids at the base of the skull. In some unicarotid birds the fetal carotid that does not persist as a definitive carotid simply regresses and becomes a ligamentous vestige or a small regional artery; in others it becomes modified as a superficial vessel of the neck (see Annot. 7).

(16) **A. occipitalis.** A branch of the internal carotid in some birds and a branch of the external carotid in others (see Annot. 21).

(17) **A. carotis cerebralis; Anastomosis intercarotica.** The cerebral carotid artery is defined as the extension of A. carotis interna past the origin of A. ophthalmica externa; i.e. its intrasphenoid segment in the carotid canal of Basis cranii and its intracranial part (Stresemann, 1927–34; Kitoh, 1962). Anastomosis intercarotica connects right and left As. carotes cerebrales by means of a transverse vessel or a side-to-side anastomosis (Wingstrand, 1951; Baumel and Gerchman, 1968).

(18) **A. sphenoidea.** Springs from the part of A. carotis cerebralis that courses in the carotid canal; leaves the base of the skull via Foramen orbitalis, at the side of the basisphenoid rostrum. In some forms R. palatinus and R. sphenomaxillaris arise as independent vessels from the cerebral carotid artery.

(19) **A. ophthalmica interna.** Arises from the segment of A. carotis cerebralis that lies at the side of the hypophysis within the Sella turcica; this artery does not persist as a substantial vessel in the adults of some avian species (Wingstrand, 1951).

(20) **A. cerebralis caudalis.** Synonymy: A. cerebri posterior.
A. choroidea ventriculi lateralis et tertii. Synonymy: A. choroidea anterior (Ariens-Kappers, 1933).
A. interhemispherica. Synonymy: A. cerebri posterior communis. A. interhemispherica is a name proposed by Baumel (1967). It is usually an unpaired artery that arises from either the left or right A. cerebralis caudalis, and is distributed to both hemispheres (Baumel, 1967).

(21) **A. basilaris.** Usually R. caudalis of either the left or right A. carotis cerebralis is prolonged as A. basilaris; its counterpart persists as a vestigial local vessel. Occasionally both caudal rami anastomose to form A. basilaris, particularly in galliform,

falconiform, and strigiform birds (Baumel and Gerchman, 1968; Kitoh, 1962). The smaller terminal branch(es) of A. basilaris anastomose with branches of A. occipitalis profunda, A. spinalis ventralis, and A. vertebralis ascendens (T. Nishida, pers. comm.) (see Annot. 11).

(22) **A. cerebellaris ventralis.** Represented on each side by two separate ventral cerebellar arteries in *Gallus* (Kitoh, 1962). In certain other species the single stem of the ventral cerebellar artery divides into rostral and caudal branches (Vitums *et al.*, 1965). See also Baumel (1967) for a discussion of the relative territories of distribution of dorsal and ventral cerebellar arteries in different species of birds.

(23) **Aa. labyrinthi.** Rostral and caudal labyrinthine arteries (Schmidt, 1964) arise directly from A. cerebellaris ventralis or A. medullaris lateralis (*Columba*) and enter the inner ear region via foramina that conduct branches of N. vestibulocochlearis. See **Osteo**. Fossa acusticus interna.

(24) **A. ophthalmica externa.** Homologue of A. stapedia of other vertebrates (Hafferl, 1933; Goodrich, 1958).
Rete mirabile ophthalmicum is also known as Rete mirabile temporale. Arterial and venous retia mirabilia ophthalmica are enmeshed with one another.

(25) **A. intramandibularis.** Substitute name for A. alveolaris inferior, a mammalian term inappropriate in modern birds, all of which lack teeth (see **Osteo**. Annot. 70 regarding teeth in fossil birds).

(26) **Aa. ciliares anteriores et posteriores.** Following usage in the human *Nomina Anatomica* (IANC, 1966) and *Nomina Anatomica Veterinaria* (ICVAN, 1973), the poles of the Bulbus oculi are arbitrarily defined as anterior and posterior even though in most birds Axis bulbi is directed rostrolaterally rather than rostrally. Aa. ciliares are named according to the pole of the eye that they enter. Other terms of direction and position with respect to the eye and orbit are: dorsal and ventral; nasal and temporal.

(27) **A. ophthalmotemporalis.** Following Wingstrand and Munk (1965) this artery is regarded as the main continuation of A. ophthalmica externa and the principal artery of supply to Bulbus oculi. In the caudomedial region of the orbit around the optic nerve a rich, plexiform system of anastomoses occurs between branches of A. ophthalmica externa, A. ophthalmica interna, and A. ethmoidalis.

(28) **Circulus arteriosus iridicus** (Oehme, 1969a). Formed by A. ciliaris posterior longa temporalis that bifurcates on reaching the peripheral margin of the iris.
Circulus arteriosus ciliaris is formed by contributions from Aa. ciliares anteriores that are rami of the several palpebral arteries. In *Columba* the termination of A. ophthalmotemporalis is a distinct A. ciliaris posterior longa nasalis; however, Oehme does not acknowledge the presence of this vessel in any of the species he studied.

(29) **A. pectinis oculi.** Birds do not possess an artery comparable to the A. centralis retinae of mammals. The Pecten oculi is a vascular body that projects from the floor of the Bulbus oculi into the chamber of the vitreous humor. According to Wingstrand and Munk (1965) the Pecten supplies nutrients to the inner retinal layers.
Rr. choroidei of the Aa. ciliares longae et breves supply the choriocapillaris and external retinal layers (Wingstrand and Munk, 1965).

(30) **A. carotis externa.** The pattern of branching of A. carotis externa varies from species to species; the branches that are listed are consistently present in one configuration or another in all species that have been studied with any degree of completeness (see Fig. 1 and Annot. 12).

(31) **Aa. auriculares.** The origin of these arteries is variable; they may arise from A. carotis externa, A. maxillaris, or A. mandibularis.

(32) **A. mandibularis.** Synonymy: A. lingualis; A. facialis externa.

(33) **A. laryngea propria.** Sometimes qualified as "superior" to distinguish it from the artery to the syrinx which is erroneously referred to as the "inferior" larynx. In the literature A. laryngea often refers to the common stem of A. esophagealis, A. trachealis descendens, Rr. pharyngeales as well as A. laryngea propria itself.

(34) **A. sublingualis.** A branch of A. lingualis in certain birds. See Annot. 32.

(35) **A. submandibularis superficialis.** Especially strong in *Gallus* which possess highly vascularized skin appendages such as wattles in the intermandibular region.

(36) **Plexus pterygoideus.** Located on both ventral and dorsal aspects of the pterygoid bone and its attached muscles in the caudal part of the floor of the orbit. The plexus is formed by anastomosing rami of A. sphenoidea, A. maxillaris, and A. palatina.

(37) **R. nasalis.** An extensive system of anastomoses connect nasal rami of A. ethmoidalis, A. maxillaris, and A. palatina.

(38) **A. palatina.** Continuation of A. maxillaris. Right and left Rr. palatini mediales flank the nasal choana, and are confluent rostral to the choana forming the unpaired R. palatinus medianus.

(39) **A. sternoclavicularis.** Synonymy: A. coracoidea (Glenny, 1951).

(40) **A. sternalis interna.** Often unpaired; arising from A. sternoclavicularis of one side it courses on the inner aspect of the sternum opposite the attachment of the carina externally.

(41) **A. coracoidea dorsalis.** Synonymy: A. acromialis. Branch of A. sternoclavicularis that parallels the dorsal aspect of Os coracoideum and extends to the caudal part of Artc. humeralis. "Acromion" refers to the point of the shoulder, and should be reserved for vessels that supply the cranial aspect of the shoulder region (see rami of A. comes n. vagi).

(42) **A. esophagotrachealis.** In galliform birds this artery springs from A. subclavia directly, and sends branches craniad. These branches supplement those from A. esophagotracheobronchialis (Glenny, 1951). In *Meleagris* A. esophagotrachealis gives off Rr. pericardii. See Annot. 8.

(43) **Truncus pectoralis.** A more meaningful term than its common synonym, A. thoracica externa, since the artery primarily supplies the breast muscles (pectus = breast L.). Both the *Nomina Anatomica* (IANC, 1966) and *Nomina Anatomica Veterinaria* (ICVAN, 1973) use "pectoral" for nerves and vessels.

(44) **A. pectoralis media.** Present in *Columba*; absent in *Gallus*.
A. cutanea thoracoabdominalis. Synonymy: A. abdominopectoralis (Neugebauer, 1845). In *Gallus* this cutaneous artery replaces A. pectoralis media; it supplies the skin of the caudal breast region and adjacent abdomen, and is the principal artery to the incubation (brood) patch. A. cutanea abdominalis (A. caudae lateralis) contributes to the supply of the incubation patch. See **Ven.** Annot. 49, 65.

(45) **A. supracoracoidea.** In some forms this artery shares a common stem with A. subscapularis. A. supracoracoideus (and vein) accompanies N. supracoracoideus into the muscle of the same name.

(46) **A. bicipitalis.** Accompanies N. bicipitalis into the biceps muscle and then into the Propatagium.

(47) **A. radialis.** Mostly supplies structures in the antebrachium; has little or no distribution in the wrist and hand.

(48) **Aa. interosseae dorsales.** Pass dorsad through the radio-ulnar interosseous space, and supply the extensor muscles and dorsal integument of the antebrachium.

(49) **A. ulnaris profunda.** A minor artery in *Gallus* that accompanies the strong V. ulnaris profunda. In other forms A. ulnaris profunda is a more substantial artery.

(50) **Lig. aortae.** The aortic attachment of this vestige of the fetal left Radix aortae is located just cranial to the origin of A. celiaca, and is often patent in adult birds.

(51) **Aa. intersegmentales** (Stresemann, 1927–34). Occur over the entire length of the vertebral column, and are named on a regional basis: Aa. intersegmentales cervicales, truncales, synsacrales, and caudales. The parent arteries of the intersegmental arteries are Aa. vertebrales ascendens et descendens, Aorta descendens, A. sacralis mediana, and A. caudae medianae (see Annot. 65).

(52) **Aa. intercostales dorsales.** A vertebralis descendens gives off the dorsal intercostal arteries to the cranialmost intercostal spaces. Aa. intercostales dorsales are the same as Rr. ventrales of Aa. intersegmentales truncales.

(53) **Aa. intercostales ventrales** are branches of A. thoracica interna.

(54) **A. esophagealis.** Supplies the terminal part of the esophagus, sharing this role with Rr. esophageales of A. esophagotracheobronchialis. A. esophagealis is a branch A. celiaca (*Gallus* and *Columba*), a direct branch of the aorta (*Buteo*), or a branch of A. proventricularis dorsalis (*Larus*, *Pteroglossus*). In *Coturnix* esophageal arteries arise from both the aorta and the celiac artery.

(55) **A. gastrica dorsalis.** A. proventricularis dorsalis is prolonged as A. gastrica dorsalis.

(56) **Aa. hepaticae.** Both A. hepatica sinistra and A. hepatica dextra may be represented by more than one vessel. In *Gallus* A. hepatica dextra sends off A. duodenojejunalis and A. jejunalis to the region of the duodenojejunal flexure of the gut. See Pavaux and Jolly (1968) and Miyaki (1973) for descriptions of the intrahepatic vessels of birds.

(57) **Aa. splenicae.** The splenic [lienal] arteries usually originate from the right ramus of A. celiaca, occasionally from the undivided trunk of A. celiaca. For variability of Aa. splenicae see Malinovský *et al.* (1973), (*Anas*); Malinovský and Višňanská (1975) (*Anser*); Fukuta *et al.* (1969) (*Gallus*) who also describe Aa. lienales accessoriae. See **Cardvas.** Intro.

(58) **A. vesica felleae.** Common synonym A. cystica. The gall bladder is present in most orders of birds, but lacking in some (see **Digest.**).

(59) **Anastomosis aa. hepaticarum.** Synonymy: R. communicans (Miyaki, 1973). The anastomotic artery connects right and left hepatic arteries, and runs along the left ramus of the right hepatic portal vein (*Columba*: *Gallus*, Miyaki, 1973).
Rr. medii supply the dorsal interlobar liver substance.

(60) **Aa. duodenales, jejunales, ileae.** The duodenum is the only differentiated part of the small intestine; "jejunal" and "ileal" in birds are arbitrary terms used in a regional sense that correspond to the condition in mammals wherein the jejunum and ileum are the intermediate and terminal segments of the small intestine (see **Digest.** Annot. 63).

(61) **Aa. ileae; Aa. ileocecales.** Ileal branches of the right ramus of A. celiaca rather than ileocecal branches of this artery are present in species of birds that lack long ceca parallelling the terminal ileum. The ileum is also vascularized by branches of the cranial and caudal mesenteric arteries (see Annot. 63).

(62) **A. pancreaticoduodenalis.** Some species of birds possess long, parallel left and right Aa. pancreaticoduodenales. Rr. pancreatici are collaterals of rami that pass to the duodenum.

(63) **A. ileocecalis.** Both A. celiaca and A. mesenterica cranialis send ileal branches to the supraduodenal loop of ileum. Anastomoses between the cranial and caudal mesenteric arterial systems are found in the ileorectal junctional region where the roots of the intestinal ceca are located.

(64) **A. marginalis intestini tenuis.** The prominent, irregular "marginal artery of the small intestine" consists of a chain of anastomosing arteries that extend along the mesenteric border of the small intestine from the duodenal ansa to the supraduodenal ansa. The marginal artery is fed proximally from the celiac system and distally from the system of the cranial mesenteric artery.

(65) **Aa. intersegmentales synsacrales.** A variable number of paired somatic branches of the aorta and its extension arise from the level of the root of A. mesenterica cranialis to the caudal end of the synsacrum. The lumbar, sacral, and caudal vertebral elements that coalesce to make up the synsacrum are not clearly defined (Boas, 1929); therefore the more general term "synsacral arteries" seems preferable to attempting to apply more specific regional terms. See Annot. 51; **Osteo.** Annot. 141; and **PNS** Annot. 38.

(66) **A. ovarica; A. testicularis.** The ovarian artery occurs on the left side only in most species of birds. Accessory ovarian (and testicular) arteries may stem directly from the aorta or other adjacent arteries (Nishida, 1964). The ovarian vessels undergo tremendous hypertrophy during egg-laying.

(67) **Aa. oviductales.** Oviductal arteries occur on the left side only, and undergo tremendous hypertrophy during egg-laying. Like most elongate viscera the oviduct acquires its blood supply from several arteries along its length. Longitudinal anastomoses occur between branches of these arteries near or on the oviduct similar to anastomoses along the intestine. See Annot. 72; and **Genit. fem.** Annot. 35–42.

The origin of the oviductal arteries varies between different species that have been studied. Individual variation within the same species is also demonstrated by these arteries. In general A. oviductalis cranialis springs from the left cranial renal artery, directly from the aorta or from A. iliaca externa. A. oviductalis media arises from the left A. ischiadica or its branch, A. renalis media. The Aa. oviductales caudales originate from the left A. iliaca interna and A. pudenda. One of the major arteries may be absent in certain individuals and replaced by branches from the next oviductal artery in the series. See **Genit. fem**. Annot. 35–42.

(68) **A. iliaca externa.** On leaving the pelvis and entering the thigh the name of this artery changes to A. femoralis. An A. iliaca communis is not present in birds; Aa. iliacae interna et externa spring independently from the aorta (see Annot. 81).

(69) **A. pubica.** Synonymy: A. epigastrica (Neugebauer, 1845); A. pelvica interna (Gadow and Selenka, 1891); A. umbilicalis (Hafferl, 1933; Nishida, 1963). A. pubica is suggested as a new term having topographic significance; i.e. the artery (vein and nerve) courses along the ventral border of the pubis and sends rami into the abdominal muscles and peritoneum. The proper umbilical artery is a branch of A. pubica that courses in the ventral extraperitoneal fat of the ventral abdominal wall to the umbilical scar. (See **Ven.** Annot. 74.)

(70) **A. coxae cranialis; A. coxae caudalis.** These are new names that replace the terms Aa. gluteae cranialis et caudalis; these arteries supply muscles and integument overlying the pre- and postacetabular parts of the ilium. The adjective "gluteal" refers to the buttock (Gr.). The preacetabular ilium does not compare at all to the mammalian gluteal region; no gluteal region is described for birds, only a "coxal region" (coxa = hip, L.). Contemporary avian myologists do not use "gluteal" for muscles of the hip region.

(71) **A. circumflexa femoris.** This term has been used erroneously for the longitudinal, spiral continuation of A. femoralis toward the knee within M. femorotibialis (Grzimek, 1963; Westpfahl, 1961). Medial and lateral circumflex femoral arteries in this list refer to transverse branches of the femoral and ischiadic arteries that "bend around" the proximal part of the femur as in mammals.

(72) **Aa. oviductales marginales.** The ventral and dorsal marginal arteries of the oviduct are channels produced by anastomoses between ascending and descending rami of major branches of the three regional arteries of the oviduct. The longitudinally oriented marginal arteries parallel the oviduct and are located within the ventral and dorsal ligaments of the oviduct, generally accompanied by satellite veins.

(73) **A. femoralis proximocaudalis.** This is a substitute name for A. profunda femoris; the latter term has been avoided to do away with comparisons with the mammalian A. profunda femoris that is a branch of the femoral artery and has a longitudinal orientation in the limb.

(74) **A. suralis.** The use of the term "sural" (sura L. = calf of the leg) for the nerves and vessels of this region of the crus simplifies some of the terminologies for vessels of the flexor compartment by limiting the number of terms using "tibialis". Branches of the sural arteries and veins accompany the superficial rami of N. tibialis to the proximal calf and overlying integument (Neugebauer, 1845).

(75) **A. poplitea.** This vessel is the continuation of A. ischiadica caudal to the knee joint; its extent is from the level of the root of A. suralis to the point of origin of A. fibularis.

(76) **A. tibialis cranialis.** This artery is the prolongation of A. poplitea distal to the origin of A. fibularis. A. tibialis cranialis leaves the flexor compartment of the crus via the distal tibiofibular interosseous foramen (Neugebauer, 1845).

(77) **Rete tibiotarsale.** Several collaterals of A. tibialis cranialis closely parallel the artery itself, and anastomose with one another (Rete tibialis) and distally with Aa. tarsales et metatarsales (Rete tarsi). See Hyrtl, 1864.

(78) **Rr. fibulares.** The superficial and deep fibular rami of A. tibialis cranialis accompany the superficial and deep fibular [peroneal] nerves.

(79) **A. metatarsalis dorsalis communis.** This artery is the prolongation of A. tibialis cranialis into the foot. See **Ven.** Annot. 70.
Aa. tarsales plantares. Transmitted via the Foramina vascularia proximalia of the Tarsometatarsus to the plantar aspect of the foot.

(80) **A. bursalis; Rr. bursocloacales.** In *Columba* the main artery of supply to the cloacal bursa is from R. caudalis of A. mesenterica caudalis. A minor share of its blood supply is from bursocloacal rami of A. pudenda (see Pintea *et al.* (1967) for the pattern in *Gallus*).

(81) **A. iliaca interna.** Synonymy: A pudenda communis (Gadow and Selenka, 1891); A. hypogastrica (Barkow, 1829). In birds and other higher vertebrates A. iliaca interna is distributed to the terminal parts of the digestive tract, reproductive tract, and urinary system. See Annot. 68.

(82) **A. pudenda.** Synonymy: A. pudenda interna.

(83) **A. caudae lateralis.** Synonymy: A. pudenda externa (Gadow and Selenka, 1891); A. musculocutanea caudae lateralis (Neugebauer, 1845). A. caudae lateralis is a simplification of the term of Neugebauer that is highly descriptive of the distribution of this artery to the dorsolateral tail region. There it supplies the bulb of the tail feathers, the muscles, and overlying integument of the tail. It has little distribution to the region of the vent (see **Ven.** Annot. 64, 65). Aa. coccygeae laterales are described by Gadow and Selenka (1891) as a strong pair of intersegmental arteries from the A. caudae medianae that supply the uropygial gland.

(84) **Aa. vertebromedullares.** Arteries to the spinal cord are derived regionally from dorsal rami of intersegmental branches of Aa. vertebrales, Aorta descendens, A. sacralis mediana, and A. caudalis mediana. Aa. vertebromedullares (Sterzi, 1903) are also known as Rr. spinales (Kitoh, 1964) and Aa. nervomedullares (Lob, 1967).

Names for branches of Aa. vertebromedullares are most after Lob (1967); Aa. vertebromedullares supply spinal nerve roots, dorsal root and sympathetic ganglia, spinal dura mater, vertebrae, and Medulla spinalis (see Annot. 21).

(85) **A. spinalis ventralis; Aa. dorsolaterales.** Longitudinal arteries that are formed by anastomoses between ascending and descending rami of Aa. radiculares ventrales et dorsales. At the foramen magnum A. spinalis ventralis communicates by way of weak anastomoses with terminal rami of A. basilaris.

(86) **Rr. marginales.** Irregular surface network of arteries (Aa. periphericae, Sterzi, 1903) derived from Aa. radiculares that send radial penetrating rami into the lateral funiculi of the spinal cord down to the gray matter. Rr. marginales supply that part of the spinal cord not served by vessels entering the ventral and dorsal sulci of the spinal cord (see Lob, 1967).

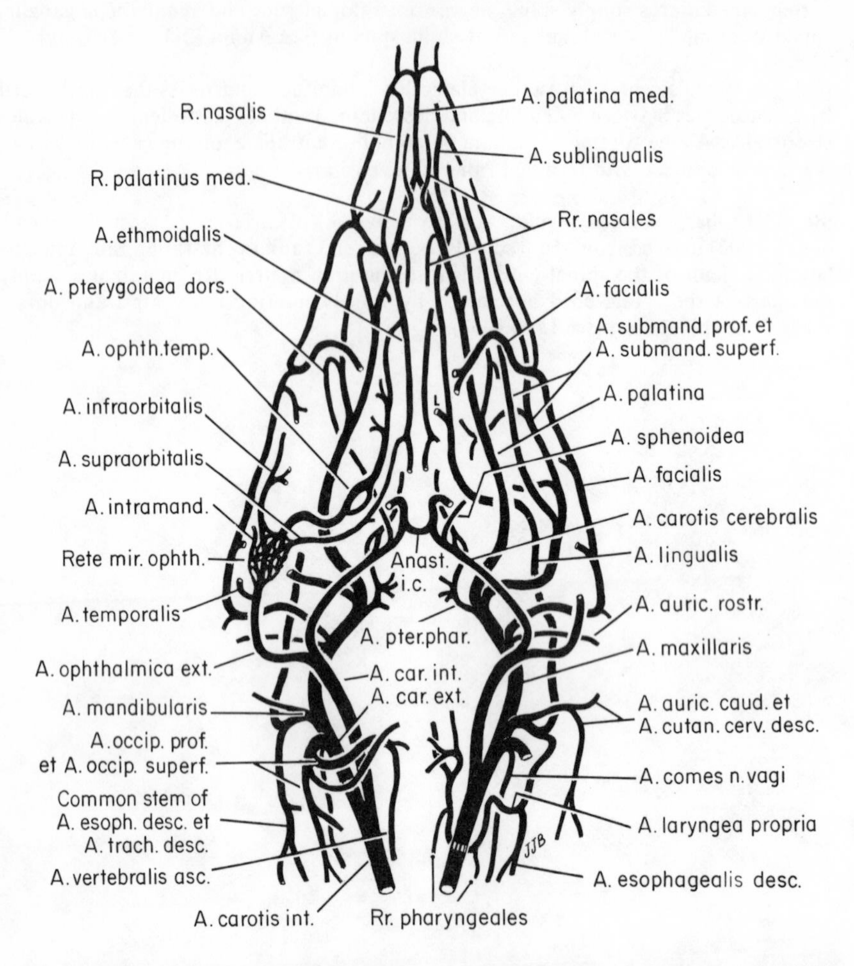

Fig. 1 Arteries of head, extracranial; *Gallus*. Dorsal view. Redrawn from Baumel (1975). The right side of the drawing emphasizes the details of the more ventrally situated arteries, while the left side features the more dorsally situated arteries.

Abbreviations: Anast. i.c., Anastomosis intercarotica; auric., auricularis; car., carotis; cutan., cutanea; esoph., esophagealis; intramand., intramandibularis; L., A. lingualis propria; occip., occipitalis; prof., profunda; pter. phar., pterygopharyngealis; Rete mir. ophth., Rete mirabile ophthalmica; submand., submandibularis; superf., superficialis; trach., trachealis.

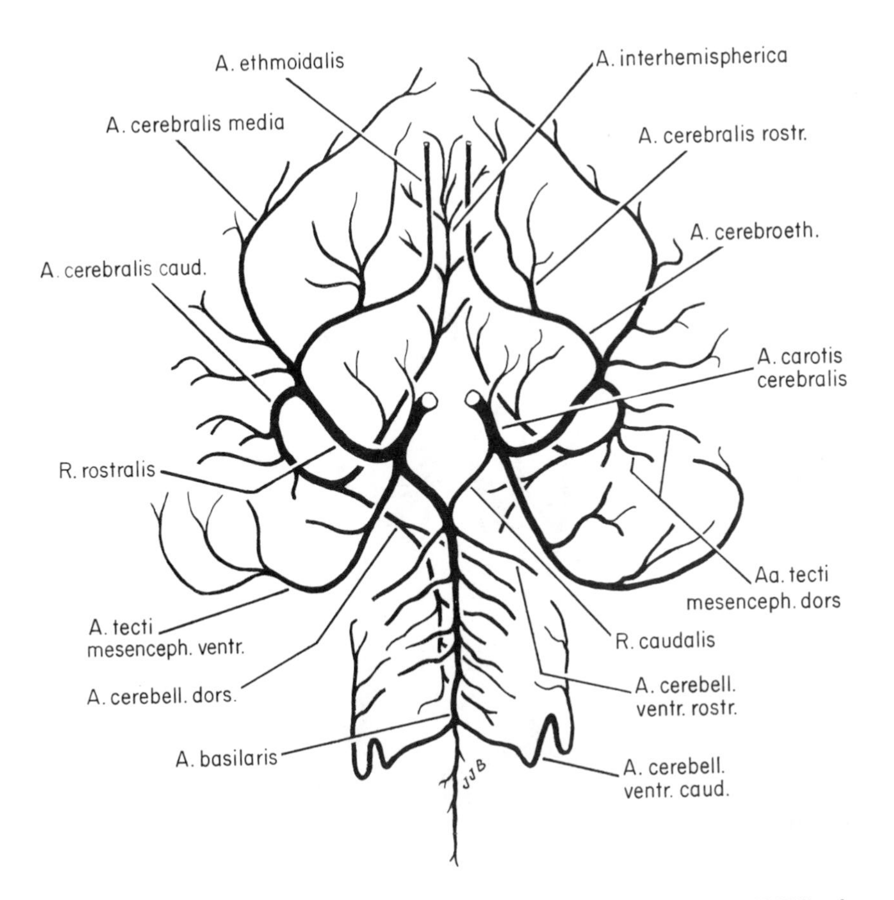

Fig. 2 Encephalic arteries; *Gallus*. Ventral view. Redrawn from Baumel (1975) after Shiina and Miyata (1932). Note the asymmetry in: (1) the roots of A. basilaris; (2) A. interhemispherica; and (3) A. cerebellaris dorsalis. See **Art.** Annot. 20, 21 for details.

Abbreviations: cerebell., cerebellaris; mesenceph., mesencephali.

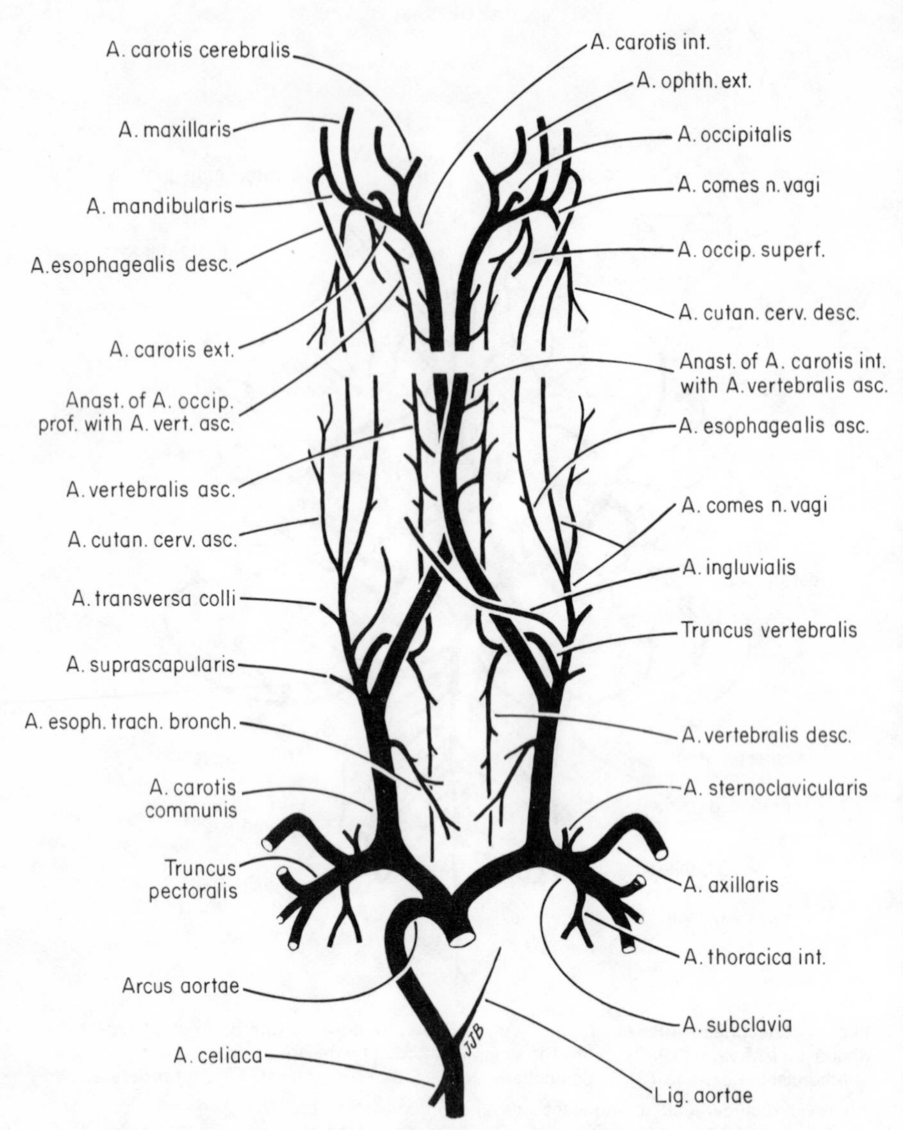

Fig. 3 Arteries of neck; *Columba livia.* Ventral view; foreshortening indicated by the white transverse band across the figure. Redrawn from Baumel (1964) and Bhaduri and Biswas (1954). A. carotis communis consists of a short trunk in the root of the neck: its principal branch is A. carotis interna. A. carotis externa branches from the cranial end of A. carotis interna in adult birds; see **Art.** Annot. 12 regarding the fetal condition.

In many bicarotid birds (see **Art.** Annot. 15) such as the pigeon, the two internal carotids converge, the left one situated proximal to the right one. They course side by side in the middle segment of the neck (Bhaduri and Biswas, 1954).

Note the hexagonal configuration produced by the brachiocephalic arteries, the common carotids and the proximal parts of the internal carotids (Bhaduri and Biswas, 1954).

Abbreviations: cutan., cutanea; cerv., cervicalis; esoph. trach. bronch., esophagotracheobronchialis; occip., occipitalis; ophth., ophthalmica; vert., vertebralis.

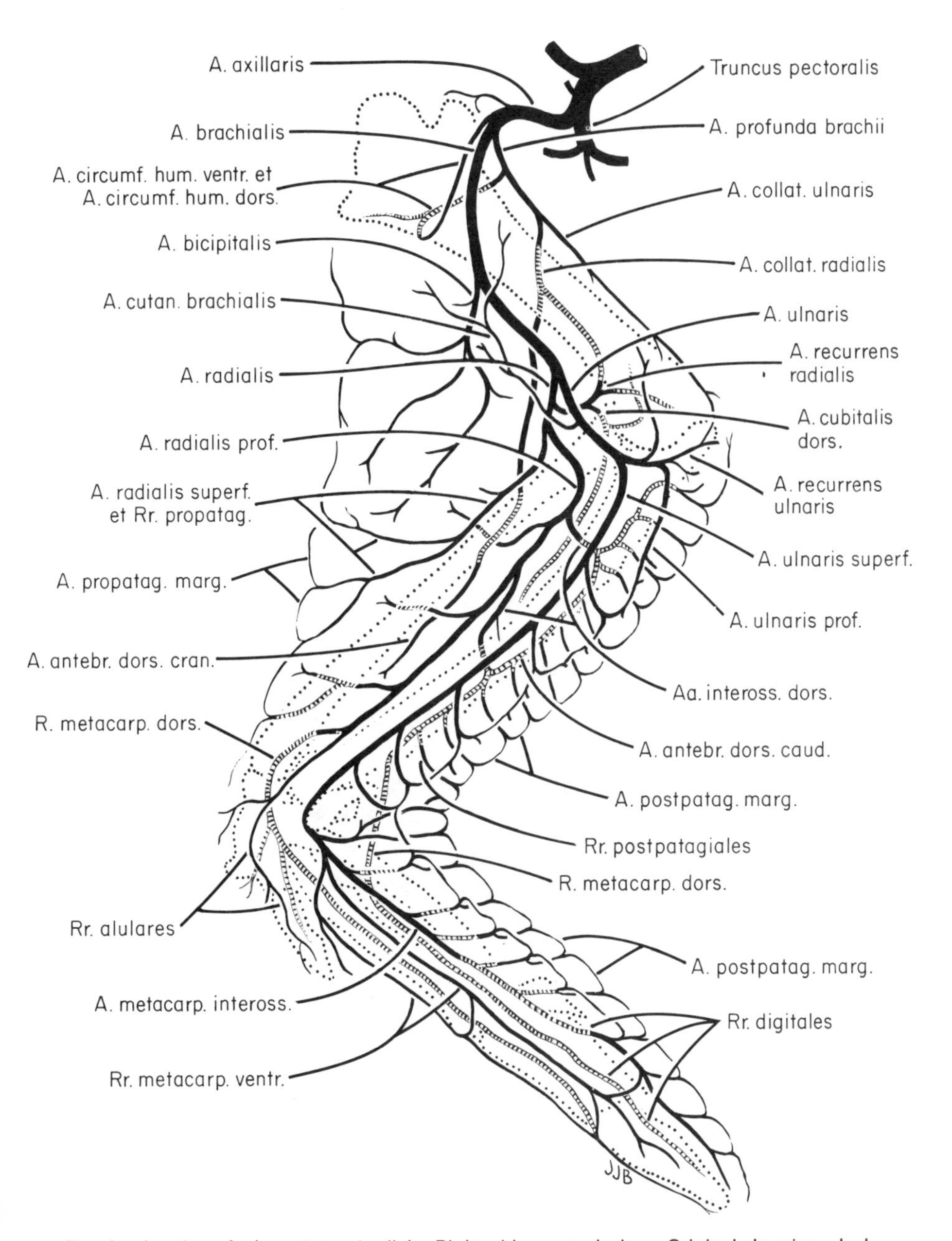

Fig. 4 Arteries of wing; *Columba livia*. Right side, ventral view. Original drawing, J. J. Baumel. Broken lines represent vessels dorsal to the skeleton.

Abbreviations: antebr., antebrachialis; circumf., circumflexa; collat., collateralis; cutan., cutanea; hum., humeri; inteross., interossea; marg., marginalis; metacarp., metacarpalis; postpatag., postpatagialis(es); propatag., propatagialis(es).

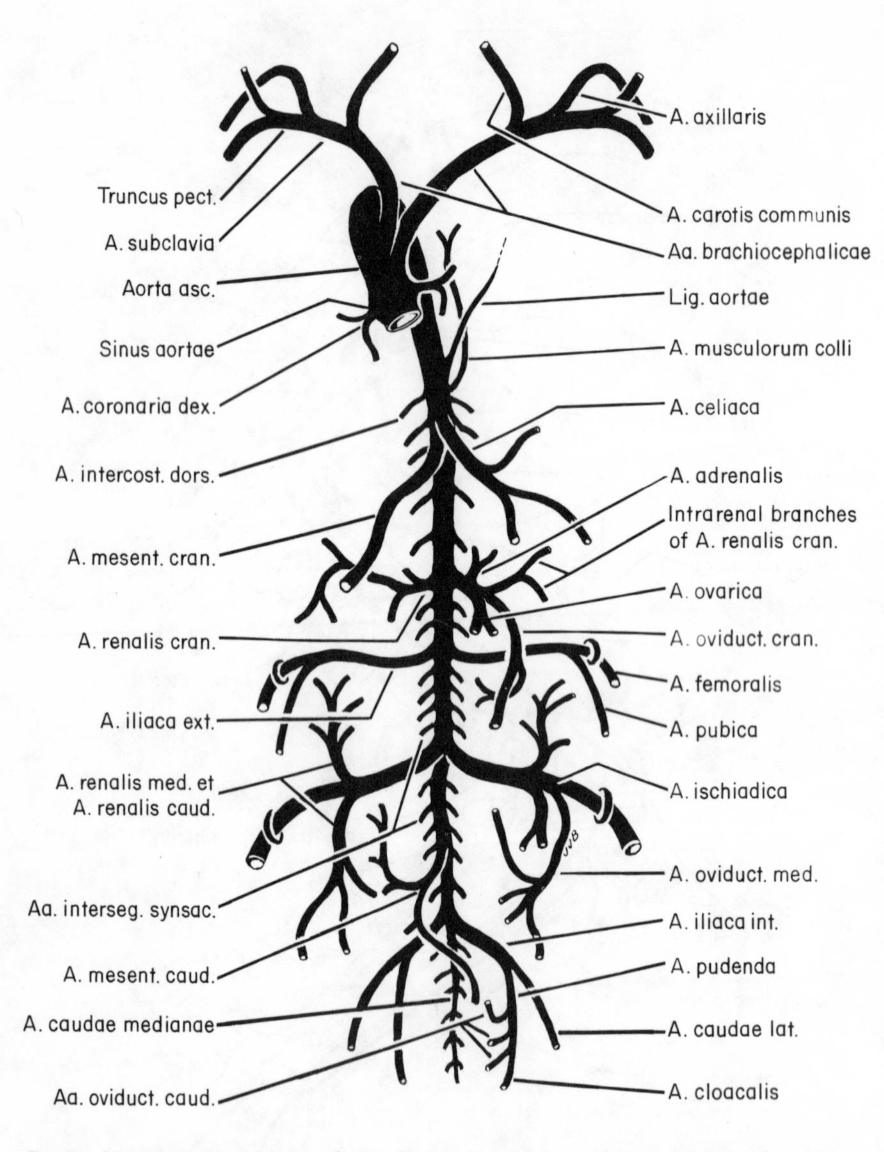

Fig. 5 Main branches of Aorta; *Gallus,* female. Ventral view. Note on left side of specimen the prominent ovarian and oviductal arteries springing from A. renalis cranialis, A. ischiadica, and A. iliaca interna (see **Art.** Annot. 66, 67, 72 for details).

Abbreviations: dex., dextra; intercost., intercostalis; interseg., intersegmentales; mesent., mesenterica; oviduct., oviductalis(es).

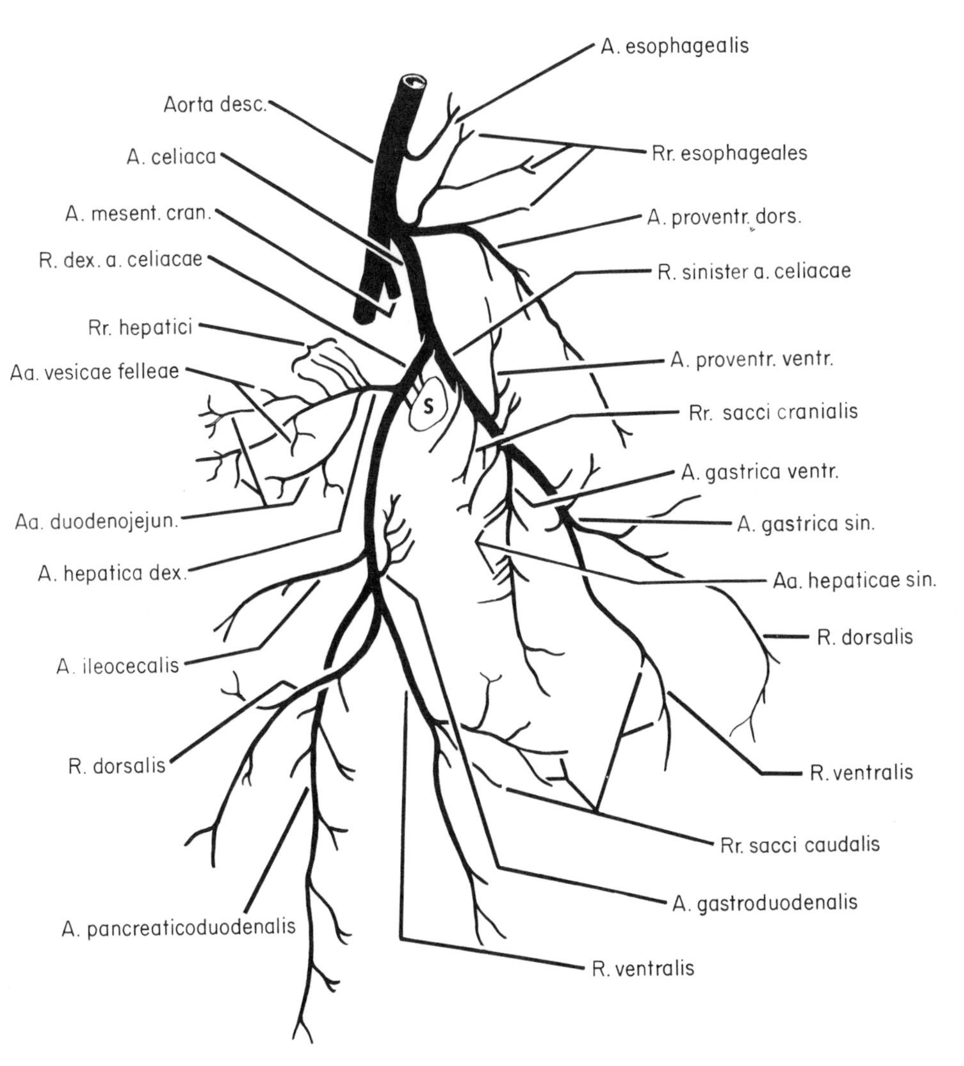

Fig. 6 Pattern of branching of A. celiaca in the goose, *Anser anser*. Ventral view. Redrawn from Malinovský and Visnanska (1975). The A. celiaca supplies the glandular and muscular stomachs, liver, pancreas, spleen and small intestine (see **Art.** Annot. 54–64).

Abbreviations: dex., dextra(er); duodenojejun., duodenojejunalis; proventr., proventricularis; sin., sinistra(e).

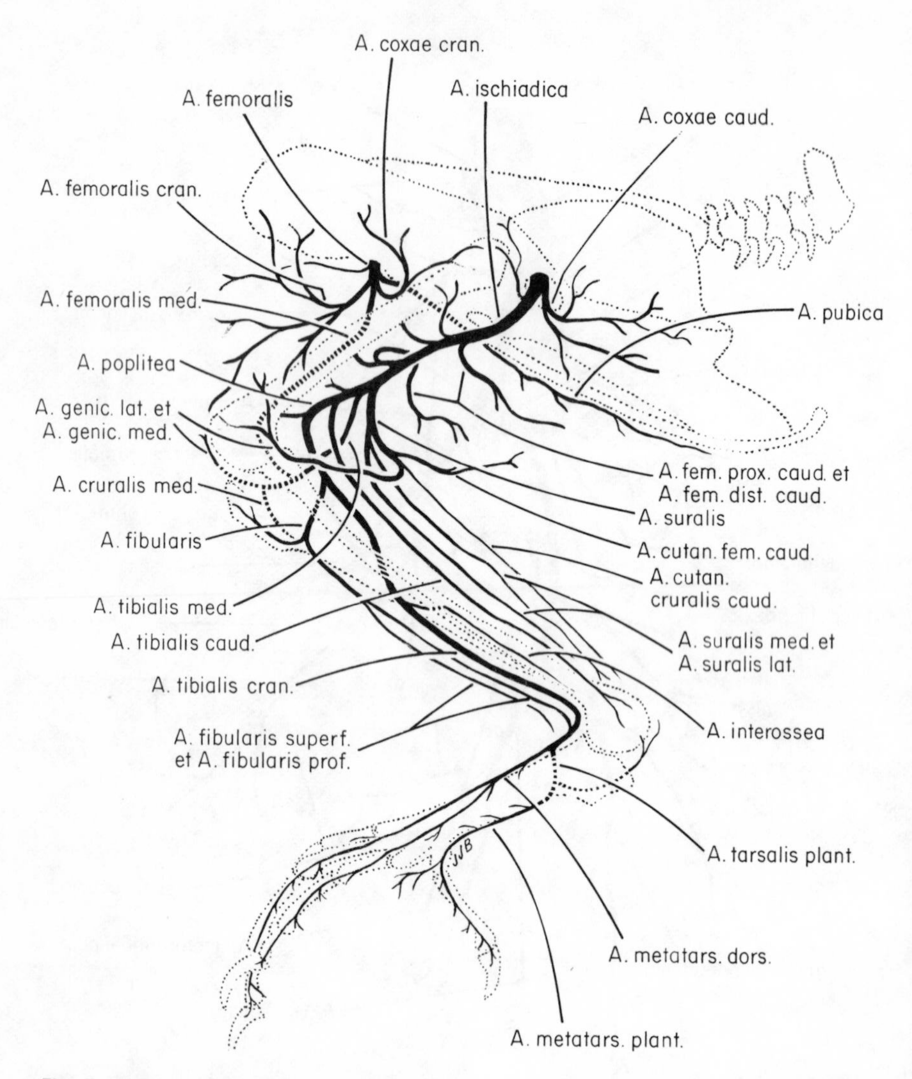

Fig. 7 Arteries of pelvic limb; *Columba livia*. Left side, lateral view. Original drawing, J. J. Baumel.

Abbreviations: cutan., cutanea; dist. caud., distocaudalis; fem., femoralis; genic., genicularis; metatars., metatarsalis; plant., plantaris; prof., profunda; prox. caud., proximocaudalis; superf., superficialis.

VENAE

VENAE PULMONALES[1]

V. pulmonalis communis
V. pulmonalis
    Rdx. cranialis[1]
    Rdx. caudomedialis
    Rdx. caudolateralis

        Vv. interparabronchiales[2]
            Vv. intraparabronchiales
                Vv. atriales
                    Venulae septales
                    Venulae intraparabronchiales

VENAE CARDIACAE[3]

(**Cor**, Fig. 1)

V. cardiaca sinistra[4]
    Pars interventricularis[4]
    Pars basilaris[4]
    Vv. conales (see **Cor.** Annot. 2)
    Vv. septales[5]
    Vv. atriales
    Vv. ventriculares
V. cardiaca circumflexa sinistra[6]
    Vv. atriales
    Vv. ventriculares
V. cardiaca dorsalis[7]
    Vv. apicis
    Vv. septales[5]
    Vv. ventriculares
Vv. cardiacae ventrales[8]
    Vv. conales (see **Cor.** Annot. 2)
    Vv. atriales
    Vv. ventriculares
Vv. cardiacae minimae[9]

VENA CAVA CRANIALIS[10]

(Fig. 3)

Vena cephalica rostralis[12] (Fig. 1)

- V. maxillaris[13] (Fig. 1)
  - V. palatina lateralis[14]
    - V. nasalis ventralis[15]
  - V. nasalis caudalis
  - V. nasalis dorsalis[15]
  - V. mandibularis interna
  - V. mandibularis externa
  - Anastomosis cum v. faciali
  - Anastomosis cum rete mirabile ophthalmico
- V. ophthalmica (Fig. 1)
  - V. ethmoidalis
    - V. nasalis medialis[15]
    - V. nasalis lateralis[15]
    - Vv. frontales[16]
      - V. frontalis superficialis
      - V. frontalis profunda
  - V. palpebralis dorsorostralis[23]
  - V. anastomotica cum sinu olfactorio[31]
  - V. ciliaris dorsalis (see rdxx. of V. facialis)
    - Vv. choroideae
- V. ophthalmotemporalis[39] (Fig. 1)
  - V. supraorbitalis[39]
  - V. infraorbitalis
  - V. ophthalmica interna
  - R. anastomoticus cum rete mirabile ophthalmico
  - Vv. musculares bulbi oculi
  - Vv. ciliares ventrales (see rdxx. of V. facilalis)
  - Vv. choroideae
  - V. pectinis oculi
  - Vv. glandulae membranae nictitantis
  - Anastomosis cum sinu cavernoso[18]
- Vv. comitantes a. maxillaris[13]
  - V. palatina medialis[14]
  - V. palatina mediana[19]
- V. submandibularis[20]
  - V. submandibularis superficialis
  - V. submandibularis profunda
- Rete venosum pterygopharyngeale
  - V. pterygopharyngealis

- V. facialis[21] (Fig. 1)
  - V. nasalis dorsalis[15]
  - V. nasalis lateralis[15]
  - Vv. frontales[16]
  - V. intermandibularis
  - Vv. palpebrales dorsorostrales[23]
    - Vv. ciliares anteriores
    - Sinus venosus sclerae[17] (see **Org. sens.** Bulbus oculi)
  - Anastomosis cum v. maxillari (see Annot. 13)
  - Vv. mandibulares caudales musculares et cutaneae
  - V. palpebralis communis (22)
    - V. palpebralis ventralis[23]
      - Vv. ciliares anteriores (see rdxx. of V. ophthalmica)
        - Sinus venosus sclerae (see **Org. sens.** Bulbus oculi)
    - V. palpebralis dorsocaudalis[23]
      - Vv. ciliares anteriores (see rdxx. of V. ophthalmica)
        - Sinus venosus sclerae[17] (see **Org. sens.** Bulbus oculi)
  - V. intramandibularis[24]
  - Vv. temporales
  - Anastomosis cum v. ophthalmica externa
- Rete venosum quadratopterygoideum
- V. mandibularis[25] (Fig. 1)
  - V. lingualis
    - V. lingualis propria
    - V. sublingualis
    - V. laryngea
    - V. pharyngealis
  - V. esophagotrachealis
    - V. esophagealis descendens
    - V. trachealis descendens
    - Vv. hyobranchiales [hyoideae]
- Rete palatopharyngeale[26]
- Anastomosis interjugularis[27] (Figs. 1, 3)
  - Vv. pharyngeales dorsales
  - Vv. musculares ventrales colli
  - V. occipitalis ventromediana[28]

Vena cephalica caudalis[29] (Figs. 1, 3)

- Sinus durae matris[30] (Figs. 1, 2)
  - Sinus olfactorius
    - V. cerebralis dorsorostralis

Vena cephalica caudalis—*continued*

Sinus sagittalis olfactorius
V. diencephali rostralis
Sinus sagittalis dorsalis[31] (Fig. 2)
- V. emissaria mediana
- V. cerebralis dorsomedialis
- V. cerebralis dorsocaudalis
- V. cerebralis interna
  - V. choroidea ventriculi lateralis (see below, Sinus cavernosus)
  - V. diencephali dorsalis
- V. cerebralis magna
  - V. cerebralis caudalis media
  - V. cerebralis caudalis

Sinus transversus
- V. diencephali lateralis

Sinus sphenotemporalis
- V. cerebralis media
- V. cerebralis rostralis
- V. corporis striati lateralis
- V. tecti mesencephali lateralis

Sinus petrosus rostralis
- V. intercruralis
- V. tecti mesencephali rostralis
- V. tecti mesencephali caudalis
- V. tecti mesencephali ventralis
- V. nervi trigemini
- V. cerebralis basilaris
- V. myelencephali basilaris
- V. cruris cerebellaris rostralis
  - V. cerebellaris rostromediana
  - V. myelencephali rostrodorsalis
  - V. cerebellaris rostroventralis

Sinus cavernosus[18]
- V. carotis cerebralis
- Vv. adenohypophysiales (see **Endoc.** Vasa sanguinea hypophysis)
- V. choroidea ventriculi tertii (see above, V. cerebralis interna)

Sinus petrosus caudalis[32]
- V. tecti mesencephali dorsalis
- V. semicircularis rostralis[32]

Sinus occipitalis (Fig. 2)
- Vv. meningeales
- Vv. cerebelli dorsales
  - V. labyrinthi dorsalis[33]

V. occipitalis dorsomediana[34]
Sinus foraminis magni [occipitalis]
Sinus fossae auriculae cerebelli[35]
V. choroidea ventriculi quarti
V. nervi glossopharyngealis
Vv. myelencephali dorsales
V. cerebellaris lateralis
V. semicircularis horizontalis[36]
V. occipitalis interna (see Annot. 28)
V. occipitalis externa[32]
V. occipitalis communis[37]
V. vertebralis ascendens[38]
V. occipitocollica[38]
V. carotis cerebralis
V. sphenoidea (see **Art.** Annot. 18)
V. palatina
V. sphenomaxillaris
V. labyrinthi ventralis[33]
V. ophthalmica externa[39] (see **Art.** Annot. 24)
V. cutanea cervicalis descendens
Rete mirabile ophthalmicum[39] (see **Art.** Annot. 24)
Anastomosis cum v. faciali[21]
V. occipitalis profunda[41]
V. auricularis rostralis[40]
V. auricularis caudalis[40]
Anastomosis cum v. faciali[21]

VENA JUGULARIS[11,27]

(Fig. 1, 3)

V. occipitalis superficialis[41]
V. cutanea cervicalis descendens (see **Cardvas.** Intro.)
V. mandibularis[25]
Vv. musculares craniales colli
Vv. cutaneae colli
V. trachealis descendens
V. esophagealis descendens
Vv. ingluviales
Rdxx. esophageales
Vv. tracheales
Vv. esophageales
V. cutanea cervicalis ascendens (see **Cardvas.** Intro.)
V. transversa colli

VENA JUGULARIS—*continued*

V. suprascapularis[42]
Vv. thyroideae
V. vertebralis[38, 43]
    V. vertebralis ascendens[43]
    V. vertebralis descendens[43]
        Vv. intercostales dorsales[44]
V. esophagotracheobronchialis[45] (see rdxx. of V. mandibularis)
V. trachealis ascendens
V. esophagealis ascendens
Vv. musculares caudales colli

SINUS VENOSUS VERTEBRALIS INTERNUS[46, 82]

(Figs. 1, 3)

Vv. intersegmentales cervicales et truncales[82]
    Vv. intercostales dorsales
Anastomoses cum vv. jugularibus[43]
Anastomosis cum v. portali renali craniali[46] (see **Urin.** Fig. 5)
Anastomosis cum v. portali renali caudali[46]
Anastomosis cum sinu foraminis magni (see Sinus durae matris)
Anastomosis cum v. occipitali interna

VENA SUBCLAVIA

(Fig. 3)

Truncus pectoralis [V. thoracica externa][47] (Fig. 3)
    V. pectoralis cranialis
    V. pectoralis media[47]
    V. pectoralis caudalis
Vv. thoracicae internae[48, 44]
    Rdx. ventralis
    Rdx. dorsalis
    V. sternalis interna
    Vv. intercostales ventrales (see **Art.** Annot. 52, 53)
V. sternoclavicularis
    V. sternalis externa
    V. clavicularis
    V. coracoidea dorsalis
V. subscapularis
    V. supracoracoidea
V. axillaris (Figs. 3, 4)
    V. cutanea thoracoabdominalis[49]

V. infrascapularis
Vv. brachiales[50] (Fig. 4)
V. bicipitalis
V. circumflexa humeri ventralis
V. cutanea brachialis
V. profunda brachii
    V. circumflexa humeri dorsalis
    V. collateralis ulnaris
    V. collateralis radialis
V. basilica (Fig. 4)
    V. radialis (Fig. 4)
        V. radialis profunda
            Vv. interosseae dorsales
                V. antebrachialis dorsalis caudalis
                    Rdxx. postpatagiales
        V. radialis superficialis
        V. cutanea propatagialis
            V. antebrachialis dorsalis cranialis
        V. cubitalis dorsalis
    V. ulnaris (Fig. 4)
        V. cubitalis ventralis
        V. ulnaris superficialis[50]
            V. ulnaris recurrens
        V. ulnaris profunda[50]
            Rdxx. postpatagiales
            Anastomosis cum v. ulnari superficiali
            Vv. carpales dorsales
            Vv. carpales ventrales
            Vv. metacarpales dorsales
            Vv. metacarpales ventrales
            V.. metacarpalis interossea
            Vv. digitales dorsales
            Vv. digitales ventrales
    V. propatagialis marginalis
    V. postpatagialis marginalis
V. proventricularis cranialis[51] (see Fig. 6; and **Cor.** Annot. 16)

VENA CAVA CAUDALIS[52]

(Fig. 5)

V. hepatica dextra[53] (see Figs, 5, 6; and Vv. portales hepaticae)
Rdx. dorsocranialis
Rdx. dorsocaudalis

VENA CAVA CAUDALIS—*continued*

Rdx. ventralis
Vv. centrales[53]
Vv. hepaticae mediae[53]
V. hepatica sinistra[53] (Figs. 5, 6)
Rdx. dorsocranialis
Rdx. dorsocaudalis
Rdx. ventralis
V. umbilicalis
V. hepatica accessoria[54]
V. adrenalis (Fig. 5)
Vv. portales adrenales[55]
Vv. ovaricae[56] (see Fig. 5; and **Genit. fem.** Vasa sanguinea ovarica)
Vv. contortae
Vv. pedunculares
Vv. intramurales
Vv. testiculares[56]
Vv. ureterodeferentiales craniales[57]
Vv. uretericae

VENA ILIACA COMMUNIS

(Fig. 5)

Valva portalis renalis (see Fig. 5; and Annot. 59)
Vv. renales craniales[58]
Rdxx. renales efferentes[58, 59]
Vv. intralobulares[58] (see **Urin.** Vasa sanguinea intrarenalia)
V. oviductalis cranialis[61] (Fig. 5)
Vv. ovaricae[56]
Vv. infundibuli
Vv. magni
V. renalis caudalis[58] (Fig. 5)
Vv. intersegmentales synsacrales[60]
Rdxx. renales efferentes
Vv. intralobulares[58] (see **Urin.** Vasa sanguinea intrarenalia)
V. oviductalis media[61] (Fig. 5)
Vv. magni
Vv. isthmi
Vv. uterinae (see **Genit. fem.**)
Vv. ureterodeferentiales mediae
Vv. uretericae
V. portalis renalis cranialis (Fig. 5)
Anastomosis cum sinu venoso vertebrali interno (see **Urin.** Fig. 5)
Rr. renales afferentes[59]

Systema portale renale[59] (see **Urin.**)

Vena iliaca interna[62] (Fig. 5)

V. pudenda[63]
Vv. ureterodeferentiales caudales
Vv. uretericae
V. oviductalis caudalis[61] (see **Art.** Annot. 67, 72; and **Genit. fem.**)
Vv. uterinae (see **Genit. fem.**)
V. vaginalis
V. oviductalis marginalis ventralis[61]
V. oviductalis marginalis dorsalis[61]
V. bursocloacalis[63] (see V. mesenterica caudalis)
Vv. bursales
Vv. cloacales (see **Cloaca** Annot. 26)
V. caudae lateralis[64]
V. cutanea abdominalis[65, 49]
V. caudae medianae
Vv. intersegmentales caudales[60]
Vv. intersegmentales caudales[60]
Anastomosis interiliaca[59, 62] (see Fig. 5; and **Urin.** Annot. 19)
Anastomosis cum v. mesenterica caudali
V. portalis renalis caudalis[62] (see **Urin.** Annot. 19)
Anastomoses cum sinu venoso vertebrali interno[46]
Vv. intersegmentales caudales et synsacrales[60]
Rr. renales afferentes[59]

Vena ischiadica[66] (Figs. 5, 7)

V. coxae caudalis[67]
V. femoralis medialis
V. circumflexa femoris medialis (see **Art.** Annot. 71)
V. circumflexa femoris lateralis[68] (see **Art.** Annot. 71)
Anastomosis cum v. femorali
V. femoralis proximocaudalis[68]
V. femoralis distocaudalis
V. nutricia femoris proximalis
V. suralis[69] (Fig. 7)
V. cutanea femoralis caudalis
V. cutanea femoralis lateralis
V. cutanea cruralis caudalis
V. suralis medialis
V. suralis lateralis
V. poplitea[66] (Fig. 7)
V. genicularis lateralis

Vena ischiadica—*continued*

V. nutricia femoris distalis
V. tibialis medialis
V. genicularis medialis
V. cruralis medialis
V. tibialis caudalis
Vv. tarsales plantares (see **Art.** Annot. 79)
Vv. metatarsales plantares[70]
V. metatarsalis plantaris superficialis[70]
Arcus venosus plantaris
Vv. pulvinarum
Vv. digitales
V. fibularis [peronea]
V. tibialis cranialis[71]
Rete tibiotarsale[72]
V. fibularis [peronea] superficialis[73]
V. fibularis [peronea] profunda[73]
V. metatarsalis dorsalis communis[70]
Vv. metatarsales dorsales
Vv. digitales

Vena iliaca externa (see Fig. 5)

V. pubica[74]
V. femoralis (see V. ischiadica)
V. coxae cranialis[67]
V. femoralis cranialis
V. cutanea femoralis cranialis
Anastomosis cum v. ischiadica[66]

SYSTEMA PORTALE HEPATICUM[75]

(Fig. 6)

V. portalis hepatica dextra[76]
V. mesenterica communis[78] (Fig. 6)
V. mesenterica cranialis[78] (Fig. 6)
Vv. jejunales
Vv. ileae
V. ileocecalis

V. marginalis intestini tenuis (see **Art.** Annot. 64)

V. mesenterica caudalis [V. coccygomesenterica][78, 62] (Figs. 5, 6)

- Rdx. cranialis
  - Vv. ileoceales
- Rdx. caudalis
  - Vv. rectales
  - V. bursalis
  - Vv. cloacales
- Anastomosis cum anastomose interiliaca[78,62] (Fig. 5)

V. proventriculosplenica[79]
- V. proventricularis dorsalis[79]
  - V. gastrica dorsalis
- V. proventricularis dextra[79]
- Vv. splenicae[79]

V. gastropancreaticoduodenalis (Fig. 5)
- V. duodenojejunalis
- V. gastica dextra
  - Vv. sacci caudalis et cranialis
  - V. pylorica
- V. pancreaticoduodenalis (Fig. 5)
  - V. ileocecalis
  - Vv. duodenales
  - Vv. pancreaticae

Rr. intrahepatici v. portalis hepaticae (Fig. 5)
- Ramus dexter v. portalis[76]
  - R. cranialis
  - R. lateralis
  - R. caudalis
- Rami medii v. portalis[76] (Fig. 5)
- Ramus sinister v. portalis[80] (Fig. 5)
  - Pars transversa[77]
  - R. cranialis
  - R. lateralis
  - R. caudalis

V. portalis hepatica sinistra[80] (Fig. 5)
- Vv. proventriculares caudales[79] (see **Cor.** Annot. 16)
- V. proventricularis ventralis[79]
- V. gastrica sinistra
  - Vv. sacci caudalis et cranialis
- V. gastica ventralis
  - V. pylorica

Vv. portales hepaticae propriae[81]

VENAE MEDULLAE SPINALIS

Sinus venosus vertebralis internus[46, 82]
- Vv. intersegmentales[82]
  - Rdxx. ventrales
  - Rdxx. dorsales
    - Vv. vertebromedullares[83]
      - Vv. radiculares ventrales[88]
        - Vv. radiculares propriae
        - V. spinalis ventralis[84]
          - Vv. sulci[85]
          - Vv. sulcocommissurales
          - Vv. marginales[86]
        - Vv. ventrolaterales[87]
      - Vv. radiculares dorsales[88]
        - Vv. radiculares propriae
        - V. spinalis dorsalis[89]
        - Vv. dorsolaterales[90]
        - Vv. fissurae[91]

ANNOTATIONS

(1) **Vv. pulmonales.** The intrapulmonary tributaries of each V. pulmonalis correspond closely to the intrapulmonary rami of A. pulmonalis. The two caudal radices and their larger radicles lie generally ventral to corresponding rami of A. pulmonalis; the cranial radix is medial to its artery (e.g. *Gallus*, *Columba*). See **Cor.** Annot. 20.

(2) **Vv. interparabronchiales.** These veins are tributaries of each of the three main radices of V. pulmonalis. Consult Radu and Radu (1971) for patterns of intrapulmonary vessels in several avian species, and Abdalla and King (1975) for angioarchitectural detail in *Gallus*.

(3) **Vv. cardiacae.** For the most part cardiac veins are not satellites of coronary arteries and their rami; veins are commonly located subepicardially (exception: Vv. septales). See remarks regarding cardiac veins in **Cardvas.** Intro. See **Cor.** Fig. 1.

(4) **V. cardiaca sinistra.** Synonymy: V. cordis magna. The left cardiac vein is not usually the largest of Vv. cordis; therefore the synonym, V. cordis magna, in birds is a misnomer. Lindsay (1967) describes two parts of this vein in *Gallus*: Pars interventricularis and Pars basilaris. Pars interventricularis occasionally continues directly as V. cardiaca circumflexa sinistra (Lindsay, 1967). See **Cor.** Fig. 1.

(5) **Vv. septales.** In *Gallus* the cranial part of Septum interventriculare is sometimes drained by left and right Trunci septales that empty into V. cardiaca sinistra and Vv. cardiacae ventrales, respectively (Lindsay, 1967).

(6) **V. cardiaca circumflexa sinistra.** At times this vein is a tributary of V. cardiaca sinistra instead of an independent vein (Lindsay, 1967).

(7) **V. cardiaca dorsalis.** Synonymy: V. cordis media. The dorsal cardiac vein usually is the largest of the cardiac veins; its cranial segment (Lindsay, 1967) lies in Sulcus interventricularis subsinuosus [dorsalis].

(8) **Vv. cardiacae ventrales.** Synonymy: Vv. cordis minores. The system of ventral cardiac veins is located mostly in the subepicardial tissues of the ventral wall of Ventriculus dexter; the three main tributaries of the system span the coronary sulcus, and empty directly into Auricula dextra via separate ostia.

(9) **Vv. cardiacae minimae.** Synonymy: Vv. luminalia; Vv. thebesii. Vv. minimae drain from the myocardium directly into the heart chambers. The foramina of these veins are abundant in the right and left atria and right ventricle; in the left ventricle they are said to occur infrequently (Uchiyama, 1929).

(10) **V. cava cranialis.** V. subclavia, V. jugularis, and Truncus pectoralis all converge to form V. cava cranialis of each side; consequently no V. brachiocephalica is present in birds.

(11) **V. jugularis** (Figs. 1, 3). In most birds V. jugularis dextra is significantly larger in caliber than V. jugularis sinistra. In general, Anastomosis interjugularis connects the two Vv. cephalicae rostrales. In certain species it is not a transverse vessel, but is obliquely disposed. In these cases it is apparent that part of the venous blood from the left side of the head is shunted to V. jugularis dextra. Neugebauer (1845) discusses the anastomosis and asymmetry of jugular veins in different avian species. Wade (1876) notes that the left jugular is atrophied and nonfunctional in several passerine species. See Annot. 27, Anast. interjugularis.

(12) **V. cephalica rostralis.** Synonymy: V. cephalica anterior; V. facialis communis. See Annot. 13.

(13) **V. maxillaris.** Synonymy: V. facialis interna. V. maxillaris does not accompany A. maxillaris; the artery courses in the roof of the pharynx just deep to the mucosa, flanked by small paired or plexiform Vv. comitantes a. maxillaris. The prominent V. maxillaris courses in the floor of the orbit, turns ventrad between the caudal wall of the orbit and Os pterygoideum, and reaches the roof of the pharynx where its name changes to V. cephalica rostralis.

(14) **V. palatina medialis; V. palatina lateralis.** The medial palatine vein drains into V. maxillaris and Vv. comitantes a. maxillaris and Rete palatopharyngeum. V. palatina lateralis is a tributary of either V. maxillaris or V. facialis in different species.

(15) **Vv. nasales.** In different species may be tributaries of V. facialis, V. ethmoidalis, or both. In *Gallus* the medial and lateral nasal veins flow together, and form a common nasal vein that drains into V. ethmoidalis. In *Columba* V. nasalis lateralis flows into the system of V. facialis.

(16) **Vv. frontales.** May flow into the system of V. facialis or that of V. ethmoidalis.

(17) **Sinus venosus sclerae.** Synonymy: Canal of Schlemm. This sinus is drained by the several Vv. ciliares anteriores. See **Org. sens.** Annot. 7, 8.

(18) **Sinus cavernosus.** Synonymy: Sinus anuli basilaris, Kaku (1959); Sinus

perihypophysialis, Hasegawa (1956). Communicates with V. ophthalmotemporalis, V. ophthalmica interna (Wingstrand, 1965), and with V. carotica cerebralis. See also Vitums *et al.* (1964).

(19) **V. palatina mediana.** This unpaired vein splits and forms the paired Vv. palatinae mediales at the rostral end of the nasal choana.

(20) **V. submandibularis.** In some species this vein is a tributary of V. mandibularis instead of a tributary of V. cephalica rostralis.

(21) **V. facialis.** Common synonymy: V. facialis externa. For the most part situated subcutaneously; its terminal part courses mesad just rostral to the external acoustic meatus, and empties into V. cephalica rostralis, the terminal part of V. cephalica caudalis, or into V. mandibularis.

(22) **V. palpebralis communis.** V. palpebralis dorsocaudalis joins V. palpebralis ventralis to form the common palpebral vein in some birds (Neugebauer, 1845).

(23) **V. palpebralis dorsorostralis.** May drain into the ethmoid system of veins as in *Gallus* or into the system of V. facialis (e.g. *Columba*). Vv. palpebrales receive ciliary radicles from the nictitating membrane, bulbar conjunctiva, and the iridial region of Bulbus oculi itself.

(24) **V. intramandibularis** runs in the mandibular canal with A. intramandibularis and the prolongation of N. mandibularis. See **Art.** Annot. 25.

(25) **V. mandibularis.** Synonymy: V. lingualis. Main vein from the region of the lower jaw. V. mandibularis empties into V. cephalica rostralis near Anastomosis interjugularis (e.g. *Gallus*), further caudally as in *Columba* and *Milvus milvus* (Neugebauer, 1845) near the junction of the rostral and caudal cephalic veins, or into V. jugularis itself.

(26) **Rete palatopharyngeale.** Synonymy: Rete mirabile basilaris (Neugebauer, 1845).

(27) **Anastomosis interjugularis.** Proposed as a substitute name for Anastomosis venarum cephalicarum anteriorum (Neugebauer, 1845); Ramus communicans pharyngicus (Bodrossy, 1938); and V. tranversa (Matsumoto, 1955). The anastomosis interjugularis actually joins the two rostral cephalic veins rather than the jugular veins themselves (see Annot. 11).

(28) **V. occipitalis ventromediana** (Richards, 1968). Considered a more descriptive term than R. anticus annuli venosi occipitalis (Neugebauer, 1845). The ventromedian occipital vein extends dorsally from the interjugular anastomosis into the median fissure between right and left columns of the ventral neck musculature. It bifurcates into right and left limbs that anastomose with the cranial ends of the vertebral veins and the internal occipital veins as the latter emerge from the atlanto-occipital interval. See Fig. 1.

(29) **V. cephalica caudalis.** Drains blood mostly from the cranial cavity, inner ear region, and suboccipital region (Neugebauer, 1845; Matsumoto, 1955; Richards, 1968). The cranial dural venous sinuses receive blood from the veins of the brain. Most of the sinuses send blood dorsocaudally to sinuses in the occipital region; the

latter flow into occipital veins and the internal vertebral sinus. V. cephalica caudalis is not well developed in *Gallus* and *Meleagris*; in these forms it appears to be partly replaced by the large V. occipitalis ventromediana (see Annot. 28).

(30) **Sinus durae matris.** Sinuses and encephalic veins named mostly according to Kaku (1959); for synonyms, see Neugebauer (1945) and Matsumoto (1955). Only the principal tributaries of the dural sinuses are listed; minor named radicles are not included. (see Fig. 2).

(31) **Sinus sagittalis dorsalis.** Synonymy: Sinus longitudinalis. In some species flows predominantly into Sinus occipitalis; in others it flows rostrad into Sinus olfactorius then via ethmoid veins into the ophthalmic veins.

(32) **Sinus petrosus caudalis.** Prolonged as V. semicircularis rostralis; the latter courses in its osseous canal along the rostral semicircular canal and emerges from the occipital region of the skull via a foramen lateral to Foramen magnum where its name changes to V. occipitalis externa.
V. semicircularis rostralis is proposed as a new term to substitute for Sinus semicircularis or Sinus petrosus posterior (Neugebauer, 1845). For details of relationships of veins that parallel the semicircular canals see Neugebauer (1845) and Ewald (1892). See **Org. sens.** Labyrinthus osseous.

(33) **Vv. labyrinthi.** Named according to Schmidt (1964). See Sinus occipitalis; V. cerebelli dorsalis.

(34) **V. occipitalis dorsomediana.** Drains into Sinus occipitalis by entering the skull through a single occipital fonticulus as in *Columba livia.* The veins are paired and situated near one another on each side of the median plane in certain birds (e.g. *Ceryle alcyon* and *Progne subis*). See **Osteo.** Annot. 87.

(35) **Sinus fossae auriculae cerebelli.** The term Sinus fossae auriculae cerebelli is a more meaningful term than Sinus foveae hemispherii cerebelli (Neugebauer, 1845; Matsumoto, 1955; Kaku, 1959. See **Osteo.** Annot. 38.

(36) **V. semicircularis horizontalis.** Synonymy: V. auris interna (Neugebauer, 1845). This vein courses in an osseous tube attached to the osseous horizontal semicircular canal, and joins V. ophthalmica externa dorsal to the external acoustic meatus.

(37) **V. occipitalis communis.** A substitute term for V. occipitalis lateralis (Neugebauer, 1845); this vessel is formed by the Vv. occipitalis interna et externa. The confluence of these veins may reasonably be considered as the commencement of V. cephalica caudalis. According to Romanoff (1960), V. occipitalis lateralis is believed to represent the true (primitive) jugular vein of embryonic development in birds.

(38) **V. vertebralis; V. occipitocollica.** The cranialmost ends of both these veins anastomose with V. occipitalis communis. V. occipitocollica receives muscular and cutaneous radicles; it communicates with V. cephalica caudalis, V. vertebralis ascendens, V. jugularis, and Sinus venosus vertebralis internus. See Neugebauer (1845). See Fig. 3.

(39) **V. ophthalmica externa.** Synonymy: V. stapedia. V. ophthalmotemporalis as well as V. supraorbitalis and V. infraorbitalis could also be treated as tributaries of V. ophthalmica externa (see V. ophthalmica). See Fig. 1.

Rete mirabile ophthalmicum. Synonymy: Rete mirabile temporale. Arterial and venous external ophthalmic retia are enmeshed with one another.

(40) **Vv. auriculares.** May be tributaries of V. facialis as in *Gallus*.

(41) **V. occipitalis profunda; V. occipitalis superficialis.** Both of these veins are companion veins of the same named arteries (see A. carotis externa).

(42) **V. suprascapularis.** Synonymy: V. cephalica humeri (Bodrossy, 1938); Szabó (1958). According to Szabó this is a particularly strong vein in *Melopsittacus*.

(43) **Vv. vertebrales ascendentes et descendentes.** These flow into V. jugularis by way of enlarged transverse anastomosing veins that also carry blood to V. jugularis from Sinus venosus vertebralis internus. See Annot. 38.

(44) **Vv. intercostales dorsales.** Two or three of the cranialmost Vv. intercostales dorsales have connections with V. vertebralis descendens and with Sinus venosus vertebralis internus. The more caudal series of dorsal intercostal veins drain into Sinus venosus vertebralis internus. See V. thoracica interna; Annot. 48.

(45) **V. esophagotracheobronchalis.** See **Art.** Annot. 8 for a discussion of the corresponding artery.

(46) **Sinus venosus vertebralis internus.** Synonymy: Sinus columnae vertebralis, (Szabó, 1958). The principal communications of this sinus are with V. jugularis at the root of the neck and with Sinus foraminis occipitalis at the Foramen magnum. Sinus venosus vertebralis internus *receives* intersegmental veins via the intervertebral foramina; it also communicates with Vv. vertebrales, Vv. portalis renalis cranialis et caudalis, V. renalis caudalis, and V. caudalis mediana. Mouchett and Cuypers (1959) have described numerous anastomoses connecting the caudal renal portal veins with the internal vertebral sinus in *Gallus*. See Fig. 3; and **Urin**. Fig. 5.

(47) **V. pectoralis communis.** Synonymy: V. thoracica externa. Lacking in some species of birds; the individual Vv. pectorales are direct tributaries of V. subclavia (*Melopsittacus*, Szabó, 1958). V. pectoralis media is not present in all species.

(48) **Vv. thoracicae internae.** Flow into V. axillaris, Vena cava cranialis, or both.

(49) **V. cutanea thoracoabdominalis** (Neugebauer, 1845). Drains the integument of the ventrolateral breast and abdomen, particularly the incubation (brood) patch where it communicates with radicles of V. cutanea abdominalis. Empties into V. axillaris rather than into pectoral veins. See Annot. 65; and **Art.** Annot. 44.

(50) **Vv. brachiales.** Paired (or plexiform) satellite veins of A. brachialis that anastomose with each other and with radial and ulnar veins in the cubital region. See Fig. 4.

V. ulnaris superficialis frequently consists of paired venae comitantes that flank the strong, subcutaneous A. ulnaris superficialis. **V. ulnaris profunda** is the strongest of the veins draining the manus, carpus, and antebrachium; it follows the weak A. ulnaris profunda. The artery and vein are located deep to M. flexor carpi ulnaris. See Fig. 4.

(51) **V. proventricularis cranialis.** Nishida and Mochizuki (1976) compare the cranial proventricular vein of *Anas* with that of *Gallus* and *Columba* ((Malinovský, 1965). Blood from the cranial segment of the proventriculus drains directly toward the heart rather than into the hepatic portal system and discharges into right atrium or left cranial vena cava. See Fig. 6.

(52) **Vena cava caudalis.** The short trunk of this vein does not parallel the vertebral column, but passes in a cranioventral direction to reach the right atrium of the heart; consequently the caudal vena cava does not receive dorsal intercostal or other intersegmental body wall veins. See **Ven.** Annot. 48, 58, 60.

(53) **Vv. hepaticae.** Intrahepatic radices of Vv. hepaticae have been named on the basis of dissections in *Gallus*. Miyaki (1973) designates the main radices of the left hepatic lobe, but not those of the right lobe. In *Anser* the larger radices of left and right hepatic veins are more numerous than in *Gallus*; they were not named by Pavaux and Jolly (1968). Vv. hepaticae mediae drain blood from Pars interlobaris of the liver; observed in *Anser*, *Gallus* and *Columba* (Pavaux and Jolly, 1968). **Vv. centrales** are tributaries of all the intrahepatic radices. See Fig. 6.

(54) **V. hepatica accessoria.** Miyaki (1973) describes one or more accessory hepatic veins that drain the cranial part of the right lobe of the liver directly into the caudal vena cava (*Gallus*).

(55) **Vv. portales adrenales.** Synonymy: Vv. adrenales afferentes. Body wall veins that drain into the veins of the adrenal gland; therefore, an adrenal portal system exists in birds (Neugebauer, 1845; Szabó, 1958; Goodchild, 1969).

(56) **Vv. ovaricae; Vv. testiculares.** May form common stems with the adrenal vein. Multiple left ovarian veins may exist; one or more ovarian veins may drain into V. oviductalis cranialis (*Gallus*); small right ovarian veins return venous blood from the vestigial right ovary.

(57) **Vv. ureterodeferentiales.** Term introduced by Nishida (1964).

(58) **Vv. renales.** Vv. renales craniales drain the cranial division of the kidney; V. renalis caudalis drains the middle and caudal divisions of the kidney. Several Vv. intralobulares of the kidney may become confluent to form one Rdx. renalis efferens or a single V. intralobularis may be a direct tributary of the proper V. renalis (Johnson *et al.*, 1972). See Fig. 5; and **Urin.** Vasa sanguinea intrarenalia.

(59) **Systema portale renale.** The presence of a functional renal portal system is a basic feature of the avian kidney. Physiologically it is of great importance in the excretion of uric acid by tubular secretion (see Johnson, 1978). Anatomically the principal veins of the system form a venous ring named the Circulus venosus portalis renalis by Kurihara and Yasuda (1975b). The components of this ring are: (1) cranial and caudal renal portal veins; (2) anastomoses of cranial renal portal veins with the internal vertebral venous sinus; and (3) the interiliac anastomosis (see **Urin.** Fig. 5). According to Akester (1967) blood from V. iliaca interna and V. mesenterica caudalis may flow craniad in V. portalis renalis caudalis directly into V. iliaca communis, thence into Vena cava caudalis. Alternatively, venous blood from the pelvic limb may be diverted from V. iliaca externa by Valva portalis renalis into V. portalis renalis

caudalis, then into V. mesenterica caudalis that leads to the hepatic portal system. Blood diverted by the renal portal vein also may flow craniad in the V. portalis renalis cranialis that communicates with Sinus venosus vertebralis internus. See Fig. 5. **Rr. renales afferentes; Rdxx. renales efferentes.** Rr. renales afferentes of Vv. portales renales may conduct venous blood into the kidney parenchyma, and mix there with arterial blood in the peritubular network of capillary sinuses (Siller, 1971); this mixed blood then drains into Rdxx. renales efferentes that empty into Vv. renales that lead into the system of V. cava caudalis. See **Urin.** Vasa sanguinea intrarenalia and Annot. 19.

(60) **Vv. intersegmentales.** In the caudal region drain on each side into V. caudae medianae and V. iliaca interna up to the level of Anastomosis interiliaca. Cranial to this level Vv. intersegmentales synsacrales empty into the trunk of V. renalis caudalis that is located just lateral to the vertebral column (*Gallus*).

(61) **Vv. oviductales marginales ventralis et dorsalis.** Consist of longitudinal anastomoses between ascending and descending radices of the cranial, middle, and caudal oviductal veins. The marginal veins of the oviduct course in the ventral and dorsal ligaments of the oviduct with companion arteries. See **Art.** Annot. 67; and **Genit. fem.** Annot. 35–44.

(62) **V. iliaca interna; V. portae renalis caudalis.** The embryonic V. iliaca interna (Miller, 1903; Hamilton, 1952; Romanoff, 1960) is derived from the posterior cardinal vein. In the adult bird the caudal renal portal vein is defined as the segment of V. iliaca interna cranial to **Anastomosis interiliaca** (postcardinal anast.). V. iliaca interna (Neugebauer, 1845) is also known in the literature as V. hypogastrica and V. pudenda communis. See Fig. 5.

(63) **V. pudenda** (Neugebauer, 1845). The drainage territory of this vein from terminal parts of the urogenital and gastrointestinal tracts corresponds closely to that of the mammalian V. pudenda interna. The term, **V. bursocloacalis,** is taken from the study of Pintea *et al.*, 1967). Most of venous drainage from cloacal bursa in *Columba* is via V. bursalis of V. mesenterica caudalis.

(64) **V. caudae lateralis.** A substitute name for V. musculocutanea caudae lateralis (Neugebauer, 1845), a term descriptive of its distribution. This vein (and its companion artery) pierces the iliocaudal membrane at the caudal edge of the bony pelvis, extends dorsocaudad, drains blood from the tail bulb, its associated muscles, and integument of the dorsum of the tail region. V. caudae lateralis does not serve to any appreciable extent the region of the vent; therefore, the term, V. pudenda externa, by which it is also called is inappropriate.

(65) **V. cutanea abdominalis.** (Bodrossy, 1938). Empties into V. caudae lateralis; drains incubation patch area of the abdominal wall in *Columba*; in *Gallus* and *Meleagris* V. cutanea abdominalis drains into V. femoralis proximocaudalis, a tributary of V. ischiadica.

(66) **V. ischiadica.** Major vein of the pelvic limb. Caudal to the knee its name changes to **V. poplitea.** Near the hip joint most of the blood from V. ischiadica is diverted to V. femoralis via Anastomosis cum v. femorali. V. ischiadica is diminished in size proximal to the anastomosis. V. ischiadica continues into the pelvis via the ilio-ischiadic foramen with A. ischiadica and terminates in V. portalis renalis caudalis. See Figs. 5, 7.

(67) **Vv. coxae.** These names are substitutes for Vv. gluteae cranialis et caudalis. See **Art.** Annot. 70.

(68) **V. femoralis proximocaudalis.** A substitute term for V. profunda femoris (see also **Ven.** Annot. 65). See **Art.** Annot. 71, for remarks relating to V. circumflexa femoris lateralis.

(69) **V. suralis.** This name is used by Neugebauer (1845) and Bodrossy (1938). Discussion of use of the term, "sural", is found in **Art.** Annot. 74. See Fig. 7.

(70) **Vv. metatarsales.** Terminology of the vessels of the foot is from the detailed, thorough study of Volmerhaus and Hegner (1963). The strong V. metatarsalis plantaris superficialis is the V. metatarsea interna s. magna of Neugebauer (1845).

(71) **V. tibialis cranialis.** Does not accompany A. tibialis cranialis through the distal tibio-fibular interosseous foramen, but runs with A. fibularis through the proximal interosseous foramen. A smaller collateral vein accompanies A. tibialis cranialis into the extensor compartment of the crus (*Columba*).

(72) **Rete tibiotarsale.** A network of anastomosing veins that parallels the cranial tibial artery, and continues as a loose plexus of veins over the tarsal and proximal metatarsal regions. See **Art.** Annot. 77.

(73) **Vv. fibulares [peroneae] superficialis** et **profunda.** With their companion arteries these veins accompany superficial and deep rami of the fibular nerve.

(74) **V. pubica.** New term for the vein that courses within the attachments of the abdominal muscles along the ventral border of the pubis; corresponds to deep circumflex iliac vein of man. For synonymy see **Art.** Annot. 69. See Fig. 5.

(75) **V. portalis hepatica.** Inasmuch as birds also possess renal and adrenal portal systems, the designation "hepatica" must be employed.

(76) **V. portalis hepatica dextra.** The short trunk of this vein divides into right and left rami that are distributed to the respective lobes of the liver (Malinovský, 1965; Miyaki, 1973). Each ramus gives off three principal branches to its lobe. See Miyaki (1973) for a different set of names for the principal intrahepatic branches of R. dexter and R. sinister of V. portalis hepatica dextra in *Gallus*. The terms used in this list fit well with the vessels depicted in Miyaki's illustrations. See Fig. 6.

(77) **Pars transversa** of Ramus sinister of the right hepatic portal vein yields numerous smaller dorsocranial rami and one or more substantial Rr. medii venae portalis into Pars interlobaris of the liver (Pavaux and Jolly, 1968; Miyaki, 1973).

(78) **Vv. mesentericae.** The synonym, V. coccygomesenterica, is firmly established in the literature and is retained as an official alternative term for V. mesenterica caudalis. The radices of this vein accompany the rami of A. mesenterica caudalis, and require the corresponding names. At its cranial end V. mesenterica caudalis joins V. mesenterica cranialis forming V. mesenterica communis. Near the junction of its cranial and caudal radices V. mesenterica caudalis is confluent with Anastomosis interiliaca. See Fig. 6.

(79) **Vv. proventriculares.** These veins join Vv. splenicae in *Buteo buteo* and *Gallus* (Malinovský, 1965), and produce a substantial tributary of V. portalis hepatica dextra; in other forms (e.g. *Columba*) proper Vv. splenicae drain directly into V. portalis hepatica dextra. See Annot. 51.

(80) **V. portalis hepatica sinistra.** A small, distinct V. portalis hepatica sinistra is formed by confluence of V. proventricularis caudalis, V. proventricularis ventralis, V. gastrica ventralis, and V. gastrica sinistra in *Buteo buteo* and *Gallus* (Malinovský, (1965). In other species (*Columba, livia Sutrnus vulgaris* and *Melopsittacus undulatus*, the gastric and proventricular veins empty individually into rami of Ramus sinister of the right hepatic portal vein. See Fig. 6.

(81) **Vv. portales hepaticae propriae** enter the left lobe of the liver independent of V. portae hepatica sinistra (Pavaux and Jolly, 1968; Yasuda, 1973).

(82) **Vv. intersegmentales.** Vv. intersegmentales of neck, trunk, and tail drain to a great extent into Sinus venosus vertebralis internus. For other connections of Vv. intersegmentales see **Ven.** Annot. 46, 52, 60. In the region of the lumbar intumescence of the spinal cord Sinus venosus vertebralis internus is lacking; in this region Vv. intersegmentales empty into Vv. renales.

(83) **Vv. vertebromedullares.** Term of Sterzi (1903). Names for tributaries of Vv. vertebromedullares follow the terminology of Lob (1967) in *Gallus*. See **Art.** Annot. 84.

(84) **V. spinalis ventralis.** At times doubled along part(s) of the its length. Both V. spinalis ventralis and V. radicularis ventralis are best developed opposite Corpus gelatinosum of the lumbar intumescence of Medulla spinalis.

(85) **Vv. sulci.** Receive drainage from the ventral commissure and most of the gray matter of the spinal cord (Lob, 1967).

(86) **Vv. marginales.** Synonymy: Vv. periphericae (Sterzi, 1903). Drain via short radial radicles into surface veins (vasocorona) that receive blood from the dorsolateral white matter of the spinal cord down to the level of the apposition of white and gray matter (Lob, 1967).

(87) **Vv. ventrolaterales.** Weakly developed longitudinal veins along the line of attachment of the ventral nerve rootlets to the spinal cord.

(88) **Vv. radiculares ventrales.** Stronger than Vv. radiculares dorsales (Lob, 1967).

(89) **V. spinalis dorsalis.** Courses in the median dorsal sulcus of the spinal cord; this vein is weaker than its ventral counterpart.

(90) **Vv. dorsolaterales.** Longitudinal veins that course along the line of attachment of the dorsal nerve rootlets to the spinal cord.

(91) **Vv. fissurae.** Drain into Vv. dorsolaterales; Vv. fissurae drain blood from the white matter located dorsal to the dorsal commissure of the gray matter of the spinal cord.

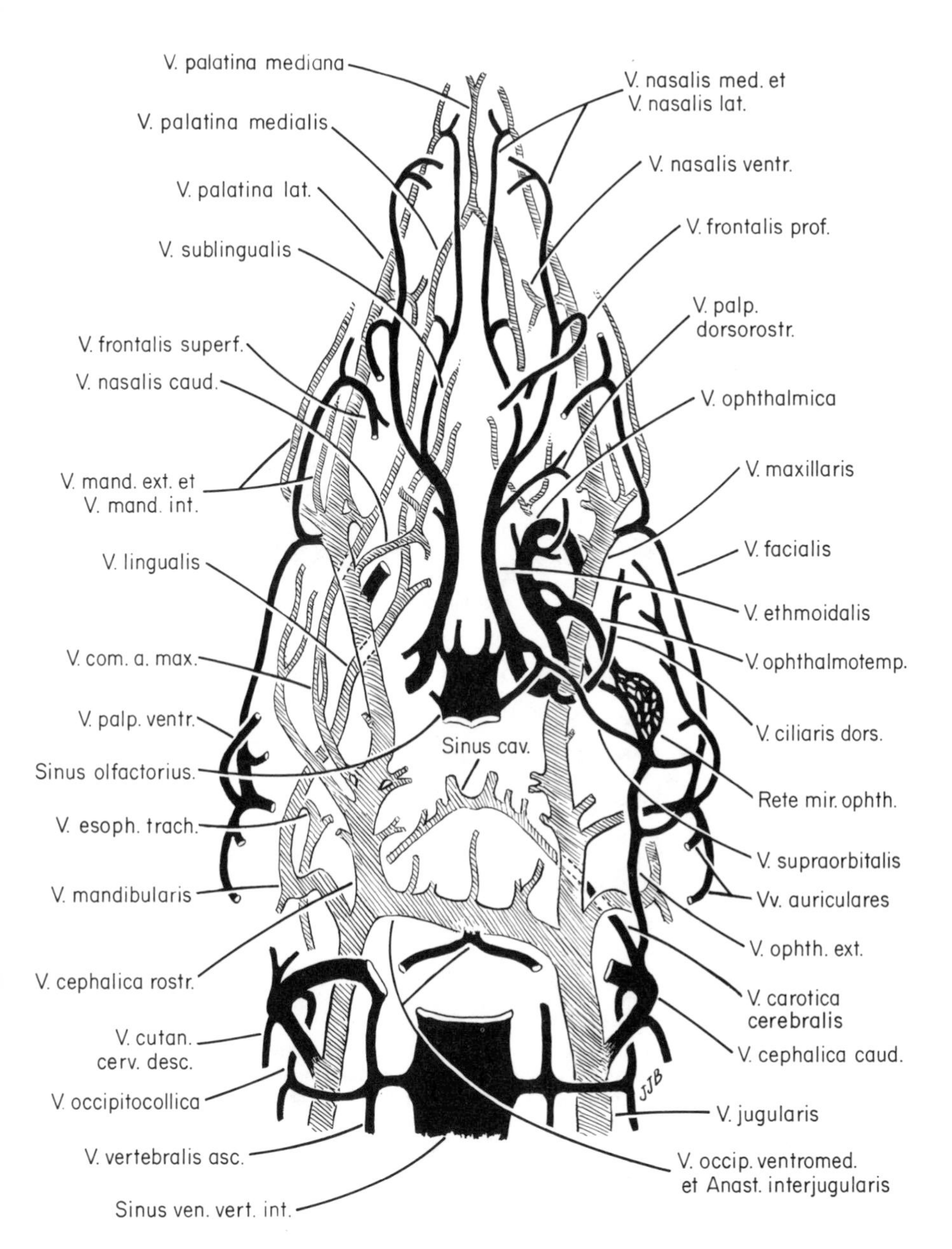

Fig. 1 Veins of head, extracranial; *Gallus*. Dorsal view. Redrawn from Baumel (1975). Black veins are located closer to the viewer (more dorsally) than the hatched veins.

Abbreviations: ophth., ophthalmica; palp., palpebralis; Rete mir. ophth., Rete mirabile ophthalmicum; Sinus cav., Sinus cavernosus; vert., vertebralis.

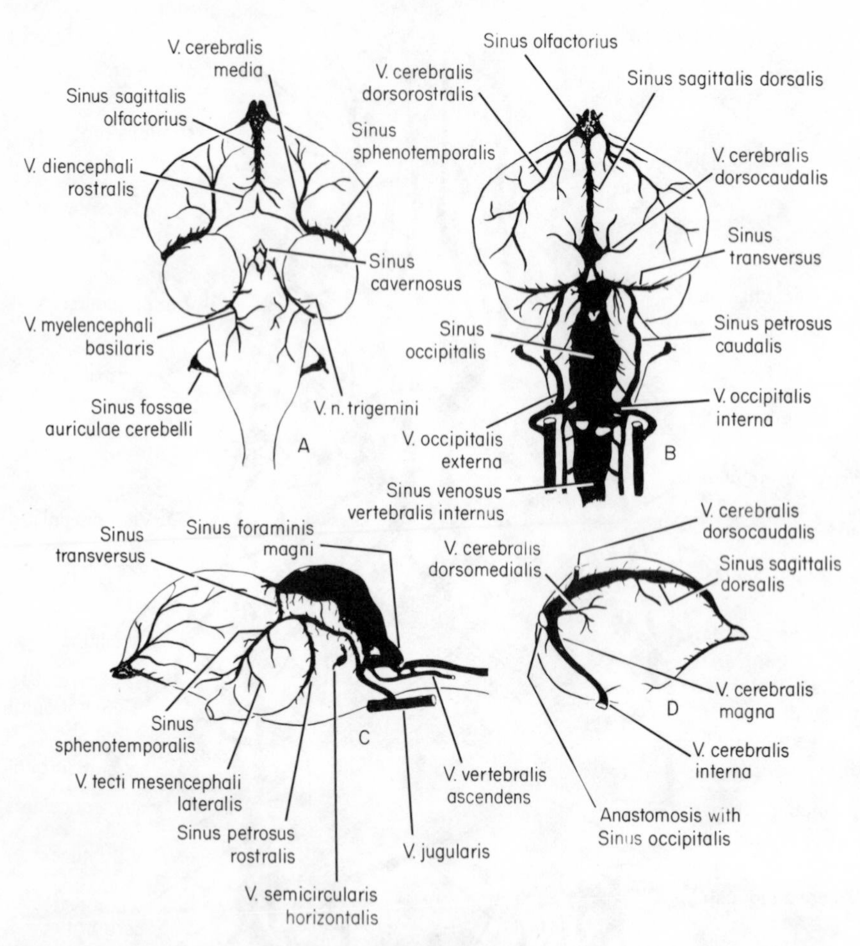

Fig. 2 Dural venous sinuses and encephalic veins; *Gallus*. Redrawn from Kaku (1959) with modifications. Most of the encephalic venous blood leaves the cranial cavity via, or in the vicinity of, the Foramen magnum where the dural sinuses empty into the jugular veins and the internal vertebral venous sinus. See Fig. 1; consult Baumel (1975) for a brief account of the dural sinuses and the veins of brain. A., ventral aspect of the brain; B., dorsal aspect of brain; C., left lateral aspect of brain; D., medial aspect of left telencephalic hemisphere.

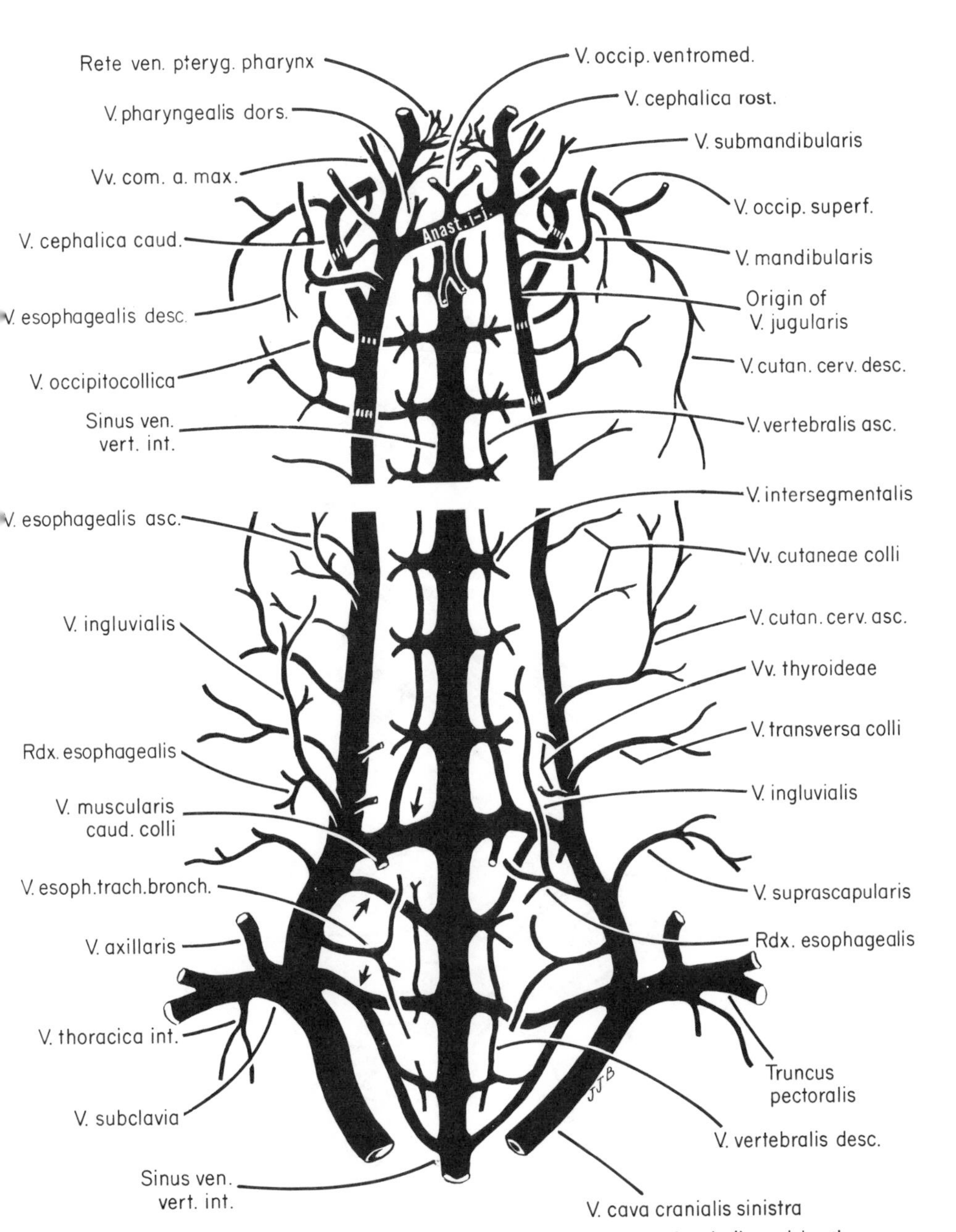

Fig. 3 Venous channels of neck; *Gallus*. Ventral view; foreshortening indicated by the white transverse band across the figure. Original drawing, J. J. Baumel. Three longitudinal venous channels drain venous blood of the head and neck. Note also the asymmetry of the jugular veins; the right jugular vein is larger than the left one. The Anastomosis interjugularis actually connects the two cranial cephalic veins, not the jugulars *per se*.

Sinus venosus vertebralis internus communicates with the cranial dural sinuses via the Foramen magnum (see Fig. 2); the internal vertebral sinus drains at the root of the neck into the jugular veins on each side via several anastomoses (see arrows). The sinus lies in the epidural space dorsal to the spinal cord (see Baumel, 1975).

Abbreviations: arrows, Anastomoses cum vv. jugularibus; Anast. i-j., Anastomosis interjugularis; cerv., cervicalis; com., comitantes; cutan., cutanea; max., maxillaris; occip., occipitalis; Rete ven. pteryg. pharyng., Rete venosum pterygopharyngeale; esoph. trach. bronch., esophagotracheobronchialis; vert., vertebralis.

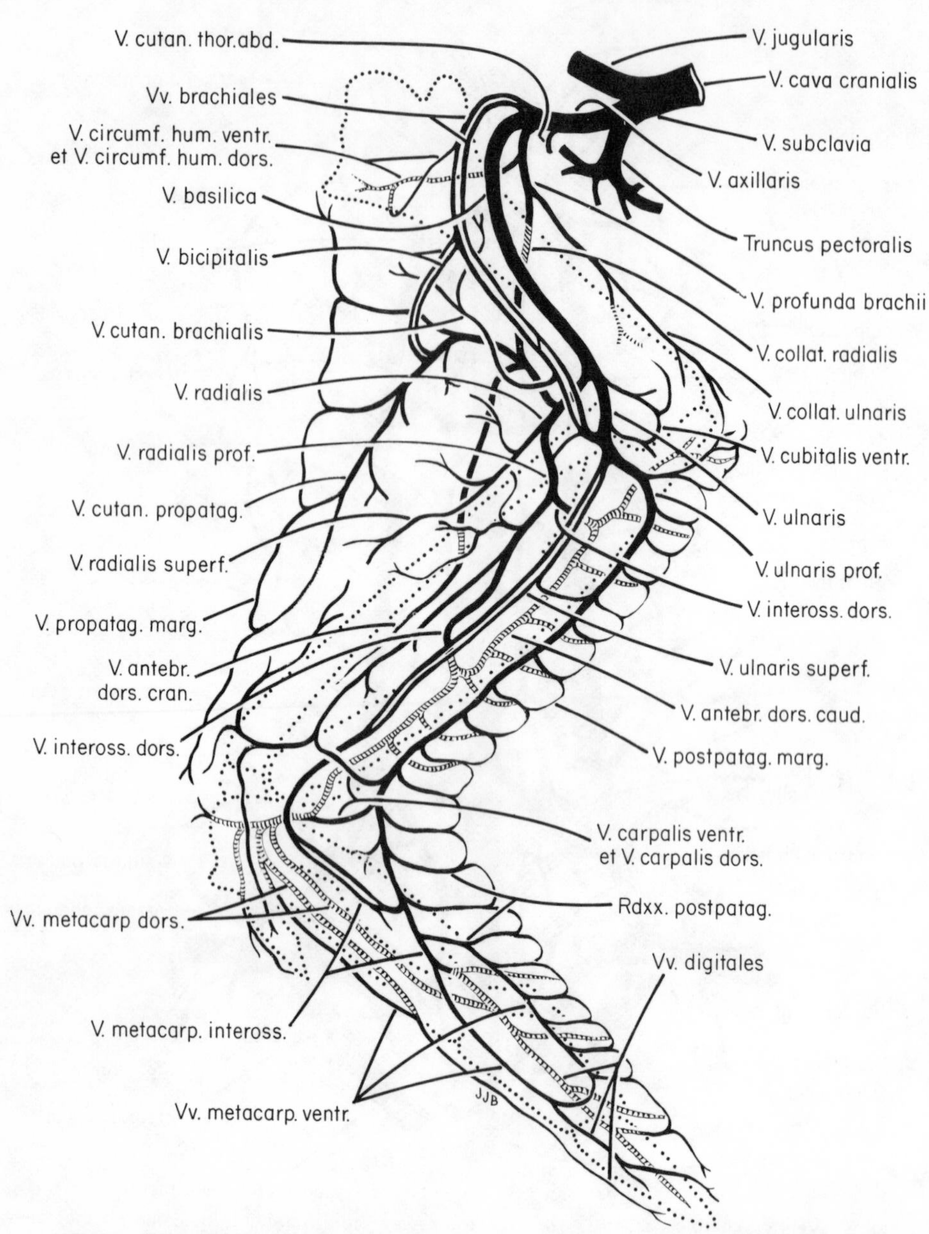

Fig. 4 Veins of wing; *Columba livia*. Right side, ventral view. Broken lines represent vessels dorsal to the skeleton. Note that: (1) V. ulnaris profunda is the strongest antebrachial vein, whereas its corresponding artery is rather weak; (2) V. basilica is the main brachial vein; the confluence of V. subclavia and V. jugularis and Truncus pectoralis produces the V. cava cranialis on each side (**Art.** Fig. 4; and Fig. 3); hence brachiocephalic veins are not present in birds.

Abbreviations: antebr., antebrachialis; circumf., circumflexa; collat., collateralis; cutan., cutanea; hum., humeri; inteross., interossea; marg., marginalis; metacarp., metacarpalis; postpatag., postpatagialis(es); propatag., propatagialis; thor. abd., thoracoabdominalis.

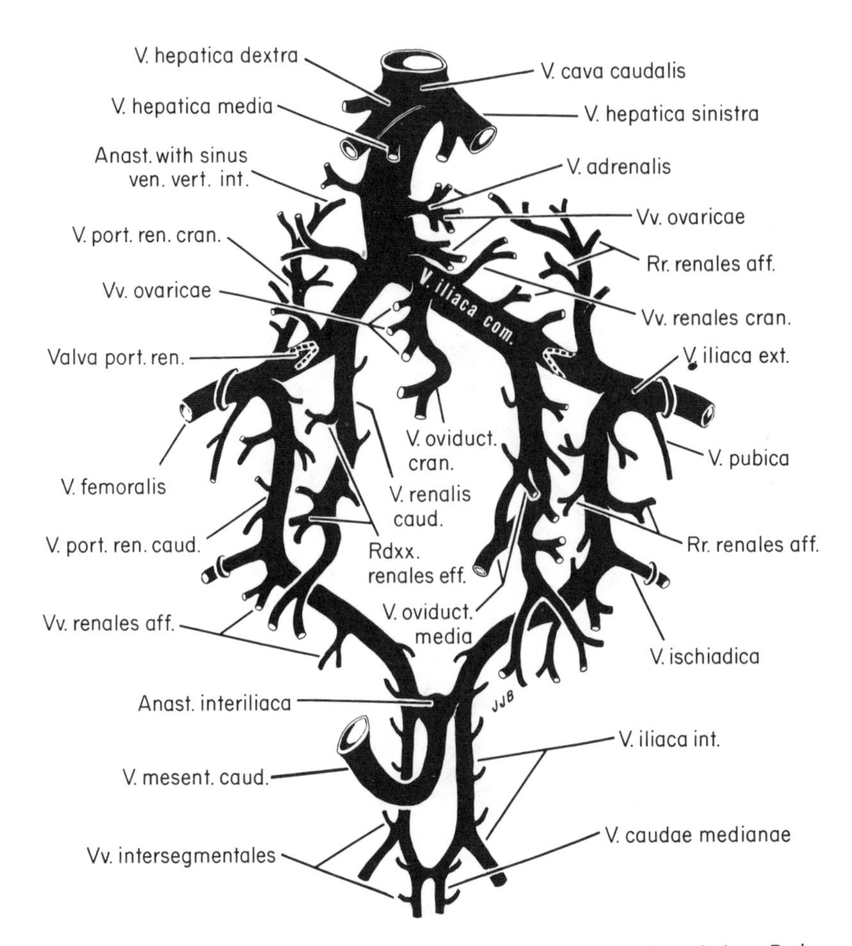

Fig. 5 Caudal vena cava and renal portal system of veins; *Gallus*. Ventral view. Redrawn from Baumel (1975). See **Ven.** Annot. 62 that deals with the development of V. iliaca interna and explains the apparent inconsistency in the naming of V. iliaca communis.

Abbreviations: Aff., afferentes; anast., anastomosis; com., communis; eff., efferentes; mesent., mesenterica; oviduct., oviductalis; port., portalis; ren., renalis; ven., venosus; vert., vertebralis.

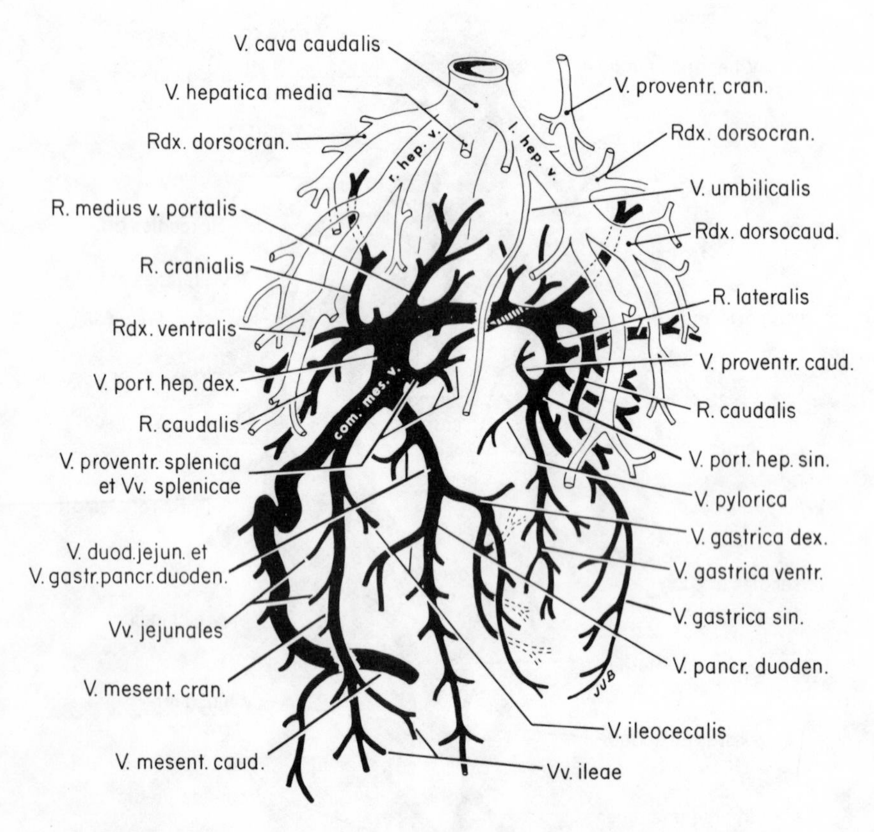

Fig. 6 Hepatic portal system and hepatic veins; *Gallus*. Ventral view. Redrawn from Baumel (1975). At upper right of drawing observe V. proventricularis cranialis which is not a tributary of Systema portale hepaticum; see **Ven.** Annot. 51 for details.

Abbreviations: com. mes. v., V. mesenterica communis; dex., dextra; duod(en)., duodeno- or duodenalis; gastr. pancr. duoden., gastopancreaticoduodenalis; hep., hepatica; jejun., jejunalis; 1. hep. v., V. hepatica sinistra; mesent., mesenterica; pancr. duoden., pancreatico-duodenalis; pot., portalis; proventr., proventricularis; sin., sinistra.

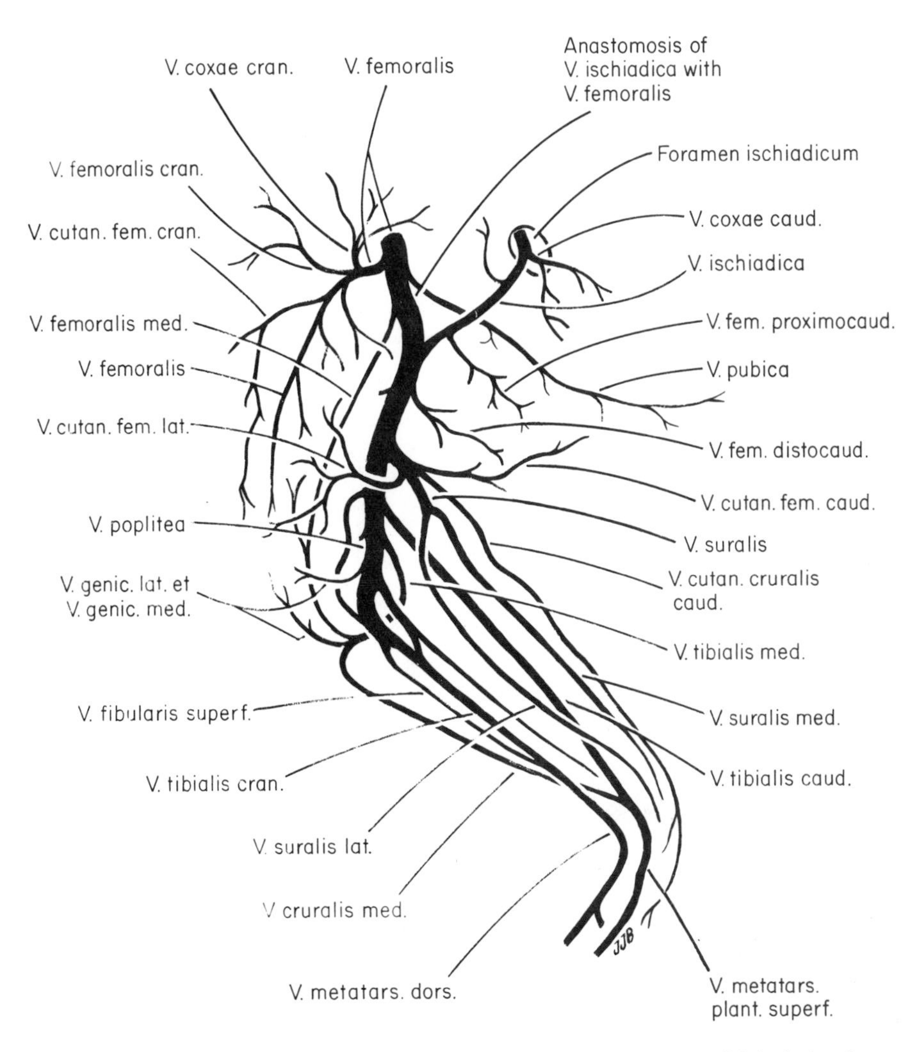

Fig. 7 Veins of pelvic limb; *Columba livia.* Left side, lateral view. Original drawing, J. J. Baumel. Although V. ischiadica is the main view of the pelvic limb, most of the blood from the territory drained by V. ischiadica enters the pelvis via V. femoralis. This is accomplished by means of a huge anastomosis connecting the two veins located on the medial side of the femur slightly distal to the hip joint.

Abbreviations: cutan., cutanea; fem., femoralis; genic., genicularis; metatars., metatarsalis; plant., plantaris; prof., profunda; superf., superficialis.

SYSTEMA LYMPHATICUM ET SPLEN

L. N. PAYNE

Houghton Poultry Research Station, Houghton, Huntingdon, Cambs PE17 2DA, England

With Contributions from Sub-Committee Members: Gy. Fehér, *University of Budapest*; J. Kaman, *University of Veterinary Medicine, Brno*; T. Miyaki, *Kanazawa University*; M. Yasuda, *Nagoya University*

TERMINI GENERALES

Vas lymphaticum
    Valvula lymphatica
Plexus lymphaticus
Cor lymphaticum[1]
Lympha
Lymphonodus [Nodus lymphaticus][2]
    Sinus centralis
Lymphonodulus solitarius [Nodulus lymphaticus solitarius][3]
Lymphonodulus aggregatus [Nodulus lymphaticus aggregatus][3]
Lymphonodulus muralis[4]

THYMUS

Lobus thymicus
    Capsula thymica
        Septum interlobulare thymicum
            Trabecula thymica
    Lobulus thymicus
        Cortex thymicus
        Medulla thymica
            Corpuscula thymica[5]

BURSA CLOACALIS[6]

Tunica serosa bursalis
Tunica muscularis bursalis
Tunica mucosa bursalis
  Plica bursalis[7]
    Folliculus lymphaticus[7]
      Cortex[7]
      Medulla[7]
    Septum interfolliculare[7]

SPLEN[8]

Tunica serosa splenica
Capsula splenica[9]
  Trabeculae splenicae[9]
Hilus splenicus
Pulpa splenica rubra
Pulpa splenica alba
  Lymphonoduli splenici

Aa. splenicae (see **Art.** Annot. 57)
  A. trabecularis[8]
    A. centralis[8]
      Arteriolae penicillares[8]
        Arteriolae terminales[8]
Vv. splenicae[8] (see **Ven.** Annot. 79)
  V. trabecularis[8]
    V. colligens[8]
      Venula splenica[9]
        Vasa sinusoidea splenica

Splen accessorius[10]

LYMPHONODULI AGGREGATI CANALIS ALIMENTARII[3, 11]

Lymphonoduli pharyngeales[12]
Lymphonoduli esophageales[12]
Lymphonoduli cecales[12]
Annuli lymphatici jejunales[13]
Annuli lymphatici ileales[13]

LYMPHONODI[2]

Lymphonodus cervicothoracicus[2]
Lymphonodus lumbaris[2]

VASA LYMPHATICA[14]

Truncus thoracoabdominalis[15] (Fig. 1)

Plexus lymphaticus celiacus
- Vas l. celiacum[16] (Fig. 1)
- Vas l. mesentericum craniale[17] (Fig. 1)
- Vas l. adrenale (Fig. 1)
- Vas l. ovaricum
- Vas l. testicularia (Fig. 1)
- Vasa l. renalia (Fig. 1)
- Vasa l. ureterodeferentialia (Fig. 1)
- Vas l. iliacum externum (Fig. 1)
- Vas l. femorale
- Vas l. iliacum internum (Fig. 1)
  - Vas l. ischiadicum (Fig. 1)
    - Vas l. popliteum
      - Vas l. tibiale craniale
      - Vas l. tibiale caudale
- Vas l. mesentericum caudale[18] (Fig. 1)
- Vas l. pudendum (Fig. 1)
  - Vas l. cloacale (Fig. 1)
- Vas l. sacrale medianum (Fig. 1)
  - Vas l. bursae cloacalis (Fig. 1)

Vas lymphaticum jugulare[19] (Fig. 1)

Vas l. caroticum commune
Vas l. vertebrale
Vas l. thyroideum (Fig. 1)
Vasa l. ingluvialia (Fig. 1)
Vasa l. esophagealia (Fig. 1)
Vas l. esophagotracheale (Fig. 1)
Vas l. cephalicum caudale
Vas l. cephalicum rostrale

Vas lymphaticum subclavium[20] (Fig. 1)

Vas l. axillare
Vas l. sternoclaviculare
Vas l. pectorale commune

Vas lymphaticum subclavium—*continued*
Vas l. brachiale
Vas l. brachiale profundum
Vas l. basilicum
Vas l. radiale
Vas l. ulnare

Vasa lymphatica thoracica interna[21] (Fig. 1)
Vasa l. pulmonalia superficialia (Fig. 1)

Vas lymphaticum cardiacum commune[22]
Vas l. cardiacum dextrum (Fig. 1)
Vas l. cardiacum sinistrum (Fig. 1)

Vas lymphaticum pulmonale commune[23] (Fig. 1)
Vas l. pulmonale profundum dextrum (Fig. 1)
Vas l. pulmonale profundum sinistrum

Vas lymphaticum proventriculare[24]

ANNOTATIONS

(1) **Cor lymphaticum.** Lymph hearts are said to occur in adult birds in various groups including certain Anserini and Larinae as well as *Casuarius* and *Rhea* (see Baum, 1930), but in *Gallus* they occur only in the embryo (Romanoff, 1960, p. 663).

(2) **Lymphonodus [Nodus lymphaticus].** Cervicothoracic and lumbar lymph nodes occur in Anseriformes and certain other birds (Jolly, 1909–10; Fürther, 1913) but not in *Gallus* (Baum, 1930). Four other nodes are mentioned in the duck by Manabe (cited by Kondo, 1937a) but are not well authenticated. So-called lymph nodes found in representatives of the Passeriformes, Coraciiformes, Charadriiformes and Galliformes by Manabe (cited by Kondo, 1937a) were thought by Kondo (1937a) to be aggregations of mural lymphoid nodules (see Annot. 4).

(3) **Lymphonodulus solitarius; Lymphonodulus aggregatus.** Solitary lymphoid nodules occur in most organs and tissues in the chicken; aggregated nodules occur in the wall of the alimentary canal and in the lacrimal ducts, in the Glandula membranae nictitantis and its duct, and in the duct of the Glandula nasalis (see Payne, 1971, p. 1022). Nodules consist of thymus-derived lymphoid tissue, and often include bursa-derived germinal centres. Numerous synonyms exist (see Payne, 1971, p. 1027).

(4) **Lymphonodulus muralis.** The mural lymphoid nodule is well-known in the wall of lymphatic vessels in *Gallus* (Kondo, 1937a; Biggs, 1957). Kondo (1937b) observed apparently aggregated mural lymphoid nodules in the walls of the lymphatic vessels of the leg of *Pelecanus* and *Phalacrocorax*.

(5) **Corpuscula thymica.** Synonymy: Hassal's corpuscles. The characteristics of these structures in the chicken have been reviewed by Hodges (1974, pp. 217–221).

(6) **Bursa cloacalis.** Synonymy: Bursa of Fabricius. See **Cloaca** Annot. 12.

(7) **Plica bursalis; Folliculus lymphaticus; Cortex; Medulla; Septum interfolliculare.** The anatomy of these structures in the chicken, a species in which the Bursa cloacalis has been the subject of intensive research, has been reviewed by Payne (1971, p. 999). When the Bursa is fully developed, its Tunica mucosa bursalis is arranged in 11 to 13 longitudinal primary folds or Plicae bursales, each of which is subdivided into 6 or 7 secondary plicae. Each Plica is generally described as consisting of numerous Folliculi lymphatici, separated by Septa interfollicularia. Each Folliculus lymphaticus possesses an outer Cortex and an inner Medulla. Separating the Cortex and Medulla is a narrow layer of epithelial cells. During natural regression of the Bursa (Payne, 1971, p. 1000), lymphocytes are lost from the Cortex and Medulla, and the epithelial layer becomes more prominent. Near the centre of the follicle, cavities begin to form which are lined by the epithelial layer, the latter now being columnar or pseudo-stratified. Many of these cavities appear to re-establish continuity with the lumen of the Bursa. The term Folliculus implies a small sac or pouch; the presence of the epithelial layer between the Cortex and Medulla, and the formation of cavities lined by this epithelium, therefore indicate that "Folliculus" rather than "Lobulus" is an appropriate term for the lymphatic components which form the wall of the Bursa cloacalis.

(8) **Splen.** Synonymy: Lien. The vascular supply and structure of the spleen have been fully described by Fukuta *et al.* (1969a, b).

(9) **Venula splenica.** Venule with a tall endothelial lining, through which lymphocytes migrate.

(10) **Splen accessorius.** One or more small accessory spleens may occur near the spleen, weighing from 4 to 45 mg in 69 of 144 chicks (Glick and Sato, 1964).

(11) **Lymphonoduli aggregati canalis alimentarii.** Lymphoid tissue occurs in the lamina propria and/or submucosa of the alimentary tract from Pharynx to Cloaca. With the exception of the Lymphonoduli pharyngeales, esophageales, and cecales, this lymphoid tissue is variable in amount and location. The term "gut associated lymphoid tissue" (GALT) has been used as a general term for both aggregated and solitary lymphoid tissue in the alimentary canal.

(12) **Lymphonoduli pharyngeales, esophageales, cecales.** Synonymy: pharyngeal, esophageal, and cecal tonsils. These are Lymphonoduli aggregati of relatively constant occurrence and relatively large size. The Lymphonoduli esophageales, however, are somewhat dubious; they were not reported by Calhoun (1932–33), but were described by Zietschmann (1911), Schauder (1923), and Kovacs (1928).

(13) **Annuli lymphatici jejunales et ileales.** Cranial and caudal bands of lymphoid tissue occur in both the jejunum and ileum of *Anas* (Leibovitz, 1968). Counterparts in *Gallus* have not been recorded.

(14) **Vasa lymphatica.** Dransfield (1944, 1945) provided the most detailed account of the distribution of lymphatic vessels in *Gallus* without naming them. Baum (1930) and Kondo (1937a) named the major vessels. (Annot. 14 cont. on p. 414).

Previously unnamed lymphatic vessels are named here according to the blood vessels which they follow or the organs which they drain. Some previously named vessels have been renamed on the same basis. Most lymphatic vessels, although named in the singular, have two or more trunks with frequent anastomoses (Dransfield, 1944, 1945). For a topographical summary, see King (1975, pp. 2010–2013).

(15) **Truncus thoracoabdominalis.** Usually right and left trunks are present, with frequent anastomoses, but a single vessel is sometimes present which may or may not bifurcate at its termination. In *Gallus* the trunk or trunks always terminate in the Vena cava cranialis (Miyaki and Yasuda, 1977). Thoracic and lumbar parts of the Truncus are designated by Baum (1930). Dransfield (1944, 1945) restricted the term thoracic duct to the lymphatics cranial to the celiac lymphatic plexus, and did not name the caudal part.

(16) **Vas l. celiacum.** This vessel receives branches from the thoracic esophagus, the proventriculus, gizzard, duodenum, ileum, ceca, spleen, pancreas and liver (Dransfield, 1944, 1945).

(17) **Vas l. mesentericum craniale.** This vessel receives branches from the jejunum, ileum and ceca, testis, ovary and the cranial part of the oviduct (Dransfield, 1944, 1945).

(18) **Vas l. mesentericum caudale.** Drains the rectum (Dransfield, 1944, 1945).

(19) **Vas lymphaticum jugulare.** Joins the V. jugularis just cranial to the junction of the latter with V. subclavia (Dransfield, 1944, 1945). Alternatively it drains into either the Truncus thoracoabdominalis or the Vena cava cranialis (Miyaki and Yasuda, 1977).

(20) **Vas lymphaticum subclavium.** Unites with the V. subclavia close to the junction of the latter with the V. jugularis (Dransfield, 1944, 1945), or it empties into the Truncus thoracoabdominalis or Vas lymphaticum jugulare (Miyaki and Yasuda, 1977).

(21) **Vasa lymphatica thoracica interna.** Accompany the Venae thoracicae internae, draining superficial lymphatics of the lung and abdominal muscles, and join the Vena cava cranialis close to its formation (Dransfield, 1944, 1945).

(22) **Vas lymphaticum cardiacum commune.** A single trunk (Fig. 1) which empties into the right Vena cava cranialis, close to its termination at the atrium, its tributaries being independent of the coronary arteries and veins (Dransfield, 1944, 1945).

(23) **Vas lymphaticum pulmonale commune.** A single trunk which joins the left Vena cava cranialis close to or in common with the Truncus thoracoabdominalis. Its tributaries travel along the left and right pulmonary veins, after draining the deep pulmonary lymphatics (Dransfield, 1944, 1945).

(24) **Vas lymphaticum proventriculare.** Follows the cranial proventricular vein (see **Ven.** Annot. 51) and joins the left Vena cava cranialis close to or in common with left Truncus thoracoabdominalis (Dransfield, 1944, 1945).

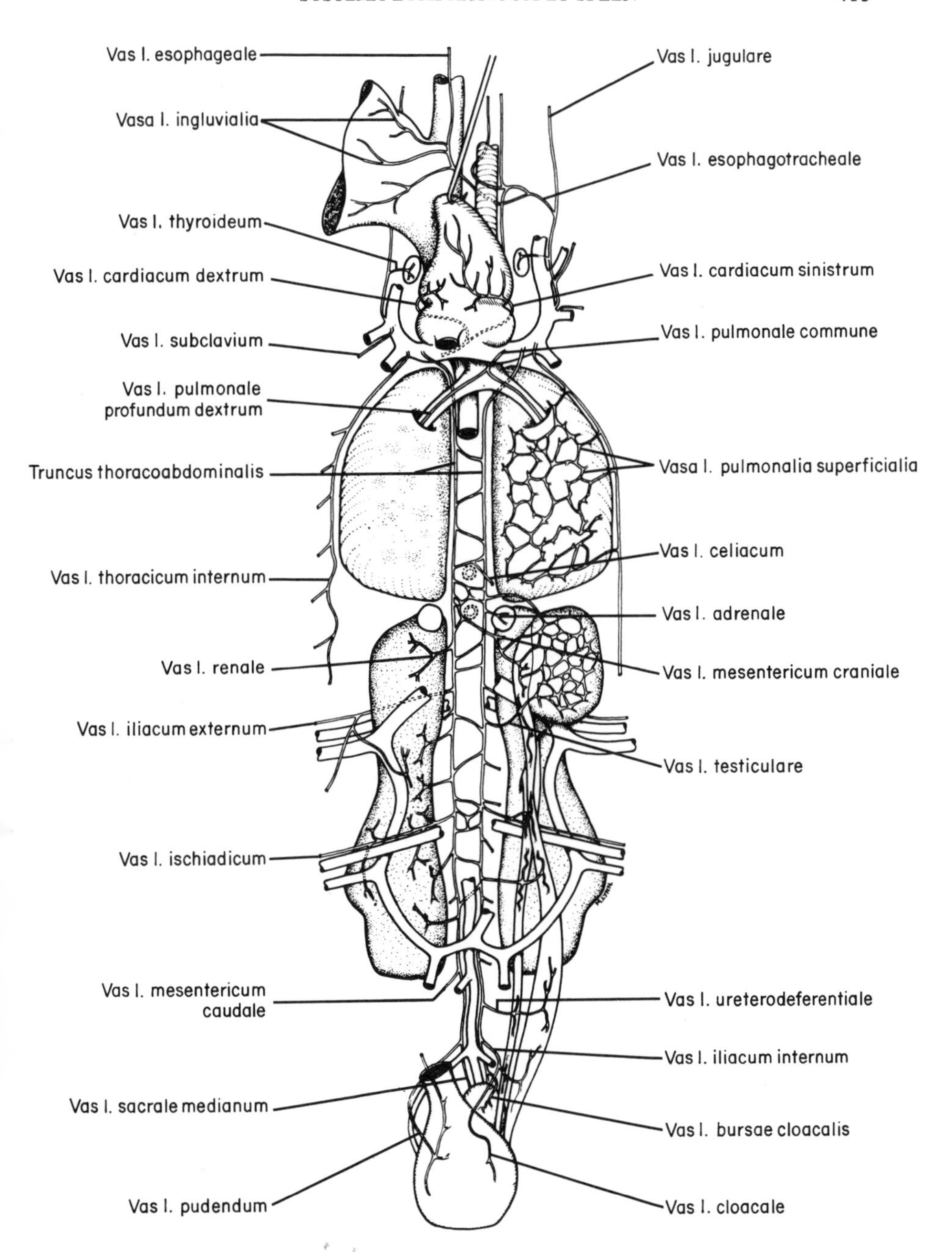

Fig. 1 Ventral view diagram of the lymphatic vessels of the trunk in *Gallus,* based on Baum (1930) and Dransfield (1944). The Vas lymphaticum cardiacum commune, shown by broken lines, is formed by union of the Vas l. cardiacum sinistrum with Vas l. cardiacum dextrum.

SYSTEMA NERVOSUM CENTRALE

JAMES E. BREAZILE

Department of Physiological Sciences
College of Veterinary Medicine
Oklahoma State University
Stillwater, Oklahoma

With contributions from Sub-Committee Members: J. J. Baumel, *Creighton University*; R. L. Boord, *University of Delaware*; J. L. Dubbeldam, *Rijksuniversiteit te Leiden*; W. Hodos, *University of Maryland*; H. J. Karten, *Massachusetts Institute of Technology*; T. J. Neary, *Creighton University*

Dr T. J. Neary deserves special mention for the indispensible help he provided in the editing of this chapter. His assistance was particularly valuable in preparing the illustrative material. The illustrations were put in their final form jointly by Dr Neary and Dr J. J. Baumel.

In their stereotaxic atlas of the brain of the pigeon, Karten and Hodos (1967) discussed the problems and historical considerations of the nomenclature of the avian brain. They noted that anatomists over the years had been unable to agree on a uniform terminology of the avian central nervous system even though a number of suggestions had been put forth for such a nomenclature. Now, a decade later, a terminology has been compiled, and is presented in this chapter. It represents agreement among the several neuroscientists of the ICAAN Sub-Committee on the Systema Nervosum Centrale as well as the other authorities who were consulted during the years that the terminology was being assembled and annotated.

By no means is this nomenclature to be thought of as immutable. Much active research is underway in avian neuroscience; as new information accumulates it will, of course, be necessary to review and revise this terminology to incorporate new findings and interpretations. Those who use this nomenclature are urged to submit suggestions for revision to the Chairman or to individual subcommittee members [Ed. (J.J.B.].

MEDULLA SPINALIS

(see Figs 2–5; and **PNS** Intro. *Spinal nerves*)

Fissura mediana [ventralis]
Funiculi medullae spinalis
    Funiculus dorsalis
    Funiculus lateralis
    Funiculus ventralis
Pars cervicalis
    Intumescentia cervicalis[1]
Pars thoracica
Pars synsacralis (see **Osteo.** Annot. 141, 145)
    Intumescentia lumbosacralis[1]
    Corpus gelatinosum[2]
    Sinus rhomboidalis
    Lobi accessorii[3]
Pars caudalis
Sulcus medianus [dorsalis]
Sulcus dorsolateralis
Sulcus ventrolateralis

SECTIONES MEDULLAE SPINALIS

(Figs. 2–5)

Canalis centralis
Commissura grisea
Commissura alba[4]
Rdx. dorsalis n. spinalis
Rdx. ventralis n. spinalis
Rdx. n. accessorii
Septum dorsale medianum
Substantia grisea[5]
    Cornu dorsale
        Apex cornus dorsalis
    Cornu ventrale
    Nuc. substantiae gelatinosae
    Nuc. dorsolateralis
    Nuc. marginalis[6]
    Nuc. proprius[7]
    Nuc. intermedius medullae spinalis[8]
    Nuc. cervicalis lateralis[9]
    Nuc. motorius
    Nuc. n. accessorii[9]
    Nuc. tr. spinalis [radicis descendentis] n. trigemini
        Pars marginalis

Pars gelatinosa
Pars magnocellularis
Substantia gelatinosa centralis
Nuc. cornucommissuralis
Substantia alba
Funiculus dorsalis
Fasciculus gracilis
Fasciculus cuneatus
Fasciculi proprii
Funiculus lateralis
Tr. spinocerebellaris ventralis[10]
Tr. spinocerebellaris dorsalis[10]
Tr. reticulospinalis lateralis
Tr. rubrospinalis
Tr. spino-olivaris
Tr. spinoreticularis
Tr. spinotectalis
Tr. spinothalamicus[11]
Fasciculus proprius
Fasciculus dorsolateralis[12]
Tr. spinalis [Rdx. descendens] n. trigemini
Funiculus ventralis
Tr. vestibulospinalis [lateralis]
Fasciculus longitudinalis medialis
Pars interstitiospinalis[13]
Pars tectospinalis[14]
Pars vestibulospinalis [ventralis]
Pars reticulospinalis [Tr. reticulospinalis medialis]
Fasciculi proprii

ENCEPHALON

RHOMBENCEPHALON

MYELENCEPHALON

MEDULLA OBLONGATA (Figs. 2, 3, 6–10)

Calamus scriptorius
Canalis centralis
Fibrae arcuatae superficiales
Fissura mediana [ventralis]
Fossa rhomboidea

MEDULLA OBLONGATA—*continued*

Funiculus dorsalis
  Fasciculus cuneatus
  Fasciculus gracilis
Funiculus ventralis
N. abducens
N. vestibulocochlearis
  Pars cochlearis
  Pars vestibularis
N. glossopharyngeus
N. vagus
N. hypoglossus
Pedunculus cerebellaris caudalis[15]
  Corpus juxtarestiforme
  Corpus restiforme
Sulcus intermedius dorsalis
Sulcus limitans
Sulcus medianus [dorsalis]
Sulcus ventrolateralis
Velum medullare caudale
Ventriculus quartus
  Obex
  Plexus choroideus ventriculi quarti
  Recessus lateralis

SECTIONES MEDULLAE OBLONGATAE

Area postrema
Commissura alba
Commissura infima[20]
Commissura grisea
Complexus olivaris caudalis[16]
  Nuc. olivaris accessorius medialis
    Pars dorsalis
    Pars intermedia
    Pars ventralis
  Nuc. olivaris accessorius dorsalis
    Pars lateralis
    Pars medialis
  Nuc. olivaris principalis
    Pars lateralis
    Pars medialis
Decussatio cochlearis dorsalis

Fibrae arcuatae internae
Fibrae arcuatae externae
Fasciculus longitudinalis medialis[17]
Fasciculus uncinatus
Funiculus dorsalis
    Fasciculus gracilis
    Fasciculus cuneatus
Funiculus lateralis
Funiculus ventralis
Lemniscus lateralis[36]
Lemniscus medialis[18]
Lemniscus spinalis
    Tr. spinotectalis[14]
    Tr. spinothalamicus
Nuc. ambiguus[30]
Nuc. angularis[19]
Nuc. centralis medullae oblongatae
    Pars dorsalis
    Pars ventralis
Nuc. cervicalis lateralis[9]
Nuc. cuneatus
Nuc. cuneatus accessorius [lateralis]
Nuc. commissuralis[20]
Nuc. gracilis
Nuc. intercalatus[21]
Nuc. intermedius medullae oblongatae[22, 8]
Nuc. intrafascicularis n. hypoglossi
Nuc. laminaris[23]
Nuc. magnocellularis cochlearis[24]
Nuc. motorius dorsalis n. vagi[25]
    Pars dorsalis
    Pars ventralis
Nuc. n. accessorii[31]
    Pars dorsalis
    Pars intermedia
    Pars ventralis
Nuc. n. glossopharyngei[25]
Nuc. n. hypoglossi[26]
Nuc. prepositus n. hypoglossi
Nuc. raphae magnae
Nuc. raphae obscurae
Nuc. raphae pallidae
Nuc. reticularis gigantocellularis

SECTIONES MEDULLAE OBLONGATAE—*continued*

Nuc. reticularis lateralis
  Subnuc. magnocellularis
  Subnuc. parvocellularis
  Subnuc. subtrigeminalis
Nuc. reticularis paragigantocellularis lateralis
Nuc. reticularis paramedianus[27]
Nuc. reticularis parvocellularis
Nuc. retroambiguus
Nuc. supraspinalis
Nuc. tr. solitarii[28]
Nuc. tr. spinalis [radicis descendentis] n. trigemini
  Subnuc. caudalis
    Pars marginalis
    Pars gelatinosa
    Pars magnocellularis
  Subnuc. interpolaris
  Subnuc. oralis
Nuc. vestibulares[29]
  Nuc. vestibularis descendens
  Nuc. vestibularis dorsolateralis
  Nuc. vestibularis dorsomedialis
  Nuc. vestibularis superior [rostralis]
  Nuc. vestibularis tangentialis
  Nuc. vestibularis ventrolateralis
Rdx. n. accessorii[31]
Rdx. n. cochlearis
Rdx. n. facialis
Rdx. n. glossopharyngei
Rdx. n. vagi
Rdx. n. vestibularis
  Pars ascendens
  Pars descendens
Tenia ventriculi quarti
Tr. arcuatus superficialis dorsalis
Tr. bulbotectalis
Tr. bulbothalamicus
Tr. laminocerebellaris
Tr. lamino-olivaris
Tr. occipitomesencephalicus[32]
Tr. olivocerebellaris
Tr. reticulospinalis lateralis
Tr. reticulospinalis ventralis

Tr. rubrospinalis
Tr. solitarius[28]
Tr. spinalis [Rdx. descendens] n. trigemini
Tr. spinocerebellaris dorsalis
Tr. spinocerebellaris ventralis
Tr. tectobulbaris dorsalis
Tr. tectobulbaris ventralis
Tr. tectospinalis[14]
Tr. vestibulocerebellaris
Tr. vestibulomesencephalicus
Tr. vestibulospinalis lateralis
Tr. vestibulospinalis ventralis

METENCEPHALON
PONS

(Figs 10–13)

Corpus trapezoideum[34]
Decussatio n. trochlearis
Fasciculus longitudinalis medialis (see Medulla oblongata)
Fasciculus uncinatus
Lemniscus lateralis[11, 36]
Lemniscus medialis[18]
Lemniscus spinalis
  Tr. spinotectalis[14]
  Tr. spinothalamicus
Nuc. centralis superior
Nuc. corporis trapezoidei
Nuc. cuneiformis
Nuc. lateralis pontis
  Pars caudalis
  Pars rostralis
Nuc. lemnisci lateralis[36]
  Pars dorsalis
  Pars lateroventralis[37]
  Pars ventralis
Nuc. loci cerulei
Nuc. medialis pontis
  Pars caudalis
  Pars rostralis
Nuc. n. abducentis
Nuc. n. abducentis accessorii[35]
Nuc. n. facialis
  Pars dorsalis[33]

PONS—*continued*

Pars intermedia
Pars ventralis
Nuc. motorius n. trigemini[38]
Pars lateralis
Pars medialis
Pars ventralis
Nuc. olivaris rostralis[23]
Nuc. papillioformis
Nuc. parabrachialis dorsalis
Nuc. parabrachialis ventralis
Nuc. raphae pontis
Nuc. reticularis parvocellularis
Nuc. reticularis pontis caudalis
Nuc. reticularis pontis rostralis
Nuc. sensorius principalis n. trigemini[39]
Pars dorsalis
Pars ventralis
Nuc. subceruleus dorsalis
Nuc. subceruleus ventralis
Nuc. tr. spinalis [radicis descendentis] n. trigemini
Rdx. ascendens n. trigemini
Rdx. mesencephalica n. trigemini
Rdx. n. abducentis
Tr. occipitomesencephalicus[32]
Tr. quintofrontalis[40]
Tr. spinalis [Rdx. descendens] n. trigemini
Tr. spinocerebellaris dorsalis
Tr. spinocerebellaris ventralis[10]
Tr. tectobulbaris dorsalis

CEREBELLUM

(Figs. 1–3)

Corpus cerebelli
Fissura prima
Fissura uvulonodularis
Lobus caudalis
Lobus flocculonodularis
Lobus rostralis
Ventriculus cerebelli
Fissurae cerebelli
Folia cerebelli
Hemispherium cerebelli

- Vinculum lingulae
- Ala lobuli centralis
- Lobulus quadrangularis
  - Pars rostralis
  - Pars caudalis
- Lobulus simplex
- Lobulus ansiformis
  - Crus rostrale
  - Crus caudale
- Lobulus paramedianus
- Auricula cerebelli
  - Paraflocculus
    - Paraflocculus dorsalis
    - Paraflocculus ventralis
  - Flocculus
    - Pedunculus flocculi

Sulci cerebelli
Vallecula cerebelli
Vermis cerebelli

Lobulus	
Lobulus I	------[Lobulus lingulae]
Lobulus IIa Lobulus IIb	------[Lobulus centralis]
Lobulus IIIa Lobulus IIIb Lobulus IVa Lobulus IVb Lobulus Va Lobulus Vb	-----[Lobulus culminis]
Lobulus VIa Lobulus VIb Lobulus VIc	-----[Declive]
Lobulus VIIa	-----[Folium vermis]
Lobulus VIIb	-----[Tuber vermis]
Lobulus VIII	-----[Lobulus pyramidis]
Lobulus IXa Lobulus IXb Lobulus IXc	-----[Lobulus uvulae]
Lobulus X	-----[Lobulus noduli]

SECTIONES CEREBELLI[41]

Arbor vitae cerebelli
Appendix falciformis
Pedunculus cerebelli caudalis[15, 41]

SECTIONES CEREBELLI—*continued*

Pedunculus cerebelli intermedius[15, 41]
Pedunculus cerebelli rostralis[15, 41]
    Tr. dentato-rubro-thalamicus
Tr. spinocerebellaris ventralis[10]
Commissura cerebellaris
Corpus medullare
Cortex cerebelli
    Stratum moleculare
    Stratum ganglionicum
    Stratum granulosum
Decussatio tr. cerebellobulbaris
Fasciculus uncinatus
Fissura precentralis
Fissura caudolateralis
Fissura prima
Fissura prepyramidalis
Fissura secunda
Fissura caudalis
Formatio quadrangularis
Lamina alba
Nuc. cerebellaris medialis [Nuc. fastigii]
Nuc. cerebellaris intermedius
Nuc. cerebellaris lateralis
Proc. cerebellovestibularis
Tr. cerebellobulbaris
Tr. laminocerebellaris
Tr. spinocerebellaris dorsalis[10]
Tr. spinocerebellaris ventralis[10]
Tr. pontocerebellaris
Tr. vestibulocerebellaris
Tr. olivocerebellaris

MESENCEPHALON

(Figs. 12–21)

Aqueductus mesencephali
Tectum mesencephali[42, 43]
    Lamina tecti
    Colliculus mesencephali[43] (Fig. 10)
    Commissura tectalis
    Brachium colliculi mesencephali[44]
    Ventriculus tecti mesencephali

Tegmentum mesencephali
- Commissura caudalis [posterior]
- N. oculomotorius
- N. trochlearis

SECTIONES MESENCEPHALI
(Figs. 12–21)

Aqueductus mesencephali
Colliculus mesencephali[43] (Fig. 10)
- Stratum fibrosum periventriculare
- Stratum griseum periventriculare
- Stratum album centrale
- Stratum griseum centrale
  - Pars profunda
  - Pars intermedia
  - Pars superficialis
- Stratum griseum et fibrosum superficiale
- Stratum opticum
- Stratum zonale

Commissura caudalis [posterior]
- Pars pretectalis
- Pars medullaris

Commissura lemnisci lateralis
Complexus isthmi[49]
- Nuc. magnocellularis isthmi
- Nuc. parvocellularis isthmi
- Nuc. isthmo-opticus[50]
- Nuc. semilunaris

Decussatio n. oculomotorii[45]
Decussatio n. trochlearis
Decussatio pedunculi cerebellaris rostralis
Decussatio tr. rubrospinalis
Decussatio tr. tectospinalis
Fasciculus longitudinalis medialis
Lemniscus lateralis[11, 36]
Lemniscus medialis[18]
Lemniscus spinalis
- Tr. spinotectalis[14]
- Tr. spinothalamicus

Nuc. annularis dorsalis
Nuc. annularis lateralis
Nuc. annularis rostralis

SECTIONES MESENCEPHALI—*continued*

Nuc. annularis ventralis
Nuc. centralis superior
Nuc. ciliaris
Nuc. commissuralis caudalis [posterior][46]
    Pars diffusa parvocellularis
    Nuc. interstitialis commissurae caudalis
Nuc. ectomamillaris[47]
Nuc. fasciculus longitudinalis medialis
Nuc. intercollicularis[48]
Nuc. interpeduncularis
Nuc. lemnisci lateralis
    Pars dorsalis
    Pars ventralis
    Pars lateroventralis
Nuc. lentiformis mesencephali
    Pars magnocellularis
    Pars parvocellularis
Nuc. linearis caudalis
Nuc. linearis intermedius
Nuc. linearis rostralis
Nuc. mesencephalicus lateralis[52]
    Pars dorsalis
Nuc. mesencephalicus n. trigemini
Nuc. mesencephalicus profundus
    Pars ventralis
    Nuc. tegmenti pedunculopontini[57]
        Pars disseminata
        Pars compacta dorsalis
        Pars compacta ventralis
Nuc. n. oculomotorii
    Pars accessoria[45]
    Pars dorsalis
    Pars ventralis
    Pars dorsolateralis
Nuc. n. trochlearis
Nuc. parageniculatus colliculi mesencephali[53]
Nuc. paragrisealis centralis mesencephali[54]
Nuc. pretectalis[55]

Pars diffusa [lateralis]
Pars medialis [principalis]
Nuc. principalis precommissuralis
Nuc. ruber
Nuc. subpretectalis
Nuc. tegmenti dorsalis
Nuc. tegmenti ventralis
Organum subcommissurale
Pedunculus cerebellaris rostralis[15]
Pars ascendens
Pars descendens
Rdx. mesencephalica n. trigemini
Rdx. n. oculomotorii
Rdx. n. trochlearis
Substantia grisea centralis
Substantia grisea et fibrosa periventricularis
Sulcus subhabenularis
Tegmentum mesencephali
Tr. bulbotectalis
Tr. bulbothalamicus
Tr. geniculotectalis
Tr. habenulointerpeduncularis
Pars lateralis
Pars medialis
Tr. interstitiospinalis
Tr. isthmomesencephalicus
Tr. isthmo-opticus
Tr. occipitomesencephalicus[32]
Tr. nuc. ectomamillaris[47, 68]
Tr. opticus marginalis
Pars lateralis
Pars medialis
Tr. pretectosubpretectalis
Tr. quintofrontalis[40]
Tr. rubrospinalis
Tr. septomesencephalicus[56]
Tr. tectobulbaris dorsalis
Tr. tectobulbaris ventralis
Tr. tectospinalis
Tr. vestibulomesencephalicus
Velum medullare rostrale

PROSENCEPHALON
DIENCEPHALON
(Figs. 16–23)

Adhesio interthalamica
Commissura caudalis [posterior]
Foramen interventriculare
Organum subcommissurale
Plexus choroideus ventriculi tertii
Recessus neurohypophysialis [infundibuli]
Ventriculus tertius

Hypothalamus (Figs. 14–23)

Chiasma opticum
Corpus mamillare
Hypophysis [Glandula pituitaria] (see **Endoc.**)
Recessus inframamillaris
Tr. opticus
Tuber cinereum
- Eminentia mediana
- Pars caudalis tuberis
- Pars parainfundibularis tuberis
- Pars rostralis tuberis
- Sulcus tuberoinfundibularis

Sectiones hypothalami (Figs. 16–23)

Decussatio supraoptica dorsalis
- Pars lateralis
- Pars medialis
- Pars ventralis

Decussatio supraoptica ventralis
- Pars dorsalis
- Pars ventralis

Decussatio tr. infundibularis
Fasciculus lateralis prosencephali[51]
- Tr. quintofrontalis[40]
- Ansa lenticularis[57]
- Tr. striohypothalamicus medialis
- Tr. striomesencephalicus
- Tr. striothalamicus dorsolateralis
- Tr. thalamostriaticus
- Tr. thalamofrontalis lateralis
- Tr. thalamofrontalis medialis
- Tr. thalamofrontalis intermedius

Fasciculus medialis prosencephali
Infundibulum

Regio preoptica [rostralis] hypothalami
- Nuc. preopticus periventricularis
- Nuc. magnocellularis preopticus
- Nuc. rostralis [anterior] hypothalami
- Nuc. decussationis supraopticae dorsalis
- Nuc. decussationis supraopticae ventralis
- Nuc. paraventricularis
- Nuc. supraopticus
- Nuc. suprachiasmaticus
- Nuc. preopticus medialis
- Nuc. preopticus lateralis

Regio medialis hypothalami
- Nuc. periventricularis hypothalami
- Area dorsalis hypothalami
- Nuc. dorsomedialis hypothalami
- Nuc. ventromedialis hypothalami
- Nuc. tuberis infundibuli

Regio caudalis hypothalami
- Nuc. intercalatus
- Nuc. mamillaris lateralis
- Nuc. mamillaris medialis
- Nuc. supramamillaris interstitialis
- Nuc. premamillaris

Regio lateralis hypothalami

Tr. hypothalamohypophysialis
- Tr. paraventriculohypophysialis
- Tr. supraopticohypophysialis
- Tr. suprachiasmaticohypophysialis

Tr. nuc. ectomamillaris[47, 68]

Tr. infundibularis

Tr. opticus

Sectiones thalami (Figs. 16–23)

Ansa lenticularis[57]

Area ventralis thalami

Nuc. opticus principalis thalami[58]
- Nuc. dorsolateralis rostralis
  - Pars lateralis
  - Pars medialis [magnocellularis]
- Nuc. lateralis rostralis [anterior]

Nuc. dorsointermedialis caudalis[65]

Nuc. dorsolateralis caudalis[11, 18, 59]

Nuc. dorsolateralis superficialis

Sectiones thalami—*continued*

Nuc. dorsomedialis thalami
  Pars caudalis [posterior]
  Pars rostralis [anterior][60]
Nuc. ectomamillaris
Nuc. entopeduncularis ventralis rostralis[61]
Nuc. externus thalami
Nuc. interstitialis magnocellularis lateralis
Nuc. interstitio-pretecto-subprectectalis
Nuc. geniculatus lateralis dorsalis
Nuc. geniculatus lateralis ventralis
Nuc. intercalatus thalami[11,18,59]
Nuc. internus dorsalis [superior]
Nuc. internus ventralis [inferior]
Nuc. magnocellularis strati grisei thalami
Nuc. parvocellularis strati grisei thalami
Nuc. caudalis intermedius
Nuc. caudoventralis thalami
Nuc. ovoidalis[62]
Nuc. postrotundus
Nuc. principalis precommissuralis[65]
Nuc. prestriaticus
Nuc. rotundus[63]
Nuc. semilunaris para-ovoidalis[64]
Nuc. spiriformis lateralis[65]
  Pars dorsalis
  Pars ventralis
Nuc. spiriformis medialis
  Pars lateralis
  Pars intermedius
  Pars medialis
Nuc. subrotundus
Nuc. superficialis magnocellularis
Nuc. superficialis parvocellularis[11,59,66]
Nuc. suprarotundus
Nuc. triangularis[67]
Nuc. tr. tectothalamici cruciati
Nuc. ventrolateralis thalami
Plexus choroideus ventriculi tertii
Stilus corporis geniculati
Tr. bulbothalamicus
Tr. isthmo-opticus[50]
Tr. nuc. ovoidalis[44]

Tr. opticus marginalis
Tr. occipitomesencephalicus[32]
Pars hypothalamica[71]
Tr. nuc. ectomamillaris[47, 68]
Tr. quintofrontalis[40]
Tr. tectorotundus [Brachium colliculi mesencephali]
Tr. tectothalamicus cruciatus
Tr. tectothalamicus dorsalis
Tr. tectothalamicus ventrolateralis
Tr. thalamofrontalis intermedius
Tr. thalamofrontalis lateralis
Tr. thalamofrontalis medialis

Epithalamus (Figs. 16–19)

Commissura habenularis
Glandula pinealis (see **Endoc.** Annot. 6)
Nuc. habenularis lateralis
Nuc. habenularis medialis
Nuc. subhabenularis lateralis
Nuc. subhabenularis medialis
Stria habenularis [medullaris]
Tr. archistriatohabenularis et precommissuralis
Tr. corticohabenularis
Tr. hypothalamohabenularis
Tr. olfactohabenularis
Tr. septohabenularis
Tr. teniae habenularis
Tr. habenulointerpeduncularis
Pars lateralis
Pars medialis

TELENCEPHALON

(see Figs. 16–27 and Table I)

Bulbus olfactorius
Ventriculus olfactorius
Fissura interhemispherica
Fissura subhemispherica[69]
Fovea limbica
Hemispherium telencephali
Polus occipitalis
Pars frontalis
Pars parietalis

TELENCEPHALON—*continued*

Eminentia sagittalis[78]
Vallecula telencephali
Tuber ventrale
Tuber ventrofrontale
Tuber ventromediale
Tuber ventrolaterale
Ventriculus lateralis
Plexus choroideus ventriculi lateralis

SECTIONES TELENCEPHALI[70]

Archistriatum[71]
Archistriatum rostrale
Archistriatum intermedium
Pars dorsalis
Pars ventralis
Archistriatum mediale
Archistriatum caudale
Nuc. teniae
Ansa lenticularis[57]
Area entorhinalis
Area paraentorhinalis
Area parahippocampalis
Pars linearis
Area pre-entorhinalis
Area prepiriformis
Area septalis
Nuc. septalis lateralis
Nuc. septalis medialis
Area temporo-parieto-occipitalis
Bulbus olfactorius (Figs. 1–3, 27)
Fila olfactoria
Lamina glomerulosa
Lamina granularis externa
Lamina molecularis externa
Lamina mitralis
Lamina molecularis interna
Lamina ependymalis
Ventriculus olfactorius
Commissura rostralis[72]
Pars bulbaris

Pars temporalis
Commissura pallii
Complexus paleostriatus[74]
Paleostriatum augmentatum[75]
Paleostriatum primitivum[76]
Nuc. intrapeduncularis[76]
Cortex prepiriformis[73]
Cortex piriformis
Ectostriatum
Fasciculus diagonalis
Fasciculus lateralis prosencephali[51] (see Sectiones Hypothalami for components)
Fasciculus medialis prosencephali
Fissura neopaleostriatica
Hippocampus
Hippocampus dorsalis
Hyperstriatum accessorium[78]
Hyperstriatum dorsale[78]
Hyperstriatum intercalatum supremum
Hyperstriatum ventrale
Pars dorsalis
Pars ventralis
Lamina frontalis superior[78]
Lamina frontalis suprema[78]
Lamina hyperstriatica
Lamina medullaris dorsalis
Lamina medullaris ventralis
Lobus parolfactorius[77]
Neostriatum
Pars caudalis
Pars rostralis
Pars intermedia lateralis
Pars intermedia medialis
Nuc. accumbens
Nuc. ansae lenticularis caudalis[79]
Nuc. ansae lenticularis rostralis[79]
Nuc. commissurae pallii
Nuc. basalis [trigeminalis prosencephali][80]
Nuc. interstitialis telencephali
Nuc. interstitialis commissurae rostralis
Nuc. interstitialis commissurae pallii
Nuc. intercalatus hyperstriati accessorii[78]
Lamina externa
Lamina interna

SECTIONES TELENCEPHALI—*continued*

Nuc. olfactorius rostralis
Nuc. teniae
Nuc. tr. diagonalis
Nuc. tr. septomesencephali
Substantia grisea periventricularis lateralis
Tr. dorso-archistriaticus
Tr. corticohabenularis
Tr. corticoseptalis
Tr. epistriaticus dorsalis
Tr. fronto-archistriaticus[81]
Tr. frontothalamicus
Tr. fronto-occipitalis
Tr. mesencephalicus ventralis
Tr. quintofrontalis[40]
Tr. occipitomesencephalicus[32]
    Pars hypothalamica[71]
Tr. olfactorius lateralis
Tr. olfactorius medialis
Tr. septomesencephalicus[56]
    Pars dorsalis
    Pars basalis
Tr. striohypothalamicus medialis
Tr. striomesencephalicus
Tr. striothalamicus dorsolateralis
Tr. thalamofrontalis lateralis
Tr. thalamofrontalis intermedius
Tr. thalamofrontalis medialis
Tr. thalamo-para-olfactorius
Tr. thalamostriaticus
Tuberculum olfactorium

MENINGES

Dura mater spinalis[82]
Dura mater encephali[82]
    Dura mater propria
    Lamina periostealis
    Plica tentorialis[83]
    Diaphragma sellae
Arachnoidea spinalis
Arachnoidea encephali
    Cavitas subarachnoidea
        Liquor cerebrospinalis
        Cisternae subarachnoideae

Pia mater spinalis
Lig. denticulatum
Lig. ventromedianum
Septum ventromedianum
Ligg. suspensoria transversa[84]
Pia mater encephali

ANNOTATIONS

(1) **Intumescentia cervicalis.** Enlargement of the region of the spinal cord from which the brachial plexus arises; usually larger than the lumbosacral intumescence, especially in flying birds. In flightless birds, especially the Ostrich, the Intumescentia lumbosacralis conspicuously surpasses the cervical intumescence in size (Kuhlenbeck, 1975). See **Osteo.** Annot. 145.

(2) **Corpus gelatinosum.** (Fig. 4) A specialized glial structure which lies within the Sinus rhomboidalis. The structure is often called the glycogen body because of its high glycogen content (Gage, 1917; Terni, 1924). The Corpus gelatinosum has been recently studied with the aid of electron microscopy by Lyser (1973) and Welsch and Wachtler (1969) who confirmed that its structural elements are glial cells containing large numbers of glycogen granules. They further demonstrated that the glial elements are innervated by nonmyelinated nerve fibers. The Corpus consists of dorsal and ventral portions connected by a constriction formed of pia mater. The ventral portion encloses the Canalis centralis of the spinal cord (Hodges, 1974).

(3) **Lobi accessorii.** Most evident in the lumbosacral levels of the spinal cord; the lobes contain neurons which appear similar to those of Nuc. marginalis. These lobes have been referred to as lobes of Lachi and as the nuclei of Hoffmann-von Kölliker (Ariëns Kappers *et al.*, 1936). See Annot. 6.

(4) **Commissura alba.** Both dorsal and ventral white commissures are present in birds. The dorsal white commissure contains collaterals of dorsal root fibers and axons of cell bodies located within the grey matter. The ventral white commissure contains axons which cross the midline to form spinoreticular, spinothalamic, and spinotectal tracts of the spinal cord.

(5) **Substantia grisea.** Leonard and Cohen (1975) and Brinkman and Martin (1973) have demonstrated that the grey matter of the spinal cord of birds exhibits a laminar organization similar to that described by Rexed for the cat (1952, 1954) and subsequently by other investigators for other mammals. It appears that the delineation of these laminae according to Leonard and Cohen (1975) more accurately depicts their cytological relationship to similar laminae in mammals. According to their delineation, there are nine laminae within the grey matter, designated laminae I through IX.

(6) **Nuc. marginalis** (Fig. 5). Lies within the white matter of the spinal cord, usually near the surface, but also represented by scattered nerve cell bodies which have been called as "paragriseal" neurons by numerous investigators. Nuc. marginalis is also known as the nucleus of Hoffmann-von Kölliker. The marginal nucleus is particularly well developed in the lumbosacral levels where it protrudes from the spinal cord as the Lobi accessorii (see Annot. 3) near the attachment of the Lig. denticulatum.

(7) **Nuc. proprius.** Synonymy: Dorsal magnocellular column of the spinal cord grey matter (Jungherr, 1969, Bolton, 1971).

(8) **Nuc. intermedius medullae spinalis** (Fig. 5). The dorsal commissural portion of Nuc. intermedius within thoracolumbar levels of the spinal cord has been demonstrated to contain autonomic preganglionic nerve cell bodies and to be an avian homologue to the intermediolateral cell column (Nuc. intermediolateralis) of mammals (Terni, 1923; Macdonald and Cohen, 1970). This nucleus has commonly been referred to as the column of Terni.

(9) **Nuc. cervicalis lateralis.** Although no literature references are available to document the presence (or homology) of Nuc. cervicalis lateralis with that of mammals, cell accumulations are present in a comparable position in birds. Whether these represent a continuation of the marginal or paragriseal cell system of the spinal cords of birds or are separate entities is not known. See Annot. 6.

(10) **Tr. spinocerebellaris ventralis; Tr. spinocerebellaris dorsalis.** The presence of dorsal and ventral spinocerebellar tracts in birds has been demonstrated by Friedlander (1898), Sanders (1929), Ariëns Kappers *et al.* (1936), Larsell (1948), and Whitlock (1952). Oscarsson *et al.* (1963) studied these tracts in the duck and verified their existence, but indicated that the dorsal spinocerebellar tract may function in a somewhat different capacity than that of mammals. No direct evidence is available concerning the presence of a rostral spinocerebellar tract in birds comparable to that of mammals.

(11) **Tr. spinothalamicus.** The spinothalamic tract of birds resembles that of mammals in: its origin from the dorsal horn of the spinal grey, its decussation at segmental levels to the contralateral spinal cord, and its contribution to the lateral funiculus (Akker, 1970, Oscarsson *et al.* (1963). Its terminations in the thalamus have been demonstrated by Karten and Revzin (1966). The spinothalamic tract terminates in Nuc. intercalatus thalami, Nuc. dorsolateralis caudalis thalami and Nuc. superficialis parvocellularis thalami (Karten and Revzin, 1966; Delius and Bennetto, 1972).

(12) **Fasciculus dorsolateralis.** Located in part in both the dorsal and lateral funiculi and corresponds to Lissaur's tract of mammals.

(13) **Pars interstitiospinalis.** Synonymy: Tr. tegmentospinalis (Kuhlenbeck, 1975).

(14) **Pars tectospinalis.** Pars tectospinalis of the Fasciculus longitudinalis medialis represents Tr. tectospinalis, and consists of both crossed and uncrossed descending fibers (Ariëns Kappers, 1936; Kuhlenbeck, 1975). See Medulla oblongata, Pons, Mesencephalon.

(15) **Pedunculus cerebellaris caudalis.** The cerebellar peduncle is commonly referred to as the Corpus restiforme (caudal cerebellar peduncle), Brachium pontis (middle cerebellar peduncle), and Brachium conjunctivum (rostral cerebellar peduncle). For purposes of simplicity, the topographical terms of rostral, middle, and caudal cerebellar peduncles have been adopted. This terminology is particularly appropriate for the caudal peduncle which consists of both Corpus restiforme and Corpus juxtarestiforme.

(16) **Complexus olivaris caudalis.** There is much variance in the literature concerning the nuclei of the caudal olivary complex in birds. The number of nuclei named by various authors range from what is presented here to only dorsal and ventral laminae (Kooy, 1915, 1917). The detailed presentation of nuclei is presented

here in order to stimulate further investigation concerning the homologies which may exist between birds and mammals. For an extensive study of the caudal olivary complex in birds see Vogt-Nilsen (1954).

(17) **Fasciculus longitudinalis medialis.** This is the oldest and most constant longitudinal fiber system in the central nervous system of vertebrates. It is large in all vertebrates and exhibits a relatively larger size in more primitive animals. The medial longitudinal fasciculus of birds appears to be homologous to that of other vertebrates (Sarnat and Netsky, 1974).

(18) **Lemniscus medialis.** The medial lemniscus of birds resembles that of mammals in: its orgin from Nuc. gracilis and Nuc. cuneatus (Akker, 1970; Friedlander, 1898), its decussation within the Medulla oblongata and its ascending within the ventromedial portion of the brain stem to terminate in the thalamic, intercalate and dorsolateral nuclei (Wallenberg, 1904; Delids and Bennetto, (1972).
Nuc. ambiguus. Synonymy: Nuc. motorius ventralis n. vagi (Bolton, 1971).

(19) **Nuc. angularis.** The angular nucleus is divisible into three parts in the pigeon brain: Pars lateralis, consisting of large cells; Pars medialis which resembles the adjacent part of Nuc. magnocellularis; and Pars ventralis which consists of small cells. The ventral part recieves afferents from both the cochlear and lagenar nerves, whereas the medial and lateral parts receive afferents from the cochlear nerve (Boord and Rasmussen, 1963). The distribution of these afferents appear to maintain a tonotopic distribution within the Nuc. angularis, similar to that observed in the cochlear nuclei of mammals. It is suggested that Pars lateralis and Pars ventralis of Nuc. angularis are homologous to the caudal division of the dorsal cochlear nucleus of mammals. The medial part of Nuc. angularis, with the medial and lateral parts of Nuc. magnocellularis appear to correspond to the ventral cochlear nucleus of mammals (Boord and Rasmussen, 1963).

(20) **Commissura infima; Nuc. commissuralis.** The latter may represent the Nuc. commissuralis tractus solitarii of Cajal.

(21) **Nuc. intercalatus.** The intercalate nucleus lies between the hypoglossal and vagus nerve nuclei and has been referred to as the dorsomedial nuclear group of the vagus nerve. It appears, however, that both a Nuc. intercalatus and a dorsomedial portion of the vagus nerve are present in birds as separate entities.

(22) **Nuc. intermedius medullae oblongatae.** The intermediate nucleus of the Medulla oblongata should not be confused with Nuc. intermedius of the spinal cord (see Annot. 8). Nuc. intermedius of the Medulla oblongata is said to contain both vagal and hypoglossal neurons in some birds, in others it is apparently entirely vagal or entirely hypoglossal in its composition.

(23) **Nuc. laminaris.** Many others have proposed that the laminar nucleus represents a primary auditory nucleus. Cochlear and lagenar fibers, however, are not directly distributed to Nuc. laminaris. Boord (1968) indicates that Nuc. laminaris does, however, receive afferents from Nuc. angularis, and suggests that this nucleus is an avian homologue of the medial nucleus of the rostral olivary complex of mammals.

(24) **Nuc. magnocellularis cochlearis.** The magnocellular nucleus can be divided into three subnuclei on the basis of cell morphology and fibrillar architecture (Boord

and Rasmussen, 1963). Two parts of the nucleus have been recognized by other investigators (Brandis, 1894; Holmes, 1903; Cajal, 1908; Craigie, 1930; and Sanders, 1929). Boord and Rasmussen (1963) demonstrated that the ventral lateral portion of the nucleus receives mixed afferents from the cochlear and lagenar components of the vestibulocochlear nerve, whereas the remainder of the Nuc. magnocellularis cochlearis receives afferents only from the cochlear components. They suggest that Nuc. magnocellularis is an avian homologue of the rostral part of the ventral cochlear nucleus of mammals. The large celled medial part of the Nuc. magnocellularis receives fibers that terminate as bulbs of Held and rise from the basal and middle thirds of the cochlea. The lateral part receives fibers from the apical third of the cochlea which terminate as bulbs of Held and as pericellular plexuses. On the basis of this architectural arrangement it is derived that the medial part of the Nuc. magnocellularis is the avian equivalent of region III of the rostroventral cochlear nucleus of Harrison and Irving (1965) and that the lateral part of the nucleus is equivalent to region II of the rostral ventral cochlear nucleus.

(25) **Nuc. motorius dorsalis n. vagi** (see Annot. 30). This nucleus has been referred to as the ventrolateral nucleus of the vagus nerve by those who classified Nuc. intercalatus as Nuc. dorsomedialis and by others. Craigie identified this as a nucleus of the glossopharyngeal nerve as well as the vagus nerve. The ventrolateral nuclear cell mass of the vagus and glossopharyngeal nerves, however, is clearly the Nuc. ambiguus of mammals. A portion of the nucleus of the glossopharyngeal nerve is, however, associated with the dorsal motor nucleus of the vagus. Several investigators have divided the nucleus of the glossopharyngeal nerve into dorsal and ventral parts. The dorsal part is associated with the vagus nucleus (Kuhlenbeck, 1975).

(26) **Nuc. n. hypoglossi.** There has been speculation that a portion of the hypoglossal nerve arises from Nuc. ambiguus since the hypoglossal nerve apparently supplies the muscle of the syrinx. Ariëns Kappers (1920), however indicates that this innervation arises from the Nuc. n. hypoglossi, and represents the avian counterpart of mammalian Ramus descendens hypoglossi which innervates the sternohyoideus muscle. See **PNS** Annot. 30.

(27) **Nuc. reticularis paramedianus.** Petrovicky (1966) indicates that the paramedian reticular nucleus is absent in the bird; it appears, however, to be present in the chicken, turkey, Java Dove (*Streptopelia "risoria"*), and pigeon sections examined in the development of this nomenclature. The stereotaxic atlas of the pigeon brain of Karten and Hodos (1967) indicates the presence of this nucleus.

(28) **Nuc. tr. solitarii.** Facial nerve projections to Tr. solitarius terminating in Nuc. tr. solitarii are commonly assumed to contain general visceral afferent fibers, but neither cutaneous exteroceptive nor special visceral afferent (taste) fibers have been demonstrated to be contained within the facial nerve (Kuhlenbeck, 1975). There are, however, terminations of facial afferents within the rostral extremity of Nuc. tr. solitarii; glossopharyngeal nerve afferents (containing most, if not all avian taste afferents to the brain); and some general visceral afferents that terminate caudal to the terminals of the facial nerve. Vagus nerve afferents enter the caudal portion of the nucleus (Kuhlenbeck, 1975).

(29) **Nuclei vestibulares.** The nomenclature of the vestibular nuclei adopted in this work is derived from Sanders (1929); for comparisons with other terminologies consult Larsell (1967) and Kuhlenbeck (1975).

(30) **Nuc. ambiguus.** Synonymy: Nuc. motorius ventralis n. vagi (Bolton, 1971).

(31) **Nuc. n. accessorii.** Huber (1936) describes a column of cells that occupy a position in the lateral part of the gray matter at about the level of the central canal and extend over the upper three or four cervical segments of the spinal cord (in the pigeon). This is known as the column of Lenhossék. Huber notes that the cells of this column are probably the cells of origin of the fibers of Lenhossék which have been regarded as the non-mammalian representatives of the spinal portion of the accessory nerve in mammals.

(32) **Tr. occipitomesencephalicus.** Zeier and Karten (1971) have demonstrated that the occipitomesencephalic tract originates exclusively from the rostral two-thirds of the Archistriatum, and is distributed ipsilaterally to the following structures: lateral part of the Nuc. spiriformis medialis, Nuc. subrotundus, Nuc. principalis precommissuralis, and lateral reticular formation, Nuc. intercollicularis, Stratum griseum centrale of Colliculus mesencephali, Locus ceruleus, Nuclei subceruleus dorsalis and ventralis and Nuc. pontis lateralis. Caudal to this level the tract is distributed bilaterally to Nuc. reticularis parvocellularis, Nuc. subtrigeminalis, Nuc. tr. spinalis. n. trigemini and Nuclei gracilis and cuneatus. The contralateral component continues into the spinal cord where it terminates within the first few cervical segments, overlapping somewhat with the termination of the septomesencephalic tract. This tract may be considered to represent the avian counterpart of Bagley's bundle in the goat (Haartsen and Verhaart, 1967) which is considered to be a variant form of the pyramidal tract of primates (Zeier and Karten, 1971).

(33) **Nuc. n. facialis, Pars dorsalis.** This nucleus innervates the M. depressor mandibulae; it is generally continuous with the medial part of the Nuc. motorius n. trigemini. This relationship has been referred to as forming a trigeminal-facial complex, but the variation in communications and branching patterns of peripheral nerves lends no real benefit to such a consideration (see Kuhlenbeck, 1975).

(34) **Corpus trapezoideum.** The trapezoid body is composed of third order axons from cells of Nuc. laminaris and second order axons from cells of the medial part of Nuc. angularis and lateral part of Nuc. magnocellularis (Boord, 1968). As these nuclei are considered to represent avian homologues of the dorsal and ventral cochlear nuclei and the medial nucleus of the rostral olivary complex of mammals, it appears that the Corpus trapezoideum of birds is homologous to that of mammals.

(35) **Nuc. n. abducentis accessorius.** This nucleus is found in numerous forms of birds as a small nuclear mass located ventrolateral to the principal motor nucleus of the abducent nerve. Kuhlenbeck, (1975) believes that it innervates the muscles of the nictitating membrane (see **PNS** Annot. 16).

(36) **Nuc. lemnisci lateralis.** The lateral lemniscus has been demonstrated to serve as the major afferent pathway to the nucleus of the lateral lemniscus and Nuc. mesencephalicus lateralis, pars dorsalis in the pigeon (Boord, 1968) and as efferents from the avian homologues of the mammalian cochlear nuclei and Nuc. laminaris, an avian counterpart of the medial nucleus of the rostral olivary complex (Cajal, 1908; Stotler, 1905). Because of these relationships the lateral lemniscus of birds can be considered to be homologous to the mammalian lateral lemniscus.

(37) **Nuc. lemnisci lateralis, Pars lateroventralis.** Boord (1968) equates Pars latero-

ventralis of the lateral lemniscus with the Nuc. ventralis lemnisci lateralis of Karten and Hodos (1967).

(38) **Nuc. motorius n. trigemini.** The relationships of the subdivisions of the motor nucleus of the trigeminal nerve to specific muscle groups are discussed by Kossaka and Hiraiwa (1905). See Annot. 33.

(39) **Nuc. sensorius principalis n. trigemini.** The size of this nucleus varies greatly in different taxonomic forms, apparently correlating with the size of the beak (Stingelin, 1961). This nucleus receives a topographically organized projection from the trigeminal ganglion and serves as the origin of the quintofrontal tract (Wallenberg, 1903; Woodburne, 1936; Zeigler and Karten, 1973). The nucleus is clearly divisible into dorsal and ventral components (Woodburne, 1936).

(40) **Tr. quintofrontalis.** This tract arises from both dorsal and ventral divisions of Nuc. sensorius principalis n. trigemini, undergoes partial decussation and terminates in Nuc. basalis of the telencephalon; it possibly represents a lemniscal pathway in birds (Wallenberg, 1903; Woodburne, 1936; and Witkovsky *et al.*, 1973).

(41) **Sectiones cerebelli.** Tr. isthmocerebellaris; Tr. cerebellomotorius; Tr. tectocerebellaris are often described in avian neuroanatomical literature. Anatomic evidence for these tracts, however, is doubtful. For this reason they have not been included as official terms in this nomenclature.

(42) **Tectum mesencephali.** Synonymy: Lobus opticus; Tectum opticum. As a gross brain feature Tectum mesencephali has traditionally been called Lobus opticus (or Tectum opticum; see below, Annot, 43). According to Cohen and Karten 1974, p. 46), who have summarized this matter, the term "optic lobe" is a misnomer, as only a limited portion of the lobe is actually related to the visual system. Since the midbrain tectum does function in other spheres of activity, it is deemed best to avoid the name with a functional connotation. Therefore, Tectum mesencephali, is recommended by all members of the Sub-Committee as the preferred name.

(43) **Colliculus mesencephali.** Synonymy: Tectum opticum. Cohen and Karten (1974, p. 46) point out that the failure to recognize the fundamental distinction between the optic lobe and the optic tectum has led to much confusion, particularly regarding the relationship of avian and mammalian brains. According to them: "Properly speaking, the optic lobe of birds is equivalent to the superior colliculus of mammals, and the optic tectum of birds is considered equivalent to the superficial cap of the superior colliculus . . . ." The superficial laminated rind constituting the surface of Tectum mesencephali contains the primary retinal input and associated interneuronal and efferent zones; this is designated as Colliculus mesencephali, which has a topographic basis for the name. See Annot. 42.

(44) **Brachium colliculi mesencephali.** The mammalian brachium of the caudal colliculus is represented in birds as the Tr. nuc. ovoidalis. This tract contains axons with cell bodies of origin in Nucleus mesencephali lateralis, Pars dorsalis, the avian counterpart of the mammalian caudal colliculus (Karten, 1967); it terminates primarily in Nuc. ovoidalis, the avian counterpart of the mammalian ventral division of the medial geniculate nucleus (Karten, 1967). Brachium colliculi mesencephali, therefore refers specifically to "tectofugal" fibers projecting from the mesencephalic colliculus to Nuc. rotundus, the avian counterpart of the mammalian Nuc. lateralis

caudalis of the thalamus (Karten and Revzin, 1966; Karten and Hodos, 1970). See Annot. 43.

(45) **Decussatio n. oculomotorii.** Some of the fibers from the ventral part of the oculomotor nuclear complex decussate proximal to emerging as the oculomotor nerve (Kuhlenbeck, 1975).
Nuc. n. oculomotorii, Pars accessoria. Synonymy: Nucleus of Edinger-Westphal.

(46) **Nuc. commissuralis caudalis [posterior].** This nucleus is distinct from Nuc. spiriformis lateralis and Nuc. spiriformis medialis which have been considered by some investigators as homologous to the nucleus of the caudal commissure of mammals.

(47) **Nuc. ectomamillaris.** Synonomy: nucleus of the basal optic tract. This nucleus appears to be homologous to the nucleus of the basal optic tract of mammals.

(48) **Nuc. intercollicularis.** This nucleus has been demonstrated to receive afferents from the spinal cord, Nuc. lateralis of the cerebellum, Colliculus mesencephali, Nuc. gracilis and Nuc. cuneatus (Karten, 1963, 1965, 1967).

(49) **Complexus isthmi.** This isthmal complex is comprised of Nuc. magnocellularis isthmi, Nuc. parvocellularis isthmi, Nuc. isthmo-opticus, and Nuc. semilunaris. It receives an intense and topographically organized input from the Colliculus mesencephali (Cohen and Karten, 1974).

(50) **Nuc. isthmo-opticus.** Synonymy: Ganglion opticum dorsale (Bellonci, 1888; Jelgersma, 1896), Nuc. opticus medialis (Perlia, 1889), Ganglion isthmi (Edinger and Wallenberg, 1899) and the nucleus of the isthmo-optic tract (Craigie, 1928) as well as the Isthmo-optic nucleus (Huber and Crosby, 1929). This nucleus has been demonstrated to give rise to a definitive efferent tract (Tractus isthmo-opticus) which projects to the retina (Cowan and Powell, 1963; Holden, 1966; Ogden, 1967; and Cowan and Wenger, 1968).

(51) **Fasciculus lateralis prosencephali.** According to Kuhlenbeck (1977) the avian lateral forebrain bundle is composed of the nine tracts listed under this term.

(52) **Nuc. mesencephalicus lateralis, Pars dorsalis:** This nucleus appears to be the avian homologue of the caudal colliculus of mammals. It has been referred to as the Ganglion laterale and has been proposed to be the pneumotaxic center of birds. The latter conclusion is unlikely as this nucleus is clearly a mesencephalic structure. The term "torus semilunaris" has been applied to this nucleus, but the term does not apply well to birds, and should be retained only for reptiles and amphibians.

(53) **Nuc. parageniculatus colliculi mesencephali.** Synonymy: Nuc. parageniculatus tecti optici and Nucleus opticus lateralis by Craigie (1930) and tectal grey by Huber and Crosby (1926).

(54) **Nuc. paragrisealis centralis mesencephali.** Synonymy: Nuc. of Darkschewitsch. As no alternate term has been applied in the literature which does not invoke the eponym, it was necessary to coin an alternate term. This nucleus is especially well developed in birds.

(55) **Nuc. pretectalis:** Kuhlenbeck (1937) identified Nuclei pretectalis lateralis, medialis, and principalis. He also identified. Nuc. areae pretectalis, pars lateralis and pars medialis. This terminology has not been widely adopted by other authors, but may be preferred as these structures are readily identified microscopically. Ariëns Kappers *et al.*, (1936) indicated that the pretectal and subprectal areas in birds are extremely well developed, and considered these structures to be intermediaries between the diencephalon and the tectum of the midbrain.

(56) **Tr. septomesencephalicus.** This tract arises from the dorsal telencephalon, predominantly from the Hyperstriatum accessorium and projects to the lateral Neostriatum, the peri-ectostriatal field, internal lamella of the ventral geniculate nucleus, pretectal nuclei, Colliculus mesencephalici, Nuc. intercalatus thalami, Nuc. spiroformis medialis, Nuc. ruber, the medial recticular formation, pontine nuclei and Nuclei cuneatus and gracilis. The fiber system continues contralaterally into the spinal cord to terminate within the dorsal horn of the spinal grey where it minimally overlaps with the terminations of the occipitomesencephalic tract (see Annot. 32; Adamo, 1967; Karten, 1971; and Zecha, 1962).

(57) **Ansa lenticularis.** Ansa lenticularis arises exclusively from the Paleostriatum primitivum—Nuc. intrapeduncularis component of the paleostriatum complex (see Annot. 76). Ansa lenticularis terminates in Nuc. ansae lenticularis rostralis, Nuc. ansae lenticularis caudalis, Nuc. dorsointermedius caudalis, Nuc. spiriformis lateralis and Nuc. tegmenti pedunculopontini. On the basis of anatomic and histochemical studies by Karten and Dubbeldam (1973) it appears that the avian Ansa lenticularis is a counterpart of that of mammals.

(58) **Nuc. opticus principalis thalami.** Nuc. dorsolateralis rostralis, Pars lateralis, Pars medialis (also known as pars magnocellularis) and Nuc. lateralis rostralis are collectively considered to be the Nuc. opticus principalis thalami (Karten *et al.*, 1973). The complex appears to be an avian homologue of the dorsal nucleus of the lateral geniculate of mammals. These nuclei represent a primary thalamic termination of the optic tract in birds, and remain quite distinct from the Nuc. geniculatus lateralis, Pars ventralis and from pretectal nuclei.

(59) **Nuc. dorsolateralis caudalis; Nuc. superficialis parvocellularis; Nuc. intercalatus thalami.** These nuclei have been demonstrated to represent thalamic relay nuclei for cutaneous sensory information in the birds (Delius and Bennetto, 1972). Nuc. dorsolateralis caudalis and Nuc. superficialis parvocellularis have been demonstrated to project to telencephalic structures, Hyperstriatum intercalatus and the rostral and medial portions of Neostriatum caudale (Erulkar, 1955; Delius and Bennetto, 1972). The homologies of these nuclei with thalamic nuclei in mammals is not clear at the present time, but the possibility is evident that at least a portion of the nuclei involved, most likely Nuc. dorsolateralis caudalis and Nuc. superficialis parvocellularis, are avian counterparts of the Nuc. ventralis posterolateralis of mammals.

(60) **Nuc. dorsomedialis, Pars rostralis.** This nucleus gives rise to a fiber system which accompanies the Fasciculus prosencepali medialis rostrally to terminate within the medial division of the Hyperstriatum dorsale and the immediately adjacent dorsomedial portion of the Hyperstriatum accessorium (Zeier and Karten, 1971). No functional significance has been alleged to these structures, but it has been speculated that Nuc. dorsomedialis rostralis may be comparable to the mammalian anterior thalamic complex, and its projection field in Hyperstriatum dorsale and Hyper-

striatum accessorium may represent a portion of the mammalian "limbic" cortex (Karten *et al.*, 1973). This speculation is strengthened by the proximity of the telencephalic region involved in this projection to the region of the avian brain which has been traditionally considered to represent the hippocampal formation of mammals (Karten *et al.*, 1973).

(61) **Nuc. entopeduncularis ventralis superior.** This nucleus has been considered to be an avian homologue of the subthalamic nucleus of mammals. It is referred to as the entopeduncular nuclear group by Craigie (1928), and is represented by nerve cell bodies which are scattered along Fasciculus prosencephali lateralis. The cells resemble those of Paleostriatum primitivum. This nucleus includes: (1) Nuc. dorsalis supraopticum, (2) bed nucleus of Tr. thalamofrontalis anterior, and (3) Nuc. parastriatus of Rendahl (1924).

(62) **Nuc. ovoidalis.** Synonymy: Nuc. B (Rendahl, 1924); Nuc. anterior ventralis (Edinger and Wallenberg, 1889). Nuc. ovoidalis represents the thalamic relay nucleus of the auditory system. Karten (1967) shows that this nucleus is the avian homologue of the ventral portion of the mammalian medial geniculate body.

(63) **Nuc. rotundus.** Anatomic, electrophysiologic and behavioral studies indicate that Nuc. rotundus represents a major thalamic relay for the visual system of birds (Cowan *et al.*, 1961; Karten and Revzin, 1966). Afferents to the nucleus arise in Tectum mesencephali and efferents project to the Ectostriatum (Karten *et al.*, 1973). This pathway is referred to as the "tectofugal pathway" for the visual system of birds (Karten *et al.*, 1973). Nuc. rotundus appears to be the avian counterpart of the Nuc. lateralis caudalis of mammals.

(64) **Nuc. semilunaris para-ovoidalis.** This nucleus lies adjacent to the ventrolateral aspect of Nuc. ovoidalis, and receives a significant number of afferent fibers from Tr. nuc. ovoidalis (Karten, 1967).

(65) **Nuc. spiriformis lateralis.** Complexus spiriformis is considered to include Nuclei spiriformis lateralis, spiriformis medialis, principalis precommissuralis, dorsointermedialis caudalis, and dorsolateralis caudalis (Karten and Dubbeldam, 1973). The afferents to this region of the thalamus include those from the spinal cord, deep cerebellar nuclei, basal ganglia homologues, the occipitomesencephalic tract, and possibly a small projection from Nuc. cuneatus and Nuc. gracilis. Each of these afferent systems project predominantly to different parts of the spiriform complex, with varying degrees of overlap. These observations suggest that the spiriform complex corresponds at least in part to the ventral tier of nuclei of the mammalian thalamus (Karten and Dubbeldam, 1973).

(66) **Nuc. superficialis parvocellularis.** Synonymy: Nuc. tr. septomesencephali, pars rostralis (Craigie, 1930).

(67) **Nuc. triangularis.** Synonymy: Nuc. tr. habenulopeduncularis (Huber and Crosby (1929) and Craigie (1930).

(68) **Tr. nuc. ectomamillaris.** Synonymy: Tr. opticus basalis.

(69) **Fissura subhemispherica.** Synonymy: Transverse cerebral fissure (Baumel, 1967). The cleft between the ventral surface of one telencephalic hemisphere and the dorsal surface of the mesencephalic tectum (see Annot. 83).

(70) **Sectiones telencephali.** In order to assist in clarifying the very confused terminology which has been applied to the major divisions of the telencephalon, Table I provides the general terms utilized by various authors.

(71) **Archistriatum.** Consideration of its cytoarchitectural organization and hodologic relationships makes it apparent that the Archistriatum is vastly more complex than has been previously assumed (Zeier and Karten, 1971, 1973). Some of the associated subdivisions of the Archistriatum are included in this nomenclature; however, it should be emphasized that the subdivisions listed are by no means all of the cytoarchitectonic components of the Archistriatum. By way of simplification of the structural relationships of the Archistriatum (based on investigations of Zeier and Karten, 1973), it appears that its caudal one-third and most medial parts (Archistriatum caudalis and Archistriatum medialis) represent "limbic" components of the Telencephalon. These regions project to the Hypothalamus via Tr. occipitomesencephalus, Pars hypothalamica, and may represent the avian counterpart of the mammalian Amygdala (Zeier and Karten, 1971) (see Annot. 56). Its rostral two-thirds (Archistriatum rostrale and Archistriatum intermedium) give rise to the nonhypothalamic component of the Tr. occipitomesencephalicus, and appear to function in a "somatic" rather than a viscero-endocrine effector mechanism. It has been proposed that this area of the Archistriatum compares well with the pericentral cortex of the goat and the sensorimotor cortex of primates (Zeier and Karten, 1971).

(72) **Commissura rostralis.** Synonymy: Commissura interarchistriatica (Ariëns Kappers *et al.,* 1936). On crossing the midline this commissure divides into two major fascicles: a relatively diffuse rostromedial branch (Pars bulbaris) and a more compact, laterally directed caudal branch (Pars temporalis). Pars bulbaris passes rostromedially to Fasciculus prosencephali lateralis, and terminates in an area ventral and lateral to Nuc. accumbens, Lobus paraolfactorius, dorsal portion of the olfactory tubercle, and limited areas of the medial edge of Paleostriatum intercalatum supremum. Pars temporalis partially terminates in the rostral one-third of the Archistriatum, and continues by two paths: a dorsolateral projection to the temporo-parieto-occipital area as it continues rostrolaterally to Hyperstriatum intercalatum supremum, and a ventrolateral projection paralleling Tr. fronto-archistriaticus to terminate in the deep layers of the piriform cortex (Zeier and Karten, 1973).

(73) **Area [Cortex] prepiriformis.** The prepiriform area [cortex] of birds does not appear to be a homologue of the olfactory tubercle of mammals.

(74) **Complexus paleostriaticus.** Several investigations have suggested that the paleostriate complex, consisting of Paleostriatum augmentatum, Paleostriatum primitivum and Nuc. intrapeduncularis, represent the equivalent of the mammalian "basal ganglia" (caudate-putamen and globus pallidus), whereas the overlying "striatal" masses of Neostriatum; Ectostriatum; Hyperstriatum ventrale, dorsale, and accessorium are representative of the mammalian neocortex (Karten, 1968, 1969; Karten and Hodos, 1970; Nauta and Karten, 1970; Zeier and Karten, 1971; and Juorio and Vogt, 1967). These assumptions have been supported by Karten and Dubbeldam (1973) utilizing histochemical and neuroanatomic methods of study. The results of these studies indicate that Lobus parolfactorius should be considered to represent the basomedial part of the head of the caudate nucleus of mammals, and thus should be as a part of the Complexus paleostriaticus.

(75) **Paleostriatum augmentatum.** This appears to represent the avian homologue of the mammalian caudate-putamen (Karten and Dubbeldam, 1973).

(76) **Paleostriatum primitivum; Nuc. intrapeduncularis.** These structures appear to correspond to the outer and inner laminae of the Globus pallidus of mammals (Karten and Dubbeldam, 1973).

(77) **Lobus parolfactorius.** The parolfactory lobe has been considered to represent a medial portion of the Paleostriatum augmentatum (Stingelin, 1956, 1958). Karten and Dubbeldam (1973), however, emphasize a clear distinction between these two structures based on cytoarchitectonics and hodologic characteristics. Lobus parolfactorius gives rise to efferents which join the Fasciculus prosencephali medialis to the lateral preoptic area and rostral hypothalamus. On the basis of their studies, these investigators concluded that the baso-medial portions of the head of the caudate nucleus of mammals correspond to the Lobus parolfactorius of birds.

(78) **Hyperstriatum accessorium** (Figs. 1, 2). Hyperstriatum accessorium makes up a major part of the gross brain feature, the **Eminentia sagittalis** or "Wulst", which is located on the dorsal aspect of the telencephalic hemisphere. The sagittal eminence may include (depending on the author one reads) the: Hyperstriatum accessorium and a variable subpial molecular layer, Lamina frontalis superior, Lamina frontalis suprema, Hyperstriatum dorsale, and Nuc. intercalatus hyperstriati accessorii. Medially the sagittal eminence grades into the parahippocampal and hippocampal areas (Karten *et al.*, 1973).

(79) **Nuc. ansae lenticularis caudalis.** Synonymy: Nucleus of the dorsal supraoptic decussation (Huber and Crosby, 1929); Nuc. entopeduncularis postero-superior (Kuhlenbeck, 1937). Nuc. ansae lenticularis caudalis corresponds to the entopeduncular nucleus of the alligator (Huber and Crosby, 1926) and a cell group believed by Papez (1935) to be comparable to the mammalian subthalamic nucleus. Powell and Cowan (1961) referred to the nucleus as the Nuc. posteroventralis and Baker-Cohen (1978) identified it as the entopeduncular nucleus. These nuclei have been referred to as the avian equivalent of the internal segment of the mammalian Globus pallidus, or Substantia nigra. Karten and Dubbeldam (1973) however, indicate that on the basis of current information these assumptions are not justified, but that Nuclei ansae lenticularis rostralis and caudalis may more correctly be considered to represent the avian counterpart of the subthalamic nucleus, nuclei of the field of Forel and other cell groups which receive afferents from the Ansa lenticularis in mammals. It should be emphasized, however, that the mammalian homologue of these nuclei is not definitive at present.

(80) **Nuc. basilis [trigeminalis prosencephali].** The synonym, Nuc. trigeminalis prosencephali, indicates that this nucleus is primarily related to the termination of ascending fibers from the principal sensory nucleus of the trigeminal nerve. It may in fact represent a thalamic relay nucleus which is somewhat remote from the thalamus proper, but is is not clear that Nuc. basalis exclusively serves this function (Wallenberg, 1903; Woodburne, 1936; and Witkovsky *et al.*, 1973). The primary term. Nuc. basalis, is retained in the present terminology on account of this contingency and the fact that this term is well established in the literature. Stingelin (1961) noted that the relative size of Nuc. basalis is proportional to the size of the principal sensory nucleus of the trigeminal nuclear complex, which in turn reflects the relative degree of beak development in various birds. Physiologic studies indicate

that the principal sensory nucleus of the trigeminal complex is linked to Nuc. basalis by way of a direct monosynaptic pathway, Tr. quintofrontalis (Witkovsky *et al.*, 1973).

(81) **Tr. fronto-archistriaticus.** This tract originates in Nuc. basalis and terminates in Archistriatum and overlying Neostriatum. Thus the term fronto-archistriaticus is not fully descriptive of its origin and termination (Zeier and Karten, 1971; Cohen and Karten, 1974).

(82) **Meninges.** Earlier workers were able to identify only a Lamina pia-arachnoidea, with little or no development of a subarachnoid space, and the Dura mater encephali in birds (see Hodges, 1974; Baumel, 1975, for reviews). More recently the studies of Böhme (*Anat. Hist. Embryol.* 3: 233–242, 1974) and that of Jones and Dolman (*J. Anat.* 128: 13–29, 1979) using light and electron microscopy have demonstrated distinct pial and arachnoid layers, the subarachnoid cavity, and arachnoid granulations in the pigeon and the chicken, equivalent to those in mammals.
Dura mater. The cranial Dura mater consists of two layers closely adherent to one another, but separable. Cranial dural venous sinuses are located in the plane between the two layers; both layers are vascularized by rami of the cerebral arteries. The internal vertebral venous sinus lies in the epidural space between the tube of spinal dura mater and the periosteum of the vertebral canal.

(83) **Plica tentorialis.** This is a transverse fold of Dura mater that lies in the Fissura subhemispherica and partially separates the mesencephalic tectum (lobe) from the ventral surface of the occipital pole of the telencephalic hemisphere on each side. The fold is attached to (and extends) the Crista tentorialis, an osseous ledge on the side wall of the cranial cavity (see **Osteo.** Fig. 3) and the **VNA** (ICVAN, 1973).

(84) **Ligg. suspensoria transversa.** Ligamentous bands of Pia mater that stretch from the Lig. denticulatum to the Lig. ventromedianum and on to the opposite Lig. denticulatum, forming a sort of hammock for the lumbosacral intumescence of the spinal cord (see Dingler, 1965).

TABLE I Nomenclature of telencephalon

Edinger, Wallenberg and Holmes, 1903	Rose, 1914	Huber and Crosby, 1929; Kappers, Huber and Crosby, 1936	Kuhlenbeck, 1938; Jones and Levi-Montalcini, 1958	Karten and Hodos, 1967	ICAAN
Cortex frontalis	B	Hyperstriatum accessorium	Nucleus diffusus dorsalis	Hyperstriatum accessorium	Hyperstriatum accessorium
Frontalmark	A	Nucleus intercalatus hyperstriati supremi	Nucleus diffusus dorsolateralis	Hyperstriatum intercalatum	Hyperstriatum intercalatum supremum
	C	Hyperstriatum dorsale	Nucleus epibasalis dorsalis,	Hyperstriatum dorsale	Hyperstriatum dorsale
		Nucelus intercalatus hyperstriati superioris	pars superior	Lamina frontalis superior	Lamina frontalis superior
Hyperstriatum	D, D_1	Hyperstriatum ventrale (dorso-ventrale) (ventro-ventrale)	Nucleus epibasalis dorsalis, pars inferior	Hyperstriatum ventrale (dorso-ventrale) (ventro-ventrale)	Hyperstriatum ventrale Pars dorsalis Pars ventralis
	G_1	Neostriatum frontale	Nucleus epibasalis centralis, pars medialis	Neostriatum frontale	Neostriatum, Pars rostralis
	G, G_2	Neostriatum intermediale	Nucleus epibasalis centralis, pars posterior	Neostriatum intermedium	Neostriatum, Pars intermedia
	L, G_3	Neostriatum caudale		Neostriatum caudale	Neostriatum, Pars caudalis

TABLE I—*continued*

Edinger, Wallenberg and Holmes, 1903	Rose, 1914	Huber and Crosby, 1929; Kappers, Huber and Crosby, 1936	Kuhlenbeck, 1938; Jones and Levi-Montalcini, 1958	Karten and Hodos, 1967	ICAAN
Parolfactory lobe				Lobus parolfactorius	Lobus parolfactorius
Mesostriatum laterale	R	Nucleus basalis	Nucleus epibasalis ventrolateralis	Nuc. basalis	Nuc. basalis
Ectostriatum	S	Ectostriatum	Nucleus epibasalis centralis accessorium	Ectostriatum	Ectostriatum
Epistriatum	K	Archistriatum	Nucleus epibasalis caudalis	Archistriatum	Archistriatum
Mesostriatum	H	Paleostriatum augmentatum	Nucleus basalis	Paleostriatum augmentatum	Paleostriatum augmentatum
Nucleus entopeduncularis	J	Paleostriatum primitivum	Nucleus entopeduncularis	Paleostriatum primitivum	Paleostriatum primitivum

General Comments. Most of the CNS figures are representative transverse sections of the brain redrawn from the Karten and Hodos' atlas of the brain of the pigeon (1967). Only two selected sections of the spinal cord have been included. The sequential figures start at the caudal end of the brain and extend rostrally. One half of the brain is shown; the sections of the cerebellum have not been included as parts of the figures of the mid-brain and the stem of the hindbrain.

Figure 3 is a parasagittal section of the brain 0·5 mm lateral to the median plane; it is marked with longitudinal scales (in mm) to be used for locating the levels of the various transverse sections. This section identifies a number of nuclei and tracts that serve as reference points.

The legends for the figures all contain the Karten–Hodos designations of the stereotaxic planes. For example, P 4·00 refers to a transverse section 4·00 mm caudal (posterior) to the 0·00 base line; A 8·50 refers to a section 8·5 mm rostral (anterior) to the base line.

Abbreviations:
Area parahipp. Area parahippocampalis
Area temp.-parieto-occip. Area temporo-parieto-occipitalis
Brach. collic. mesen. Brachium colliculi mesencephali
Collic. mesen. Colliculus mesencephali
Commiss. tect. Commissura tectalis
Complexus oliv. caud. Complexus olivaris caudalis
Decuss. n. IV Decussatio n. trochlearis
Decuss. supraopt. Decussatio supraoptica
Fasc. diagonalis Fasciculus diagonalis
Fasc. lat. pros. Fasciculus lateralis prosencephali
Fasc. longit. med. Fasciculus longitudinalis medialis
Hyperstr. intercal. Hyperstriatum intercalatum
Lam. front. Lamina frontalis
Lam. hyperstr. Lamina hyperstriatica
Nuc. centr. medullae obl. Nuc. centralis medullae oblongatae
Nuc. centr. sup. Nuc. centralis superior
Nuc. crbell lat. Nuc. cerebellaris lateralis
Nuc. commiss. pallii Nuc. commissurae pallii
Nuc. cun. access. Nuc. cuneatus accessorius
Nuc. dors. intermed. caud. Nuc. dorsointermedialis caudalis
Nuc. dorsolat. caud. Nuc. dorsolateralis caudalis
Nuc. dorsomed. thal. Nuc. dorsomedialis thalami
Nuc. ectomam. Nuc. ectomamillaris
Nuc. genic. lat. Nuc. geniculatus lateralis
Nuc. hab. Nuc. habenularis
Nuc. int. dors. Nuc. internus dorsalis
Nuc. int. ventr. Nuc. internus ventralis
Nuc. intercal. hyperstr. access. Nuc. intercalatus hyperstriati accessorii
Nuc. intercal. thal. Nuc. intercalatus thalami
Nuc. intermed. medullae spin. Nuc. intermedius medullae spinalis
Nuc. interpedunc. Nuc. interpeduncularis
Nuc. interstit. commiss. caud. Nuc. interstitialis commissurae caudalis
Nuc. lat. pont. Nuc. lateralis pontis
Nuc. lemn. lat. Nuc. lemnisci lateralis
Nuc. lentif. mesen. Nuc. lentiformis mesencephali
Nuc. lin. caud. Nuc. linearis caudalis
Nuc. loc. cer. Nuc. loci cerulei
Nuc. magnocell. isthmi Nuc. magnocellularis isthmi
Nuc. magnocell. preopt. Nuc. magnocellularis preopticus
Nuc. magnocell. strati gris. thal. Nuc. magnocellularis strati grisei thalami
Nuc. mam. Nuc. mamillaris
Nuc. magnocell. coch. Nuc. magnocellularis cochlearis

Nuc. med. pont. Nuc. medialis pontis
Nuc. mesen. lat. dors. Nuc. mesencephalicus lateralis, Pars dorsalis
Nuc. mesen. prof. Nuc. mesencephalicus profundus
Nuc. motor. dors. n. X Nuc. motorius dorsalis n. vagi
Nuc. motor. n. V Nuc. motorius n. trigemini
Nuc. n. III Nuc. n. oculomotorii
Nuc. n. IV Nuc. n. trochlearis
Nuc. n. VI Nuc. n. abducentis
Nuc. n. VI access. Nuc. n. abducentis accessorii
Nuc. n. VII Nuc. n. facialis
Nuc. n. IX Nuc. n. glossopharyngei
Nuc. n. XII Nuc. n. hypoglossi
Nuc. oliv. rostr. Nuc. olivaris rostralis
Nuc. opt. princip. thal. Nuc. opticus principalis thalami
Nuc. paragris. centr. mesen. Nuc. paragrisealis centralis mesencephali
Nuc. parvocell. isthmi Nuc. parvocellularis isthmi
Nuc. periventric. hypothal. Nuc. periventricularis hypothalami
Nuc. retic. gigantocell. Nuc. reticularis gigantocellularis
Nuc. retic. lat. Nuc. reticularis lateralis
Nuc. retic. parvocell. Nuc. reticularis parvocellularis
Nuc. retic. paragigant. lat. Nuc. reticularis paragigantocellularis lateralis
Nuc. retic. paramed. Nuc. reticularis paramedianus
Nuc. retic. pont. caud. Nuc. reticularis pontis caudalis
Nuc. sens. princip. n. V Nuc. sensorius principalis n. trigemini
Nuc. semilun. Nuc. semilunaris
Nuc. spirif. Nuc. spiriformis
Nuc. spin. crbell, dors. Nuc. spinocerebellaris dorsalis
Nuc. subcer. ventr. Nuc. subceruleus ventralis
Nuc. subhab. Nuc. subhabenularis
Nuc. subpretect. Nuc. subpretectalis
Nuc. subst. gel. Nuc. substantiae gelatinosae
Nuc. superf. parvocell. Nuc. superficialis parvocellularis
Nuc. teg. dors. Nuc. tegmenti dorsalis
Nuc. teg. pedunc. pont. Nuc. tegmenti pedunculopontini
Nuc. teg. ventr. Nuc. tegmenti ventralis
Nuc. tr. spin. n. V Nuc. tr. spinalis n. trigemini
Nuc. tuberis infund. Nuc. tuberis infundibuli
Nuc. vestib. dors. lat. Nuc. vestibularis dorsolateralis
Nuc. vestib. desc. Nuc. vestibularis descendens
Nuc. vestib. dors. med. Nuc. vestibularis dorsomedialis
Nuc. vestib. sup. Nuc. vestibularis superior
Nuc. vestib. tang. Nuc. vestibularis tangentialis
Nuc. vestib. ventr. lat. Nuc. vestibularis ventrolateralis
Paleostr. aug. Paleostriatum augmentatum
Paleostr. prim. Paleostriatum primitivum
Pedunc. crbell. caud + intermed. Pedunculus cerebelli caudalis + Pedunculus cerebelli intermedius
Pedunc. crbell. rostr. Pedunculus cerebelli rostralis
Rdx. n. III Rdx. n. oculomotorii
Rdx. n. VI Rdx. n. abducentis
Reg. lat. hypothal. Regio lateralis hypothalami
Subst. gris. fibr. periventric. Substantia grisea et fibrosa periventricularis
Tr. corticohab. Tr. corticohabenularis
Tr. dorso-archistr. Tr. dorso-archistriaticus
Tr. hab. interpedunc. Tr. habenulointerpeduncularis
Tr. isthm. opt. Tr. isthmo-opticus.
Tr. lam. oliv. Tr. lamino-olivaris

Abbreviations—*continued* on p. 454.

Tr. nuc. ectomam. Tr. nuc. ectomamillaris
Tr. occip. mesen. Tr. occipitomesencephalicus
Tr. opt. marg. lat. Tr. opticus marginalis, Pars lateralis
Tr. opt. marg. med. Tr. opticus marginalis, Pars medialis
Tr. sept. mesen. Tr. septomesencephalicus
Tr. spin. crbell. dors. Tr. spinocerebellaris dorsalis
Tr. spin. n. V. Tr. spinalis n. trigemini
Tr. thalamostr. Tr. thalamostriaticus
Tr. vestib. mesen. Tr. vestibulomesencephalicus
Ventric. tect. mesen. Ventriculus tecti mesencephali

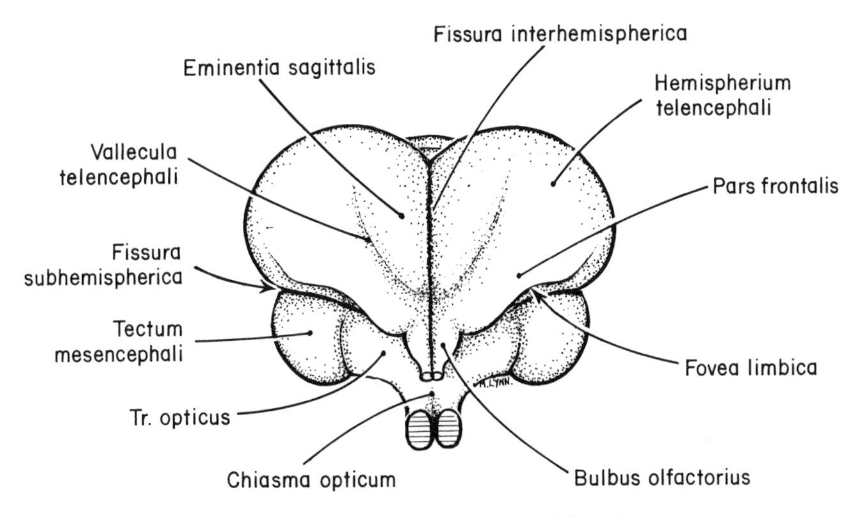

Fig. 1 Frontal aspect of the brain of the pigeon.

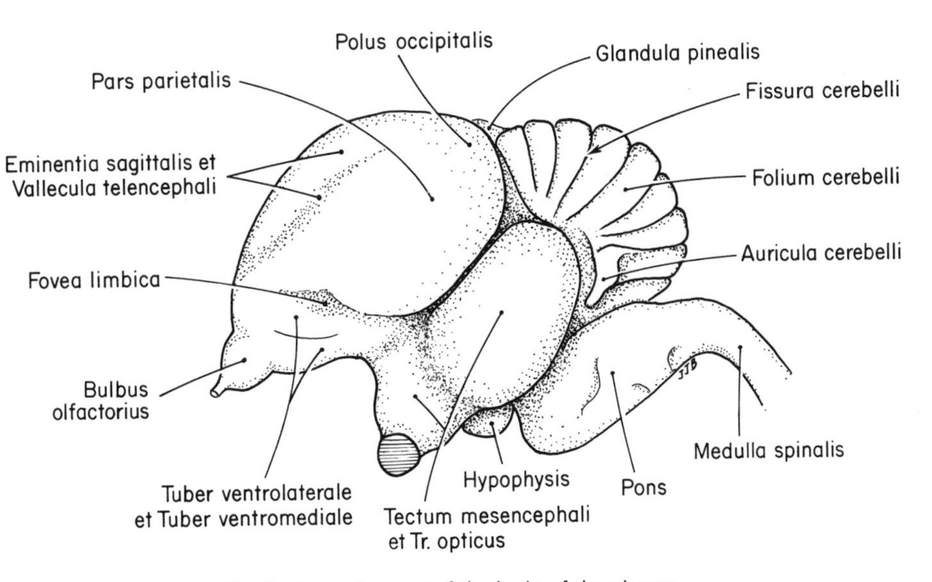

Fig. 2 Lateral aspect of the brain of the pigeon.

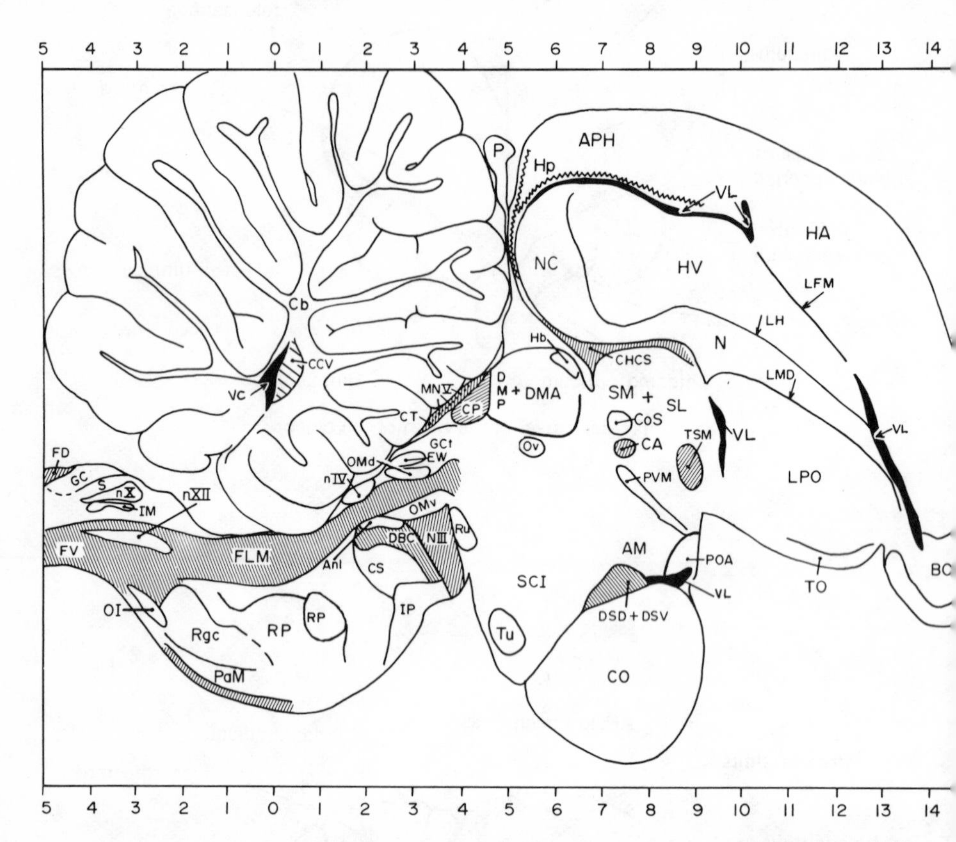

Fig. 3 Parasagittal section of the pigeon brain (0·5 mm from the median plane). The planes of the transverse sections of the brain are indicated by perpendicular lines connecting the longitudinal scales (in mm) at top and bottom of the figure. Note the zero base line through the middle of the cerebellum. All transverse sections rostral to the base line are designated as A (anterior), e.g. A 5·50, etc.; those sections caudal to the base line as P (posterior), e.g. P 4·00. From Karten and Hodos (1967).

Abbreviations:

Anl,	Nuc. annularis
APH,	Area parahippocampalis
BO,	Bulbus olfactorius
CA,	Commissura rostralis
Cb,	Cerebellum
CC,	Commissura cerebellaris
CHCS,	Tr. corticohabenularis + Tr. corticoseptalis
CO,	Chiasma opticum
COS,	Commissura pallii
CP,	Commissura caudalis
CS,	Nuc. centralis superior
CT,	Commissura tectalis
DBC,	Decussatio pedunculi cerebelli rostralis
DMP + DMA,	Nuc. dorsomedialis thalami, Pars caudalis + Pars rostralis
DSD + DSV,	Decussatio supraoptica dorsalis + Decussatio supraoptica ventralis
EW,	Nuc. n. oculomotorii, Pars accessoria (Nuc. of Edinger–Westphal)
FD,	Funiculus dorsalis
FLM,	Fasciculus longitudinalis medialis
FV,	Funiculus ventralis
GC,	Nuc. gracilis + Nuc. cuneatus
GCt,	Substantia grisea centralis
HA,	Hyperstriatum accessorium
Hb,	Nuc. habenularis medialis
Hp,	Hippocampus
HV,	Hyperstriatum ventrale
IM,	Nuc. intermedius medullae oblongatae
IP,	Nuc. interpeduncularis
LFM,	Lamina frontalis suprema
LH,	Lamina hyperstriatica
LMD,	Lamina medullaris dorsalis
LPO,	Lobus parolfactorius
MN V,	Nuc. mesencephali n. trigemini
N,	Neostriatum
N III,	N. oculomotorius
n IV,	Nuc. n. trochlearis
n X,	Nuc. motorius dorsalis n. vagi
n XII,	Nuc. n. hypoglossi
OI,	Complexus olivaris caudalis
OM,	Nuc. n. oculomotorii
OV,	Nuc. ovoidalis
P,	Glandula pinealis
PaM,	Nuc. paramedianus
POA,	Regio preoptica
PVM,	Nuc. magnocellularis preopticus
Rgc,	Nuc. reticularis gigantocellularis
RP,	Nuc. reticularis pontis caudalis + Nuc. reticularis pontis rostralis
Ru,	Nuc. ruber
S,	Nuc. tr. solitarii
SM + SL,	Area septalis
TO,	Tuberculum olfactorium
TSM,	Tr. septomesencephalicus
Tu,	Nuc. tuberis
Vc,	Ventriculus cerebelli
VL,	Ventriculus lateralis

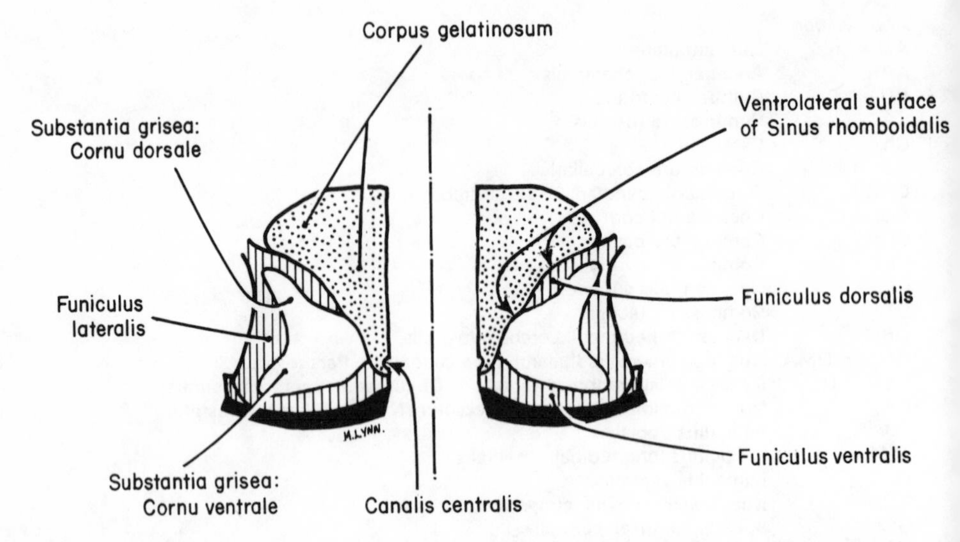

Fig. 4 Transverse section of the avian spinal cord (Pars synsacralis). The section is located at the level of the widest part of Sinus rhomboidalis in the lumbosacral enlargement of the cord. In this region the dorsal parts of the avian spinal cord are not united, producing an elongate, diamond-shaped fossa that contains the gelatinous body (see Annot. 2). Note that the central canal is situated within the gelatinous body.

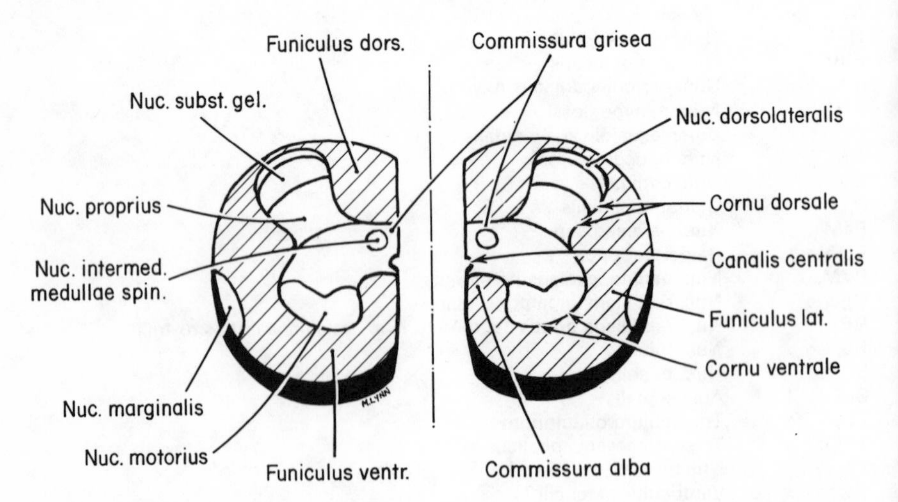

Fig. 5 Generalized transverse section of the thoracic spinal cord of the bird. Nuc. intermedius medullae oblongatae is commonly referred to in the literature as the Nucleus or Column of Terni.

Figs 6–8 Transverse sections of the Medulla oblongata of the pigeon. Note that sections of the Cerebellum are not included in these figures. Levels of sections (see Fig. 3): Fig. 6 (P 4·00); Fig. 7 (P 2·75); Fig. 8 (P 1·50). Redrawn from Karten and Hodos (1967).

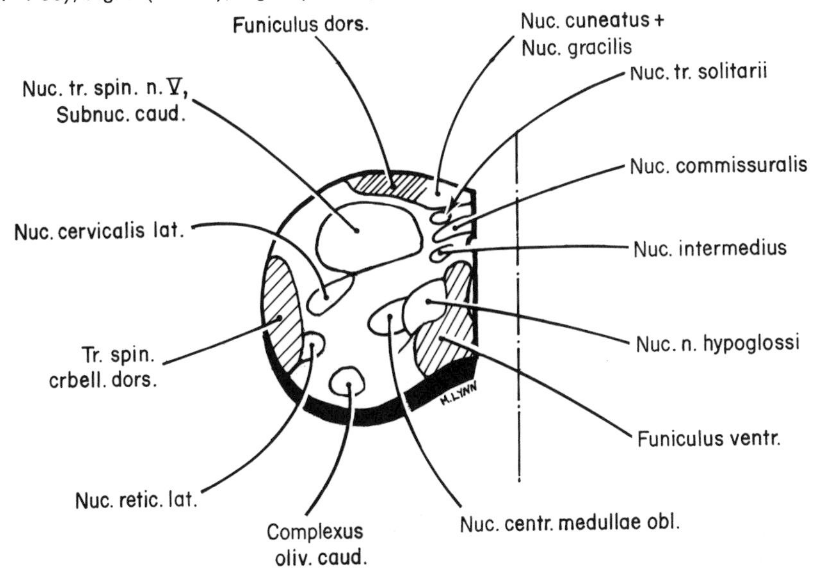

Fig. 6

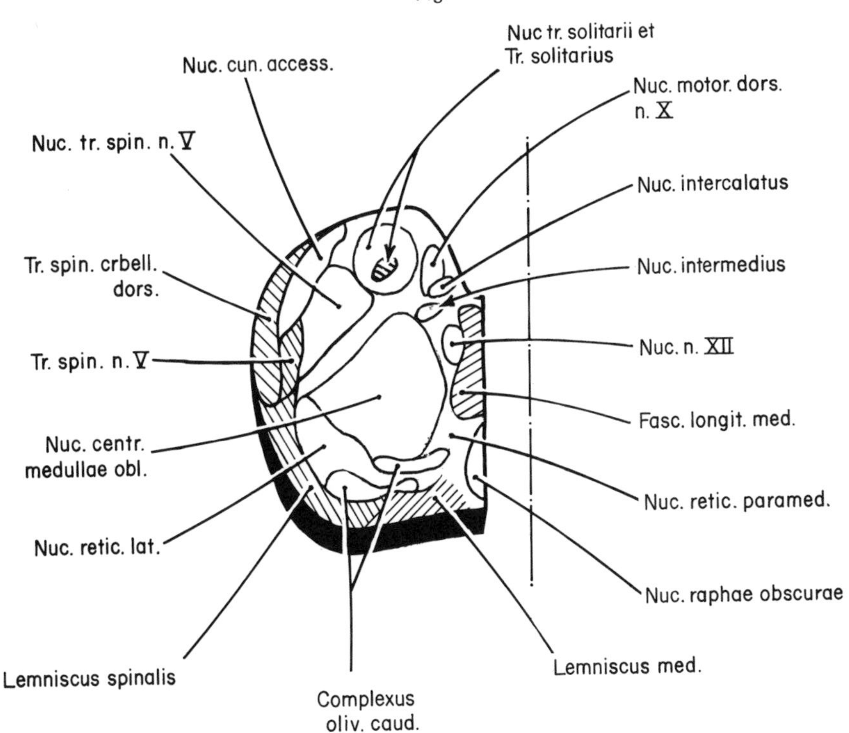

Fig. 7

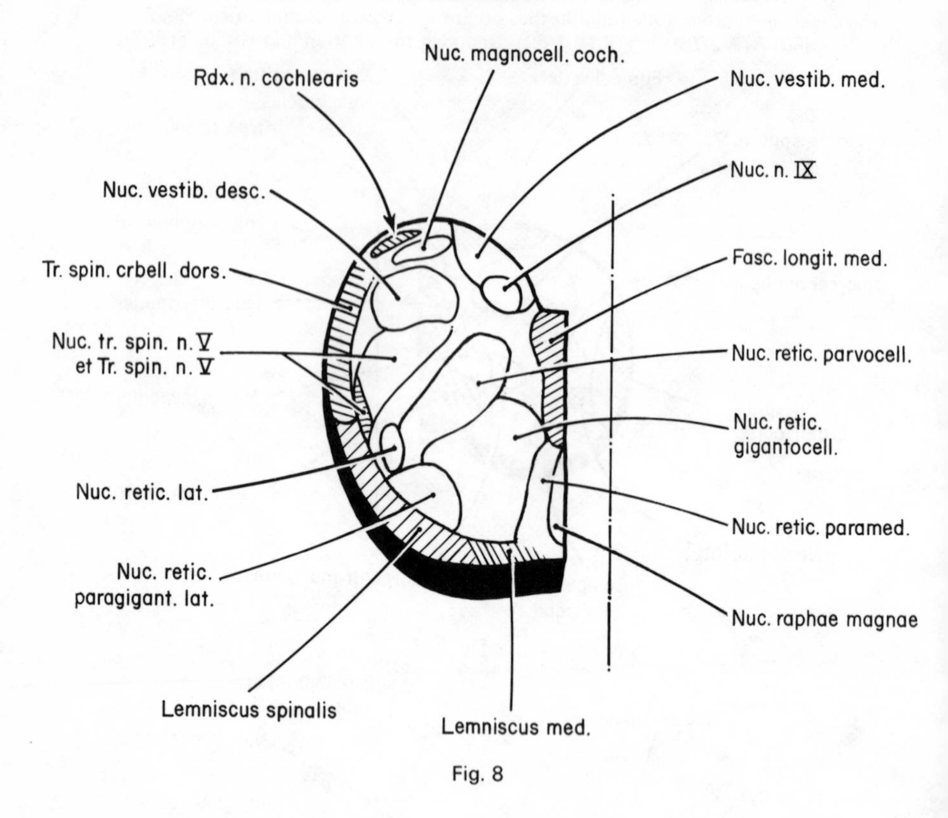

Fig. 8

Figs 9–11 Transverse sections of the Pons and rostral part of the Medulla oblongata of the pigeon. Note that sections of the Cerebellum are not included in the figures. Levels of sections (see Fig. 3): Fig. 9 (P 0·50); Fig. 10 (AP 0·00); Fig. 11 (A 0·75). Redrawn from Karten and Hodos (1967).

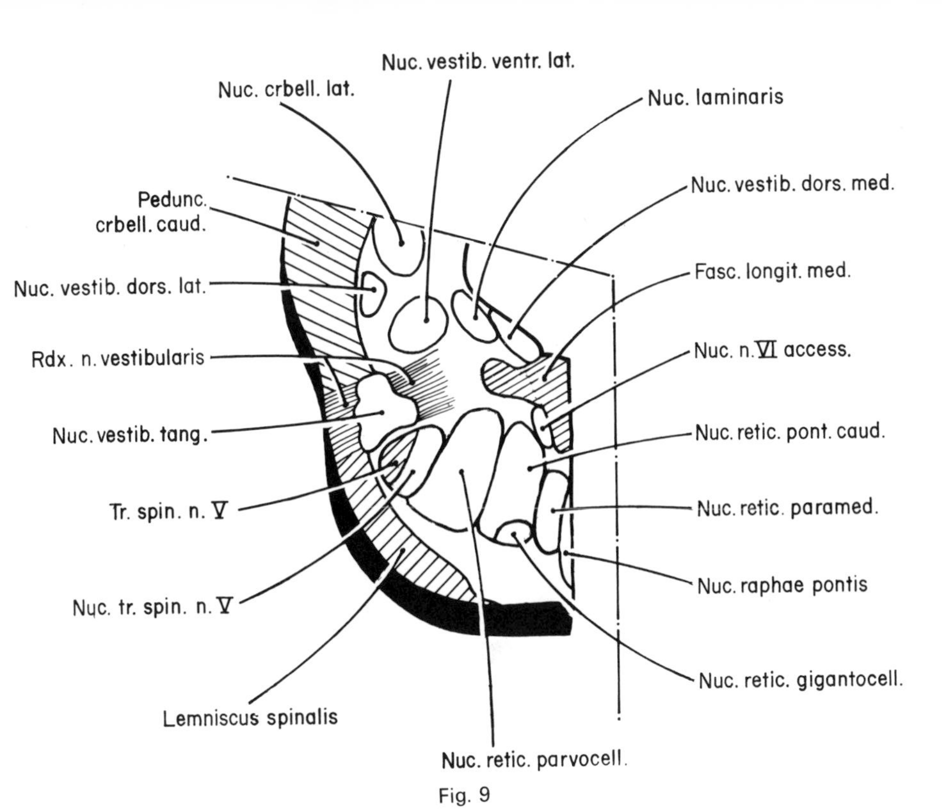

Fig. 9

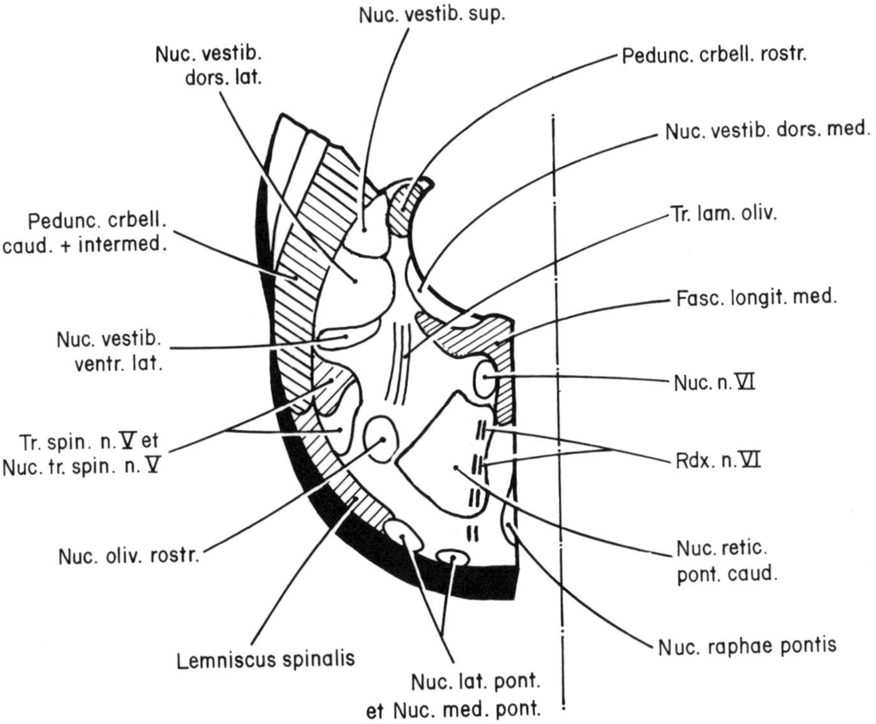

Nuc. vestib. sup.
Nuc. vestib. dors. lat.
Pedunc. crbell. rostr.
Nuc. vestib. dors. med.
Tr. lam. oliv.
Pedunc. crbell. caud. + intermed.
Fasc. longit. med.
Nuc. vestib. ventr. lat.
Nuc. n. VI
Tr. spin. n. V et Nuc. tr. spin. n. V
Rdx. n. VI
Nuc. retic. pont. caud.
Nuc. oliv. rostr.
Nuc. raphae pontis
Lemniscus spinalis
Nuc. lat. pont. et Nuc. med. pont.

Fig. 10

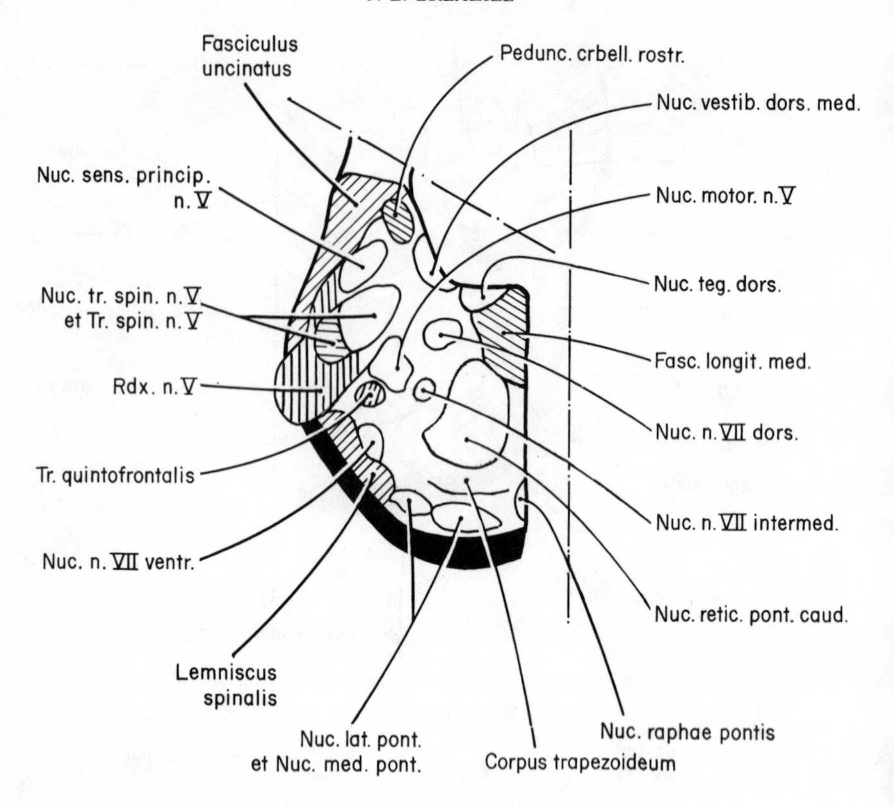

Fig. 11

Figs 12–15 Transverse sections of the Mesencephalon and rostral part of the Pons in the pigeon. Note that sections of the Cerebellum are not included in the figures. Levels of sections (see Fig. 3): Fig. 12 (A 1·50); Fig. 13 (A 2·50); Fig. 14 (A 3·25); Fig. 14 (A 4·25). Redrawn from Karten and Hodos (1967).

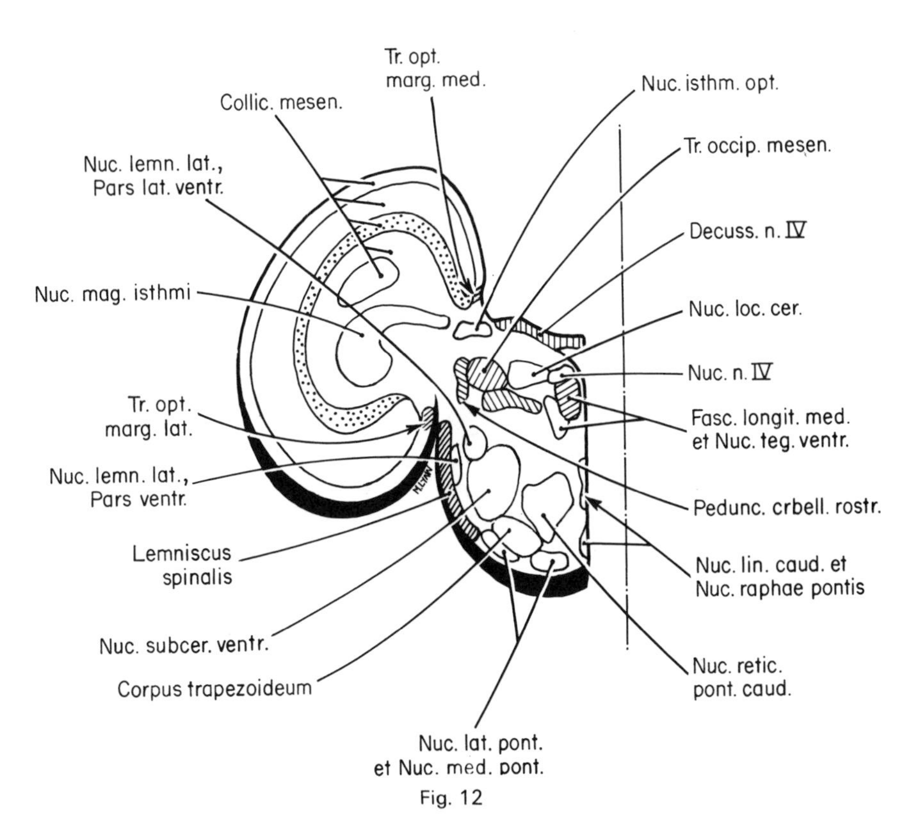
Tr. opt.
marg. med.
Collic. mesen.
Nuc. isthm. opt.
Tr. occip. mesen.
Nuc. lemn. lat.,
Pars lat. ventr.
Decuss. n. IV
Nuc. mag. isthmi
Nuc. loc. cer.
Nuc. n. IV
Tr. opt.
marg. lat.
Fasc. longit. med.
et Nuc. teg. ventr.
Nuc. lemn. lat.,
Pars ventr.
Pedunc. crbell. rostr.
Lemniscus
spinalis
Nuc. lin. caud. et
Nuc. raphae pontis
Nuc. subcer. ventr.
Nuc. retic.
pont. caud.
Corpus trapezoideum
Nuc. lat. pont.
et Nuc. med. pont.

Fig. 12

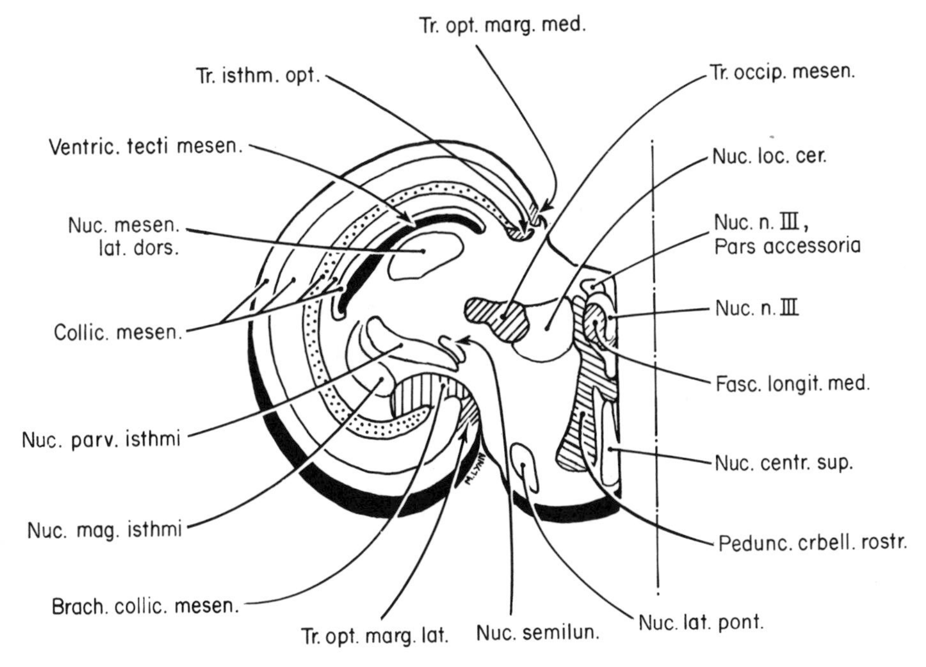
Tr. opt. marg. med.
Tr. isthm. opt.
Tr. occip. mesen.
Ventric. tecti mesen.
Nuc. loc. cer.
Nuc. mesen.
lat. dors.
Nuc. n. III,
Pars accessoria
Nuc. n. III
Collic. mesen.
Fasc. longit. med.
Nuc. parv. isthmi
Nuc. centr. sup.
Nuc. mag. isthmi
Pedunc. crbell. rostr.
Brach. collic. mesen.
Tr. opt. marg. lat.
Nuc. semilun.
Nuc. lat. pont.

Fig. 13

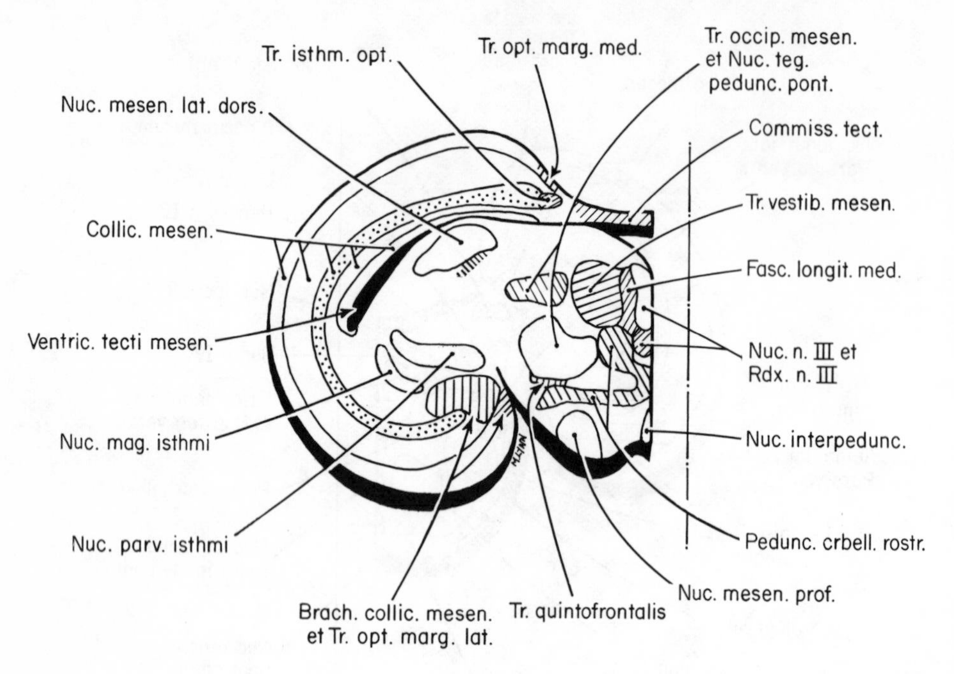
Tr. isthm. opt.
Tr. opt. marg. med.
Tr. occip. mesen.
et Nuc. teg.
pedunc. pont.
Nuc. mesen. lat. dors.
Commiss. tect.
Tr. vestib. mesen.
Collic. mesen.
Fasc. longit. med.
Ventric. tecti mesen.
Nuc. n. III et
Rdx. n. III
Nuc. mag. isthmi
Nuc. interpedunc.
Nuc. parv. isthmi
Pedunc. crbell. rostr.
Nuc. mesen. prof.
Brach. collic. mesen.
et Tr. opt. marg. lat.
Tr. quintofrontalis
M. LYNN

Fig. 14

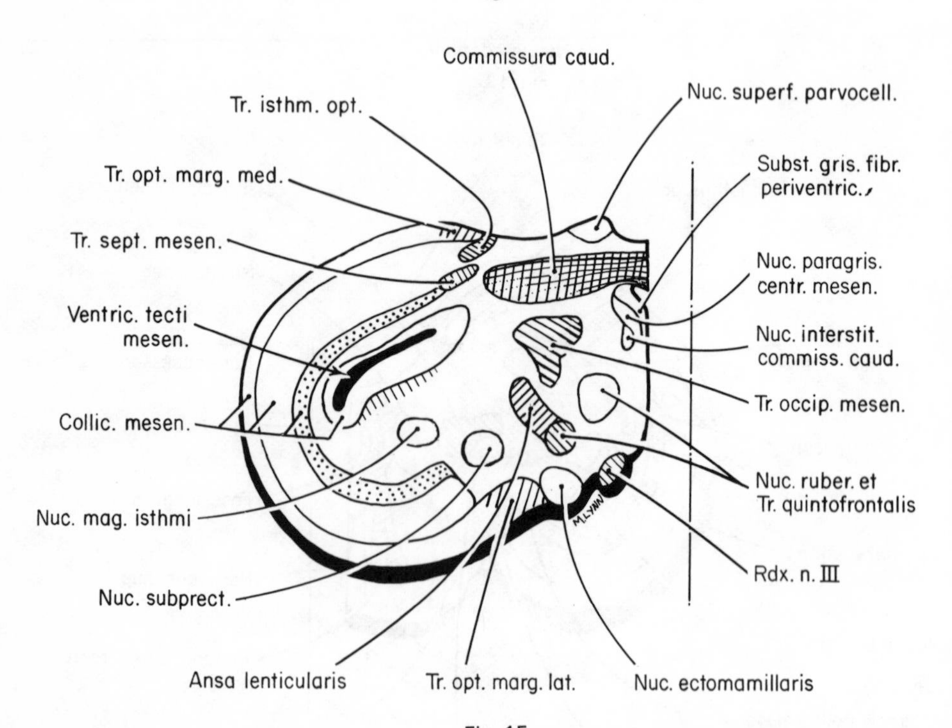
Commissura caud.
Tr. isthm. opt.
Nuc. superf. parvocell.
Tr. opt. marg. med.
Subst. gris. fibr.
periventric.
Tr. sept. mesen.
Nuc. paragris.
centr. mesen.
Ventric. tecti
mesen.
Nuc. interstit.
commiss. caud.
Tr. occip. mesen.
Collic. mesen.
Nuc. ruber. et
Tr. quintofrontalis
Nuc. mag. isthmi
Rdx. n. III
Nuc. subprect.
Ansa lenticularis
Tr. opt. marg. lat.
Nuc. ectomamillaris
M. LYNN

Fig. 15

Figs 16–21 Transverse sections of the caudal part of the Telencephalon, the caudal part of the Diencephalon, and the Mesencephalon of the pigeon. Levels of sections (see Fig. 3): Fig. 16 (A 5·00); Fig. 17 (A 5·25); Fig. 18 (A 5·50); Fig. 19 (A 5·75); Fig. 20 (A 6·00); Fig. 21 (A 6·25). Redrawn from Karten and Hodos (1967).

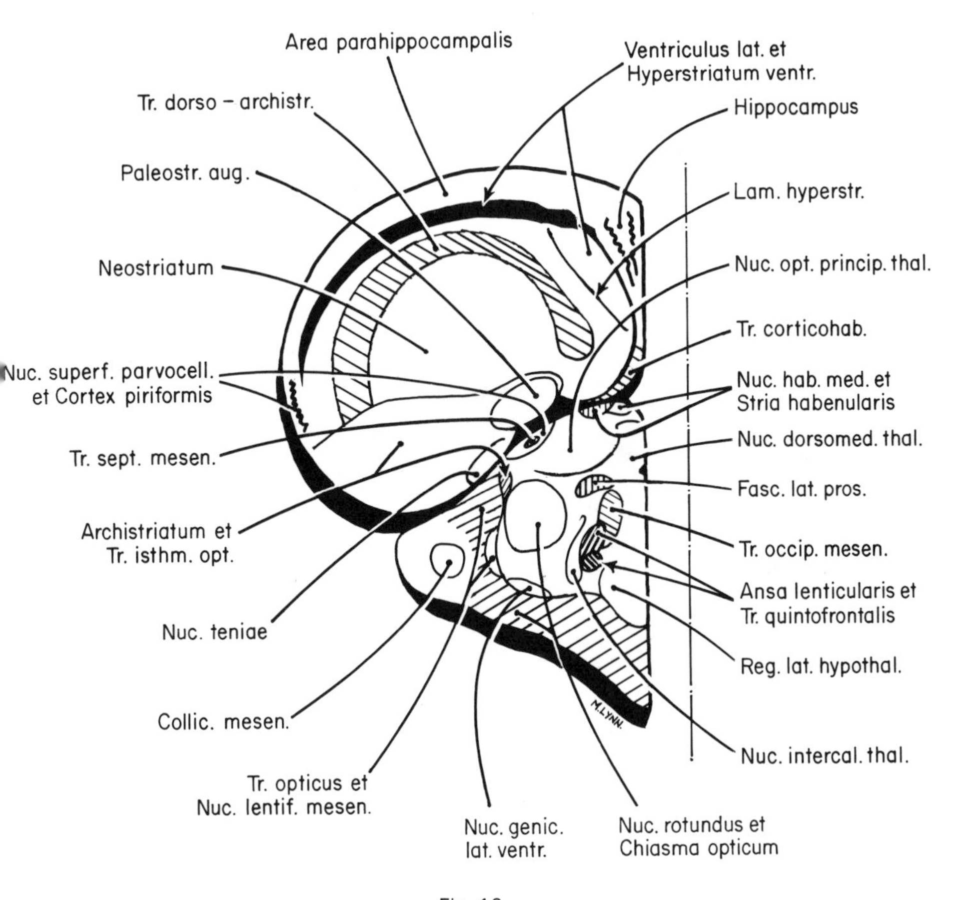

Fig. 16

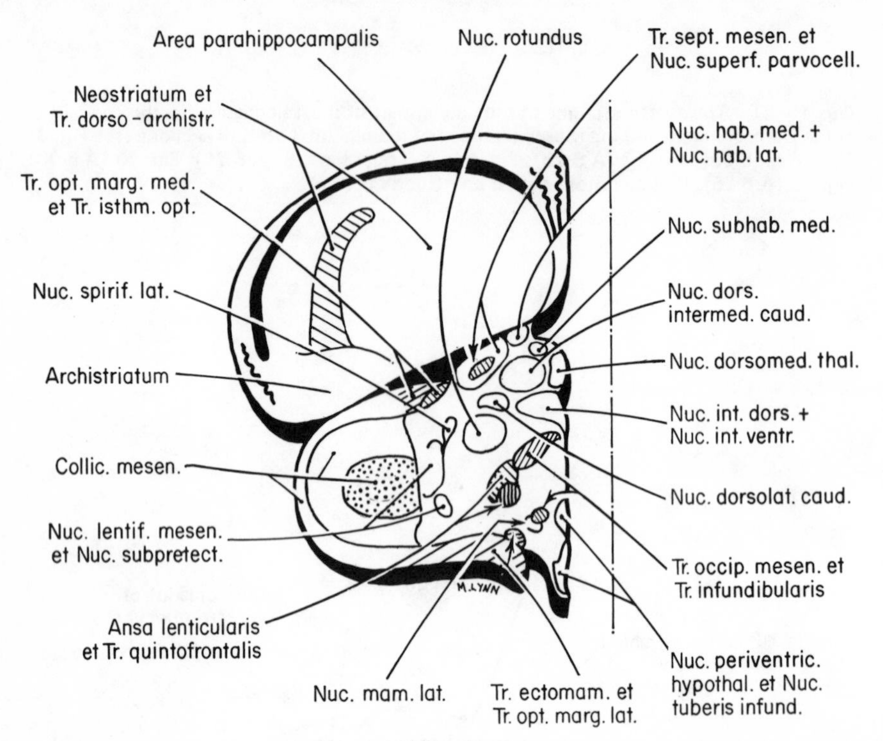
Area parahippocampalis
Nuc. rotundus
Tr. sept. mesen. et
Nuc. superf. parvocell.
Neostriatum et
Tr. dorso-archistr.
Nuc. hab. med. +
Nuc. hab. lat.
Tr. opt. marg. med.
et Tr. isthm. opt.
Nuc. subhab. med.
Nuc. spirif. lat.
Nuc. dors.
intermed. caud.
Archistriatum
Nuc. dorsomed. thal.
Nuc. int. dors. +
Nuc. int. ventr.
Collic. mesen.
Nuc. dorsolat. caud.
Nuc. lentif. mesen.
et Nuc. subpretect.
Tr. occip. mesen. et
Tr. infundibularis
M. LYNN
Ansa lenticularis
et Tr. quintofrontalis
Nuc. periventric.
hypothal. et Nuc.
tuberis infund.
Nuc. mam. lat.
Tr. ectomam. et
Tr. opt. marg. lat.

Fig. 17

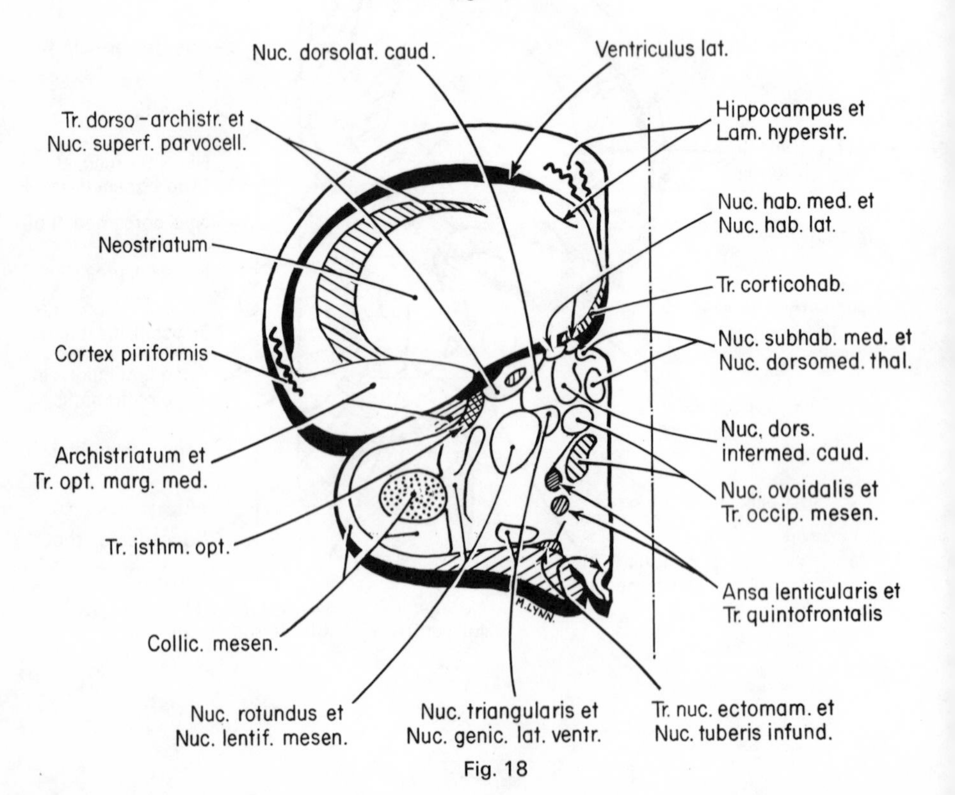
Nuc. dorsolat. caud.
Ventriculus lat.
Tr. dorso-archistr. et
Nuc. superf. parvocell.
Hippocampus et
Lam. hyperstr.
Nuc. hab. med. et
Nuc. hab. lat.
Neostriatum
Tr. corticohab.
Cortex piriformis
Nuc. subhab. med. et
Nuc. dorsomed. thal.
Nuc. dors.
intermed. caud.
Archistriatum et
Tr. opt. marg. med.
Nuc. ovoidalis et
Tr. occip. mesen.
Tr. isthm. opt.
Ansa lenticularis et
Tr. quintofrontalis
M. LYNN
Collic. mesen.
Nuc. rotundus et
Nuc. lentif. mesen.
Nuc. triangularis et
Nuc. genic. lat. ventr.
Tr. nuc. ectomam. et
Nuc. tuberis infund.

Fig. 18

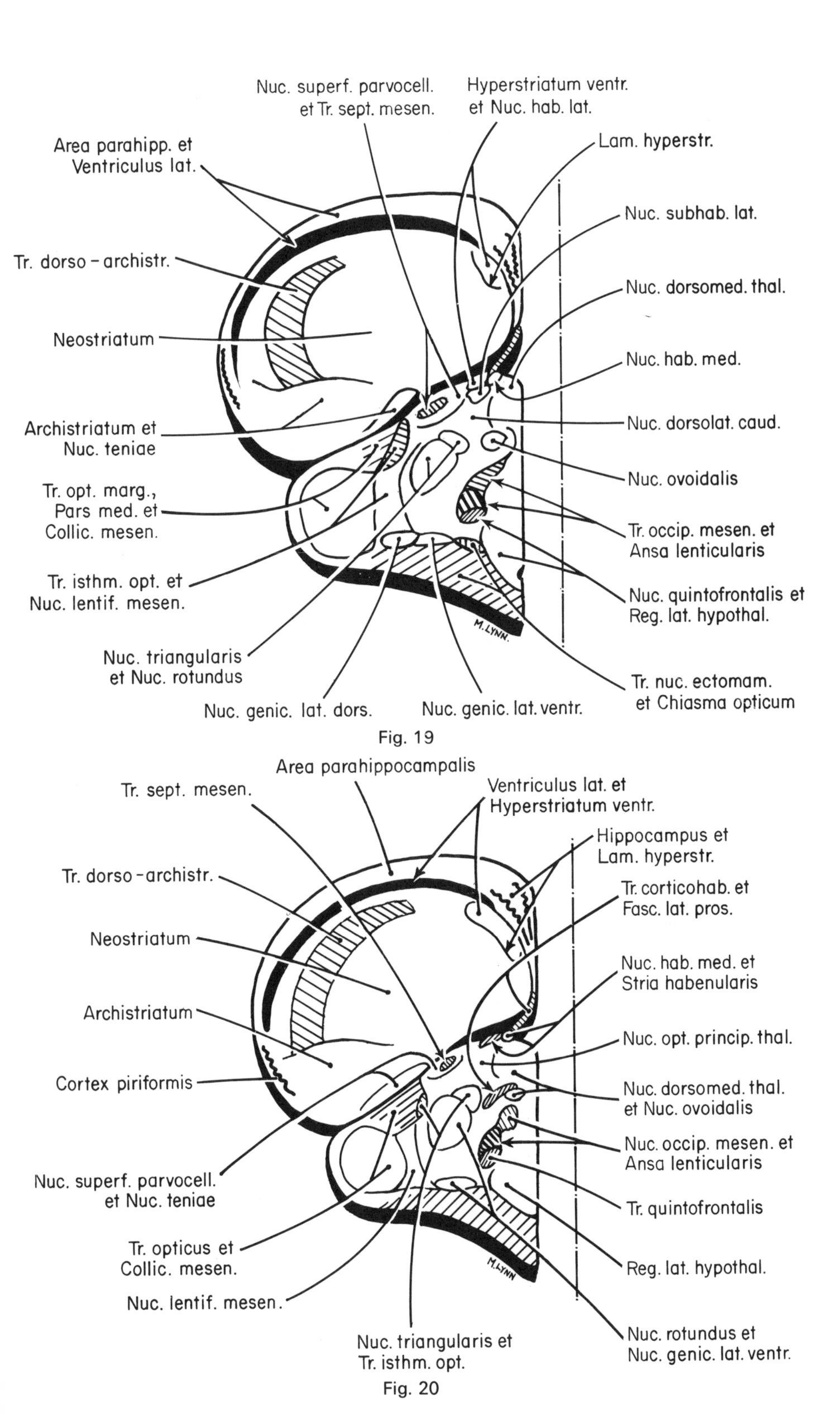
Nuc. superf. parvocell.
et Tr. sept. mesen.
Hyperstriatum ventr.
et Nuc. hab. lat.
Area parahipp. et
Ventriculus lat.
Lam. hyperstr.
Nuc. subhab. lat.
Tr. dorso-archistr.
Nuc. dorsomed. thal.
Neostriatum
Nuc. hab. med.
Archistriatum et
Nuc. teniae
Nuc. dorsolat. caud.
Nuc. ovoidalis
Tr. opt. marg.,
Pars med. et
Collic. mesen.
Tr. occip. mesen. et
Ansa lenticularis
Tr. isthm. opt. et
Nuc. lentif. mesen.
Nuc. quintofrontalis et
Reg. lat. hypothal.
M. LYNN.
Nuc. triangularis
et Nuc. rotundus
Tr. nuc. ectomam.
et Chiasma opticum
Nuc. genic. lat. dors.
Nuc. genic. lat. ventr.
Fig. 19

Area parahippocampalis
Tr. sept. mesen.
Ventriculus lat. et
Hyperstriatum ventr.
Hippocampus et
Lam. hyperstr.
Tr. dorso-archistr.
Tr. corticohab. et
Fasc. lat. pros.
Neostriatum
Nuc. hab. med. et
Stria habenularis
Archistriatum
Nuc. opt. princip. thal.
Cortex piriformis
Nuc. dorsomed. thal.
et Nuc. ovoidalis
Nuc. occip. mesen. et
Ansa lenticularis
Nuc. superf. parvocell.
et Nuc. teniae
Tr. quintofrontalis
Tr. opticus et
Collic. mesen.
M. LYNN
Reg. lat. hypothal.
Nuc. lentif. mesen.
Nuc. rotundus et
Nuc. genic. lat. ventr.
Nuc. triangularis et
Tr. isthm. opt.
Fig. 20

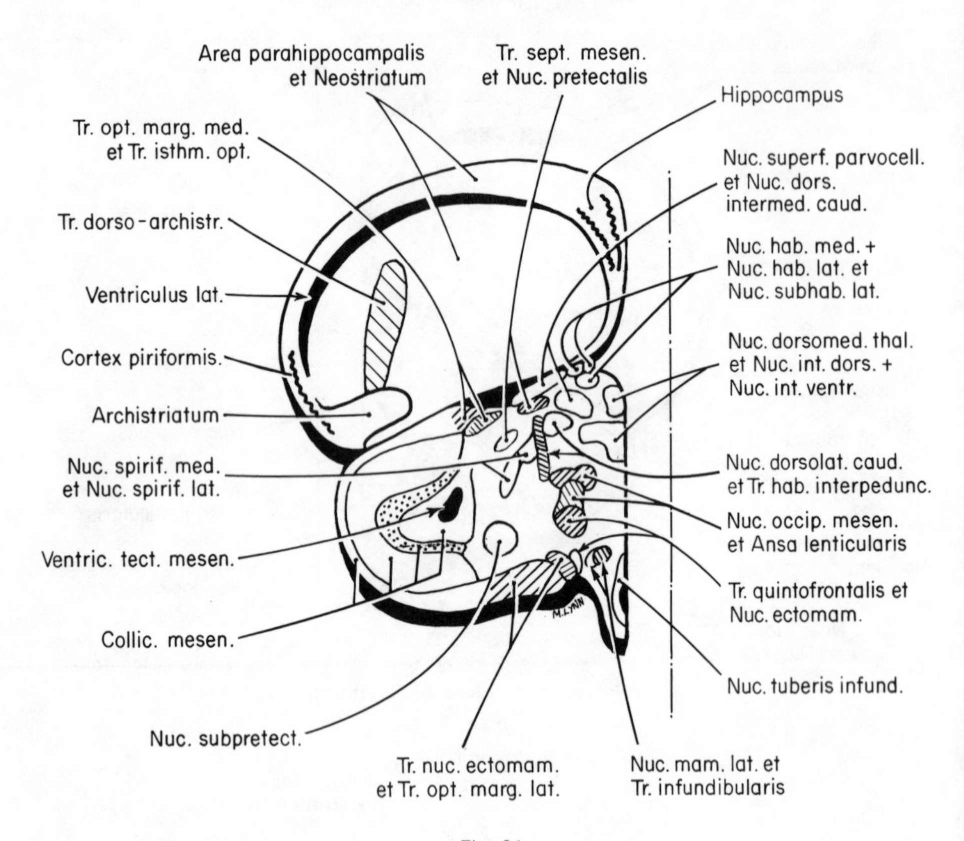

Fig. 21

Figs 22–24 Transverse sections of the caudal part of the Telencephalon and the rostral part of the Diencephalon of the pigeon. Levels of sections (see Fig. 3): Fig. 22 (A 7·25); Fig. 23 (A 7·75); Fig. 24 (A 8·75). Redrawn from Karten and Hodos (1967).

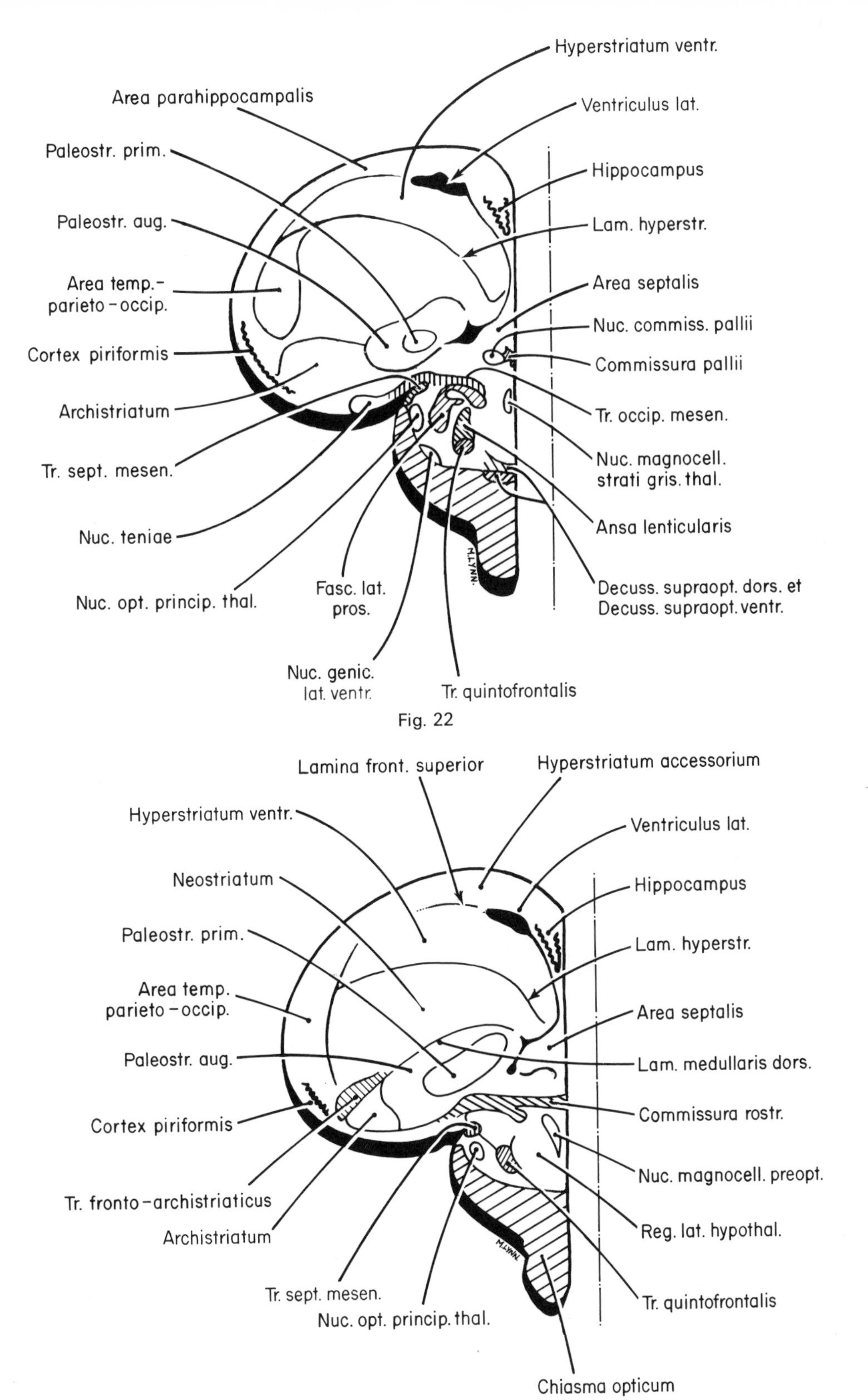
Hyperstriatum ventr.
Area parahippocampalis
Ventriculus lat.
Paleostr. prim.
Hippocampus
Paleostr. aug.
Lam. hyperstr.
Area temp.-
parieto-occip.
Area septalis
Nuc. commiss. pallii
Cortex piriformis
Commissura pallii
Archistriatum
Tr. occip. mesen.
Tr. sept. mesen.
Nuc. magnocell.
strati gris. thal.
Nuc. teniae
Ansa lenticularis
Nuc. opt. princip. thal.
Fasc. lat.
pros.
Decuss. supraopt. dors. et
Decuss. supraopt. ventr.
Nuc. genic.
lat. ventr.
Tr. quintofrontalis

Fig. 22

Lamina front. superior
Hyperstriatum accessorium
Hyperstriatum ventr.
Ventriculus lat.
Neostriatum
Hippocampus
Paleostr. prim.
Lam. hyperstr.
Area temp.
parieto-occip.
Area septalis
Paleostr. aug.
Lam. medullaris dors.
Cortex piriformis
Commissura rostr.
Nuc. magnocell. preopt.
Tr. fronto-archistriaticus
Archistriatum
Reg. lat. hypothal.
Tr. sept. mesen.
Nuc. opt. princip. thal.
Tr. quintofrontalis
Chiasma opticum

Fig. 23

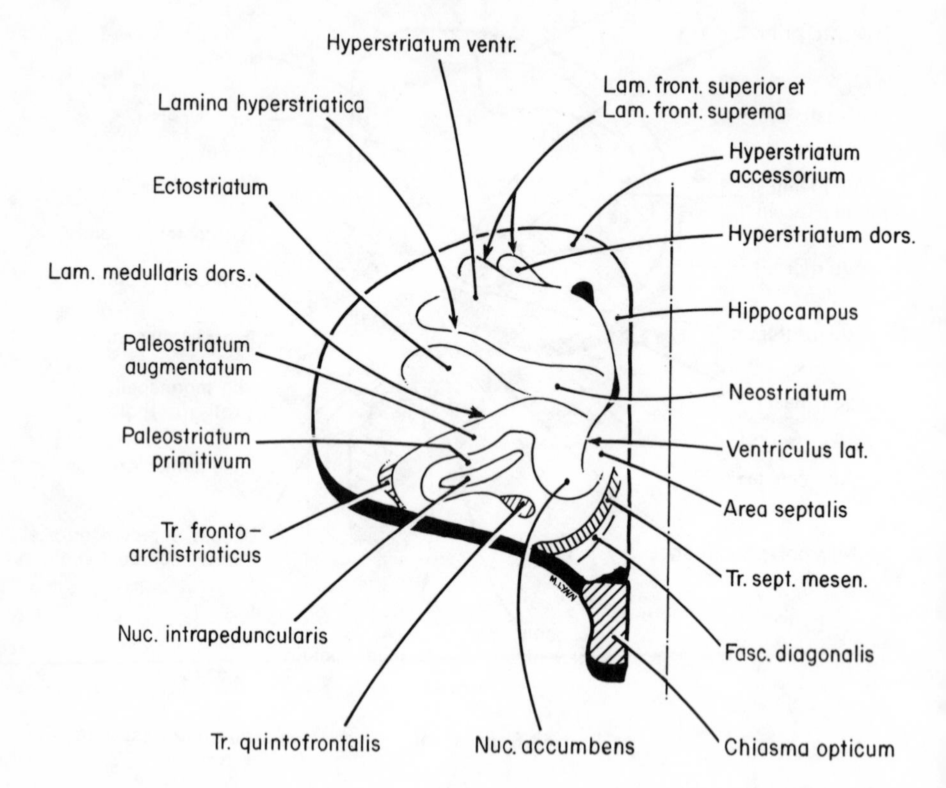

Fig. 24

Figs 25–27 Transverse sections of the rostral part of the Telencephalon of the pigeon. Levels of sections (see Fig. 3). Fig. 25 (A 10·50); Fig. 26 (A 11·50); Fig. 27 (A 13·75). Redrawn from Karten and Hodos (1967).

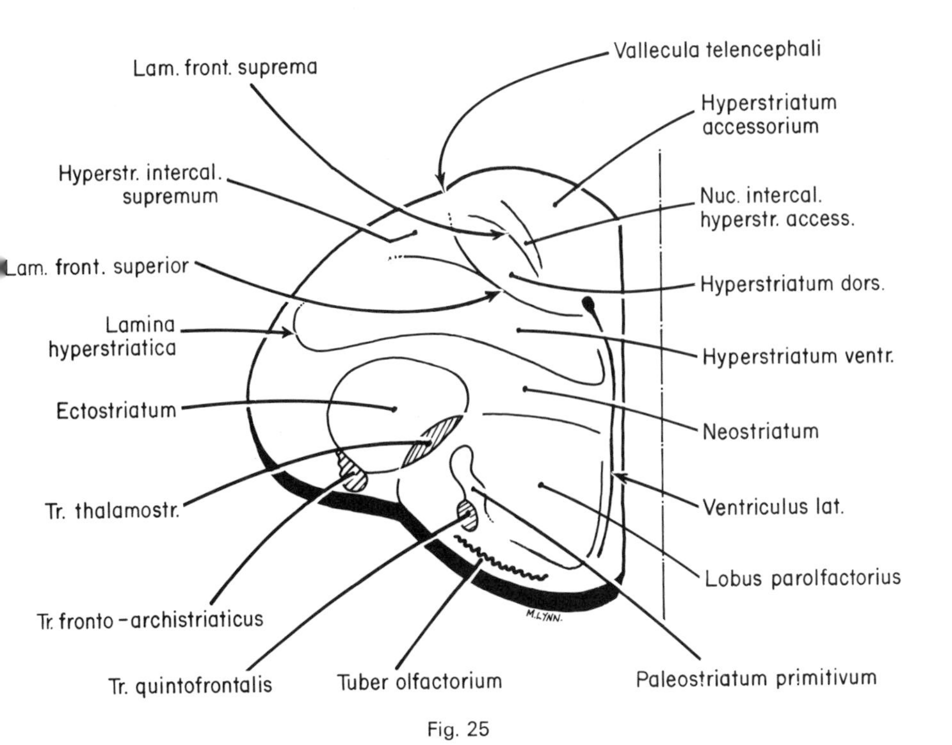
Lam. front. suprema
Vallecula telencephali
Hyperstriatum accessorium
Hyperstr. intercal. supremum
Nuc. intercal. hyperstr. access.
Lam. front. superior
Hyperstriatum dors.
Lamina hyperstriatica
Hyperstriatum ventr.
Ectostriatum
Neostriatum
Tr. thalamostr.
Ventriculus lat.
Lobus parolfactorius
Tr. fronto-archistriaticus
M.LYNN.
Tr. quintofrontalis
Tuber olfactorium
Paleostriatum primitivum

Fig. 25

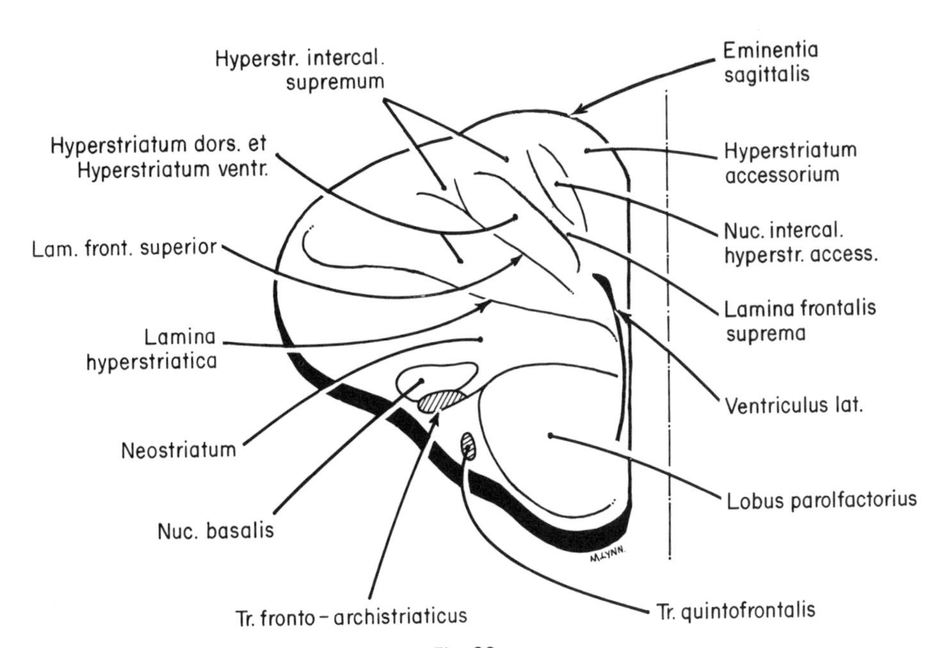
Hyperstr. intercal. supremum
Eminentia sagittalis
Hyperstriatum dors. et Hyperstriatum ventr.
Hyperstriatum accessorium
Lam. front. superior
Nuc. intercal. hyperstr. access.
Lamina frontalis suprema
Lamina hyperstriatica
Ventriculus lat.
Neostriatum
Lobus parolfactorius
Nuc. basalis
M.LYNN.
Tr. fronto-archistriaticus
Tr. quintofrontalis

Fig. 26

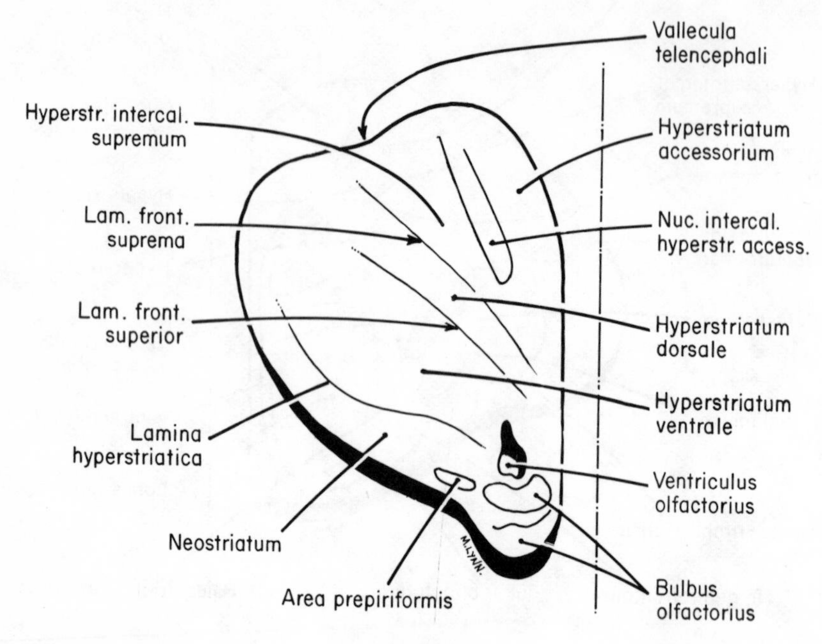
Vallecula
telencephali
Hyperstr. intercal.
supremum
Hyperstriatum
accessorium
Lam. front.
suprema
Nuc. intercal.
hyperstr. access.
Lam. front.
superior
Hyperstriatum
dorsale
Hyperstriatum
ventrale
Lamina
hyperstriatica
Ventriculus
olfactorius
Neostriatum
Area prepiriformis
Bulbus
olfactorius
M. LYNN.

Fig. 27

SYSTEMA NERVOSUM PERIPHERIALE

JAMES E. BREAZILE

Department of Physiological Sciences, Oklahoma State University, Stillwater, Oklahoma 74074, USA

MIKIO YASUDA

Department of Veterinary Anatomy, Nagoys University, Nagoya 464, Japan

With contributions from Sub-Committee Members: J. J. Baumel, *Creighton University*; T. Bennett, *University of Nottingham*; J. L. Dubbeldam, *Rijksuniversiteit te Leiden*; S. L. Freedman, *University of Vermont*; A. B. Gilbert, *Poultry Research Centre, Edinburgh*; L. Malinovský, *University J. E. Purkynev, Brno*; A. Waluszewska-Bubień, *School of Agriculture, Wroclaw*; T. Watanabe, *Nagoya University*

Dr T. Bennett served as the original Chairman of the Sub-Committee responsible for the development of the autonomic nervous system terminology, but he was compelled to relinquish this activity for personal reasons. His preparation of the initial list of terms provided the foundation for the subsequent development of the definitive nomenclature of the avian autonomic nervous system. Dr Bennett's place was assumed by Dr J. E. Breazile. Dr M. Yasuda was chairman of the Sub-Committee for the development of the somatic component of the peripheral nervous system terminology, and he was instrumental in the compilation of this material. In the later stages of the ICAAN project it was determined that the peripheral nervous system could best be presented as an entity, combining both the somatic and visceral parts into one unified nomenclature. Thus the Sub-Committee on Systema Nervosum Peripheriale came to be jointly chaired by Drs Breazile and Yasuda. The illustrations for this chapter, original ones and those redrawn from other sources, were rendered by Dr J. Baumel. [Ed. (J.J.B.)].

Cranial nerves. Considerable confusion remains concerning the course and distribution of the cranial nerves of birds. Much of this confusion results from interspecific variability, but some is due to a lack of definitive descriptive investigation. An example of this confusion is evidenced by the relative distribution of cranial nerves IX–XII that supply the larynx, pharynx, trachea, esophagus, and the subcutaneous muscle sheets of the neck. The situation is worsened by the extensive communications between these nerves.

Spinal nerves. The cranial nerve of birds are organized similarly to those of mammals, thus numbering and naming of these nerves are easily managed. On the other hand, the spinal nerves vary in number over a relatively wide range on account of the variable number of vertebrae in different birds. Therefore, homologies are difficult to determine (see Boas, 1933). Differentiation of the exact boundaries between the vertebrae of different regions (e.g. cervical and thoracic, lumbar and sacral) is problematical. The nerves of the pelvic region are named collectively as Nn. synsacrales which obviates the problem of identifying the vertebrae incorporated in the Synsacrum. One may still speak of the cervical or thoracic spinal nerves in a regional sense; however, in the transitional zone between two vertebral regions it is risky to ascribe a particular spinal nerve to one regional series or to an adjacent one (see **Osteo.** Annot. 112, 139–141). A safer and more precise method of designating spinal nerves is to number them, starting at the base of the cranium and proceeding caudally (see Huber, 1936; Baumel, 1975).

Cutaneous and muscular nerves. Relatively few named, cutaneous nerves are included in this nomenclature. Consult Yasuda (1964) for a detailed presentation of the cutaneous nerves of *Gallus*. In most cases the general term "Rr. musculares" is provided, rather than named branches to individual muscles. In numerous instances named nerves are, in fact, muscular branches (e.g. N. supracoracoideus).

Fibular—peroneal. In order to put the peripheral nerve terminology into agreement with osteological terminology (Fibula), "fibular" is used as the primary term instead of "peroneal" for reference to the branch of the ischiadic (sciatic) nerve. This usage also pertains for the muscles and vessels of the pelvic limb, and is a departure from that of the human *Nomina Anatomica* (IANC, 1966, 1977) and the *Nomina Anatomica Veterinaria* (ICVAN, 1973).

Connexus—Ramus communicans. Except for the "Rami communicantes" that join spinal nerves to the Truncus paravertebralis, the term "Connexus" is employed instead of the Ramus communicans of other anatomical nomenclatures. Thus confusion between these specifically designated communicating rami and other nerve communications is avoided. One should be aware that for each listing of a connexus there will be another listing of the connexus with the names of the structures involved reversed (e.g. Ganglion ciliare: Connexus cum n. ophthalmico; N. ophthalmicus: Connexus cum g. ciliari). The term "anastomosis" is not used at all in reference to communications between nerves, since the term literally applies to junctions of tubular structures.

Visceral nerves. The autonomic nervous system is often classified as a purely visceral efferent (motor) system. In the present terminology Pars visceralis of Systema Nervosum Peripheriale is not the equivalent of the autonomic nervous system, since the nerves making up Pars visceralis contain both visceral afferent and visceral efferent neuronal processes. The terminology of Pars visceralis makes no reference to the autonomic "sympathetic" and "parasympathetic" divisions nor to cholinergic and adrenergic systems, inasmuch as these have functional rather than anatomical connotations. The names of the neural elements are organized into craniosacral and thoracolumbar divisions for purposes of convenience.

The term Plexus subvertebralis is synonymous with the mammalian non-official terms, Plexus prevertebralis or Plexus collateralis. This terminology does not assume homology, but is somewhat more topographically descriptive, since "prevertebral" is based on the upright anatomical position of man, "subvertebral" being more appropriate in the quadrupedal anatomical position of the bird (see **Gen. Intro.** Terms of orientation; and **Topog. Anat.**) Neither the *Nomina Anatomica Veterinaria* (ICVAN, 1973) nor the 4th edition of the human *Nomina Anatomica* (IANC, 1977) has an inclusive term equivalent to the avian Plexus subvertebralis; however, the latter does provide the general regional terms: Plexus aorticus thoracicus, Plexus aorticus abdominalis, and Plexus hypogastricus superior.

The distribution of N. glossopharyngeus and N. vagus is visceral for the most part; therefore, these nerve are treated in detail in Pars visceralis, and not at all in Pars somatica.

Each major named autonomic ganglion of the head is listed with the cranial nerve which provides the so-called motor root of the ganglion. This practice has also been adopted in the 4th Edition of *Nomina Anatomica* (IANC, 1977), but the 3rd Edition of *Nomina Anatomica* (IANC, 1966) and *Nomina Anatomica Veterinaria* (ICVAN, 1973) are inconsistent in this regard.

PARS SOMATICA

NERVI CRANIALES

N. olfactorius[1]
- R. dorsalis
- R. ventralis

N. opticus[2] (Fig. 2)
- Chiasma opticum

N. oculomotorius (see Fig. 2; and Pars visceralis)
- R. dorsalis
  - Rr. musculares
- R. ventralis
  - R. ganglionicus ciliaris[3]
  - Rr. musculares

Ganglion ciliare[4] (Fig. 2)
Connexus cum n. abducenti
Connexus cum n. oculomotorio[3]
Connexus cum n. ophthalmico
Nn. choroidales[5]
N. iridociliaris[5]
Rdx. ganglionica[6]
Rdx. ophthalmica[6]
Plexus annularis ciliaris[7]
Rr. ciliares[7]
Plexus annularis iridicus[7]
Rr. iridici[7]

N. trochlearis (see Fig. 2: and **Osteo.** Fig. 3)
R. muscularis
Connexus cum n. ophthalmico

N. trigeminus (Fig. 1)
Rdx. motoria
Rdx. sensoria
Ganglion trigeminale[8]

N. ophthalmicus[9] (Figs. 1, 2)
Connexus cum n. trochleari
Rdx. iridociliaris[6] (see N. oculomotorius)
Connexus cum g. ethmoidali[10]
Connexus cum g. ciliari
Rr. glandulae nasalis[10] (see Pars visceralis, N. palatinus)
R. lateralis
Rr. glandulae nasalis
Rr. frontales
Rr. cristales
Rr. palpebrales rostrodorsales
Rr. nasales interni
R. medialis
Rr. nasales interni
R. premaxillaris dorsalis
Rr. rostri maxillaris[11]
R. premaxillaris ventralis
Rr. palatini
Rr. rostri maxillaris[11]
Rr. glandularum palati

N. maxillaris (Fig. 1)
Connexus cum n. faciali[21]
N. supraorbitalis[9]
Rr. cristales
Rr. frontales
Rr. auriculares
Rr. glandulares lacrimales
R. palpebralis caudodorsalis
R. palpebralis caudoventralis
N. infraorbitalis
Rr. palpebrales rostroventrales
N. nasopalatinus
R. palatinus
R. nasalis
Rr. nasales externi
Rr. nasales interni
Connexus cum g. sphenopalatino[12]
Connexus cum g. ethmoidali
N. mandibularis (Fig. 1)
Rr. musculares[13]
R. pterygoideus[13]
R. externus[13]
Rr. artc. quadratomandibularis
R. anguli oris
Rr. glandulares
Rr. palatini
Rr. cutanei
R. mandibularis externus
R. intermandibularis[14]
R. muscularis
Rr. cutanei
Connexus cum chorda tympani[20]
R. sublingualis[15]
Rr. glandulares
Rr. rostri mandibularis[11]
N. intramandibularis (see **Art.** Annot. 25; and **Ven.** Annot. 24)

N. abducens (Fig. 2)
Connexus cum g. ciliari
R. dorsalis[16]
R. ventralis[16]

N. facialis (see Fig. 1; and Pars visceralis for additional terms)
- Geniculum n. facialis
- Ganglion geniculatum[17]
- N. palatinus [R. rostralis][18] (see Pars visceralis)
- Ganglion sphenopalatinum (see Pars visceralis)
- Ganglion ethmoidale (see Pars visceralis)
- Chorda tympani[19] (see Pars visceralis)
- N. hyomandibularis [R. caudalis]
  - R. auricularis
  - N. depressoris mandibularis[22]
  - R. hyoideus[23]
  - R. cervicalis[23] (see **Myol.** Annot. 5)
  - Connexus cum g. cervicali craniali (Fig. 3)
  - Connexus cum n. glossopharyngeo (Fig. 3)
  - Connexus cum n. cervicali

N. vestibulocochlearis[24] (see **Org. sens.**)
- Pars vestibularis
  - Ganglion vestibulare[25]
    - R. rostralis
      - N. ampullaris rostralis
      - N. ampullaris horizontalis [lateralis]
      - N. utriculus
    - R. caudalis
      - N. ampullaris caudalis
        - N. maculae neglectae
      - N. saccularis
- Pars cochlearis
  - Ganglion cochleare
    - N. cochlearis
  - Ganglion lagenare
    - N. lagenaris

N. glossopharyngeus (see Fig. 3; and Pars visceralis for terminology)

N. vagus (see Figs. 3, 8; and Pars visceralis for terminology)

N. accessorius (Fig. 3)
- Radiculae craniales
- Radiculae spinales
- Connexus cum n. vago
- R. externus[26] (see N. vagus)

N. hypoglossus (Fig. 3)
Rdx. rostralis
Rdx. caudalis
Connexus cum g. cervicali craniali
Connexus cum n. cervicali primo[27]

N. hypoglossocervicalis[27] (Fig. 3)
Connexus cum n. vago
Connexus cum n. glossopharyngeo
Connexus cum n. cervicali secundo
R. cervicalis descendens[29]
R. laryngolingualis [rostralis]
Rr. musculares
R. trachealis [caudalis]
Rr. musculares[29, 59]
R. syringealis[30]

NERVI SPINALES[31]

(see **PNS** Intro.)

Fila radicularia
Rdx. dorsalis
Ganglion radiculare dorsale [Ganglion spinale]
Rdx. ventralis
Ramus dorsalis
Ramus ventralis
Rami communicantes (see **PNS** Intro.)
Ramus meningeus

Nn. cervicales (see Figs. 3, 4; and **PNS** Intro.)
Rr. dorsales
Rr. musculares
Nn. cutanei cervicales dorsales[31]
Plexus cervicalis dorsalis[32]
Rr. ventrales
Rr. musculares
Nn. cutanei cervicales ventrales[31]
Connexus cum n. faciali[33]
N. cutaneus colli[33]
Connexi cum n. hypoglosso[28] (Fig. 3)

Plexus brachialis[34] (Figs 4, 8)
- Rr. musculares
- Plexus brachialis accessorius[35]
  - N. cutaneus omalis
- Rdxx. plexus
- Trunci plexus
- Fasciculi plexus[36]
  - Fasciculus dorsalis
    - N. anconealis
  - Fasciculus ventralis
- N. subscapularis
- N. subcoracoscapularis
- N. axillaris
  - Rr. musculares
  - N. cutaneus axillaris
- N. radialis (Figs 4, 5)
  - Rr. musculares
  - N. cutaneus brachialis dorsalis
  - N. propatagialis dorsalis
  - N. cutaneus antebrachialis dorsalis
  - R. superficialis
    - Rr. postpatagiales
  - R. profundus
    - Rr. carpales dorsales
    - R. alularis
    - Nn. metacarpales dorsales
      - Rr. postpatagiales
      - Rr. digitales
- N. supracoracoideus
- N. pectoralis (Fig. 4)
  - N. pectoralis cranialis
    - Rr. musculares
    - Nn. cutanei pectorales
  - N. pectoralis caudalis
    - Rr. musculares
    - Nn. cutanei abdominales
- N. bicipitalis
  - Rr. propatagiales
- N. medianoulnaris (Fig. 5)
  - N. cutaneus brachialis ventralis
  - N. medianus
    - Rr. musculares
    - N. propatagialis ventralis

R. superficialis
R. profundus
Rr. carpales ventrales
R. alularis
Nn. metacarpales ventrales
Rr. postpatagiales
Rr. digitales
N. cutaneus antebrachialis ventralis
N. ulnaris
Rr. musculares
N. cutaneus cubiti
R. caudalis
R. cranialis
Rr. postpatagiales
Rr. metacarpales ventrales
Rr. digitales

Nn. thoracici (see **PNS** Intro.)
Rr. dorsales
Rr. ventrales
Nn. intercostales
Rr. cutanei
Rr. musculares[37]

Nn. synsacrales[38] (see Fig. 6) (see **PNS** Intro.)

Plexus lumbosacralis[39]
Rdxx. plexus
Trunci plexus

Plexus lumbaris (Figs 6, 8)
N. cutaneus femoralis lateralis
N. cutaneus femoralis medialis[42]
N. cutaneus cruralis cranialis
N. pubicus[43]
N. femoralis (Figs 6, 7)
N. cutaneus femoralis cranialis
N. coxalis cranialis (see **Art.** Annot. 70)
Rr. musculares[40]
N. obturatorius
R. obturatorius medialis
R. obturatorius lateralis
Plexus sacralis (Figs 6, 8)
N. coxalis caudalis (see **Art.** Annot. 70)

Plexus lumbosacralis—*continued*

N. cutaneus femoralis caudalis
Rr. musculares[40]
Connexus caudalis[41]
N. ischiadicus (Figs 6, 7)
N. tibialis
N. suralis lateralis
N. suralis medialis
N. interosseus (Fig. 7)
N. cutaneus suralis
N. plantaris medialis[44]
N. metatarsalis plantaris
N. parafibularis [paraperoneus][45]
N. plantaris lateralis[46]
N. metatarsalis plantaris
R. digitalis
Rr. musculares[40]
N. fibularis [peroneus]
N. fibularis superficialis
Nn. metatarsales dorsales
Rr. digitales
N. fibularis profundus
Nn. metatarsales dorsales
Rr. digitales

Plexus pudendus (Fig. 6)

N. lateralis caudae
Rr. venti
Rr. musculares
Rr. cutanei
Connexus caudalis[41] (see Plexus sacralis)
N. intermedius caudae (Fig. 6)
N. pudendus[62] (see Pars visceralis for other branches)
Rr. musculares
Rr. proctodeales
Rr. cutanei

Nn. caudales (see **PNS** Intro.) (Fig. 6)

Plexus caudalis[47]

N. medialis caudae
N. bulbi rectricium[47]
Rr. musculares[48]
Rr. cutanei
Rr. glandulae uropygialis

PARS VISCERALIS[49]
(see **PNS** Intro.)

DIVISIO CRANIOSACRALIS

N. oculomotorius (see Fig. 2; and Pars somatica)
- Ganglion ciliare[4]
    - Connexus cum n. oculomotorio
    - Connexus cum n. ophthalmico
    - Connexus cum n. abducenti
    - N. iridociliaris[5]
        - Rdx. ganglionica[6]
        - Rdx. ophthalmica[6]
        - Plexus annularis ciliaris[7]
            - Rr. ciliares[7]
        - Plexus annularis iridicus[7]
            - Rr. iridici[7]

N. facialis (see Fig. 1; and Pars somatica)
- Ganglion geniculatum[17]
- Chorda tympani[19, 20]
    - Connexus cum n. mandibulari
    - Ganglia mandibularia[20]
        - Rr. glandulares (see **Digest.**)
    - Connexus cum n. maxillari[21]
        - Rr. glandulares lacrimales (see N. maxillaris)
- N. palatinus [R. rostralis][18] (see N. hyomandibularis)
    - R. ventralis
        - Connexus cum g. sphenopalatino
        - Ganglion sphenopalatinum[50]
            - Rr. glandulae membranae nictitantis[18]
            - Connexus cum n. nasopalatino[18]
    - R. dorsalis
        - Connexus cum g. ethmoidali
        - Ganglion ethmoidale[50]
            - Connexus cum n. ophthalmico
            - Rr. glandulae nasalis[18]
            - Connexus cum g. sphenopalatino

N. glossopharyngeus (Fig. 3)

- Ganglion proximale[51]
- Ganglion distale[51]
  - Connexus cum n. hyomandibulari
  - Connexus cum g. cervicali craniali
- N. subcaroticus [precaroticus][57]
- N. lingualis
  - Rr. pharyngeales
  - Rr. mm. hyobranchialium[51]
  - Rr. glandulares
  - Rr. gustatorii
- Connexus vagoglossopharyngealis[52] (Fig. 3)
- N. laryngopharyngealis
  - N. largyngealis
    - Rr. musculares[53]
    - Rr. tracheales
  - N. pharyngealis
    - Rr. glandulares[53]
    - Rr. pharyngeales
    - Rr. esophageales
  - Connexus cum n. hypoglossocervicali
- N. esophagealis descendens[54]
  - Rr. esophageales
  - Rr. ingluviales
  - Rr. tracheales
  - Connexus cum n. vago

N. vagus[61] (Figs. 3, 8)

- Ganglion proximale[55]
  - Connexus cum n. accessorio
- Connexus cum g. cervicali craniali
- R. externus[26]
- Connexus vagoglossopharyngealis[52] (Fig. 3)
- Connexus cum n. hypoglossocervicali
- Rr. laryngeales
- Rr. pharyngeales
- Rr. tracheales
- Rr. thymici
- Ganglion distale[55]
  - Rr. glandulares[56]
  - Rr. glomi carotici[57]
- N. cardiacus cranialis[58]

N. recurrens
Rr. bronchiales
Rr. esophageales
N. pulmoesophagealis[68]
R. pulmonalis
R. esophagealis
R. descendens
Rr. esophageales
Connexi cum plexo celiaco
R. ascendens
Rr. esophageales
Rr. ingluviales
Rr. tracheales
Rr. musculorum tracheae[59]
Rr. pulmonales[68]
Nn. cardiaci caudales[58, 67]
Rr. septi obliqui[60]
Truncus communis n. vagi[61] (Fig. 8)
Rr. proventriculares
Rr. viscerales abdominales [Nn. gastrici][61]
Rr. proventriculares
Connexi cum plexo celiaco (Fig. 8)
Rr. ventriculares
Rr. pylorici
Rr. duodenales
Rr. pancreatici
Rr. hepatici

N. pudendus[62] (see Fig 6, 8; and Pars somatica)
Rr. ureterales
Rr. oviductales
Rr. ductus deferentis
Nn. cloacales (see **Cloaca** Annot. 27)
Rr. corporis vascularis paracloacalis
Rr. bursae cloacalis

Ganglion rectale[63] (see **Cloaca** Annot. 27)

Ganglia cloacalia (see **Cloaca** Annot. 27)

N. intestinalis[64] (see Divisio thoracolumbaris) (Fig. 8)
Ganglia n. intestinalis[64]

DIVISIO THORACOLUMBARIS

Truncus paravertebralis (see **PNS** Intro. and Annot. 72) (Figs 6, 8)

Ganglia paravertebralia (Figs 3, 6, 8)
Rr. communicantes (see **PNS** Intro.)
Connexi interganglionici
Ansae connexorum interganglionicorum

Plexus subvertebralis (see **PNS** Intro.) (Figs 3, 8)

Ganglia subvertebralia

Truncus paravertebralis cervicalis [N. vertebralis][65]

Ganglion cervicale craniale[65] (Fig. 3)
N. ophthalmicus externus
Connexus cum n. trigemino
N. caroticus externus
N. caroticus cerebralis [internus]
Connexus cum n. faciali
Connexus cum n. glossopharyngeo
Connexus cum n. vago
Connexus cum n. hypoglosso
Truncus subvertebralis [N. caroticus cervicalis][66]
Plexus subvertebralis[66] (see Annot. 57)

Truncus paravertebralis thoracicus

N. cardiacus[67] (Fig. 8)
Rr. pulmonales[68]
Plexus pulmonalis
Nn. splanchnici thoracici
Plexus subvertebralis thoracicus
Ganglion celiacum
Ganglia mesenterica cranialia
Plexus celiacus[69] (Fig. 8)
Plexus splenicus
Plexus hepaticus
Plexus pancreaticoduodenalis
Plexus proventricularis
Plexus gastricus
Plexus mesentericus cranialis[69]

Truncus paravertebralis synsacralis (Figs 6, 8)

Nn. splanchnici synsacrales
Plexus subvertebralis synsacralis
Plexus aorticus
Plexus adrenalis[70]
Ganglia adrenalia
N. hepaticus[71]
Plexus mesentericus caudalis[69]
Plexus iliacus internus

Truncus paravertebralis caudalis[72]

Nn. splanchnici caudales
Ganglia impares[72]
Plexus pelvici

N. intestinalis[64] (see N. vagus) (Fig. 8)

Ganglia n. intestinalis[64]

ANNOTATIONS

(1) **N. olfactorius.** Consult Baumel (1975) and Watanabe and Yasuda (1968) for descriptions of the peripheral course and relationships of the olfactory nerve. The latter authors refer to R. dorsalis as R. externus and to R. ventralis as R. internus.

(2) **N. opticus.** See O'Flaherty (1971) for a study of the fiber size distribution of the optic nerve of the duck (*Anas*).

(3) **R. ganglionicus ciliaris.** Synonymy: Rdx. brevis. This branch is also known as the oculomotor root of the ciliary ganglion [Connexus cum n. oculomotorii].

(4) **G. ciliare.** The paper of Oehme (1968) reviews the literature pertaining to the avian ciliary ganglion and thoroughly treats the connections and "branches" of the ganglion as well as its components in several species of Corvidae.

(5) **N. iridociliaris.** Synonymy: N. ciliaris major (Cords, 1904); N. ciliaris longus. See Annot. 6, 7. **Nn. choroidales.** Synonymy: Nn. ciliares breves.

Oehme (1958) discusses the nomenclatural difficulties in applying mammalian names to the nerves "branching" from G. ciliare that pass to the Bulbus oculi. The terms, N. iridociliaris and Nn. choroidales, were proposed by Oehme (1968). Watanabe *et al.* (1967) described the formation and distribution of these nerves in *Gallus*. There is some uncertainty concerning whether this nerve supplies autonomic fibers to the iridial muscle of birds (Isomura, 1973; Oehme, 1969; Holtzmann, 1896; Schwalbe, 1879), but it has been demonstrated to supply the fundus and the ciliary body (Oehme, 1968).

(6) **Rdx. ganglionica; Rdx. ophthalmica.** N. iridociliaris is formed by the union of two roots, one from N. ophthalmicus (often consisting of multiple rootlets) and the other from G. ciliare. Oehme (1968) contends that Rdx. ophthalmica contributes sensory and sympathetic nerve fibers and that Rdx. ganglionica contributes postganglionic parasympathetic fibers to Bulbus oculi. See Watanabe *et al.* (1967) for synonymy. See also Fig. 2.

(7) **Plexus annularis ciliaris; Plexus annularis iridicus** (Oehme, 1968). Synonymy: Plexus ciliaris (Watanabe *et al.*, 1967). N. iridociliaris courses rostrally within the bulb between the scleral and choroidal layers and ramifies into several branches. At the periphery of the ciliary muscle the branches form a circular plexus within the muscle. One or more rami of the ciliary annular plexus form the iridial ring of nerves which in turn gives rise to Rami iridici that supply the iris (Oehme, 1968). See Fig. 2.

(8) **G. trigeminale.** Synonymy: G. semilunare. The trigeminal ganglion of some birds (e.g. *Gallus, Anser*) is partially divided into a smaller ophthalmic part and a larger maxillomandibular part. The extent of this division is reflected in the configuration of the Fossa ganglii trigemini of the interior of the floor of the cranial cavity in the dried skull (see **Osteo.**).

(9) **N. ophthalmicus.** N. ophthalmicus is the principal afferent nerve of the upper jaw (Rostrum maxillare). In the older literature this nerve is known as N. ophthalmicus profundus. This terminology is confusing since in the older literature N. ophthalmicus superficialis refers to a branch of N. maxillaris (N. supraorbitalis).

(10) **Rr. glandulae nasalis.** Secretomotor nerve fibers to Gl. nasalis (salt gland) course via Connexus cum g. ethmoidali to N. ophthalmicus and are distributed via glandular rami (see N. facialis). The innervation of Gl. nasalis is treated by Ash *et al.* (1969).

(11) **Rr. rostri.** Afferent rami of N. ophthalmicus and N. mandibularis carry impulses centrally from the various sensory corpuscles and discs deep to the Rhamphotheca. These endings have been extensively studied in different birds (see Malinovský and Zemanek, 1969, for a concise review of the literature; for other citations see J. Schwartzkopff, 1973). See **Osteo.** Annot. 41.

(12) **Connexus cum g. sphenopalatino.** Synonymy: R. communicans cum n. palatino of N. facialis (see Pars visceralis).

(13) **Rr. musculares.** These rami of N. mandibularis supply the various parts of the jaw muscles. Certain of the Rr. musculares have been given specific names by Barnikol (1953): R. pterygoideus and R. externus. See **Myol.** Annot. 16, 17.

(14) **R. intermandibularis.** Synonymy: R. circumflexus. Innervates skin between the two mandibular rami as well as Mm. intermandibulares (see **Myol.** Annot. 26).

(15) **R. sublingualis** (Cords, 1904). This branch of N. mandibularis has been referred to as N. or R. lingualis in the literature. The ramus does not have a lingual distribution proper, but is distributed to the mucosa of the floor of the oral cavity (see N. glossopharyngeus and N. hypoglossus).

(16) **R. dorsalis; R. ventralis.** R. dorsalis of N. abducens corresponds to the combined dorsolateral and ascending rami of Baumel (1975), and supplies M. rectus lateralis and M. quadratus membranae nictitans. R. ventralis corresponds to Baumel's R. descendens, and is distributed to M. pyramidalis of the Membrana nictitans.

(17) **G. geniculatum.** In a study of the sensory ganglion of N. facialis, Yntema and Hammond (1954) recognized a G. radicis and a G. geniculi, the former arising from the neural crest and the latter of placodal origin. There does not appear to be a report of the G. radicis in adult birds in the literature.

(18) **N. palatinus.** Synonymy: R. rostralis; N. nasopalatinus (Bonsdorf, 1852). This ramus of N. facialis is considered by Santamaria-Arnaiz (1962) to be the homologue of the greater superficial petrosal or Vidian nerve of mammals. The nerve conducts parasympathetic fibers to G. ethmoidale and G. sphenopalatinum (see Pars visceralis): The ultimate distribution of this pathway presumably includes the Gl. nasalis, Gl. membranae nictitantis, Gl. lacrimalis, the palatine glands and glands of the nasal mucosa. These last communications are effected through Connexus cum n. nasopalatino.

(19) **Chorda tympani.** Smith (1904–05) describes two configurations of the Chorda tympani. In *Columba* and *Sturnus vulgaris,* the Chorda springs from N. hyomandibularis near its exits from Basis cranii, enters the caudal portion of the tympanic cavity, crosses dorsal to Cartilago extracolumellaris, traverses the cavity with Lig. columellosquamosum (see **Org. sens.**), and leaves the cavity near the articulation of Quadratum with the squamous and otic bones. Crompton (1953) notes the same relationships of Chorda tympani in *Spheniscus demersus.* In *Gallus,* on the other hand, the Chorda tympani arises from the region of G. geniculatum and arches above the tympanic cavity within the canal of the external ophthalmic vessels, entering the orbit with these vessels.

(20) **Connexus cum n. mandibulari; Gg. mandibularia; Connexus cum Chorda tympani.** Cords (1904) reviewed the earlier literature regarding the existence and connections of the avian Chorda tympani. Cords and more recently Hsieh (1951) indicate that the Chorda joins either the intramandibular segment of N. mandibularis or its R. sublingualis. See Fig. 1.

(21) **Connexus cum n. faciali; Connexus cum n. maxillari.** The connection of the facial nerve with the maxillary nerve is the so-called lesser superficial petrosal nerve of Hsieh (1951). He describes it as forming a common stem with the Chorda tympani. The nerve becomes distinct from the Chorda within the canal of the external ophthalmic vessels, communicates with the perivascular sympathetic nerves, and joins the maxillary nerve just medial to the Rete mirabile ophthalmicum. According to Hsieh some of the facial nerve fibers that join N. maxillaris are distributed to the lacrimal gland (see Annot. 18). If, however, this connection is to be considered an homologue of the mammalian lesser superficial nerve, its origin must be from the N. glossopharyngeus.

(22) **N. depressoris mandibularis.** Synonymy: N. digastricus.

(23) **R. hyoideus.** This ramus innervates M. stylohyoideus and M. serpihyoideus. (see **Myol.** Annot. 27, 28).

(24) **N. vestibulocochlearis.** Terminology of the branches of this cranial nerve are based primarily on the works of Ewald (1892) and Boord (1969).

(25) **G. vestibulare.** This ganglion occupies the Fossa acustica interna of the caudal fossa of the cranial cavity.

(26) **R. externus.** The external ramus of N. accessorius is distributed to the craniolateral portion of the cucullaris muscle (Baumel, 1975). It appears to be a branch of the vagus nerve, but consists of fibers from N. accessorius that join and travel with the vagus nerve for a distance (Baumel, 1975).

(27) **N. hypoglossocervicalis.** This nerve is formed by the combination of the ventral elements of the hypoglossal nerve and the first cervical nerve.

(28) **Connexi cum n. hypoglosso.** Synonymy: plexus hypoglossocervicalis and plexus cervicalis (Nishi, 1938).

(29) **R. cervicalis descendens.** The source of these nerve fibers is unknown but are most likely from the N. vagus and/or cervical nerves. This branch of the hypoglossocervical nerve supplies M. sternotrachealis. See Annot. 59. See also **Myol.** Annot. 37 and Table 1.

(30) **R. syringealis.** The actual source of the nerve fibers in this branch of the hypoglossocervical nerve is unknown, and might be cervical or vagal. Nottebohm and Nottebohm (1976) contend that hypoglossal nerve fibers proper supply the syringeal musculature (principally in passeriforms).

(31) **Nn. spinales.** The naming of spinal nerves is somewhat difficult because of the problem of distinguishing between regions of vertebrae (see **PNS** Intro.) See Yasuda (1964) for a detailed terminology of cutaneous nerves of *Gallus*.

(32) **Plexus cervicalis dorsalis.** Synonymy: Plexus suboccipitalis (Nishi, 1938). The term used here is proposed by Yasuda (1964).

(33) **N. cutaneus colli.** This nerve is formed by a communication of N. facialis and the ventral rami of the third cervical nerve (Yasuda, 1964).

(34) **Plexus brachialis.** The terminal branches of this plexus have been classed into the following four divisions by Fürbringer (1879): (1) Mn. thoracici superiores; (2) Nn. brachiales superiores; (3). Nn. brachiales inferiores and (4) Nn. thoracici inferiores. These terms are used by Buri (1900), Baumel (1958), and Yasuda (1960). On the other hand the present nomenclature emphasizes the intrinsic organization of the plexus by the use of the terms, Radices, Trunci and Fasciculi (see Baumel 1958; 1975).

(35) **Plexus brachialis accessorius.** Synonymy: Nebenplexus (Fürbringer, 1879). This plexus is separable into two entities in the martin (*Progne subis*) (Baumel, 1958).

(36) **Fasciculi plexus.** Nerve cords formed by junction of divisions of the trunks of the plexus.

(37) **Rr. musculares.** These branches include those to intercostal muscles and to M. costoseptales. For details consult deWet *et al.* (1967) (see also **Myol.** Annot. 61).

(38) **Nn. synsacrales.** These nerves include the caudal thoracic, lumbar, sacral and cranialmost caudal nerves, which are associated with the Synsacrum. See **Osteo.** Annot. 141.

(39) **Plexus lumbosacralis.** This term does not indicate all the spinal nerves that contribute to the plexus, but is commonly employed. Boas (1933) used the term in the most definitive work available on the description of this plexus. Yasuda (1961) described the plexus as consisting of a plexus cruralis, plexus sacralis (ischiadicus) and plexus pudendus. The 25th spinal nerve corresponds to N. furcalis and the 30th to N. bigeminus in *Gallus* (Baumel, 1975).

(40) **Rr. musculares.** The muscular branches of the femoral nerve are distributed primarily to the extensor muscles of the knee. The direct muscular branches of the sacral plexus, exclusive of tibial and fibular nerve distribution, supply primarily the M. iliotibialis and M. iliofibularis.

(41) **Connexus caudalis.** Synonymy: R. communicans caudalis (Buchholz, 1959–60). This nerve connects one of the muscular rami supplying the flexor muscle group of the thigh with a cutaneous ramus of N. lateralis caudae (see Plexus pudendus and Baumel, 1975, p. 2053).

(42) **N. cutaneus femoralis medialis.** Synonymy: N. saphenus. See Buchholz (1959–60) and Yasuda (1964) for the distribution of this nerve.

(43) **N. pubicus.** Synonymy: N. ilioinguinalis (Buchholz, 1959–60). Arises from the lumbar plexus or the intrapelvic part of the medial femoral cutaneous nerve, parallels the pubis, and sends muscular rami to abdominal muscles (Baumel, 1975).

(44) **N. plantaris medialis.** This nerve ends in the integument of the medial aspect of the ankle joint, and does not extend into the foot. The proximal part of the nerve corresponds to R. tibialis profundus and the distal part corresponds to R. cutaneus tarsalis medialis of Buchholz (1959–60).

(45) **N. parafibularis [paraperoneus]** Holmes (1963) called this nerve N. paraperoneus; Yasuda (1961) referred to it as the tibial plantar nerve, and Buchholz (1959–60) referred to it as the tertiary fibular nerve. This branch of the tibial nerve accompanies N. fibularis through the Ansa m. iliofibularis and extends into the foot as N. plantaris lateralis (see Annot. 46 and Fig. 7).

(46) **N. plantaris lateralis.** Synonymy: N. metatarsalis profundus plantaris (Buchholz, 1959–60). See Annot. 45.

(47) **N. bulbi rectricium.** This branch of the caudal plexus innervates the bulb of the rectrices. See **Myol.** Annot. 67; and Baumel (1975, p. 2053).

(48) **Rr. musculares.** The caudal plexus supplies dorsal and ventral axial muscles of the tail, M. bulbi rectricium, the dorsal integument of the tail and uropygial gland (Baumel, 1975, p. 2053).

(49) **Pars visceralis.** This portion of the peripheral nervous system includes both visceral efferent and visceral afferent components. The cell bodies of the visceral afferent nerve fibers are located in cranial nerve ganglia and dorsal root ganglia of spinal nerves. See **PNS** Intro.

(50) **G. sphenopalatinum.** Synonymy: Ventral pterygopalatine ganglion. For connections of this ganglion see Schrader (1970); Watanabe and Yasuda (1970); and Baumel (1975).
G. ethmoidale. Synonymy: G. orbitonasale; dorsal pterygopalatine ganglion. For connections of this ganglion see Schrader (1970); Watanabe and Yasuda (1970); and Baumel (1975).

(51) **G. proximale.** Synonymy: G. jugulare; G. radicis. In general this ganglion of N. glossopharyngeus is closely related or actually joined to G. proximale of N. vagus in a fossa of the base of the skull.
G. distale. The distal ganglion of N. glossopharyngeus is closely adjacent to the G. cervicale craniale.
Rr. mm. hyobranchialium. The rami of the N. lingualis of the glossopharyngeal nerve innervate some of the hyobranchial muscles (e.g. M. serpihyoideus).

(52) **Connexus vagoglossopharyngealis.** Synonymy: Anastomosis of Staderini. The variability of this connection in *Gallus* is described by Waluszewska-Bubień (1972). Similar variability occurs in *Anas*, in which the connection may be represented by two or three nerves.

(53) **Rr. musculares.** The laryngeal branch of the glossopharyngeal nerve innervates Mm. constrictor et dilator glottidis. See **Resp.** Annot. 30.
Rr. glandulares. These rami of the laryngeal nerve apparently contain secretomotor fibers to the several small glands associated with the larynx.

(54) **N. esophagealis descendens.** The fibers that make up this nerve to the Esophagus and the Ingluvies are very likely of vagal origin (Baumel, 1975, p. 2034; Watanabe, 1968).

(55) **G. proximale.** Synonymy: G. jugulare; G. radicis. See Annot. 41.
G. distale. Synonymy: G. trunci; G. courvreuri; G. thoracicum; G. nodosum.

(56) **R. glandulares.** These rami of G. distale innervate the ultimobranchial, parathyroid and thyroid glands.

(57) **Rr. glomi carotici.** The source of the fibers in these branches is uncertain. They may be glossopharyngeal or vagal or a combination of these. It is possible that these nerve fibers reach this level through the subcarotid nerve (Terni, 1929). According to Terni **N. subcaroticus [precaroticus]** arises from the glossopharyngeal nerve just below the base of the skull, and courses with the cervical part of the internal carotid artery to the root of the neck, and communicates with the Plexus subvertebralis (see Baumel, 1975, p. 2057).

(58) **N. cardiacus cranialis; Nn. cardiaci caudales.** The cranial cardiac nerves contain predominantly afferent nerve fibers from the heart, whereas the caudal cardiac nerves contain the only significant "parasympathetic" branches to the heart (Fedde *et al.* 1963).

(59) **Rr. musculorum tracheae.** There is some doubt concerning whether these branches of N. vagus contain components of N. glossopharyngeus and N. hypoglossus as well (see Annot. 30).

(60) **Rr. septi obliqui.** These branches provide partial innervation to the muscle of the oblique septum (see Annot. 37; and **Pericar.** Annot. 2).

(61) **Truncus communis n. vagi.** Synonymy: N. vagus impar. The right and left vagi unite on the Proventriculus to form this nerve; there appears to be some interchange of nerve fibers in this junction (Watanabe, 1960). A common vagal trunk is not always formed in the chicken (Waluszewska-Bubień, 1972). When present the common trunk divides at the caudal extremity of the Proventriculus into right and left vagus nerves or **Rr. viscerales abdominales** (synonymy: Nn. gastrici). See Baumel (1975, pp. 2035–36). See Fig. 8.

(62) **N. pudendus.** Synonymy: N. pelvini (Freedman and Sturkie, 1963). **N. intermedius caudae.** Synonymy: N. cutaneus caudae (see Baumel, 1975). This branch of the pudendal plexus is distributed mostly to the ventral tail muscles (see Fig. 6).

(63) **G. rectalis.** Synonymy: G. coli (Watanabe, 1972) (see **Cloaca** Annot. 27).

(64) **N. intestinalis.** For review of the structure and proposed function of this nerve see Hsieh (1951); Watanabe (1972); Bennett (1974) and Pusstilnik (1937). N. intestinalis courses near the mesenteric border of the gut from the end of the rectum to the duodenum; numerous ganglia occur along its length.

(65) **Truncus paravertebralis cervicalis [N. vertebralis]; G. cervicale craniale.** The cervical paravertebral trunk represents the avian counterpart of the cervical sympathetic trunk of mammals. In birds this nerve passes through the Foramina transversaria from the thoracic inlet to the cranial cervical ganglion (Thebault, 1898). The cranial cervical ganglion of the bird is formed by fusion of the first two cervical ganglia. Individual ganglia are associated with the remainder of the cervical nerves (Hsieh, 1951).

(66) **Plexus subvertebralis.** Synonymy: Common carotid plexus (Hsieh, 1951). See Intro. visceral nerves and Annot. 57.
Truncus subvertebralis [N. caroticus cervicalis]. Synonymy: retrocarotid trunk (Terni, 1929). The trunk arises from the cranial cervical ganglion, and courses caudally along the cervical part of the internal carotid artery in the Sulcus caroticus (see **Osteo.** Annot. 121). See also N. subcaroticus and Annot. 57.

(67) **N. cardiacus.** This is the so-called sympathetic cardiac nerve. This nerve along with other cardiac nerves have been described in detail by Hsieh (1951). The plexuses formed by these nerves are described by Ssinelnikow (1928); Thebault (1898) and Hsieh (1951) (see Annot. 58).

(68) **N. pulmoesophagalis; Rr. pulmonales.** For a description of these structures see King and Molony (1971); McLelland and Abdalla (1972); Fedde *et al.* (1963) and Fedde (1970).

(69) **Plexus celiacus; Plexus mesentericus cranialis et caudalis.** For a description of these structures see Watanabe and Paik (1973) and Baumel (1975).

(70) **Plexus adrenalis Gg. adrenalia.** See Hsieh (1951) and Freedman (1968). See Fig. 8.

(71) **N. hepaticus.** The distribution of nerves to the liver are described by Staderini (1889), Cords (1904), Marage (1889), Thebault (1898) and Hsieh (1951). The hepatic nerve is derived from a ganglion in the right adrenal plexus and from the cranial mesenteric plexus (Hsieh, 1951).

(72) **Gg. impares.** In the tail region where the paravertebral trunks converge and form a single trunk, several unpaired ganglia occur along the length of the trunk.

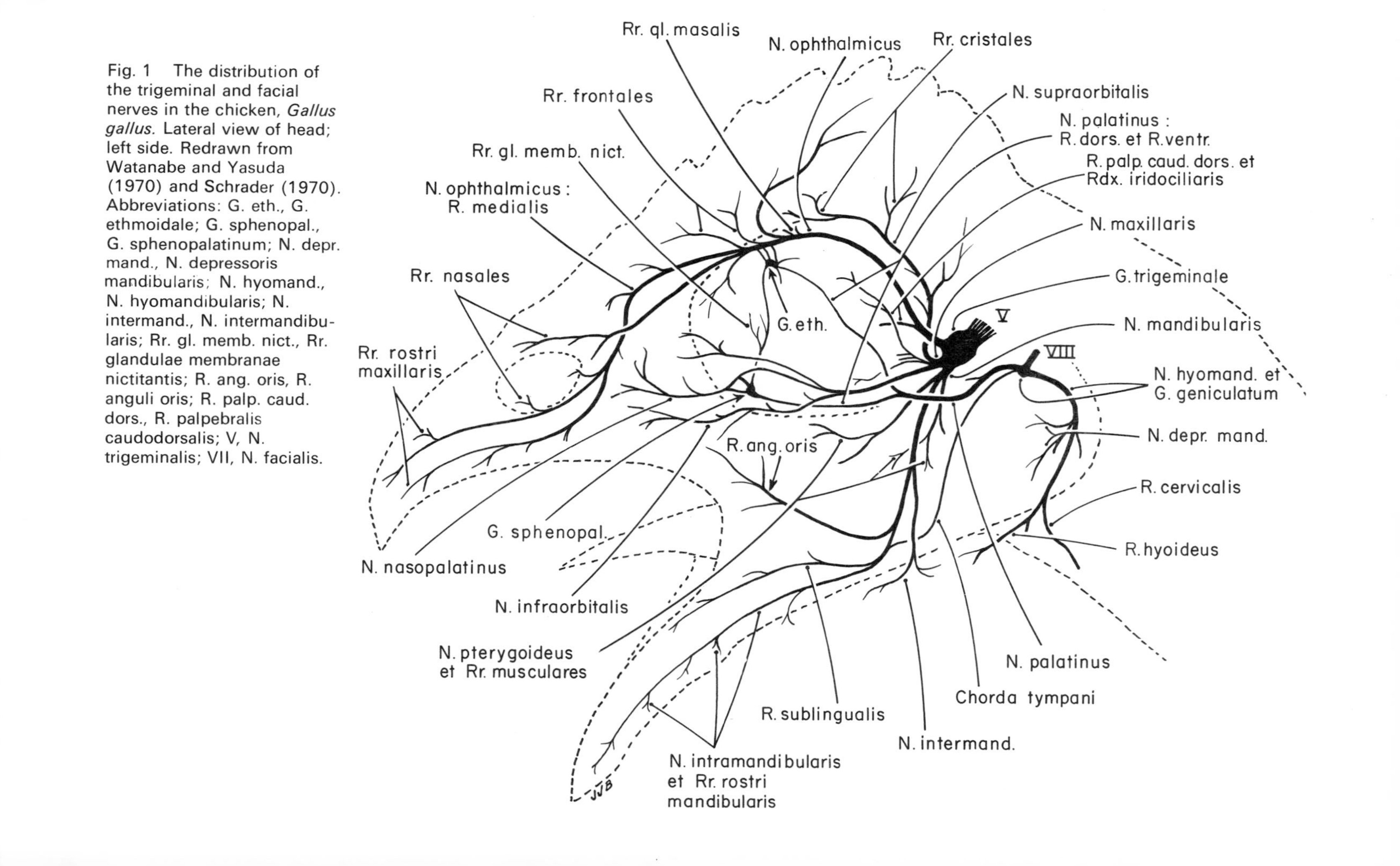

Fig. 1 The distribution of the trigeminal and facial nerves in the chicken, *Gallus gallus.* Lateral view of head; left side. Redrawn from Watanabe and Yasuda (1970) and Schrader (1970). Abbreviations: G. eth., G. ethmoidale; G. sphenopal., G. sphenopalatinum; N. depr. mand., N. depressoris mandibularis; N. hyomand., N. hyomandibularis; N. intermand., N. intermandibularis; Rr. gl. memb. nict., Rr. glandulae membranae nictitantis; R. ang. oris, R. anguli oris; R. palp. caud. dors., R. palpebralis caudodorsalis; V, N. trigeminalis; VII, N. facialis.

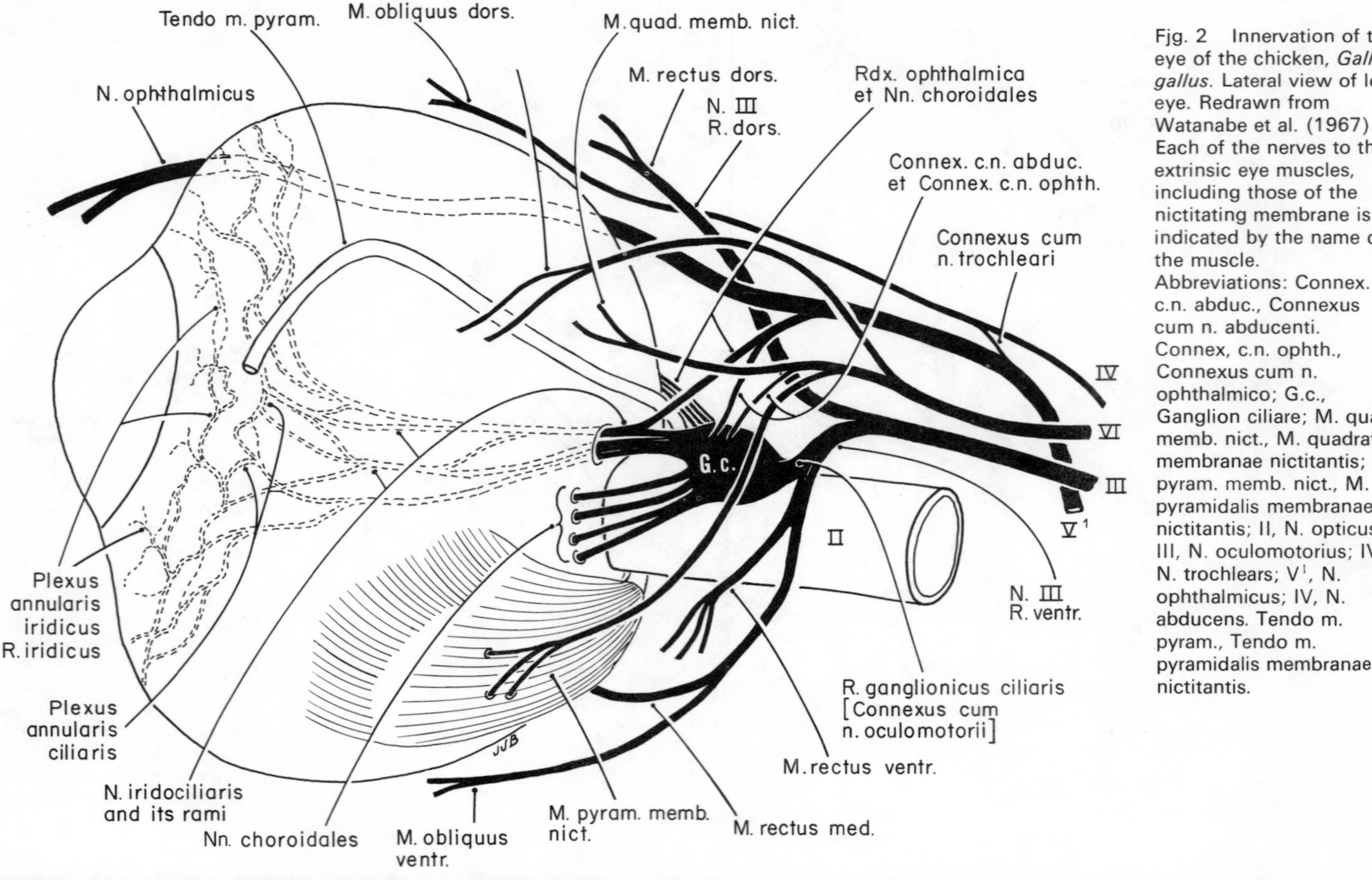

Fjg. 2 Innervation of the eye of the chicken, *Gallus gallus*. Lateral view of left eye. Redrawn from Watanabe et al. (1967). Each of the nerves to the extrinsic eye muscles, including those of the nictitating membrane is indicated by the name of the muscle. Abbreviations: Connex. c.n. abduc., Connexus cum n. abducenti. Connex, c.n. ophth., Connexus cum n. ophthalmico; G.c., Ganglion ciliare; M. quad. memb. nict., M. quadratus membranae nictitantis; M. pyram. memb. nict., M. pyramidalis membranae nictitantis; II, N. opticus; III, N. oculomotorius; IV, N. trochlears; V[1], N. ophthalmicus; IV, N. abducens. Tendo m. pyram., Tendo m. pyramidalis membranae nictitantis.

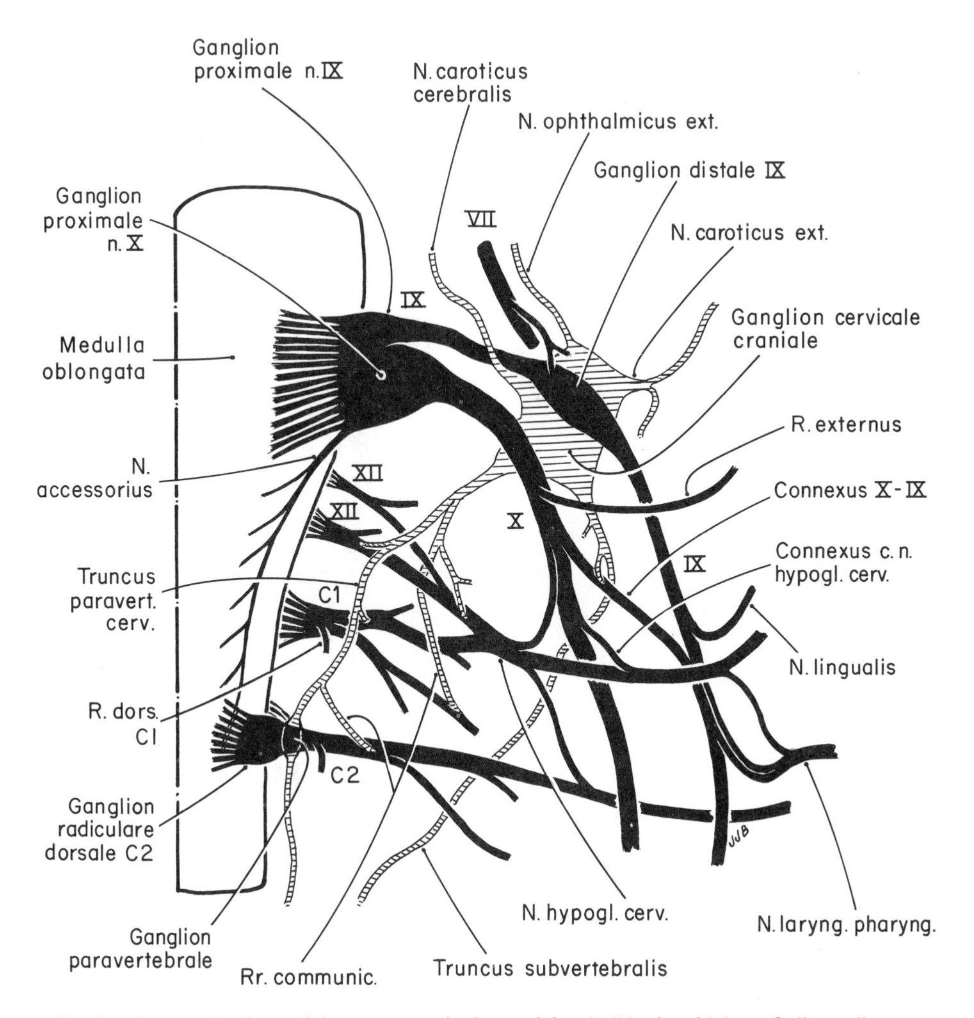

Fig. 3 Communications of the nerves at the base of the skull in the chicken, *Gallus gallus*. Dorsal aspect, right side, of the upper cervical spinal cord and Medulla oblongata. Redrawn from Baumel (1975). Plexiform communications between the last four cranial nerves, the upper cervical nerves and the cranial end of the paravertebral autonomic nerve trunk occur intracranially and immediately below the base of the skull.
Note: 1) the three nerves springing from the upper end of G. cervicale craniale accompany arteries of similar names; 2) the proximal ganglia of the cranial nerves IX and X are fused with one another and occupy a fossa in the base of the skull; 3) the first cervical nerve differs from most of the other spinal nerves in lacking a dorsal root ganglion and by having no paravertebral ganglion attached to it; 4) the extensive connections between the first cervical nerve and the nerves XII and X.
Abbreviations: C1, C2, Nn. cervicales 1 et 2; Connexus X-IX, Connexus vagoglossopharyngealis; Connexus c. n. hypogl. cerv., Connexus cum n. hypoglossocervicali; N. hypogl. cerv., N. hypoglossocervicalis; N. laryng. pharyng., N. laryngopharyngealis; Rr. communic., Rr. communicantes; Truncus paravert. cerv., Truncus paravertebralis cervicalis; IX, N. glossopharyngealis; X, n. vagus; N. XII, N. hypoglossus; VII, N. facialis.

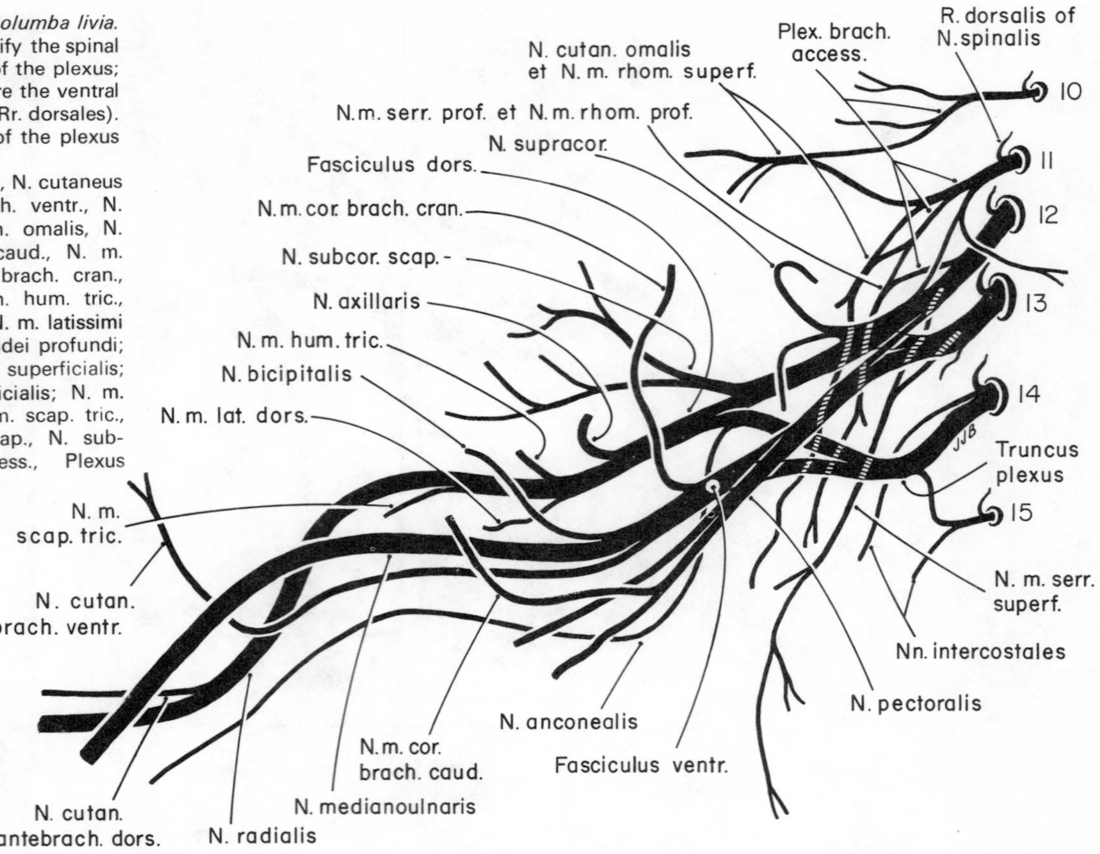

Fig. 4 Brachial plexus of the pigeon, *Columba livia*. Ventral view, right side. Numerals identify the spinal nerves that contribute to the formation of the plexus; the roots (Rdxx. plexus) of the plexus are the ventral rami of the spinal nerves (see the smaller Rr. dorsales). The proximal half (viewer's right half) of the plexus is located within the thoracic cavity.

Abbreviations: N. cutan. antebrach. dors., N. cutaneus antebrachialis dorsalis; N. cutan. brach. ventr., N. cutaneus brachialis ventralis; N. cutan. omalis, N. cutaneus omalis; N. m. cor. brach. caud., N. m. coracobrachialis caudalis; N. m. cor. brach. cran., N. m. coracobrachialis cranialis; N. m. hum. tric., N. m. humerotricipitis; N. m. lat. dors., N. m. latissimi dorsi; N. m. rhom. prof., N. m. rhomboidei profundi; N. m. rhom. superf., N. m. rhomboidei superficialis; N. m. serr. superf., N. m. serrati superficialis; N. m. serr. prof., N. m. serrati profundi; N. m. scap. tric., N. m. scapulotricipitis; N. subcor. scap., N. subcoracoscapularis; Plex. brach. access., Plexus brachialis accessorius.

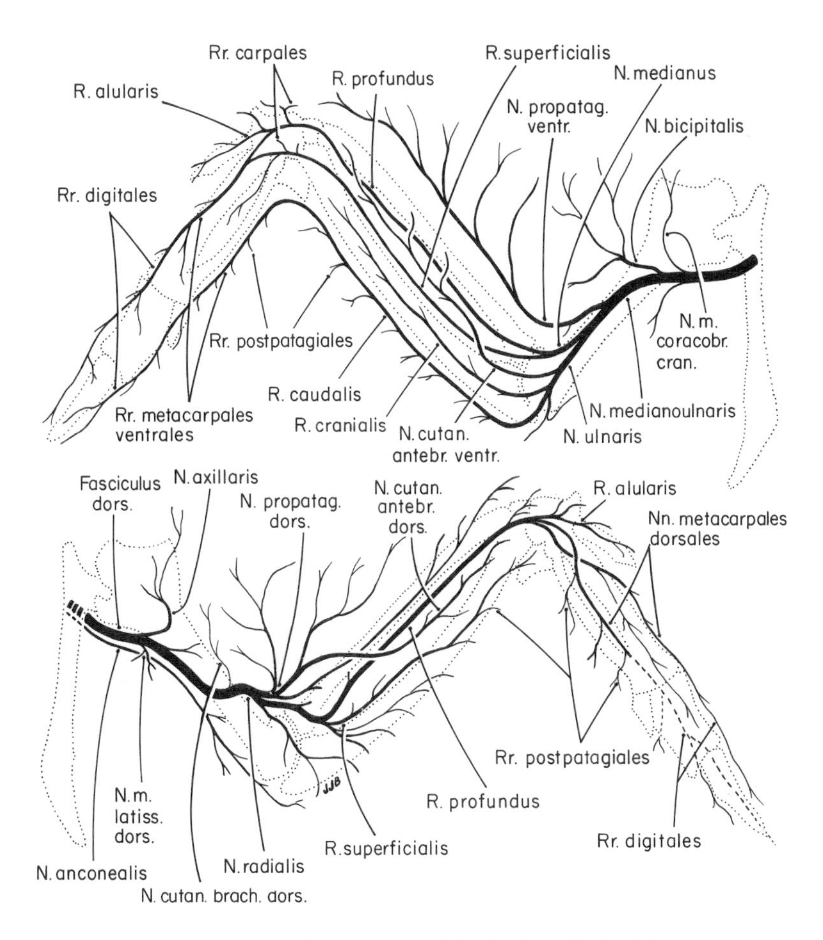

Fig. 5 Innervation of the wing of the pigeon, *Columba livia*. Right side. A, ventral nerves; B, dorsal nerves.
Note: 1) only a few of the muscular rami are identified; 2) the propatagial and postpatagial nerves are mostly cutaneous in their distribution; the postpatagial rami innervate the follicles of the large flight feathers; 3) see Fig. 4 for the origins of the radial and medianoulnar nerves from the brachial plexus.
Abbreviations: N. cutan. antebr., N. cutaneus antebrachialis; N. cutan. brach., N. cutaneus brachialis; N. m. coracobr. cran., N. m. coracobrachialis cranialis; N. m. latiss. dors., N. m. latissimi dorsi; N. propatag., N. propatagialis.

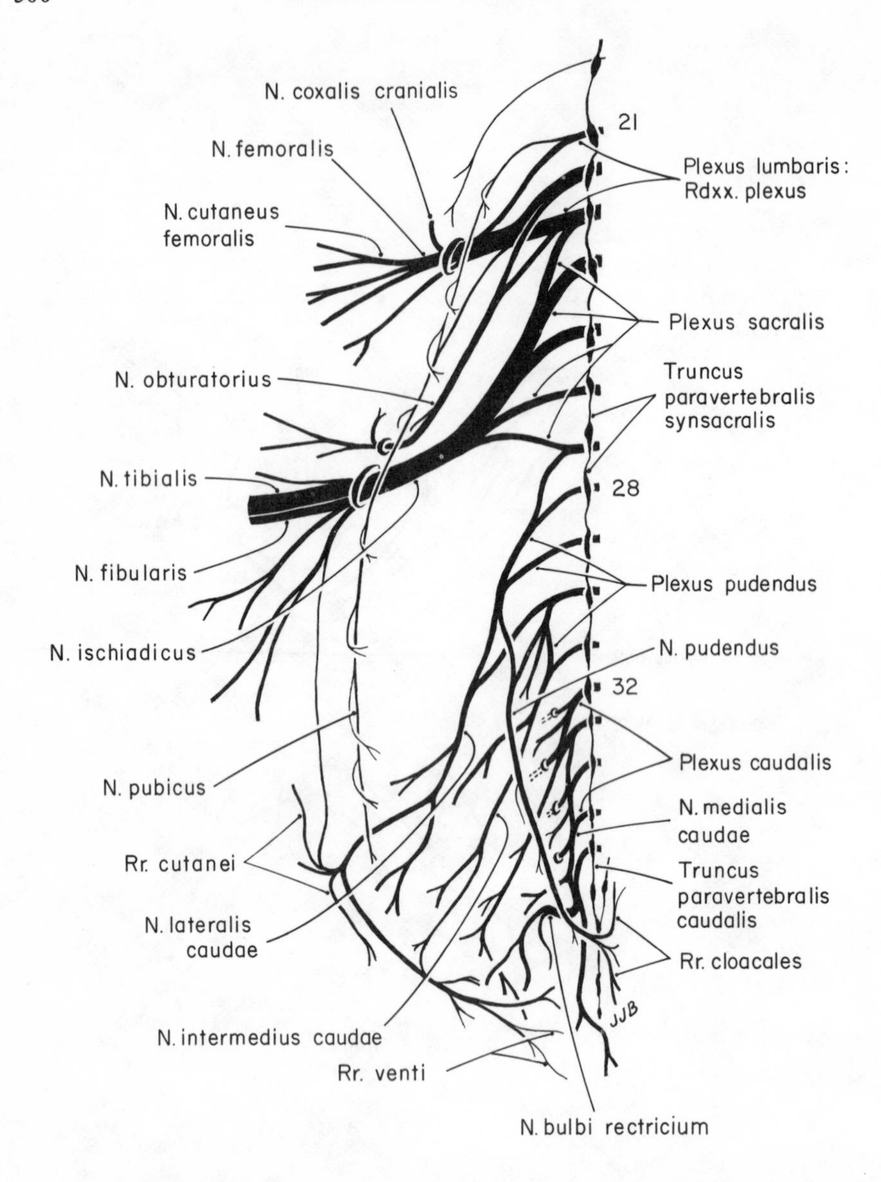

Fig. 6 Lumbosacral, pudendal, and caudal plexuses of the pigeon, *Columba livia*. Ventral view, right side. Numerals indicate the spinal nerves involved in formation of the plexuses. Note: 1) N. pudendus accompanies the ureter; this nerve provides part of the visceral innervation of the Ureter, Ductus deferens, Oviductus, Cloaca and Rectum; 2) N. pubicus courses along the ventral aspect of the Pubis within the abdominal muscles which it innervates; 3) no visceral rami of the paravertebral trunk are indicated. The paravertebral ganglia mostly are attached to the ventral rami of the spinal nerves as drawn; 4) usually the Plexus lumbaris is formed by the ventral rami of spinal nerves 21–23; the Plexus sacralis is formed by nerves 23–27; the Plexus pudendus is formed by nerves 27–31.

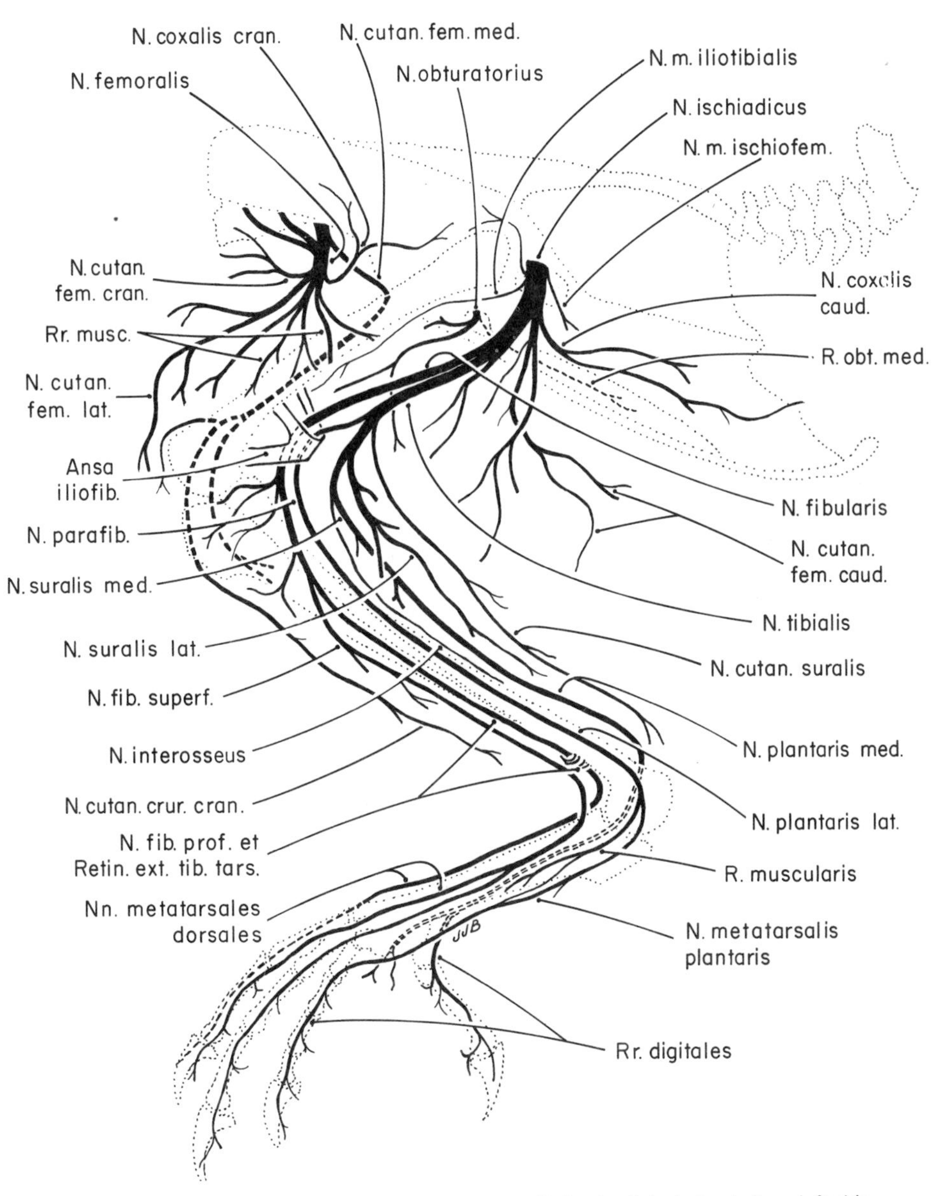

Fig. 7 Innervation of the pelvic limb of the pigeon *Columba livia*. Lateral view, left side. Note: 1) most muscular nerves are unnamed; 2) N. plantaris medialis is the prolongation of N. suralis medialis; 3) N. plantaris lateralis is the continuation of N. parafibularis. N. parafibularis is actually adherent to N. fibularis as the two traverse the Ansa m. iliofibularis; 4) see *Art.* Annot. 74 and *Ven.* Annot. 69 concerning the sural nerves and related vessels. Abbreviations: Ansa iliofib., Ansa m. iliofibularis; N. cutan. fem., N. cutaneus femoralis; N. cutan. crur. cran., N. cutaneus cruralis cranialis, N. fib. prof., N. fibularis profundus; N. fib. superf., N. fibularis superficialis; N. m. ischiofem., N. m. ischiofemoralis; N. parafib., N. parafibularis; R. musc., R. muscularis; R. obt. med. R. obturatorius medialis; Retin. ext. tib. tars., Retinaculum extensorium tibiotarsi.

Fig. 8 Thoracoabdominal visceral nerves of the chicken, *Gallus gallus*. Redrawn from Baumel (1975) after Stiemens (1934). Ventrolateral view of the spinal nerves and paravertebral trunk of the left side showing the various splanchnic nerves forming the subvertebral plexuses (see Chap. Intro.) on the aorta and on the roots of its visceral branches. Note: 1) the connection between the vagus nerve and the celiac plexus and the relation of the pudendal nerve to the Plexus pelvicus; 2) the rectal, ileal, and jejunal segments of N. intestinalis; 3) N. cardiacus provides pulmonary rami (not shown) as well as cardiac rami; 4) the loops (ansae) of the paravertebral trunk in the cervical, thoracic, and caudal regions; 5) the Gg. impares which represent the fusion of the right and left paravertebral trunks in the caudal region.
Abbreviations: A, Gl. adrenalis; C, Cecum; D. Duodenum; H, Hepar, I, Ileum; J, Jejunum; K, Ren; P, Proventriculus; Plex. mesent. caud., Plexus mesentericus caudalis; Plex. pancr. duod., Plexus pancreaticoduodenalis; Plex. subvert. synsac., Plexus subvertebralis synsacralis; R, Rectum; S, Splen; Truncus paravert. synsac., Truncus paravertebralis synsacralis; V, Ventriculus.

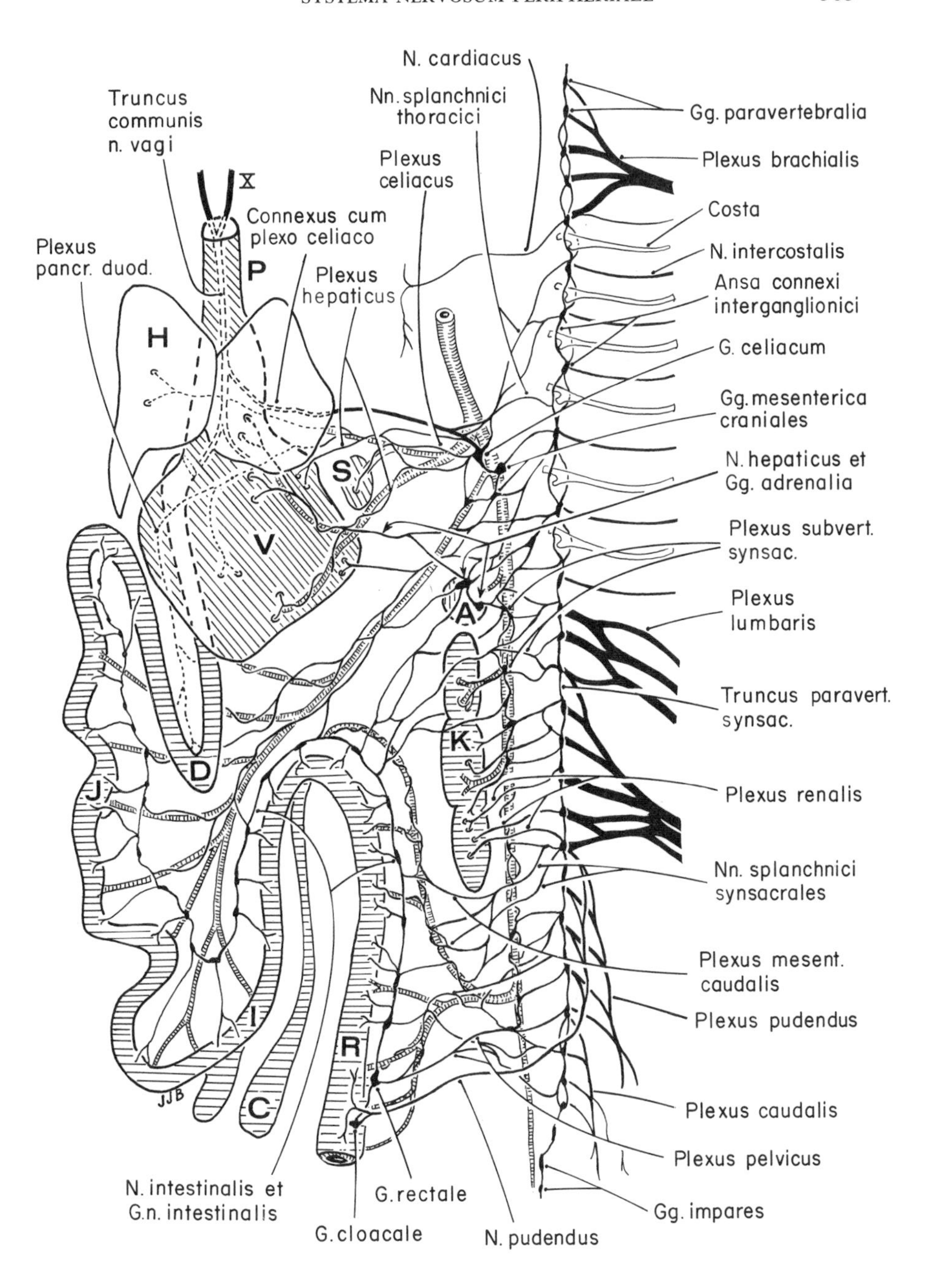
N. cardiacus
Truncus communis n. vagi
Nn. splanchnici thoracici
X
Plexus celiacus
Connexus cum plexo celiaco
Plexus pancr. duod.
P
Plexus hepaticus
H
S
V
A
K
D
J
I
R
C
JJB
Gg. paravertebralia
Plexus brachialis
Costa
N. intercostalis
Ansa connexi interganglionici
G. celiacum
Gg. mesenterica craniales
N. hepaticus et Gg. adrenalia
Plexus subvert. synsac.
Plexus lumbaris
Truncus paravert. synsac.
Plexus renalis
Nn. splanchnici synsacrales
Plexus mesent. caudalis
Plexus pudendus
Plexus caudalis
Plexus pelvicus
Gg. impares
N. intestinalis et G.n. intestinalis
G. rectale
G. cloacale
N. pudendus

ORGANA SENSORIA

HOWARD E. EVANS

Department of Anatomy, College of Veterinary Medicine, Cornell University, Ithaca, New York 14853, USA

With contributions from Sub-Committee Members: B. G. Bang, *Johns Hopkins University*; R. L. Boord, *University of Delaware*; R. B. Chiasson, *University of Arizona*; V. Ilyichev, *Academy of Sciences, Moscow*; J. McLelland, *University of Edinburgh*; R. Pearson, *University of Liverpool*; P. A. L. Wight, *Poultry Research Centre, Edinburgh*

For their roles in the preparation of the sense organ terminology, particular gratitude is extended to Dr John McLelland and Ronald Pearson who submitted the first draft in early 1973. Their efforts on this draft provided impetus for the development of the preliminary list of terms. As the work progressed, Drs Valery Ilyichev and Robert Boord were especially helpful with the ear terminology. Dr Robert Chaisson supplied histological sections of the eye from which Fig. 2 was drawn by artist, Lewis Sadler. In addition to contributions from Sub-Committee Members, helpful comments came from Drs R. D. Klemm, J. Schwartzkopff, and both Drs D. and M. Webster. Drs J. J. Baumel and A. S. King made many valuable suggestions and corrections on several drafts of the lists of terms.

Our knowledge of the structure and function of the sense organs of birds is rather incomplete although much present day research concerns the use of various sense organs in orientation and navigation. Birds utilize polarized light, stellar patterns, magnetic coordinates, barometric pressure, and infrasound as well as the better known sensory modalities of sight, hearing, heat, touch, taste, and olfaction. The structure of sense organs in birds is different from that of the mammal or the reptile and requires its own descriptive terms.

The eye of the bird is distinctive in many of its features. It varies greatly in shape from round to tubular and the axis of the eye may be directed rostrally (owl) laterally (finch), or caudally (woodcock). On account of this variation of the axis of the bulb, terms of position, direction, and movement of the bulb are especially defined. Remaining consistent with the *Nomina Anatomica Veterinaria* (ICVAN, 1973) for the eye, the poles of the bulb are "anterior" and "posterior". The part of the eye nearest the beak (Rostrum) is referred to as "nasal"; that nearest the ear is known as "temporal". See **Art.** Annot. 26.

The lens of the eye has an annular pad surrounding a central core whose refractive properties are influenced by this unique arrangement. The ciliary processes are in contact with the lens or attached to it, and appear to affect it directly. The ciliary muscle is attached to the inner surface of the scleral ring, and inserts at the Sinus venosus sclerae; it appears capable of dilating the filtration angle of the cilioscleral sinus. At the point where the optic nerve exits from the bulb, the choroid and retina project into the vitreous body as a pleated, pigmented, highly vascular Pecten which probably plays a nutritive role although over thirty functions have been ascribed to it. See Wingstrand and Munk (1965); Tucker (1975).

The ear of the bird, which can be isolated *in situ* without great difficulty, is a model of simplicity with several distinctive features which are not present in mammals: a canal which connects the Scala tympani with the Scala vestibuli at the base of the Cochlea; the Lagena at the end of the cochlear duct; a single ear ossicle with its processes attached to the tympanic membrane; etc. In the arrangement of terms for the ear, the structures of the membranous labyrinth have been separated from the structures of the osseous labyrinth in order to clarify their relationships. For example the Scala vestibuli and Scala tympani are creations of the osseous labyrinth whereas the Ductus cochlearis and Lagena are the terminal portions of the membranous labyrinth. Just as the semicircular "ducts" signify the membranous component versus the semicircular "canals" for the osseous portion, so should other "ducts" and "canals" agree. Thus the so-called Ductus brevis which joins the basal portions of the Scala tympani and the Scala vestibuli is herein named the Canalis interscalaris basalis while the connection of the two scalae at the apex (the Helicotrema) is called the Canalis interscalaris apicalis. The latter is an appropriate term since in the bird the connection between the two scalae has length rather than being a simple hole (trema).

OCULUS [ORGANUM VISUS]

(Fig. 1)

Bulbus oculi

Polus anterior (see **Term. sit.** Annot. 1; and **Gen. Intro.** Terms of orientation)

Polus posterior (see **Term. sit.** Annot. 1; and **Gen. Intro.** Terms of orientation)

Equator[1]

Axis bulbi [opticus][2]

Axis visuale [visus][3]

Tunica fibrosa bulbi
- Sclera
  - Annulus ossicularis sclerae[4] (Fig. 3)
    - Ossiculum sclerale
  - Ossicula posteriora sclerae[5]
    - Os nervi optici[6]
  - Sinus cilioscleralis[7] (Fig. 2)
  - Sinus venosus sclerae[8] (Fig. 2)
  - Lamina cartilaginea sclerae
- Cornea
  - Annulus conjunctivalis[9]
  - Limbus cornealis
  - Facies anterior
  - Facies posterior
  - Epithelium anterius cornealis
  - Lamina limitans anterior[10]
  - Substantia propria cornealis[11]
  - Lamina limitans posterior[12]
  - Epithelium posterius cornealis

Tunica vasculosa bulbi[13]
- Choroidea
  - Lamina suprachoroidea [fusca]
  - Spatium perichoroideum[14]
  - Lamina vasculosa
  - Lamina choroidocapillaris
  - Lamina basalis
- Corpus ciliare
  - Musculus ciliaris[15] (Fig. 2)
    - M. cornealis anterior[16]
      - Fibrae radiales[17]
    - M. cornealis posterior
  - Corona ciliaris
  - Processus ciliares[18]
  - Plicae ciliares
- Iris
  - Margo pupillaris
    - Granula iridica
  - Margo ciliaris[19]
  - Facies anterior
  - Facies posterior
  - Annulus iridicus[20]
  - Plicae iridicae[21]

Tunica vasculosa bulbi—*continued*

- Pupilla iridicae[21]
- Pupilla
- M. sphincter pupillae[22]
- M. dilator pupillae[22]
- Stroma iridicum
- Tapetum lucidum iridicum[23]
  - Corpus iridocytorum (Fig. 2)
    - Iridocyti[23]
- Epithelium anterius iridis
- Stratum pigmentum iridis
- Angulus iridocornealis
  - Reticulum trabeculare [Lig. pectinatum][24] (Fig. 2)
  - Spatia anguli iridocornealis[25]
- Circulus arteriosus iridicus (see **Art.** Annot. 28)
- Circulus arteriosus ciliaris (see **Art.** Annot. 28)

Tunica nervosa bulbi

- Retina (see **CNS** Annot. 42, 43)
  - Pars optica retinae
  - Ora serrata
  - Pars ciliaris retinae
  - Pars iridica retinae
  - Pecten oculi[26] (see Fig. 1; and **Art.** Annot. 29)
    - Pons pectinis
    - Lamina basalis pectinis
  - Area centralis rotunda[27]
  - Area centralis horizontalis[28]
    - Fovea centralis[29]
  - Area temporalis[30]
    - Fovea temporalis[31]
  - Stratum pigmentosum retinae[32]
    - Pars optica
    - Pars ciliaris
    - Pars iridica
  - Strata nervosa retinae
    - Stratum neuroepitheliale[33]
    - Stratum bipolare
    - Stratum ganglionicum
  - Nervus opticus
- Camera anterior bulbi (Fig. 1)
  - Angulus iridocornealis (see Sclera)
  - Humor aquosus

Camera posterior bulbi
    Humor aquosus
Camera vitrea bulbi
    Corpus vitreum
        Membrana vitrea
        Stroma vitreum
        Humor vitreus
    Fundus oculi[33]
Lens (Fig. 2)
    Capsula lentis
    Fibrae lentis
    Substantia lentis
    Corpus centrale lentis
    Pulvinus annularis lentis[34]
    Vesicula lentis
    Aqua vesiculae lentis
    Polus anterior
    Polus posterior
    Facies anterior
    Facies posterior
    Axis lentis
    Equator lentis
    Radii lentis
Zonula ciliaris[35] (Fig. 2)
    Fibrae zonulares
    Spatia zonularia

ORGANA ACCESSORIA OCULI

Musculi bulbi (see Membrana nictitans; and **PNS** Annot. 16)

M. rectus dorsalis
M. rectus ventralis
M. rectus medialis
M. rectus lateralis
M. obliquus dorsalis (see **Myol.** Annot. 14)
M. obliquus ventralis
M. quadratus membranae nictitantis (see **Myol.** Annot. 15)
    Vagina fibrosa tendinis [Trochlea]
M. pyramidalis membranae nictitantis (see **Myol.** Annot. 15)
    Tendo m. pyramidalis[36]

Vagina bulbi

Spatium episclerale

Palpebrae (see **Topog. anat.** Regio orbitalis)

Palpebra dorsalis[37]
Palpebra ventralis[37]
Facies cutanea
Facies conjunctivalis
Rima palpebrarum
Commissura temporalis palpebrarum
Commissura nasalis palpebrarum
Angulus oculi temporalis
Angulus oculi nasalis
Limbus palpebralis
  Plicae marginales[38]
Cilia palpebralia (see **Integ.** Annot. 40)
Tarsus[39]
M. levator palpebrae dorsalis (see **Myol.** Annot. 16)
M. depressor palpebrae ventralis (see **Myol.** Annot. 16)
M. orbicularis palpebrarum (see **Myol.** Annot. 16)
Membrana nictitans [Palpebra tertia][40] (see Musculi bulbi)
  Plica marginalis[41]

Conjunctiva

Tunica conjunctiva bulbi[9]
Tunica conjunctiva palpebrarum
Tunica conjunctiva membranae nictitantis
Saccus conjunctivalis
  Fornix conjunctivae dorsalis
  Fornix conjunctivae ventralis
  Fornix conjunctivae membranae nictitantis

Apparatus lacrimalis (Fig. 1b)

Gl. lacrimalis[42]
  Ductus gl. lacrimalis
Gl. membranae nictitantis[43] (Fig. 1b)
  Ductus gl. membranae nictitantis
Canaliculi lacrimales
Ostia canaliculi lacrimales[41]
Ductus nasolacrimalis

AURIS [ORGANUM VESTIBULOCOCHLEARE]
(Fig. 3)

AURIS INTERNA

LABYRINTHUS MEMBRANACEUS

Organum vestibulare (see **PNS** N. vestibulocochlearis)
- Endolympha
- Ductus endolymphaticus
- Saccus endolymphaticus
- Utriculus
- Macula utriculi
  - Statoconia
  - Membrana statoconiorum
  - Neuroepithelium
- Crista [Papilla] neglecta[44]
  - Neuroepithelium
- Ductus semicirculares (Fig. 3)
  - Ductus semicircularis rostralis [anterior]
  - Ductus semicircularis caudalis [posterior]
  - Ductus semicircularis horizontalis [lateralis]
  - Crus membranaceum commune
  - Epithelium ductus semicircularis
- Ampullae membranaceae
  - Ampulla membranacea rostralis [anterior]
  - Ampulla membranacea caudalis [posterior]
  - Ampulla membranacea horizontalis [lateralis]
    - Crista ampullaris
      - Neuroepithelium
    - Planum semilunatum[45]
    - Cupula
      - Septum cruciatum[46]
  - Ductus utriculosaccularis (Fig. 4)
  - Sacculus
    - Macula sacculi
      - Statoconia
      - Membrana statoconiorum
      - Neuroepithelium
  - Ductus sacculocochlearis [reuniens][47] (Fig. 4)

Organum cochleare (see **PNS** N. vestibulocochlearis)

Ductus cochlearis[47]

Tegmentum vasculosum[48]

Membrana tectoria

Papilla [Crista] basilaris

Membrana basilaris

Neuroepithelium

Lagena[49] (Fig. 4)

Macula lagenae

Statoconia

Membrana statoconiorum

Neuroepithelium

LABYRINTHUS OSSEUS

Perilympha

Spatium perilymphaticum

Cochlea[56]

Apex cochleae (Fig. 4)

Basis cochleae

Scala vestibuli (Fig. 4)

Fossa scalae vestibuli[50]

Cisterna scalae vestibuli[51]

Scala tympani (Fig. 4)

Recessus scalae tympani

Canalis interscalaris basalis [Ductus brevis][52] (Fig. 4)

Canalis interscalaris apicalis [Helicotrema][53] (Fig. 4)

Canaliculus perilymphaticus[54]

Vestibulum

Canales semicirculares ossei

Canalis semicircularis rostralis [anterior]

Canalis semicircularis caudalis [posterior]

Canalis semicircularis horizontalis [lateralis]

Crus osseum commune[55] (Fig. 3)

Ampullae osseae

Ampulla ossea rostralis [anterior]

Ampulla ossea caudalis [posterior]

Ampulla ossea horizontalis [lateralis]

AURIS MEDIA

Cavitas tympanica (see **Osteo**. Annot. 21)
Ostium tympanicum tubae pharyngotympanicae (see **Osteo.** Annot. 94, 98; and **Digest.** Annot. 19)
Foramen m. columellae
Recessus antevestibularis (see **Osteo.** Annot. 22)
Fenestra vestibularis
Lig. annulare columellae[57]
Fenestra cochlearis (Figs. 3, 4)
Membrana tympanica secundaria
Membrana tympanica
Margo fibroelasticus[58]
Sinus pneumaticus marginalis (Fig. 3)
Columella[59] (Fig. 3)
Scapus columellae[60]
Basis columellae[61]
Lig. columellosquamosum[62]
Cartilago extracolumellaris[63]
Processus caudalis
Processus ventralis
Processus rostralis
Musculus columellae[64] (Fig. 3)
Organum paratympanicum[65]

AURIS EXTERNA[66]

Meatus acusticus externus (see **Osteo.** Annot. 19, 21)
Porus acusticus externus
Plica cavernosa[67]
Glandulae meatus acustici externi
Pennae auriculares[68]
Tectrices auriculares rostrales
Tectrices auriculares caudales
Operculum auris[69]

ORGANUM OLFACTORIUM
(see **Resp.** Annot. 15)

Concha nasalis caudalis
Tunica mucosa nasi
Regio olfactoria
Neuroepithelium

ORGANUM GUSTATORIUM
(see **PNS** N. lingualis)

Gemma gustatoria[70]
Porus gustatorius

ORGANA SENSORIA ACCESSORIA[71]

Terminationes nervosae liberae[72]
Corpuscula nervosa terminalia
  Corpuscula nervosa capsulata
    Corpusculum lamellosum avium[73]
    Corpusculum bicellulare[74]
  Corpusculum nervosum acapsulatum
    Meniscus tactus[75]

ANNOTATIONS

(1) **Equator.** The greatest circumference of the eyeball. This is a valid term for birds with a spherical or globose eyeball (*Hirundo, Corvus*) but less meaningful for a flat (*Gallus, Cygnus*) or tubular eyeball (*Bubo, Strix*) (Walls, 1942).

(2) **Axis bulbi** [Axis opticus]. A line through the central point of the cornea and the lens which thus passes from the anterior to the posterior pole of the Bulbus oculi. This term avoids the Latin and Greek combination of Axis (L.) and opticus (G.).

(3) **Axis visuale [visus]**. A line passing through the center of the lens and the Fovea centralis, assumed to be the line of most acute vision. If there is no fovea, or the fovea lies more centrally, the Axis bulbi and Axis visuale coincide (Walls, 1942).

(4) **Annulus ossicularis sclerae.** There are usually 10 to 18 overlapping ossicles forming a ring within the sclera anterior to the equator. Birds with tubular eyes (owls) have a funnel-shaped ring formed by concave scleral ossicles. See Curtis and Miller (1938) and Nelson (1942).

(5) **Ossicula posteriora sclerae.** Ossifications in the sclera posterior to the equator distinct from the scleral ring.

(6) **Os nervi optici.** An ossification in the cartilage surrounding the optic nerve, also referred to as "Gemminger's ossicle" and "os opticus" (Tiemeier, 1950, 1953).

(7) **Sinus ciliosceleralis** (Fig. 2). A posterior extension of the anterior chamber via a choroidal meshwork at the base of the iris. The venous sinus of the sclera (canal of Schlemm) lies in or borders this region.

(8) **Sinus venosus sclerae.** A venous annulus (canal of Schlemm) near the corneoscleral junction which drains aqueous humor from the cilioscleral sinus. There may

be two venous sinuses of the sclera, depending on the course of an annular artery which sometimes divides the chamber.

(9) **Annulus conjunctivalis.** A ring marking the narrow zone of transition where the bulbar conjunctiva becomes continuous with the corneal epithelium. The location of the annulus corresponds to the corneal limbus (see Fig. 2).

(10) **Lamina limitans anterior.** A fibrillated membrane underlying the corneal epithelium also referred to as Bowman's membrane.

(11) **Substantia propria cornealis.** Connective tissue that separates the anterior from the posterior Lamina limitans.

(12) **Lamina limitans posterior.** A homogeneous membrane which is situated, when present, at the posterior surface of the Substantia propria. This is very thin in many birds; it is also known as Descemet's membrane.

(13) **Tunica vasculosa bulbi.** This vascular layer which includes the choroid and iris is sometimes referred to as the uveal tract.

(14) **Spatium perichoroideum.** The choroid is thickest in the fundus and has a prominent capillary bed. According to Walls (1942, p. 645), there is a thick sinusoidal structure between the vascular bed and the thin pigmented Lamina suprachoroidea. In prepared slides this area appears rather empty. Traversing this thick open layer, there are connective tissue cords and columns which often contain muscle cells. In the flicker (*Colaptes*) the choroid is not empty-looking but contains a thick mass of mucoid tissue which has probably been developed to prevent detachment of the retina during "woodpecking".

(15) **Musculus ciliaris.** Synonymy: M. sclerocornealis. The ciliary muscle of birds is attached to the inner surface of the sclera and extends from the scleral cartilage (optic cup) across the width of the scleral ring, to the corneoscleral junction where it ends in close association with the Sinus venosus sclerae. The muscle may appear as one or be divided into an anterior and posterior portion.

(16) **M. cornealis anterior.** Synonomy: Crampton's muscle. The anterior corneal muscle represents the corneal end of the reptilian ciliary muscle. It originates on the scleral ring and inserts on the limbus of the cornea. The M. cornealis anterior most directly alters the contour of the cornea. It is largest in hawks and owls, smallest in aquatic birds and absent in the cormorant (*Phalacrocorax*).
M. cornealis posterior. Synonymy: Brücke's muscle. This muscle originates on the thin scleral sheet which forms the anchorage of the trabecular retinaculum and represents the scleral end of the reptilian ciliary muscle. The M. cornealis posterior is large in cormorants and the gannet (*Morus* (Walls, 1942).

(17) **Fibrae radiales.** Synonymy: Müller's muscle. These fiber bundles are present in some birds as subdivisions of the M. cornealis anterior or the M. cornealis posterior with fibers radial in orientation (Walls 1942: 279, 616). Müller's muscle according to Rochon-Duvigneaud (1950, p. 228) consists of radiating fibers which pass between Crampton's and Brucke's muscles and does not merit a designation as a muscle.

(18) **Processus ciliares.** Unlike mammals, the ciliary processes of birds are numerous and irregular, occupying the entire ciliary zone. The ciliary processes are often fused with one another and attach directly to the lens capsule.

(19) **Margo ciliaris.** The border of the iris and ciliary body at the iridocorneal angle.

(20) **Annulus iridicus.** The free portion of the iris consisting of a pupillary ring and a ciliary ring.

(21) **Plicae iridicae.** Folds or indentations on the margin of the pupil.

(22) **M. sphincter pupillae; M. dilator pupillae.** There have been various opinions expressed as to the presence, absence, proportions, or morphology of sphincter and dilator muscles of the bird's iris. Although Oehme (1969b) reported smooth muscle fibers in both the sphincter and dilator muscles, Pilar and Vaughan (1971) say these smooth muscle fibers represent a very minor portion of the total muscle mass, and do not contribute significantly to the contracture tension.

(23) **Tapetum lucidum iridicum.** This structure consists of refractive cells (iridocytes) in the iris of Columbiformes which are visible in histological sections under transmitted or polarized light. Chiasson and Ferris (1968) described two types of cells in the Inca Dove *Scardafella inca*: cells with large reflecting platelets scattered in the superficial layer of the iris and deeper cells with smaller platelets forming a more discrete iridocyte body.

(24) **Reticulum trabeculare [Lig. pectinatum].** The Reticulum trabeculare consists of a trabecular meshwork in the iridocorneal angle connecting the base of the iris to the cornea. Wychgram (1912) has shown that during accommodation when the sphincter of the iris contracts, the trabeculae are tensed.

(25) **Spatia anguli iridocornealis.** Synonymy: Spaces of Fontana. These interstices between the trabeculae of the reticulum in the iridocorneal angle of the anterior chamber allow for drainage into the Sinus venosus sclerae.

(26) **Pecten oculi.** This projection from the retina into the vitreous body is conical and vaned in the Ostrich, rhea, and tinamou. It is usually thin and pleated in other birds. The kiwi (*Apteryx*) has a reptilian-like pecten similar to the conus papillaris of lizards. Wood (1917) illustrated the gross appearance in many species. Wingstrand and Munk (1965) concluded that the pecten is a nutritive organ necessary for the maintenance of the inner retinal layers. Dieterich and Pfautsch (1973) support the nutritive role of the pecten. Brach (1975) suggests that the pecten is related primarily to intraocular pH regulation. There are several different explanations of how the pecten functions. The capillaries of the pecten form an extensive anastomotic network and their ultrastructure suggests that there is active transcellular transport through the capillary endothelial cells. The pecten has numerous pigment cells which give it a jet black appearance. It is attached to the Fundus oculi at the exit of the optic nerve. See Tucker (1975) concerning the mechanical significance of the pecten.

(27) **Area centralis rotunda.** This is a small, round mound of high cone density and a more or less 1:1 ratio of cone cells to ganglion cells.

(28) **Area centralis horizontalis.** A ribbon-like area replacing the rotund area in species of open habitat such as the shore birds (e.g. Charadriiformes). During the alert posture, the head is held with the beak pointing skyward so that this area is close to the horizontal (see Duijm, 1958; and Walls, 1942, p. 188). See also Pearson (1972, pp. 297–300, Tables 39, 40) for summaries of species which demonstrate presence or absence of ribbon-like, horizontal central areas of the retina.

(29) **Fovea centralis.** The central fovea is a depression with steeply sloping sides in the Area centralis. Better developed than the temporal fovea in all birds except eagles and swifts (*Apus*) (Pumphrey, 1961).

(30) **Area temporalis.** Additional area in Alcedinidae (kingfishers) and Trochilidae (hummingbirds) in the temporal portion of the retina.

(31) **Fovea temporalis.** Depression in the Area temporalis of the kingfishers and hummingbirds. Owls have only the temporal fovea (Walls, 1942) presumably because a central fovea would be of little use in a tubular eye.

(32) **Stratum pigmentosum retinae.** As in mammals but penetrating further between the rods and cones (Pearson, 1972).

(33) **Stratum neuroepitheliale.** Same as S. photosensorium of *Nomina Histologica* (IANC, 1975). See King-Smith (1971) for details of the retinal structure.
Fundus oculi. The portion of the interior of the eyeball around the posterior pole; that part farthest removed from the opening or pupil.

(34) **Pulvinus annularis lentis.** Synonymy: Ringwulst, bourrelet, or annular pad of the lens. An encircling band separated by a lenticular chamber from the central area of the lens. The ciliary processes contact this annular pad. In swifts (*Apus*), the annular pad is one-half of the area of a median section; it is small in flightless birds such as the kiwi (*Apteryx*) and Ostrich (*Struthio*).

(35) **Zonula ciliaris.** The zonular fibers pass from the ciliary processes, folds, and adjacent areas to the lens capsule. Owing the proximity and contact of some anterior ciliary processes with the lens, the zonular fibers in this region are short or fused in the contact zone.

(36) **Tendo m. pyramidalis.** Synonymy: Arcus tendineus nervi optici. This is the tendon of the M. pyramidalis attached to the Membrana nictitans.

(37) **Palpebra dorsalis; Palpebra ventralis.** The upper lid is short and thick whereas the lower lid is longer, thinner and very movable. Closure of the eye is due mainly to the movement of the lower lid. The lower lid has a fibrous tarsus. When the lids are closed, they meet well above the pupil. In most birds the lids are wide open before hatching, but in altricial birds, the lids remain closed for a time after hatching (Walls, 1942). See **Myol.** Annot. 16; and **Topog. anat.** Annot. 24.

(38) **Plicae marginales.** The margin of the eyelid is thickened and segmented into a variable number of folds, 17 to 19 in the House Sparrow (*Passer*) to allow for stretching and closure. "When the lids are closed, the junction does not form a straight line but a slight curve upward." (Slonaker, 1918, p. 355).

(39) **Tarsus.** According to Walls (1942, p. 424) only the lower lid has a fibrous tarsus.

(40) **Membrana nictitans [Palpebra tertia].** The nictitating membrane of owls and dippers appears cloudy or white because its inner surface is covered by a specialized epithelium which improves its cleansing action. It is a protective, translucent membrane pulled by muscles from the nasal to the temporal commissure. In mammals, the Plica semilunaris is a rudiment of the nictitating membrane. Glands supplying fluid and mucus to the lids and cornea are the lacrimal glands and the gland of the nictitating membrane. For a detailed description of the structure, tendons, muscles and histology of the nictitating membrane of the chicken, see Simić and Jablan-Plantič (1959). See also Slonaker (1918, 1922) and Stibbe (1928).

(41) **Plica marginalis.** This marginal plait on the anterior surface of the nictitating membrane functions to draw fluids and detritus from the conjunctival space, between the nictitating membrane and the lids to the lacrimal ostia at the nasal commissure of the lids (Slonaker, 1918, p. 362).

(42) **Glandula lacrimalis.** The small lacrimal gland is located in the caudolateral wall of the orbit, having a single duct (Slonaker, 1918) or multiple ducts (McLelland, 1975) that open inside the lower lid.

(43) **Glandula membranae nictitantis.** Synonomy: Harder's gland; Gl. lacrimalis: Gl. palpebrae tertiae; and Gl. lacrimalis accessorius. This large gland located within the orbit restromedial to the eyeball empties its copious secretion via a duct into the conjunctival sac between the eyeball and nictitating membrane. It is the major tear gland of birds rather than an accessory gland. See Miller *et al.* (1971) for a discussion of the role of this gland as a source of antibody cells.

(44) **Crista [Papilla] neglecta.** A small sensory area on the floor of the Utriculus, often present in mammals as a part of the Crista ampullaris caudalis (see Jorgensen, 1970).

(45) **Planum semilunatum.** A half-moon shaped cellular zone surrounding the Crista ampullaris on the side-walls of the ampulla in the pigeon (Dohlman, 1964).

(46) **Septum cruciatum.** Synonomy: Eminentia cruciata; Septum cruciforme. A horizontal fold on the Cupula within the ampullae of the rostral and caudal semicircular ducts (Igarashi and Yoshinobu, 1966).

(47) **Ductus cochlearis.** Synonomy: Scala media. An outgrowth of the Sacculus, filled with endolymph, and still connected to it by the Ductus sacculocochlearis. The blind end of the cochlear duct is the Lagena. On each side of the cochlear duct are perilymphatic spaces: the Scala vestibuli and Scala tympani. See Amerlinck (1923), Schwartzkopff and Winter (1960), and Jorgensen (1970).

(48) **Tegmentum vasculosum.** A thick, folded epithelium extending into the cochlear duct. In life the Tegmentum vasculosum compresses the Scala vestibuli so as to occlude the lumen (Schwartzkopff, 1973; and Amerlinck, 1923).

(49) **Lagena.** The slightly expanded; blind end of the avian cochlear duct. See Jorgensen (1970).

(50) **Fossa scalae vestibuli.** An enlargement of the perilymphatic space at the distal end of the Scala vestibuli.

(51) **Cisterna scalae vestibuli.** A small perilymphatic space in the region of the vestibular window (Schwartzkopff, 1968).

(52) **Canalis interscalaris basalis [Ductus brevis].** A connection at the base of the cochlea between the Scala vestibuli and the Scala tympani (DeBurlet, 1934; Schwartzkopff, 1973). See **Org. sens.** Intro.

(53) **Canalis interscalaris apicalis [Helicotrema].** Synonomy: Ductus scalae tympani. This is the apical connection between the Scala vestibuli and the Scala tympani (Fig. 4) at the apex of the cochlea. The term "Ductus" should only be used for the membranous labyrinth. See **Org. sens.** Intro.

(54) **Canaliculus perilymphaticus.** Synonomy: Aqueductus cochleae; Canaliculus cochleae. A connection between the Scala tympani and the subarachnoid space. In man the so-called perilymphatic duct is not a true duct but rather a canaliculus containing connective tissue and fluid (Anson and Donaldson, 1973).

(55) **Crus osseum commune.** The confluence of the rostral and caudal semicircular canals; see Fig. 2.

(56) **Cochlea.** A thin walled, slightly curved osseous tube enclosing the Cochlear duct and its terminal Lagena. The cochlea is essentially a straight tube, not a spiral as in mammals.

(57) **Ligamentum annulare columellae.** A ring of fibrous tissue attaching the foot-plate of the Columella to the rim of the vestibular window.

(58) **Margo fibroelasticus.** This fibrous ring is the thickened margin of the tympanic membrane. In its rostrovental attachment area, it includes an air sinus (Pohlman, 1921, p. 239), the Sinus pneumaticus marginalis.

(59) **Columella.** The only auditory ossicle of the bird. It is of hyoid arch origin and extends from the tympanic membrane to the vestibular window. At the tympanic or distal end there is a tripod-shaped extracolumellar cartilage, and at the vestibular or proximal end it possesses an expanded basis or footplate occluding the window (Crompton, 1953; Frank and Smit, 1976).

(60) **Scapus columellae.** The ossified shaft or body of the Columella; also called Pars stapedialis or Stapes.

(61) **Basis columellae.** The footplate or expanded proximal end of the Columella. In some owls "the internal surface of the footplate forms a remarkable vesicular protrusion" (Schwartzkopff, 1955). Another term for the footplate is the Clipeolus (Ilyichev, 1972).

(62) **Lig. columellosquamosum.** Synonomy: Platner's ligament. Pohlman (1921, p. 247) describes and figures it in the chicken as an elastic band between the Proc. rostralis of the extracolumellar cartilage and the tympanic margin at the quadrato-squamosal articulation.

(63) **Cartilago extracolumellaris.** A tripod-like cartilage of the Columella attached to the tympanic membrane. Also referred to as the Pars extrastapedialis of the Columella.
Proc. caudalis; Proc. ventralis; Proc. rostralis. Synonomy: Proc. caudalis, Proc. suprastapedialis; Proc. ventralis, Proc. infrastapedialis; Proc. rostralis, Proc. extrastapedialis. These parts of the cartilaginous extracolumella anchor the Columella to the tympanic membrane. See Fig. 3.

(64) **Musculus columellae.** Synonymy: M. tensor tympani; M. stapedius; M. occipito-tympanicus (Edgeworth, 1935). Innervated by the N. facialis (Pohlman, 1921), it is a derivative of second branchial arch musculature, and is therefore homologous to the stapedius muscle of mammals. This new term clearly associates the muscle with the columella, the only middle ear ossicle in the bird.

(65) **Organum paratympanicum.** A vesicular remnant of pharyngeal pouches noted by Vitali (1912) and reinvestigated by Federici (1927). The organ is of variable form, embedded in connective tissue caudo-dorsal to the quadrato-prootic articulation. It is also called the Organ of Vitali (Maderson and Jaskoll, 1976; Romanoff, 1960).

(66) **Auris externa.** The external ear of birds may have specialized feathers or ear flaps associated with it. Diving birds have the narrowest lumen, while parrots, passerine birds, and falconiform birds have the widest ear funnels. The external ears are asymmetrical in several genera of owls and the skull bones are modified accordingly. (Schwartzkopff, 1973; Ilyichev, 1961; Kelso, 1940). See **Osteo.** Annot. 19, 21.

(67) **Plica cavernosa.** A semicircular erectile fold deep in the external acoustic meatus close to the tympanic membrane. It is firmly attached at the upper and lower bony borders of the external acoustic meatus but its base is separated by a bursa from the wall of the meatus. Pohlman (1921) illustrates the fold in one of his figures and states that Wurm first described it in 1885.

(68) **Pennae auriculares.** Feathers associated with the opening to the external ear. In some birds the rostral shielding feathers are rather open-spaced whereas the caudal sound reflecting feathers are very dense. In diving birds the feathers may overlap and cover the meatus (Ilyichev, 1961; Lucas and Stettenheim, 1972, p. 99). (See Annot. 66).

(69) **Operculum auris.** The skin-fold covering the external acoustic meatus of some owls (e.g. *Asio otus*). According to Schwartzkopff (1973) there may be "not only posterior flaps but also erectable anterior flaps". Both have specialized feathers which can serve either to close or enlarge the ear opening. See **Topog. anat.** Annot. 6.

(70) **Gemma gustatoria.** There are few taste buds on the base of the tongue in most birds although parrots are said to have greater numbers than other species. (Lindenmaier and Kare, 1959; Moore and Elliott, 1946; Gentle, 1971).

(71) **Organa sensoria accessoria.** Various types of sensory endings have been described, some of which may serve specific modalities such as touch, temperature, atmospheric pressure, vibration, chemical reception, etc. (Grzycki, 1973; Malinovský, 1967; Schwartzkopff, 1973; Zweers *et al.*, 1977).

(72) **Terminationes nervosae liberae.** Free nerve endings are found in various tissues of the body and some of these may be associated with general or specific chemical sensations.

(73) **Corpusculum lamellosum avium.** This new term was devised in order to avoid the eponym, "Herbst". These corpuscles consist of a lamellated inner and outer core around a central axon. They are found in joint capsules and in the bill of ducks. It has been demonstrated that this lamellar corpuscle is different from the Vater-Pacinian corpuscle of mammals (Malinovský, 1967; Malinovský and Zemanek, 1969; Munger, 1971; Zweers *et al.*, 1977).

(74) **Corpusculum bicellulare.** A new term to avoid the eponym "Grandry". The "Grandry" corpuscle of avian anatomical literature is composed of two tactile cells with a nerve disc between them enclosed in a capsule deep in the corium. These corpuscles are commonly found in the beak of ducks (see **Osteo.** Annot. 41). Typically the corpuscles are pictured and described as bicellular but Malinovský (1967), quoting the literature, reported among different kinds of birds a variability of 1 to 8 cells. See Zweers *et al.* (1977).

(75) **Meniscus tactus.** A new term in place of the eponym "Merkel"; also used in the *Nomina Histologica* (IANC, 1977). The "Merkel" corpuscle consists of a cup-shaped terminal network with a nerve fiber entering into intimate contact with a special tactile cell. It is situated more superficially than the others and lacks a distinct capsule.

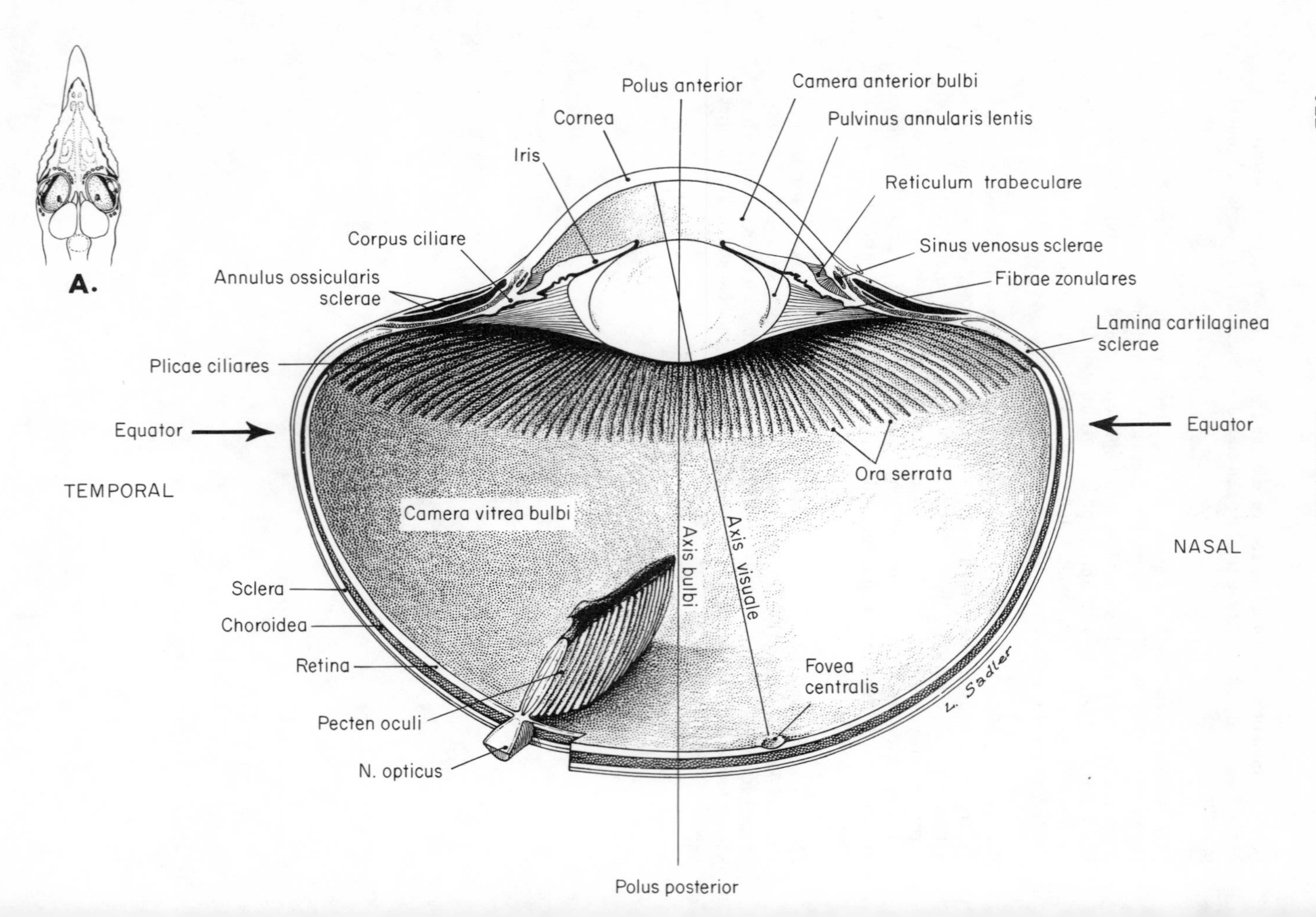
A.
Polus anterior
Camera anterior bulbi
Cornea
Pulvinus annularis lentis
Iris
Reticulum trabeculare
Corpus ciliare
Sinus venosus sclerae
Annulus ossicularis sclerae
Fibrae zonulares
Lamina cartilaginea sclerae
Plicae ciliares
Equator
Equator
Ora serrata
TEMPORAL
Camera vitrea bulbi
Axis bulbi
Axis visuale
NASAL
Sclera
Choroidea
Retina
Fovea centralis
L. Sadler
Pecten oculi
N. opticus
Polus posterior

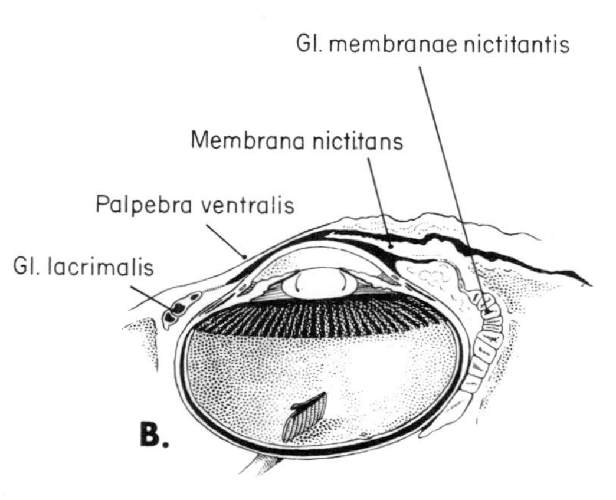

Fig. 1 Interior of the eye; *Gallus*. The eye is sectioned in the dorsal (horizontal) plane; the view is of the interior of the ventral half of the bulb.

Inset A (upper left corner) is a section through the head in the dorsal plane which shows the position of the eyes within the orbits. Inset B (lower right corner) is a section through the eye and ventral lid in the dorsal plane which demonstrates some of the accessory structures of the eye.

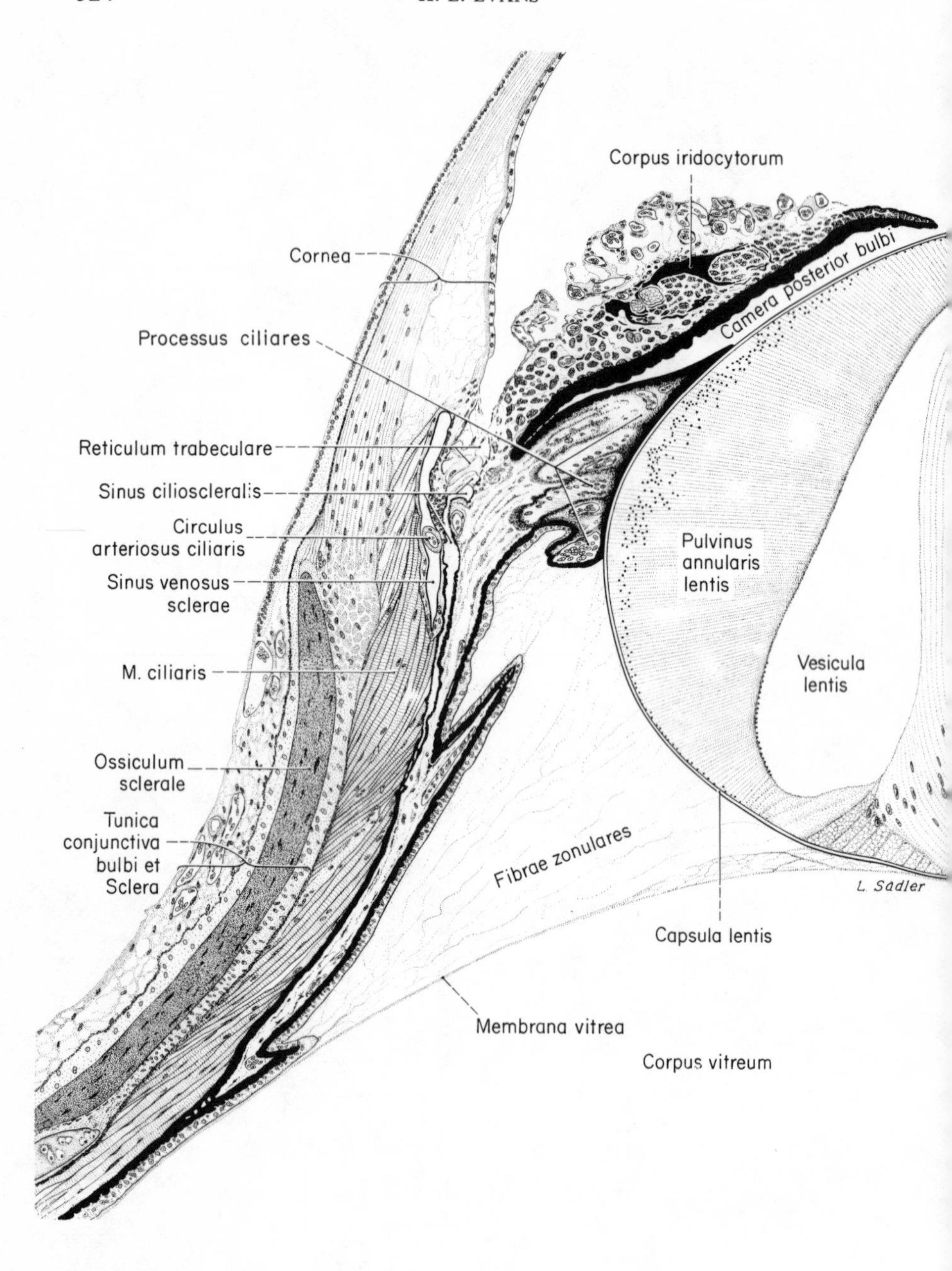

Fig. 2 Part of a meridional section of the eye through the corneoscleral junction, ciliary region, iris, and lens; Inca Dove (*Scardafella inca*). Drawn from histological slides, courtesy of R. B. Chaisson.

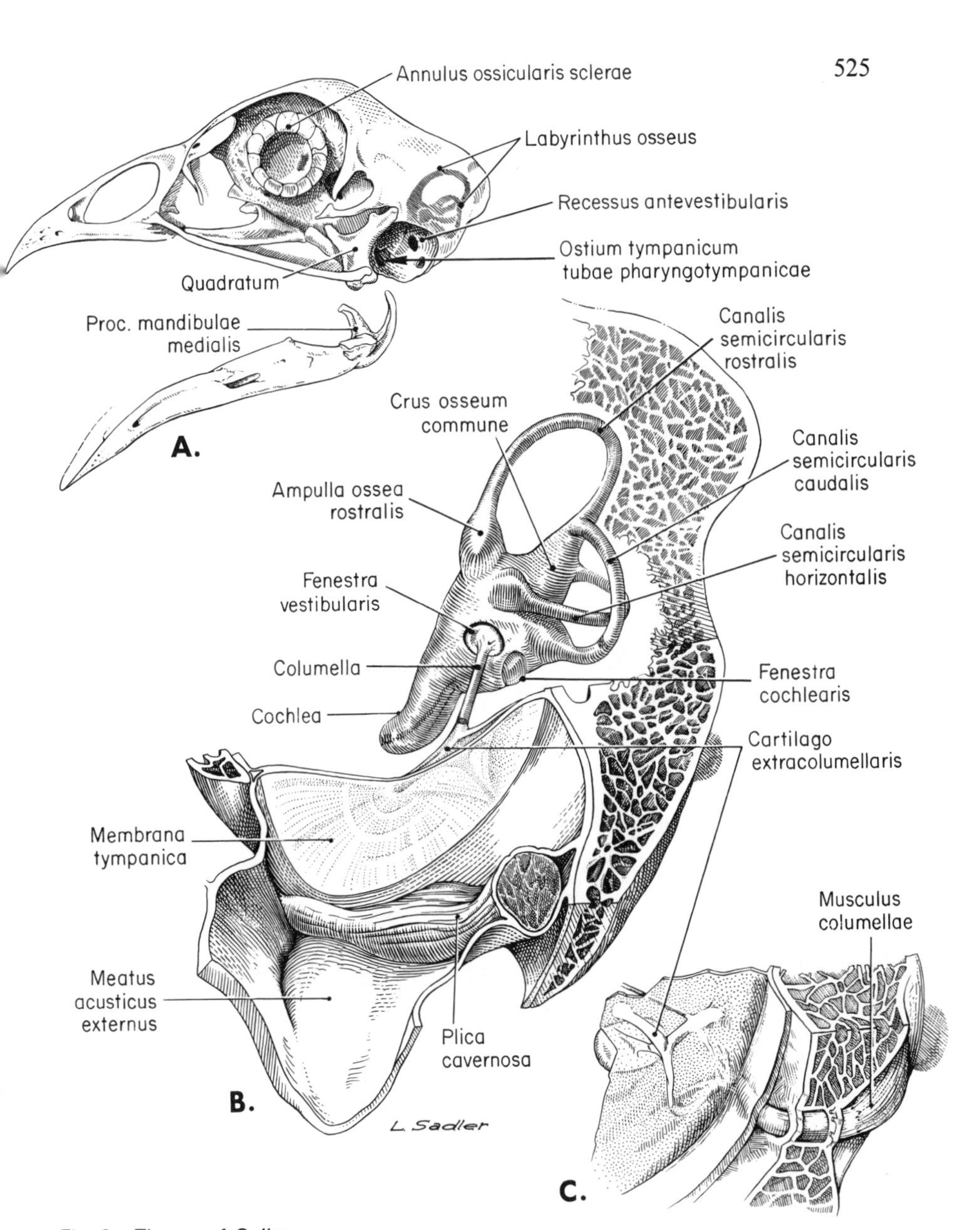

Fig. 3 The ear of *Gallus.*

A. Lateral view of the skull showing the ear region. Membrana tympanica and the Columella are removed; the osseous labyrinth is shown in transparency.

B. Dorsolateral view of the left ear, sculptured to show the osseous labyrinth and the floor of the external acoustic meatus.

C. Lateral view of the caudoventral quadrant of the ear, showing the processes of the extracolumellar cartilages through the tympanic membrane and the tendon of M. columellae entering the tympanic cavity via a foramen (see Fig. 3a, not labelled) in its caudoventral wall.

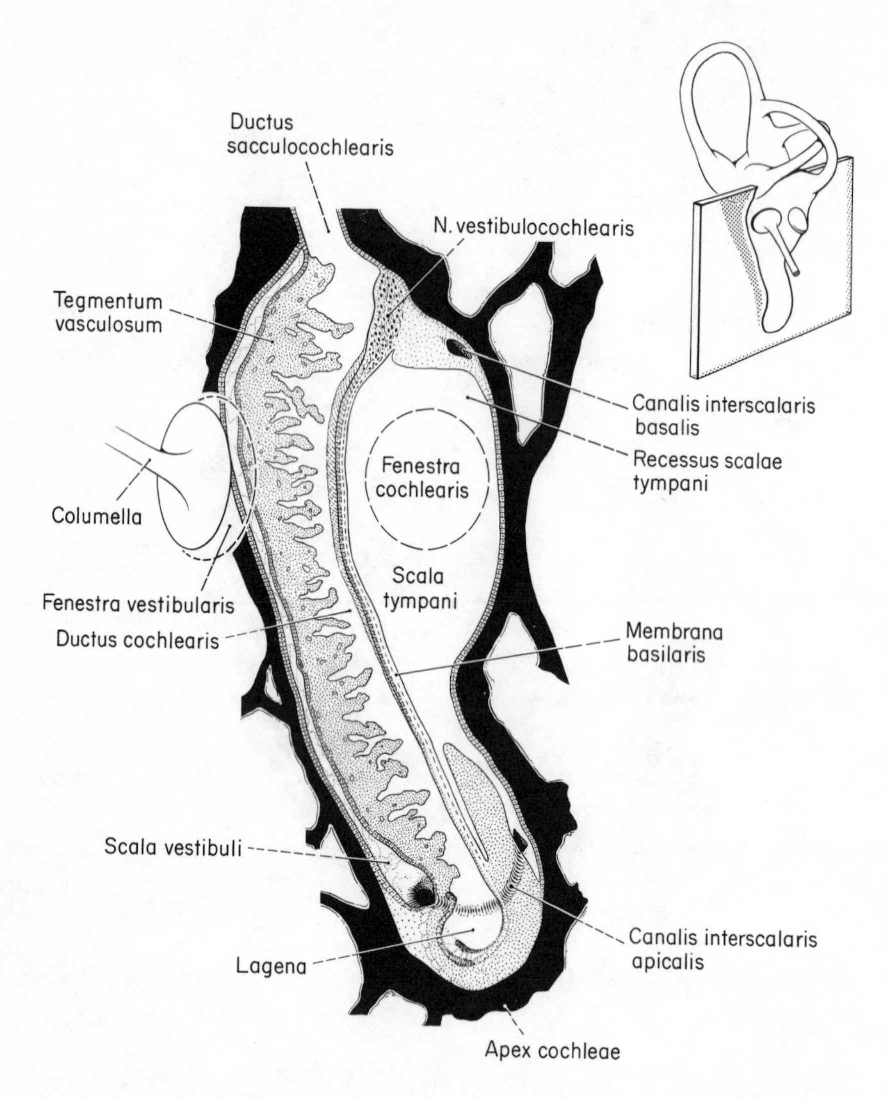

Fig. 4 Schematic representation of the avian cochlea. The inset at the upper right corner shows the plane of the section. Notice that the Scala vestibuli in birds is greatly reduced by contrast with the Scala tympani (see Annot. 48).

TAXONOMIC LIST

(Scientific and Common Names)

RICHARD L. ZUSI

Division of Birds, Smithsonian Institution,
United States National Museum,
Washington, D.C. 20560, USA

This list contains common names, generic names, subfamilies and tribes found in the text and refers each to its family (sometimes subfamily) or order. Families in brackets are known only from fossils. Terms with standardized endings that are readily recognized to family or order are omitted (e.g. gruid = Gruidae; ciconiiform = Ciconiiformes). Common group names that apply to more than one family (e.g. blackbird, sparrow) are omitted and can only be identified from the original reference. Those names that refer to chordate taxa other than Aves are referred to their class or subphylum.

Acryllium	Numididae
Aix	Anatidae
albatross	Diomedeidae
Alca	Alcidae
Alcedo	Alcedinidae
Ammodramus	Emberizidae, Emberizinae
Amphioxus	Cephalochordata
Anas	Anatidae
Anatinae	Anatidae
Anatini	Anatidae
Andean Condor	Cathartidae
Anhinga	Anhingidae
anhinga	Anhingidae
Anser	Anatidae
Anseranas	Anatidae
Anserinae	Anatidae
Anthracothorax	Trochilidae

Aprosmictus	Psittacidae
Apteryx	Apterygidae
Apus	Apodidae
Ara	Psittacidae
Aramus	Aramidae
Archaeopteryx	[Archeopterygidae]
Ardea	Ardeidae
Asio	Strigidae
auk	Alcidae
avocet	Recurvirostridae
Aythya	Anatidae
Aythyini	Anatidae
Balaeniceps	Balaenicipitidae
barbet	Capitonidae
Blackbird	Turdidae
Black Vulture	Cathartidae
Blue Coua	Cuculidae
booby	Sulidae
Botaurus	Ardeidae
bowerbird	Ptilonorhynchidae
Branta	Anatidae
broadbill	Eurylaimidae
Brotogeris	Psittacidae
Bubo	Strigidae
Bucephala	Anatidae
Buceros	Bucerotidae
Bucorvus	Bucerotidae
Budgerigar	Psittacidae
bustard	Otidae
Buteo	Accipitridae
Cacatua	Psittaciformes, Cacatuidae
Cairina	Anatidae
Calidris	Scolopacidae
Caprimulgus	Caprimulgidae
cassowary	Casuariidae
Casuarius	Casuariidae
Cathartes	Cathartidae
Centrocercus	Phasianidae, Tetraoninae
Cepphus	Alcidae
Cercyle	Alcedinidae
Chamaeza	Formicariidae

Charadrii	Charadriiformes
Charadrius	Charadriidae
Chauna	Anhimidae
chicken	Phasianidae, Phasianinae
Choriotis	Otidae
Chrysolophus	Phasianidae, Phasianinae
Ciconia	Ciconiidae
Circus	Accipitridae
Coccyzus	Cuculidae
cockatoo	Psittaciformes, Cacatuidae
Colaptes	Picidae
Columba	Columbidae
condor	Cathartidae
Coracina	Campephagidae
Coragyps	Cathartidae
cormorant	Phalacrocoracidae
Corvus	Corvidae
Coturnix	Phasianidae, Phasianinae
crane	Gruidae
Crax	Cracidae
Crocodilia	Reptilia
Crotophaga	Cuculidae
crow	Corvidae
Crowned Crane	Gruidae
Crowned Pigeon	Columbidae
Crypturus	Tinamidae
Crypturellus	Tinamidae
cuckoo	Cuculidae
curassow	Cracidae
Cyanocitta	Corvidae
Cygnus	Anatidae
darter	Anhingidae
Delichon	Hirundinidae
Dendrocygna	Anatidae
dipper	Cinclidae
dove	Columbidae
Dromaius	Dromiceidae
duck	Anatidae
eagle	Accipitridae
Egretta	Ardeidae
emu	Dromiceidae

Eos	Psittaciformes, Loriidae
Eudocimus	Threskiornithidae
Falco	Falconidae
flamingo	Phoenicopteridae
flicker	Picidae
fowl	Phasianidae, Phasianinae
francolin	Phasianidae, Phasianinae
Fregata	Fregatidae
frigate-bird	Fregatidae
frogmouth	Podargidae
Fulica	Rallidae
Gallinago	Scolopacidae
gallinule	Rallidae
Gallus	Phasianidae, Phasianinae
gannet	Sulidae
Gavia	Gaviidae
Geococcyx	Cuculidae
glossy cuckoo	Cuculidae
goatsucker	Caprimulgidae
goose	Anatidae
Great Horned Owl	Strigidae
grebe	Podicipedidae
grouse	Phasianidae, Tetraoninae
Grus	Gruidae
guineafowl	Numididae
gull	Laridae
Haliaeetus	Accipitridae
hawk	Accipitridae
heron	Ardeidae
Hesperornis	[Hesperornithidae]
Hirundo	Hirundinidae
Hoatzin	Opisthocomidae
hornbill	Bucerotidae
Horned Screamer	Anhimidae
House Sparrow	Ploceidae
House Wren	Troglodytidae
hummingbird	Trochilidae
Hydrophasianus	Jacanidae
Ichthyornis	[Ichthyornithidae]
Inca Dove	Columbidae

jacana	Jacanidae
Jackdaw	Corvidae
Java Dove	Columbidae
jay	Corvidae
Jynx	Picidae
Kakapo	Psittacidae
kingfisher	Alcedinidae
kiwi	Apterygidae
Lanius	Laniidae
Leptoptilos	Ciconiidae
Large Owlet-frogmouth	Podargidae
Lari	Charadriiformes
Larus	Laridae
limpkin	Aramidae
loon	Gaviidae
Lophoceros	Bucerotidae
Lorius	Psittaciformes, Loriidae
lyrebird	Menuridae
magpie	Corvidae
Manucodia	Paradisaeidae
Marabou Stork	Ciconiidae
Melanitta	Anatidae
Melanopareia	Rhynocryptidae
Meleagris	Phasianidae, Meleagridinae
Melopsittacus	Psittacidae
merganser	Anatidae
Mergini	Anatidae
Mergus	Anatidae
Merops	Meropidae
Morus	Sulidae
Nestor	Psittacidae
Northern Jacana	Jacanidae
Numenius	Scolopacidae
Nyctibius	Nyctibiidae
Nycticorax	Ardeidae
Old World vulture	Accipitridae
Opisthocomus	Opisthocomidae
oscine	Passeriformes
osprey	Pandionidae
Ostrich	Struthionidae

Otis	Otidae
owl	Strigiformes
Oxyura	Anatidae
Oxyurini	Anatidae
Pagodroma	Procellariidae
Painted Stork	Ciconiidae
Papuan Hornbill	Bucerotidae
parrot	Psittaciformes
partridge	Phasianidae
Passer	Ploceidae
passerine	Passeriformes
Pavo	Phasianidae, Phasianinae
peacock-pheasant	Phasianidae, Phasianinae
pea fowl	Phasianidae, Phasianinae
Pedioecetes	Phasianidae, Tetraoninae
Pelecanus	Pelecanidae
pelican	Pelecanidae
penguin	Spheniscidae
Perdix	Phasianidae, Phasianinae
Perisoreus	Corvidae
petrel	Procellariidae
Phalacrocorax	Phalacrocoracidae
phalarope	Scolopacidae
Phasianini	Phasianidae, Phasianinae
Phasianus	Phasianidae, Phasianinae
pheasant	Phasianidae, Phasianinae
Phoenicopterus	Phoenicopteridae
Phonygammus	Paradisaeidae
Pica	Corvidae
Picoides	Picidae
Picus	Picidae
pigeon	Columbidae
Platalea	Threskiornithidae
plover	Charadriidae
Pluvialis	Charadriidae
Podiceps	Podicipedidae
Podilymbus	Podicipedidae
Progne	Hirundinidae
Psittacula	Psittacidae
psittacine	Psittaciformes
Pteroglossus	Ramphastidae
puffin	Alcidae
Pygoscelis	Spheniscidae

quail	Phasianidae
rail	Rallidae
Rallus	Rallidae
Rhamphastos	Ramphastidae
ratite	Struthioniformes, Rheiformes, Casuariiformes, Apterygiformes
Rhea	Rheidae
Roadrunner	Cuculidae
Rostratula	Rostratulidae
Rosy Finch	Fringillidae, Carduelinae
Rynchops	Rynchopidae
Sagittarius	Sagittariidae
sandgrouse	Pteroclididae
sandpiper	Scolopacidae
Scopus	Scopidae
shorebird	Charadriiformes, Charadrii
snipe	Scolopacidae
Spheniscus	Spheniscidae
spoonbill	Threskiornithidae
starling	Sturnidae
Steatornis	Steatornithidae
Sterna	Laridae
stilt	Recurvirostridae
stork	Ciconiidae
Strix	Strigidae
Struthio	Struthionidae
Sturnus	Sturnidae
Sula	Sulidae
swallow	Hirundinidae
swan	Anatidae
swift	Apodidae
Tadorna	Anatidae
Teleostei	Osteichthyes
Tephrodornis	Campephagidae
Thalasseus	Laridae
tinamou	Tinamidae
toucan	Ramphastidae
touraco	Musophagidae
Trochilus	Trochilidae
Trogon	Trogonidae

trogon	Trogonidae
trumpeter	Psophiidae
turkey	Phasianidae, Meleagridinae
Turnix	Turnicidae
Vanellus	Charadriidae
Vini	Psittaciformes, Loriidae
vulture	Cathartidae
woodcock	Scolopacidae
woodcreeper	Dendrocolaptidae
woodpecker	Picidae
wryneck	Picidae
Zonotrichia	Emberizidae, Emberizinae

REFERENCES

Abdalla, M. A. and King, A. S. (1975). The functional anatomy of the pulmonary circulation of the domestic fowl. *Resp. Physiol.* **23**: 267–290.

Abdalla, M. A. and King, A. S. (1976). The functional anatomy of the bronchial circulation of the domestic fowl. *J. Anat.* **121**: 537–550.

Abdel-Magied, E. M. and King, A. S. (1978). The topographical anatomy and blood supply of the carotid body region of the domestic fowl. *J. Anat.* (*Lond.*) (in press).

Abraham, A. and Stammer, A. (1966). Über die Structur und die Innervierung der Augenmuskeln der Vögel unter Berücksichtigung des Ganglion ciliare. *Acta Biol.* (Hungary) **12**: 87–118.

Adamo, N. J. (1967). Connections of efferent fibers from hyperstriatal areas in chicken, raven, and African love-bird. *J. Comp. Neurol.* **131**: 337–356.

Adams, W. E. (1937). A contribution to the anatomy of the avian heart as seen in the Kiwi (*Apteryx australis*) and the yellow-crested penguin (*Megadyptes antipodum*). *Proc. zool. Soc. Lond.* Ser. B, **107**: 417–441.

Aitken, R. N. C. (1966). Postovulatory development of ovarian follicles in the domestic fowl. *Res. vet. Sci.* **7**: 138–142.

Aitken, R. N. C. (1971). The oviduct. *In* "The Physiology and Biochemistry of the Domestic Fowl" (D. J. Bell and B. M. Freeman, eds), Vol. 3. Academic Press, London.

Aitken, R. N. C. and Johnston, H. S. (1963). Observations on the fine structure of the infundibulum of the avian oviduct. *J. Anat.* (*Lond.*) **97**: 87–99.

Akester, A. R. (1960). The comparative anatomy of the respiratory pathways in the domestic fowl (*Gallus domesticus*), pigeon (*Columba livia*), and domestic duck (*Anas platyrhyncha*). *J. Anat.* (*Lond.*) **94**: 487–505.

Akester, A. R. (1967). Renal portal shunts in the kidney of the domestic fowl. *J. Anat.* (*Lond.*) **101**: 569–594.

Akester, A. R. (1971). The heart. *In* "Physiology and Biochemistry of the Domestic Fowl" (D. J. Bell and B. M. Freeman, eds), Vol. 2. Academic Press, London.

Akester, A. R., Pomeroy, D. E. and Purton, M. D. (1973). Subcutaneous air pouches in the Marabou stork. (*Leptoptilos crumeniferus*). *J. Zool.* (*Lond.*) **170**: 493–499.

Akker, L. M. Van Den (1970). "An Anatomical Outline of the Spinal Cord of the Pigeon". Van Gorcum, Netherlands.

Alix, E. (1874). "Essai on l'appareil locomoteur des Oiseaux". Masson, Paris.

Alverdes, I. (1924). Der Nebenhoden des Haussperlings. *Z. mikrosk.-anat. Forsch.* **1**: 202–227.

American Poultry Association. (1966). "The American Standard of Perfection" (1940 Edn). American Poultry Association, Davenport, Iowa.

American Poultry Association (1974). "The American Standard of Perfection". Jacob North Printing Co., Lincoln, Nebraska.

Amerlinck, A. (1923). Contribution à l'étude de la Membrane de Reissner et de l'epithelium de revêtement du canal cochleaire des oiseaux. *Arch. Biol.* **33**: 301–328.

Ames, P. L. (1971). The morphology of the syrinx in passerine birds. *Bull. Peabody Mus. Nat. Hist.* (Yale Univ.) **37**: 1–194.

Ames, P. L. (1975). The application of syringeal morphology to the classification of the Old World Insect Eaters (Muscicapidae). *Bonn zool. Beitr.* **26**: 107–134.

Anson, B. J. and Donaldson, J. A. (1973). "Surgical anatomy of the temporal bone and ear", 2nd Edn. Saunders, Philadelphia.
Antony, M. (1920). Über die Speicheldrüsen der Vögel. *Zool. Jb.* **41**: 547–660.
Ariëns-Kappers, C. U. (1933). The forebrain arteries in plagiostomes, reptiles, birds, Wirbeltiere und des Menschen." Bohn, Haarlem.
Ariëns-Kappers, C. U. (1933). The forebrain arteries in plagiostomes, reptiles, birds, and monotremes. *Proc. Kon. Akad. Wetensch.* Amsterdam **36**: 52–62.
Ariëns Kappers, C. U., Huber, G. C. and Crosby, E. C. (1936). "The Comparative Anatomy of the Nervous System of Vertebrates Including Man". Macmillan, New York.
Ash, R. W., Pearce, J. W. and Silver, A. (1969). An investigation of the nerve supply to the salt gland of the duck. *Quart. J. Exp. Physiol., Cog. Med. Sci.* **54**: 281–295.
Assenmacher, I. (1953). Étude anatomique du système artériel cervicocéphalique chez l'oiseau. *Arch. Anat. Histol. Embryol.* **35**: 181–202.
Assenmacher, I. (1973). The peripheral endocrine glands. *In* "Avian Biology" (D. S. Farner and J. R. King, eds), Vol. 3. Academic Press, London.
Bailey, R. E. (1953). Accessory reproductive organs of male fringillid birds: seasonal variations and response to various sex hormones. *Anat. Rec.* **115**: 1–19.
Baker-Cohen, D. F. (1968). Comparative enzyme histochemical observations on sub-mammalian brains. *Ergebn. Anat. Entwgesch.* **40**: 1–70.
Ballmann, P. (1969a). Die Vögel aus der altburdigalen Spaltenfüllung von Wintershof (West) bei Eichstätt in Bayern. *Zitteliana* (München) **1**: 5–60.
Ballmann, P. (1969b). Les oiseaux miocènes de la Grive-Saint-Alban (Isère). *Géobios*, **2**: 157–204.
Bang, B. G. (1971). Functional anatomy of the olfactory system in 23 orders of birds. *Acta Anat.* **79** (suppl. 58): 1–76.
Barkow, H. (1829). Anatomisch-physiologische Untersuchungen, vorzüglich über das Schlagadersystem der Vögel. *Arch. Anat. Physiol.* 1829: 306–496.
Barkow, H. C. L. (1856). "Syndesmologie der Vögel", pp. 1–41. Königlichen University, Breslau.
Barnett, C. H. (1954a). A comparison of the human knee and avian ankle. *J. Anat.* **88**: 59–70.
Barnett, C. H. (1954b). The structure and function of fibrocartilages within vertebrate joints. *J. Anat.* **88**: 363–368.
Barnikol, A. (1953). Zur Morphologie des Nervus Trigeminus der Vögel unter besonderer Berücksichtigung der Accipitres, Cathartidae, Striges, und Anseriformes. *Z. wiss. Zool.* **157**: 285–332.
Bartels, M. (1925). Über die Gegend des Dieters und Bechterewskernes der Vögeln. *Z. ges. Anat.* **77**: 726–784.
Bas, C. (1954–55). On the relation between the masticatory muscles and the surface of the skull in *Ardea cinerea* (L.). Parts I, II, III. *Proc. Koninkl. Nederl. Akad. v. Wetensch.* (Amsterdam), Series C. Part 1—**57**: 678–685 (1954); Parts II and III—**58**: 101–108, 109–113 (1955).
Bates, G. L. (1918). The reversed under wing-coverts of birds and their modifications, as exemplified in the birds of West Africa. *Ibis* (Ser. 10) **6**: 529–583.
Batojeva, S. Ts. and Batojev, Ts. Zh. (1972). On the anatomy of the pancreas of domestic birds. *Arkh. Anat. Histol. Embriol.* **63**: 105–108.
Baum, H. (1930). Das Lymphgefässsystem des Huhnes. *Z. Anat. EntwGesch.* **93**: 1–34.
Baumel, J. J. (1958). Variation in the brachial plexus of *Progne subis. Acta Anat.* **34**: 1–34.
Baumel, J. J. (1964). Vertebral-dorsal carotid artery interrelationships in the pigeon and other birds. *Anat. Anz.* **114**: 113–130.

Baumel, J. J. (1967). The characteristic asymmetrical distribution of the posterior cerebral artery of birds. *Acta Anat.* **67**: 523–549.

Baumel, J. J. and Gerchman, L. (1968). The avian intercarotid anastomosis and its homologue in other vertebrates. *Amer. J. Anat.* **122**: 1–18.

Baumel, J. J. (1971). Morphology of the tail apparatus in the pigeon (*Columba livia*). XIX World Veterinary Congress (Mexico City) **3**: 849 (Abstract).

Baumel, J. J. (1975). Heart and blood vessels. Aves. *In* "Sisson and Grossman's The Anatomy of Domestic Animals" (R. Getty, ed.), Vol. 2, 5th Edn. Saunders, Philadelphia.

Baumel, J. J. (1975). Aves Nervous System. *In* "Sisson and Grossman's The Anatomy of the Domestic Animals" (R. Getty, ed.), Vol. 2, 5th Edn. Saunders, Philadelphia.

Beddard, F. E. (1884). A contribution to the anatomy of *Scopus umbretta. Proc. zool. Soc. Lond.* 1884: 543–553.

Beddard, F. E. (1885). Notes on the visceral anatomy of birds. I. On the so-called omentum. *Proc. zool. Soc. Lond.* 1885: 836–844.

Beddard, F. E. (1885). On the structural characters and classification of the cuckoos. *Proc. zool. Soc. Lond.* 1885: 168–187.

Beddard, F. E. (1888). Notes on the visceral anatomy of birds. II. On the respiratory organs in certain diving birds. *Proc. zool. Soc. London* 1888: 252–258.

Beddard, F. E. (1896). On the oblique septa ("diaphragm" of Owen) in the passerines and in some other birds. *Proc. zool. Soc. London.* 1896: 225–231.

Beddard, F. E. (1898). "The structure and classification of birds". Longmans, Green, London.

Beddard, F. E. (1911). On the alimentary tract of certain birds and on the mesenteric relations of the intestinal loops. *Proc. zool. Soc. London.* **1**: 47–93.

Beer, C. G. (1964). Incubation. *In* "A New Dictionary of Birds", (A. L. Thomson, ed.). McGraw-Hill, New York.

Bellairs, A., d'A. and Jenkin, C. R. (1960). The skeleton of birds. *In*: "Biology and Comparative Physiology of Birds." (A. J. Marshall, ed.), Vol. 1. Academic Press, New York and London.

Bellairs, R. (1965). The relationship between oocyte and follicle in the hen's ovary as shown by electron microscopy. *J. Embryol. exp. Morph.* **13**: 215–233.

Bellairs, R., Harkness, M. and Harkness, R. D. (1963). The vitelline membrane of the hen's egg: a chemical and electron microscopical study. *J. Ultrastruct. Res.* **8**: 339–359.

Bellonci, J. (1888) Über die zentrale Endigung des Nervus opticus bei der Vertebraten. *Z. Wiss. Zool.* **47**: 1–45.

Bennett, T. (1974). Peripheral and autonomic nervous systems. *In* "Avian Biology" (D. S. Farner and J. R. King, eds), Vol IV. Academic Press, London.

Bennett, T. and Malmfors, T. (1970). The adrenergic nervous system of the domestic fowl. (*Gallus domesticus*) (L.)). *Z. Zellforsch.* **106**: 22–50.

Benoit, J. (1950). Organes uro-genitaux. *In* "Traité de Zoologie" (P.-P. Grassé, ed.), Vol. 15. Masson, Paris.

Benoit, J. (1950) Les glandes endocrines. *In* "Traité de Zoologie" (P.-P. Grassé, ed.), Vol. 15. Masson, Paris.

Benoit, J. and Assenmacher, I. (1953). Rapport entre la stimulation sexuelle prèhypophysaire et al neurosécretion chez l'oiseau. *Arch. Anat. micr. Morph. exp.* (Paris) **42**: 334–386.

Berger, A. J. (1952). The comparative functional morphology of the pelvic appendage in three genera of Cuculidae. *Amer. Midland. Nat.* **47**: 513–605.

Berger, A. J. (1953). The pterylosis of *Coua caerulea. Wilson Bull.* **65**: 12–17.

Berger, A. J. (1955). On the anatomy and relationship of glossy cuckoos of the genera *Chrysococcyx*, *Lamphomorpha*, and *Chalcites*. *Proc. U.S. Nat. Mus.* **103**: 585–597.

Berger, A. J. (1956). The expansor secundariorum muscle, with special reference to passerine birds. *J. Morph.* **99**: 137–168.

Berger, A. J. (1957). On the anatomy and relationships of *Fregilupus varius*, an extinct starling from the Mascarene Islands. *Bull. Am. Mus. Nat. Hist.* **113**: 225–272.

Berger, A. J. (1960). The musculature. *In* "Biology and Comparative Physiology of Birds" (A. J. Marshall, ed.), Vol. 1. Academic Press, New York and London.

Berger, A. J. (1960). Some anatomical characters of the Cuculidae and the Musophagidae. *Wilson Bull.* **72**: 60–105.

Berger, A. J. (1961). "Bird Study". Dover, New York.

Berger, A. J. (1966). The musculature. *In* "Avian Myology" (J. C. George and A. J. Berger, eds). Academic Press, New York and London.

Berger, A. J. (1968). Appendicular myology of Kirtland's Warbler. *Auk* **85**: 594–616.

Berger, A. J. (1969). Appendicular myology of passerine birds. *Wilson Bull.* **81**: 220–223.

Bhaduri, J. L. and Biswas, B. (1945). The main cervical and thoracic arteries of birds. Series I. Coraciiformes. Part 1. *Nat. Inst. Sci. India*, **11**: 236–245.

Bhaduri, J. L., Biswas, B. and Das, S. K. (1957). The arterial system of the domestic pigeon (*Columba livia* Gmelin). *Anat. Anz.* **104**: 1–14.

Bhaduri, J. L., De, A. and Biswas, B. (1965). The main throacic and cervical arteries of the flamingo, *Phoenicopterus roseus* Pallas. *Proc. Zool. Soc.* (Calcutta) **18**: 167–172.

Bickford, A. A. (1965). A fully formed and functional right oviduct in a single comb White Leghorn pullet. *Avian Dis.* **9**: 464–470.

Biggs, P. M. (1957). The association of lymphoid tissue with the lymph vessels in the domestic chicken. *Acta Anat.* **29**: 36–47.

Bignon, F. (1889). Contribution à l'étude de la pneumaticité chez les oiseaux. Les cellules aeriennes cervico-cephaliques des oiseaux et leurs rapports avec les os de la tête. *Mem. Soc. Zool. France* **2**: 260–320.

Bittner, H. (1924). Die Sektion des Hausgeflügels und der Versuchssingvögel. *Berl. tierträrzt. Wschr.* **40**: 99–101.

Bittner, H. (1925). Beitrag zur topographischen Anatomie der Eingeweide des Huhnes. *Z. Morph. Ökol. Tiere* **2**: 785–793.

Bland-Sutton, J. (1897). "Ligaments, Their Nature and Morphology". Blakiston, Philadelphia.

Bleicher, M. and Legait, E. (1932). De l'homologie de la cavité accessorie souscutaneé des oiseaux avec le sinus de mammillares. *C.R. Assoc. Anat.* **27**: 43–51.

Blom, L. (1973). Ridge pattern and surface ultrastructure of the ocidućal mucosa of the hen (*Gallus domesticus*). *Kgl. danske Vidensk Selsk. Skr., Biol.* **20**: 3–15.

Boas, J. E. V. (1929) Biologisch-anatomische Studien über den Hals der Vögel, *D. Kgl. Danske Vidensk. Selsk. Skrifter, naturvidensk. og Mathem. Afd.* Series 9, Vol. **1**, (3): 105–222.

Boas, J. E. V. (1933). Kreuzbein, Becken, und Plexus lumbosacralis der Vögel. *D. Kgl. Danske Vidensk. Selsk. Skrifter, naturvidensk. og Mathem. Afd.* Series 9, Vol. **5**: 1–59, 15 pls.

Bobr, L. W., Lorenz, F. W. and Ogasawara, F. X. (1964). Residence sites of spermatozoa in fowl oviducts. *J. Reprod. Fert.* **8**: 39–47.

Bock, W. J. (1960). Secondary articulation of the avian mandible. *Auk* **77**: 19–55.

Bock, W. J. (1962). The pneumatic fossa of the humerus in the Passeres. *Auk* **79**: 425–443.

Bock, W. J. (1963). The cranial evidence for ratite affinities. *Proc. XIII Internatl. Cong*: 39–54.

Bock, W. J. (1964). Kinetics of the avian skull. *J. Morph.* **114**: 1–52.
Bock, W. J. (1968). Mechanics of one- and two-joint muscles. *Amer. Mus. Novit.* **2319**: 1–45.
Bock, W. J. (1972). Morphology of the tongue apparatus of *Ciridops anna* (Drepanididae). *Ibis* **114**: 61–78.
Bock, W. J. (1977). Morphology of the passerine larynx. *Wilson Bull.* (in press).
Bock, W. J. Balda, R. P. and Vander Wall, S. B. (1973). Morphology of the sublingual pouch and tongue musculature in Clark's Nutcracker. *Auk* **90**: 491–519.
Bock, W. J. and Hikida, R. S. (1968). An analysis of twitch and tonus fibers in the hatching muscle. *Condor* **70**: 211–222.
Bock, W. J. and Hikida, R. S. (1969). Turgidity and function of the hatching muscle. *Amer. Midl. Nat.* **81**: 99–106.
Bock, W. J. and McEvey, A. (1969). The radius and relationship of owls. *Wilson Bull.* **81**: 55–68.
Bock, W. J. and Morioka, H. (1971). Morphology and evolution of the ectethmoid-mandibular articulation in the Melpiphagidae (Aves). *J. Morph.* **135**: 13–50.
Bock, W. J. and Morony, J. (1972). Snap-closing jaw ligaments in flycatchers. *Amer. Zool.* **12**: 729–730.
Bock, W. J. and Shear, C. R. (1972). A staining method for gross dissection of vertebrate muscle. *Anat. Anz.* **130**: 222–227.
Bodrossy, L. (1938). "Das Venensystem der Hausvogel". Inaug. Diss. Budapest. [in Hungarian, German summary].
Böhm, A. A. and von Davidoff, M. (1910). "A Textbook of Histology including Microscopic Technic" (G. D. Huber, ed.), 2nd Edn. Saunders, Philadelphia.
Böker, H. (1929). Flugvermögen und Kropf bei *Opisthocomus cristatus* und *Stringops habroptilus*. *Morph. Jb.* **63**: 152–207.
Bolton, T. B. (1971). The structure of the nervous system. *In* "Physiology and Biochemistry of the Domestic Fowl" (D. J. Bell and B. M. Freeman, eds). Academic Press, London.
Boord, R. L. (1968). Ascending projections of the primary cochlear nuclei and nucleus laminaris in the pigeon. *J. Comp. Neurol.* **133**: 523–542.
Boord, R. L. (1969). The anatomy of the avian auditory system. *Ann. N.Y. Acad. Sci.* **167**: 186–198.
Boord, R. L. and Rasmussen, G. L. (1963). Projection of the cochlear and lagenar nerves on the cochlear nuclei of the pigeon. *J. Comp. Neurol.* **120**: 463–475.
Boulton, R. (1927). Ptilosis of the House Wren (*Troglodytes aedon aedon*). *Auk* **44**: 387–414.
Boyd, J. D. and Hamilton, W. J. (1952). Cleavage, early development and implantation of the Egg. *In* "Marshall's Physiology of Reproduction" (A. S. Parkes, ed.), Vol. 2, 3rd Edn. Longmans Green, London.
Brach, V. (1975). The effect of intraocular ablation of the pecten oculi of the chicken. *Invest. Ophth.* **14**: 166–168.
Bradley, O. C. (1960). "The Structure of the Fowl" (T. Grahame, ed.), 4th Edn. Oliver and Boyd, Edinburgh.
Brandis, F. (1894). Untersuchungen über das Gehirn der Vogel. Ursprung der Nerven der Medulla oblongata. *Wilhelm Roux Arch. Entwickl. Organismen*, **43**: 96–116.
Braun, E. J. and Dantzler, W. H. (1972). Function of mammalian-type and reptilian-type nephrons in kidney of desert quail. *Amer. J. Physiol.* **22**: 617–629.
Brinkman, R., and Martin, A. H. (1973). A cytoarchitectonic study of the spinal cord of the domestic fowl *Gallus gallus domesticus*. 1. Brachial region. *Brain Res.* **56**: 43–62.
Brodkorb, P. (1971). Origin and evolution of birds. *In* "Avian Biology" (D. S. Farner and J. R. King, eds), Vol. I. Academic Press, New York and London.

Broman, I. (1942). Über die Embryonalentwicklung der Enten-Syrinx. *Anat. Anz.* **93**: 241–251.

Brooks, W. S. and Garrett, S. E. (1970). The mechanism of pipping in birds. *Auk* **87**: 458–466.

Buchholz, V. (1959–60). Beitrag zu makroskopischen Anatomie des Armgeflechtes under der Beckennerven beim Haushuhn (*Gallus domesticus*). *Wiss. Z. Humboldt-Univ. Berlin, Math.-Nat. R.* **9**: 515–594.

Buckley, G. A. and Wheater, L. E. (1968). The isolated expansor secundariorum: a smooth muscle preparation from the wing of a domestic fowl. *J. Pharm. Pharmacol.* 20 (Suppl.): 114s.

Budras, K. D. (1972). Das Epoophoron der Henne und die Transformation seiner Epithelzellen in Interrenal- und Interstitialzellen. *Ergebn. Anat. Entwickl. Gesch.* **46**: 1–70.

Budras, K. D. and Sauer, T. (1975). Morphology of the epididymis of the cock (*Gallus domesticus*) and its effect upon the steroid sex hormone synthesis. 1. Ontogenesis, morphology and distribution of the epididymis. *Anat. Embryol.* **148**: 175–196.

Bühler, P. (1970). Schädelmorphologie und Kiefermechanik der Caprimulgidae (Aves). *Z. Morph. Tiere* **66**: 337–399.

Bulliard, H. (1926). La brosse du dindon (*Meleagris gallopavo*). Étude morphologique. *Assoc. Anat. Bull.* (1926): 132–145.

Buri, R. O. (1900). Zur Anatomie des Flügels von *Micropus melba* und einigen anderen Coracornithes, zugleich Beitrag zur Kenntnis der systematischen Stellung der Cypselidae. *Jena Z. Med.* **33**: 361–610.

Burt, W. H. (1929). Pterylography of certain North American woodpeckers. *Univ. Calif. Pubs. Zool.* **30**: 427–442.

Burt, W. H. (1930). Adaptive modifications in the woodpeckers. *Univ. Calif. Publ. Zool.* **32**: 455–524.

Burton, P. J. K. (1971). Some observations on the splenius capitis muscle of birds. *Ibis* **113**: 19–28.

Burton, P. J. K. (1974a). "Feeding and the feeding apparatus in waders". *Brit. Mus. Nat. Hist. Publ.* 719. London.

Burton, P. J. K. (1974b). Anatomy of head and neck in the Huia (*Heteralocha acutirotris*) with comparative notes on other Callaeidae. *Bull. Brit. Mus. Nat. Hist.* **27**: 1–48.

Burton, P. J. K. (1974c). Jaw and tongue features in Psittaciformes and other orders with special reference to the anatomy of the Toothbilled Pigeon (*Didunculus strigirostris*). *J. Zool.* (Lond.) **174**: 255–276.

Butler, G. W. (1889). On the subdivision of the body-cavity in lizards, crocodiles and birds. *Proc. zool. Soc. Lond.* 1889: 452–474.

Cadow, G. (1933). Magen und Darm der Fruchttauben. *J. Ornith.* **81**: 236–252.

Cajal, S. Ramon y. (1908). Les ganglions terminaux du nerf acoustique des oiseaux. *Trab. Lab. Invest. Biol.* (Univ. Madrid) **6**: 195–225.

Calhoun, M. L. (1932–33). The microscopic anatomy of the digestive tract of *Gallus domesticus. Iowa J. Science* **7**: 261–304.

Calhoun, M. H. (1954). "Microscopic Anatomy of the Digestive System". Iowa State University Press, Ames.

Campana, A. (1875). "Recherches d'anatomie, de physiologie et d'organogénie pour la détermination des lois de la genèse et de l'évolution des espèces animales. I. Mémoire: Physiologie de la respiration chez les oiseaux, anatomie de l'appareil pneumatique-pulmonaire, des faux diaphragmes, des séreuses et de l'intestin chez le poulet". Masson, Paris.

Cane, A. K. and Spearman, R. I. C. (1967). A histochemical study of keratinization in domestic fowl (*Gallus gallus*). *J. Zool.* (*Lond.*) **153**: 337–352.
Cazin, M. (1887). Recherches anatomiques histologiques et embryologiques sur l'appareil gastrique des oiseaux. *Ann. Sci. nat. Zool.* **4**: 177–323.
Chandler, A. C. (1916). A study of the structure of feathers, with reference to their taxonomic significance. *Univ. Calif. Pub. Zool.* **13**: 243–446.
Chiasson, R. B. and Ferris, W. R. (1968). The iris and associated structures of the Inca Dove (*Scardafella inca*). *Amer. Zool.* **8**: 818.
Chiodi, V. and Bortolami, R. (1967). "The Conducting System of the Vertebrate Heart". Edizioni Claderini, Bologna.
Cholodkowsky, N. (1892). Zur Kenntnis der Speicheldrüsen der Vögel. *Zool. Anz.* **15**: 250–254.
Clara, M. (1929). Bau und Entwicklung des sogenannten Fettgewebes beim Vogel. *Z. mikrosk.-anat. Forsch.* **19**: 32–113.
Clara, M. (1934). Über den Bau des Magendarmkanals bei der Amseln (Turdidae). *Z. Anat. EntwGesch.* **102**: 718–771.
Clark, L. F., Rahn, H. and Martin, M. D. (1942). Seasonal and sexual dimorphic variations in the so-called "air sac" region of the Sage Grouse. *Bull. Wyoming Game Fish Dep.* **2**: 13–27.
Clench, M. H. (1970). Variability in body pterylosis, with special reference to the genus *Passer*. *Auk* **87**: 650–691.
Coe, M. J. (1960). Inflation of the neck pouch of the Marabou Stork. *Nature* **188**: 598.
Cohen, D. H., and Karten, H. J. (1974). The structural organization of avian brain: An overview, *In* "Birds Brain and Behavior" (I. J. Goodman and M. W. Schein, eds). Academic Press, New York.
Cohen, H. and Davies, S. (1937). The development of the cerebro-spinal fluid spaces and choroid plexuses in the chick. *J. Anat.* **72**: 23–53.
Coil, W. H. and Wetherbee, D. K. (1959). Observations on the cloacal gland of the Eurasian Quail, *Coturnix coturnix*. *Ohio J. Sci.* **59**: 268–270.
Cole, F. J. (1944). "A History of Comparative Anatomy from Aristotle to the Eighteenth Century". Macmillan, London. [Reprinted in 1975 by Dover, New York.]
Compton, L. (1938). The pterylosis of the Falconiformes with special attention to the taxonomic position of the osprey. *Univ. Calif. Pub. Zool.* **42**: 173–212.
Cords, E. (1904). Beiträge zur Lehre vom Kopfnervensystem der Vögel. *Anat. Hefte* **26**: 49–100.
Correa, W. M., Silva, G. H., Kawabe, L. T. and Carneiro, A. C. (1969). The supracloacal chromolipoid body of the Black Vulture (*Coragyps atratus*): anatomical, histological and histochemical considerations. *Can. J. comp. Med.* **33**: 160–163.
Coues, E. (1890). "Handbook of Field and General Ornithology". Macmillan, London.
Coues, E. (1903). "Key to North American Birds", 5th Edn, Vols I and II. Dana Estes & Co., Boston.
Coues, E. (1927). "Key to North American Birds". Page Co., Boston.
Cover, M. S. (1953). The gross and microscopic anatomy of the respiratory system of the turkey: I. The nasal cavity and infraorbital sinus *Am. J. Vet. Res.* **14**: 113–117.
Cowan, W. M., Adamson, L., and Powell, T. P. S. (1961). An experimental study of the avian visual system. *J. Anat.* **95**: 545–563.
Cowan, W. M. and Powell, T. P. S. (1963). Centrifugal fibers in the avian visual system. *Proc. Roy. Soc. B* **158**: 232–252.
Cowan, W. M. and Wenger, E. (1968). The development of the nucleus of origin of centrifugal fibers to the retina in the chick. *J. Comp. Neurol.* **133**: 207–240.

Cornselius, C. (1925). Morphologie, Histologie, und Embryologie des Muskelmagens der Vögel. *Morph. Jb.* **54**: 507–559.

Cracraft, J. (1971). The functional morphology of the hind limb of the domestic pigeon, *Columba livia. Bull. Amer. Mus. Nat. Hist.* **144**: 171–268.

Craigie, E. H. (1928). Obervations on the brain of the hummingbird (*Chrysolampis mosquitus* Linn. and *Chlorostibon caribaeus* Lawr.). *J. Comp. Neurol.* **45**: 377–481.

Craigie, E. H. (1930). The brain of the kiwi (*Apteryx australis*). *J. Comp. Neurol.* **49**: 223–357.

Cralley, J. C. (1965). "The vascular anatomy of the starling, *Sturnus vulgaris* Linnaeus". Ph.D. Diss., Anatomy, University Illinois. [University Microfilms, Ann Arbor, Michigan, 66–4161.]

Crompton, A. W. (1953). The development of the chondrocranium of *Spheniscus demersus* with special reference to the columella auris of birds. *Acta Zool.* **34**: 71–146.

Curtis, E. L. and Miller, R. C. (1938). The sclerotic ring in North American birds. *Auk* **55**: 225–243.

Curtis, M. R. (1910). The ligaments of the oviduct of the domestic fowl. *Bull. Maine agric. Exptl. Sta.* No. **176**: 1–20.

Dahl, E. (1970). Studies of the fine structure of ovarian interstitial tissue. *Z. Zellforsch. mikrosk. Anat.* **109**: 227–244.

Davenport, H. A. (1966). Introduction and topographic anatomy. *In* "Morris' Human Anatomy" (B. J. Anson, ed.), 12th Edn. Blakiston Division of McGraw-Hill, New York.

Davids, J. A. G. (1952). Étude sur les attaches au crâne des muscles de la tête et du cou chez *Anas platyrhyncha platyrhyncha L.* II, III. *Proc. Koninkl. Nederl. Akad. Wetensch.* (*Amsterdam*) *Series C*, **55**: 525–533; 534–540.

Davidson, M. F., Draper, M. H. and Leonard, E. M. (1968). Structure and function of the oviduct of the laying hen. *J. Physiol. Lond.* **196**: 9–10 P.

Davies, F. (1930). The conducting system of the bird's heart. *J. Anat.* **64**: 129–147.

DeBurlet, H. M. (1934). Vergleichende Anatomie des statoakustischen Organs. a) Die inner Ohrsphäre. Vol. 2, part 2, pp. 1293–1380. b) Die mittlere Ohrsphäre. Vol. 2, pp. 1381–1432. *In* "Handbuch der vergleichenden Anatomie der Wirbeltiere" (L. Bolk, E. Göppert, E. Kallius and W. Lubosch, eds). Urban & Schwarzenberg, Berlin and Vienna.

De Kock, J. M. (1955). The cranial morphology of *Sturnus vulgaris vulgaris* Linnaeus. *Ann. Univ. Stellenbosch* **31** (Sect. A, No. 3): 153–175.

De Kock, L. L. (1958). On the carotid body of certain birds. *Acta anat.* **35**: 161–178.

De Kock, L. L. (1959). The carotid body system of the higher vertebrates. *Acta anat.* **37**: 265–279.

Delacour, J. (1964). "The Waterfowl of the World". Country Life Ltd., London.

Delius, J. D. and Bennetto, K. (1972). Cutaneous sensory projections to the avian forebrain. *Brain Res.* **37**: 205–221.

deWet, P., Fedde, M. R. and Kitchell, R. L. (1967). Innervation of the respiratory muscles of *Gallus domesticus. J. Morph.* **123**: 17–34.

Dieterich, C. E. and Pfautsch, M. (1973). Fine structure observations of the pecten oculi capillaries of the chicken. Freeze-etching, scanning and transmission electron microscopic investigations. *Z. Zellforsch.* **146**: 473–489.

Dingler, E. C. (1965). Einbau des Ruckenmarks in Wirbelkanal bei Vögeln. *Anat. Anz.* **115** (Suppl.): 71–84.

Dohlman, G. F. (1964). Secretion and absorption of endolymph. *Ann. Otol. Rhin. Laryng.* **73**: 708–723.

Donath, T. and Crawford, G. N. C. (1969). "Anatomical Dictionary with Nomenclatures and Explanatory Notes", 1st English Edn. Pergamon Press, Oxford and London.

Dransfield, J. W. (1944). "The lymphatic system of the domestic fowl". M.V.Sc. Thesis, Liverpool University.

Dransfield, J. W. (1945). The lymphatic system of the domestic fowl. *Br. vet. J.* **101**: 171–179.

Draper, M. H., Davidskn, M. F. and Wyburn, G. M. (1972). The fine structure of the fibrous membrane forming region of the isthmus of the oviduct of *Gallus domesticus*. *Quart. J. exp. Physiol.* **57**: 297–309.

Draper, M. H., Johnston, H. S. and Wyburn, G. M. (1968). The fine structure of the oviduct of the laying hen. *J. Physiol.* Lond. **196**: 7–8 P.

Drimmelen, G. C. van. (1946). Sperm nests in the oviduct of the domestic hen. *J.S. Afr. vet. med. Ass.* **17**: 42–52.

Dubale, M. S. and Rawal, U. M. (1965). A morphological study of the cranial muscles associated with the feeding habit of *Psittacula krameri* (*Scopoli*). *Pavo* **3**: 1–13.

Dubale, M. S. (1969). The jaw muscles of some Indian birds. *Proc. Nat. Acad. Sci. India* **39** (B); I and II: 201–212.

Duijm, M. (1951). On the head posture in birds and its relation to some anatomical features. I, II. *Proc. Koninkl. Nederl. Akad. Wetensch.* (Amsterdam) Series C **54**: 202–211; 261–271.

Duijm, M. (1958). On the position of a ribbon-like central area in the eyes of some birds. *Arch. neerl. Zool.* **13**: 128–145.

Dullemeijer, P. (1951). The correlation between muscle system and skull structure in *Phalacrocorax carbo sinensis* (Shaw and Nodder). I, II, III *Proc. Koninkl. Nederl. Akad. Wetensch.* (Amsterdam), Series C, **54**: 247–259; 400–404; 533–536.

Duncker, H.-R. (1971). The lung air sac system of birds. *Ergebn. Anat. Entw. Gesch.* **45** (6): 1–171.

Duncker, H.-R. (1972). Structure of avian lungs. *Respir. Physiol.* **14**: 44–63.

Dwight, J., Jr. (1900). The sequence of plumages and moults of the passerine birds of New York. *Ann. N.Y. Acad. Sci.* **13**: 73–360.

Dziuk, H. E. and Duke, G. E. (1972). Cineradiographic studies of gastric motility in turkeys. *Amer. J. Physiol.* **222**: 159–166.

Edgeworth, F. H. (1935). "The Cranial Muscles of Vertebrates". Cambridge University Press, London.

Edinger, L. and Wallenberg, A. (1899). Untersuchungen über das Gehirn der Tauben. *Anat. Anz.* **15**: 245–271.

Edington, G. H. and Miller, A. E. (1941). The avian ulna: its quill knobs. *Proc. Roy. Soc. Edinburgh*, **B61**: 138–148.

Eiselen, G. (1939). Untersuchungen über den Bau und die Entstehung von Schmalzkielen bei Tauben. *Z. wiss. Zool.* **152**: 409–438.

Ellis, C. J. and Thome, P. (1975). Syringeal histology. IV. Parulidae: Ovenbird, *Seiurus aurocapillus*, and Mourning Warbler, *Oporonis philadelphia*. *Iowa State J. Res.* **50**: 153–172.

Ellis, L. C. (1976). The endocrine role of the pineal gland. *Amer. Zool.* **16**: 3–101.

Engels, W. L. (1938). Tongue musculature of passerine birds. *Auk* **55**: 642–650.

Erulkar, S. D. (1955). Tactile and auditory areas in the brain of the pigeon. *J. Comp. Neurol.* **103**: 421–458.

Evans, H. E. (1969). Anatomy of the Budgerigar. *In* "Diseases of Cage and Aviary Birds" (M. Petrak, ed.). Lea and Febiger, Philadelphia.

Ewald, J. R. (1892). "Physiologische untersuchungen über das Endorgan des Nervus octavus". J. R. Bergmann, Wiesbaden.

Fahrenholz, C. (1937). Drüsen der Mundhohle. *In* "Handbuch der vergleichenden Anatomie der Wirbeltiere" (L. Bolk, E. Göppert, E. Kallius and W. Lubosch, eds), Vol. 3. Urban & Schwarzenberg, Berlin and Vienna.

Fedde, M. R. (1970). Peripheral control of avian respiration. *Fed. Proc.* **29**: 1664–1673.

Fedde, M. R., Burger, R. E. and Kitchell, R. L. (1963). Localization of vagal afferents involved in the maintenance of normal avian respiration. *Poult. Sci.* **42**: 1224–1236.

Fedde, M. R., Burger, R. E. and Kitchell, R. L. (1964). Anatomic and electro-myographic studies of the costo-pulmonary muscles in the cock. *Poult. Sci.* **43**: 1177–1184.

Feder, F.-H. (1969). Beitrag zur makroskopischen und mikroskopischen Anatomie des Verdauungsapparats beim Wellensittich (*Melopsittacus undulatus*). *Anat. Anz.* **125**: 233–255.

Feder, F.-H. (1972). Zur mikroskopischen Anatomie des Verdauungsapparats beim Nandu (*Rhea americana*). *Anat. Anz.* **132**: 250–265.

Federici, F. (1927). Über die innervation des von Vitali entdecklen Sinnesorganes im Mittelohr der Vögel paratympanisches Organ. *Anat. Anz.* **62**: 241–254.

Fehér, Gy. and Fancsi, T. (1971). Vergleichende Morphologie der Bauchspeicheldrüsen von Hausvögeln. *Acta vet. hung.* **21**: 141–164.

Fehér, Gy. and Gyürü, F. (1971). Data on the postembryonal changes of the yolk sac in the domestic fowl. I. Postembryonal changes of the yolk-sac in chickens. *Magy. Allatorv. Lap.* **26**: 353–360.

Feinstein, B. (1962). Additional cases of bilobated kidneys in the hornbills. *Auk* **79**: 709–711.

Feldotto, A. (1929). Die Harnkanälchen des Huhnes. *Z. mikrosk.-anat. Forsch.* **17**: 353–370.

Fenna, L. and Boag, D. A. (1974). Adaptive significance of the caeca in Japanese Quail and Spruce Grouse (Galliformes). *Can. J. Zool.* **52**: 1577–1584.

Fischer, G. (1905). Vergleichend anatomische Untersuchungen über den Bronchialbaum der Vögel. *Zoologica* **19**: 1–45.

Fisher, H. I. (1940). The occurrence of vestigial claws on the wings of birds. *Amer. Midl. Nat.* **23**: 234–243.

Fisher, H. I. (1945). Flying ability and the anterior intermuscular line on the coracoid. *Auk* **62**: 125–129.

Fisher, H. I. (1946). Adaptations and comparative anatomy of the locomotor apparatus of New World vultures. *Amer. Midl. Nat.* **35**: 545–727.

Fisher, H. I. (1966). Hatching and the hatching muscle in some North American ducks. *Trans. Ill. State Acad. Sci.* **59**: 305–325.

Fisher, H. I. and Goodman, D. C. (1955). The myology of the Whooping Crane, *Grus americana. Ill. Biol. Monogr.* **24**: 1–127.

Foelix, R. F. (1970). Vergleichend-morphologische Untersuchungen an den Speicheldrüsen körnerfressender Singvögel. *Zool. Jb. Anat.* **87**: 523–587.

Forbes, W. A. (1877). On the bursa Fabricii in birds. *Proc. zool. Soc. Lond.* 1877: 304–318.

Forbes, W. A. (1881). On the conformation of the thoracic end of the trachea in the "ratite" birds. *Proc. zool. Soc. Lond.* 1881: pp. 778–788.

Forbes, W. A. (1882). On some points on the anatomy of the Indian Darter (*Plotus melanogaster*), and on the mechanism of the neck of darters (*Plotus*) in connection with their habits. *Proc. zool. Soc. Lond.* 1882: 208–212.

Forsyth, D. (1908). The comparative anatomy, gross and minute, of the thyroid and parathyroid glands in mammals and birds. II. Aves. *J. Anat. Physiol.* **42**: 302–319.

Francis, E. T. B. (1964). The excretory system. *In* "A New Dictionary of Birds" (A. L. Thomson, ed.). Nelson, London.

Frank, G. H. and Smit, A. L. (1976). The morphogenesis of the avian columella auris with special reference to *Struthio camelus*. *Zool. Africana* **11**: 159–182.

Freedman, S. L. (1968). The innervation of the suprarenal gland of the fowl (*Gallus domesticus*). *Acta Anat.* **69**: 18–25.

Freedman, S. L. and Sturkie, P. D. (1963). Blood vessels of the chicken's uterus (shell gland). *Amer. J. Anat.* **113**: 1–7.

Freedman, S. L. and Sturkie, P. D. (1963). Extrinsic nerves of the chicken's uterus (shell gland). *Anat. Rec.* **147**: 431–437.

Freund, L. (1926). Das äussere Ohr der Sauropsiden. *Zool. Anz.* **66**: 319–325.

Frewein, J. (1967). Die Gelenkräume, Schleimbeutel und Sehnenscheiden an den Zehen des Haushuhnes. *Zbl. Veterinärmed.* **14-A**: 129–136.

Friedlander, A. (1898). Untersuchungen über das Ruckenmark und das Kleinhirn der Vogel. *Neurol. Zbl.* **17**: 397–409.

Fritschi, E. (1926). Beiträge zur Strumafrage beim Huhn. Virchow's *Arch. path. Anat. Physiol.* **260**: 422–435.

Fuchs, A. (1955). On the correlation between the skull structure and the muscles in the male *Phasianus colchicus* L. VI. Some remarks on a number of ligaments and other connective tissue connections. *Proc. Koninkl. Nederl. Akad. v. Wetensch.* (Amsterdam) Series C, **58**: 114–120.

Fujihara, N., Nishiyama, H. and Nakashima, N. (1976). Studies on the accessory reproductive organs in the drake. 2. Macroscopic and microscopic observations on the cloaca of the drake with special reference to the ejeculatory groove region. *Poult. Sci.* **55**: 927–935.

Fujii, S. (1963). Histological and histochemical studies on the oviduct of the domestic fowl with special reference to the region of uterovaginal junction. *Arch. histol. jap.* **23**: 447–459.

Fujii, S. (1975). Scanning electron microscopical observation on the mucosal epithelium of hen's oviduct with special reference to the transport mechanism of spermatozoa through the oviduct. *J. Fac. Fish. Anim. Husb. Hiroshima Univ.* **14**: 1–13.

Fujuii, S., Tamura, T. and Kunisaki, H. (1965). Histochemical study of mucopolysaccharides in goblet cells of the chicken oviduct. *J. Fac. Fish. Anim. Husb. Hiroshima Univ.* **6**: 25–35.

Fujioka, T. (1959). On the origins and insertions of the muscles of the thoracic limb in the fowl. *Jap. J. Vet. Sci.* **21**: 85–95.

Fujioka, T. (1962). On the origins and insertions of the muscles of the pelvic limb in the fowl. *Jap. J. Vet. Sci.* **24**: 183–199 [in Japanese].

Fujioka, T. (1963). On the origins and insertions of the muscles of the head and neck in fowl. Part I. Muscles of the head. *Jap. J. Vet. Sci.* **25**: 207–226.

Fukuta, K., Nishida, T. and Yasuda, M. (1969a). Blood vascular system of the spleen in the fowl. *Jap. J. vet. Sci.* **31**: 179–185 [in Japanese].

Fukuta, K., Nishida, T. and Yasuda, M. (1969b). Structure and distribution of the fine blood vascular system in the spleen. *Jap. J. vet. Sci.* **31**: 303–311 [in Japanese].

Fürbringer, M. (1879). Zur Lehre von den Umbildungen der Nervenplexus. *Morph. Jb.* **5**: 324–394.

Fürbinger, M. (1886). Über Deutung und Nomenklatur der Muskulatur des Vogelflugels. *Morph. Jb.* **11**: 121–125.

Fürbinger, M. (1888). "Untersuchungen zur Morphologie und Systematik der Vögel", I. Specieller Theil. T. J. Van Holkema, Amsterdam.

Fürbringer, M. (1902). Zur vergleichenden Anatomie des Brustschulterapparates und der Schultermuskeln. V. Teil. Vögel. *Jena. Z. Naturw.* **36** (N.F. 29): 289–736.

Fürther, H. (1913). Beiträge zur Kentniss der Vogellymphknoten. *Jena. Z. Med. Naturw.* **50**: 359–410.

Gadow, H. (1879a). Versuch einer vergleichende Anatomie des Verdauungssystems der Vögel. I. Thiel. *Jena Z. Naturw.* **13**: 97–171.

Gadow, H. (1879b). Versuch einer vergleichende Anatomie des Verdauungssystems der Vögel. II. Theil. *Jena Z. Naturw.* **13**: 339–403.

Gadow, H. (1887). Remarks on the cloaca and on the copulatory organs of the Amniota. *Phil. Trans. R. Soc. Ser. B* **178**: 5–37.

Gadow, H. (1889). On the taxonomic value of the intestinal convolutions in birds. *Proc. zool. Soc. Lond.* 1889: 303–315.

Gadow, H. (1891). Crop and sternum of *Opisthocomus cristatus*: a contribution to the question of the correlation of organs and the inheritance of acquired characters. *Proc. R. Irish Acad. Series III* **11**: 147–154.

Gadow, H. (1896). Cloaca. *In* "A Dictionary of Birds" (A. Newton, ed.). Black, London.

Gadow, H. (1896). Syrinx. *In* "A Dictionary of Birds" (A. Newton, ed.). Black, London.

Gadow, H. and Selenka, E. (1891). Vögel: I. Anatomischer Theil. *In* "Bronn's Klassen und Ordnungen des Thier-Reichs", Bd. 6 (4). C. F. Winter, Leipzig.

Gage, S. H. (1917). Glycogen in the nervous system of vertebrates. *J. Comp. Neurol.* **27**: 451–456.

Gans, C. and Bock, W. J. (1965). The functional significance of muscle architecture—a theoretical analysis. *Ergeb. Anat. Entw. Gesch.* **38**: 115–142.

Gardner, L. L. (1926). The adaptive modifications and the taxonomic value of the tongue in birds. *Proc. U.S. natl. Mus.* **67**: Art. 19.

Garrod, A. H. (1872). On the mechanism of the gizzard in birds. *Proc. zool. Soc. Lond.* 1872: 525–529.

Garrod, A. H. (1873). On some points in the anatomy of *Steatornis*. *Proc. zool. Soc. Lond.* 1873: 526–535.

Garrod, A. H. (1874a). On the "showing-off" of the Australian Bustard (*Eupoditis australis*). *Proc. zool. Soc. Lond.* 1874: 471–474.

Garrod, A. H. (1874b). Further notes on the mechanism of "showing-off" in the bustards. *Proc. Soc. Lond.* 1874: 102–105.

Garrod, A. H. (1876). Notes on the anatomy of *Plotus anhinga*. *Proc. zool. Soc. Lond.* 1876: 335–345.

Garrod, A. H. (1878a). Note on the gizzard and other organs of *Carpophaga latrans*. *Proc. zool. Soc. Lond.* 1878: 102–105.

Garrod, A. H. (1878b). Notes on points in the anatomy of Levaillant's darter (*Plotus levaillanti*). *Proc. zool. Soc. Lond.* 1878: 679–681.

Garrod, A. H. (1879). On the conformation of the thoracic extremity of the trachea in the Class Aves. Part I. The Gallinae. *Proc. zool. Soc. Lond.* 1879: 354–380.

Gasch, F. R. (1888), Beiträge zur vergleichenden Anatomie des Herzens der Vögel und Reptilien. *Arch. Naturgesch.* **54**: 119–152.

Gaunt, A. S., Gaunt, S. L. L. and Hector, D. H. (1976). Mechanics of the syrinx in *Gallus gallus*. I. A comparison of pressure events in chickens to those in oscines. *Condor* **78**: 208–223.

Gegenbauer, C. (1871). Beiträge zur Kenntnis des Beckens der Vögel. *Jena Z. Naturw.* **6**: 157–220.

Gentle, M. J. (1971). The lingual taste buds of *Gallus domesticus*. *Brit. Poult. Sci.* **12**: 245–248.

George, J. C. and Berger, A. J. (1966). "Avian Myology". Academic Press, New York.

George, W. G. (1962). The classification of the Olive Warbler, *Peucedramus taeniatus*. *Amer. Mus. Novit.* **2103**: 1–41.

Gerhardt, U. (1933). Kloake und Begattungsorgane. *In* 'Handbuch der vergleichenden Anatomie der Wirbeltiere" (L. Bolk, E. Göppert, E. Kallius and W. Lubosch, eds.), Vol. 6. Urban & Schwarzenberg, Berlin and Vienna.

Gerisch, D. (1971). "Die Bronchi atriales in der Lunge des Haushuhnes (*Gallus gallus domesticus* L.)—Ein Beitrag zur Morphologie und Nomenklatur". Inaug. Diss., Hanover.

Gertner, M. (1969). Vascular system of the oviduct of the goose. *Agrartud. Egyet. mezögtud. Kar. Közl. Godollo*: 73–82 [in Hungarian].

Ghetie, V. and Atanasui, I. (1962) Die Myologie des Zungenbeinaufhängeapparates und der Zunge beim Hühner- und Wassergeflugel. *Rev. Biol.* (Rumania) **7**: 85–94.

Ghetie, V., Chitescu, St., Cotofan, V. and Hillebrand, A. (1976). "Atlas De Anatomie A Păsărilor Domestice". Editura Acad. Republicii Socialiste România, Bucharest.

Gilbert, A. B. (1965). Innervation of the ovarian follicle of the domestic hen *Quart. J. exp. Physiol.* **50**: 437–445.

Gilbert, A. B. (1969). Innervation of the ovary of the domestic hen. *Quart. J. exp. Physiol.* **54**: 404–411.

Gilbert, A. B. (1971a). The ovary. *In* "The Physiology and Biochemistry of the Domestic Fowl" (D. J. Bell and B. M. Freeman, eds.), Vol. 3. Academic Press, London.

Gilbert, A. B. (1971b). The egg: its physical and chemical aspects. *In* "The Physiology and Biochemistry of the Domestic fowl" (D. J. Bell and B. M. Freeman, eds.), Vol. 3. Academic Press, London.

Gilbert, A. B. (1971c). The endocrine ovary. *In* "The Physiology and Biochemistry of the Domestic Fowl" (D. J. Bell and B. M. Freeman, eds.), Vol. 3. Academic Press, London.

Gingerich, P. D. (1972). A new partial mandible of *Ichthyornis*. *Condor* **74**: 471–473.

Giroud, A. and Leblond, C. P. (1951). The keratinization of epidermis and its derivatives especially the hair, as shown by X-ray diffraction and histochemical studies. *Ann. N.Y. Acad. Sci.* **53**: 613–626.

Glenny, F. H. (1951). A systematic study of the main arteries in the region of the heart. Aves XII. Galliformes, Part 1. *Ohio J. Sci.* **51**: 47–54.

Glenny, F. H. (1955). Modifications of pattern in the aortic arch system of birds and their phylogenetic significance. *Proc. U.S. Natl. Mus.* **104**: 525–621.

Glenny, F. H. and Friedmann, H. (1954). Reduction of the clavicles in the Mesoenatidae, with some remarks concerning the relationship of the clavicle to flight-function in birds. *Ohio J. Sci.* **54**: 111–113.

Glick, B. and Sato, K. (1964). Accessory spleens in the chicken. *Poult. Sci.* **43**: 1610–1612.

Glimstedt, G. (1942). "Über Morphogenese, Histogenese und Bau der Gehörgangdrüsen bei einigen Vögeln". Gleerupska University, Bokhandeln, Lund.

Gobreil, R. E. (1970). Arterial system of the Herring Gull (*Larus argentatus*). *J. Zool. Lond.* **160**: 337–354.

Goedbloed, E. (1958). The condylus occipitalis in birds. I, II, III. *Proc. Koninkl. Nederl. Akad. Wetensch.* (Amsterdam) Series C, **61**, 36–47; 48–58; 59–65.

Gomot, L. (1958). Interaction ectoderme-mesoderme dans la formation des invaginations uropygiennes des oiseaux. *J. Embryol. exp. Morph.* **6**: 162–170.

Goodchild, W. M. (1956). "Biological aspects of the urinary system of *Gallus domesticus* with particular reference to the anatomy of the ureter". M.Sc. Thesis, University of Bristol, England.

Goodchild, W. M. (1969). The venous system of the adrenal glands of *Gallus domesticus*. *Brit. Poult. Sci.* **10**: 183–185.

Goodman, D. C. and Fisher, H. I. (1962). "Functional anatomy of the feeding apparatus in waterfowl. Aves: Anatidae". Southern Illinois University Press, Carbondale.

Goodrich, E. S. (1930). "Studies on the Structure and Development of Vertebrates". Macmillan, London. [Reprinted 1958, Dover, New York.]

Goodrich, E. S. (1958). "Studies on the Structure and Development of Vertebrates." Dover Publ., New York. [Reprint of original 1930 Edn.]

Göppert, E. (1903). Die Bedeutung der Zunge für den sekundären Gaumen und den Ductus naso-pharyngeus. *Morph. Jb.* **31**: 331–359.

Gorham, F. W. and Ivy, A. C. (1938). General function of the gall-bladder from the evolutionary standpoint. *Field Mus. Natur. Hist. Publ., Zool. Ser.* **22**: 159–213.

Grau, H. (1943). Anatomie der Hausvögel. *In* "Ellenberger and Baum's Handbuch der vergleichenden Anatomie der Haustiere" (O. E. Zietzschmann, E. Ackerknecht and H. Grau, eds), 18th Edn. Springer, Berlin.

Gray, A. A. (1926). An aqueduct in the bird's labrynth not previously recorded and its evolutionary significance. *Proc. Roy. Soc. Med.* **19**: 41–46.

Gray, J. C. (1937). The anatomy of the male genital ducts in the fowl. *J. Morph.* **60**: 393–405.

Green, J. D. (1951). The comparative anatomy of the hypophysis, with special reference to its blood supply and innervation. *Amer. J. Anat.* **88**: 225–311.

Greenewalt, C. H. (1969). How birds sing. *Sci. Amer.* **221**: 126–139.

Greenlee, T. K., Jr., Beckham, C. and Pike, D. (1975). A fine structural study of the development of the chick flexor digital tendon: a model for synovial sheathed tendon healing. *Amer. J. Anat.* **143**: 303–314.

Groebbels, F. (1932). "Der Vogel. Bau, Funktion, Lebenserscheinung, Einpassung", Vol. 1. Borntraeger, Berlin.

Gross, W. B. (1964a). Voice production in the chicken. *Poult. Sci.* **43**: 1005–1008.

Grzycki, S. (1973). Variability and structure of tactile corpuscles in the bird's tongues. *Morph. Jb.* **119**: 427–433.

Guzsal, E. (1966). Histological studies on the mature and postovulation ovarian follicle of fowl. *Acta. vet. Acad. sci. hung.* **16**: 37–44.

Guzsal, E. (1968). Histochemical study of goblet cells of the hen's oviduct. *Acad. vet. Acad. sci. hung.* **18**: 251–256.

Guzsal, E. (1974). Erection apparatus of the copulatory organ of ganders and drakes. *Acta vet. Acad. Sci. hung.* **24**: 361–373.

Haartsen, A. B. and Verhaart, W. J. C. (1967). Cortical projections to brain stem and spinal cord in the goat by way of the pyramidal tract and bundle of Bagley. *J. Comp. Neurol.* **18**: 189–201.

Haecker, V. (1900). "Der Gesang der Vögel, seine anatomischen und biologischen Grundlagen". Fischer, Jena.

Hafferl, A. (1933). Gefässsystem. V. Das Arteriensystem. *In* "Handbuch der vergleichenden Anatomie der Wirbeltiere" (L. Bolk, E. Göppert, E. Kallius and W. Lubosch, eds). Urban & Schwarzenberg, Berlin and Vienna.

Haines, R. W. (1942). The tetrapod knee joint. *J. Anat.* **76**: 270–301.

Hamilton, H. L. (1952). "Lillie's Development of the Chick", 3rd Edn. Henry Holt, New York.

Hanke, B. (1957). Zur Histologie des Ösophagus der Tinamidae. *Bonn. zool. Beitr.* **8**: 1–4.

Harrison, J. M. (1964). Moult. *In* "A New Dictionary of Birds" (A. L. Thomson, ed.). Nelson, London.

Harrison, J. M. and Irving, R. (1965). The anterior ventral cochlear nucleus. *J. Comp. Neurol.* **124**: 15–42.

Harvey, E. B., Kaiser, H. E. and Rosenberg, L. E. (1968). "An Atlas of the Domestic Turkey (*Meleagris gallopavo*). Myology and Osteology. U.S. Atomic Energy Commission, Division Biological Medicine. U.S. Government Printing Office, Washington, D.C.

Hasegawa, K. (1956). On the vascular supply of hypophysis and of hypothalamus in domestic fowl. *Fukuoka Acta med.* **47**: 89–98 [in Japanese].

Hasse, C. (1871). "Zur Morphologie des Labyrinths der Vögel". *Anat. Studien* (herausg. von Hasse), 2 Heft, Nr. VI.

Hayes, V. E. and Hikida, R. S. (1976). Naturally-occurring degeneration in chick muscle development: ultrastructure of the M. complexus. *J. Anat.* **122**: 67–76.

Hazelwood, R. A. (1973). The avian endocrine pancreas. *Amer. Zool.* **13**: 699–709.

Hedonius, I. (1892). Chemische Untersuchung der hornartigen Schicht des Muskelmagens der Vögel. *Skand. Arch. Physiol.* **3**: 244–252.

Heidrich, H. (1908). Die Mund- und Schlundkopfhöhle der Vögel und ihre Drüsen. *Morph. Jb.* **37**: 10–69.

Hellman, B. and Hellerström, C. (1960). The islets of Langerhans in ducks and chickens with special reference to their argyrophil reaction. *Z. Zellforsch. mikr. Anat.* **52**: 278–290.

Helm, A. F. (1884). Über die Hautmuskeln der Vögel, ihre Beziehungen zur Federfluren and ihre Funktionen. *J. Ornith.* **32**: 321–379.

Helmi, C. and Cracraft. J. (1977). The growth patterns of three hindlimb muscles in the chicken. *J. Anat.* **123**: 615–635.

Hikida, R. S. (1972). The structure of the sarcotubular system in avian muscle. *Amer. J. Anat.* **134**: 481–496.

Hikida, R. S. and Bock, W. J. (1974). Analysis of fiber types in the pigeon's metapatagialis muscle. I. Histochemistry, end plates and ultrastructure. *Tissue and Cell* **6**: 411–430.

Hikida, R. S. and Bock, W. J. (1976). Analysis of fiber types in the pigeon's metapatagialis muscle. II. Effects of denervation. *Tissue and Cell* **8**: 259–276.

Hill, K. J. (1971). The structure of the alimentary tract. *In* "Physiology and Biochemistry of the Domestic Fowl" (D. J. Bell and B. M. Freeman, eds), Vol. 1. Academic Press, London.

Hill, O. (1964). Syrinx. *In* "A New Dictionary of Birds" (A. L. Thompson, ed.). Nelson, London.

Hodges, R. D. (1965). The blood supply to the avian oviduct, with special reference to the shell gland. *J. Anat.* **99**: 485–506.

Hodges, R. D. (1970). The structure of the fowl's ultimobranchial gland. *Ann. Biol. anim. Biochem. Biophys.* (series 2) **10**: 255–279.

Hodges, R. D. (1974). "The Histology of the Fowl". Academic Press, London.

Hofer, H. (1945). Untersuchungen über den Bau des Vogelschädels, besonders über den der Spechte und Steisshühner. *Zool. Jb.* (Abt. Anat. v. Ontogenie d. Tiere) **69**: 1–158.

Hofer, H. (1949). Die Gaumenlücken der Vögel. *Acta Zool.* **30**: 209–248.

Hofer, H. (1950). Zur Morphologie der Keifermuskulatur der Vögel. *Zool. Jb.* (*Anat.*) **70**: 427–600.

Hofer, H. (1955). Neuere Untersuchungen zur Kopfmorphologie der Vögel. *Acta XI Congr. Int. Orn. 1954*: 104–137.

Hoff, K. M. (1966). "A comparative study of the appendicular muscles of Strigiformes and Caprimulgiformes". Ph.D. Thesis, Washington State University, Pullman.

Hoffman, K. B. and Pregl, F. (1907). Über Koilin. Hoppe-Seyler's *Z. Physiol. Chem.* **52**: 448–471.

Holden, A. L. (1966). An investigation of the centrifugal pathway to the pigeon retina. *J. Physiol.* **186**: 133P.

Holmes, E. B. (1962). The terminology of the short extensor muscles to the third toe in birds. *Auk* **79**: 485–488.

Holmes, E. B. (1963). Variation in the muscles and nerves of the leg in two genera of grouse (*Tympanuchus* and *Pedioecetes*). *U. Kans. Publ. Mus. Nat. Hist.* **12**: 363–474.

Holmes, G. (1903). On the comparative anatomy of the nervus acusticus. *Trans. Roy. Irish Acad.* (Dublin) section B, **32**: 101–144.

Holmgren, N. (1955). Studies on the phylogeny of birds. *Acta Zool.* **36**: 243–328.

Holtzman, H. (1896). Untersuchungen uber Ciliärganglion und Ciliärnerven *Morph. Arbeit.* **6**: 114–142.

Honess, R. F. and Allred, W. J. (1942). Structure and function of the neck muscles in inflation and deflation of the oesophagus in the Sage Grouse. *Bull. Wyoming Game Fish Dep.* **2**: 1–12.

Howard, H. (1929). The avifauna of Emeryville Shellmound. *Univ. Calif. Publ. Zool.* **32**: 301–394.

Howell, A. B. (1938). Muscles of the avian hip and thigh. *Auk* **55**: 71–81.

Hsieh, T. M. (1951). "The sympathetic and parasympathetic nervous system of the fowl". Ph.D. Diss., University of Edinburgh.

Huber, G. C. (1917). On the morphology of the renal tubules of vertebrates. *Anat. Rec.* **13**: 305–339.

Huber, G. C. and Crosby, E. C. (1926). On thalamic and tectal nuclei and fiber paths in the brain of the American alligator. *J. Comp. Neurol.* **40**: 97–227.

Huber, G. C. and Crosby, E. C. (1929). The nuclei and fiber paths of the avian diencephalon, with consideration of telencephalic and certain mesencephalic centres and connections. *J. Comp. Neurol.* **48**: 1–225.

Huber, J. F. (1936). Nerve roots and nuclear groups in the spinal cord of the pigeon. *J. comp. Neurol.* **65**: 43–91.

Hudson, G. E. (1937). Studies on the muscles of the pelvic appendage in birds. *Amer. Midl. Nat.* **18**: 1–108.

Hudson, G. E. and Lanzillotti, P. J. (1955). Gross anatomy of the wing muscles in the family Corvidae *Amer. Midl. Nat.* **53**: 1–41.

Hudson, G. E., Lanziliotti, P. J. and Edwards, G. D. (1959). Muscles of the pelvic limb in galliform birds. *Amer. Midl. Nat.* **61**: 1–67.

Hudson, G. E. and Lanzillotti, P. J. (1964). Muscles of the pectoral limb in galliform birds. *Amer. Midl. Nat.* **71**: 1–113.

Hudson, G. E., Chen Wang, S. Y. and Provost, E. E. (1965). Ontogeny of the supernumerary sesamoids in the leg muscles of the Ring-necked Pheasant. *Auk* **82**: 427–437.

Hudson, G. E., Parker, R. A., Vanden Berge, J. and Lanzillotti, P. J. (1966). A numerical analysis of the modifications of the appendicular muscles in various genera of gallinaceous birds. *Amer. Midl. Nat.* **76**: 1–73.

Hudson, G. E. Hoff, K. M., Vanden Berge, J. and Trivette, E. C. (1969). A numerical study of the wing and leg muscles of Lari and Alcae. *Ibis* **111**: 459–524.

Hudson, G. E., Schreiweis, D. O. and Chen Wang, S. Y. (1972). A numerical study of the wing and leg muscles of tinamous (Tinamidae). *Northwest Sci.* **46**: 207–255.

Hughes, A. F. W. (1933–34). On the development of the blood vessels in the head of the chick. *Phil. Trans Roy. Soc. London* Series B, **224**: 75–129.

Humphrey, P. S. and Clark, G. A.., Jr. (1961). Pterylosis of the Mallard Duck. *Condor* **63**: 365–385.

Humphrey, P. S. and Clark, G. A., Jr. (1964). The anatomy of waterfowl. *In* "The Waterfowl of the World" (J. Delacour, ed.), Vol. 4. Country Life Ltd., London.

Huxley, T. H. (1877). "A Manual of the Anatomy of Vertebrated Animals". London.

Huxley, T. H. (1882). Respiratory organs of *Apteryx. Proc. zool. Soc. Lond.* 1882 560–569.

Hyrtl, J. (1864). Neue Wundernetze und Geflechte bei Vögeln und Säugethieren. *Denkschr. Mathem.-naturwiss.* (Wien) **22**: 113–147.

IANC (International Anatomical Nomenclature Committee) (1966). "Nomina Anatomica", 3rd Edn. Excerpta Medica Foundation, Amsterdam.

IANC (International Anatomical Nomenclature Committee) (1975). "Nomina Histologica", 2nd Edn. Revision presented for approval at Tokyo International Congress of Anatomists.

IANC (International Anatomical Nomenclature Committee) (1977). "Nomina Anatomica", 4th Edn. Excerpta Medica Foundation, Amsterdam. [This edition also contains "Nomina Histologica" and "Nomina Embryologica."]

ICVAN (International Committee on Veterinary Anatomical Nomenclature) (1973). "Nomina Anatomica Veterinaria", 2nd Edn. Internatl. Comm. Vet. Anat. Nomencl., Vienna.

Igarashi, H. and Yoshinobu, T. (1966). Comparative observations of the eminentia cruciata in birds and mammals. *Anat. Rec.* **115**: 269–278.

Ilyichev, V. D. (1961). Morphological and functional details of the external ear in crepuscular and nocturnal birds. *Dok. Biol. Sci. Sect.* **137**: 253–256.

Ilyichev, V. D. (1972). "Bio-acoustics of Birds". Moscow University Press.

Isomura, G. (1973). A nerve originating from the superior cervical ganglion in the fowl. *Anat. Anz.* **133**: 82–89.

Jacobshagen, E. (1937). Mittel- und Enddarm. *In* "Handbuch der vergleichende Anatomie der Wirbeltiere" (L. Bolk, E. Göppert, E. Kallius and W. Lubosch, eds), Vol. 3. Urban & Schwarzenberg, Berlin and Vienna.

Jäger, G. (1857). Das Os humero-scapulare der Vögel. *SitzBer. Akad. Wiss. Wien* **23**: 387–423.

Jäger, G. (1858). Das Wirbelkörpergelenk der Vögel. *SitzBer. Akad. Wiss. Wien* **23**: 527–564.

Jelgersma, G. (1896). De verbindingen van den groote hersenen by de vogels met de oculomotoriuskern. *Nederlandsche vereeniging van Psychiatrie Festbundel*, pp. 241–250.

Johnsgard, P. A. (1961). "Tracheal anatomy of the Anatidae and its taxonomic significance". Report Wildfowl Trust, Slimbridge, Glos., U.K.

Johnson, O. W. (1968). Some morphological features of avian kidneys. *Auk* **85**: 216–228.

Johnson, O. W. (1974). Relative thickness of the renal medulla in birds. *J. Morph.* **142**: 277–284.

Johnson, O. W. (1978). Urinary system. *In* 'Form and Function in Birds" (A. S. King and J. McLelland, eds). Academic Press, London.

Johnson, O. W. and Mugaas, J. N. (1970). Some histological features of avian kidneys. *Amer. J. Anat.* **127**: 423–436.

Johnson, O. W. and Ohmart, R. D. (1973). Some features of water economy and kidney microstructure in the Large-billed Savannah Sparrow (*Passerculus sandwichensis rostratus*). *Physiol. Zool.* **46**: 276–284.

Johnson, O. W., Phipps, G. L. and Mugaas, J. N. (1972). Injection studies of cortical and medullary organization in the avian kidney. *J. Morph.* **136**: 181–190.

Johnson, O. W. and Skadhauge, E. (1975). Structural and functional correlations in the kidneys and observations of colon and cloacal morphology in certain Australian birds. *J. Anat.* **120**: 495–505.

Johnston, H. S., Aitken, R. N. C. and Wyburn, G. M. (1963). The fine structure of the uterus of the domestic fowl. *J. Anat.* **97**: 333–334.

Jollie, M. T. (1957). The head skeleton of the chicken and remarks on the anatomy of this region in other birds. *J. Morph.* **100**: 389–436.

Jollie, M. T. (1958). Comments on the phylogeny and skull of the Passeriformes. *Auk* **75**: 26–35.

Jolly, J. (1909–10). Recherches sur les ganglions lymphatiques des oiseaux. *Arch. Anat. micr. Morph. exp.* **11**: 179–290.

Jolly, J. (1915). La bourse de Fabricius et les organes lympho-épithéliaux. *Arch. anat. microsc. Morph.* **16**: 363–547.

Jolly, J. (1923). "Traité Technique D'Hématologie". Maloine, Paris.

Jones, D. R. (1969). Afferent vagal activity related to respiratory and cardiac cycles. *Comp. Biochem. Physiol.* **28**, 961–965.

Jorgensen, J. M. (1970). On the structure of the macula lagenae in birds with some notes on the avian maculae utriculi and sacculi. *Vidensk. Medd. Dan. Naturh. Foren.* **133**: 121–147.

Juillet, A. (1912). Recherches antomiques, embryologiques, histologiques et comparatives sur le poumon des oiseaux. *Arch. Zool. Exp. Gén.* **9**: 207–371.

Jungherr, E. (1943). Nasal histopathology and liver storage in subtotal vitamin A deficiency of chickens. *Bull. Storrs. Agric. Exp. Stn.* No. **250**: 1–36.

Jungherr, E. L. (1969). "The neuroanatomy of the domestic fowl (*Gallus domesticus*)", *Avian Diseases*, Special Issue.

Juorio, A. V. and Vogt, M. (1967). Monoamines and their metabolites in the avian brain. *J. Physiol.* **189**: 489–518.

Kaku, K. (1959). On the vacular supply in the brain of the domestic fowl. *Fukuoka Acta Medica* **50**: 4293–4306.

Kallius, E. (1905). Beiträge zur Entwicklung der Zunge. II Teil. Vögel (*Anas boschas* L.; *Passer domesticus* L.) *Anat. Hefte* (Abt. 1): 309–586.

Kar, A. B. (1947). Studies on the ligaments of the oviduct in the domestic fowl. *Anat. Rec.*: 175–192.

Karten, H. J. (1963). Ascending pathways from the spinal cord of the pigeon. *Proc. XVI internatl. Congr. Zool.* (*Wash.*) **2**: 23 (abstract).

Karten, H. J. (1965). Projections of the optic tectum of the pigeon (*Columba livia*). *Anat. Rec.* **148**: 297–298. (abstract).

Karten, H. J. (1967). The organization of the ascending auditory pathway in the pigeon (*Columba livia*). I. Diencephalic projections of the inferior colliculus (nucleus mesencephali lateralis, pars dorsalis). *Brain Res.* **6**: 409–427.

Karten, H. J. (1971). Efferent projections of the Wulst of the owl. *Anat. Rec.* **169** (abstract).

Karten, H. J. and Dubbeldam, J. L. (1973). The organization and projections of the paleostriatal complex in the pigeon (*Columba livia*). *J. Comp. Neurol.* **148**: 61–90.

Karten, H. J. and Hodos, W. (1967). "A Stereotaxic Atlas of the Brain of the Pigeon". Johns Hopkins Press, Baltimore.

Karten, H. J. and Hodos, W. (1970). Telencephalic projections of the nucleus rotundus in the pigeon (*Columba livia*). *J. Comp. Neurol.* **140**: 35–52.

Karten, H. J., Hodos, W., Nauta, W. J. H. and Revzin, A. M. (1973). Neural connections of the "visual wulst" of the avian telencephalon. Exerimental studies in the pigeon (*Columba livia*) and owl (*Speotyto cunicularia*). *J. Comp. Neurol.* **150**: 253–278.

Karten, H. J. and Revzin, M. (1966). The afferent connections of the nucleus rotundus in the pigeon. *Brain Res.* **2**: 368–377.

Kelso, L. (1940). Variation of the external ear-opening in the Strigidae. *Wilson Bull.* **52**: 24–29.

Kenneth, J. H. (1966). "Henderson's Dictionary of Biological Terms", 8th Edn. Van Nostrand, Princeton, N.J. and New York.

Kern, A. (1926). Das Vogelherz. Untersuchungen an *Gallus domesticus* Briss. *Morph. Jb.* **56**: 264–315.

Kern, D. (1963). Die Topographie der Eingeweide der Körperhohle des Haushuhnes (*Gallus domesticus*) unter besonderer Berücksichtigung der Serosa- und Gekröseverhaltnisse. Inaug. Diss., Universität Giessen.

Kesteven, H. L. (1925). The parabasal canal and nerve foramina and canals in the bird skull. *J. Roy. Soc. New South Wales* **59**: 108–123.

Khanna, R. K. (1957). On the presence of specialized connecting (conducting) tissue in the heart of the Indian Rock-Pigeon *Columba livia intermedia* Strick. *J. Animal Morph. Physiol.* (India) **4**: 33–42.

King, A. S. (1956). The structure and function of the respiratory pathways of *Gallus domesticus*. *Vet. Rec.* **68**: 544–547.

King, A. S. (1966). Structural and functional aspects of the avian lungs and air sacs. *In* "International Review of General and Experimental Zoology" (W. J. L. Felts and R. J. Harrison, eds), Vol. II. Academic Press, New York.

King, A. S. (1975). Aves Urogenital System. *In* "Sisson and Grossman's The Anatomy of the Domestic Anaimals" (R. Getty, ed.), Vol. 2, 5th Edn. Saunders, Philadelphia.

King, A. S. (1975). Aves respiratory system. *In* "Sisson and Grossman's The Anatomy of the Domestic Animals" (R. Getty, ed.), Vol. 2, 5th Edn. Saunders, Philadelphia.

King, A. S. (1980). The Phallus. *In* "Form and Function in Birds", (A. S. King and J. McLelland, eds.), Vol. 2. Academic Press, London.

King, A. S. (1975). Aves, lymphatic system. *In* "Sisson and Grossman's The Anatomy of the Domestic Anaimals" (R. Getty, ed.), Vol. 2, 5th Edn. Saunders, Philadelphia.

King, A. S. and Atherton, J. D. (1970). The identity of the air sacs of the turkey (*Meleagro gallopavo*). *Acta Anat.* **77**: 78–91.

King, A. S. and Cowie, A. F. (1969). The functional anatomy of the bronchial muscle of the bird. *J. Anat.* (*Lond.*) **105**: 323–336.

King, A. S. and McLelland, J. (1975). "Outlines of Avian Anatomy". Baillière Tindall, London.

King, A. S. and Molony, V. (1971). The anatomy of respiration. *In* "Physiology and Biochemistry of the Domestic Fowl" (D. J. Bell and B. M. Freeman, eds.), Vol. 1. Academic Press, London.

King-Smith, P. E. (1971). Special senses. *In* "Physiology and Biochemistry of the Domestic Fowl" (D. J. Bell and B. M. Freeman, eds), Vol. II. Academic Press, London.

Kinsky, F. C. (1971). The consistent presence of paired ovaries in the Kiwi (*Apteryx*) with some discussion of this condition in other birds. *J. Ornith.* **112**: 334–357.

Kitoh, J. (1962). Observations on the arteries with their anastomoses in and around the brain in the fowl. *Jap. J. vet. Sci.* **24**: 141–150 [in Japanese].

Kitoh, J. (1964). Arterial supply of the spinal cord. *Jap. J. Vet. Sci.* **26**: 169–175 [in Japanese].

Klemm, R. D. (1969). "Comparative myology of the hind limb of procellariiform birds". *Southern Ill. Univ. Monogr.*, Sci. Ser. **2**: 1–269.

Klemm, R. D., Knight, C. E. and Stein, S. (1973). Gross and microscopic morphology of the Glandula proctodealis (Foam Gland) of *Coturnix c. japonica* (Aves). *J. Morph.* **141**: 171–184.

Knight, C. E. (1970). "The anatomy of the structures involved in the erection-dilution mechanism in the male domestic fowl". Ph.D. thesis, Michigan State University.

Köditz, W. (1925). Über die Syrinx einiger Clamatores und ausländischer Oscines. *Z. wiss. Zool.* **126**: 70–144.

Kolda, J. and Komárek, Vl. (1958). "Anatomie Domacich Ptaku". Ceskoslovenska Akad. Zemedělskych ved., Praha.

Komárek, Vl. (1958) Krajiny těla husy a Kura [Regiones corporis der Gans und des Huhnes], Acta univ. agric. sylvicul. (Brno) **6**: 1–19.

Komárek, Vl. (1969). Die männliche Kloake unserer Entenvögel. *Anat. Anz.* **124**: 434–442.

Komárek, Vl. (1970). Vertebra avia. *Sci. Agric. Bohemoslovaca.* **2**: 35–49.

Komárek, V. (1970). The cloaca of the turkey-cock and of the cock. *Acta vet. Brno* **39**: 227–234.

Komárek, V. (1971). The female cloaca of anseriform and galliform birds. *Acta vet. Brno* **40**: 13–22.

Komárek, V. and Marvan, F. (1969). Beitrag zur mikroskopischen Anatomie des Kopulationsorganes der Entenvogel. *Anat. Anz.* **124**: 467–476.

Komárek, V. and Prochazkova, E. (1970). Growth and differentiation of the ovarian follicles in the postnatal development of the chicken. *Acta Vet. Brno* **39**: 11–16.

Kondo, M. (1937a). Die lymphatischen Gebilde im Lymphgefässsystem des Huhnes. *Folia Anat. Jap.* **15**: 309–325.

Kondo, M. (1937b). Die lymphatische Gebilde im Lymphgefässsystem der verschiedenen Vogelarten. *Folia Anat. Jap.* **15**: 329–348.

Koonz, C. H., Strandine, E. J. and Gray, R. E. (1963). A study of factors responsible for keel blisters in poultry. (Abstract) *Poult. Sci.* **42**: 1281.

Kooy, F. H. (1915). De phylogenese van de oliva inferior. *Ned. Tijdss. voor Geneesk.* **51**: 2533–2536.

Kooy, F. H. (1917). The inferior olive in vertebrates. *Folia Neurobiol.* **10**: 205–369.

Kosaka, K. and Hiraiwa, K. (1905). Über die Facialiskerne beim Huhn. *Jb. Psychiat.* **25**: 57–69.

Kose, W. (1904). Über die Carotisdrüse und das chromaffine Gewebe der Vögel. *Anat. Anz.* **25**: 609–617.

Kovacs, J. (1928). Comparative histologic studies upon the oral and anal parts of the intestine of the domestic fowl. *Közleményck az osszehasontító élet-és kórtan köréböl.* **21**: 400–405. Abstract in *Jahresbericht vet. medizu.* **1**: 56.

Krause, R. (1922). "Mikroskopische Anatomie der wirbeltiere in Einzeldarstellungen II. Vögel und Reptilien". DeGruyter, Berlin.

Kudo, N., Sugimura, M. and Yamano, S. (1975). Anatomical studies of corpus paracloacalis vascularis in cocks. *Jap. J. vet. Res.* **23**: 1–10.

Kuhlenbeck, H. (1937). The ontogenetic development of the diencephalic centres in a bird's brain (chick) and comparison with reptilian and mammalian diencephalon. *J. Comp. Neurol.* **66**: 23–75.

Kuhlenbeck, H. (1975). "The Central Nervous System of Vertebrates", Vol. 4: Spinal Cord and Deuterencephalon. Karger, New York.

Kuhlenbeck, H. (1977). "The Central Nervous System of Vertebrates. Vol. 5, Pt. 1. Derivatives of the Prosencephalon: Diencephalon and Telencephalon. Karger, Basel.

Kuhlenbeck, H. and Haymaker, W. (1949). The derivatives of the hypothalamus in the human brain; their relation to the extrapyramidal and autonomic systems. *Milit. Surg.* **105**: 26–52.

Kurihara, S. and Yasuda, M. (1975). Morphological study of the kidney in the fowl. I. Arterial system. *Jap. J. Vet. Sci.* **37**: 29–47.

Kurochkin, E. N. (1968). "Locomotion and morphology of the pelvic extremities in swimming and diving birds". *Acad. Sci. Ukrainian SSR, Inst. Zool., Moscow*, 18 pp. [in Russian].

Kuroda, N. (1954). On some osteological and anatomical characters of Japanese Alcidae (Aves). *Jap. J. Zool.* **11**: 311–327.

Kuroda, N. (1954). "On the Classification and Phylogeny of the Order Tubinares, particularly the Shearwaters (*Puffinus*)", 1–779. Publ. by author.

Kuroda, N. (1960). On the pectoral muscles of birds. *Misc. Reports Yamashina Inst. Ornith, Zool.* **2**: 50–59.

Kuroda, N. (1961a). A note on the pectoral muscles of birds. *Auk* **78**: 261–263.

Kuroda, N. (1961b). Analysis of three adaptive body forms in the Steganopodes, with notes on pectoral muscles. *Misc. Reports Yamashina Inst. Ornith. Zool.* **3**: 54–66.

Kuroda, N. (1962). On the cervical muscles of birds. *Misc. Reports Yamashina Inst. Ornith. Zool.* **3**: 189–211.

Lake, P. E. (1971). The male in reproduction. *In* "Physiology and Biochemistry of the Domestic Fowl" (D. J. Bell and B. M. Freeman, eds.). Vol. 3. Academic Press, London.

Lake, P. E., Smith, W. and Young, D. (1968). The ultrastructure of the ejaculated fowl spermatozoon. *Quart. J. exp. Physiol.* **53**: 356–366.

Lakjer, T. (1926). "Studien über die Trigeminus—versorgte Kaumuskulatur der Sauropsiden". C. A. Reitzel, Copenhagen.

Lambrecht, K. (1914). Morphologie des Mittelhandknochens Os metacarpi der Vögel. *Aquila, Jb.* **1, 21**: 53–84.

Lambrecht. K. (1933). Literatur des Skelettssystems. *In* "Handbuch der Palaeonithologie". Gebrüder Borntraeger, Berlin.

Lange, B. (1931). Integument. II. Integument der Sauropsiden. *In* "Handbuch der vergleichenden Anatomie der Wirbeltiere" (L. Bolk, E. Göppert, E. Kallius and W. Lubosch, eds), Vol. 1. Urban & Schwartzenberg, Berlin and Vienna.

Langley, J. N. (1904). On the sympathetic system of birds, and on the muscles which move the feathers. *J. Physiol.* **30**: 221–252.

Larsell, O. (1948). The development and subdivisions of the cerebellum of birds. *J. Comp. Neurol.* **89**: 123–189.

Larsell, O. (1967). The Comparative Anatomy and Histology of the Cerebellum from Myxinoids through Birds. Univ. of Minnesota Press, Minneapolis.

Larson, L. M. (1930). Osteology of the California Road-Runner Recent and Pleistocene. *Univ. Cal. Publ. Zool.* **32**: 409–428.

Larsson, L.-I., Sundler, F., Håkanson, R., Rehfeld, J. F. and Stadil, F. (1974). Distribution and properties of gastrin cells in the gastrointestinal tract of chicken. *Cell Tiss. Res.* **154**: 409–421.

Lasiewski, R. C. (1972). Respiratory function in birds. *In:* "Avian Biology" (D. S. Farner and J. R. King, eds), Vol. 2. Academic Press, London.

Lebedinsky, N. G. (1920). Beiträge zur Morphologie und Entwicklungsgeschichte des Unterkiefers der Vögel. *Verh. naturf. Ges. Basel* **31**: 39–112.

Lebedinsky, N. C. (1921). Zur Syndesmologie der Vögel. *Anat. Anz.* **54**: 8–15.

Leblond, C. P. (1951). Histological structure of hair, with a brief comparison to other epidermal appendages and epidermis itself. *Ann. N.Y. Acad. Sci.* **53**: 464–475.

Leblond, C. P., Greulich, R. C. and Pereira, J. P. M. (1964). Relationship of cell formation and cell migration in the renewal of stratified squamous epithelia. *In* "Advances in Biology of Skin" (W. Montagna and R. E. Billingham, eds), Vol. 5. Macmillan, New York.

Leibovitz, L. (1968). Weyonella Philiplevinei, N.S.P., a coccidial organism of the White Pekin Duck. *Avian Dis.* **12**: 670–681.

Leonard, R. B., and Cohen, D. H. (1975). A cytoarchitectonic analysis of the spinal cord of the pigeon (*Columba livia*). *J. Comp. Neurol.* **163**: 159–180.

Levi, W. M. (1957). "The Pigeon". Levi Publ. Co., Sumter, S. C.

Liebe, W. (1914). Die männliches Begattungsorgan der Hausente. *Jena Z. Naturw.* **51**: 627–696.

Liebelt, R. A. and Eastlick, H. L. (1954). The organlike nature of the subcutaneous fat bodies in the chicken. *Poult. Sci.* **33**: 169–179.

Lillie, F. R. (1952). "Development of the Chick" (revised by H. L. Hamilton), 3rd Edn. Holt, Rinehart & Winston, New York.

Lindenmaier, P. and Kare, M. R. (1959). The taste end organs of the chicken. *Poult. Sci.* **38**: 545–550.

Lindgren, H. (1868). Über den Bau der Vogelnieren. *Z. rationelle Med.* **33**: 15–35.

Lindsay, F. E. F. (1967). The cardiac veins of *Gallus domesticus*. *J. Anat.* **101**: 555–568.

Lindsay, F. E. F. and Smith, H. J. (1965). Coronary arteries of *Gallus domesticus*. *Amer. J. Anat.* **116**: 301–314.

Lob, G. (1967). Untersuchungen am Huhn über die Blutgefässe von Rückenmark und Corpus gelatinosum. *Morph. Jb.* **110**: 316–358.

Lockner, F. R. and Youngren, O. M. (1976). Functional syringeal anatomy of the Mallard. I. In situ electromyograms during ESB elicited calling. *Auk* **93**: 324–342.

Locy, W. A. and Larsell, O. (1916a). The embryology of the bird's lung. Part 1. *Amer. J. Anat.* **19**: 447–504.

Locy, W. A. and Larsell, O. (1961b). The embryology of the bird's lung. Part 2. *Amer. J. Anat.* **20**: 1–44.

Lorenz, F. W. (1964). Recent research on fertility and artificial insemination of domestic birds. Vth Cong. Int. Ripro. Anim. Fecond. Art., Vol. 4.

Lucas, A. M. (1968). Lipoid secretion in the avian epidermis. *Anat. Rec.* **160**: 386–387.

Lucas, A. M. (1970). Avian functional anatomic problems. *Fed. Proc.* **29**: 1641–1648.

Lucas, A. M. (1975). Common integument. *In* "Sisson and Grossman's The Anatomy of Domestic Animals" (R. Getty, ed.), 5th Edn. Saunders, Philadelphia.

Lucas, A. M. and Stettenheim, P. R. (1965). Avian Anatomy. *In* "Diseases of Poultry" (H. E. Biester and L. H. Schwarte, eds), Chap. 1. Iowa State University Press, Ames.

Lucas, A. M. and Stettenheim, P. R. (1972). "Avian Anatomy. Integument." Agriculture Handbook 362. U.S. Dept. Agric., U.S. Government Printing Office, Washington, D.C.

Lucas, F. A. (1897). The tongue of birds. *Rep. U.S. nat. Mus.* 1897: 1003–1020.

Lyser, K. M. (1973). The fine structure of the glycogen body of the chicken. *Acta Anat.* **85**: 533–549.

Macdonald, R. L. and Cohen, D. H. (1970). Cells of origin of sympathetic pre- and postganglionic cardioacceleratory fibers in the pigeon. *J. Comp. Neurol.* **40**: 343–358.

Maderson, P. F. A. and Jaskoll, T. (1976). A long-ignored sensory structure in the avian middle ear. *Amer. Zool.* **16**: 200.

Magnan, A. (1912). Essai de morphologie stomacale en fonction du régime alimentaire ches les oiseaux. *Ann. Sci. nat. Zool.* **15**: 1–41.

Magnus, H. (1870). Untersuchungen über den Bau knöchernen Vogelkopfes. *Z. Zool.* **21**: 1–108.

Malinovský, L. (1962). Contribution to the anatomy of the vegetative nervous system in the neck and thorax of the domestic pigeon. *Acta. Anat.* **50**: 326–347.

Malinovský, L. (1965). Contribution to the comparative anatomy of the vessels in the abdominal part of the body cavity in birds. III. Nomenclature of branches of the a. coeliaca and of tributaries of the v. portae. *Folia Morph.* **13**: 252–264.

Malinovský, L. (1967). Die Nervenendkörperchen in der Haut von Vögeln und ihre Variabilität. *Z. mikr. -Anat. Forsch.* **77**: 279–303.

Malinovský, L. and Zemanek, R. (1969). Sensory corpuscles in the beak skin of the domestic pigeon. *Folia Morph.* **17**: 241–250.

Malinovský, L., Visnanska, M. and Roubal, P. (1973). Branching of A. coeliaca in some domestic birds. I. Domestic duck (*Anas platyrhynchos f. domestica*). *Scripta Medica* (Brno) **46**: 325–336.

Malinovský, L. and Visnanska, M. (1975). Branching of the coeliac artery in some domestic birds. II. The domestic goose. *Folia Morph.* **23**: 128–135.

Marage, R. (1889). Anatomie descriptive du sympathique chez les oiseaux. *Ann. Sci. Natur. Zool.* **7**: 1–72.

Marples, B. J. (1932). The structure and development of the nasal glands of birds. *Proc. zool. Soc. Lond.* **2**: 829–844.

Marshall, A. J. (1961). Reproduction. *In* "Biology and Comparative Physiology of Birds" (A. J. Marshall, ed.), Vol. 2. Academic Press, New York and London.

Marshall, A. J. (1962). "Parker and Haswell's Textbook of Zoology", 7th Edn. Macmillan, London.

Marshall, E. K. (1934). The comparative physiology of the kidney in relation to theories of renal secretion. *Physiol. Rev.* **14**: 133–159.
Martin, R. (1904). "Die vergleichenden Osteologie der Columbiformes". Inaug. Diss., University of Basel.
Marvan, F. (1969). Postnatal development of the male genital tract of the *Gallus domesticus*. *Anat. Anz.* **124**: 443–462.
Marvan, F. and Těšik, I. (1970). Comparative anatomical study of the tongue of fowl, turkey and guinea fowl. *Acta vet. Brno.* **39**: 235–243.
Matoltsy, A. G. (1962). Mechanism of keratinization. *In* "Fundamentals of Keratinization" (E. O. Butcher and R. F. Sognnaes, eds). *Amer. Assoc. Adv. Sci. Pub.* **70**, Washington, D.C.
Matoltsy, A. G. (1969). Keratinization of the avian epidermis. An ultrastructural study of the newborn chick skin. *J. Ultrastruct. Res.* **29**: 438–458.
Matoltsy, A. G. and Huszar, T. (1972). Keratinization of the reptilian epidermis: an ultrastructural study of the turtle skin. *J. Ultrastruct. Res.* **38**: 87–101.
Matsumoto, C. (1955). The venous sinuses of the dura mater of the brain and the veins in the bird kind. *J. Kurume Med. Assoc.* **18**: 765–797 [in Japanese].
Matthews, L. H. (1949). The origin of stomach oil in the petrels, with comparative observations on the avian proventriculus. *Ibis* **91**: 373–392.
McFarland, L. Z., Warner, R. L., Wilson, W. O. and Mather, F. B. (1968). The cloacal gland complex of the Japanese Quail. *Experientia* **24**: 941–943.
McIntosh, J. R. and Porter, K. R. (1967). Microtubules in the spermatids of the domestic fowl. *J. Cell Biol.* **35**: 153–173.
McLelland, J. (1965). The anatomy of the rings and muscles of the trachea of *Gallus domesticus*. *J. Anat.* (*Lond.*) **99**: 651–656.
McLelland, J. (1968). The hyoid muscles of *Gallus gallus*. *Acta Anat.* **69**: 81–86.
McLelland, J. (1975a). Aves Digestive System. *In* "Sisson and Grossman's The Anatomy of the Domestic Animals" (R. Getty, ed.). Vol. 2, 5th Edn. Saunders, Philadelphia.
McLelland, J. (1975b). Sense organs, Aves. *In* "Sisson and Grossman's The Anatomy of the Domestic Animals" (R. Getty, ed.), Vol. 2, 5th Edn. Saunders, Philadelphia.
McLelland, J. and Abdalla, A. B. (1972). The gross anatomy of the nerve supply to the lungs of *Gallus domesticus*. *Anat. Anz.* **131**: 320–328.
McLelland, J. and King, A. S. (1970). The gross anatomy of the peritoneal coelomic cavities of *Gallus domesticus*. *Anat. Anz.* **127**: 480–490.
McLelland, J. and King, A. S. (1975). Aves. Celomic cavities and mesenteries. *In* "Sisson and Grossman's The Anatomy of the Domestic Animals" (R. Getty, ed.), Vol. 2, 5th Edn. Chap. 62. Saunders, Philadelphia.
McLeod, W. M. and Wagers, R. P. (1939). The respiratory system of the chicken. *J. Amer. Vet. Med. Assoc.* **95**: 59–70.
Mayr, E. (1931). Die Syrinx einiger Singvögel aus Neu-Guinea. *J. Ornith.* **79**: 333–337.
Menaker, M. and Oksche, A. (1974). The avian pineal organ. *In* "Avian Biology" (D. S. Farner and J. R. King, eds), Vol. 4. Academic Press, London.
Middleton, A. L. A. (1972). The structure and possible function of the avian seminal sac. *Condor* **74**: 185–190.
Mikami, S. I., Oksche, A., Farner, D. S. and Vitums, A. (1970). Fine structure of the vessels of the hypophysial portal system of the White-Crowned Sparrow, *Zonotrichia leucophrys gambelii*. *Z. Zellforsch. mikr. Anat.* **106**: 155–170.
Mikami, S. I. and Ono, K. (1962). Glucogen deficiency induced by extirpation of alpha islets of the fowl pancreas. *Endocr.* **71**: 464–473.
Miller, A. H. (1934). The vocal apparatus of some North American owls. *Condor* **36**: 204–213.

Miller, A. H. (1941). The buccal food-carrying pouches of the Rosy Finch. *Condor* **43**: 72–73.
Miller, A. M. (1903). The development of the postcaval vein in birds. *Amer. J. Anat.* **2**: 283–298.
Miller, A. P., Sato, K. and Glick, B. (1971). The chicken lacrimal gland, gland of Harder, caecal tonsil, and accessory spleens as sources of antibody-producing cells. *Cell Immunol.* **2**: 142–152.
Miller, W. DeW. (1924). Further notes on ptilosis. *Bull. Amer. Nat. Hist.* **50**: 305–331.
Milne-Edwards, A. (1865). Observations sur l'appareil respiratoire de quelques oiseaux. *Ann. Sci. nat.* **3**: 137–142.
Milne-Edwards, A. (1867–71). "Recherches anatomiques et paléontologiques pour servir a l'histoire des oiseaux fossiles de la France", Vols 1, 2. Masson et Fils, Paris.
Miskimen, M. (1951). Sound production in passerine birds. *Auk* **68**: 493–504.
Miskimen, M. (1963). The syrinx in certain tyrant flycathers. *Auk* **80**: 156–165.
Mitchell, P. C. (1895). On the anatomy of *Chauna chavaria*. *Proc. zool. Soc. Lond.* 1895: 350–358.
Mitchell, P. C. (1901). On the intestinal tract of birds; with remarks on the valuation and nomenclature of zoological charcters. *Trans. Linn. Soc. Lond.* **8**: 173–275.
Mitchell, P. C. (1913). The peroneal muscles in birds. *Proc. zool. Soc. Lond.* 1913: 1039–1072.
Mivart, St. G. (1895). On the hyoid bone of certain parrots. *Proc. zool. Soc. Lond.* 1895: 162–174.
Miyaki, T. (1973). The hepatic lobule and its relation to the distribution of blood vessels and bile ducts in the fowl. *Jap. J. vet. Sci.* **35**: 403–410.
Miyaki, T. and Yasuda, M. (1977). On the thoracic duct and the lumbar lymphatic vessel in the fowl. *Jap. J. vet. Sci.* **39**: 559–570.
Möllendorff, W. von (1930). "Handbuch der mikroskopischen Anatomie des Menschen", Vol. VII, Part 1. Springer, Berlin.
Montagna, W. (1945). A re-investigation of the development of the wing of the fowl. *J. Morph.* **76**: 87–113.
Moore, C. A. and Elliott, R. (1946). Numerical and regional distribution of taste buds on the tongue of the bird. *J. Comp. Neurol.* **84**: 119–131.
Morejohn, G. V. (1966). Variations of the syrinx of the fowl. *Poult. Sci.* **45**: 33–39.
Morgan, W. and Adams, A. (1959). Identification of two oviducts in live hens. *Poult. Sci.* **38**: 861–864.
Morony, J. J., Jr.; Bock, W. J. and Farrand, J., Jr. (1975). "Reference List of Birds of the World". Dept. Ornithology, American Museum of Natural History, New York.
Moser, E. (1906). Die Haut des Vogels. *In* "Handbuch der Vergleichenden Anatomie der Haustiere" (W. Ellenberger, ed.). Julius Springer, Vienna.
Mouchett, R. and Cuypers, Y. (1959). Étude de la vascularisation du rein de coq. *Arch. Biol.* **69**: 577–590.
Mudge, G. P. (1903). On the morphology of the tongue of parrots, with a classification of the order, based upon the structure of the tongue. *Trans. zool. Soc. Lond.* **16**: 211–278.
Muller, B. (1907). The air sacs of the pigeon. *Smithson. Misc. Coll.* **50**: 365–414.
Müller, H. J. (1961). Über strukturelle Ähnlichkeiten der Ohr- und Occipitalregion bei Vögeln und Säugern. *Zool. Anz.* **166**: 391–402.
Müller, J. (1847). "On certain variations in the vocal organs of the Passeres that have hitherto escaped notice". (Trans. from the German by F. J. Bell). Clarendon Press, Oxford.
Müller, S. (1927). Zur Morphologie des Oberflachenreliefs der Rumpfdarmsschleimhaut bei den Vögeln. *Jena Z. Naturw.* **58**: 533–606.

Munger, B. L. (1971). Patterns of organization of peripheral sensory receptors. *In "Handbook of Sensory Physiology" (W. R. Lowenstein, ed.), Vol. I. Springer, Berlin.*

Muratori, G. (1934). Contributo istologico all' innervazione della zone arteriosa glomo carotidea. *Arch. Ital. Anat. Embriol.* **33**: 421–442.

Murie, J. (1867). On the tracheal pouch of the Emu (*Dromaeus novae-hollandiae* Vieill.). *Proc. zool. Soc. Lond.* 1867: 405–415.

Murie, J. (1868). Observations concerning the presence and function of the gular pouch in *Otis kori* and *Otis australis*. *Proc. zool. Soc. Lond.* 1868: 471–477.

Murie, J. (1869). Note on the gular pouch of *Otis tarda*. *Proc. zool. Soc. Lond.* 1869: 140–142.

Murrillo-Ferrol, N. L. (1967). The development of the carotid body in *Gallus domesticus*. *Acta Anat.* **68**: 102–126.

Myers, J. A. (1917). Studies of the syrinx of *Gallus domesticus*. *J. Morph.* **29**: 165–215.

Nagelschmidt, L. (1939). Die Langerhansschen Inseln der Bauchspeicheldrüse bei den Vögeln. *J. Morph. mikr. Anat.* **45**: 200–255.

Naik, D. R. and Dominic, C. J. (1963). The intestinal caeca as a criterion in avian taxonomy. *Proc. 50th Ind. Sci. Congr. 1962, Part III*, p. 533.

Naik, D. R. and Dominic, C. J. (1969). A study of the intestinal caeca of some Indian birds. *Proc. 56th Ind. Sci. Congr. 1969, Part III*, pp. 473–474.

Nalbandov, A. V. and James, M. F. (1949). The blood vascular system of the chicken ovary. *Amer. J. Anat.* **85**: 347–367.

Narbaitz, R. and de Robertis, E. M. (1968). Postnatal evolution of steroidogenic cells in the chick ovary. *Histochemie* **15**: 187–193.

Nauck, E. Th. (1938). Extremitätenskelett der Tetrapoden. *In* "Handbuch der vergleichenden Anatomie der Wirbeltiere" (L. Bolk, E. Göppert, E. Kallius and W. Lubosch, eds), Bd. V. Urban & Schwarzenberg, Berlin and Vienna.

Nauta, W. J. H. and Karten, H. J. (1970). A general profile of the vertebrate brain, with sidelights on the ancestry of the cerebral cortex. *In* "The Neuro-Sciences, Second Study Program" (F. O. Schmitt, ed.). Rockefeller University Press, New York.

Nelson, N. M. (1942). The sclerotic plates of the white leghorn chicken. *Anat. Rec.* **84**(3): 295–306.

Neugebauer, L. A. (1845). Systema venosum avium cum eo mammalium et imprimis hominis collatum. *Nova Acta Acad. Leopoldino-Carolinae Naturae Curiosum* **21**: 517–698.

Newton, A. (1896). "A Dictionary of Birds". Black, London.

Newton, I. (1967). The adaptive radiation and feeding ecology of some British finches. *Ibis* **109**: 33–98.

Niethammer, G. (1933). Anatomisch-Histologische und Physiologische Untersuchungen über die Kropfbildungen der Vögel. *Z. wiss. Zool.*, Abt. A **144**: 12–101.

Niethammer, G. (1966). Sexual dimorphismus am Öesophagus von *Rostratula*. *J. Ornith.* **107**: 201–204.

Nishi, S. (1938). II. Muskeln des Rumpfes. III. Muskeln des Kopfes, Parietale Musculatur. *In* "Handbuch der vergleichenden Anatomie der Wirbeltiere" (L. Bolk, E. Göppert, E. Kallius and W. Lubosch eds), Bd. 5. Urban & Schwarzenberg, Berlin and Vienna.

Nishida, T. (1960). On the blood vascular system of the thoracic limb in the fowl. Part 1. The artery. *Jap. J. Vet. Sci.* **22**: 223–231 [in Japanese].

Nishida, T. (1963). The blood vascular system of the hind limb in the fowl. Part 1. The artery. *Jap. J. Vet. Sci.* **24**: 93–106 [in Japanese].

Nishida, T. (1964). Blood vascular system of the male reproductive organs. *Jap. J. vet. Sci.* **26**: 211–221 [in Japanese].

Nishida, T. and Mochizuki, K. (1976). The venous system of the proventriculus of duck (*Anas domesticus*). *Jap. J. vet. Sci.* **38**: 255–262.

Nishida, T., Paik, Y. and Yasuda, M. (1969). Blood vascular supply of the glandular stomach (ventriculus glandularis) and the muscular stomach (ventriculus muscularis). *Jap. J. vet. Sci.* **31**: 51–70 [in Japanese].

Nishiyama, H. (1950). Studies on the physiology of reproduction in the male fowl. 1. On the accessory organs of the phallus. *Sci. Bull. Fac. Agric. Kyushu Univ.* **12**: 27–36.

Nishiyama, H. (1955). Studies on the accessory reproductive organs in the cock. *J. Fac. Agric. Kyushu Univ.* **10**: 277–305.

Nitzsch, C. L. (1867). "Nitzsch's Pterylography" (translated into English by W. S. Dallas; P. O. Slater, ed.). Roy. Society, London.

Nonidez, J. F. and Goodale, H. D. (1927). Histological studies on the endocrines of chickens deprived of ultraviolet light. I. Parathyroids. *Amer. J. Anat.* **38**: 319–347.

Nottebohm, F. and Nottebohm, M. E. (1976). Left hypoglossal dominance in the control of canary and White-crowned Sparrow song. *J. comp. Physiol.* **108**: 171–192.

Oakes, B. W. and Bialkower, B. (1977). Biomechanical and ultrastructural studies on the elastic wing tendon from the domestic fowl. *J. Anat.* **123**: 369–387.

Oehme, H. (1968). Das Ganglion ciliare der Rabenvögel (Corvidae) *Anat. Anz.* **123**: 261–277.

Oehme, H. (1969a). Blutgefässe und Bindgewebe der Vogeliris. *Morph. Jb.* **113**: 555–589.

Oehme, H. (1969b). Der Bewegungsapparat der Vogeliris (Eine vergleichende morphologisch-funktionelle Untersuchung). *Zool. Jb. Anat.* **86**: 96–128.

O'Flaherty, J. J. (1971). The optic nerve of the mallard duck: Fiber diameter, frequency distribution and physiological properties. *J. Comp. Neurol.* **143**: 17–24.

Ogawa, M. and Sokabe, H. (1971). The macula densa site of avian kidney. *Z. Zellforsch. mikrosk. Anat.* **120**: 29–36.

Ogden, T. E. (1967). On the function of efferent retinal fibers. *In* "Structure and Function of Inhibitory Neuronal Mechanisms". Wenner Gren Center International Symposia **7**: 89–109. Pergamon Press, London.

Oksche, A. (1964). The fine structure of the neurosecretory system of birds in relation to its functional aspects. *Proc. 2nd Intern. Congr. Endocrin.*, London 1964: 167–171.

Oksche, A. and Farner, D. S. (1974). Neurohistological studies of the hypothalamo-hypophysial system of *Zonotrichia leucophrys gambelii* (Aves, Passeriformes). *Adv. in Anat. Embryol. Cell Biol.* Fasc. 4 **48**: 1–136.

Oksche, A., Wilson, W. O. and Farner, D. S. (1964). The hypothalmic neuro-secretory system of *Coturnix coturnix japonica. Z. Zellforsch mikr. Anat.* **61**: 688–709.

Oliveira, A. (1958). Contribuicãe para o estudo anatomico da arteria celiaca e sua distribucão no *Gallus domesticus. Veterinaria* **12**: 1–22.

Oliveira, A. (1959). Contribuicão para o estudo anatomico das afluentes e confluentes do distrito venoso portal no *Gallus gallus domesticus. Veterinaria* **13**: 43-78.

Oribe, T. (1968). On the distribution of blood vessels of the mature follicle. *Jap. J. Zootech. Sci.* **39**: 228–234.

Osborne, D. R. (1968). "The functional anatomy of the skin muscles in Phasianinae". Ph.D. Thesis, Michigan State University, East Lansing.

Oscarsson, O., Rosen, I., and Uddenberg, N. (1963). Organization of ascending tracts in the spinal cord of the duck. *Acta Physiol. Scand.* **59**: 143–153.

Owen, R. (1842). Monograph on *Apteryx australis*, including its myology. *Proc. zool. Soc. Lond.* 1842: 22–41.

Owen, R. (1866). "On the anatomy of vertebrates", Vol. 2. Longmans Green, London.

Paik, Y. K., Fujioka, T. and Yasuda, M. (1974). Division of pancreatic lobes and distribution of pancreatic ducts. *Jap. J. vet. Sci.* **36**: 213–229.

Paik, Y. K., Nishida, T., and Yasuda, M. (1969). The blood vascular system of the pancreas in the fowl [in Japanese]. *Jap. J. Vet. Sci.* **31**: 241–251.
Palmgren, P. (1949). Zur biologischen Anatomie der Halsmuskulatur der Singvogel. *In* "Ornithologie als Biologische Wissenschaft" (E. Mayr and E. Schuz, eds). Carl Winter, Heidelberg.
Papez, J. W. (1935). Thalamus of turtles and thalamic evolution. *J. Comp. Neurol.* **61**: 433–475.
Parker, W. K. (1888). On the vertebral chain of birds. *Proc. Roy. Soc.* **43**: 465–482.
Patzelt, V. (1936). Der Darm. *In* "Handbuch der microskopischen Anatomie des Menschen" (W. von Möllendorf, ed.), Vol. 5, Part III. Springer-Verlag, Berlin.
Pavaux, C. and Jolly, A. (1968). Note sur la structure vasculo-canaliculaire du foie des oiseaux domestiques. *Rev. Méd. Vét.* **119**: 445–466.
Payne, D. C. and King, A. S. (1959). Is there a vestibule in the lung of *Gallus domesticus*? *J. Anat. (Lond.)* **93**: 577.
Payne, D. C. and King, A. S. (1960). The lung of *G. domesticus*: secondary bronchi. *J. Anat. (Lond.)* **94**: 292–293.
Payne, L. N. (1971). The lymphoid system. *In* "Physiology and Biochemistry of the Domestic Fowl" (D. J. Bell and B. M. Freeman, eds). Academic Press, London.
Peaker, M. and Linzell, J. L. (1975). "Salt Glands in Birds and Reptiles". Monographs of the Physiological Society No. 32. Cambridge University Press, Cambridge.
Pearson, R. (1972). "The Avian Brain". Academic Press, London and New York.
Pelham, R. W., Ralph, C. L. and Campbell, I. M. (1972). Mass spectral identification of melantonin in blood. *Biochem. Biophys. Res. Commun.* **46**: 1236–1242.
Pelissier, M. (1923). L'appareil ligamentaire des rémiges des oiseaux. *Arch. d'Anat. Hist. Embryol.* (Strasbourg) **2**: 307–341.
Perlia, R. (1898). Über ein Opticuscentrum beim Huhne. *Arch. f. Ophth.* **35**: 20–24.
Pernkopf, E. (1930). Beiträge zur vergleichenden Anatomie des Vertebratenmagens. *Z. Anat. EntwGesch.* **91**: 329–390.
Pernkopf, E. and Lehner, J. (1937). Vorderdarm. Vergleichende Beschreibung des Vorderdarm bei den einzelnen Klassen der Kranioten. *In* "Handbuch der vergleichende Anatomie der Wirbeltiere" (L. Bolk, E. Göppert, E. Kallius and W. Lubosch, eds), Vol. 3. Urban & Schwarzenberg, Berlin and Vienna.
Petit, M. (1933). Péritoine et cavité péritoneale chez les oiseaux. *Rev. vet. J. Med. vet.* **85**: 376–382.
Petren, T. (1926). Die Coronärarterien des Vogelherzens. *Morph. Jb.* **56**: 239–249.
Petrovicky, P. (1966). Reticular formation of the pigeon. *Folia Morph.* **14**: 334–346.
Petry, G. (1951). Über die Formen und die Verteilungen elastisch-muskuloser Verbindungen in der Haustaube. *Morph. Jahrb.* **91**: 511–535.
Pilar, G. and Vaughan, P. C. (1971). Ultrastructure and contractures of the pigeon iris striated muscle. *J. Physiol.* **219**: 253–266.
Pilz, H. (1937). Artmerkmale am Darmkanal des Hausgeflügels (Gans, Ente, Huhn, Taube). *Morph. Jb.* **79**: 275–304.
Pintea, V., Constantinescu, G. M. and Radu, C. (1967). Vascular and nervous supply of bursa of Fabricius in the hen. *Acta Vet. Acad. Sci. Hung.* **17**: 263–268.
Pintea, V. and Rizkalla, W. (1967). Lympho-epithelial and glomic structures in the upper wall of the cloaca in the hen. *Acta vet. Acad. Scient. Hung.* **17**: 249–255.
Pitelka, F. A. (1945). Pterylography, molt, and age determinations of American jays of the genus *Aphelocoma*. *Condor* **47**: 229–261.
Pitman, C. R. S. (1964). Eggs, natural history of. *In* "New Dictionary of Birds" (A. L. Thomson, ed.). Nelson, London.
Plate, L. H. (1918). Über Drüsen und Lymphknoten in der Ohrfalte der Truthenne und des Auerhahns. *Arch. mikr. Anat.* **9**: 208–217.

Pohlman, A. G. (1921). The position and functional interpretation of the elastic ligaments in the middle-ear region of *Gallus*. *J. Morph.* **35**: 229–262.

Poole, M. (1909). The development of the subdivisions of the pleuroperitoneal cavity in birds. *Proc. zool. Soc. Lond.* **77**: 210–235.

Porta, A. (1908). I muscoli caudale e anali nei generi *Pavo* e *Meleagris*. *Zool. Anz.* **33**: 116–120.

Portmann, A. (1950). Les organes respiratoires. *In* "Traité de Zoologie" (P.-P. Grasse, ed.), Vol. 15: Oiseaux. Masson, Paris.

Portmann, A. (1950). Squelette. *In* "Traité de Zoologie" (P.-P. Grasse, ed.), Vol. 15: Oiseaux. Masson, Paris.

Portmann, A. (1961). Sensory organs: Part I. Skin, Taste and Olfaction. Part II. Equilibrium. *In* "Biology and Comparative Physiology of Birds" (A. J. Marshall, ed.), Vol. 2. Academic Press, New York and London.

Prakash, R. (1956). The heart and its conducting system in the common Indian fowl. *Proc. Nat. Inst. Sci. India* **22**: 22–27.

Prakash, R., Bhatnagar, S. P. and Yousuf, N. (1960). The development of the conducting tissue in the heart of chicken embryos. *Anat. Rec.* **136**: 322 (abstract).

Preuss, F. and Rautenfeld, D. G. von (1974). Umstrittenes zur Anatomie der Bursa cloacae, der Papilla vaginalis und des Phallus femininus beim Huhn. *Berl. Münch Tierärztl. Wschr.* **87**: 456–458.

Prochazkova, E. and Komárek, V. (1970). Growth of the zona vasculosa and zona parenchymatosa in postnatal development of the ovary in the chicken. *Acta Vet. Brno* **39**: 3–10.

Pumphrey, R. J. (1961). Sensory organs: *In* "Biology and Comparative Physiology of Birds" (A. J. Marshall, ed.), Vol. 2. Academic Press, New York and London.

Pusstilnik, E. (1937). Zum Problem der Innervation der Beckenorgane der Wirbeltiere. *Anat. Anz.* **84**: 106–112.

Quay, W. B. (1967). Comparative survey of the anal glands of birds. *Auk* **84**: 379–389.

Quay, W. B. (1970). The significance of the pineal. *In* "Hormones and Environment" (G. K. Benson and J. C. Phillips, eds). *Mem. Soc. Endocr.* No. **18**: 423–445.

Quiring, D. P. (1933–34). The development of the sino-atrial region of the chicken heart. *J. Morph.* **55**: 81–118.

Quitzow, H. (1970). "Die Bronchen der Hühnerlunge". Inaug. Diss., Freien University Berlin.

Radu, C. and Radu, L. (1971). Le dispotif vasculaire du poumon chez les oiseaux domestiques (coq, dindon, oie, canard). *Rev. Méd. vét.* **122**: 1219–1226.

Radu, C. (1975). Les fosses renales des oiseaux domestiques (*Gallus domesticus*, *Meleagris gallopavo*, *Anser domesticus* et *Anas platyrhynchos*. *Anat. Histol. Embryol.* **4**: 10–23.

Raikow, R. (1970). Evolution of diving adaptations in the stifftail ducks *Univ. Calif. Publ. Zool.* **94**: 1–52.

Raikow, R. (1975). The evolutionary reappearance of ancestral muscles as developmental anomalies in two species of birds. *Condor* **77**: 514–517.

Raikow, R. (1976). Pelvic appendage myology of the Hawaiian honeycreepers (Drepanididae). *Auk* **93**: 774–792.

Raikow, R. (1977). Pectoral appendage myology of the Hawaiian honeycreepers (Drepanididae). *Auk* **94**: 331–342.

Rautenfeld, D. B. von (1973). "Zur Form und Funktion des Kopulations-organes beim Haushuhn (*Gallus domesticus*)." Inaug. Diss., Freien University, Berlin.

Rautenfeld, D. B. von, Preuss, F. and Fricke, W. (1974). Neue Daten zur Erektion und Reposition des Erpelphallus. *Prakt. Tierarzt* **10**: 553–556.

Rawal, U. M. (1970). A comparative account of the lingual myology of some birds. *Proc. Indian Acad. Sci.* **71**: 36–46.

Rawal, U. M. (1971). Adaptations for food getting in the House Swift. *Proc. Indian Acad. Sci.* **73**: 224–235.

Rawal, V. M. and Bhatt, P. L. (1973). Myology of feeding apparatus of common grey hornbill, *Tockus birostris* Scopoli. *Vidya (B. Sciences)* **16**: 77–85.

Rawles, M. E. (1960). The integumentary system. *In* "Biology and Comparative Physiology of Birds" (A. J. Marshall, ed.), Vol. 1. Academic Press, New York and London.

Rendahl, H. (1924). Embryologische und morphologische Studien über das Zwischenhirn beim Huhn. *Acta Zool.* **5**: 241–344.

Renggli, F. (1967). Vergleichend anatomische Untersuchungen über die Kleinhirn und Vestibulariskerne der Vogel. *Rev. Suisse Zool.* **74**: 701–778.

Reviers, de M. (1971a). Le développement testiculaire chez le Coq. I. Croissance des testicules et développement des tubes séminiferes. *Annls. Biol. anim. Biochim. Biophys.* **11**: 519–530.

Reviers, de M. (1971b). Développement testiculaire chez le Coq. II. Morphologie de l'epithelium séminifere et établissement de la spermatogénese. *Annls. Biol. anim. Biochim. Biophys.* **11**: 531–546.

Revzin, A. M. (1967). Unit responses to visual stimuli in the nucleus rotundus of the pigeon (*Columbia livia*). *Fed. Proc.* **26**: 656 (abstract).

Revzin, A. M. (1969). A special visual projection area in the hyperstriatum of the pigeon (*Columbia livia*). *Brain Res.* **15**: 246–249.

Revzin, A. M. and Karten, H. J. (1966). Rostral projections of the optic tectum and the nucleus rotundus in the pigeon. *Brain Res.* **3**: 264–276.

Rexed, B. (1952). The cytoarchitectonic organization of the spinal cord in the cat. *J. Comp. Neurol.* **96**: 415–495.

Rexed, B. (1954). A cytoarchitectonic atlas of the spinal cord in the cat. *J. Comp. Neurol.* **100**: 297–379.

Richards, L. P. and Bock, W. J. (1973). Functional anatomy and adaptive evolution of the feeding apparatus of the Hawaiian honeycreeper genus *Loxops* (Drepaniidae). *A.O.U. Ornith. Monogr.* **15**: 1–173.

Richards, S. A. (1967). Anatomy of the arteries of the head in the domestic fowl. *J. Zool. (Lond.)* **152**: 221–234.

Richards, S. A. (1968). Anatomy of the veins of the head in the domestic fowl. *J. Zool. (Lond.)* **154**: 223–234.

Richardson, K. C. (1935). The secretory phenomena in the oviduct of the fowl, including the process of shell formation examined by microincineration technique. *Phil. Trans. Roy. Soc.* **225B**: 149–195.

Rigdon, R. H. and Frolich, J. (1970). The heart of the duck. *Zbl. Veterinärmed.* (Reihe A) **17**: 85–94.

Robb, J. S. (1965). "Comparative Basic Cardiology". Grune & Stratton, New York.

Robin, Ch. & L. Chabry (1884). Note sur les organes elastiques de l'aile des oiseaux. *J. Anat. Physiol.* (Paris) **20**: 291–316.

Rochon-Duvigneaud, A. (1950). Les yeux et la vision. *In* "Traité de Zoologie", Vol. 15: Oiseaux, Masson, Paris.

Romanoff, A. L. and Romanoff, A. J. (1949). "The Avian Egg". Wiley, New York.

Romanoff, A. L. (1960). "The Avian Embryo". Macmillan, New York.

Romer, A. S. (1927). The development of the thigh musculature of the chick. *J. Morph. Morph. Physiol.* **43**: 347–385.

Romer, A. S. (1962). "The Vertebrate Body", 3rd Edn. Saunders, Philadelphia.

Rooth, J. (1953). On the correlation between the jaw muscles and the structure of the skull in *Columba palumbus L. 1. Koninkl. Nederl. Akad. v. Wetensch* (Amsterdam) Series C, **56**, 251–264.

Röse, C. (1890). Beiträge zur vergleichenden Anatomie des Herzens der Wirbelthiere. *Morph. Jb.* **16**: 26–96.

Rothwell, B. and Tingari, M. D. (1973). The ultrastructure of the boundary tissue of the seminiferous tubule in the testis of the domestic fowl (*Gallus domesticus*). *J. Anat.* **114**: 321–328.

Rüppell, W. (1933). Physiologie und Akustik der Vogelstimme. *J. Ornith.* **81**: 433–542.

Saito, I. (1966). Comparative anatomical studies of the oral organs of poultry. IV. Macroscopical observations of the salivary glands. *Bull. Fac. Agric. Univ. Miyazaki* **12**: 110–120.

Salt, R. (1954). The structure of the cloacal protuberance of the Vesper Sparrow (*Pooecetes gramineus*) and certain other passerine birds. *Auk* **71**: 64–73.

Salt, G. W. and Zeuthen, E. (1960). The respiratory system. *In* "Biology and Comparative Physiology of Birds" (A. J. Marshall, ed.), Vol. I, Chap. 10. Academic Press, New York and London.

Sanders, E. B. (1929). A consideration of certain bulbar, midbrain and cerebellar centres and fiber tracts in birds. *J. Comp. Neurol.* **49**: 155–222.

Sandoval, J. (1963). Estudio sobre la anatomía comparada y functional del esqueleto cefálico de la gallina y su morfogenesis. *Anales de Anat.* **12**: 283–359.

Sandoval, J. (1964). La organizacion de la cavidad nasal de la gallina con especial referencia a su desarrollo olfactorio. *Anales de Anat.* **13**: 249–258.

Santamaría-Arnáiz, P. (1962). Untersuchungen über die parasympathischen Kopfganglien beim Sperling (*Passer domesticus*). *Morph. Jb.* **103**: 85–107.

Sappey, P. (1847). "Recherches sur l'appareil respiratoire des oiseaux". Baillière, Paris.

Sarnat, H. B. and Netsky, M. G. (1974). "Evolution of the Nervous System". Oxford University Press, London.

Schauder, W. (1923). Anatomie der Hausvogel. *In* "Martin's Lehrbuch der Anatomie der Haustiere". Schickhardt & Ebner, Stuttgart.

Schepelmann, E. (1906). Über die gestaltende Wirkung verschiedener Ernährung auf die Organa der Gans, insbesondere über die funktionelle Anpassung an die Nahrung. *Arch. EntwMech. Org.* **21**: 500–595.

Scheschin, J. (1925). "Beitrag zur Histologie der Vogelschilddrüse". Inaug. Diss.. Wien.

Schmidt, R. S. (1964). Blood supply of pigeon inner ear. *J. comp. Neurol.* **123**: 187–204.

Schneider, A. (1931). Über den Kopfanhang des Truthuhns (*Meleagris gallopavo* L.), *J. Ornith.* **79**: 236–255.

Schrader, E. (1970). "Die Topographie der Kopfnerven vom Huhn". Inaug. Diss., Freien University, Berlin.

Schulze, F. E. (1908). Die Lungen des afrikanischen Strausses. *Sitzber. Preuss. Akad. Wiss.*: 416–431.

Schulze, F. E. (1910). "Uber die Luftsäcke der Vögel". *Verhandl. 8th Internat. Zool. Kongr. Graz*: 446–482.

Schumacher, S. (1919). Der Bürzeldocht. *Anat. Anz.* **52**: 291–301.

Schummer, A. (1973). *In* "Lehrbuch der Anatomie der Haustiere" (R. Nickel, A. Schummer and E. Seiferle, eds), Vol. 5. Parey, Berlin.

Schüz, E. (1927). Beitrag zur Kenntnis der Puderbildung bei den Vögeln. *J. Ornith.* **75**: 86–223.

Schwalbe, G. (1879). Das Ganglion oculomotorii, ein Beitrag zur vergleichenden Anatomie der Kopfnerven. *Jena. Z. Naturwiss.* **13**: 173–268.

Schwartzkopff, J. (1955). On the hearing of birds. *Auk* **72**: 340 347.

Schwartzkopff, J. (1968). Hearing in birds. *Ciba Symposium*: 4–59.

Schwartzkopff, J. (1973). Mechanoreception. *In* "Avian Biology" (D. S. Farner and J. R. King, eds). Vol. III. Academic Press, London.

Schwartzkopff, J. and Winter, P. (1960). Zur Anatomie der Vogel-Cochlea unter naturlichen Bedingungen. *Biol. Zbl.* **79**: 607–625.

Seichert, V. and Rychter, Z. (1972). Vascularization of developing anterior limb of the chick embryo. II. *Folia Morph.* **20**: 352–361.

Selby, C. C. (1955). An electron microscope study of the epidermis of mammalian skin in thin sections. I. Dermoepidermal junction and basal cell layer. *J. Biophys. Biochem. Cytol.* **1**: 429–444.

Sengel, P. (1976). "Morphogenesis of Skin." Cambridge University Press. London and New York.

Sell, J. (1959). Incidence of persistent right oviducts in the chicken. *Poult. Sci.* **38**: 33–35.

Setterwall, C. G. (1901). Studier öfver Syrinx hos Polymyoda Passeres. *Akad. Lund. Afhandl.* 128 pp.

Shah, R. V. and Menon, G. K. (1972—Publ. 1975). Histochemical studies on pigeon definitive feathers during post-hatching, induced and regenerative modes of development: III. Lipids, lipase and β-hydroxy butyrate dehydrogenase. *Pavo* **10**: 30–42.

Shah, R. V., Menon, G. K., Desai, J. H. and Jani, M. B. (1977). Feather loss from capital tracts of Painted Storks related to growth and maturity: 1. Histophysiological changes and lipoid secretion in the integument. *J. Anim. Morphol. Physiol.* **24**: 99–107.

Shaner, R. F. (1923). On the muscular architecture of the vertebrate ventricle. *J. Anat.* **58**: 59–70.

Sharp, P. J. and Follett, B. K. (1969). The blood supply to the pituitary and basal hypothalamus in the Japanese quail (*Coturnix coturnix japonica*). *J. Anat.* **104**: 227–232.

Shiina, J. and Miyata, D. (1932). Studies on the cerebral arteries of birds. I. The arterial supply on the brain surface of some kinds of birds. *Acta anat. nipp.* **5**: 13–38.

Shufeldt, R. W. (1890). The Myology of the Raven (*Corvus corax sinuatus*)". Macmillan, London.

Shufeldt, R. W. (1909). "Osteology of birds". *New York State Museum Bull.* **130**: University of State of N.Y., Albany.

Sick, H. (1937). Morphologisch-funktionelle Untersuchungen über die Feinstruktur der Vogelfeder. *J. Ornith.* **85**: 206–372.

Siller, W. G. (1971). Structure of the kidney. *In* "Physiology and Biochemistry of the Domestic Fowl" (D. J. Bell and B. M. Freeman, eds), Vol. 1. Academic Press, London.

Siller, W. G. and Hindle, R. M. (1969). The arterial blood supply to the kidney of the fowl. *J. Anat.* **104**: 117–135.

Simić, V. and Andrejevic, V. (1963). Morphologie und Topographie der Brustmuskeln bei den Hausphasianiden und der Taube. *Morph. Jb.* **104**: 546–560.

Simić, V. and Andrejevic, V. (1964). Morphologie und Topographic der Brustmuskeln bei den Hausschwimmvögeln. *Morph. Jb.* **106**: 480–490.

Simić, V. and Jablan-Pantič, O. (1959). Morphologischer Beitrag über den Mechanismus des dritten Augenlids bei den Hausvögeln. *Anat. Anz.* **106**: 76–85.

Simić, V. and Janković, N. (1959). Ein Beitrag zur Kenntnis der Morphologie und Topographie der Leber beim Hausflügel und der Taube. *Acta vet.* (Beogr.) **9**: 7–34.

Simons, P. C. M. (1971). "Ultrastructure of the Hen Eggshell and its Physiological Interpretation". *Central Institute for Poultry Research 'Het Spelderholt'*, Communication no. 175. Beekbergen, Netherlands.

Singh, R. M. and Dominic, C. J. (1970). Distribution of the portal vessels of the avian pituitary in relation to the median eminence and the pars distalis. *Experientia* **26**: 962–964.

Slonaker, J. R. (1918). A physiological study of the anatomy of the eye and its accessory parts of the English Sparrow (*Passer domesticus*). *J. Morph.* **31**: 351–459.

Slonaker, J. R. (1922). The development of the eye and its accessory parts in the English Sparrow (*Passer domesticus*) *J. Morph.* **35**: 263–357.

Smith, G. (1904–05). The middle ear and columella of birds. *Quart. J. Micros. Sci.* (Lond.) **48**: 11–22.

Smith, H. W. (1956). "Principles of Renal Physiology". Oxford University Press, New York.

Smith, P. H. (1974). Pancreatic islets of the Coturnix Quail. A light and electron microscope study with special reference to the islet organ of the splenic lobe. *Anat. Rec.* **178**: 567–586.

Smith, R. B. (1971). Observations on nerve cells in human, mammalian, and avian cardiac ventricles. *Anat. Anz.* **129**: 436–444.

Spanner, R. (1925). Der Pfortaderkreislauf in der Vogelniere. *Morph. Jb.* **54**: 560–696.

Spatz, H., Doepen, R. and Gaupp, V. (1948). Zur Anatomie des Infundibulum und des Tuber cinereum beim Kanichen. *Dtsch. Z. Nervenheilk* **159**: 229–268.

Spearman, R. I. C. (1966). The keratinization of epidermal scales, feathers and hairs. *Biol. Rev.* **41**: 59–96.

Spearman, R. I. C. (1969). The epidermis and feather follicles of the King Penguin (*Aptenodytes patagonica*) (Aves). *Z. Morph. Tiere* **64**: 361–372.

Sperber, I. (1944). Studies of the mammalian kidney. *Zool. Bidr. Uppsala* **22**: 252–431.

Sperber, I. (1960). Excretion. *In* "Biology and Comparative Physiology of Birds" (A. J. Marshall, ed.), Vol. 1. Academic Press, New York and London.

Ssinelnikow, R. (1928). Die Herznerven der Vogel. *Z. Anat. Entwickl.* **86**: 540–562.

Staderini, R. (1889). Sopra la distribuzione dei nervi glosso-faringeo vago e ipoglosso in alcuni rettili ed uccelli. *Att. Accad. Fisiocrit. Siena* **1**: 585–599.

Starck, D. (1955). Die endokraniale Morphologie der Ratiten, besonders der Apterygidae und Dinornithidae. *Morph. Jb.* **96**: 14–72.

Starck, D. and Barnikol, A. (1954). Beiträge zur Morphologie der Trigeminusmuskulatur der Vögel. *Morph. Jb.* **94**: 1–64.

Stein, R. C. (1968). Modulation in bird sounds. *Auk* **85**: 229–243.

Steinbacher, G. (1935). Funktionell-anatomische Untersuchungen an Vogelfüssen mit Wendezehen und Rückzehen. *J. Ornith.* **83**: 214–282.

Steiner, H. (1917). "Das Problem der Diastataxie des Vogelflügels". Gustav Fischer, Jena.

Steiner, H. (1922). Die ontogenetische und phylogenetische Entwicklung des Vogelflügelskelettes. *Acta Zool.* **3**: 307–359.

Steiner, H. (1938). Der *Archaeopteryx*-Schwanz der Vogelembryonen. *Vjschr. Naturf. Ges. Zürich Beibl.* 30, Festschr. Karl Hescheler **83**: 279–300.

Stellbogen, E. (1930). Über das äussere mittlere Ohr des Waldkauzes (*Syrnium aluco* L.). *Z. Morph. Ökol. Tiere* **19**: 686–731.

Stephan, B. (1970). Eutaxie, Diastataxie, und andere Probleme der Befiederung des Vogelflügels. *Mitt. Zool. Mus. Berlin* **46**: 339–437.

Sterzi, G. (1903). I vasi sanguigni della midolla spinale degle uccelli. *Arch. Ital. Anat. Embriol.* **2**: 216–236.

Stettenheim, P. R. (1959). "Adaptations for underwater swimming in the common murre (*Uria aalge*)". *Ph.D. Diss., University of Michigan.*

Stettenheim. P. (1972). The integument of birds. In "Avian Biology" (D. S. Farner and J. R. King, eds), Vol. II. Academic Press, London.

Stettenheim, P. (1974). The bristles of birds. *Living Bird*, pp. 201–234, *Ann. Cornell Lab. Ornith.*, Cornell University.

Stibbe, E. P. (1928). A comparative study of the nictitating membrane of birds and mammals. *J. Anat.* **62**: 159–176.

Stingelin, W. (1961). Grossenunterschiede des sensiblen Trigeminuskerns bei verschiedenen Vögeln. *Rev. Suisse Zool.* **68**: 247–251.

Stoeckel, M. E. and Porte, A. (1970). A comparative electron microscopic study of the fowl, pigeon and the turtle-dove of the C cells localized in the ultimobranchial body and the thyroid. *In* "Calcitonin 1969" (S. Taylor and G. Foster, ed). Heineman, London.

Stoll, R. and Maraud, R. (1955). Sur la constitution de l'épididyme du coq. *C.R. Séanc. Soc. Biol.* **149**: 687–689.

Stolpe, M. (1932). Physiologisch-anatomische Untersuchungen über die hintere Extremität der Vögel. *J. Ornith.* **80**: 161–247.

Stork, H. J. (1972). Zur Entwicklung pneumatischer Räume im Neurocranium der Vögel (Aves). *Z. Morph. Tiere* **73**: 81–94.

Stotler, W. A. (1905). Further studies of the bulbar acoustic centres. *Anat. Rec.* **10**: (abstract): 349.

Stresemann, E. (1927–34). Sauropsida: Aves, Bd. 7(2). *In* "Handbuch der Zoologie" (W. Kükenthal and T. Krumbach, eds). de Gruyter, Berlin and Leipzig.

Sturkie, P. D. (1976). "Avian Physiology", 3rd Edn. Springer, New York and Berlin.

Sugimura, M., Kudo, N. and Yamano, S. (1975). Fine structure of corpus paracloacalis vascularis in cocks. *Jap. J. vet. Res.* **23**: 11–16.

Sullivan, G. E. (1962). Anatomy and embryology of the wing musculature of the domestic fowl (*Gallus*). *Austral. J. Zool.*, **10**: 458–518.

Surface, F. M. (1912). The histology of the oviduct of the domestic hen. *Bull. Maine agric. Exptl. Sta.* **206**: 397–430.

Suschkin, P. P. (1899). Zur Morphologie des Vogelskelets. 1. Schädel von *Tinnunculus*. *Nov. Mem. Soc. Imperiale Nat. de Moscou.* XVI- 1–163.

Susi, F. R. (1969). Keratinization in the mucosa of the ventral surface of the chicken tongue. *J. Anat.* **105**: 477–486.

Swenander, G. (1899). Beiträge zur Kenntnis des Kropfes der Vögel. *Zool. Anz.* **22**: 140–142.

Swenander, G. (1902). Studien über den Bau des Schlundes und des Magens der Vögel. *K. norsk. Vidensk. Selsk. Skr.* **6**: 1–240.

Sy, M. (1936). Functionell-anatomische Untersuchungen am Vogelflügel. *J. Ornith.* **84**: 199–296.

Szabō, L. (1958). A. hullámos papagáj (*Melopsittacus undulatus*) érrendszere. [The vascular system of the Australian Lovebird (*Melopsittacus undulatus*).] Thesis: Anatomy, Veterinary Medicine, Budapest.

Tamura, T. and Fujii, S. (1967). Studies on the cloacal gland of the quail. I. Macroscopical and microscopical observations. *Japan J. poult. Sci.* **4**: 187–193.

Taylor, T. G., Simkiss, K. and Stringer, D. A. (1971). The skeleton: its structure and metabolism. *In*: "Physiology and Biochemistry of the Domestic Fowl" (D. J. Bell and B. M. Freeman, eds). Academic Press, London and New York.

Tcheng, K. T., Fu, S. K. and Chen, T. Y. (1963a). On the vasculature of the aortic bodies in birds. *Scientia Sinica* **12**: 339–345.

Tcheng, K. T., Fu, S. K. and Chen, T. Y. (1963b). Supracardial encapsulated receptors of the aorta and pulmonary artery in birds. *Scientia Sinica* **12**: 73–81.

Technau, G. (1936). Die Nasendrüse der Vögel zugleich ein Beitrag zur Morphologie der Nasenhöhle. *J. Ornith.* **84**: 511–617.

Terni, T. (1923). Richerche anatomiche sul sistema nervosa autonomia degli uccelli. *Arch. Ital. Anat. Embriol.* **20**: 433–510.

Terni, T. (1924). Richerche sulla considetta sostanza gelatinosa (corpo glicogenico) del midollo longo-sacrale degli uccelli. *Arch. Ital. Anat. Embriol.* **21**: 55–86.

Terni, T. (1929). Recherches morphologiques sur le sympathique cervical des oiseaux et sur l'innervation autonome de quelques organes glandulaires du cou. *C.R. Assoc. Anat.* **24**: 473–480.

Thebault, V. (1898). Étude des rapports qui existent entre les systemes pneumogastrique et sympathique chez les oiseaux. *Ann. Sci. Natur. Zool.* **6**: 1–243.

Thomson, A. L. (ed.) (1964). "A New Dictionary of Birds". Nelson, London.

Tiemeier, O. W. (1950). The os opticus of birds. *J. Morph.* **86**: 25–46.

Tiemeier, O. W. (1953). The embryogeny of the os opticus in the English Sparrow, *Passer domesticus. Trans. Kansas Acad. Sci.* **56**: 440–448.

Tingari, M. D. (1971). On the structure of the epididymal region and ductus deferens of the domestic fowl (*Gallus domesticus*). *J. Anat.* **109**: 425–435.

Tingari, M. D. (1973). Observations on the fine structure of spermatozoa in the testis and excurrent ducts of the male fowl, *Gallus domesticus. J. Reprod. Fert.* **34**: 255–265.

Tixier-Vidal, A. and Follett, B. K. (1973). The adenohypophysis. *In* "Avian Biology" (D. S. Farner and J. R. King, eds). Vol. 3. Academic Press, London.

Tracuic, E. (1967). L'anatomie microscopique de l'épididyme chez *Sterna hirundo. Anat. Anz.* **121**: 381–386.

Tracuic, E. (1969). La structure de l'epididyme de *Coloeus monedula* (Aves, Corvidae). *Anat. Anz.* **125**: 49–67.

Trotter, M. and Peterson, R. R. (1966). Arthrology. *In* "Morris' Human Anatomy" (B. J. Anson, ed.), 12th Edn. McGraw-Hill, New York.

Tucker, R. (1975). The surface of the pectin oculi in the pigeon. *Cell Tiss. Res.* **157**: 457–465.

Turkewitsch, B. G. (1936). Die Abhängigkeit der anatomischen Struktur des knöchernen inneren Ohres von der Körperlage der Vögel. *Zool. Jb., Abt. Anat. Ont.* **61**: 121–138.

Uchiyama, T. (1928). Zur Frage der Vv. minimae thebesii und der Sinusoide beim Hühnerherzen. *Morph. Jahrb.* **60**: 196–322.

Vallancien, B. (1963), Comparative anatomy and physiology of the auditory organ in vertebrates. *In* "Acoustic Behaviour of Animals" (R. G. Busnel, ed.). Elsevier, Amsterdam.

Vanden Berge, J. C. (1970). A comparative study of the appendicular musculature of the Order Ciconiiformes. *Amer. Midl. Nat.* **84**: 289–364.

Vanden Berge, J. C. (1975). Aves myology. *In* "Sisson and Grossman's The Anatomy of the Domestic Animals" (R. Getty, ed.), 5th Edn. Saunders, Philadelphia.

Vanden Berge, J. C. (1976). M. iliotibialis medialis and a review of the M. iliotibialis complex in flamingos. *Auk* **93**: 429–433.

van Oort, E. D. (1905). Beitrag zur Osteologie des Vogelschwanzes. *Tijdschr. nederl. dierk. Ver.* **9**: 1–144.

Van Tyne, J. and Berger, A. J. (1959). "Fundamentals of Ornithology". John Wiley, New York.

Van Tyne, J. and Berger, A. J. (1976). "Fundamentals of Ornithology", 2nd Edn. John Wiley, New York.

Vitali, G. (1912). Di un interessante derivato dell' ectoderma della prima fessura branchiale nervoso di sonso nell' orecchio medio degli uccelli. *Anat. Anz.* **40**: 631–639.

Vitums, A., Mikami, S.-I., Oksche, A. and Farner, D. S. (1964). Vascularization of the hypothalamo-hypophysial complex in the White-Crowned Sparrow, *Zonotrichia leucophrys gambelii. Z. Zellforsch.* **64**: 541–569.

Vitums, A., Mikami, S-I. and Farner, D. S. (1965). Arterial blood supply to the brain of the White-crowned sparrow, *Zonotrichia leucophrys gambelii. Anat. Anz.* **116**: 309–326.

Vogt-Nilsen, L. (1954). The inferior olive in birds, a comparative morphological study. *J. Comp. Neurol.* **101**: 447–481.

Voitkevich, A. A. (1966). "The Feathers and Plumage of Birds". Sidgwick and Jackson, London.

Vollmerhaus, B. and Hegner, D. (1963). Korrosionsanatomische Untersuchungen am Blutgefässsystem der Hühnerfusses. *Morph. Jb.* **105**: 139–184.

Vos, H. J. (1934). Über den Weg der Atemluft in der Entenlunge. *Z. Vergl. Physiol.* **21**: 552–578.

Wade, C. H. (1876). Notes on the venous system of birds. *J. Linn. Soc. Zool.* **12**: 531–535.

Wallenberg, A. (1903). Der Ursprung des Tractus isthmo-striatus (oder bulbostriatus) der Taube. *Neurol. Zbl.* **22**: 98–101.

Wallenberg, A. (1904). Neue Untersuchungen über den Hirnstamm der Taube. *Anat. Anz.* **24**: 357–369.

Walls, G. L. (1942). "The Vertebrate Eye and Its Adaptive Radiation". Cranbrook Institute of Science, Bloomfield Hills, Michigan.

Waluszewska-Bubień, A. (1972). Topography of the vagus nerve in the domestic hen (*Gallus Gallus F. Domestica L.*) *Zool. Poloniae* **22**: 5–42.

Warner, R. W. (1971). The structural basis of the organ of voice in the genera *Anas* and *Aythya* (Aves). *J. Zool.* (*Lond.*) **164**: 197–207.

Warner, R. W. (1972a). The syrinx in family Columbidae. *J. Zool.* (*Lond.*) **166**: 385–390.

Warner, R. W. (1972b). The anatomy of the syrinx in passerine birds. *J. Zool.* (*Lond.*) **168**: 381–393.

Warwick, R. (1978). The future of *Nomina Anatomica*—a personal view. *Anat. Rec.* **190**: 1–4.

Watanabe, T. (1960). On the peripheral courses of the vagus nerve in the fowl. *Jap. J. Vet. Sci.* **22**: 145–154.

Watanabe, T. (1964). Peripheral courses of the hypoglossal, accessory and glossopharyngeal nerves. *Jap. J. Vet. Sci.* **26**: 249–258.

Watanabe, T. (1968). A study of retrograde degeneration in the vagal nuclei of the fowl. *Jap. J. Vet. Sci.* **30**: 331–340.

Watanabe, T. (1972). Sympathetic nervous system of the fowl. Part 2. Nervus intestinalis. *Jap. J. vet. Sci.* **34**: 303–313.

Watanabe, T., Isomura, J. and Yasuda, M. (1967). Distribution of nerves in the oculomotor and ciliary muscles. *Jap. J. Vet. Sci.* **29**: 151–158.

Watanabe, T. and Paik, Y. K. (1973). Sympathetic nervous system of the fowl. Part 3. Plexus celiacus and plexus mesentericus cranialis. *Jap. J. Vet. Sci.* **35**: 389–401.

Watanabe, T. and Yasuda, M. (1968). Peripheral course of the olfactory nerve in the fowl. *Jap. J. Vet. Sci.* **30**: 275–279.

Watanabe, T. and Yasuda, M. (1970). Peripheral course of the trigeminal nerve. *Jap. J. Vet. Sci.* **32**: 43–57 [in Japanese].

Watson, M. (1883). "Voyage of H.M.S. Challenger: Zoology" (C. W. Thomson and J. Murray, eds), Vol. 7. Longmans, London.

Watzka, M. (1933). Vergleichende Untersuchungen über den ultimobranchialen Körper. *Z. mikr. anat. Forsch.* **34**: 485–533.

Webster, H. D. (1948). The right oviduct in chickens. *Amer. vet. Med. assoc.* **112**: 221–223.

Welsch, U. and Wachtler, K. (1969). Feinbau des Glykogenkorpers im Ruckenmark der Taube. *Z. Zellforsch.* **97**: 160–169.

Welty, J. C. (1962). "The Life of Birds". Saunders, Philadelphia.

Westpfahl, U. (1961). Das Arteriensystem das Haushuhnes (*Gallus domesticus*). *Wiss. Z. Humboldt-Univ. Berlin, Math.-Nat. R.* **10**: 93–124.

Wetmore, A. (1918). A note on the tracheal air-sac in the Ruddy Duck. *Condor* **20**: 19–20.

Weymouth, R. D., Lasiewski, R. C. and Berger, A. J. (1964). The tongue apparatus in hummingbirds. *Acta anat.* **58**: 252–270.

White, S. S. (1970). "The larynx of "*Gallus domesticus*". Ph.D. Thesis, University of Liverpool, England.

White, S. S. (1975). The larynx. *In* "Sisson and Grossman's The Anatomy of Domestic Animals" (R. Getty, ed.), Vol. 2, 5th Edn. Saunders, Philadelphia.

White, S. S. and Chubb, J. C. (1967). The muscles and movements of the larynx of *G. domesticus. J. Anat.* (*Lond.*) **102**: 575.

Whitlock, D. G. (1952). A neurohistological and neurophysiological study of afferent fiber tracts and receptive areas of the avian cerebellum. *J. Comp. Neurol.* **97**: 567–635.

Williamson, J. H. (1965). Cystic remnants of the right Müllerian duct and egg production in two strains of White Leghorns. *Poult. Sci.* **44**: 321–324.

Wingstrand, K. G. (1951). "The Structure and Development of the Avian Pituitary". C. W. K. Gleerup, Lund.

Wingstrand, K. G. and Munk, O. (1965). The pectin oculi of the pigeon with particular regard to its function. *Biol. Skr. Danske Vid. Selsk.* **14**: 1–64.

Wingstand, K. G. (1966). Comparative anatomy of the hypophysis. *In* "The Pituitary Gland" (G. W. Harris and B. T. Donovan, eds), Vol. 1. Butterworths, London.

Win[illegible], H. (1958). Persistent right oviducts in fowls including an account of the his[illegible]ology of the fowl's normal oviduct. *Aust. vet. J.* **34**: 140–147.

Witkovsky, P., Zeigler, H. P. and Silver, R. (1973). A single-unit analysis of the nucleus basalis in the pigeon. *J. Comp. Neurol.* **147**: 119–128.

Witschi, E. (1961). Sex and secondary sexual characters. *In* "Biology and Comparative Physiology of Birds" (A. J. Marshall, ed.), Vol. 2. Academic Press, New York and London.

Wolf, L. (1967). "Das Herz der Vogel", Inaug. Diss., Humboldt-University, Berlin.

Wolfson, A. (1952). The cloacal protuberance—a means for determining breeding condition in live male passerines. *Bird-Banding* **23**: 159–165.

Wolfson, A. (1954). Sperm storage at lower than body temperature outside the body cavity in some passerine birds. *Science* **120**: 68–71.

Wood, C. A. (1917). "The Fundus Oculi of Birds Especially as Viewed by the Ophthalmoscope". Lakeside Press, Chicago.

Wood, C. A. (1924). The Polynesian Fruit Pigeon, *Globicera pacifica*, its food and digestive apparatus. *Auk* **41**: 433–438.

Woodburne, R. T. (1936). A phylogenetic consideration of the primary and secondary centers and connections of the trigeminal complex in a series of vertebrates. *J. Comp. Keurol.* **65**: 403–501.

Wunderlich, L. (1884). Beiträge zur vergleichenden Anatomie und Entwickelungsgeschichte des unteren Kehlkopfes der Vogel. *Nova. Acta der Kgl. Leop.-Carol. Deutschen Akad. der Naturforscher* **48** (1): 1–80.

Wyburn, G. M., Aitken, R. N. C. and Johnston, H. S. (1965). The ultrastructure of the zona radiata of the ovarian follicle of the domestic fowl. *J. Anat.* **99**: 469–506.
Wyburn, G. M. and Baillie, A. H. (1966). Some observations on the fine structure and histochemistry of the ovarian follicle of the fowl. *In* "Physiology of the Domestic Fowl" (C. Horton-Smith and E. C. Amoroso, eds). Oliver and Boyd, Edinburgh.
Wychgram, E. (1912). Über das Ligamentum pectinatum im Vogelauge. Vorlaufige Mittelung. *Arch. vergl. Ophthalmol.* **3**: 22–29.
Yarrell, W. (1833). On the organs of voice in birds. *Trans. Linn. Soc. Lond.* **16**: 305–321.
Yasuda, M. (1960). On the nervous supply of the thoracic limb in the fowl. *Jap. J. Vet. Sci.* **22**: 89–101.
Yasuda, M. (1961). On the nervous supply of the hind limb in the fowl *Jap. J. Vet. Sci.* **23**: 145–155.
Yasuda, M. (1964). Distribution of cutaneous nerves in the fowl. *Jap. J. Vet. Sci.* **26**: 241–254.
Yntema, C. L. and Hammond, W. S. (1954). The origin of intrinsic ganglia of trunk viscera from vagal neural crest in the chick embryo. *J. Comp. Neur.* **101**: 515–542.
Yousuf, N. (1965). The conducting system of the heart of the House Sparrow, *Passer domesticus indicus*. *Anat. Rec.* **152**: 235–250.
Yudin, K. A. (1961). On the mechanism of the jaw in Charadriiformes, Procellarii-formes, and some other birds. *Trudy Zool. Inst. Leningrad* **29**: 257–302 [in Russian].
Yudin, K. A. (1965). The phylogeny and classification of the Charadriiformes. *In* "Fauna of U.S.S.R.", Vol. 2: Birds, part 1 (New Series No. 91). U.S.S.R. Academy of Sciences Zoological Institute [in Russian].
Zecha, A. (1962). The "pyramidal tract" and other telencephalic efferents in birds. *Acta morph. neerlando-scandinavica* **5**: 194–195.
Zeek, P. M. (1951). Double trachea in penguins and sea lions. *Anat. Rec.* **111**: 327–344.
Zeier, H. and Karten, H. J. (1971). The archistriatum of the pigeon: Organization of afferent and efferent connections. *Brain Res.* **31**: 313–326.
Zeier, H. J. and Karten, H. J. (1973). Connections of the anterior commissure in the pigeon (*Columba livia*). *J. Comp. Neurol.* **150**: 201–216.
Zeigler, H. P. and Karten, H. J. (1973). Brain mechanisms and feeding behavior in the pigeon (*Columba livia*). I. Quinto-frontal structures. *J. Comp. Neurol.* **152**: 59–82.
Zietzschmann, O. (1911). Der Verdauungsapparat der Vögel. *In* "Handbuch der vergleichenden mikroscopischen Anatomie der Haustiere" (W. Ellenberger, ed.), Bd. 3, pp. 377–416. Paul Parey, Berlin.
Ziswiler, V. (1965). Zur Kenntnis des Samenöffnens und der Struktur des hörnernen Gaumens bei Körnerfressenden Oscines. *J. Ornith.* **106**: 1–48.
Ziswiler, V. (1967). Vergleichend morphologische Untersuchungen am Verdauung-strakt körnerfressender Singvögel zur Ablärung ihrer systematischen Stellung. *Zool. Jb., Abt. Syst.* **94**: 427–520.
Ziswiler, V. and Farner, D. S. (1962). Digestion and the digestive system. *In* "Avian Biology" (D. S. Farner and J. R. King, eds), Vol. 2. Academic Press, London.
Zusi, R. L. (1959). The role of the depressor mandibulae in certain passerine birds. *Auk* **76**: 537–539.
Zusi, R. L. (1962). "Structural Adaptations of the Head and Neck in the Black Skimmer, *Rynchops nigra*". Publ. No. 3, Nuttall Ornithological Club, Cambridge, Mass.
Zusi, R. L. (1967). The role of the depressor mandibulae muscle in kinesis of the avian skull. *Proc. U.S. Natl. Mus.* **123**:(3607): 1–28.
Zusi, R. L. (1974). An interpretation of skull structure in penguins. *In* "Biology of Penguins" (B. Stonehous, ed.). Macmillan, New York.

Zusi, R. L. and Storer, R. W. (1969). Osteology and myology of the head and neck of the pied-billed grebes (*Pobilymbus*). *Misc. Publ. Mus. Zool. Univ. of Michigan, No.* **139**: 1–49.

Zweers, G. A. (1971). "A Stereotaxic Atlas of the Brainstem of the Mallard (Anas platyrhnchos L.)". Diss. University of Leiden.

Zweers, G. A. (1974). Structure, movement and myography of the feeding apparatus of the Mallard (*Anas platyrhynchos L.*). *Neth. J. Zool.* **24**: 323–467.

Zweers, G. A., Gerritsen, A. F. Ch. and van Kranenburg-Voogd, P. J. (1977). Mechanics of feeding of the mallard (*Anas platyrhynchos*; Aves, Anseriformes). *Contrib. Vert. Evol.* **3**: 1–110.

GENERAL INDEX

This Index was prepared by Dr J. J. Baumel with the assistance of Fr. Edward Sharp, S. J., of the Computer Center, Creighton University.

Key to Index: page numbers in Roman type indicate an item in the list of terms; page numbers in italic indicate an item in the text of the annotations or introductory material; page numbers in bold type indicate an item in an illustration or its legend. See pages xv and xvi for abbreviations.

B

C

H

I

Membrana testae interna, 319, *326*, **330**

N

Q

R

S

U

V

W

Y

Z